Encyclopedia of Communities of Practice in Information and Knowledge Management

Elayne Coakes
University of Westminster, UK

Steve Clarke
University of Hull, UK

Idea Group
REFERENCE

IDEA GROUP REFERENCE
Hershey · London · Melbourne · Singapore

Acquisitions Editor: Renée Davies
Development Editor: Kristin Roth
Senior Managing Editor: Amanda Appicello
Managing Editor: Jennifer Neidig
Copy Editors: Julie LeBlanc, Shanelle Ramelb, Sue VanderHook and Jennifer Young
Typesetters: Diane Huskinson, Sara Reed and Larissa Zearfoss
Support Staff: Michelle Potter
Cover Design: Lisa Tosheff
Printed at: Yurchak Printing Inc.

Published in the United States of America by
        Idea Group Reference (an imprint of Idea Group Inc.)
        701 E. Chocolate Avenue, Suite 200
        Hershey PA 17033
        Tel: 717-533-8845
        Fax: 717-533-8661
        E-mail: cust@idea-group.com
        Web site: http://www.idea-group-ref.com

and in the United Kingdom by
        Idea Group Reference (an imprint of Idea Group Inc.)
        3 Henrietta Street
        Covent Garden
        London WC2E 8LU
        Tel: 44 20 7240 0856
        Fax: 44 20 7379 3313
        Web site: http://www.eurospan.co.uk

Library of Congress Cataloging-in-Publication Data

Encyclopedia of communities of practice in information and knowledge management / Elayne Coakes and Steve Clarke, editors.
    p. cm.
 Includes bibliographical references and index.
 Summary: "This encyclopedia will give readers insight on how other organizations have tackled the necessary means of sharing knowledge across communities and functions"--Provided by publisher.
 ISBN 1-59140-556-4 (hc) -- ISBN 1-59140-558-0 (ebook)
 1. Organizational learning. 2. Knowledge management. 3. Social networks. 4. Social learning. 5. Information society. I. Title: Communities of practice in information and knowledge management. II. Coakes, Elayne, 1950- III. Clarke, Steve, 1950-
    HD58.82.E53 2006
    658.4'038--dc22
                        2005013816

British Cataloguing in Publication Data
A Cataloguing in Publication record for this book is available from the British Library.

All work contributed to this encyclopedia is new, previously-unpublished material. Each article is assigned to at least 2-3 expert reviewers and is subject to a blind, peer review by these reviewers. The views expressed in this encyclopedia are those of the authors, but not necessarily of the publisher.

Editorial Advisory Board

List of Contributors

Contents

Table of Contents

Table of Contents

Contents
by Category

1. Generic Aspects of CoPs

Generic Aspects of CoPs

Classification and Critique of CoPs

CoPs and Formal Workgroups

CoPs and Networking

The Strategic Advantages of CoPs

2. CoPs and the Business Environment

CoPs and Virtual Communities

Developing Organizational Strategies for CoPs

Role of CoPs within Complexity

Role of CoPs within the Business Environment

Role of CoPs within the Public Environment

3. Organizational Aspects of CoPs

Organizational Culture and CoPs

Organizational Change Elements of Establishing, Facilitating and Supporting CoPs

CoPs and Organizational Development - Ethics and Values?

Measuring the Output of CoPs

4. Virtual Teams and the Role of Communities

Distinguishing Between Work Groups, Teams, Knowledge Networks and Communities

Virtual Teaming

5. The Role of Knowledge Management

Virtual Knowledge Communities

Knowledge Communities and Issues

The Meaning of Knowledge

6. Enabling Technology

Software and Hardware for Community Work Support

Where Does Knowledge Management Software Fit?

7. The Philosophy Theory of CoPs/KM

Narrative Inquiry and CoPs

What Organizational Development Theory Can Contribute to Our Understanding of CoPs

What Sociotechnical Theory Can Contribute To Our Understanding of CoPs

Social Aspects and Issues of CoPs

Psychoanalysis, Organizations, and Communities

Psychological Aspects and Issues

Social Philosophy and CoPs

Foreword

I am delighted to write the Foreword to this new, creative, and innovative encyclopedia on Communities of Practice (CoPs) in information and knowledge management (KM). The global importance of knowledge management has been recognized only recently, and KM is still in its infancy. It is, nevertheless, a complex and challenging area of enquiry. This encyclopedia is therefore especially welcome at a time when recognition that the creation, retention, and dissemination of knowledge are key to organizational survival and competitiveness.

By addressing a major aspect of knowledge management (CoPs), the authors help provide understanding where there has previously been mystery. Although this is one (major) area of KM, the authors have been careful to take a holistic approach to the issues, and have avoided the temptation of a reductionist approach. This makes the encyclopedia comprehensive in scope, and appropriate for students, managers, and academics. The encyclopedia is highly readable, highly enjoyable, and full of superb insights that may be used and adapted to many different settings.

The notion of 'Communities of Practice' has arisen alongside the development of knowledge management. As with KM, CoPs have existed far longer than they have been recognized, but their importance for the public and private sectors, and for other organizations, is only just beginning to be understood. The appropriate use of technology to support such communities is key to their success, but it is the word *appropriate* that must be emphasized. Unlike many texts that purport to address KM, but which really deal solely with information technology, this encyclopedia addresses the key organizational and communications issues that are the real challenges facing us in the 21st century. Technology is an enabler, but the drivers for action must come from the needs of the community, be it a public organization, a large company, a charity, an SME, or a group of people who share interests and directions, or who want to develop those interests in a cooperative fashion. If technology becomes the driver those needs become subsumed, and all too often it is apparent that systems are serving themselves, rather than the communities they should be serving.

The *Encyclopedia of Communities of Practice in Information and Knowledge Management* contributors provide a wonderful collection of material that will both interest, and serve as reference, for anyone who wants to think about the world in which we now live, and who want to improve that world.

Professor Brian Lehaney

Preface

OVERVIEW

Communities of Practice (CoPs) are developing in importance for the organizational world. In the field of knowledge management there is increasing importance being placed on the social aspects of knowledge and how it can be managed, as opposed to how technology can be utilized. Technology, it is now being argued as a supportive mechanism rather than a driver for the management of knowledge. People, it is also maintained, prefer to share their knowledge on a face-to-face basis rather than through electronically mediated means. Once the knowledge is shared in a tacit manner, there remains the issue of how it may be shared through technology, as information to be accessed as required.

Much has been made of the difficulties associated with turning so-called tacit knowledge into explicit. One way of characterizing this problem is to see explicit knowledge as recordable (on paper, computer disk, etc.), and tacit as inherently difficult to record and hence to share through such media as computers. A possible solution to this is to see sharing tacit knowledge as a process, achieved through human interaction, rather than as simple content.

CoPs can contribute to this by connecting people so that they can collaborate and share their tacit (personal) knowledge about a particular work context or practice. Thus railway engineers will meet to discuss issues such as how to devise good signalling; midwives will meet to discuss best practice in drugs for childbirth and so on. As Wenger (2001) argues they are people who "share an interest in a domain of human endeavour and engage in a process of collective learning that creates bonds between them" (p. 2).

Here, in the *Encyclopedia of Communities of Practice in Information and Knowledge Management*, for the first time, we combine discussions of how CoPs can assist organizations, both voluntarily and privately funded, from both practitioners and academics. These discussions come from a wide variety of industrial sectors and from across the world.

COPs IN THE ORGANIZATIONAL WORLD

There are a number of definitions of CoPs available in literature. However, Etienne Wenger is credited as being the person who has most developed the concept. Thus, here we offer his definition taken from his study of supporting technologies:

Communities of practice are a specific kind of community. They are focused on a domain of knowledge and over time accumulate expertise in this domain. They develop their shared practice by interacting around problems, solutions, and insights, and building a common store of knowledge. (2001, p. 1)

There are a many ways of defining a community, including: the domain, the practice, and the community itself.

In the organizational world these communities are very much focused on expertise, and are intended as social structures for sharing practice and practical knowledge. It is therefore embedded in the CoP concept that CoPs have the ability to cut across departmental or even organizational boundaries, and can provide learning (and teaching) opportunities to all levels of staff, of all ages and experience, in an informal manner. Communities, once established, may outlast the organization in which they were created, and may even grow across time to encompass practitioners in very diverse organizations. Communities need to be re-informed in their practice by regular meetings and engaging in joint activities. Through these activities they re-enforce their social bonds.

In gathering and collating the material for this encyclopedia, we and our Editorial Advisory Board have found the contributions naturally falling into seven categories, and consequently a Table of Contents by Category is included in this encyclopedia and detailed in the following section.

ORGANIZATION OF THE ENCYCLOPEDIA

In this encyclopedia we have more than 100 entries related to the topic area of Communities (of Practice), including both practical examples and theoretical discussions: this is the largest collection to date of articles in this field.

This encyclopedia is organized into categories consisting of related articles. It is organized in a manner that will make your search for specific information easier and quicker.

In addition, a comprehensive index is included at the end of the encyclopedia to help you find cross-referenced articles easily and quickly.

The seven major categories also include sub-categories. All are detailed below:

Category 1: Generic Aspects of CoPs

In this category the sub-categories covered are:

Classification and Critique of CoPs
CoPs and Formal Workgroups
CoPs and Networking
The Strategic Advantages of CoPs
Story-Telling within CoPs and Knowledge Transfer
Language (and Symbol) Development in CoPs

The category begins with a general introduction to the concept of communities of practice by Coakes and Clarke and is followed by six sub-categories: Classification and Critique of CoPs; CoPs and Formal Workgroups; CoPs and Networking; The Strategic Advantages of CoPs; Story-Telling within CoPs and Knowledge Transfer; and Language (and Symbol) Development in CoPs.

Examples of the types of contributions found in this encyclopedia can be seen from the unusual entries by Tunç Medeni. In Category 1, Medeni discusses Yaren talks. Yaren talks are an example of medieval tradition relating to craftsmen's guilds still operating but in a modified form in modern Turkey. Medeni also provides us with some fascinating photographs of these talks in action illustrating the costumes worn by the participants and discusses not only the story-telling that comprise the main activity of these talks but also the punishments that might be imposed on those who transgress its rules. Modern CoPs rarely have punishments for transgressors. His second unusual contribution is to be found in Category 4.

A very important sub-category of this category is that concerning language and symbol development within communities and here we look at two articles. Ahmad and Al-Sayed discuss how language within a medical community has evolved and von Wartburg considers the part that metaphors play within figurative speech and as part of the socialization process for members for the community.

Category 2: CoPs and the Business Environment

In this category the sub-categories covered are:

CoPs and Virtual Communities
Developing Organizational Strategies for CoPs
Role of CoPs within Complexity
Role of CoPs within the Business Environment
Role of CoPs within the Public Environment
CoPs and Competitive Advantage
Role of CoPs in Supporting Economic Development
The Role of Knowledge Management in CoPs and Supply Chains

A number of issues, or themes, are apparent in these articles. One considered by Teigland in her articles with Schenkel, relates to economic development and regional innovation systems. These systems are networks of organizations, institutions and individuals within which there is the creation and exploitation of innovation. These communities comprise not only for-profit firms but also academic institutions, policy and government authorities, large and small firms that are spatially contiguous and develop 'local' capital through their activities. A related article by Mason and Castleman looks at SMEs in regional clusters and the value of virtual CoPs for promoting innovation and knowledge sharing.

Category 3: Organizational Aspects of CoPs

In this category the sub-categories covered are:

Organizational Culture and CoPs
Organizational Change Elements of Establishing, Facilitating and Supporting CoPs
CoPs and Organizational Development - Ethics and Values?
Measuring the Output of CoPs
Inter-Organizational Communities
Using Communities to Support Research
Using CoPs for Organizational Learning
CoPs and the Development of Best Practices
Leadership Issues within CoPs
Collective Learning within CoPs
CoPs and Their Life-Cycle
CoPs and Project Management
How Are Social and Community Links Captured and Supported in CoPs?

This category reviews the internal and organisational aspects that affect and are affected by CoPs. The 13 entries within the section reflect the importance of these issues.

The idea that CoPs can enhance the development of intellectual capital and the local economy is considered by Pyke, and by Bellarby and Orange. Bellarby and Orange look at the voluntary sector as does Walker. In the latter study, community spaces, it is argued, have long been in existence where voluntary sector workers engage in discourse and informal learning.

Category 4: Virtual Teams and the Role of Communities

In this category the sub-categories covered are:

Distinguishing Between Work Groups, Teams, Knowledge Networks and Communities
Virtual Teaming
Vortals (Communities Operating Via an Electronic Network Rather Than in Contiguous Space)
Teamwork Issues in Virtual Teams

Here we see the second unusual entry offered by Medeni relating to a virtual community of fantasy game-players. The world of Wold has been developed and researched for sometime and here we see some insights into how this world operates and how the community has developed.

Category 5: The Role of Knowledge Management

In this category the sub-categories covered are:

Knowledge Sharing
Issues in Knowledge Sharing
Knowledge Communities
Virtual Knowledge Communities
Knowledge Communities and Issues
The Meaning of Knowledge

Here our contributors look at the role of knowledge management within CoPs, issues such as: how do communities share knowledge? What is tacit knowledge within CoPs and how it is considered?

We learn from Rodriguez-Elias et al. how to develop knowledge management tools to support knowledge flows. We also find in Zappvigna's article how systemic functional linguistics can assist us to discover the relationship between doing, meaning and saying. The article from Chen et al. provides an overview of the inter-organizational knowledge transfer and its related literature, and present a proposed inter-organisational knowledge transfer process model based on theoretical and empirical studies.

This category also looks at some very interesting articles discussing malpractice in CoPs and the issues involved in knowledge sharing, as knowledge management theory often leaves us with the impression that knowledge can be as easily managed, like products and commodities, which may well not be the case.

Category 6: Enabling Technology

In this category the sub-categories covered are:

Software and Hardware for Community Work Support
Where Does Knowledge Management Software Fit?
Tools - Repositories, Modelling, Scenario Development, and Analysis (Etc.) To Support and Capture CoP Activities

Category 6 is concerned with the technology support for communities and in its three sub-categories look at software and hardware, especially knowledge management software and also specific software types such as data warehouse and their role in knowledge sharing.

We have a number of articles discussing what technology is needed to support CoPs. In one article by Coakes, we see a generic discussion of what facilities CoPs need to function and how technology can supply (some) of these functions and facilities. In an article by Chua, we learn about three tools for educational communities: portals, course management systems, and videoconferencing, which can be used to create and sustain communities of practice, and provide value-added services to participants in an interactive environment. The article by Dotsika explores the advantages and pitfalls of supporting 'computerised' versions of these communities, reviews a number of existing software tools and looks into emerging technologies considering their role and appropriateness. In Ruhi we look at a best practices model for utilizing these technologies.

Category 7: The Philosophy Theory of CoPs/KM

Finally, in this category the sub-categories covered are:

Narrative Inquiry and CoPs
What Organizational Development Theory Can Contribute To Our Understanding of CoPs
What Sociotechnical Theory Can Contribute To Our Understanding of CoPs
Social Aspects and Issues of CoPs
Psychoanalysis, Organizations and Communities
Psychological Aspects and Issues
Social Philosophy and CoPs

Philosophy and theory in relation to communities of practice is well represented with seven entries. The theories we consider include narrative inquiry; sociotechnical, social theory and social philosophy. We also look at psychoanalysis and psychological aspects and issues within communities.

Three very interesting contributions here are made by Nobre. In her first article, Nobre argues that the dominant stream of management theory is still largely influenced by the command and control paradigm developed over a century ago and that there is a growing awareness of the dangers of assuming a reductive and limited view of organisational complexity. Indeed Grieves agrees and additionally comments that the use of organisational development theory enables organisations to achieve effectiveness through careful analysis and diagnostic techniques as well as through carefully considered intervention strategies. In her second article, Nobre explores the many hidden dimensions of human actions within the organisational environment and considers the practice of the theory of psychodynamics and the role of consultants engaging with a client organization. Her final contribution argues that the growth in importance of communities within organisational settings is a sign of a change in paradigm.

Finally, Clarke offers the view that CoPs are in essence social groupings, and that something is therefore to be gained by considering the contribution of social theory and philosophy to the domain. Some of the current research and practice informed by critical social theory is used here to shed light on the issues.

CONCLUSION AND CONTRIBUTION

The *Encyclopedia of Communities of Practice in Information and Knowledge Management*, we believe, will become the leading reference source for dynamic and innovative research in the field of CoPs for information and knowledge management. With the ever-increasing interest in knowledge management, this volume provides a comprehensive, critical and descriptive examination of all facets of CoPs in information and knowledge management in societies and organizations.

This encyclopedia contains numerous research contributions from leading scholars from all over the world on all aspects of communities of practice in information and knowledge management, with comprehensive coverage of each specific topic, highlighting recent and future trends and describing the latest advances. It also contains a compendium of key terms, definitions and explanations of concepts, processes and acronyms, and thousands of comprehensive references on existing literature and research on communities of practice in information and knowledge management.

We hope you enjoy reading it as much as we have enjoyed the challenge of collating the 100 plus entries.

Elayne Coakes
Steve Clarke

REFERENCES

Wenger, E. (2001) *Supporting communities of practice: A survey of community-oriented technologies, Version 1.3.* Available from *http://www.ewenger.com/tech*

About the Editors

Elayne Coakes is senior lecturer in BIM at the University of Westminster, UK. She is active in promoting and researching the sociotechnical view of information systems, in particular researching the inter-relationship between humans, information and knowledge (and learning) and the systems through which they are utilised and managed within organisations.

Her PhD (from Brunel) is in the field of information systems strategy.

She is the author and editor of a number of books and articles relating to sociotechnical thinking and KM and is currently interested in communities of practice. She is an associate editor for *OR Insight* and on the Editorial Board of the *International Journal of Knowledge Management*.

Steve Clarke received a BSc in economics from The University of Kingston Upon Hull, an MBA from the Putteridge Bury Management Centre, The University of Luton, and a PhD in human centred approaches to information systems development from Brunel University — all in the UK. He is professor of information systems and director of research for the University of Hull Business School, UK.

Steve has extensive experience in management systems and information systems consultancy and research, focusing primarily on the identification and satisfaction of user needs and issues connected with knowledge management. His research interests include: social theory and information systems practice, strategic planning, and the impact of user involvement in the development of management systems. Major current research is focused on approaches informed by critical social theory.

An Adaptive Multi-Agent Environment

Rejane Pinheiro
University of Fortaleza, Brazil

Elizabeth Furtado
University of Fortaleza, Brazil

INTRODUCTION

This article aims to develop a new environment of collaborative learning, by taking into account the criteria of construction of knowledge by the apprentices and the adaptative management of that knowledge by artificial agents. The multi-agent technology has been chosen due to the possibility of having artificial agents with internal decision processes to help students in the construction of their own projects and enabling learning objects available in accordance with the cognitive characteristics of the students and of their group. In this multi-agent system, exchanges of messages between the agents can occur so that they can perform theirs tasks in the best possible way.

BACKGROUND

In CSCL (computer-supported collaborative learning) environments, collaborative learning happens mainly when groups of students have as a common objective the resolution of a certain problem (Santos, 2003). In this article, environments of collaborative learning including TelEduc (Rocha, 2002), AulaNet (Lucena & Fuks, 2000), AME-A (Pereira, D'Amico & Geyer, 1998), AdaptWeb (Freitas et al., 2002), and WebSaber (Santos, 2003) were studied, and it was observed that the criteria related to the adaptation in presenting the didactic material for the students, along with the possibility of the learners to act as authors in the construction of these didactic materials, are characteristics that have not been found in these analyzed educational environments.

According to Bannan-Ritland, Dabbagh, and Murphy (2000), most of the systems of learning objects that use didactic basis are made up of the behaviorist or cognitive kind, where the apprentices continue to be receivers of information that will be contained in the software. These researchers assert that educational environments, using learning objects through a constructivist approach, have not yet been developed.

The same authors mention that the environments of learning objects that use objectivist theories allow the users only to receive instrumental content, when it is finished, and do not give the apprentices any options so to create educational material (Bannan-Ritland et al., 2000). Therefore, those environments restrict the possibility of the apprentices to be an author or co-author on the development of the objects, making active participation of the apprentices impossible.

The reconstruction of digital resources, such as text and video/audio components, allows the users to adapt, rebuild, or reconfigure those media in their own representation of meaning, while it does not happen when the apprentice receives an instructional content developed by the teacher (Wiley, 2000).

The adaptation of the media or of the learning objects is related to the way of showing these objects on the interface of the system. Thus, we will use the concept of personalized learning, which consists of using the software technology to make the learning objects available, according to the profiles of the apprentices.

THE ENVIRONMENT PROPOSED

This educational environment is formed of two sides: the pedagogical side and the technological one. The pedagogical side consists of the use of two learning theories: the Genetic Epistemology of Jean Piaget (1975) will be used to construct the students' model (profile) to make the adaptation of the interface possible, and the socio-cultural theory of Vygotsky (1998) will be used for the construction of the group's model (profile) according to the relationship of the students for the construction of their learning objects. The technological side that was already mentioned above is constituted by the multi-agent technology.

This educational environment uses the socio-constructivist approach analyzed by Dillenbourg, Baker, Blaye, and O'Malley (1994) which focuses on the individual in a social interaction context and is based on the two learning theories mentioned above.

The learning process will include constructed educational projects, which will be developed by groups of students working face-to-face and at a distance. The students will be able to construct their own didactic material through learning objects that will be available in the environment. Wiley (2000) defines a learning object as any entity, digital or non-digital, that can be used, reused, or referenced during technology-supported learning.

This educational environment uses a playful locale where the learners construct their knowledge and discover the solutions for their problems with the construction of toys and the use of games. In the playful locale, the learners have the help of the teachers, who instead of imposing the information will assist them so that they can discover how to solve the tasks that will appear in implementing the learning process.

In the considered environment, the students will use the available learning objects to construct their stages in order to reach the objectives sought, and will also have the help of the facilitator and artificial agents. So, the learning objects will function as toys or games that stimulate the imagination of the students and will enable them in accordance with their level of cognitive development. These objects will be developed by the students in groups or alone.

This educational environment is formed by human agents who are the students and the facilitator, and by three types of artificial agents: Accompanying Agent, Student's Profile Agent, and Group's Profile Agent.

These virtual agents are based on probabilities and utilities, and the process of decision is made through the use of Bayesians Nets, defined by Jensen and Lauritzen (1999) as a set of variables and lines directed between the variables. Each variable has a set of finite states and a probabilist value whose result depends on the parents' variables; this dependence is called conditional probability.

The base of knowledge of these agents—based on uncertainty that will be used in this research—uses Bayesians Nets to calculate the probabilities of the relevant variables to one determined situation and uses Decision Nets to calculate the utility of these variables.

The Student's Profile Agent is responsible for the cognitive characteristics of the students based on Piaget's theory of the learner's development stages. They can be in the following levels: novice, intermediary, experienced. The base of knowledge of this virtual agent will store the variables that correspond to the cognitive characteristics of the students and when this agent were requested to inform which the best profile to insert student is, it will have to cover the Bayesian Net to search for this result.

The Group's Profile Agent is responsible for the interaction of the learners of a group and for the interaction of the groups between themselves. The base of knowledge of this agent will have variables that will be defined based on the theory of Vygotsky, where mediation between the individuals and zone of proximal development is used.

The function of the Accompanying Agent is to assist the learners in implementing the learning process that will occur in the development of the educational projects. The base of knowledge of this agent will contain learning objects already constructed by other learners.

The architecture of this educational environment can be seen in Figure 1; the interactions that occur between the human and virtual agents in this system are numbered and correspond with the following actions:

- **[1]:** The student requests the help of the Accompanying Agent, and he answers to this requirement.
- **[2] and [3]:** The Accompanying Agent requests the learning objects that will be available in the interface of the environment.
- **[4.1] and [4.2]:** The facilitator has access to the data bases of learning objects for possible updates that it needs to make.
- **[5]:** Interactions between learners or groups of students and the facilitator can be performed.
- **[6.1] and [6.2]:** The Accompanying Agent and the Facilitator request results about the

Figure 1. Architecture of the educational environment

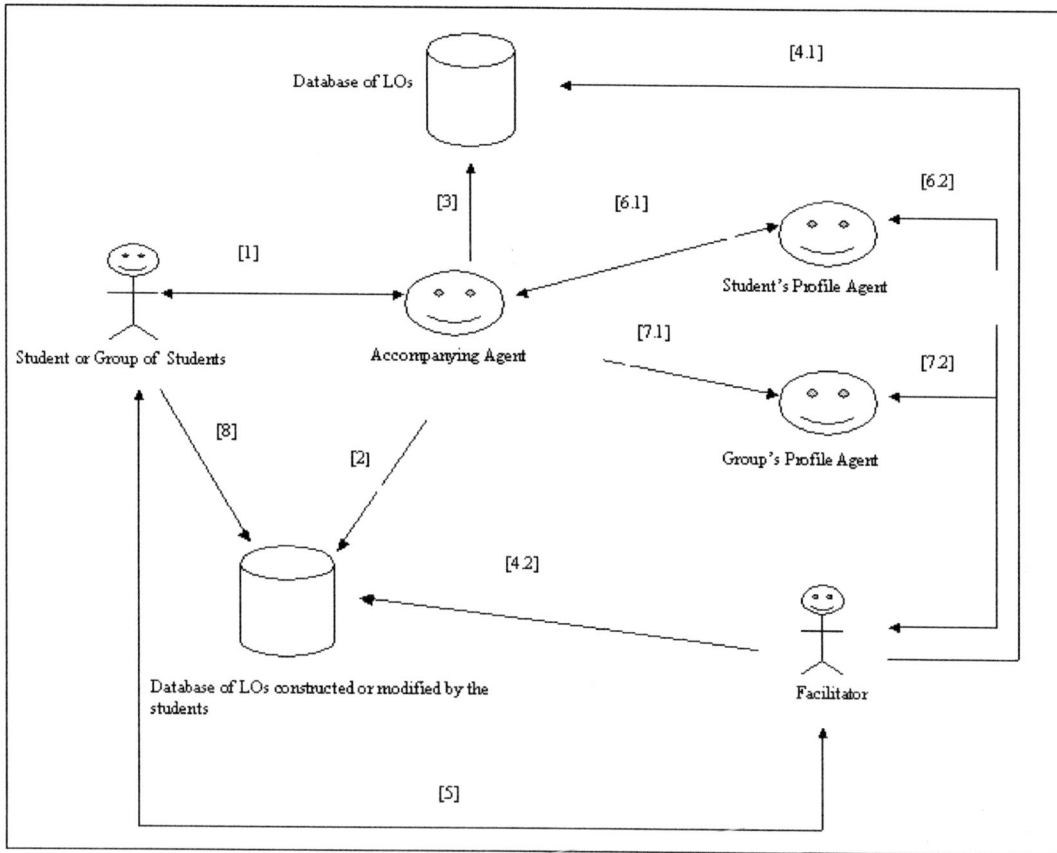

cognitive characteristics of the students to the Student's Profile Agent, who returns the answer to them.

• **[7.1] and [7.2]:** The Accompanying Agent and the Facilitator request results about the interactions of the students to the Group's Profile Agent, who returns the answer to them.

• [8]: The students or groups of students record the didactic material (media that was developed by them) in the database that has learning objects constructed or modified by the students.

An educational project begins when the Facilitator chooses the subject that will be worked on by the users. Before informing the users of the chosen subject, the Facilitator verifies if the Accompanying Agent has information about this subject in its base of knowledge. If it does not, the Facilitator must decide if the project continues without the aid of this agent. The Facilitator selects an archive of simulation more adjusted to the subject that will be worked on in the project and sends a message informing the users of

his choice. Situations of challenge are presented to the users, and they have to construct learning objects to develop the task that the simulator is requesting.

When the students were constructing learning objects, they could ask for help from the Accompanying Agent, or this agent can perceive that the students need help. This artificial agent will supply an adaptive aid so that the user can modify the learning object and continue constructing the didactic material. The Accompanying Agent decides how he is going to help the users through the use of the Bayesians Nets that will be constructed in the beginning of the project and when the Student's Profile or Group's Profile is modified.

This environment has two types of adaptation: the type of aid given to the learners, and the method of presentation of learning objects in the interface of the system. The form to carry through the adaptive aid to the learners was already specified above. The presentation of learning objects is related to which media will be enabled for use for the groups of

Figure 2. Bayesian Net related to the topic of algorithms

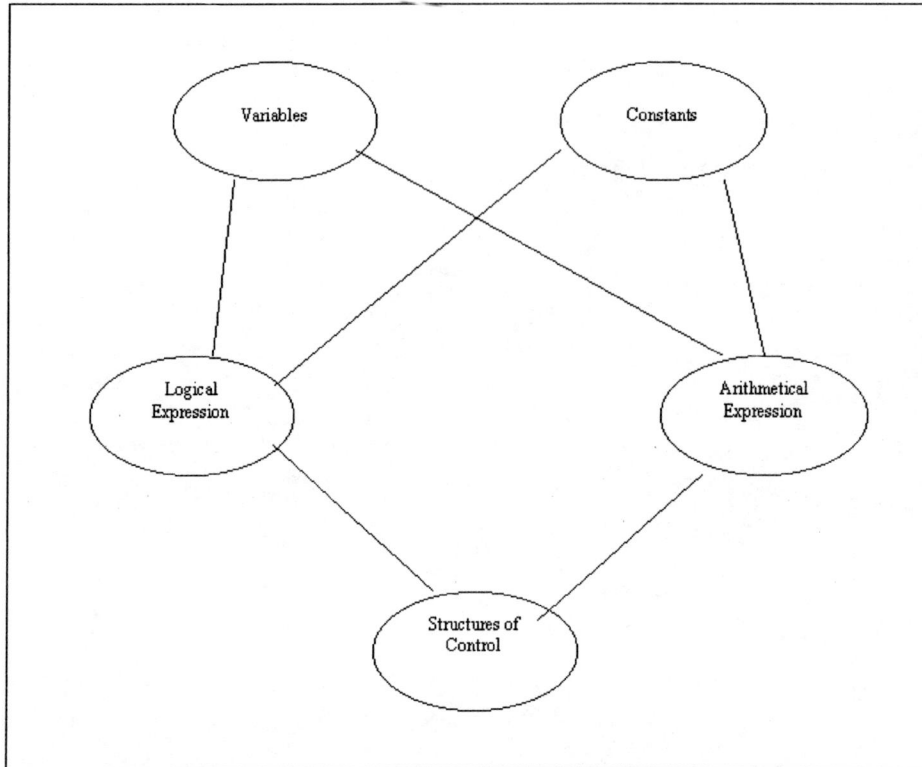

learners; this decision will be made by the Student's Profile Agent analyzing the cognitive characteristics of these students.

When the users finish their projects that correspond to the construction of the learning objects, they will discuss the generated projects in order to try to unify the learning objects built.

The development of a generic project in the considered environment was described. Now we are going to present the development of a specific project that corresponds to the learning of algorithms.

To study algorithms, learners must have knowledge of the following topics: constants, variables, arithmetical and logical expressions, and structures of control (sequential, selection, and repetition). Soon, the Accompanying Agent must request information to the Student's Profile Agent about the level of knowledge of the student on this subject.

In accordance with the level of knowledge of the student, the Accompanying Agent will search in the Bayesian Net and will indicate the suitable didactic material (learning object) to be used by the student.

In Figure 2, a Bayesian Net is shown. It is formed by the topics of knowledge that contain learning objects associated to them, and these objects represent didactic materials that are stored in a database.

CONCLUSION

In this article, we presented an environment of cooperative learning that has as its main characteristics: the adaptation in presenting learning objects, and the possibility of the students acting as authors in the construction of these learning objects with the help of the teacher and of the Accompanying Agent. This artificial agent will assist the students in their learning process, having had each student's profile as a basis for this help.

The key points of the development of this environment of collaborative learning are: to provide the construction of knowledge by means of technology to the apprentices, and to permit the management of this knowledge through the artificial agents and through the facilitator.

The base of knowledge of the artificial agents who compose the architecture of this environment is constituted by Bayesians Nets, which have interrelated variables that will make it possible for the virtual agents to make the best decisions to adapt enabling learning objects in the interface and assist the students in the environment.

REFERENCES

Andrade, A., Jaques, P., Vicari, R., Bordini, R., & Jung, J. (2000). Uma proposta de modelo computacional de aprendizagem à distância baseada na concepção sócio-interacionista de Vygotsky. *Proceedings of the Workshop de Ambientes de Aprendizagem Baseados em Agentes; Simpósio Brasileiro de Informática na Educação*, SBIE, Maceió.

Bannan-Ritland, B., Dabbagh, N., & Murphy, K. (2000). Connecting learning objects to instructional design theory: A definition, a metaphor, and a taxonomy. Retrieved from *http://www.reusability.org/read/chapters/bannan-ritland.doc*

Dillenbourg, P., Baker, M., Blaye, A., & O'Malley, C. (1994). The evolution of research on collaborative learning. Retrieved from *http://tecfa.unige.ch/staf/staf-f/deiaco/doII4/resume_article.doc*

Freitas, V., Marçal, V.P., Gasparini, I., Amaral, M., Proença, M. Jr., Brunetto, M.C., Pimenta, M., Ribeiro, C.P., Lima, J.V., & Oliveira, J.P. (2002). AdaptWeb: An adaptive Web-based courseware. Retrieved from *www.inf.ufrgs.br/~tapejara/electra/docs/02_icte_badajoz.pdf*

Jensen, F., & Lauritzen, S. (1999). Probabilistic networks. Retrieved from *http://citeseer.nj.nec.com/jensen97probabilistic.html*

Lucena, C., & Fuks, H. (2000). *Professores e aprendizes na Web: A educação na era da Internet.* Rio de Janeiro: Clube do Futuro.

Pereira, A.S., D'Amico, C., & Geyer, C. (1998). Gerenciamento do conhecimento no ambiente AME-A. Retrieved from *www.inf.ufrgs.br/~adriana/publicacoes.htm*

Piaget, J. (1975). *Seis estudos de psicologia.* Rio de Janeiro: Forense University.

Rocha, H.V. (2002). Projeto TelEduc: Pesquisa e desenvolvimento de tecnologia para educação à distância. Retrieved from *http://teleduc.nied.unicamp.br/pagina/publicacoes/figura premio_abed2002.pdf*

Santos, N. (2003). WebSaber: Um ambiente para a aprendizagem cooperativa baseada na resolução de problemas. Retrieved from *lsm.dei.uc.pt/ribie/docfiles/txt200372911256Websaber.pdf*

Schank, R., & Cleary, C. (1995). *Engines for education.*

Vygotsky, L.S. (1998). *A formação social da mente: O desenvolvimento dos processos psicológicos superiores.* São Paulo: Martins Fontes.

Wiley, D.A. (2000). Connecting learning objects to instructional design theory: A definition, a metaphor, and a taxonomy. Retrieved from *http://www.reusability.org/read/chapters/wiley.doc*

Wooldridge, M., & Jennings, N. (1997). Intelligent agents: Theory and practice. Retrieved from *http://www.doc.mmu.ac.uk/staff/mike/ker95/ker95-html.html*

Aspects and Issues of Communities of (Mal)practice

Jon Pemberton
Northumbria University, UK

Brenda Stalker
Northumbria University, UK

INTRODUCTION

The concept of *community of practice (CoP)* is now embedded within all areas of public- and private-sector organisations, although the term has different connotations dependent on its context. It is not a new concept, and it could be argued that many of the developments in this area are evolutionary rather than cutting-edge innovations. Informal groupings have always existed, but in the quest to harness and develop knowledge and 'add value' to organisations, the CoP has been embraced and developed, as various strands of management practice have fused and merged. Typically, these incorporate knowledge management, management strategy, complex adaptive systems, and latterly, knowledge ecology. Whether they exist as a social gathering or technological network, the sharing of expertise and the creation of new knowledge, often tacit in nature, is a central tenet of a CoP's existence (Lave & Wenger, 1991). There are clear parallels with organisational learning and the knowledge-centric organisation, and few would dispute the potential benefits that CoPs can bestow on the individuals making up these communities and the organisations that these CoPs reside in (Wenger & Snyder, 2000; McDermott, 2002).

BACKGROUND

Unlike other articles in this book that trumpet the successes, benefits, and reach of CoPs within organisations, this article discusses the possible negative consequences of communities of practice, both from an individual and organisational perspective. It is not intended to denigrate the value of such groups, but merely to flag potential pitfalls and problems associated with CoPs. This discussion is based not only on published material, but first-hand experience and knowledge of a research-based CoP within higher education.

A number of themes are examined in this article, with discussion centring on why these might have a negative impact, or even lead to inferior practices in the workplace. It should also be noted at this juncture that not all of the issues raised here apply to all groups—it is highly dependent on context, function areas, and so forth.

Time is of the Essence

The emergence of a CoP in any organisation is dependent on a number of factors, not least:

- the context and focus of the group,
- the individual initiators of the community,
- whether it is an offshoot of other formal or informal organisational groupings,
- whether the technological infrastructure exists to support online discussion boards and real-time meetings, and
- through chance meetings of individuals with similar interests and motivation.

Initially, they begin life as a relatively small grouping of individuals or online participants, with membership cascading as word filters through an organisation or additional individuals are invited to participate in the group. Several meetings or online forums usually take place before membership achieves an equilibrium, and the core group is then established. The value attached by individuals to these meetings is a critical success factor, and there is evidence to suggest that this value can dissipate over time, leading to the demise of the CoP (Wenger, McDermott & Snyder, 2002).

The early stages of any CoP are critical since this is where trust is established between participants and assessments made as to the potential value of these groups by their members (Ardichvilli, Page & Wentling, 2003). Provided this transition is relatively smooth, the CoP may exist for several months or years, but this is highly dependent on the motivation of its members and its internal management. Over time, however, interest can subside—it is, after all, a voluntary commitment and not formally tied to job enhancement or progression. The departure from an organisation of key participants, typically founders or organisers, may also lead to the disintegration of the community. This is certainly corroborated in the authors' case, where membership of their research-based CoP, established in 2002, started in double figures; over time this has diminished, and meetings now generally consist of a core of five regular participants.

Time is also a critical success factor in terms of online communication, as the posting of questions and responses in a written format is a vastly more lengthy process than verbal communication. In this sense, face-to-face CoPs have a distinct advantage over discussion/bulletin board-type communities.

Follow the Leader

Communities are, by definition, groups of like-minded individuals keen to share existing knowledge and practice, and create new knowledge in the process. For CoPs to function effectively, internal leadership and coordination must also be present (Wenger et al., 2002). Leaderless communities seldom survive as groups fragment and momentum is lost.

To ensure the issues discussed have the support of the community, a careful balance between guidance and authority is needed, so that the views of the 'leader' are not solely reflected in the group. This issue should not be underestimated where managers act as CoP coordinators or leaders, as managers tend to command and control; for CoPs to function effectively, it is critical that new skills of brokerage and translation are developed (Brown & Duguid, 1998).

An added complication arises where organisations seek to ensure cultural conformity to a specific organisational identity (Moore & Sonsino, 2003).

Managers typically seek to impose this upon CoPs which may be at variance with the self-regulation enjoyed by CoP members. In this situation, creativity and innovation may suffer as a consequence. Furthermore, CoPs are motivated by a communicative *Habermasian,* rather than an instrumental logic that is driven by deliverables, and seeks to alter the traditional perspectives of managerial control within CoPs (O'Donnell et al., 2003). In essence, managers can foster the cultural context for facilitating CoPs, but should step back by allowing members to negotiate their own norms and agree on their own boundaries.

In the authors' own CoP, the freedom to explore new ideas and set its own agenda, free from the shackles of organisational missives, has been achieved by the commitment of its members and facilitated by a coordinator acting as a 'leader' for the purposes of organising meetings. During meetings, however, equal status is afforded to all participants, and the individual personalities of the members are such that the CoP functions effectively by virtue of the creativity and freedom it bestows on its participants.

Outside in

The emergence of a CoP within an organisation is typically down to an individual, or group of individuals, and its formulation is not a result of management intervention—its existence lies outside the formal organisational structure (Wenger et al., 2002). Depending on the type of organisation, a CoP may emerge without management awareness or, for that matter, other employees not part of this community. Gradually, over time, awareness of the CoP emerges, often through members feeding solutions to problems into the formal problem-solving processes of the organisation (Wenger, 1998). Where these solutions are perceived as beneficial, greater scrutiny of the CoP sometimes ensues, with managers and employees questioning:

• why they themselves are not part of the group;
• whether it is consuming organisational resources that other employees do not enjoy;
• whether members of the group appear to have an advantage, in some sense, over non-members?

In many cases, these concerns do not surface and the CoP is seen as team-based structure adding value to the organisation's performance. Where they do, resentment can arise via jealousy or a misplaced mistrust, as well as the fact that certain organisational members were not invited to participate in the group (Wenger et al., 2002). Where this happens with individuals at management level, covert or explicit actions may be taken to limit the ability of the group to exist by making physical and organisational resources less available for the CoP.

The negative effects described here are not a direct consequence of the CoP itself, but the perceptions of it from non-members. For this reason, it is important that the processes governing the setting up of a CoP, and its subsequent existence, are both transparent and inclusive.

Throughout the existence of the authors' own CoP, new members, both internal and external to the organisation, have joined the group and, intermittently, attended meetings, appearing and disappearing accordingly. Where a particular issue is felt to be of value to members outside the group, for example a new writing opportunity or the emergence of a conference that might have a wider appeal, this information has been communicated to non-CoP members.

Dominant Forces

Bringing together a variety of individuals, often with different organisational status and professional expertise, is not without problems. For example, power-distance relationships are potentially divisive in this situation, especially where members do not feel free to express themselves or are inhibited by the presence of more senior organisational members.

An interesting comparison exists with executive judgement or expert juries, where planning takes place at the highest level within an organisation. It is well documented that individuals who are perceived to carry weight in terms of status and authority often dominate the discussion, and other colleagues merely agree with them, despite their own personal beliefs and experience (Madridakis & Wheelwright, 1989). This 'political' situation can be mirrored in CoPs where members have different roles and status in an organisation. In certain situations, this manifests itself in poor decisions being made and, in extreme cases, the break up of these groups where participants feel their contribution is marginalised.

Evidence for such statements arises from the research-based CoP to which the authors belong. A senior member of staff with an interest in research requested to attend a CoP meeting and effectively 'hijacked' the meeting with personal opinions, clearly at variance with the core beliefs of the group. On the senior staffer's departure, it emerged that the existing members felt very uncomfortable with the ideas put forward, and there was widespread consensus that this was not a direction the CoP wanted to explore. In the end, no confrontation was needed, as the member of staff in question did not attend any further meetings.

Port in a Storm

CoPs can act as a welcome refuge for its members, without organisational constraints influencing behaviours to any great extent. Such localised empowerment is an attractive feature of these groupings. Yet, the potential energy generated in these innovative communities is often drained and dissipated by members' excursions into the chaotic and political milieu of organisational life. In extreme cases, this refuge is both a source and container of anxiety, creating organisational spaces that are dysfunctional for both individuals and organisations alike. Emotional containment is also fundamental to the functioning of the group and often allows people to stand back from the daily pressure of their working environment, permitting reflection and experimentation with new ways of thinking and organising. Thus, although a certain amount of insularity is inevitable and necessary, emotional support should be provided, along with encouragement to channel energies outwards, particularly in environments of turbulent change (Nicolini, Sher, Childerstone & Gorli, 2003).

The feeling of providing a 'protected space' is a feature of the authors' own research CoP, where dialogue and disclosure are implicitly subject to the 'Chatham House Rule'. Participants have reflected upon the sensation of speaking openly in the group, without the obligatory self-regulation that sanctions their contributions elsewhere. Such dialogue has been identified by one member with having the freedom to express 'deviant' thoughts that he or she would not normally feel comfortable sharing with others in the organisation. Interestingly, this expression of feeling 'deviant' has served as a source of light amusement within the group, as if releasing a 'safety valve', rather than a catalyst for descending into a group cathartic moment.

Clever Devils

Wenger et al. (2002) highlight a number of problems related to CoPs, including imperialism, narcissism, and factionalism. These are related in many ways, stemming from the belief that the CoP, by nature of its specialised expertise, is in some way 'superior' to other parts of the organisation. This arrogance can result in dogmatism both from within and outside the community. Internally, this manifests itself by factions developing with particular viewpoints that may produce a volatile setting at variance with the aims of the group. Equally, an imbalance in perceived intellectual status or expertise amongst participants can create tensions and resentment, especially if accompanied by over-confidence. As an example, in a research-based CoP, new researchers can feel marginalised by established researchers, depending on the dynamics of the group. Over time, this may dissipate or intensify, the latter typically leading to departure of members from the community.

In the authors' experience, within their CoP, such concerns have been very real and particularly felt by new individuals joining the group, but initially left unspoken. However, once articulated and discussed, the highlighting of these concerns serves to make the 'rules of engagement' explicit within the group and encourages participation from all group members. Over time, these concerns appear to have disappeared as the more inexperienced researchers have grown in confidence, realising that their role is as valid as the so-called more experienced researchers.

Another aspect of CoPs relates to the ownership of *intellectual property* generated within the community. Unlike codified organisational knowledge in the public domain that has an inherent quality control, often through peer review, there are no such filters for shared personal knowledge. For this reason, O'Donnell et al. (2003) argue that information presented to the group requires scrutiny. Ironically, however, individuals who vigorously interrogate the veracity or

validity of other members' contributions may inhibit the confidence of members to disclose and participate in a critical examination of practice. Once again, the badges of status—including experience, organisational standing, or recognised expertise—clearly have the potential to impact on relationships and influence knowledge sharing within a CoP. A fine balance is needed, but when the environment is conducive, the theoretical origins of CoPs, drawing on Vygotskian ideas of *proximal development*, demonstrate how the expert members of CoPs may scaffold the learning of less experienced members (Nicolini et al., 2003).

One-Way Street

A typical CoP centres on the interaction process whereby a holistic view of the range of complex problems and situations is developed, facilitating the integration of a diverse body of knowledge within organisations and, as a consequence, developing knowledge and verifying best practices (Bhatt, 2001).

The notion of best practice has been embraced throughout all sectors of business, commerce, and industry, particularly in the context of benchmarking for example (Camp, 1995). Much has been written in this area of the benefits of benchmarking best practice, but this has the potential to stifle imagination, creativity, and vision in the desire to conform to perceived accepted norms of best practice (Pemberton, Stonehouse & Yarrow, 2001).

There are clear parallels with CoPs from the point of view of a grouping of like-minded individuals with common interests who may lose sight of the 'bigger picture' by adopting a blinkered and narrow view of life (Wenger et al., 2002). In essence, this approach is then perceived as best practice in the eyes of the CoP, whereupon creativity can be compromised and lead to 'bad' practice, possibly at variance with other organisational processes and procedures.

This may be a problem in the short term, but outside influences or a recognition by CoP members when relating their experiences to the organisation are usually enough to ensure that a re-alignment of the CoP's views and practices takes place. In the authors' own CoP, the openness of the group and the willingness of its members to participate fully in the organisation's activities outside the CoP have ensured that this has not happened—the benefits of CoP membership has been seen in terms of new collaborations with individuals within and outside the CoP, enhanced publication possibilities, and a better understanding of inter-subject research within the organisation.

In extreme cases, where CoP practice deviates so greatly from organisational practice, individuals can become unsettled and begin to question their role in the CoP, or in the worst-case scenario, question their value to the organisation. There are clearly strong links to the personalities within the group, internal leadership, and the longevity of the CoP in terms of the direction and development of shared 'good' or 'bad' practice.

CONCLUSION

To established supporters and advocates of CoPs, many of the issues discussed in this article will make uncomfortable reading, but to others, will strike accord. This is because CoPs exist in many forms, and while there is commonality of issues, the nature of some CoPs will make some of the themes discussed here highly relevant.

Like any grouping of individuals, be they physical or virtual, there is a responsibility on the CoP members to ensure that their input, and the subsequent consensus that emerges, supports not only their own ends, but complements the organisation's operations and strategy. On this score, and as corroborated by the contributors to this book, CoPs have made, and will continue to make, a valuable contribution to exchange, creation, and diffusion of knowledge within public- and private-sector organisations.

That said, any effective mechanism that allows individuals and groupings to prosper in terms of job capabilities and performance should always be reviewed and questioned from time to time. CoPs are no different in this respect. Thus, while this encyclopaedia generally celebrates the undisputable benefits and applications of CoPs in modern society, keeping sight of some of the issues raised in this article may help to ensure that the pitfalls are avoided, thereby guaranteeing that CoPs have a beneficial and positive effect on knowledge sharing and creation within all manner of organisations.

REFERENCES

Ardichvilli, A., Page, V., & Wentling, T. (2003). Motivation and barriers to participation in virtual knowledge-sharing communities of practice. *Journal of Knowledge Management*, 7(1), 64-77.

Bhatt, G.D. (2001). Knowledge management in organisations: Examining the interaction between technologies, techniques and the people. *Journal of Knowledge Management*, 5(1), 68-75.

Brown, J.S., & Duguid, P. (1998). Organizing knowledge. *California Management Review*, 40(3), 90-111.

Camp, R.C. (1995). *Business process benchmarking: Finding and implementing best practices*. Milwaukee, WI: ASQC Quality Press.

Lave, J., & Wenger, E. (1991). *Situated learning: Legitimate peripheral participation*. Cambridge, MA: Cambridge University Press.

Madridakis, S., & Wheelwright, S.C. (1989). *Forecasting methods for management*. New York: John Wiley & Sons.

McDermott, R. (2002). Measuring the impact of communities: How to draw meaning from measures of communities of practice. *Knowledge Management Review*, 5(2), 26-29.

Moore, J., & Sonsino, S. (2003). *Leadership unplugged: The new renaissance of value propositions*. London: Palgrave Macmillan.

Nicolini, D., Sher, M., Childerstone, S., & Gorli, M. (2003, June). In search of the 'structure that reflects': Promoting organizational reflection in a UK Health Authority. *Proceedings of the Organizational Learning and Knowledge 5th International Conference*, Lancaster University, UK.

O' Donnell, D., Porter, G., McGuire, D., Garavan, T., Heffernan, M., & Cleary, P. (2003). Creating intellectual capital: A Habermasian community of practice (CoP) introduction. *Journal of European Industrial Training*, 27(2-4), 80-87.

Pemberton, J., Stonehouse, G., & Yarrow, D. (2001). Benchmarking and the role of organizational learning in developing competitive advantage. *Knowledge and Process Management, 8*(2), 123-135.

Wenger, E. (1998). Communities of practice. Learning as a social system. *Systems Thinker*. Retrieved June 25, 2004, from *http://www.co-i-l.com/coil/knowledge-garden/cop/lss.shtml*

Wenger, E.C., McDermott, R., & Snyder, W.M. (2002). *Cultivating communities of practice.* Boston: Harvard Business School Press.

Wenger, E.C., & Snyder, W.M. (2000). Communities of practice: The organisational frontier. *Harvard Business Review,* (January/February), 139-145.

KEY TERMS

Benchmarking: An improvement process in which a company measures its performance against that of 'best in class' companies, determines how those companies achieved their performance levels, and uses the information to improve its own performance.

Best Practice: A superior method or innovative practice that contributes to the improved performance of an organisation, usually recognized as 'best' by other peer organisations.

Chatham House Rule: A code introduced in 1927 at Chatham House, London, and now is internationally recognised as a guarantee of anonymity for contributors within a meeting in order to promote free 'off-the-record' discussion.

Expert Jury: A method based of forecasting and planning using the expertise of a panel of the firm's executives.

Habermasian Logic: The learning process of the human species takes place through the accumulation of both technical and moral-practical knowledge within social interactions yielding a 'logic of growing insight'.

Intellectual Property: A general term for intangible property rights that are a result of intellectual effort.

Power-Distance: The extent to which less powerful members of institutions and organisations accept that power is distributed unequally.

Proximal Development: The distance between the actual developmental level as determined by independent problem solving and the level of potential development as determined through problem solving under guidance or in collaboration with more capable peers.

Boundaries in Communities

José Córdoba
University of Hull, UK

INTRODUCTION AND BACKGROUND

This article suggests a way of complementing the notion of boundary objects from communities of practice to enable learning: That of extending the notion of boundary objects to account also for boundary people. There are some people whose participation in a community could provide benefits for them and the community. Although it has been suggested that in a community of practice there are different types of membership, little is mentioned about how learning could be fostered by developing *inclusive* membership. This could be a way of bringing relevant experience to the attention of a community.

BOUNDARIES AND MEMBERSHIP

In a community of practice, there are two main elements that constitute learning: experience and competence (Wenger, 1998). A community can be seen as a recurrent encounter between people who share interests with this permanency generating their competence, participation, and own identity. The community feeds itself from the experience of its members, including newcomers.

According to Wenger, McDermott, and Snyder (2002), cultivation of communities of practice requires establishing first a domain of competence, something that members care about. Nurturing this requires organizing activities of a community and roles for participants. It also requires establishing ways of dealing with contingencies (i.e., conflict). The result of this will be generating knowledge, which can be explicit (i.e., documents).

Although in the theory of communities of practice, it is acknowledged that communities have boundaries that define who is in and who is not part of it, there is very little guidance on how communities can deal with the resulting exclusion of individuals. It is assumed that members share interests that lead them to become part of a community and to define their engagement. Individual motivation is a condition for the formation of communities of practice, and the theory's main thrust is to provide guidance for the adequate development (or nurturing) of communities. An issue that remains unexplored is how to facilitate inclusion of those whose interest is (or might not be) developed to belong to a community, but who could greatly benefit from participating in it as well as benefiting the community.

CRITIQUE ON BOUNDARIES: BRINGING PEOPLE TO A COMMUNITY

The notion of what constitutes a boundary has been explored in management science, more particularly, in the literature of critical systems thinking (Midgley, 2000; Ulrich, 1983). A boundary is a social construction by which knowledge and people to be considered relevant in a situation are defined (Churchman, 1968). This notion presents a similarity with that of a boundary object of the theory of communities of practice (mentioned elsewhere in this encyclopedia). A boundary object helps people from a community to communicate with the rest of the world and to coordinate activities (Wenger, 1998).

This concept of an object could be extended to account for people who may be excluded from participation in a community of practice. Therefore, the idea of boundary people can be put forward. Midgley (1992) suggests that in any situation, reflection on people *and* issues which become marginalized from any decision could help those deciding to foster inclusion and participation. In a community, this type of reflection could also help members define their identity by acknowledging who they are and what they do, or who they could *become*. Often, Wenger (1998) argues, defining what and who constitutes a community helps individuals to define their own identity.

Non-participation and marginality are two issues that are accounted for in the theory of communities of practice. The first refers to a non-intensive engagement (i.e., when people are new to a community). The second refers to situations where there are barriers for people to become full members of a community. This situation may be problematic for the development of a community. In this aspect, reflection on boundaries and marginalization of both objects and people could help potential participants and community members identify issues that need to be addressed to facilitate inclusion and learning.

Midgley (1992) suggests that the definition of a boundary brings value judgments about what and who is to be included and marginalized from decisions. These value judgments could be subject to debate to enable a community to debate on possibilities of including some peripheral and marginalized members and their experience as a core element of their practice. The following questions could help a community to reflect on issues of inclusion and marginalization:

- Who is to be included within this community?
- What can constitute knowledge within the community?
- What and whose value judgments are supporting the above definitions?
- What and who is to be marginalized from activities? Why?
- From the above questions, what barriers for inclusion and learning could be identified?

CONCLUSION

In this article, a perspective to facilitate inclusion in a community of practice has been developed. This perspective takes the notion of a boundary object and extends it to account for the possible existence of people in the margins of boundaries whose participation in a community of practice could bring benefits for learning. In the dynamics of a community, it is inevitable some people and their knowledge could be marginalized. Reflecting on the implications of maintaining their marginality or avoiding it by including them into community's activities could be seen positively as a way of fostering learning and competency.

REFERENCES

Churchman, C. W. (1968). *The systems approach.* New York: Delacorte Press.

Midgley, G. (1992). The sacred and profane in critical systems thinking. *Systems Practice, 5,* 5-16.

Midgley, G. (2000). *Systemic intervention: Philosophy, methodology and practice.* New York: Kluwer Academic/Plenum.

Ulrich, W. (1983). *Critical heuristics of social planning: A new approach to practical philosophy.* Berne: Haupt.

Wenger, E. (1998). *Communities of practice: Learning, meaning and identity.* Cambridge: Cambridge University Press.

Wenger, E., McDermott, R., & Snyder, W. M. (2002). *A guide to managing knowledge: Cultivating communities of practice.* Boston: Harvard Business School Press.

KEY TERMS

Boundary: A social construction that defines knowledge and people to be included (or benefited) from a decision.

Marginality: A situation resulting from the definition of a boundary over what is to be included in the mainstream of activities of a community of practice. Marginality can be defined in terms of non-participation in a community.

Peripherality: A form of participation in a community of practice by which individuals take less intense membership. This can be seen as a form of becoming a newcoming member of the community. This situation should not be seen as free of conflicts. Instead, it can be an opportunity for a community to develop their practice.

Value Judgement: An assessment that could reveal the values that support a community and its members, who or what is to be included and who or what is to be marginalized.

Building a Dynamic Model of Community Knowledge Sharing

Geoffrey A. Walker
University of Northumbria, UK

INTRODUCTION

Many case studies have been undertaken about how informal, sponsored, and supported communities of practice operate within private and public sector organizations. To date, however, no examination has been made of how informal communities of practice operate within the third sector, the sector of community, and voluntary organizations. The third sector has a long history of using community space, in various forms, either physical or notional, to engage individuals in discourse and informal learning. The rise of the network society has added value to this process by allowing active individuals to personalize networks through the use of technologies which enhance communication. The third sector is now demonstrating that individuals and groups are seeking to create open access knowledge-sharing spaces which attempt to combine face-to-face networks with computer-mediated communications to support informal learning between community development practitioners.

This article examines the role of Sunderland Community Development Network in the creation of informal communities of practice. It pays particular attention to three key areas:

1. **Community space:** How core, active, peripheral, and transactional community spaces within third sector partnerships create an ebb and flow of informal communities of practice.
2. **Personalized networking:** How issue-based activity, inside and outside communities, can lead to the rapid appearance and disappearance of informal communities of practice.
3. **Knowledge-sharing space:** How core members of a third sector organization can create a dynamic model of roles within informal communities of practice capable of impacting upon

processes of governance beyond the organization.

BACKGROUND: SUNDERLAND AND THE COMMUNITY DEVELOPMENT NETWORK

Sunderland is a new city in the North East of England with a population of 300,000. Toward the end of the last century, it suffered adversely from the post-industrialization process. Both shipbuilding (ships had been built on the River Wear for over 1,500 years) and coal mining (Monkwearmouth Colliery was one of the largest deep mines in Europe) went into terminal decline. The dawn of the new millennium, however, has witnessed an economic, social, and cultural renaissance in the city. Sunderland's Nissan car plant is now the largest in the UK with 12,000 employees. Sunderland University has a new riverside campus adjacent to a thriving marina and an emerging shellfish industry. Sunderland Football Club has a new arena (built on the former site of Monkwearmouth Colliery), boldly titled "The Stadium of Light", and there is an award-winning museum and winter gardens in the heart of the city center.

Sunderland Community Development Network (SCDN) forms the neighborhood-based component of the city's renaissance and is open to community groups, community networks, voluntary sector organizations, volunteers and residents who are, or want to be, active in their communities.

The aims of SCDN are to link together neighborhood renewal (Social Exclusion Unit 2000) areas of the city in communities of practice; maximize the power of communities to shape the future of the city; provide a decision-making and discussion forum for communities; provide effective, meaningful, and coordinated representation at all levels of the city council's Local Strategic Partnership (LSP); and provide a structure of accountability for community representation and the communication of information. The concept of partnership working in this manner was first suggested in a document produced by the Neighbourhood Renewal Unit (2001). In summary, SCDN aims to capture, store, and transfer the wide range of knowledge contained within Sunderland's community-based organizations and make this knowledge accessible to other sectors.

THE EMERGENCE OF SCDN

SCDN has been emerging as a meta-network since September 2001 under the innovatory leadership of VOICES. VOICES was originally established as Sunderland Voluntary Sector Partnership (VSP) in 1994, and since September 1994, has played an active role on the City of Sunderland Partnership (CoSP). Three community development workers were appointed in May 1998 to develop networks in areas where there was no existing infrastructure and to build the community and voluntary sector in the city. In 2000, the VSP gained charitable company status in the name of Sunderland Voluntary and Community Sector Partnership. The official launch of the new company was held in October 2000 to coincide with the signing of the local compact between the CoSP and the voluntary and community sector. The name VOICES was adopted to reflect the role of the VSP in ensuring local people's needs, views, and opinions are integral to the decision-making processes of policy makers at local, regional, and national levels.

The core group of VOICES has many years of experience of community development activity, stretching back to the 1970s, long before the introduction of the Internet and other network technologies. Some members of the core group have taken readily to e-mail and other network technologies while others struggle with it. All, however, are very skilled face-to-face networkers and demonstrate a high level of trust in the communities they support.

The meta-network provides a range of knowledge-sharing platforms through which dialogue can flow, both formally and informally. These platforms include formal strategy meetings, informal lunches, events and residential conferences, and seminars as well as sharing documents and discussion via e-mail and the Internet. Key informants constantly refer to the informal dialogue, which takes place before, after, and around meetings. The informal sharing of knowledge is seen to lie at the hub of the collective learning and knowledge-sharing process, which takes place within the meta-network. Access to knowledge is sought in a seamless way by combining face-to-face informality with document sharing and the use of e-mail, firmly grounded in the needs of communities. Knowledge is also accessed via the mobile telephone and text messaging which adds value to the use of other technologies. A high level of trust is placed upon individuals with key skills and competencies within the network, as containers and carriers of knowledge on community development.

SCDN has been debating, for more than 2 years, the importance of legitimizing peripheral participation (Lave & Wenger, 1997) within the network and the LSP. Legitimate peripheral participation provides a way to speak about the relations between newbies, veterans, activities, identities (Wenger, 1998), communities of knowledge (Brown & Duguid, 1991), and practice. It is concerned with the process by which newcomers become part of a community of practice (Wenger et al., 2002). As a result of this debate, a model has been devised which aims to provide a means of legitimizing peripheral participation within it.

In this model, members of the network are divided into a tripartite framework of community development responsibilities within each of the 12 themed and 6 area-based neighborhood renewal groups of the LSP, as follows:

1. **Capacity-builder:** With previous partnership-working experience, well-developed informal and formal meeting skills and knowledge of decision-making structures.
2. **Mentor:** With experience of representation or other partnership working.
3. **Learner:** With experience of meetings at a neighborhood level but no previous representation experience.

Clearly, individuals tend to exhibit all of these roles to a greater or lesser degree. In terms of the meta-network, however, these three roles form a dynamic learning framework for the community participants within the 18 working groups of the LSP. It is clear that this tripartite framework creates a dynamic model for developing informal communities of practice as the three roles constantly combine and disperse leaving critical masses of knowledge which can be accessed in a number of ways through:

1. The manipulation of the spaces where communities are formed.
2. The establishment of personalized networks.
3. The creation of knowledge-sharing spaces.

COMMUNITY SPACE

The traditional gathering place of community activists for centuries has been the village hall, community center, or their physical equivalent: the place of democratic engagement and dialogue on issues affecting the community. Community activists often put forward the view that it is possible to create an equitable community space, both mental and physical, where the views of individuals and groups can be freely exchanged in a form of true participatory democracy. Such a belief can be seen as an extension of the concept of agora where the creation of a level playing field, by definition, leads to engagement in the free expression of ideas, opinions, and innovation.

Does such a shared mental and physical community space exist, however, when the barriers to effective use of place, space, and cyberspace are manifold?

Several commentators have grappled with the concept of community space. They have revealed a complexity, which goes far beyond that manifest in village halls and community centers. In order to understand this complexity, the following concepts are now examined in turn: community space, liminal space, reproduction of space, defensible space, the space of flows, and the semiotics of global space.

Wenger (1998) talks of community space in which groups operate. The facilitators, innovators, and leaders occupy the core space. Active, interested individuals inhabit the active space. Interested individuals who are not necessarily active occupy peripheral space, and the transactional space is where partnerships are forged. This paradigm suggests the existence of four distinct community spaces. It does not, however, explain how groups apparently move with ease from one space to another or alternatively occupy several spaces simultaneously. For example, individuals may well occupy core space in one group, active space in

another, and so on. SCDN's tripartite framework means that the dynamic roles cut across the boundaries of community space.

By introducing the concept of liminal space, we can envisage how individuals and groups journey between the spaces outlined above. Liminal space, an anthropological term, refers to the limbo which an individual inhabits while performing a rite of passage between one space and another. A physical example of this space is the Aboriginal "Walkabout" where teenage aborigines must spend time alone surviving in the outback prior to acceptance as an adult member of the group. A comparison can be made here with the concept of a lurker in an electronic environment or a learner within SCDN. Lurking in an electronic environment would be considered a form of situated learning by Lave and Wenger (1997) and, as such, a legitimate form of peripheral participation. Adding the concept of liminal space to the paradigm creates a new dynamic, which does, at least, appear to go some way toward illustrating how individuals and groups occupy several spaces simultaneously.

Puttnam (2000) refers to the bridging and bonding of social capital within communities. Social capital is created either by forming bridges between communities or bonding communities where they share common characteristics. It is, therefore, legitimate to suggest that social capital is formed in liminal space.

Lefebvre's (1991) discourse on the relationship between mental and physical space highlights not only the production of community space but also the reproduction of this space:

The problematic of space, which subsumes the problems of the urban sphere ... and of everyday life, has displaced the problematic of industrialisation. It has not, however, destroyed that earlier set of problems: the social relationships that obtained previously still obtain; the new problem is, precisely, the problem of their reproduction. (p. 89)

In physical terms, former British Prime Minister, Winston Churchill, went some way to expressing the relationship when he said:

There is no doubt whatever about the influence of architecture and structure upon human character. We shape our buildings and afterwards our buildings shape us.

SCDN, like many other networks, has experimented extensively with variations in physical space in order to facilitate knowledge sharing. However, are there human and psychological constructs which influence individual and group behavior in community spaces?

Building upon the idea of human structures, Goffman (1959) derived the concept of defensible space, the cognitive space between individuals where they form opinions and assumptions of others. In physical space, we can visibly assess people's changing opinions through human interaction, which is supported by body language. In cyberspace, however, where body language can play a different part, defensible space becomes the space of legitimate peripheral participation. Discourse and dialogue in cyberspace can often be viewed as significantly more reflective than that which takes place in physical space. The roles of capacity-builder, mentor, and learner assist discourse and dialogue through enabling conversations on who is learning what, from whom, and the impact of this upon the network.

Castells (1989) argues that access to flows of information and resources is the key to participation in the networked society. He refers to a subtle interaction between physically colocated resources and virtual information-based resources. He calls this space the space of flows. He suggests a further dimension to community space. The space of flows is the personal space, which individuals manipulate, in and around the groups they populate. They create this space by constructing complex problem-solving, personal, social networks. These networks manipulate information and resources on a personal level through a complex web of digital technologies and face-to-face interactions.

Due to the constant and rapid evolution of community space within networked society, SCDN has attempted to create dynamic issue and area-based thematic communities of practice which accommodate the informality of the relationships created. Each member of an issue or area-based group has to relate to other members of the groups in terms of their ability to act as capacity-builder, mentor, and learner. This interaction leads to semiotic relationships between communities of practice with high levels of synergy capable of rapid transformation and dissolution around a particular theme or issue. Such interaction also relies on high levels of personal interaction within networks and meta-networks.

PERSONALIZED NETWORKING

Human networks are hugely complex phenomena. We are only just beginning to understand the implications of understanding networks:

Today we increasingly recognize that nothing happens in isolation. Most events and phenomena are connected, caused by, and interacting with a huge number of other pieces of a complex universal puzzle. We have come to see that we live in a small world, where everything is linked to everything else. We are witnessing a revolution in the making as scientists from all different disciplines discover that complexity has a strict architecture. We have come to grasp the importance of networks. (Barabasi, 2003, p. 7)

In this small world, individuals and groups are as likely to reach out around the globe for knowledge as they are to visit their next door neighbor in search of information (Watts, 2003).

Given this complexity, how do we provide a platform for a networked community?

While face-to-face contact is paramount within SCDN and often cannot be replicated in electronic systems, the constraint of time and space on active individuals has led to the network accepting that the community is not necessarily located in a fixed space. The idea of community being with you wherever you are is a welcome and reassuring idea associated with the trust, strengths, and connections needed for effective networking. As a result, SCDN has begun to use ICT to add value to human systems in a personalized manner, often referred to as personalized networking.

Research into personalized networking (Wellman, 2001a, b) has shown that knowledge transfer and the idea of communities of practice (groups of people that create, share, and exchange knowledge) is relative to situated learning (how useful the knowledge is within a particular situation or toward a particular end). This requires multi-faceted means of creating dialogue where meaning flows through individuals and groups.

Initially, however, the quality of the dialogue may not be as important as the process of democratic engagement, as it is often about allowing people to explore new ideas and discarding those that are not fit for purpose. Evidence also suggests that network identity can also emerge through consensual agreement on what is community-based knowledge in the emerging dialogue.

The division of responsibility into capacity-builder, mentor, and learner creates dynamic spaces within personalized networks where knowledge can be shared informally.

Personalized networks appear to vary not only with regard to the skills and experience of capacity-builders, mentors, and learners but also in relation to where the networker is located within community space. It would be relatively easy to map personalized networks if community spaces were mutually exclusive and static. However, such spaces are mutually reliant and dynamic; as such, they are capable of potential highly complex topologies of personalized networks.

Social network analysis tools prove difficult to deploy in such a complex context. A high reliance on subjective and qualitative analysis is needed to understand the complexity of personalized networks within meta-networks. SCDN has attempted to create matrices of cross-cutting usage of technologies such as e-mail, Web, and text-messaging within the tripartite framework. This has proved difficult to progress in a collaborative computer-mediated environment, and progress has been limited to face-to-face workshops.

FUTURE: KNOWLEDGE-SHARING SPACE

In partnership with Sunderland City Council's E-government Unit, SCDN is developing an appropriate architecture for a community technology, into which its collective knowledge can be filtered and codified (http://www.sunderlandcommunitynetwork.org.uk). Data, information, and knowledge is drawn from a wide range of cross-cutting sources emanating from core groups, activists, peripheral groups, and transactional partners at varying levels. Taxonomies and topologies are created which are dynamic and organic, developed through user-defined language in detailed consultations with network members. For example, knowledge is coded as theme-based, issue related, or network representation related and, in turn, validated through dynamic use by members. All types of knowledge are upheld as equally valid. As more and more people search and use the network's knowledge, the more common definition naturally surfaces according to the emerging dialogue. The key is to build intelligence into analysis of the use of language in the dialogue that emerges.

From the outset of the project, it was clear that a cultural shift was required to get beyond data and information and move toward knowledge sharing among network members. Such cultural problems are widely recognized by academics and practitioners, as most individuals and groups within organizations are comfortable dealing with hard facts and figures rather than soft outcomes as a starting point. This marks the first phase of development of cultural shift and is only useful in the network's thinking if it is accompanied by a roadmap toward appropriate and effective management of knowledge in the long term.

SCDN's first stage of developing a knowledge base consists of compiling information on the actors within the network. This is the point at which the architecture of shared learning space is structured through recognizing the interaction of actors with the emerging architecture. Such structuration (the structuring of social relations across time and space), however, must contain the flux, which allows actors within the network to customize their own personalized networking structures. Understanding the degree of flexibility that actors need to interpret their personalized networks is paramount.

The second stage is referred to as profiling the key skills and experiences which members bring to the network. Profiling is a mix of knowledge supplied by professional community development workers and network members themselves about the network itself. This extends to a need to determine performance according to both external and internal transactional criteria with other partners. The profiling stage begins the process of monitoring and evaluating the level of participation and reification within the dialogue that populates the shared learning space.

The final stage is tooling of network members to meet the increasing demands of personalized networking. This is the stage at which members' skill gaps are identified and filled. As noted previously, SCDN has divided participants in the shared learning space into three key roles: capacity-builders, mentors, and learners. Each actor-role compliments the other around a particular theme such as, health, diversity, and community safety.

Although paper-based forms of communication and telephone calls may not be easily codified, the idea of a meta-network means that these conversations are likely to become embedded within an online application provided that the dialogue is ongoing, regular, and frequent. This is dependent on the overall utility of the knowledge base that can only be determined through the level of usage. Part of the work of Sunderland City Council's E-government Unit is to allow 60 members of the network access to a portable computer and to the Sunderland E-government Web site. In the short term, these people will be able to use the community-based Web portal to see what they might expect in an online environment run by and for themselves. If this enthusiasm is cascaded throughout the network and access is widespread, most members could be accessing their knowledge base most days to add value to their personalized networking.

CONCLUSION

This article has examined SCDN's role in the creation of informal communities of practice. In particular, it has analyzed the part played by the tripartite framework of capacity-builder, learner, and mentor in progressing flexible and dynamic communities of practitioners. The tripartite framework has been examined within the context of community space, personalized networking, and knowledge-sharing spaces. Outcomes to date, as to the robustness of this dynamic model of community knowledge-sharing, are positive; however, the model has not yet reached a significant level of maturity, and the possibility of long-term success remains uncertain.

While the article offers a positive and transferable model of the creation of informal communities of practice, the development of the model has been subject to a number of barriers:

1. The time spent gaining agreement on the model (almost 18 months) in the meta-network. For people who are unfamiliar with informal, dynamic, and flexible working relationships, the

model appears simultaneously complex and radical.

2. Agreeing knowledge-sharing protocols with transactional partners where the shared vision did not appear to be as advanced as that of SCDN. The LSP has a wide range of partners, all from different sectors of the economy and all appear to be at different levels of skills and experience in partnership working.

3. The protracted discussions on the creation of a critical mass for the development of the model within the meta-network was hampered by turnover in key personnel. It was recognized from the outset that champions of the model would play a key role in the creation of this critical mass. The skills acquired by the champions in the dynamic working environment led to their rapid progression to roles within other networks and organizations with a loss of skills and experience to SCDN.

4. The lack of an education program on the tripartite framework for newbies which makes significant connections with veterans. This has developed on an ad hoc basis. There is now a recognition of the need for a strategy which connects skills and knowledge which is evident in the work with Sunderland City Council's E-government Unit.

REFERENCES

Barabasi, A. L. (2002). *Linked: The new science of networks.* Cambridge, MA: Perseus.

Brown, J. S., & Duguid, P. (1991). Organisational learning and communities of practice: Towards a unified view of working, learning and innovation. *Organization Science, 2,* 40-57.

Castells, M. (1989). *The informational city: Information technology, economic restructuring and the urban-regional process.* Oxford: Blackwell.

Goffman, E. (1959). *The presentation of self in everyday life.* London: Penguin.

Lave, J., & Wenger, E. (1997). *Situated learning: Legitimate peripheral participation.* Cambridge: Cambridge University Press.

Lefebvre, H. (1991). *The production of space.* Oxford: Blackwell.

Neighbourhood Renewal Unit. (2001). *Local strategic partnership guidance.* London: HMSO.

Putnam, R. D. (2000). *Bowling alone: The collapse and revival of American community.* New York: Simon & Schuster.

Social Exclusion Unit. (2000). *A national strategy for neighbourhood renewal.* London: HMSO.

Watts, D. J. (2003). *Six degrees: The science of a connected age.* London: Norton.

Wellman, B. (2001a). Physical place and cyberplace: The rise of personalised networking. *International Journal of Urban and Regional Research, 25.*

Wellman, B. (2001b). Computer networks as social networks. *Science, 293,* 2031-2034.

Wenger, E. (1998). *Communities of practice.* Cambridge: University Press.

Wenger, E., et al. (2002). *Cultivating communities of practice.* Boston: Harvard Business School Press.

KEY TERMS

Community: Any group of people with a shared set of values or beliefs. The term is inclusive of geographic communities, communities of interest, communities of identity, and communities of practice.

Community Development: Is concerned with the relationship between social and economic development, building a capacity for local cooperation, self-help, and the use of expertise and methods drawn from outside the local community.

Community Space: A space, real, virtual, or a combination of both where a sense of community is created.

Knowledge: The information, customs, and techniques contained within an organization about customers, products, and services which is contained within people's minds or filed in analog or digital format.

Knowledge Management: A systematic attempt to use knowledge within an organization to improve overall performance.

Networking: Activities that enable individuals, groups, or organizations to interact with each other in social formations which enhance communication and create new opportunities.

Personalized Networking: A type of networking where a range of technologies are used to add value to a human network.

A Classification of Communities of Practice

Norm Archer
McMaster University, Canada

INTRODUCTION

Communities of practice have been in existence since the days when individual craftsmen got together to share ideas and issues. Eventually, these developed into craft guilds and finally into professional associations. But more specifically, focused communities of practice have recently begun to attract a great deal of attention in the business community because they provide a way for strategically growing and managing knowledge as an asset (Grant, 1996; Nonaka & Takeuchi, 1995; Powell, 1998). The increasing complexity in products, services, and processes requires more specialization and collaboration between workers. However, orchestrating the involvement of disparate groups that work on complex projects requires finding a balance between differentiation, when teams work separately, and integration, when groups meet to exchange knowledge. For example, development projects usually benefit when expertise is drawn from diverse sources, including potential users, where the interests, skills, and formal and tacit knowledge of the different groups can be drawn together by skillful project managers (Garrety, Robertson & Badham, 2004). By responding to new economic pressures for rapid transformation, communities of practice can help improve knowledge exchange in critical areas, so organizations can maintain or improve their competitive positions.

The growth of interest in communities of practice has resulted in their spread into several classifications of modern organizations, all of which must share knowledge and learning to thrive. How effectively communities of practice perform in these different environments is of great interest, and, in order to study them in detail, we suggest classifying them according to the structure of the organizations they serve. We have been able to identity four such classifications: internal communities of practice, communities of practice in network organizations, formal networks of practice, and self-organizing networks of practice. Among these four classifications are charac-teristics of particular interest, especially when successful practices exhibited in one classification can be replicated in others. This article outlines the characteristics of each classification, explores their differences and similarities, and summarizes the findings from a review of the literature. The objective of this article is to encourage the migration of successful ideas for knowledge transfer and learning among the different classifications.

BACKGROUND

As the realization grows that knowledge is a critical business resource with a pivotal role in the marketplace, knowledge management, transfer and learning are attracting a great deal of attention in today's organizations (Kraatz, 1998; Nonaka & Takeuchi, 1995; Nooteboom, 2000; Norman, 2002; Parise & Henderson, 2001; Powell, 1998). Knowledge management is related to the wider field of management in the context of overlapping and synergistic relationships in such activities as learning and innovation, benchmarking and best practice, strategy, culture, and performance measurement (Martin, 2000). While knowledge can exist in both tacit and explicit forms, the embodied expertise that exists in the tacit form may be the most valuable, especially if it is difficult for competitors to replicate. However, tacit knowledge is often difficult, if not impossible, to transform into written form, often making it necessary to transmit to others in the form of stories, coaching, or apprenticeship (Lam, 1997; Leonard & Sensiper, 1998; Nonaka & Takeuchi, 1995). Explicit knowledge is knowledge that exists in documents, software, hardware, and other instruments (Zack, 1999). It is more easily transmitted to others, but, for the same reason, it is more difficult to safeguard from unauthorized use.

Certain knowledge management problems arise out of the difficulty of current management paradigms to manage intangible/tacit knowledge, as compared to tangible/explicit knowledge. The latter may be supported by extended information resource management approaches, but the former has overlapping and synergistic relationships with such personalized activities as learning, innovation (Bogenrieder & Nooteboom, 2004), and benchmarking and best practices (Bardach, 2003). Such activities need not be confined within an organization, and they can cross organizational, international, and cultural boundaries with attendant transmission of knowledge of both types (Inkpen & Dinur, 1998).

Communities of practice are an organized way of implementing knowledge management, learning, and transfer. With appropriate support, motivation, and coordination, these communities can create both codification and personalization channels to distribute knowledge and support learning within and among organizations, and among individuals both internal and external to any particular organization. However, the value attributed to knowledge that gives an organization a competitive advantage will inhibit its sharing with other organizations, unless there are formal agreements relating to how and what knowledge and information is to be shared. There are a variety of motivations for professional participation in communities of practice, including tangible returns, intangible returns, and community interaction (Wasko & Faraj, 2000). However, harnessing technological innovation through communities of practice is a major organizational application (Persaud, Kumar & Kumar, 2001), potentially leading to competitive advantage (Liedtka, 1999). Communities of practice have been used widely for brokering a variety of knowledge within organizations (Burnett, Brookes-Rooney & Keogh, 2002; Gongla & Rizzuto, 2001; Saint-Onge & Wallace, 2003; Wenger, McDermott & Snyder, 2002).

Communities of practice need to have a defined objective and scope in order to succeed. Wenger et al. (2002) indicate the three most important elements to be domain, community, and practice. All these elements must be developed together in a carefully balanced manner. All grow dynamically and interact in various ways. The key is to extract the maximum benefit for the community membership, so all members are motivated to contribute and participate fully. How communities of practice are organized, evolve, managed, and how they impact the individuals included in the communities depend upon the nature of the community of practice and upon the nature of the organizations it touches. Every community of prac-

tice must have dynamic and committed leadership and objectives that are seen as important to its members. These elements may exist during the start-up of a successful community of practice, but if any of them fades or disappears, the initiative is likely to fail.

CLASSIFICATIONS OF COMMUNITIES OF PRACTICE

Communities of practice can exist in four classifications that we have been able to identify: (1) entirely within individual organizations, (2) spanning organizations that are linked through mergers, acquisitions, or by formal business partnerships (network organizations), (3) formal networks that span organizations but are not part of other formal relationships, and (4) self-organizing networks of individuals with ad hoc relationships and no formal ties. We explore each in more detail below.

Internal Communities of Practice

Internal communities of practice add value to organizations in a number of ways (Wenger & Snyder, 2000) such as: (1) helping to drive strategy, (2) starting new lines of business, (3) solving problems quickly, (4) transferring best practices, (5) developing professional skills, and (6) helping companies to recruit and retain talent. They complement activities of other organizational networks, including formal work groups, project teams, and informal networks (Wenger & Snyder, 2000). Communities of practice are used extensively in some larger organizations. For example, in the Global Services organization of IBM, there are over 60 communities of practice (Gongla & Rizzuto, 2001) with a total of 20,000 members in most of the countries it serves. These communities handle explicit knowledge or intellectual capital (gathering, evaluating, structuring, and disseminating knowledge shared among community peers), adopt a set of common roles for managing knowledge, and provide opportunities for sharing tacit knowledge among community members.

Organizations that use such communities systematically may have a support team of consultants to provide coaching for community leaders, educational activities to raise awareness and skills, facilitation services, communication with management, and coordination across the initiative (Wenger, 2004). P&G also uses communities of practice extensively (Sakkab, 2002) with 20 in place over a wide range of disciplines. Each is sponsored by an R&D vice president, and their purpose is to promote cross-fertilization and diffusion of expertise. Activities include problem solving through e-mail conferences, knowledge sharing through live seminars and Web sites, engagement of expert practitioners both internally and externally, and communication tools for knowledge diffusion throughout the organization. Communities of practice at P&G play a key role in identification, development and deployment of new research methodology, and problem solving on specific projects.

Communities of Practice in Network Organizations

A network organization is a relationship among independent organizations (Powell, 1990). Such networks have been growing rapidly in number and scope in recent years with the majority of business organizations now belonging to at least one such network (for example, a supply chain is a network organization). Member organizations in a network work in close and continuous cooperation on projects or processes involving partnerships, common products and/or services, and even a common strategy. Reasons for building such networks include faster time to market, ability to concentrate on core competencies, increase in competencies due to networking with business partners, and the need to guarantee availability of resources and materials. In addition, risk and cost mitigation is a motivation for forming network organizations for the purpose of research and development (Cologhirou, Hondroyiannis & Vonortas, 2003). Formal agreements that allow for explicit exchanges of product and service knowledge are traditionally required and are usually an adjunct to the main collaboration agreement among the network members.

In solving problems in today's environment, it is becoming increasingly important to cross boundaries, either within the organization or to unconnected organizations for fresh insights. For a community of practice, an important question involves deciding what other organizations should be connected. Networks of practice make it easy for inter-organizational exchanges to occur, and shared practices provide channels to share knowledge efficiently (Brown & Duguid,

1991). Learning and knowledge exchange through networks focuses on the inter-organizational network as a resource generator to enhance learning. Powell, Koput, and Smith-Doer (1996) suggest that the locus of innovation in an industry that is both complex and expanding with sources of expertise widely dispersed will be found in inter-organizational networks of learning rather than within individual firms.

In a network organization, the sharing of intellectual property, such as inventions, product design knowledge, and the like, are specifically encouraged through the network agreement, aided by knowledge transfer and learning through channels such as communities of practice. An example is the Toyota manufacturing company that maintains very close relations with, and insists on a high level of knowledge exchange among, its network organization partners (primarily suppliers) (Dyer & Nobeoka, 2000). Within such networks, there is always the risk of knowledge leakage (Sampson, 2004) to competitors. But some network organizations, such as the Visa and Mastercard networks, have memberships made up of competing banking organizations for the purpose of sharing operations and related knowledge.

Networks of Practice

Brown and Duguid (1991) refer to extra-organizational communities of practice as networks of practice. A network of practice is an open activity system focused on work practice, and it may exist primarily through electronic communication. It is similar to a community of practice in that it is a social space where individuals working on similar problems help each other and share perspectives about their practice. However, in a network of practice, people working within occupations or having similar interests congregate to engage in knowledge exchange about the problems and issues that are common to their occupational community and shared practice.

Communities of practice can be extended and augmented to span organizational boundaries and provide external sources of innovation even from competitors in some cases (von Hippel, 1988). These can bring together divergent and complementary views that contribute to organizational knowledge and innovation. However, as compared to the previously discussed communities of practice in network organizations, networks of practice are stand-alone and not adjunct to more general agreements and contracts. Moreover, they are likely to focus on business processes, strategies, and management, and less on products and services. Sharing ideas and experiences about strategies, management processes, and procedures are often acceptable in networks of practice, and these are usually covered by blanket agreements that assign intellectual property rights to the network and not to members of individual organizations. We differentiate between two such networks: (1) formal networks of practice and (2) self-organizing networks of practice.

Formal Networks of Practice

A formal network of practice has a membership that is controlled by fees and/or acceptance through some central authority that also assists in organizing, facilitating, and supporting member communications, events, and discussion topics. This is similar to a professional, business, or non-profit association, although these are not as focused as a community of practice and could also be classified as affinity networks (Kaplan, 2002). However, a network of practice has a focus on specific work issues and strategies of immediate importance to the membership, and it may in fact become an adjunct to an affinity network. An example of an affinity network is purchasing managers, members of an association who may form networks of practice where they communicate on a regular basis on strategies, practices, opportunities, and innovations. Other networks of practice are stand-alone such as the Open Group[1], an association of information technology companies that works to develop common software and system interoperability standards to achieve "Boundaryless Information Flow through global interoperability in a secure, reliable and timely manner". Although composed of competitors, this group develops standards and best practices that promote their common goals and certifies products that meet their standards. Similar associations exist for standards development and certification in other industries. Associations for open systems, such as Linux and Unix, also fall in this category, where product innovation is the objective, and legal agreements cover ownership and use of intellectual property.

Self-Organizing Networks of Practice

A self-organizing network of practice is a loosely organized and informal network that has no central management authority or sponsor, membership is voluntary, and there is little explicit commitment. Members may choose to join or leave as they wish. Most such networks operate virtually, so communication strategy is primarily based on knowledge codification. An example is Usenet groups. In a study of such groups, Faraj and Wasko (2001) found that obligation and not trust was a predictor of knowledge contribution, but individuals acquiring knowledge from the network trusted knowledge provided by others. People participate in such networks due to their affiliation with a profession rather than an organization. Results support general findings from communities of practice that individuals do not participate due to a need to socialize but are motivated by a need to engage in working, learning, and innovating (Brown & Duguid, 1991). Another example of such a network was the establishment of an informal network by a group of companies after the 9/11 disasters to communicate with each other in times of crisis (D'Amico, 2002).

FINDINGS

In order to contrast the differences in characteristics among the four community of practice classifications, the literature was reviewed in detail. Findings from the review are summarized in Table 1 with references to published examples and cases in the last row. We found that some benefits and problems are common to all classifications, but some characteristics vary more among particular classifications.

Major differences of note among the classifications were seen in characteristics such as:

- The type of knowledge transferred and the desired objective or outcome (more management and professional skill orientation in formal and self-organizing networks, compared with product and service orientation in internal and network organizations).
- Funding (internally for internal communities of practice; shared in network organizations and

Table 1. Characteristics of communities of practice

| Characteristic | Community of Practice | | | |
	Internal	*Network Organization*	*Formal Network*	*Self-Organizing Network*
Type of Knowledge	Product, service (technical), management skills, processes	Product, service (technical), management skills, processes	Management skills, processes; operational, product knowledge	Management skills, processes; operational, product knowledge
Desired Objective or Outcome	Innovations in products, services, improved management practices	Innovations in products, services, improved management practices	Improved management practices, products, services	Improved products, services, management practices
Funding	Internal	Shared	Shared	Voluntary
Intellectual Property	Internal	Shared by formal agreement	Controlled by the network	Shared by agreement
Management	Internal	Managed jointly as component of organizational agreement	Externally managed	Externally managed
Professional Expertise	Internal	Shared by formal agreement	Shared by agreement	No agreement
Dispute Resolution	Internal management	Legally resolved	Withdrawal	Withdrawal
Potential Knowledge Contribution	Unlimited from internal sources with need to know	Limited by formal agreement	Determined by members; No min. or max. limit	Determined by members; No min. or max. limit
Common Benefits	Developing and sharing formal best practices, learning and sharing tacit and explicit knowledge, benchmarking, innovations in management, operations, and processes			
Potential Gain — Shared Knowledge of:	Innovations in products, services	Innovations in products, services, shared access to IP	Innovations in management practices, innovations in products, services	Innovations in practices, understanding and innovating products, services
Common Problems	Unpredictable payback, initiating and maintaining interest, building and maintaining trust, encouraging steady flow of information and knowledge among participants, divergence of objectives, lack of common participant language (natural and/or professional), ensuring payback to all participants			
Potential Problems	Reorganization may be required to improve knowledge sharing and learning	Limitations of formal agreement	Ensuring knowledge contributions from all members	Unknown value of knowledge communicated; Hard to reach contributors
Remediation of Operational Problems	Attention from moderator or manager(s)	Attention from moderator, manager(s), or legal resort	Attention from moderator	Targeted attention from membership
Some Published Examples	IBM (Gongla & Rizzuto, 2001), AMS (Wenger & Snyder, 2000), P&G (Sakkab, 2002)	Toyota (Dyer & Nobeoka, 2000), Biotech firms (Oliver, 2001) Sematech (Davenport, 1997)	Critical Emergency Ops Link (D'Amico, 2002), ASAE Futures Scan (Mason, 2001)	Usenet groups (Faraj & Wasko, 2001), Democracy Online (Cashel, 2002)

formal networks of practice; voluntary in self-organizing networks).

- Intellectual property (not an issue for internal communities of practice—based on need to know; shared by formal agreement in network organizations; controlled by common agreement in formal networks; and often shared by informal agreement in self-organizing networks).

- Dispute resolution (for internal communities of practice, handled internally; for network organi-

zations, legally resolved due to contractual relationships; resolved by withdrawal in formal networks and self-organizing networks).

- Potential knowledge contribution (unlimited from internal sources with need to know for internal communities of practice; limited by formal agreement in network organizations; unlimited in formal and self-organizing networks).

- Professional expertise (internal for internal communities of practice; shared by formal agreement in network organizations; shared by agreement in formal networks; no formal sharing agreement in self-organizing networks).
- Potential problems include maintaining continuing interest and contributions from the membership in all cases but, specifically, for internal communities of practice, reorganization may be required to improve effectiveness from time to time; limitations of the formal agreement may constrain community of practice contributions in network organizations; the unknown value of knowledge communicated—members may not know each other very well—and difficulty in reaching contributors in self-organizing networks.
- Remediation of operational problems requires attention of the moderator or responsible managers for internal and network organizations; moderator attention in formal networks; targeted attention from the membership in self-organizing networks.

FUTURE TRENDS

There has been significant growth in the number of network organizations in their various forms, due to a variety of influences. These influences include a competitive environment in which business firms, governments, and non-profit organizations take advantage of the expertise of other organizations to provide synergies with their own to design, develop, manufacture, market, and distribute products and services. This is clearly a trend that is likely to continue, and communities of practice to encourage learning and share knowledge within and among firms are an important aspect of the success in such endeavors. Similarly, both formal and self-organizing networks are providing better ways for managers and professionals to share knowledge within, among, and outside organizations. All of these forms of communities of practice will continue to grow in order to encourage the application of knowledge through sharing and collaborative work (Barth, 2004).

CONCLUSION

We began this review with the belief that there would be differences in the ways that knowledge would be shared among the classifications of community of practice we identified. In the end, we found that there were both similarities and differences in their characteristics. There are broad similarities across the four in both common benefits and problems. Differences appear in types of knowledge transferred, funding, how intellectual property issues are handled, how disputes are resolved, potential for knowledge sharing, how professional expertise is shared, the types of problems that may crop up from time to time, and how operational problems are remediated.

Of particular interest is that the emphasis in networks of practice appears to be on the propagation of improved management strategies and practices, although this can also be an objective of some internal communities of practice (Leahy, 2002) and for some network organizations. Even when competitors are in the same community of practice (as is often the case in networks of practice), management practice diffusion does not seem to have the tight constraints on intellectual property associated with product and service innovations, which are usually treated more as competitive advantages. Trust among members does not seem to be an important issue in the case of sharing management strategies and practices, provided that appropriate legal contracts are in place. One can speculate that sharing of management strategies, processes, and procedures among organizations is usually not perceived as a direct threat to the organization sharing it, since such sharing benefits the entire community, and no single organization is likely to achieve a competitive advantage because of it. The same effect is observed when competitors work together to develop best practices and benchmarks (Hinton, Francis & Holloway, 2000).

REFERENCES

Bardach, E. (2003). Creating compendia of "best practice". *Journal of Policy Analysis and Management, 22*(4), 661.

Barth, S. (2004, February). Three thousand communities of practice. *KM World, 13*, 20.

Bogenrieder, I., & Nooteboom, B. (2004). Learning groups: What types are there? A theoretical analysis and an empirical study in a consultancy firm. *Organization Studies, 25*(2), 287.

Brown, J. S., & Duguid, P. (1991). Organizational learning and communities-of-practice: Toward a unified view of working, learning and innovation. *Organization Science, 2*(1), 40-57.

Burnett, S., Brookes-Rooney, A., & Keogh, W. (2002). Brokering knowledge in organizational networks: The SPN approach. *Knowledge and Process Management, 9*(1), 1-11.

Cashel, J. (2002). Interview with Steven Clift. Retrieved May 5, 2005, from *http://www.publicus.net*

Cologhirou, Y., Hondroyiannis, G., & Vonortas, N. S. (2003). The performance of research partnerships. *Managerial and Decision Economics, 24*(2/3), 85-99.

D'Amico, E. (2002). E-business. *Chemical Week, 164*(38), 22-24.

Davenport, T. (1997). Secrets of successful knowledge management. *Knowledge Inc.*

Dyer, J. H., & Nobeoka, K. (2000). Creating and managing a high-performance knowledge-sharing network: The Toyota case. *Strategic Management Journal, 21*(3), 345-367.

Faraj, S., & Wasko, M. M. (2001). *The web of knowledge: An investigation of knowledge exchange in networks of practice*. Cambridge, MA: Open Source Research Community and MIT.

Garrety, K., Robertson, P. L., & Badham, R. (2004). Integrating communities of practice in technology development projects. *International Journal of Project Management, 22*, 351-358.

Gongla, P., & Rizzuto, C. R. (2001). Evolving communities of practice: IBM Global Services experience. *IBM Systems Journal, 40*(4), 842-862.

Grant, R. M. (1996). Toward a knowledge-based theory of the firm. *Strategic Management Journal, 17*, 109-122.

Hinton, M., Francis, G., & Holloway, J. (2000). Best practice benchmarking in the UK. *Benchmarking, 7*(1), 52-61.

Inkpen, A. C., & Dinur, A. (1998). Knowledge management processes and international joint ventures. *Organization Science, 9*(4), 454-468.

Kaplan, S. (2002). *Models for group and organizational collaboration*. Walnut Creek, CA: iCohere.

Kraatz, M. S. (1998). Learning by association? Interorganizational networks and adaptation to environmental change. *Academy of Management Journal, 41*(6), 621.

Lam, A. (1997). Embedded firms, embedded knowledge: Problems of collaboration and knowledge transfer in global cooperative ventures. *Organizational Studies, 18*(6), 973-997.

Leahy, T. (2002). Extracting diamonds in the rough. Retrieved May 5, 2005, from *http://www.businessfinancemag.com*

Leonard, D., & Sensiper, S. (1998). The role of tacit knowledge in group innovation. *California Management Review, 40*(3), 112-132.

Liedtka, J. (1999). Linking competitive advantage with communities of practice. *Journal of Management Inquiry, 8*(1), 5-16.

Martin, B. (2000). Knowledge management within the context of management: An evolving relationship. *Singapore Management Review, 22*(2), 17-36.

Mason, M. (2001). Creating a community. *Association Management, 53*(8), 82-83.

Nonaka, I., & Takeuchi, H. (1995). *The knowledge creating company*. New York: Oxford University Press.

Nooteboom, B. (2000). Institutions and forms of coordination in innovation systems. *Organizational Studies, 21*(5), 915-939.

Norman, P. M. (2002). Protecting knowledge in strategic alliances: Resource and relational characteristics. *Journal of High Technology Management Research, 13*, 177-202.

Oliver, A. L. (2001). Strategic alliances and the learning life-cycle of biotechnology firms. *Organization Studies, 22*(3), 467-497.

Parise, S., & Henderson, J. C. (2001). Knowledge resource exchange in strategic alliances. *IBM Systems Journal, 40*(4), 908-924.

Persaud, A., Kumar, U., & Kumar, V. (2001). Harnessing scientific and technological knowledge for the rapid deployment of global innovations. *Engineering Management Journal, 13*(1), 12-18.

Powell, W. W. (1990). Neither market nor hierarchy: Network forms of organization. *Research in Organizational Behavior, 12*, 295-336.

Powell, W. W. (1998). Learning from collaboration: Knowledge and networks in the biotechnology and pharmaceutical industries. *California Management Review, 40*(3), 228-241.

Powell, W. W., Koput, K. W., & Smith-Doer, L. (1996). Interorganizational collaboration and the locus of innovation: Networks of learning in biotechnology. *Administrative Science Quarterly, 41*(1), 116-145.

Saint-Onge, H., & Wallace, D. (2003). *Leveraging communities of practice for strategic advantage.* Amsterdam: Butterworth-Heineman.

Sakkab, N. Y. (2002). Connect & develop complements research & develop at P&G. *Research Technology Management, 45*(2), 38-45.

Sampson, R. C. (2004). Organizational choice in R&D alliances: Knowledge-based and transaction cost perspectives. *Managerial and Decision Economics, 25*, 421-436.

von Hippel, E. (1988). Cooperation between rivals: The informal trading of technical know-how. In E. von Hippel (Ed.), *The sources of innovation* (pp. 76-92). Oxford: Oxford University Press.

Wasko, M. M., & Faraj, S. (2000). "It is what one does": Why people participate and help others in electronic communities of practice. *Journal of Strategic Information Systems, 9*, 155-173.

Wenger, E. (2004, January-February). Knowledge management as a doughnut: Shaping your knowledge strategy through communities of practice. *Ivey Business Journal Online, 1.*

Wenger, E., McDermott, R., & Snyder, W. M. (2002). *Cultivating communities of practice: A guide to managing knowledge.* Boston: Harvard Business School Press.

Wenger, E., & Snyder, W. M. (2000). Communities of practice: The organizational frontier. *Harvard Business Review, 78*(1), 139-146.

Zack, M. H. (1999). Managing codified knowledge. *Sloan Management Review, 40*(4), 45-58.

KEY TERMS

Affinity Network: Groups of people who are drawn together, based on one or more shared personal attributes. Their activities are highly relationship oriented and typically include networking, mentoring, and representing a collective voice in both organizational and external community affairs.

Benchmarking: An improvement process in which a company measures its performance against that of best in class companies, determines how those companies achieved their performance levels, and uses the information to improve its own performance. The subjects that can be benchmarked include strategies, operations, processes, and procedures.

Best Practice: Superior performance within a function independent of industry, leadership, management, or operational methods or approaches that lead to exceptional performance; best practice is a relative term and usually indicates innovative or interesting business practices which have been identified as contributing to improved performance at leading companies. Best practice exercises routinely employ a variety of strategies to facilitate knowledge sharing and the creation of knowledge content in pursuit of enhanced customer service and ultimately customer loyalty.

Explicit Knowledge: Knowledge that can easily be collected, organized, and transferred through digital means.

Formal Network of Practice: A network of practice that has a membership controlled by fees and/or acceptance through some central authority that also assists in organizing, facilitating, and supporting member communications, events, and discussion topics.

Intellectual Capital: Everything that is known within an organization as exemplified in knowledge itself, in ideas and competencies, and in systems and processes.

Knowledge Management: The processes necessary to capture, codify, and transfer knowledge across the organization to achieve competitive advantage.

Knowledge Codification: The process of amalgamating individual knowledge in organizations, putting it in some relatively cohesive context, usually in a central repository, and making it available to organizational members.

Knowledge Personalization: Recognizes the difficulty in codifying knowledge, especially that which is tacit, and relies on face-to-face interaction, dialogue, and mentoring to transfer knowledge.

Network Organization: A network of several independent organizations. Members of a network work in close and continuous cooperation, involving partnerships, common products, and/or services and even a common strategy.

Self-Organizing Network of Practice: A loosely organized and informal network that has no central management authority or sponsor, membership is voluntary, and there is little explicit commitment.

Tacit Knowledge: Knowledge that is personal, context-specific, and hard to formalize and communicate.

ENDNOTE

[1]   See The Open Group's Web site at http://www.opengroup.org/

C

29

Collective Learning within CoPs

P.A.C. Smith
The Leadership Alliance Inc., Canada

INTRODUCTION

Like it or not, CoPs are awash in assumptions, and we presume validity at our peril in organizational contexts that are increasingly complex and ambiguous. If we wish to successfully address issues via CoPs, it is critical that members continually, individually, and jointly question their suppositions, evolve fresh questions out of their ignorance, and share relevant knowledge. Although CoPs clearly have the potential to do this, in the author's experience, little attention is paid by CoP members to the processes of either individual or collective learning that would facilitate achieving such ends.

The ability to think things through and debrief experiences at non-trivial personal and contextual levels is increasingly recognized as essential to effective learning in all situations, including CoPs. Action Learning (AL) is a well-proven individual, collective, and organizational development philosophy (McGill & Brookbank, 2004) that provides a sound setting for such reflective inquiry. Its application in CoP settings seems to be largely undocumented or untried.

BACKGROUND

AL, in its traditional form, originated more than 50 years ago as a means to improve UK coal production (Revans, 1982) and has become widely practiced worldwide (Marquardt, 1999). AL involves working on real problems, focusing on learning, and actually implementing solutions. It is based on Revans' notion that effective learning requires us to both question what is known and explore what is unknown (L = P + Q). There is general acceptance today that AL is a form of learning through experience, "by doing", where the task environment is the classroom and the task the vehicle (IFAL—Canada, 1998).

AL programs are typically based on the following tenets:

- Participants are tackling real problems (no "right" answer) in real time.
- Participants meet intermittently in small learning groups (AL Sets).
- Problems are relevant to a participant's own workplace realities.
- A supportive collaborative learning process is followed.
- The group process is based on reflection, questioning, conjecture, and refutation.
- Participants take action between meetings to try to resolve their issues and return at later sessions with progress reports, learning, and so forth on which to base further AL.

AL has an "elicitive" framework, intended to draw out, capture, and build on "what is", rather than operate in a detached, analytical, and rational world of "what should be". It is well known that experience is a very untrustworthy teacher, since most of the time we have experiences from which we never learn. AL seeks to throw a net around slippery experiences and capture them as learning, that is, as replicable behavior in similar contexts and as a source of questions in differing contexts (Smith & Peters, 1997).

By promoting reflection and insightful questioning with perceptive partners in situations where solutions are not always obvious and by leaving responsibility for implementation of the solution in the participant's hands, the individual makes sense of an experience by conceptualizing it and generalizing the replicable points; plans for future actions are based on the learning gathered. In this way, an AL group provides a "safe practice field" where the participants' mental models and future actions are shaped and reshaped in continual developmental cycles.

FUTURE TRENDS

The relevance of the above methodological tenets to CoPs should be clear to CoP practitioners, and its application is straightforward. The author uses a style of AL described by McLaughlin (1998). This approach is based on the counseling approach pioneered by Gaunt (1991) whereby participants negotiate for time to explore an issue. This model is favored over the more familiar project model advocated by Revans (1982) because of its relevance to CoP settings; for example, individuals are encouraged to define their own areas of interest/concerns and work in-depth with these issues, thus building increased capacity for ownership and insight. In addition, the author disregards the stable-group membership criteria advocated in project model practice and successfully applies AL processes in "whomever turns up" CoP settings; this provides the kind of attendance freedom typically associated with CoP sessions.

CONCLUSION

It is the author's hope that CoP practitioners will give serious consideration to more generally adopting AL processes as a means to better identify and encourage individual and collective learning in CoP programs. Such a flexible and powerful methodology deserves its place in the CoP lexicon of tools.

REFERENCES

Gaunt, R. (1991). *Personal and group development for managers: An integrated approach through action learning.* Essex, UK: Longmans.

International Foundation for Action Learning-Canada. (1998). Retrieved June 10, 2005, *http://www.tlainc.com/ifalc.htm*

Marquardt, M. J. (1999). *Action learning in action: Transforming problems and people for world-class organizational learning.* Palo Alto, CA: Davies-Black

McGill, I., & Brookbank, A. (2004). *The action learning handbook.* London: Routledge Falmer.

McLaughlin, M. (1998). Action learning. *Counselling At Work, 22,* 9-11

Revans, R. W. (1982). *The origins and growth of action learning.* London: Chartwell-Bratt.

Smith, P. A. C., & Peters, V. J. (1997). Action learning is worth a closer look. *Business Quarterly, 62*(1), 30-37.

KEY TERMS

Action Learning: A collaborative but challenging group process of cyclic inquiry that facilitates insight in an individual group member facing an important real-life problem such that (s)he may take reasoned action to resolve her/his problem, and the individual and other group members learn through the overall process.

AL Set: The name given to an AL group.

L = P + Q: AL is based on the radical concept that significant learning (L) only results when based on "routine knowledge in use" (P) and "questioning insight" (Q) brought together through a process of personal and communal reflection that integrates research on what is best practice or obscure with practical action to resolve a problem.

Communities and Evaluation of E-Government Services

José Córdoba
University of Hull, UK

INTRODUCTION AND BACKGROUND

In this article, I would like to reflect on a potential contribution of the theory of communities of practice to the evaluation of e-government services. Up to date, the adoption of e-government in local, regional, and national institutions could be characterized as a process guided by a need to improve efficiency and speed in the delivery of services to citizens. However, this might have implications to different groups of citizens who might like to use services, in other words, to become users. The theory of communities of practice could provide a framework by which e-government services (and, more particularly, e-government Web sites) could be assessed with a view of facilitating participation and inclusion.

E-GOVERNMENT

Broadly speaking, e-government can be defined as a set of activities supported by information systems with the aim to improve relationships between government institutions and citizens (Heichlinger, 2004). E-government implies the use of technology to enhance access to and delivery of government services to benefit citizens, business partners, and employees. Worldwide, technological advances combined with an emerging interest in citizen participation have led institutions to embark in implementing information services for citizens online. Services include general information on the use of services, facilities for online payment, specialist advice, and news.

The aim of e-government is to enhance public participation in decision making. Worldwide varieties of e-government Web sites have been set up, providing services and information at different levels (local, regional, or national). One example of a national e-government Web site can be seen at directgov (http://www.directgov.org.uk). On this Web site, there are different sources of information, which aim at providing support for different groups, including disabled, unemployed, and the elderly. The information is also organized around common themes (i.e., caring for someone, living abroad, etc.). An example of a local e-government Web site is that of Hull City Council (http://www.hullcc.gov.uk/). On this Web site, individuals can get up to date about new services, find job opportunities, make payments on existing services, and gain an overview of what the city council can do for them.

It is difficult to assess how distinct e-government Web sites are from commercial Web sites. Both types offer information and transactions to potential customers. In fact, one emerging issue of concern is how e-government services can contribute to develop integral approaches to e-government, by which the goals of inclusion and participation can be assessed and developed.

COMMUNITIES OF PRACTICE TO ASSESS E-GOVERNMENT INITIATIVES

A perspective on how to achieve the above goals can be developed by using the theory of communities of practice. The theory states that communities are groups of individuals which interact to pursue shared enterprises (Wenger, 1998). Embedded in this notion is a process of *learning* by which individuals gain competence through participating, continuously exchanging experiences and negotiating the meanings of what they see as their practice.

According to Wenger (1998), learning is not only about knowing but living meaningfully, developing a satisfying identity, and altogether being human. Learn-

C

ing implies a careful balance between engagement and experience. This means that people should have competence to interact with others, but also be able to bring new experiences to share. This ensures that a community's learning is adequately oriented in relation to what happens in society. A component that enables this learning to happen is that of "boundary objects" (Wenger, 1998). These elements help communities coordinate activities and interact with the rest of the world. For individuals, knowledge of boundary objects could constitute a condition of membership of the community.

Following the above, a possibility to evaluate the contribution of e-government Web sites could be developed. E-government Web sites could be assessed in terms of:

- How they enable emerging communities of citizens to interact and learn from new experiences;
- How people can share information; and
- How people can gain competence to enter into groups that may help them feel inclusive and participatory.

A way of assessing the appropriateness of e-government Web sites toward these aspects could be to explore how certain boundary objects are managed, for example, how *online applications* could be seen as an object that communicates two different communities (citizens and government officers). Around this object, some additional support could be given to both communities, including:

- Awareness raising about its relevance
- Procedures to fill applications and processes
- Experiences from both users and processors
- Knowledge about content.

This support to manage boundary objects is about ensuring *participation* in communities as much as use of these objects (Wenger, 1998). Although general, these aspects could inform further research on the evaluation of e-government Web sites. They could also open up opportunities for discussion with the aim to improve the delivery of services to citizens.

CONCLUSION AND FUTURE TRENDS

In this article, it has been argued that some general ideas of the theory of communities of practice could provide insights to evaluate e-government initiatives. This could apply in particular to existing e-government Web sites at the national and local level. It is not assumed that e-government contributes (or should) to foster the development of communities. Often, communities emerge to allow people to learn, and to *become*. Recognizing which groups of citizens could be supported with e-government services and enabling them to interact via appropriate boundary objects could lead those responsible for e-government initiatives to improve participation of citizens in government affairs.

REFERENCES

Heichlinger, A. (2004). *eGovernment in Europe's regions: Approaches and progress in IST strategy, organisation and services, and the role of regional actors*. Maastricht, The Netherlands: European Institute of Public Administration.

Wenger, E. (1998). *Communities of practice: Learning, meaning and identity*. Cambridge: Cambridge University Press.

KEY TERMS

Boundary Object: An element that enables community members to participate together and communicate with the outside of the community.

Competence: A description of the membership of a community of practice according to the degree of learning about practice in the community.

E-Government: Set of activities that aim to improve relationships between government institutions and citizens with the help of information systems and technologies.

Evaluation: A process of finding the value of information services or products according to the needs of their consumers or users.

Government: The art of managing relationships between citizens to ensure their welfare and in general their personal, economic and social development.

Participation: The act of taking part in or sharing the development of something. Participation also implies critical thinking about that which is being shared or developed.

Service: A facility provided to the public in general to ensure their wealth and care.

Users: Anyone with the willingness and capacity to access online e-goverment services provided for him/her and his/her community peers. Peers could be people with the same interests or similar concerns.

Communities of Implementation

Duncan Shaw
Aston Business School, UK

Brad Baker
Aston Business School, UK

John S. Edwards
Aston Business School, UK

INTRODUCTION

The concept of communities of practice (CoPs) has rapidly gained ground in fields such as knowledge management and organisational learning since it was first identified by Lave and Wenger (1991) and Brown and Duguid (1991). In this article, we consider a related concept that we have entitled "communities of implementation."

Communities of implementation (CoIs) are similar to communities of practice in that they offer an opportunity for a collection of individuals to support each other and share knowledge in a dynamic environment and on a topic in which they share interest. In addition, and to differentiate them from CoPs, a community of implementation extends the responsibilities of a CoP by having as its focus the implementation of a programme of change. This may well extend to designing the change programme. Thus, whereas a main purpose of a CoP is to satisfy "a real need to know what each other knows" (Skyrme, 1999) in an informal way, we argue that a main purpose of a community of implementation is to "pool individual knowledge (including contacts and ways of getting things done) to stimulate collective enthusiasm in order to take more informed purposeful *action* for which the members are responsible." Individual and collective responsibility and accountability for successfully implementing the actions/change programme is a key feature of a community of implementation. Without these pressures the members might lower the priority of implementation, allowing competing priorities to dominate their attention and resources. Without responsibility and accountability, the result is likely to be (at best) an organisation which has not begun a change programme, or (at worst) an organisation which is stuck halfway through another failing initiative.

To achieve these additional objectives beyond those of a CoP, the CoI needs to provide heightened support to its members. In fact often the members will collectively strategise the development and implementation of the change programme they are leading in the organisation.

Other concepts similar to CoPs have appeared in the literature, for example "communities of knowing" (Boland & Tenkasi, 1995), but none have a specific focus on implementation. Perhaps the closest example of a CoI, as suggested by our definition, is reported by Karsten, Lyytinen, Hurskainen, and Koskelainen (2001) who describe a CoP in a paper machinery manufacturer which seems to have the necessary focus on implementation.

The theoretical aspects of this article will explore the relationship between CoPs and CoIs, and the needs for different arrangements for a CoI. The practical aspect of this article will consist of a report on a case study of a CoI that was successful in its implementation of a programme of change that aimed to improve its organisation's knowledge management activities. Over two years the CoI implemented a suite of complementary actions across the organisation. These actions transformed the organisation and moved it towards achieving its 'core values' and overall objectives. The article will explore: the activities that formed and gelled the community, the role of the community in the implementation of actions, and experiences from key members of this community on its success and potential improvements.

BACKGROUND

Communities of practice are "groups of people informally bound together by shared expertise and passion for a joint enterprise" (Wenger & Snyder, 2000). Kulkarni, Stough, and Haynes (2000) identify various aspects that are typical of CoPs:

- Emphasis on learning.
- Group formation tends to be spontaneous.
- Direction comes from a set of shared problems, professional and/or social problems.
- The role of members is to act as sounding boards for new ideas and help each other learn.

CoPs may be formed within a single organisation, or have cross-organisational or indeed non-organisational membership (e.g., a community group or political-interest group). The literature tends to concentrate on CoPs within a single organisation and CoPs associated with a professional grouping.

As mentioned above, a community of implementation is a form of CoP, but with the distinguishing feature that *its focus is the implementation of a programme of change*. This brings many differences between CoIs and CoPs which the following paragraphs will highlight.

The philosophical differences between CoPs and CoIs require them to be arranged differently. This includes having accountability for outcomes, and more formal arrangements for structure and reporting to allow progress on actions to be monitored and reviewed. Thus, a community of implementation might include fewer members than a CoP (partly to heighten the importance of each individual in taking responsibility for leading actions) and have a less fluid membership (to ensure consistency and joint agreement on actions). The fact that the social processes that lead to the inclusion of some in a group or community lead equally to the exclusion of others (Marshall & Brady, 2001) is also even more of an issue for a CoI than for other forms of CoPs.

It follows, therefore, that the selection of CoI group members is very different to the spontaneous formation of a CoP. CoI members are selected perhaps because they:

- Have complementary knowledge which is of value to the design and/or implementation of the actions. The aim is not to be "bound together by shared expertise" because CoIs do not want shared knowledge. Instead they need individuals to bring complementary knowledge that adds to the pool of knowledge in the CoI.
- Represent a department which is not already represented and to which the CoI needs access (perhaps for informational or resource reasons). Complementary knowledge and resources are likely to be found across departments in the organisation, rather than being dominated by/housed in a single department.
- Represent a stakeholder group that could facilitate, or hinder, implementation. Often union representatives (or respected individuals) are involved in change programmes because they ensure that the actions are seen as legitimate by their constituents, and can become active in 'selling' the actions to their constituents.
- Have authority to commit resources to the initiative. CoIs are not just a collection of people who are interested in change, but have no authority to carry it out. Members of the CoI must have the authority to make decisions and implement actions; otherwise, dissatisfaction with barriers presented by top managers might lower motivation and impede progress.

These philosophical differences and features of membership begin to blur the boundaries between the formal and the informal responsibilities of the group. CoIs may indeed have a formal membership and responsibility for the implementation of the programme and/or its design to be successful. This may require top management resource mobilisation rather than them simply acknowledging its importance/existence.

Many of the features described above are similar to those advocated for a good project team in the literature on business processes (Hammer & Champy, 1993). What distinguishes a CoI from project teams in general should be the commitment of the individuals in it: the key words are "passion" and "enterprise," as in the Wenger and Snyder definition. Also, the potential lack of a hierarchy in a cross-department CoI could almost remove the usual line-management responsibility and reporting duties found in project teams. Instead, responsibility and accountability will be to the group, its chair, and the community that is experiencing the implementation. In some cases the group may be almost powerless to enforce the prioritisation of their action over the priorities that compete for a CoI member's attention. Only the individual's own desire to make progress on an action—or the feeling that they are being viewed negatively as a result of their lack of progress, or that they are hindering others making progress—might force their action.

There are various ways of facilitating a community of implementation. Many of the approaches can be taken from the organisation of CoPs, for example, facilitated meetings, discussion/bulletin boards, and e-mail lists (Skyrme, 1999). Approaches can also be taken from strategic planning—to give the action-orientation needed to such a group, for example, problem structuring workshops (Rosenhead & Mingers, 2001).

A CASE STUDY OF A CoI

ConsumProt is a non-statutory regulatory organisation charged with consumer protection and raising standards in its industry. ConsumProt was tasked by both the government (through the Treasury) and its industry to maintain the levels of consumer protection and the momentum for raising standards.

When ConsumProt was set up, the possibility that an official government body would in time replace it was acknowledged, and the original expectation was that the total lifespan of the organisation would be between five and ten years. In fact, it has turned out to be approximately six years. Given that its operational environment required consultation of actions with stakeholders, ConsumProt was set up with a strong emphasis on people. This also reflected the management style of its executive, who adopted what may be classed as a 'transformational leadership' approach, encouraging empowerment and freedom of action amongst its professional workforce. Its internal management and policymaking processes were based on seven 'Core Values': Make a Difference, Innovation, Teamwork, Support, Integrity, Learning, and Enjoyment. In its prime, ConsumProt had 65-70 employees. Six months before closure, this diminished to 45 employees.

At the time when this CoI came into being, the date of the transfer of responsibility to the govern-

ment body was already known. The objective of the CoI was to identify and evaluate a knowledge management strategy which would initiate or improve processes and tasks in order to: continue effective operations for a defined period, close the operation of the organisation more effectively, and prepare the process for possible handover to the government body. Given the constraints on the organisation, ConsumProt would only consider actions which could be implemented quickly within existing resources, and which offered results within six months.

The organisation is split into two main operational departments, A and B, with a range of other supporting departments. Tensions in the organisation were high, both across and within A and B. Taking their toll on the employees were the combined pressures of a declining workforce that was not being replaced, the prospect of redundancy, uncertainty over when redundancy might happen, a high workload, and a culture of 'us and them' between A and B.

Key events in the formation and continuation of the CoI have been a series of three problem-structuring, strategy-making group workshops (see Figure 1) conducted by "outsiders" from the Aston Business School who acted as workshop facilitators. It is not uncommon for a CoP to arise from a workshop: this was the experience in Unilever (Huysman & de Wit, 2000; von Krogh, Nonaka, & Aben, 2001), one of the best-known examples of an organisation that successfully encourages CoPs. The timeline of the major events in ConsumProt's CoI is as follows:

Figure 1. The Aston Group evaluating progress during the third workshop

- **First workshop:** June 2002—to form the group and agree on initial actions for better knowledge management.
- **Formation and naming of the CoI:** known as the 'Aston Group', after the involvement and facilitation of researchers from Aston Business School. The CoI consisted of eight individuals representing departments A and B and support departments. Group members were nominated by ConsumProt's executive (senior management team) and included members of this senior team.
- **Second workshop:** September 2002—to consolidate the group's understanding of actions and develop a path for their implementation.
- **Identification of projects and make progress on their implementation:** a significant amount of work by the members of the Aston Group went into implementing the actions (which are reviewed below).
- **Third workshop:** October 2003—to evaluate the progress of the CoI and agree on final actions to closure.

The first workshop was a scoping exercise for the knowledge management strategy. The aims of the workshop were: (1) to understand what knowledge ConsumProt needs to harness to improve its business; (2) to develop effective processes to harness knowledge; and (3) to consider how these processes should be evaluated.

The outputs from the first workshop were a group understanding of:

- The knowledge or sources of knowledge ConsumProt used or needed to inform its business, such as its market, media, constituent firms, and so forth.
- The processes currently used to harness and utilise that knowledge.
- What should perhaps be done to harness that knowledge more effectively.
- An evaluation of how good ConsumProt was in harnessing and using knowledge.

The group members returned to their workplace and, over a series of meetings (both informal discussion and formal group meetings), reflected on the session's outputs and next steps. It was agreed that

an excellent opportunity existed to implement the actions developed in the first workshop to create significant enhancements within the organisation. These actions aimed to help it maintain its activities in the defined period remaining and also prepare for an effective closure. The group therefore sought additional help and facilitation from the Aston Business School, with the aim of making the outputs more relevant to the workplace.

Three months after the first workshop, a second workshop took place. The aims of this workshop were: (1) to explore how we can improve the sharing of information within the organisation; and (2) to develop an action or implementation plan of three to four achievable 'next steps' which can be completed within six months.

Early in the workshop it was decided to concentrate on four topics for improvement. These were staff development, team building, communication of roles and functions, and formal continuous process improvement. Three of these originated from the organisation's prioritisation of staff retention, while the fourth explicitly centred on another core concern, process efficiency.

For each topic the group designed a programme of actions which the members felt would make significant progress in this area. For each programme of actions, the group assigned two or three CoI members who would take responsibility for further design and implementation of the actions, and for reporting progress back to the CoI.

These subgroups of managers were responsible for actions within their assigned areas. Monthly meetings of the wider CoI group ensured consistency of actions and maintained the overall momentum of implementation. The group established its own process for reporting to ConsumProt's executive (including the chief executive) and its board. It is believed that these processes helped foster an attitude of personal and group accountability, which was a significant factor in the overall success of the CoI's 'project'.

Six months after the workshop, the work of the CoI was described by the organisation's internal staff newsletter as "an internal initiative with the aim of improving the way information and knowledge is shared and used within ConsumProt...the [senior] management team fully support[s] these initiatives."

The newsletter described seven components of the initiative, so-called 'projects' that had crystallised from the programmes of actions mentioned above. These were:

1. **Contingency Planning:** Ensuring knowledge and skills of people leaving the organisation are not lost.
2. **Retention Strategy:** Focusing on staff retention and preparing people for the future.
3. **Build an open and supportive environment:** This will encourage teamwork.
4. **Implement Continuous Improvement:** Focusing on using staff knowledge to improve our processes.
5. **Enhance everyone's knowledge:** Enhance knowledge of each other's roles and impact on the organisation.
6. **Connection:** Create a sense that we all belong to the same organisation.
7. **Software:** Make better use of existing software.

The projects were described as very interlinked, as progress on one project often led to progress on one (or more) other(s). There was also considerable overlap of the projects with the organisation's seven Core Values, mentioned earlier.

EVALUATION OF THE IMPACT OF THIS CoI

In the third workshop the CoI reviewed the progress made, again with facilitation from the Aston Business School.

The group's objective assessment was that the implementation of actions had progressed well, even though a diminishing workforce was creating additional issues and the organisation's lifespan was ever shorter. The effort made in establishing the project group and taking it forward—including holding monthly group meetings—had certainly been justified in terms of improving the effectiveness of the organisation through internal communication and process enhancements.

Between 85-90% of the group's original objectives were reviewed as having been "already met."

Examples of the group's work (and the projects which addressed them) include:

- Conducting a skills audit of IT and software user skills, with tailored training sessions then developed, using in-house expertise identified, to address any gaps highlighted. [Projects 1, 2, 3, 6, 7]
- Introducing and implementing the formal documentation of work processes and job rotation and training in key areas. This reduced the risk to the business of key skills and knowledge being lost with the anticipated turnover of personnel. [Projects 1, 2, 5]
- Implementing training in presentation skills to address a development need raised by a large number of team members. [Projects 2, 3, 6, 7]
- Establishing informal focus groups of staff at the 'delivery end' of the process to review and rationalise certain business processes using a 'lessons learned' approach. [Projects 3, 4, 6]
- Conducting 'awareness' training of job roles and responsibilities across functions. [Projects 1, 3, 5, 6]
- Establishing cross-functional communications meetings to address issues as they arise. An example is the handling of new applications for registration with ConsumProt. In order to facilitate and improve this key work process, the organisation has introduced new processes for sharing knowledge, identified parameters for investigation work, and agreed on new service standards between departments A and B. [Projects 1, 3, 4, 5, 6]
- Improving 'softer' communication—for example, through greater cross-functional social activities and the use of ConsumProt's internal newsletter to highlight personal profiles and experiences of individual team members. These have worked well in terms of enhancing relationships and internal communication. [Projects 3, 5, 6]

Additionally, all members of the project group have developed skills and awareness of the issues around the successful application of strategic knowledge management. No fewer than 35 skill development areas have been identified including: listening skills, being forced to confront uncomfortable issues, letting go of 'old' ways of doing things, and project planning.

WHAT MADE THIS CoI WORK SO WELL

A number of features of the ConsumProt CoI contributed to its success. These included the action plan, the organisation of the CoI, motivation to act, and the management of the organisation.

The Action Plan

* **A Realistic Action Plan:** Gave members confidence that progress was in the right direction and actions contributed to the overall effort towards moving in that direction. The plan included 'short-term' actions which were designed to immediately deliver business and personal benefits—to build motivation and momentum. Anticipated completion of 'long-term' actions was still only six months away, which gave them urgency and did not allow months of inactivity.
* **A Focus on the People in the Organisation:** A healthy proportion of the actions concentrated on the staff and improving their working environment, improving their employability, relieving pressures, and so forth. Consequently, the effects of the actions could be seen on their friends and colleagues, and on themselves, rather than on some distant process involving someone they would never meet.

The Organisation of the CoI

* **Monthly Meetings of the CoI:** Strengthened the feeling that the group was a community with a shared vision and which could be relied upon to make progress. Meetings allowed members to meet socially, collaborate, share, celebrate success, raise problems, and acted as a deadline for genuine progress. This heightened personal accountability to the CoI, as regular lack of progress could not be hidden.
* **Formal Minutes of Meetings:** Recorded progress on actions and agreement for future

action, read by (not just sent to) the executive and available to employees throughout the organisation.
* **Effective Leadership by an Accessible CoI Champion:** Led the group in terms of completing actions to time and specification, administrative tasks (organising meetings, writing minutes, being available for consultation), and liaising effectively with the executive.
* **Effective Monitoring of the Progress on Actions:** Through the meeting and its minutes. Monitoring acted as a motivation to make progress and increased the perception of accountability.
* **Doubling Up People on Actions:** Aimed to support struggling individuals by having a colleague to work with, and to make joint progress.
* **Confronted Difficult Issues Early On:** As discussed above, there were many organisational internal matters which might have created conflict in the CoI and slowed progress. These were confronted and resolved early on, initially in the second workshop and through subsequent CoI meetings.

Motivation to Act

* **A Realisation That the Status Quo Could Not Continue:** The organisation had to change in preparation for closure. The CoI saw that it could lead this change or be engulfed by it. This acted as a tremendous initial motivation to participate.
* **Motivation in the CoI:** Early on in particular, some very motivated individuals infected others with their will to take action and created a culture where progress between meetings was expected. This reinforced responsibility to act and motivated the group to take the initiative.
* **The Response of the People:** A flood of staff took advantage of the opportunities newly open to them. This motivated the CoI to make more progress, and rapidly. This staff reaction might have been due to the CoI putting effort in the right places by doing things that were of use to the staff, not things that only made management look good.

- **Recognition and Reward:** The Aston Group was viewed as a successful group in ConsumProt. Internal newsletters publicised their success and future activities. Responses from the executive and other staff members, who were experiencing the effects of their effort, were positive motivators. The positive feelings inside the group heightened the desire for more success.

Management of the Organisation

- **Direct Two-Way Verbal Reporting Between the CoI and the Executive:** Heightened the feeling of accountability to the executive, as delivery and reaction was given/experienced first hand by most members, not just relayed through the CoI members who were also part of the executive.
- **Genuine Support from the Executive:** Which embraced the CoI with genuine enthusiasm, resources, guidance, and giving them the authority to act.

In summary, the Aston Group project has resulted in significant enhancements to ConsumProt's internal communications, information flow, and work processes across all functions—and particularly between departments A and B. This has meant that information relating to its activities has been more readily captured and shared, and the overall project has enabled ConsumProt to remain fully effective in its operations despite the impact of change and uncertainty experienced in the period to closure—for example through facilitating ongoing reviews of processes and ensuring skills gaps could be readily filled after changes in personnel.

FUTURE TRENDS

It would be worthwhile to see if the concept of a CoI can be extended beyond a single organisation, for example to a professional group or a set of organisations connected in some form of supply chain, as with other types of CoP. Inter-organisational CoIs may also overlap with the concept of strategic communities, as explained by Kodama (2003), in connection with knowledge-creating networks.

A particularly interesting area of research might be analyses of the personalities that should contribute to an effective CoI. This would aim to support managers as they construct a group which will enable progress to be made, motivation to be sustained, and slick political manoeuvring to implement actions.

CONCLUSION

The concept of initiating a community of implementation requires considerations beyond initiating a community of practice. The emphasis on making change happen in the organisation is a key differentiator. Further differentiation stems from the additional attention to selecting the right membership of complementary knowledge, resources, and access to constituent groups. Recognition of the need for change at the highest levels of the organisation and issues of CoI member selection could indicate a top-down approach to CoI initiation. However, this may be initially unbalanced by the freedom which the established CoI requires to implement actions rapidly, rather than getting bogged down in bureaucracy which can sap enthusiasm. Quick wins and rapid payback can bolster enthusiasm and create positive momentum.

REFERENCES

Boland, R.J., & Tenkasi, R.V. (1995). Perspective making and perspective taking in communities of knowing. *Organization Science, 6*(4), 350-372.

Brown, J.S., & Duguid, P. (1991). Organizational learning and communities-of practice: Toward a unified view of working, learning and innovation. *Organization Science, 2,* 40-57.

Hammer, M., & Champy, J. (1993). *Reengineering the corporation: A manifesto for business revolution*. London: Nicholas Brealey.

Huysman, M., & de Wit, D. (2000). Knowledge management in practice. *Proceedings of KMAC2000,* Aston University, UK.

Karsten, H., Lyytinen, K., Hurskainen, M., & Koskelainen, T. (2001). Crossing boundaries and conscripting participation: Representing and integrating knowledge in a paper machinery project. *European Journal of Information Systems, 10*(2), 89-98.

Kodama, M. (2003). Knowledge creation through the synthesizing capability of networked strategic communities: Case study on new product development in Japan. *Knowledge Management Research & Practice, 1*(2), 77-85.

Kulkarni, R.G., Stough, R.R., & Haynes, K.E. (2000). Towards modelling of communities of practice (CoPs)—a Hebbian learning approach to organizational learning. *Technological Forecasting and Social Change, 64*(1), 71-83.

Lave, J., & Wenger, E.C. (1991). *Situated learning: Legitimate peripheral participation*. New York: Cambridge University Press.

Marshall, N., & Brady, T. (2001). Knowledge management and the politics of knowledge: Illustrations from complex products and systems. *European Journal of Information Systems, 10*(2), 99-112.

Rosenhead, J. & Mingers, J. (2001). *Rational analysis for a problematic world revisited*. Chichester, UK: John Wiley & Sons.

Skyrme, D.J. (1999). *Knowledge networking: Creating the collaborative enterprise*. Oxford: Butterworth-Heinemann.

von Krogh, G., Nonaka, I., & Aben, M. (2001). Making the most of your company's knowledge: A strategic framework. *Long Range Planning, 34*(4), 421-439.

Wenger, E.C., & Snyder, W.M. (2000). Communities of practice: The organizational frontier. *Harvard Business Review,* (January-February), 139-145.

KEY TERMS

Action Plan: A portfolio of complementary activities which aim to have an effective, and the desired, effect on the organisation when implemented.

Collective Responsibility: The group is sanctioned with, and accepts responsibility for, creating change in the organisation. Each member helps and supports other members to make progress on implementing actions.

Community of Implementation: A group whose purpose is to pool individual knowledge (including contacts and ways of getting things done) to stimulate collective enthusiasm in order to take more informed purposeful action, for which the members are responsible.

Facilitated Meeting: A group of people getting together to explore the issues with the help of a facilitator. The facilitator brings a methodology of facilitation which provides process support and content management. Process support heightens the effectiveness of relational behaviours in the group (e.g., everyone getting airtime) and feeling free to share controversial ideas. Content support enables the mass of complexity shared during the meeting to be made sense of.

Problem Structuring Methods: The workshops were run using a problem structuring method approach. Therefore, instead of the participants talking about the issues in an unstructured format, the facilitator modelled the group's knowledge in the form of mind maps. The facilitator led the participants through a process enabling them to share, synthesise, and learn from structuring group knowledge.

Taking Responsibility: Responsibility drives the desire to make progress on implementing actions. The individual's own desire to make progress on an action—or the feeling that they are being viewed negatively as a result of their lack of progress, or that they are hindering others making progress—might force their action.

Communities of Learners in Paleography and ICT

Antonio Cartelli
University of Cassino, Italy

INTRODUCTION AND BACKGROUND: KNOWLEDGE AND LEARNING—FROM INDIVIDUAL TO SOCIAL STUDIES

Mankind studied and analyzed knowledge and learning since its first history and two main ways of thinking imposed very early: idealism, interpreting reality as the construction of human mind, and empiricism, looking at knowledge as the effect of the human-reality interaction. Recently three ways of interpreting thinking and knowledge intervened in changing the above perspective: *relativism* (it is impossible to objectively, universally, and absolutely know), *critical theory* (knowledge is mediated by social, political, cultural, economical, ethnical, and gender agents), and *constructivism* (knowledge is built by individuals and groups, and it is socially and experientially founded).

Among the above theories, constructivism played a great role in interpreting both individual and social learning and had a great influence on hypotheses explaining knowledge construction and evolution in communities, including communities of practice. The bases for today's constructivist theories can be found in many studies. Dewey (1949), for example, was the first scientist looking at the teaching-learning process in a pragmatic way. The inquiry was for Dewey the essential element of the subject-reality interaction; the experimental method had to guide teachers' work and students' learning, and at the basis of the knowledge process, there had to be the theory of research. Individuals' knowledge was continuously developing from common sense (traditions, popular misconceptions, etc.) to scientific knowledge. Main consequences of Dewey's educational project were activism with school-laboratories and active schools.

Dewey's ideas were collected and amplified by Kilpatrick, who introduced the project as a general

method of learning (i.e., problem-finding had to be used together with problem-solving in everyday teaching).

The hypotheses of Dewey and Kilpatrick were born in North America, but soon spread in Europe, where they found a rich soil and differentiated in at least two threads. Binet, Decroly, and Claparède privileged the psychological aspects of activism; on the contrary, Freinet and Freinet favored its social aspects (Varisco, 2002). "Modern School" was the name that Freinet and Freinet gave to their educational project; they hypothesized the creation of a cooperative school within which the social techniques and practices—like typography, correspondence, and cooperative catalogues—had a special relevance (their experiences had counterparts in many countries, and the case of don Milani in Italy is just an example for them).

CONSTRUCTIVISM AND INDIVIDUAL KNOWLEDGE CONSTRUCTION

Piaget and Ausubel, who are usually considered precursors of constructivism, hypothesized an active role of the individual in the cognitive process. Piaget (1971) suggested the theory of genetic epistemology to interpret the philo-ontogenetical evolution of the subject and stated that learning is the result of a continuous process of assimilation and settlement. Ausubel (1990), on the other hand, suggested two main types of learning: the mechanical and the meaningful learning, both depending on previous knowledge and on the ways the subjects build new knowledge—that is, there is a meaningful learning when: (a) the topic to be learned is logically meaningful; (b) the subject has special knowledge elements (subsumers) making easier the insertion of new knowledge in the reference frame of previous knowledge; and (c) the subject is willing to correlate what hc/she is learning with what he/she already knows, in other words he/she is motivated to learn.

Strictly speaking, for Piaget and Ausubel, if the subject has an active role in the cognitive process, social and cultural interactions have less or no relevance. The scientist who recognized the importance of the historical-cultural matrix into the philo-ontogenetical development of knowledge is Vygotskij; he went over the development-learning dichotomy and hypothesized a relationship between spontaneous learning and reactive learning, or in other words, between spontaneous ideas and scientific explanations. Vygotskij started from the hypothesis that spontaneous learning (due to experience) happens before school learning (which is social) and stated that education was effective if: (a) it anticipated an individual's development, and (b) it filled the ZPD (Zone of Proximal Development). When a subject acts socially in the solution of a problem that he/she is not able to autonomously solve, then he/she gets hold of new cognitive instruments (Vygotskij, 1980). Leont'ev (1977), disciple and successor of Vygotskij, introduced the idea of activity—under a well-defined form, structure, and condition, all depending on social interactions—as an action mediated by purposes; the activity substitutes the words as early knowledge units and early structural elements of human knowledge.

SOCIAL CONSTRUCTIVISM AND COMMUNITIES OF LEARNERS

In the 1980s, in the cultural-contextual psychology area, many scholars analyzed cognitive and learning practices outside the school context. The activity theory of Leont'ev found application in many studies (i.e., cultural anthropology research) and produced the situated-cultural (sometimes called situationist) approach to learning, which explicitly applied to communities of practice.

Regarding the communities of practice, the Laboratory of Comparative Human Cognition (LCHC, 1982) and Cole (1996) introduced the context in the analysis of learning experiences and hypothesized the presence of a shared elaboration system, connecting the individual learning experience to the corresponding performance by means of special schemas, in contrast with the contemporary idea of

a unique and absolute cognitive style, emerging from the culture the subject belongs to. One of the most relevant aspects of the situated-cultural approach to learning was represented by the concept of membership. Lave and Wenger (1991) analyzed membership and especially LPP (Legitimate Peripheral Participation), and stated that all members of a community had the same rights and were legitimated in participating to all resources and practices of the community. Further studies on the communities of practice led Wenger (1998) to his theory of social learning essentially based on the idea of identity; it consists of identification and negotiability between a subject and a community, and fulfills in different modes of belonging: engagement, imagination, and alignment.

Regarding learners' communities, often identified with school classes and groups of students, many studies focused on the analysis of the differences existing between in-school and extra-school learning. Brown and Campione (1994), on the other hand, defined the elements marking a community of students, or what they called CoL (community of learners). A CoL is made by students, teachers, tutors, and experts, who are organized in a community within which previous knowledge is analyzed, verified, and discussed, and new knowledge and theories are built. Soon after, the same authors modified their idea of CoLs and proposed the concept of FCL (Fostering Communities of Learners) (Brown & Campione, 1996).

The above ideas were adopted by many scientists in recent studies where computers and information and communication technology (ICT) were used to support learning, and as an example, the experiences of Scardamalia and Bereiter (1996) and Linn and Hsi (2000) are recalled here.

ICT AND THE PALEOGRAPHERS' COMMUNITY: A CASE STUDY

The experiences described in this article (made by the author in cooperation with M. Palma, Professor of Latin Paleography at the University of Cassino) are a good example of the changes ICT can induce in traditional and well-settled human activities. They are based on the use of the Internet and especially of the Web for the creation of communities of study and research, and modified the approach that scholars and students had with manuscripts and printed matters.

First of all it has to be noted that dating and localizing a medieval script, as well as identifying a scribe, have always been paleographer's essential tasks—that is, a paleographer has to answer the following typical questions: "Who, when, where wrote a charter or a manuscript between late antiquity and the invention of printing?" Before photography and other reproduction technologies, each paleographer had to manually examine charters in the libraries they were kept in, and his/her work was mostly made while being alone. With the Internet, a virtual repository of images and charters' reproduction was made available and a new set of communication instruments could be used by scientists for sharing ideas.

The Web site of *Didactic Materials for Latin Paleography* is the first experience of a new way of teaching this discipline. Within this site the didactic materials of the course of Latin paleography and especially two kinds of documents were made available:

1. Plates reproducing pages of ancient manuscripts in the different scripts adopted in the Middle Ages (Beneventan, Caroline, Gothic, etc.), together with the transcription of their texts; and
2. Texts freely extracted and/or translated from printed or electronic documents, or made available by the authors and collected in the various sections: Codicology, Cataloging, Preservation, Palimpsests, History of Paleography, and so forth.

A further experience concerned the creation of a dynamic Web site interfaced with a RDBMS (Relational Data Base Management System): *Women and Written Culture in the Middle Ages* (Cartelli, Miglio & Palma, 2001). Its main aim was to systematize the data emerging from the research on women copyists while leading to an instrument that could help scholars finding new elements for further studies. The data stored in the database concerned women copyists and the manuscripts they wrote, up to the fifteenth century. A main feature of the Web site is the presence of two different sections: the former one being operated only by the editors (who can insert, modify, delete the stored data, thus ensuring the

scientific validity of the reported information); the latter one at everyone's disposal, letting users obtain the list of all women and manuscripts in the database, or letting them make queries concerning women and manuscripts with specific qualifications.

The *Open Catalogue of Manuscripts* (Cartelli & Palma, 2003b) is a more complex and articulated information system. It is devoted to the management of the documentary information in a library, and is based on the use of the Internet and especially of the Web for storing and retrieving that information. Its main aim is to direct research and popularization interests on ancient manuscripts and charters, or in other words, to give the manuscript curators the instruments to recover the function they had in the erudition times, thus giving impulse to research and study in the manuscripts' study (a function they progressively lost for the increasing of bureaucratic tasks).

The last experience concerns another information system called BMB online (*Bibliografia dei Manoscritti Beneventani—Bibliography of Beneventan Manuscripts*) (Cartelli & Palma, 2004). Differently allowed people can store within the system the quotations of Beneventan manuscripts so that general users can freely query them. Persons entrusted with the task of collecting the quotations of the manuscripts are grouped into three categories:

1. Contributors who can write, modify, and delete bibliographic data;
2. Scientific administrators who can manage all data and write, modify, and certify bibliographic materials; and
3. System administrators who are allowed to do all operations, including the modification or deletion of certified data.

The access to certified bibliographic materials is possible to everyone according to several different criteria: (a) by author; (b) by manuscript; (c) by contributor; and (d) by one or more words or part of them concerning the title, location, or bibliographical abstract of a given publication.

It has to be noted that the system includes also a closed communication subsystem represented by an electronic blackboard granting an easy exchange of messages among contributors involved in the collection of bibliographic data.

ICT Influence on Paleographers' Community

In what follows, the analysis of the ICT influence on the creation of a community of study and research is reported:

1. The experience carried out with the site of the didactic materials is mostly unique not only for the systematic nature of the plates and for the presence of their transcriptions, but also for the documents reported among the texts; many of them are in fact papers concerning recent research topics, produced for special events (i.e., mostly conferences) and are made available by the authors for didactic purposes. In such a way students attending the Paleography course are instantaneously led to the leading themes of the most recent research and to the debates of the paleographic community.

2. Students attending the Paleography course used the materials reported in the site of *Women Copyists,* and were involved in the description of manuscripts and in the collection of plates. This work—the analysis of the different data stored in the system and the discussions the students had—led them to distinguish the different hands of the women and their way of writing manuscripts. A relevant role was played in this experience from the different expertise each student had in common with colleagues.

3. The Martyrology of Arpino as a single manuscript kept in the church devoted to Our Lady in Arpino (a small town in Central Italy) had the suitable features for the creation of an *Open Catalogue* (Cartelli & Palma, 2003a) and was used for teaching as follows:

    • Some make-up courses focused on the Martyrology were designed for students showing History and Latin gaps in their basic knowledge, since the manuscript was used as a chronicle from the fourteenth to sixteenth century, and the historical events reported there had a counterpart in relevant events of that period.

    • Students were directly involved in producing the Web pages of the Martyrology (with the acquisition and editing of texts and images), in the description of the manuscript, and in the transcription of the plates into text.

4. Finally, the students attending the Latin Paleography course were invited to become contributors of BMB online and were asked to produce bibliographic materials. The discussions they had with administrators, professors, and among themselves, along with the use they made of the electronic blackboard and of the e-mail services for the exchange of messages, and the chance to work in small groups on the same problems, helped them very much in developing and deepening the skills they needed in their everyday work, and suggested further elements of investigation to scholars.

The main result emerging from the analysis of the above data seems to confirm what has been found on communities of learners (Brown & Campione, 1994; Lave & Wenger, 1991)—that is, the raising of individual features and the improvement of group experiences (mainly due to the use of the ICT).

Regarding the students, it has to be noted that professors observed better results than the ones they obtained in traditional courses—they had better scores and developed better skills. Furthermore, some results never observed before were detected in the following skills: working in a group, easing of complex tasks (thanks to the help that each student could have from companions), and raising of the individuals' peculiarities within the community.

As for the scholars, the main features observed until now concern the sharing of information on their work, the suggestions they give to young students, and the discussions they started on the hypotheses concerning scripting styles, scribes, single manuscripts, and so forth.

CONCLUSION

Together with the notes reported at the end of the above paragraph, some remarks on the effects of ICT on students attending the course of paleography can be drawn:

1. People involved in the above experiences developed computing skills greater than the ones they could obtain in traditional computing literacy courses.
2. ICT supported the carrying out of a community of study and research in paleography, and students, researchers, and professors were members of this community.
3. The students were immersed in a metacognitive environment, submitted to cognitive apprenticeship strategies (a special version of traditional apprenticeship) and involved in the discussion and evaluation of the procedures they took part in—that is, they experimented with an experience of meaningful learning (Varisco, 2002).

FUTURE TRENDS

A further remark is needed, in the author opinion, for the IS planned and carried out in the above experiences. The above systems implement the knowledge that people working every day on the Middle Ages have to share and externalize that knowledge, so leading to the sharing of the best practices into the community.

How much this remark leads to an identification of communities of practice with CoLs is too early to say, but in the special case of paleographers, it seems possible.

REFERENCES

Ausubel, D.P. (1990). *Educazione e processi cognitive*. Milan: Franco Angeli.

Brown, A.L., & Campione, J. (1994). Guided discovery in a community of learners. In K. McGilly (Ed.), *Classroom lesson: Integrating cognitive theory and classroom practice* (pp. 229-270). Cambridge, MA: MIT Press.

Brown, A.L., & Campione, J. (1996). Psychological theory and the design of innovative learning environments: On procedure, principles and systems. In L. Schaube & R. Glaser (Eds.), *Innovation in learning* (pp. 289-375). Mahwah, NJ: Lawrence Erlbaum.

Cartelli, A., Miglio, L., & Palma, M. (2001). New technologics and new paradigms in historical research. *Informing Science (The International Journal of an Emerging Discipline), 4*(2), 61-66.

Cartelli, A., & Palma, M. (2003a). Il Martirologio di Arpino come oggetto di ricerca e strumento didattico. *Tecnologie Didattiche* (a cura dell'Istituto per le Tecnologie Didattiche del CNR), *28*(1), 65-72.

Cartelli, A., & Palma, M. (2003b). The open catalogue of manuscripts between paleographic research and didactic application. In M. Khosrow-Pour (Ed.), *Proceedings of the IRMA 2003 Conference: "Information Technology & Organization: Trends, Issues, Challenges and Solutions"* (pp. 51-54). Hershey, PA: Idea Group Publishing.

Cartelli, A., & Palma, M. (2004). BMB online: An information system for paleographic and didactic research. In M. Khosrow-Pour (Ed.), *Proceedings of the IRMA 2004 Conference: "Innovation Through Information Technology"* (pp. 45-47). Hershey, PA: Idea Group Publishing.

Cole, M. (1996). *Cultural psychology*. Cambridge, UK: Belknap.

Dewey, J. (1949). *Democrazia ed educazione*. Florence: La Nuova Italia.

Lave, J., & Wenger, E. (1991). *Situated learning. Legitimate peripheral participation*. New York: Cambridge University Press.

LCHC. (1982). Culture and intelligence. In R.J. Sternberg (Ed.), *Handbook of human intelligence*. New York: Cambridge University Press.

Leont'ev, A.N. (1977). *Attività, coscienza e personalità*. Florence: Giunti Barbera.

Linn, M.C., & His, S. (2000). *Computers, teachers, peers. Science learning partners*. Mahwah, NJ: Lawrence Erlbaum.

Piaget, J. (1971). *L'epistemologia genetica*. Bari: Laterza.

Scardamalia, M., & Bereiter, C. (1996). Engaging students in a knowledge society. *Educational Leadership, 54*(3), 6-10.

Varisco, B.M. (2002). *Costruttivismo socio-culturale.* Rome: Carocci.

Vygotskij, L.S. (1980). *Il processo cognitivo.* Turin: Boringhieri.

Wenger, E. (1998). *Communities of practice. Learning, meaning and identity.* New York: Cambridge University Press.

KEY TERMS

CoL (Community of Learners): A community of students, teachers, tutors, and experts marked by the presence of the following elements: (1) multiple ZPDs (the ones of the subjects in the CoLs), (2) legitimated peripheral participation (the respect of the differences and peculiarities existing among the various subjects in the community), (3) distributed expertise, (4) reciprocal teaching, peer tutoring and various scaffoldings etc. In this community previous knowledge is analyzed, verified and discussed and new knowledge and theories are built.

CSCL (Computer-Supported Collaborative Learning): This is usually based on special tools (knowledge forum is an example of this kind) which can create electronic or virtual environments, improving collaborative learning by means of computer networks. Main ideas they are based on include: (1) intentional learning (based on motivation to learn), (2) involvement in a process of expertise development, and (3) looking at the group as a community building new knowledge.

FCL (Fostering Communities of Learners): Main features of FCL with respect to CoLs are represented by reflection and discussion. The reflection is based on three main activities: research, sharing of information, and fair jobs. Discussions and speech have the main aim of stimulating auto-criticism and auto-reflective thinking in these communities.

LPP (Legitimate Peripheral Participation): States that all members of a community (also the less expert or more peripheral to it) have the same rights and are legitimated in accessing all resources and participating in all practices of the community. It is strongly based on Vygotskji's ZPD concept.

Open Catalogue of Manuscripts: An information system made of five sections (to be intended in a flexible manner, i.e., depending on the available resources and the different solutions that a library will adopt): (a) the first section contains documents illustrating the history of the library and its manuscripts; (b) the bibliography ordered by shelfmark and, eventually, alphabetically and chronologically, is housed in the second section; (c) the descriptions of the manuscripts (i.e., previous printed catalogues or ancient handwritten catalogues suitably digitized) and new descriptions are in the third section; (d) the fourth section is devoted to the images reproducing the highest number of manuscripts in the library (potentially all); and (e) the fifth and last section is a communication subsystem including electronic blackboards, chats, forums, and special Web solutions granting the easier acquisition, writing, and editing of texts.

Scripting Style: In the Middle Ages, and especially from the eighth to fifteenth century (before Gutenberg's invention of printing), different writing styles were used from copyists for the reproduction of ancient manuscripts. Beneventan, Caroline, Gothic, and Humanistic are four of the most important and widely used scripts in that period.

ZPD (Zone of Proximal Development): The individual learning areas marked by the distance between the skill and the knowledge a subject has in a given field, and the same kind of skill and knowledge of a more clever member in the community.

Communities of Practice and Critical Social Theory

Steve Clarke
University of Hull, UK

INTRODUCTION AND BACKGROUND

In philosophical terms, a key issue of communities of practice (CoPs) can be located within one of the key philosophical debates. The need for CoPs is traceable to the inadequacy in certain contexts of the so-called scientific or problem-solving method, which treats problems as independent of the people engaged on them. Examples of this can be drawn from the management domains of information systems development, project management, planning, and many others. In information systems development, for example, the whole basis of traditional systems analysis and design requires such an approach. In essence, in undertaking problem solving, the world is viewed as though it is made up of hard, tangible objects, which exist independently of human perception and about which knowledge may be accumulated by making the objects themselves the focus of our study. A more human-centered approach would, by contrast, see the world as interpreted through human perceptions: the reason why the problem cannot be solved is precisely because it lacks the objective reality required for problem solving. In taking this perspective, it may or may not be accepted that there exists a real world "out there", but in any event, the position adopted is that our world can be known only through the perceptions of human participants.

This question of objective reality is one with which philosophers have struggled for at least 2,500 years, and an understanding of it is essential to determining the need for, and purpose of, CoPs. The next section therefore discusses some of the philosophical issues relevant to the subjective-objective debate: a search for what, in these terms, it is possible for us to know and how we might know it.

A FOUNDATION IN KANTIAN CRITICAL PHILOSOPHY

Kant's critical problem, as first formulated in the letter to Herz (February 21, 1772) (Gardner, 1999, pp. 28-29), concerns the nature of objective reality. Prior to Kant, all philosophical schema took objective reality as a given and sought to explain how it was that we could have knowledge of this reality. If this were taken as definitive, it is easy to see how we might build (empirical) knowledge in the way suggested by Locke (1632-1704): that we are born with a "tabula rasa", or blank slate, on which impressions are formed through experience. This explains the pre-Kantian debate of reason vs. experience as the source of our knowledge: the rationalist view was that, by reason alone, we are able to formulate universally valid truths (for example, around such issues as God and immortality); empiricists, by contrast, see experience as the only valid source of knowledge.

Kant's insight and unique contribution was to bring together rationalism and empiricism in his new critical transcendental philosophy, the basis of which is his Copernican Revolution in philosophy. Loosely stated, this says that objective reality may be taken as existing, but that, as human beings, we have access to this only through our senses: we therefore see this objectivity not as it is but as we subjectively construct it. Unlike Berkeley (1685-1753), Kant does not claim that objects *exist* only in our subjective constructions, merely that this is the only way in which *we can know them*: objects necessarily conform to our mode of cognition.

For this to be so, Kant's philosophy has to contain *a priori* elements: there has to be an object-enabling structure in our cognition to which objective reality can conform and thereby make objects possible for us. This is what lies at the heart of Kant's *Transcendental Idealism*.

- While objects may exist (be "empirically real"), for us, they can be accessed only through their appearances (they are "transcendentally ideal").
- Our cognition does not conform in some way to empirical reality, rather this objectivity should be seen as conforming to our modes of cognition. In this way, we construct our objective world.
- Objects of cognition must conform to our sense experience. So, in this sense, knowledge is sensible, or the result of experience.
- These objects must conform to the object-enabling structures of human cognition. The resultant transcendental knowledge is (at least) one stage removed from objective reality, and is, according to Kant, governed by *a priori* concepts within human understanding.

This brief review of some key philosophical ideas has led neatly back to the subjective-objective debate. Seen from a Kantian perspective, we simply have no access to objective reality. (Interestingly, and again quite uniquely, Kant did not maintain there to be no objective reality; on the contrary, he argued that there must be real objects, or we would be in the ludicrous position of having perceptions of a world, but there being nothing to give rise to those perceptions.)

What objects may be in themselves, and apart from all this receptivity of our sensibility, remains completely unknown to us. We know nothing but our mode of perceiving them—a mode which is peculiar to us….Even if we could bring our intuition to the highest degree of clearness, we should not thereby come any nearer to the constitution of objects in themselves. (Kant, 1787, p. 82)

In summary:

1. Objectivity is conceivable only from the perspective of a thinking subject.

2. Central to Kantian philosophy is the question of how it is possible for subject and object to be so joined—what conditions must apply in order that this might be so?
3. In the Transcendental Deduction, Kant argues that subject and object make each other possible: neither one could be represented without the other.
4. All of this rests on their being: (a) a world of objects which is unknowable to us and (b) *a priori* concepts in understanding which enable representation of this world of objects.

A WAY FORWARD THROUGH CRITICAL THEORY

Theoretically, this philosophical position leads to a grounding in those theories relevant to human understanding and interaction, which are to be found in the social and cognitive domains. Given that in CoPs we are seeking a pluralistic, human perspective, those theories which best explain social interaction might be seen as especially relevant. Drawing again on the stream of social inquiry emanating from Kantian philosophy, this leads through the critical social theory of the early 20th century Frankfurt School to contemporary social theorists such as Foucault and Habermas (see, for example, Habermas, 1971, 1987).

Key concerns within Habermasian critical social theory are issues such as social inclusion, participation, and a view of how we ought to undertake intervention in social domains, all of which are fundamental to the functioning of CoPs. Habermas (1971, 1976, 1987) follows Kant in arguing that reliable knowledge is possible only when science assumes its rightful place as one of the accomplishments of reason. While the achievements of scientific study are not disputed, the problem perceived through the route followed by Kant and Habermas is that the methods of science which have grown out of modernity are effectively self referential: that scientific study sets up rules and then tests itself against its own rules is a procedure which has given considerable advances to modern society, but to regard this as representing all knowledge is mistaken. Habermas refers to the worst excesses of this as scientism: that we must identify knowledge with science.

Habermas further argues that the scientistic (positivist) community is unable to perceive self-reflection as part of its process, and that such reflection must be built into an understanding of knowledge. As with Kant, Habermas' challenge is whether knowledge is reducible to the properties of an objective world, leading him to a definition of knowledge which is based on perception but only in accordance with *a priori* concepts that the knowing subject brings to the act of perception. Since the knowing subject is a social subject, all knowledge is mediated by social action and experience, leading to Habermas' grounding certain theories in communicative interaction.

In the study of CoPs, this leads us to the following problems:

1. Accepting all human actions as mediated through subjective understanding leads to the possibility of a pluralist basis for CoPs.
2. There is no longer a dichotomy between subject and object.
3. The difficulty now left to resolve is essentially a practical one, of how to incorporate these ideas into a pluralistic foundation for CoPs.

From the position of viewing *all* human interactions with the so-called objective world as perceptual and subject to the *a priori* understanding that we, as human actors, bring to the act of perception, we begin to see CoPs as part of our normal social interaction. Research into problem analysis within the domain of management science, where communicative action theory has been used to further develop the concepts, is helpful in making sense of this. The ability to communicate by use of language is something that human beings bring to the world by nature of their existence: that is to say, it is not developed empirically but is *a priori*. To the extent that any theoretical position can be grounded on such an *a priori* ability, such a position may be seen as fundamental to us as communicative human actors.

To the extent that communication, at least partially, may be oriented toward mutual understanding, it might be argued as the foundation of CoPs, insofar as all such analysis is seen (after Kant) as perceptual. In these terms, CoPs never relate directly to the properties of an objective world but can be defined both objectively and according to the *a priori* concepts that the knowing subject brings to the act of perception. This knowing subject, being social, mediates all knowledge through social action and experience: subject and object are linked in the acts of cognition and social interaction.

In essence, then, it is argued that our difficulties disappear once a scientific basis for our thinking is denied. This echoes Habermas' view that science should be seen as just one form of knowledge, which in any case is simply a convenient human perception of how the world works. Now, all human endeavor becomes mediated through subjective understanding, leading to the possibility of a basis for CoPs in the universal characteristics of language. The difficulty is now essentially a practical one, of how to incorporate these ideas into CoP practice.

Within Habermas' (1976, 1987) theory of communicative action is presented a universal theory of language. Oliga (1996) summarizes this from its basis in locutionary, illocutionary, and perlocutionary speech acts, based on Austin (1975). Locutionary speech acts are concerned with saying something in a meaningful form which can be understood and are effectively a necessary precondition for communication. Perlocution is concerned with communication "strategically oriented toward individual success over [an] opponent" (Oliga, 1996, p. 246). According to Habermas (1984), only illocutionary speech acts "count as communicative action." The logic of this should not be lost in relation to the objectives of this research in relation to CoPs. The remaining tasks may be summarized as:

1. CoPs are an enactment of the perspective that objective reality is questionable and that we see our world according to our own views and perceptions (Oliga, 1996).
2. To the extent that these views are communicated by language, communicative action theory can help with the process.
3. The communication undertaken, both as spoken and as documented, can be tested for its locutionary (i.e., it should be meaningful and understandable) and its perlocutionary content, that is, it should *not* be concerned with "influencing the decisions of a rational opponent" (Oliga, 1996).
4. The primary test is then to deconstruct illocutionary speech, which is oriented toward understanding.

Illocutionary speech acts are oriented toward three fundamental validity claims: truth, rightness, and sincerity. What is most compelling about this theory, however, is that all three validity claims are communicatively mediated. This viewpoint is most radically seen in respect of the truth claim, where it is proposed that such a claim results not from the content of descriptive statements, but from the Wittgensteinian approach casting them as arising in language games which are linked to culture: truth claims are socially contextual and are therefore to be assessed not by reference to fact, but by reference to communication. Rightness is about norms of behavior, which are culturally relevant and are therefore to be determined by reference to that which is acceptable to those involved and affected in the system of concern as a cultural group. Finally, sincerity is about the speaker's internal world: his/her internal subjectivity.

These ideas can now be taken forward to provide an approach to CoPs which is theoretically grounded and closer to that which is experienced in (communicative) action.

AN APPLICATION FRAMEWORK FOR CoPs

It is now possible to design an application framework containing these concepts; revisiting CoPs from a philosophical and theoretical perspective dictates that an interventionist must always pay heed to the following:

- All problem analysis is perceptual; any approach to it must therefore be conducted through the views and opinions of participants, since only through these can objectivity be seen.
- An explicitly critical perspective must be maintained with a particular focus on normative ("ought") positions to counter factual ("is") claims.
- Critique must be applied to both content and the material conditions (norms and values) within which the content is set.
- Communicative action should be used as the social medium through which values are judged.

In order to make judgments about communicative action, it is necessary to first record and then deconstruct the communications that have taken place. Recording is less problematic than might be imagined: while it is not uncommon for group sessions to be recorded and transcribed, this is only one of the ways in which conversations can be documented. Recent work by Alford (see Future Trends), investigating virtual tourism, used the Internet as the medium through which conversations were conducted and enabled not only transcription but also sound and vision recording. From this and other data, the process of deconstructing the conversations has been started with, at its core, the need to determine communication as:

- Meaningful and understandable to all concerned.
- Not oriented toward coercively influencing the decisions of others.
- Truthful: the "ought" test.
- "Right": acceptable to those involved and affected as a cultural group.
- Sincere: related to the speaker's internal subjectivity.

FUTURE TRENDS

An essential precursor to applying the framework is the recognition that (after Kant and Habermas) problem analysis is not possible without participant involvement and that CoPs are a medium for achieving this.

Making judgments about communicative action is more complex, but there are some existing guidelines to help with this. There are, of course, numerous ways in which conversations can be both set up and recorded, some of which have already been mentioned earlier in this entry. Technology is proving particularly helpful in this respect, and, perhaps unsurprisingly, the domain of information systems is one in which significant progress has been made. Lyytinen (1992), for example, cites conferencing technology, which could encourage discursive activity, and information technology, which would allow the anonymous submission of "radical change proposals", while Kemmis (2001, p. 100) argues there to be considerable potential for IT to create "communicative spaces" in which communicative action can take place.

Research carried out in the field of critical qualitative research (Carspecken, 1996; Carspecken & Apple, 1992; Forester, 1992) provides guidance toward a framework for analyzing the validity claims raised during communicative action. A detailed analysis of this is beyond the scope of this article, but these and other ideas (some examples of which are given below) are now being applied to develop techniques which adhere to the guidelines provided by communicative action theory.

Validity claims relating to truth are concerned with defining perceptions of reality. Participants all have access to the "world of truth" but will have different interpretations of it. While this may to some seem esoteric, nowhere is this more important than in the often highly pragmatic domain of information systems. Frequently, in a systems development exercise, truth will be colonized by a powerful group, and issues important to the success of the development will be ignored (see, for example, Bentley, Clarke & Lehaney, 2004; Clarke, Lehaney & Evans, 2004).

Rightness is about the cultural acceptability of the claimant's position. One of the most powerful ways to challenge these claims is by a critically normative approach. Ulrich's (1983) Critical Systems Heuristics is helpful here: asking "ought" questions about an "is" position is something worth perfecting.

Sincerity relates to the subjective position of the claimant. Carspecken (1996) suggests ways in which the researcher can check this sincerity: checking recorded interviews for discrepancies and asking the interviewee to explain them; comparing what a person says with what they do and seeking clarification; or showing the person a summary of your reconstruction and ask them to comment on its accuracy.

In the Centre for Systems Studies at Hull, these ideas are used to inform research into information systems. One such development is headed by Paul Drake (see, for example, Clarke & Drake, 2002; Drake & Clarke, 2001), a research student attached to the center. Paul is applying Habermas' systems/lifeworld concept and theory of communicative action to develop a deeper understanding of information security. Similarly, Philip Alford is looking at

communicative action as a means of deconstructing conversations in the development of virtual tourism. Phil's conversations are collected mostly through Internet-based discussion forums, and the tests for communicative competence are proving to be of considerable value in understanding a way forward for this highly dynamic domain.

CONCLUSION

Seeing CoPs from a perspective based in critical theory gives us a new perspective. From this theoretical and philosophical basis can be derived a view of CoPs as based fundamentally on the perceptions of those involved in and affected by the system of concern. Any truly pluralist method must embrace this and must therefore pursue an approach which takes subject and object each to be a condition of possibility for the other.

Following this stream of thought, CoPs enable us to define our world both objectively and according to the *a priori* concepts that the knowing subject brings to the act of perception. This knowing subject, being social, mediates all knowledge through social action and experience: subject and object are linked in the acts of cognition and social interaction.

From this research project, a framework has been developed for implementing these ideas, based on theories of communicative action drawn from Austin and Habermas, and an approach to how this might be implemented in practice has been outlined. To the extent that CoPs involve communication through language, evidence of this communication can be gathered, and, from the documentation, communicative validity can be tested. All of this provides a wider framework within which methodological application can be undertaken.

REFERENCES

Austin, J. L. (1975). *How to do things with words.* Cambridge, MA: Harvard University Press.

Bentley, Y., Clarke, S., & Lehaney, B. (2004). A critical approach to the investigation of a UK university's information system. *Proceedings of the European and Mediterranean Conference on Information Systems: EMCIS 2004.* Hôtel

Golden Tulip Carthage Tunis, Avenue de la Promenade, B.P 606.

Carspecken, P. F. (1996). *Critical ethnography in educational research*. New York: Routledge.

Carspecken, P. F., & Apple, M. (1992). Critical qualitative research: Theory, methodology, and practice. In J. Preissle (Ed.), *The handbook of qualitative research in education* (pp. 507-553). San Diego: Academic Press.

Clarke, S., & Drake, P. (2002). A social perspective on information security: Theoretically grounding the domain. In S. A. Clarke, et al. (Ed.), *Socio-technical and human cognition elements of information systems*. Hershey, PA: Idea Group Publishing.

Clarke, S. A., Lehaney, B., & Evans, H. (2004). Human issues and computer interaction: A study of a UK police call centre. In A. Sarmento (Ed.), *Issues of human computer interaction* (pp. 291-320). Hershey, PA: IRM Press.

Drake, P., & Clarke, S. A. (2001, May 20-23). Information security: A technical or human domain? In M. Khosrow-Pour (Ed.), *Managing information technology in a global economy, 2001 Information Resources Management Association Int. Conf.*, Toronto, Canada (pp. 467-471). Hershey, PA: Idea Group Publishing.

Forester, J. (1992). *Critical ethnography: On fieldwork in a Habermasian way*. In H. Willmott (Ed.), *Critical management studies* (pp. 46-65). London: Sage.

Gardner, S. (1999). *Kant and the critique of pure reason*. London: Routledge.

Habermas, J. (1971). *Knowledge and human interests*. Boston: Beacon Press.

Habermas, J. (1976). *On systematically distorted communication. Inquiry, 13*, 205-218.

Habermas, J. (1984). *The theory of communicative action*. Cambridge, UK: Polity Press.

Habermas, J. (1987). *Lifeworld and system: A critique of functionalist reason* (Vol. 2). Boston: Beacon Press.

Kemmis, S. (2001). Exploring the relevance of critical theory for action research: Emancipatory action research in the steps of Jurgen Habermas. In P. Reason & H. Bradbury (Eds.), *Handbook of action research* (pp. 91-102). London: Sage.

Lyytinen, K. (1992). Information systems and critical theory. In M. Alvesson & H. Willmott (Eds.), *Critical management studies* (pp. 159-180). London: Sage.

Oliga, J. C. (1996). *Power, ideology, and control*. New York: Plenum.

Ulrich, W. (1983). *Critical heuristics of social planning: A new approach to practical philosophy*. Berne: Haupt.

KEY TERMS

Critical Theory: The branch of social theory, grounded on Kant and pursued by the Frankfurt School. The best known contemporary critical theorist is Jurgen Habermas (1929-).

Objective Reality: Essentially, the view that there exists, independently of human perception, an objective world, and we are able to gain knowledge of that world by reference to these objects.

Problem Solving: An approach which treats all problems as independent of human viewpoints and, effectively, as solvable using scientific or pseudo-scientific methods.

Problem Structuring: The recognition that certain problems cannot be solved in the above sense but need to be made sense of through debate.

Rationalism: The concept that we may gain knowledge purely through the operation of thought and rational analysis.

Empiricism: The concept that all knowledge comes from experience.

Transcendental Idealism: The thesis, first put forward by Kant, that while objects may exist (be "empirically real"), for us, they can be accessed only through their appearances (they are "transcendentally ideal"). Our cognition does not conform in some way to empirical reality, rather this "objectivity" should be seen as conforming to the object-enabling structures of human cognition. In this way, we "construct" our objective world.

Communities of Practice and Organizational Development for Ethics and Values

Jim Grieves
University of Hull, UK

INTRODUCTION AND BACKGROUND

Ethics is the study of moral issues and choices. In organizations, such a study inevitably involves consideration of decision-making practices and interpersonal relationships. This in turn may require the investigation of complex combinations of influences which include personality characteristics, values, and moral principles as well as organizational mechanisms and the cultural climate that rewards and reinforces ethical or unethical behavioral practices. Organizations ignore ethical issues at their peril as we know from recent examples of:

- past claims of brutality, poor wages, and 15-hour days in the Asian sweatshops run by Adidas, Nike and GAP,
- banks that rate their customers by the size of their accounts,
- the race for commercial control by private firms, universities, and charities claiming exclusive development rights over natural processes in the human body and patents sought by organizations, overwhelmingly from rich countries, on hundreds of thousands of animal and plant genes, including those in staple crops such as rice and wheat,
- a lack of people management skills and supervision which was said to be responsible for the falsification of some important quality control data of an experimental mixed plutonium and uranium fuel at the Sellafield nuclear reprocessing scandal which led to cancelled orders and the resignation of its chief executive.

We can all think of other examples that have hit the headlines to indicate that modern business management must recognize its responsibility to provide an ethical framework to guide action. This is the case in respect to human resources policy, health and safety policy, marketing policy, operations management, and environmental management.

Ethical policymaking has become the watchword for both national and local government. Ethics is now taught in the police force in order to be proactive and combat discrimination. Concern is now expressed in all forms of decision making from genetic modification of foods and the patenting of human organs to the ethical decisions of pharmaceutical companies or the marketing dilemmas of global corporations. Despite these developments, we continue to find many examples of decision makers making bad ethical decisions and people who blow the whistle on many of those actions. On the positive side, we have seen how so called green organizations have proved that ethics and profit are not incompatible goals.

COMMUNITIES OF PRACTICE AND SOCIAL RELATIONSHIPS

While many of these issues will engage people at the organizational level, communities of practice need to be aware of ethical issues particularly in relation to social relationships. This is because compromises have often to be made in relation to decision making as well as in the production of products and the provision of services. We can therefore distinguish between deliberate practices which include activities to deceive others such as consumers, employees, or colleagues and stakeholders from actions which are not premeditated to deceive but do, nevertheless, contravene what we might call ethical standards.

There are also many practices that may not be legally defined as unethical but which may result from collusion between subordinates and others who hold positions of power. Such examples are often not perceived as controversial and tend to be rationalized by means of situational expediency. These are related to the five main sources of power articulated by

French and Raven (1968) and involves the relative perceptions of the manager and subordinate relationship. The examples below indicate how this can occur when an individual makes decisions that are informed by perceptions and situational circumstances that constrain reflective judgment.

1. **Reward power** is seen to legitimize actions when a subordinate perceives the manager has the ability and resources to obtain rewards for compliance with directives. These often take the form of pay, promotion, praise, recognition, and the granting of various privileges. While this is quite a natural process, it can give rise to conflicts of interest when the motives of a subordinate are informed by personal gain, and those of the manager seek to achieve instrumental objectives.

2. **Coercive power** may not legitimize the actions to conform in the eyes of subordinates, but it does explain how collusion is sometimes related to perceived fear of punishment. This may, of course, be extremely subtle since the perception of punishment may be related to desired personal objectives or rewards such as promotion or an increase in pay. The abuse of power occurs when power holders can exercise power to the extent that subordinates fear that non-compliance may lead to the allocation of undesirable tasks or to lost opportunities to progress their careers.

3. **Legitimate power** reflects the assumptions of subordinates that a power holder as manager or supervisor has a right to expect compliance with a particular course of action. This is fairly typical of the position power that exists within bureaucratic structures. Unless subordinates are extremely knowledgeable about their own rights in relation to a particular manager's legitimate right to command obedience, they are likely to be drawn along by the situation.

4. **Referent power** occurs when a particular manager exercises influence because of charismatic reasons or because of personal attributes perceived to be desirable by subordinates. In situations where a conflict of interest may occur, collusion in a course of action may result be-

cause subordinates may be over-zealous in their pursuit of particular objectives while failing to reflect on the consequences of their actions. This may often be the motive for *group-think*, the consequences of which may be disastrous for an organization.

5. **Expert power** occurs when a leader is perceived to have a special knowledge, expertise, or degree of competence in a given area. In such cases, subordinates and, indeed, other stakeholders are likely to defer to the expertise of a particular individual. Thus, alternative judgments and information can be overlooked.

Because individuals often seek to achieve organizational objectives, their tendency to ignore conflicting value systems can create value dilemmas and ethical conflicts of interest. Many people do report conflict of interest in their work. Examples often cited in relation to overt practices include bribes, gifts, slush funds, concealing information from customers, shareholders, or, more generally, from the market place, engaging in price-fixing, and so on. We can regard these examples as institutionalized practices, but there are also examples where ethical problems emerge because of workplace pressures to achieve results.

How people come to rationalize their judgments is partly explained by the exercise of power but is also informed by the belief that actions are not illegal or unethical. This is illustrated by Gellerman (1986) who argues that there are four commonly held rationalizations that lead to ethical misconduct:

1. The belief that the activity is within reasonable ethical and legal limits—that it is not "really" illegal or immoral.

2. A belief that the activity is in the individual's or the corporation's best interests—that the individual would somehow be expected to undertake the activity.

3. A believe that the activity is "safe" because it will never be found out when publicized—the classic crime-and-punishment issue of discovery.

4. A belief that, because the activity helps the company, the company will condone it and even protect the person who engages in it.

LESSONS FROM ORGANIZATIONAL DEVELOPMENT

While ethical issues in organizations can be addressed from a variety of perspectives and raise complex theoretical issues, it would be more instructive for the reader to consider whether a community of practice should address ethical issues separately from the procedures and guidelines identified within their own organizations. Where guidelines and ethical codes for practice exist, as they do in many organizations, then members of a CoP will be obliged to be guided by them. In this respect, employees are likely to be informed by organizational values, rules, and guidelines or by professional codes of practice. However, as a general principle, it is recommended that members of a CoP consider their own activities in relation to the extent to which interventions will be a consequence of their actions.

Some lessons from organizational development (OD) may assist CoP members in meeting their objectives. The first point to be made is that OD is informed by its own value system, and OD consultants accept that they are bound by humanistic and democratic values. These are seen to be essential to building trust and collaboration within an organization. Nevertheless, there are some difficulties here.

The first major dilemma for an OD consultant is the extent to which the pursuit of humanistic values is contradicted or compromised by the desire to achieve organizational effectiveness. As Cummings and Worley (1997) argue, "more practitioners are experiencing situations in which there is conflict between employees' needs for greater meaning and the organisation's need for more effective and efficient use of its resources" (p. 57). As a result, it is important to identify any areas of potential concern at the point of agreeing to a contract with the client system. It should be clear, therefore, that any contract must make it transparent that organizational efficiency and effectiveness will depend upon an open and democratic concern for improvement through the organization's employees. And this, of course, has not always been the case where more programmed approaches to change management (for example, TQM, BPR) have sometimes adopted more instrumental and formulaic approaches to their interventions.

The second dilemma is related to the value conflict that OD practitioners are likely to face in relation to the different perspectives of stakeholder groups. Whereas traditional OD tended to adopt a relatively naive functionalist perspective, contemporary OD practitioners/consultants are much more likely to be aware of the plurality of interests operating within an organization. This inevitably means that different stakeholders will need to be consulted and their views explored in order to arrive at a workable intervention strategy. Each of these potential value conflicts may arise when consultants are not clear about their roles. In other words, role conflict and role ambiguity can give rise to value dilemmas.

A third dilemma relates to technical ability. In other words, a change agent who fails to act with sensitivity to the needs of the client system and who fails to possess sufficient knowledge and skill of underpinning behavioral issues is acting unethically. Thus, the ability to analyze a problem situation and to diagnose a potential solution requires an awareness of the variety of intervention strategies appropriate to the nature of the problem identified.

CONCLUSION AND FUTURE TRENDS: ETHICAL GUIDELINES FOR COMMUNITIES OF PRACTICE

Communities of practice might therefore consider whether they have a helping role in relation to a particular client system. If they do, then they will need to question the following:

1. The needs of the client system in its widest possible sense which, of course, may include all stakeholders and/or different employee groups.
2. The extent to which compromises are possible in relation to organizational efficiency and effectiveness vis-a-vis humanistic values.
3. The clarity of purpose or remit of the CoP in order to achieve its defined objectives.
4. The extent to which members of the CoP possess sufficient knowledge and internal skills in relation to (a) intrapersonal skills; (b) interpersonal skills; (c) consultation skills; (d) knowledge of underpinning organizational behavior.

A useful set of guidelines has been provided by the Human Systems Development Consortium (HSDC). These are stated in-depth in Cummings and Worley (1997) and refer to four main areas that should be addressed. These are (1) responsibility for professional development and competence; (2) responsibility to clients and significant others; (3) responsibility to the profession; (4) social responsibility. The reader is advised to read these issues in more depth and consider their relevance to any CoP activities they are currently undertaking.

As we noted previously, an organization's reward system can compound the problem of ethical dilemmas caused by the pressure for results. Like organizations, a CoP should consider rules for its own ethical climate. A basic framework should include the following:

1. Act as a role model by demonstrating positive attitudes and behaviors that signal the importance of ethical conduct.
2. Develop a code of ethics which (a) is distributed to everyone; (b) refers to specific practices and ethical dilemmas likely to be encountered; (c) rewards compliance and penalizes non-compliance.
3. Where necessary, provide ethics training designed to identify solutions to potential problems.
4. Create mechanisms such as audits to deal with ethical issues.

REFERENCES

Cummings, T. G., & Worley, C. G. (1997). *Organizational development and change* (6th ed.). Cincinnati: Southwestern College.

French, J. R. P., & Raven, B. (1968). The bases of social power. In D. Cartwright & A. F. Zander (Eds.), *Group dynamics: Research and theory* (3rd ed.). Harper and Row.

Gellerman, S. W. (1986, July-August). Why "good" managers make bad ethical choices. *Harvard Business Review*.

KEY TERMS

Code of Ethics: A codes of ethics can have a positive impact if it satisfies four criteria: (1) they are *distributed* to every employee; (2) they are firmly *supported* by top management; (3) they refer to *specific* practices and ethical dilemmas likely to be encountered by target employees; (4) they are evenly *enforced* with rewards for compliance and strict penalties for non-compliance.

Ethical Behavior: Receives greater attention today. This is partly due to reported cases of questionable or potentially unethical behavior and the associated dysfunctions that emerge. Because ethics involves the study of moral issues and choices, it is concerned with moral implications springing from virtually every decision. As a result, managers are challenged to set the standards and act as role models for other employees.

Ethical Climate: Indicates whether an organization has a conscience. The more ethical the perceived culture of an organization, the less likely it is that unethical decision making will occur.

Humanistic Values: OD promotes humanistic values through empowerment. That is, by articulating values designed to facilitate visioning, organizational learning, and problem solving in the interests of a collaborative management. Values are seen to be central to promoting trust, collaboration, and openness occur.

Moral Behavior: The goal for managers should be to rely on moral principles, so their decisions are *principled, appropriate,* and *defensible.* Empowerment has come to imply a moral commitment at an emotional level within organizations. Seen this way, many management trends see themselves as exemplars of this moral attempt to do the right thing.

Reinforcing Ethical Behavior: The frequency of reinforcement is seen as a crucial factor in maintaining standards. Rewarding ethical conduct and punishing unethical behavior is critical to the success of ethical practice.

Structural Mechanisms: In order to ensure that ethical decisions are practiced routinely, it is important to create positions and structural mechanisms that reinforce ethical behavior. Without structural mechanisms, a code of ethics will be forgotten. An example of such a mechanism is the social or ethical audit designed to assess whether the organization is practicing what it states in its formal guideline.

C

Communities of Practice and other Organizational Groups

Eli Hustad
Agder University College, Norway

Bjørn Erik Munkvold
Agder University College, Norway

INTRODUCTION

A general challenge in communities of practice (CoP) research and practice is how this concept can be distinguished from related terms such as project teams, workgroups, and knowledge networks. What criteria determine whether a group qualifies as a CoP? While these different concepts share several common characteristics, there are also important distinguishing features. Acknowledging these differences is important when assessing which former knowledge and research streams to build upon, and for increasing the level of precision in CoP research. In this article we provide a brief comparison of related terms, based on a set of distinguishing dimensions.

DEFINITIONS

A general definition of CoPs is: "Groups of people who share a concern, a set of problems, or passion about a topic, and who deepen their knowledge and expertise in this area by interacting on an ongoing basis" (Wenger, McDermott & Snyder, 2002). However, being able to bring out the distinguishing characteristics of CoPs requires a more fine-grained perspective. Table 1 presents a comparative analysis of CoPs and three related terms, using the following dimensions: purpose, membership, degree of formality, time frame, management, and role of ICT.

DISCUSSION

The emergent, self-organizing characteristic of CoPs based upon voluntary membership and participation is in contrast to using formal controls to support knowledge exchange, such as contractual obligation, organizational hierarchies, or mandated rules. Instead CoPs promote knowledge flows along lines of practice through informal social networks on a continuous basis.

Knowledge networks extend beyond the concept of communities of practice, and they are often acknowledged by management to increase innovation and organizational efficiency (Büchel & Raub, 2002). In addition knowledge networks are more visible in the organization than CoPs which exist beside the organizational structure. The organization could achieve more benefits from knowledge networks since they are more acknowledged than CoPs in respect of allocated resources and time to participate frequently. In organizations where knowledge networks are acknowledged, one aim is to link different knowledge networks together to a constellation of networks by applying boundary practices and knowledge brokers to consciously ensure organizational learning.

The choice to participate in a CoP is reciprocal, in that the community chooses their members, and members of a community choose whether to participate or not. This property seems to be the most exceptional compared to the other organizational groups.

Finally, both CoPs and different categories of knowledge networks differ from project teams that are formally mandated, deadline- and goal-oriented, and from workgroups that are formal organizational entities that build upon job descriptions and task performance (Hackman, 1990).

Table 1. Comparing different organizational groups

	Community of Practice	Knowledge Network	Workgroup	Team
Purpose	Organizational learning Share and build knowledge about a common passion through a joint enterprise, mutual engagement, and shared repertoire	Similar as CoP, but more goal oriented, innovative	Deliver a service or a product, Fulfill organizational objectives	Accomplish a specified task according to project goals
Membership	Self-selected assignment Voluntary participation	Either self-selected or more managed membership	Mandated from job descriptions and organizational hierarchy	Team members selected by management
Degree of Formality	Low, informal	Low to medium	High	High
Time Frame	Long-time voluntary membership	Either long-time voluntary membership or decided by management	Permanent, as long as the organization structure is stable	Participating part or full time as long as the project lasts
Management	Self-organizing groups, invisible to the formal organizational structure	Varies from self-organizing groups towards more managed and mandated groups Allocated time and resources to participate	Reporting to management of department	Reporting to team manager
Role of ICT	Support creation and sustainment of distributed communities, Choice of ICT—ease of use, efficiency	Support creation and sustainment of distributed communities Choice of ICT—ease of use, efficiency Linking different knowledge networks together—to a constellation of different networks, implementing boundary practices through ICT initiated by management	Distributed workgroups dependent on ICT for interaction purposes	Virtual teams dependent on ICT for creating a shared space, and for coordinating and performing common tasks

CONCLUSION AND FUTURE AREAS OF RESEARCH

Workgroups, teams, CoPs, and knowledge networks have both similarities and differences. Workgroups and teams are more similar in terms of their formal structure, objectives, imposed participation, and directions from management. This is in contrast to CoPs and knowledge networks, where objectives are more fluid and often emerge during the participation process.

Research on organizational groups is a topic of increasing interest for both scholars and practitioners. However, the boundaries between these groups are not clear-cut, and it is a need for further development of concepts and definitions that are more commonly accepted and universally valid in these research streams.

REFERENCES

Büchel, B., & Raub, S. (2002). Building knowledge-creating value networks. *European Management Journal, 20*(6), 587-596.

Hackman, J.R. (1990). *Groups that work (and those that don't): Creating conditions for effective teamwork.* San Francisco: Jossey-Bass.

Wenger, E., McDermott, R., & Snyder, W.M. (2002). *Cultivating communities of practice.* Boston: Harvard Business School Press.

KEY TERMS

Knowledge Network: Organizational members who share a strong interest in a particular topic and interact frequently to share and create new knowledge, for example, new solutions to business problems, a new technology, or a new business. Often initiated and supported by management to build both individual and organizational capabilities.

Team: A small cross-functional group with complementary skills that is responsible for a time-specific project, with a set of performance goals, and approach for which members hold themselves mutually accountable.

Virtual Team: In addition to a common purpose as a team, these groups are geographically dispersed with no or a moderate level of physical proximity, sharing a common virtual space where they collaborate by means of ICT to fulfill the goal of the project.

Workgroup: People, usually from the same organizational unit, with a shared responsibility for a product or service.

Communities of Practice and Technology Support

Elayne Coakes
University of Westminster, UK

INTRODUCTION

In order to operate successfully, communities of practice (CoPs) require a number of resources and facilities made readily available to them. These facilities can come in both physical and virtual forms. In this article we look at these resources and facility requirements for success, and review the possibilities for technology support (software offerings) that can provide the virtual aspects of these facilities.

BACKGROUND

Here it is argued that there are six main resources or facilities that CoPs require in order to operate. These are:

1. a space to meet;
2. a place to store ideas;
3. a memory of activities;
4. a record of members and their interests;
5. a means of communication among CoP member; and
6. ways to share tacit knowledge.

A Space to Meet

In order to fulfil its function, a CoP needs a place for members to meet on a regular basis. This might be at a pre-arranged time or on an ad hoc basis. This space needs to be easily reached by all members, private to the CoP members, and accessible by invitation only. What occurs and is said within this space should be made known only to other members of the CoP, unless they agree otherwise.

Physically this space could be a room booked out for CoP meetings as required, but technologically this could also be provided online through software that permits discussion groups, e-forums, threaded discussions, online chat-rooms for instant communication, and virtual meeting rooms.

IDEA SHARING

Communities need to be able to share the ideas that they have generated in their discussions. For instance, if they are engineers discussing a maintenance problem, a number of community members may have suggestions as to how to solve the problem. At the very least this information can be stored in members' memories; however, it might be felt appropriate to keep a record of suggestions. In a physical meeting space, this could be done by a scribe noting down the main points of the discussion in an informal manner, a formal report that can be later circulated among CoP members, or a tape or video recording of the meeting that can be stored and accessed later. If the latter, the recording could be stored on a multimedia database, as could the reports in a document archive. Virtual discussions of course are easily stored in discussion threads and best practice databases that are generated and extracted from these discussions.

Activity Memory

A memory of activities has a number of concepts in common with the storage of generated ideas. It is however a more generic concept, as activities will encompass the ideas, but will also include a record of suggestions for future activities. These activities need not be restricted to conversation. They may include guest speakers to update members on a professional matter; training, whether conducted by a CoP member or an external invitee; problem-solving forums; surveys; seminars; attendance at external events; and so on. A diary of events/activities is therefore required, both past and future, as well as a means of recording what happens at these activities. Again, this could be provided informally or formally through a scribe or CoP facilitator (if one is attached to the community) and circulated in paper or electronic format. E-mail is the base technology that could be used to notify members of events and to circulate records of activities, but electronic shared diaries may also be useful. Databases storing content and documents, virtual presentations, webinars, and possibly also online courses may also be useful.

Member and Expertise Record

Obviously each community needs a way of identifying who is and is not a member of that community. In the times of guilds, members wore identifying badges or the equivalent of uniforms as external verifiers of their membership of a skilled band of artisans. Modern CoPs are unlikely to provide such obvious identifiers. Rather, a list of members and criteria for how to become a member of the CoP will be kept by a designated record-keeper. It may be that entry to the community only requires that a candidate express interest in order to be invited in, in which case membership records are of little value and may not be physically kept. If, however, membership is restricted according to set rules and potential entrants must pass the equivalent of an entrance test, or may need to be sponsored through a voting process, then more formal records may be required. These records can be kept physically in documents or through invitation to join a virtual community whereby, for example, passwords will only be issued to verifiable members. The virtual community will then share a directory of members. Once a record of members is kept, it is then easy to store profiles of members' expertise rather than relying on the memory that Jane is an expert on anti-viral drugs and so on. Member profiles, once stored on a database, provide the community with not only a pool of searchable expertise, but also with the ability to link members with similar interests and thus enhance the social networking aspect of the community.

In addition, once expertise is stored virtually in a database, individual members can enhance their profiles by linking to their own records or reports, articles, Web pages, weblogs, and so forth that can

provide additional expert content and enhance the 'library' storage of ideas.

Communication

The simplest form of communication is face-to-face conversation and a physical space this is easy and convenient. Interestingly, this type of communication can now be facilitated by technology both in a high-technology and low-technology format. The high-technology format is videoconferencing, with all its requirements for well-supported technical assistance and resources, and the low-technology version is one that can be utilised by any home PC user—the Web cam and a telephone. With broadband Internet access and Windows Messenger, it is possible to speak on a telephone (or chat through text, if that is preferred) and see the person you are talking to, albeit the image is often not clear and not necessarily synchronised to speech. When synchronous communication is required, the telephone can also be used for conferencing with several people, and online chat rooms can also be used (or even Windows Messenger on its own). E-mail, of course, will provide asynchronous communication, as can discussion threads.

Tacit Knowledge Sharing

Tacit knowledge, by its very nature, becomes information once externalised into a form that can be shared through technology. Thus tacit knowledge can only be shared when the physical space is also shared. The physical space can be supported, as described above, by telephones and camera link-ups, but essentially it is a face-to-face activity. Tacit knowledge is frequently shared through storytelling or 'How I solved the X problem'. It can also be shared through mentoring and the action of 'sitting by Nelly'. Hands-on training with an expert is often recommended for tacit knowledge acquisition.

Comment

Although it is possible to externalise tacit knowledge and record it for instance through a recorded (video and tape) Q&A session with an expert that can be then stored in a database, without synchronous interaction, those viewing the Q&A are in receipt of information that the recipient then needs to transform into knowledge through their understanding of the context.

These then are the main resources and facilities that a CoP requires for support. The marketplace provides a number of technical offerings for these resources as indicated above. See the following sites and software:

- **Software:** Community Zero; iCohere; Communispace; Tomoye (Simplify); enable2; Livelink; Sitescape Enterprise Forum 7.1; Business Workspaces (Vignette application portal); Sigma Connect; Groove; Community Software; Plumtree
- **Web Sites:** www.icohere.com; www.enable.com; www.opentext.com; www.rumius.net; www.builda community.com

In addition, the software can provide:

- searching facilities for the various stored records and content;
- role-based permissions for community activities;
- FAQ databases;
- process and workflow management where communities work together on projects (though it is arguable in this circumstance whether the community is still a CoP or has become a project team);
- white boards for virtual meetings;
- audit trails, notifications of document updating, self-governance voting and policy tools, taxonomies; and
- support for sub-communities.

FUTURE TRENDS AND CONCLUSION

It is evident that many communities can operate with minimal resources and facilities; however, there are now a growing number of software suppliers who assist the operations and support community activities with various technology tools.

As communities grow within organisations, many may develop intra-organisational connections and networks. Virtual communities are growing and developing, and technology can assist in this. We will no doubt see more tools and facilities developing, especially through enterprise portals, as the value of CoPs becomes even more evident to organisations.

KEY TERMS

Asynchronous: Out of synchronicity. Conversations with time lags, as in e-mail.

Chat Room: An area where synchronous, text-based, online conversation can take place. Sometimes conversations can be conducted ad hoc, or they can be scheduled for a specific time and topic.

E-Forum: Another name for a forum, which is an area on a Web site where you can read and post messages on a particular topic, allowing debate.

Synchronous: At the same time. Live conversation.

Threaded Discussion: When a computer-assisted discussion takes place (in a forum for instance), the discussions are grouped together under the main point and related replies. This is asynchronous activity.

Webinars: Web-enabled (virtual) seminars.

Communities of Practice and the Development of Best Practices

Miles G. Nicholls
RMIT University, Australia

INTRODUCTION

Communities of practice and the development of best practices have a particularly strong base in an industrial setting where the intellectual capital—or more correctly, the tacit knowledge—is a 'craft' bordering on 'alchemy'. The concept of 'craft' tacit knowledge in this context relates to industrial processes where the operation is often based on a *body of individuals' experience* and is not able to be determined or analysed in a scientific or repeatable manner. Some examples of industries where these processes exist include aluminium smelting and float glass manufacturing. In both of these industries, a large proportion of the production processes rely on factory floor operators utilizing 'craft' (tacit knowledge) in the pursuit of best practice. These types of situations see many individuals involved in the manufacturing process sharing a community interest, and seeking the determination of best practice as a challenge and a means of enhancing personal and group pride. Best practice is used here in both a general and a mathematical sense, since there are no deterministic solution algorithms that can be used for solving certain aspects of the processes described below.

BACKGROUND

Many mathematical models of aluminium smelting reduction cells have been developed (e.g., Grjotheim, Krohn, Malinovsky & Thonstad, 1982); however, they are usually of a macro nature and assume a given efficiency of production (i.e., the efficiency with which electric current converts the raw materials into aluminium, termed 'current efficiency'). In reality, some current is 'lost' due to the nature of the process and the way in which the reduction cells (pots) are handled by the smelter floor operators.

The actions of the operators can affect the current efficiency of a particular pot for many days in uncertain ways.

An example of this is the manner in which 'anode effects' are handled. An anode effect is effectively where a wave of molten aluminium is started in the pot (as a result of gas bubbles and magnetic fields) that frequently spills out from the ends of the pots onto the pot room floor. The manner of treating this varies according to the experience and accumulated community of practice tacit knowledge. The operators, technicians, and other scientific people form very close-knit communities of practice with respect to handling these types of occurrences. In reality, the current efficiency cannot be actually calculated, only estimated. Thus, there are parts of the smelting process that are very 'soft', while others relating to the electrochemistry are quite 'hard'. Research undertaken by Rodrigo (1998) provided for the first time a method for determining current efficiency with a higher degree of accuracy than had previously been seen. This approach used Petri nets and other non-deterministic techniques in a mixed-mode modelling approach. Urpani (1997) has encapsulated the 'craft' or 'alchemy' aspect associated with how operators handle the pots on a day-to-day basis, and attempted—through an object-oriented methodology—to determine a common 'best practice' for the operation of reduction cells. However, while valuable information came from this research, together with an increased level of understanding, the attempt to define 'best practice' failed. Consequently, there is still a very strong and robust community of practice operating in the pot rooms of aluminium smelters in an attempt to achieve this elusive best practice, which is very much in-house as the improvement of pots' efficiency (i.e., increases in the current efficiency) means big increases in profitability.

In the float glass manufacturing industry, a similar situation exists to that in aluminium smelters. The production process for glass is not difficult per se. In fact the process is many hundreds of years old. It is, however, only relatively recently that float glass production has been utilized, rather than 'drawn' glass. Float glass production requires the molten glass from the furnace to be spread across a bed of molten tin. The manner in which this is achieved, and the way in which the glass moves along the float glass 'tank', determines the smoothness of the glass (i.e., the absence of bubbles, ripples and lines, etc.). This process of moving the glass along and keeping the molten tin bed as smooth as possible is the 'craft' aspect of the process. As in the case of aluminium smelting, communities of practice from across the spectrum of people within a glass manufacturing company are in existence, striving for best practice and achieving an operating procedure that will yield the ideal float glass. Again, it is not a scientific approach that is used, more the craft (or artisan) approach.

FUTURE AND CONCLUSION

The very elusiveness of the attainment of best practice and the 'craft' nature of the tacit knowledge in industries, such as the above examples, ensure existence, strength, and continuity of communities of practice. The determination of best practice using scientific means in these types of industries is still a very long way away, suggesting that there will be ongoing communities of practice for some time to come and a strong need for the same. Knowledge (albeit intuitive knowledge) difficult to quantify and codify is shared and perpetuates the essential alchemy of the industrial process.

REFERENCES

Grjotheim, K., Krohn, C., Malinovsky, K., & Thonstad, J. (1982). *Aluminium Electrolysis—Fundamentals of the Hall-Heroult Process*. Düsseldorf: Aluminium-Verlag.

Rodrigo, H. D. (1998). *The establishment of a predictive model for current efficiency in reduction cells*. PhD Thesis, Swinburne University of Technology, Australia.

Urpani, D. (1996). *Knowledge acquisition from real world data*. PhD Thesis, Swinburne University of Technology, Australia.

KEY TERMS

Best Practice: An explicit recognition of the fact that 'optimization' techniques and the goal of obtaining specific objective function maximisation or minimisation is inapplicable in the context. Best practice in the end is determined by the stakeholders and the producers, and may involve many subjective criteria.

Craft/Alchemy: The intuitive and holistic grasp of a body of knowledge or skill relating to complex processes, often without the basis of rational explanation.

Current Efficiency: The percentage of the electrical current (drawn into the reduction cell) that is utilized in the conversion of raw materials (essentially alumina and aluminium fluoride) into the end product, aluminium. The remaining percentage is lost due to complex reactions in the production process and the physical nature of reduction cells.

Communities of Practice as Facilitators of Knowledge Exchange

Scott Paquette
University of Toronto, Canada

INTRODUCTION

For knowledge to create value in an organization, whether tacit or explicit, it must have the ability to be shared among employees. This intentional (or in some instances unintentional) flow of knowledge can become the driver for organizational learning. When examining knowledge sharing, it is important to consider the context in which the knowledge is developed, as the community in which the individual is learning can affect any knowledge that is created. Organizational learning is impacted by individuals, groups, and the organization as a whole, and how these three levels are linked by social processes (Crossan, Lane & White, 1999). However, it is very difficult to create the right social environment to produce optimum knowledge sharing and learning. Sharing knowledge is an 'unnatural act', and therefore firms must strive to create the right environment and means to assist employees in overcoming knowledge flow barriers (Ruppel & Harrington, 2001).

Previous research has identified communities of practice as a hub for sharing knowledge within an organization (Brown & Duguid, 1991; Ellis, 1998; Hildreth & Kimble, 1999). The ability of a community of practice to create a friendly environment for individuals with similar interests and problems to discuss a common subject matter encourages the transfer and creation of new knowledge. Practitioners with similar work experiences tend to be drawn to communities, and from this a common purpose to share knowledge and experience arises (Wenger, 1998). Blackler (1995) argues that the creation and deployment of knowledge is inseparable from activity, and different contexts manifest in the form of knowledge boundaries. A community of practice can help individuals remove this boundary through the creation of a common context that links different experiential knowledge in an environment suited for knowledge exchange.

BACKGROUND

Communities of practice bring value to individuals and organizations by allowing for the acquisition of knowledge that supports practice within a role or responsibility. Brown and Duguid (1998) distinguish between two types of knowledge: (1) "know-what" or topical knowledge, and (2) "know-how" or knowledge derived from experience and action. They define "know-how" as the ability for an individual to take his or her "know-what" knowledge and put it into practice.

Other perspectives focus on the knowledgeability of action (Orlikowski & Yates, 1994). Here the verb *knowing* is stressed, rather than the noun *knowledge*. The emphasis on the interactive requirement for individual learning rather than the passive receipt of knowledge is a perspective that fits well with communities of practice. The use of the verb participation, a requirement for membership within a community of practice, also suggests that knowledge is created and shared from participation in experience and active membership within a community. An individual's ability to know is inseparable from practice and context.

Communities of practice follow the logic that knowledge cannot be separated by practice, as what is learned is highly dependent on the context where the learning takes place (Hayes & Walsham, 2001). The concept of *legitimate peripheral participation* (LPP) is derived from this notion, as it postulates that members who are allowed the opportunity to fully participate in community activities begin to behave as community members, or as practitioners. It is through this membership that knowledge can be shared with the rest of the community. Learning within a community is situated, as it occurs through people interacting in context. The learner's situated perspective, including physical and social context, become an important aspect in their learning and interaction with the community (Lave & Wenger, 1991).

In some cases, a familiar context or environment becomes a crucial factor in a practitioner's ability to deal with unfamiliar, unstructured problems (Tyre & von Hippel, 1997). These members must have access to the periphery of the practice, which allows for either observation or participation in the practice that eventually contributes to their decision to join the community. The term periphery is not used in the geographical sense, but as the degree of involvement an individual may have with the community. Their participation must eventually become legitimized (though not in the formal sense), in order to empower the participants to participate in learning and personal development.

Knowledge is situated within these communities through the situated learning curriculum that is unique to each community of practice. Newcomers can access this curriculum to gain the common knowledge resident in the community as a first step towards full participation. However, learning is an improvised practice, and eventually the participant must go beyond this notion of structure and curriculum to acquire knowledge. Therefore, participation in any community where knowledge exists can be defined as the act of learning (Lave & Wenger, 1991).

Communities of practice are able to assist an individual with this knowledge conversion as long as the participants are situated within the same community. The transfer of knowledge across communities becomes more challenging due to the "sticky" nature of knowledge. As knowledge is situated within a particular context, the removal from this context may distort its value or meaning. Various means of overcoming this obstacle have been proposed. Boland and Tenaski (1995) propose the use of communication forums that span multiple communities, while both Star (1989) and Carlile (2002) support the use of boundary objects.

Facilitating Knowledge Sharing

Lesser and Storck (2001) examined communities of practice by identifying their influence on a firm's social capital. Social capital, or "the sum of actual and potential resources embedded within, available through, and derived from the network of relationships possessed by an individual or social unit " (p. 833), emphasizes the value of a cohesive group in

organizational learning. This value can clearly be seen through examining the three dimensions of social capital: the structural dimension, the relational dimension, and the cognitive dimension. The following considers each dimension and the related factors that encourage knowledge flow and learning.

The *structural dimension* refers to the ease of which individuals can make connections with other similar practitioners. It identifies the processes, resources, and tools the community creates in order to augment and encourage social interactions. These may be in the form of physical resources, such as systems, or intangible resources such as face-to-face meetings and communities of practice.

Communities of practice can bring many structural benefits to an organization's knowledge sharing initiatives. They promote the use of IT tools in knowledge sharing, which can stimulate the use of this infrastructure and create a well-networked organization by use of the provided resources. Distributed cognitive theory addresses how learning in such a collaborative environment takes place. It defines a person's *horizon of observation* as the portion of the workspace that a participant can observe or monitor. Technologies designed for communities and knowledge sharing expand a member's horizon of observation, allowing for the identification of different knowledge sources that can contribute to the learning within the community. These technologies typically incorporate tools such as recorder tools, forums, local memory storage, and other knowledge collection aids in order to increase the spread of knowledge (Eales, 2003).

The community can act as a boundary-spanning object for geographical barriers through its distributed social nature and its ability to successfully use global IT resources. By allowing communities to work or partner with other company functions, they can become the facilitator for knowledge transfer, and encourage these functions to develop new knowledge. This situates them in the role of the educator in organizational learning. When implementing a knowledge strategy, the leadership within an organization can employ communities of practice to communicate the vision of a knowledge organization, set knowledge-related priorities and funding levels, facilitate communication that crosses business unit boundaries, encourage employee participation, and ensure alignment of company-wide systems and policies (Wenger, McDermott & Snyder, 2002).

The second dimension of social capital is the *relational dimension*. Here, the interpersonal relationships and activities of interaction come together to create a community that is not only willing, but also trusting and caring to share information with others.

Useful or valuable knowledge in organizations is often developed not by specialists or the people known for their subject matter expertise who are detached from the problem, but those who can operationalize the problem based on their work and who stand to benefit directly from the solution. These people are usually members of the community of practice where the solution is discovered (Brown & Duguid, 2001). By possessing a common goal for community (such as problem solving), relational capital is formed.

One method of determining how communities of practice share knowledge is by examining why people participate in these communities. A participant's motivation and justification for involvement will shed light on how knowledge is transferred. Wasko and Faraj (2000) view knowledge with the perspective that it is embedded not only in individuals and organizations, but also communities. Therefore knowledge can be managed as a public good, which is defined as "a commodity that can be provided only if group members contribute something towards its provision: however, all persons may use it" (p. 156). Organizations can be conceptualized as a group of overlapping communities that treat knowledge as a public good. The role that these communities play in knowledge flow throughout the organization is crucial.

Employees' motivation to exchange knowledge is impacted by their view that the decision to participate in such communities is either primarily economic and motivated by self-interest, or non-economic and motivated by community interest and moral obligation. This perspective can determine what stimulates an employee to share knowledge in a community of practice. Some participants become involved to generate tangible returns, such as access to useful information and expertise, answers to specific operational questions, and personal gain. They find that information received is up to date, compared to other sources such as company manuals or other information sources. In some cases, an individual can receive personal gain, including enhanced standing in the profession, a better reputation, or even to generate personal business. On the other hand, a community of practice can be a source of intangible benefits to its members, in the form of intrinsic rewards. Certain people find participation challenging, refreshing, and a means to refine their thinking in order to develop new insights. These individuals enjoy learning and sharing with others, and become confident with their expertise. They can also enhance their own personal learning through exposure to a variety of viewpoints from around the world. These are people that believe helping people is "the right thing to do" (Wasko & Faraj, 2000).

The final dimension of social capital is the *cognitive dimension*. This dimension provides a common context that allows for the efficient transfer of knowledge between individuals. It provides not only a common language, but addresses acronyms, subtleties, and underlying assumptions that are common to the daily operations of members. It can provide taxonomies for the classification of knowledge, and means that allow for knowledge transfer.

Knowledge flows most efficiently when seekers and experts are considered members of the same community and thus share the same values, norms, processes, and narratives. Furthermore, this flow is supported through making knowledge available that is deemed useful, timely, and helpful to the community (Wasko & Faraj, 2000). However, Pan and Leidner (2003) argue that organizations need to provide multiple channels of communication to support diverse knowledge sharing needs and preferences. The boundary of a community of practice should be dynamic and include other functions, people, and external sources. This generates the requirement to consider issues such as the need for a shared context, language, and culture, which can be nourished through motivation to share knowledge with individuals from different communities or the expansion of existing groups.

Many people have argued that larger organizations do not have the structure or capability of producing continuous and valuable innovation. However, with an organization of communities as described in Brown and Duguid (2001), large organizations supported and recognized by the larger community can develop these smaller, specialized communities. This potentially develops the capability of producing new knowledge, whether individually or in conjunction with other overlapping communities. Thus, larger organizations that are 'reflectively structured' are well positioned to be both highly innovative and capable of dealing with high degrees of change.

Challenges for Communities Sharing Knowledge

In many cases, the simple establishment of a community of practice will not contribute towards knowledge sharing. Underlying factors in either its design or the individuals participating can block members from interacting and sharing their knowledge. Participants will not contribute to the knowledge sharing within a community of practice for many reasons. For example, if they are not comfortable with their level of expertise, they become the victim of attacks on their ideas and opinions, or become overwhelmed with too much information being circulated. As well, a community that provides knowledge that is not useful or not interesting to its members creates the concern that participation is too time consuming and not a valuable use of time or resources (Wasko & Faraj, 2000).

In some organizations, employees may not feel the organizational climate provides a safe or desirable forum to share their valuable knowledge. Communities of practice can provide *safe enclaves* from organizational social-political pressures and encourage further knowledge sharing. Safe enclaves are characterized by shared electronic and non-electronic social spaces that allow for underlying views to be expressed. When communities are used for political purposes, it has been found the participation from members is very limited. It is common for managers and those removed from the community to attempt to influence or govern these communities from a distance, which also negatively influences participation. Genuine participation only occurs when the use of technology does not mirror the career or financial reward structure, or the control activities of senior management (Hayes & Walsham, 2001).

FUTURE TRENDS

A recent trend in the knowledge management field is to look beyond a firm's external boundaries for new sources of knowledge. Knowledge management strategies are encompassing not only the focal organization, but its partners, suppliers, and customers. Orga-

nizations will receive further knowledge benefits from communities of practice as the communities encompass individuals and knowledge assets located outside the organization. This partnership in knowledge sharing, with the community as the base of the relationship, will infuse new knowledge into the community and expand its knowledge creation capabilities.

Communities of practice are increasing their functional contributions within organizations by closer aligning with corporate strategies. A community that has grown within an organization can be formalized to contribute to the operations of the business, while serving its members in the original intended fashion. For example, instead of forming a product development team, an organization can utilize an existing community of practice to identify and involve the most appropriate people who possess the relative knowledge and skills. These formalized communities of creation have advantages over non-formal communities as they receive management recognition and support, priority in resource allocation, and increased recognition for their members' contributions. Eventually, these formalized communities will be the product or system development team, rather than an ad hoc assembly of staff.

As communities of practice become more fully integrated into job functions and business processes, they will become visibly integrated within the organization. As an organization is a community consisting of smaller communities, the boundary between communities and the formal structure of the organization will become seamless, and each new department or working group will have traits similar to communities already found in organizations. The social benefits of these communities working for a common goal under a common community structure will impacts the organization's social capital, its employee productivity, and its success in the marketplace.

CONCLUSION

Communities of practice have the capability to grow an organization's social capital through the increase in knowledge sharing that naturally occurs within these communities. By connecting individuals with similar experiences and interests, creating relationships between individuals and groups who may have not had the opportunity to meet through the formal structure of the organization, and providing a common context that encourages people to share their knowledge, a formal or informal community of practice can create the foundation for successful knowledge sharing within an organization.

Communities of practice can result in the following benefits to an organization and its knowledge strategy (Wenger et al., 2002):

1. create new business opportunities by developing internal expertise and relationships with an organization's customer base, resulting in the conversion of insights into new products;
2. reconstitute expertise that can become lost in a dynamic organization, and create a method of locating such expertise;
3. enable companies to compete on talent, then for talent—by becoming known as a home for experts that encourages the development of skills and expertise by employees; and
4. capitalize on the participation in multi-organizational communities of practice—by extending the firm's knowledge resources beyond its traditional boundaries.

Also noted by Brown and Duguid (2001), communities can design, develop, and maintain significant repositories for the storage and dissemination of knowledge throughout the organization. Although these repositories may be technical in nature or located in the individuals who hold the knowledge, the community becomes an identified source for knowledge on a particular subject matter. As organizations recognize the importance of supporting and maintaining communities of practice, they will experience increased knowledge capabilities and business success.

REFERENCES

Blackler, F. (1995). Knowledge, knowledge work, and organizations: An overview and interpretation. *Organization Studies, 16*(6), 1021-1046.

Boland, R.J., & Tenkasi, R. (1995). Perspective making and perspective taking in communities of knowing. *Organization Science, 6*(4), 350-372.

Brown, J. F., & Duguid, P. (1998). Organizing knowledge. *California Management Review, 40*(3), 90-111.

Brown, J. S., & Duguid, P. (1991). Organizational learning and communities of practice: Toward a unified view of working, learning and innovation. *Organization Science, 2*(1), 40-57.

Brown, J. S., & Duguid, P. (2001). Knowledge and organization: A social-practice perspective. *Organization Science, 12*(2), 198-213.

Carlile, P.R. (2002). A pragmatic view of knowledge and boundaries: Boundary objects in new product development. *Organization Science, 13*(4), 442-455.

Crossan, M.M., Lane, H.W., & White, R.E. (1999). An organizational learning framework: From intuition to institution. *Academy of Management Review, 24*(3), 522-537.

Eales, R.T.J. (2003). Supporting information communities of practice within organizations. In M. Ackerman, V. Pipek & V. Wulf (Eds.), *Sharing expertise: Beyond knowledge management* (pp. 275-295). Cambridge, MA: The MIT Press.

Ellis, S. (1998). Buckman Laboratories Learning Centre. *Journal of Knowledge Management, 1*(3), 189-196.

Hayes, N., & Walsham, G. (2001). Participation in groupware-mediated communities of practice: A social-political analysis of knowledge working. *Information and Organization, 11*, 263-288.

Hildreth, P., & Kimble, C. (1999). Communities of practice in the international environment. *Proceedings of the 2nd Workshop on Understanding Work and Designing Artefacts: Design for Collaboration*, King's Manor, University of York, UK.

Lave, J., & Wenger, E. (1991). *Situated learning: Legitimate peripheral participation*. Cambridge: Cambridge University Press.

Lesser, E.L., & Storck, J. (2001). Communities of practice and organizational performance. *IBM Systems Journal, 40*(4), 831-841.

Orlikowski, W.J., & Yates, J. (1994). Genre repertoire: The structuring of communitive practices in organizations. *Administrative Science Quarterly, 39*(4), 541-574.

Pan, S.L., & Leidner, D.E. (2003). Bridging communities of practice with information technology in pursuit of global knowledge sharing. *Journal of Strategic Information Systems, 12*(1), 71-88.

Ruppel, C.P., & Harrington, S.J. (2001). Sharing knowledge through intranets: A study of organizational culture and intranet implementation. *IEEE Transactions on Professional Communication, 44*(1), 37-52.

Star, S.L. (1989). The structure of ill-structured solutions: Heterogeneous problem-solving, boundary objects and distributed artificial intelligence. In M. Huhns & L. Gasser (Eds.), *Distributed artificial intelligence* (vol. 2, pp. 37-54). Menlo Park, CA: Morgan Kauffman.

Tyre, M.J., & von Hippel, E. (1997). The situated nature of adaptive learning in organizations. *Organization Science, 8*(1), 71-83.

Wasko, M.M., & Faraj, S. (2000). "It is what one does": Why people participate and help others in electronic communities of practice. *Journal of Strategic Information Systems, 9*, 155-173.

Wenger, E. (1998). *Communities of practice: Learning, meaning, and identity*. Cambridge: Cambridge University Press.

Wenger, E., McDermott, R., & Snyder, W.M. (2002). *Cultivating communities of practice*. Boston: Harvard Business School Press.

KEY TERMS

Boundary Objects: As knowledge crosses three forms of boundaries (syntactic, semantic, and pragmatic), certain objects assist this knowledge flow. They can be repositories, standard forms, objects and models, and maps of boundaries. They are both concrete and abstract objects.

Communities of Creation: A community of practice where members mainly focus on the sharing and generation of new knowledge for the purposes of creating new ideas, practices, and artifacts (or products). They can be legitimized through involvement

in a company-sponsored product development effort, or may be informal through various practitioners with similar experience and knowledge meeting, and new innovations arise from this interaction.

Horizon of Observation: Based in distributed cognitive theory, it is the portion of the workspace that a participant can observe or monitor. It addresses how learning in a collaborative environment takes place, and how technologies expand the horizon by allowing for the identification of different knowledge sources that can contribute to learning within communities.

Knowledge Sharing: The intentional (and often unintentional) flow of knowledge between individuals, groups, and organizations. The goal is to provide valuable internal knowledge in exchange for external knowledge, which is often combined via organizational learning to create new and meaningful knowledge. Sharing implies the flow is two-way through process, structural, or social means.

Legitimate Peripheral Participation: When non-members are allowed the opportunity to fully participate in community activities, and begin to behave as community members. It is through this membership that knowledge can be shared with the rest of the community. These individuals must have access to the periphery of the practice, which allows for either observation or participation in the practice, and eventually their participation becomes legitimized. Periphery is not used in the geographical sense, but as the degree of involvement an individual may have with the community.

Safe Enclaves: A social area or community that shelters individuals from organizational social-political pressures to encourage knowledge sharing. They sometimes alter the organizational climate to provide a safe or desirable forum to share knowledge. Often they encompass shared electronic and non-electronic social spaces that allow for underlying or contrary views to be expressed.

Social Capital: The sum of actual and potential resources embedded within, available through, and derived from the network of relationships possessed by an individual or social unit. The value of a cohesive group in organizational learning is emphasized. It is composed of three dimensions which include the structural dimension, the relational dimension, and the cognitive dimension.

Communities of Practice for Organisational Learning

Andrew Wenn
Victoria University, Australia

INTRODUCTION

The term *community of practice* (CoP) arises out of the work of Lave and Wenger (1991) and Wenger (1998) and refers to the way groups of individuals interact and engage in "the sustained pursuit of a shared enterprise" (Wenger, 1998, p. 45). It is the activities of the members of these groups both individually and collectively, the construction of and practices at a local level that allows them "to meet the demands of the institution" (Wenger, 1998, p. 46) which they work for. For the CoP, learning occurs as a form of social practice.

BACKGROUND

Learning

For Wenger, learning does not occur in isolation. He proposes, in his social theory of learning, that we should understand learning as arising from social participation, and it involves four components:

1. meaning as a way of discussing our life experiences in relation to the world—learning as experience;
2. practice is a way of talking about the activities and guiding principles of what we do—learning as doing;
3. community is about the "social configurations" which organizations require of us, and, through our participation, we become recognized as competent—learning as belonging; and
4. identity is the way we talk about how learning changes us at the personal level—learning as becoming. (Wenger, 1998, p. 5)

Understanding that learning is a social process is particularly important when we wish to see how CoPs can facilitate organisational learning.

When we consider organisational learning, we must move from the individual to the "knowledge flows" that occur between individuals and the "contexts shared by individuals and groups" (Nidumolu, Subramani & Aldrich, 2001, p. 116). To be successful, an organisation must "know what it knows" (Wenger, 1998, p. 8); thus, it is necessary to understand how knowledge flows across boundaries between communities and between communities and the organisation. It is also a matter of creating, nurturing, and sustaining these flows (Wenger & Snyder, 2000; Wenn & Burgess, in press).

Boundaries and Boundary Encounters

Communities of practice have boundaries. These boundaries serve to separate different communities and are often only revealed when we realise what learning is required to move from one CoP to another or from the CoP to the larger organisation. What is important for our purposes is that these boundaries are not impermeable; a community cannot exist in total isolation to the rest of the world; there are entities that serve as boundary objects (Star & Griesemer, 1989) that are able to move between the different communities and "coordinate the perspectives of various constituencies for some purpose" (Wenger, 1998, p. 106).

Examples of boundary objects are documents (such as research publications, memos, reports, e-mails, spreadsheets, forms), terms, concepts, people, and other artifacts that are capable of communicating between the community and the organisation, creating connections between them.

Wenger (1998, pp. 112-114) identifies three types of boundary encounters (Figure 1). These can be meetings, conversations, and visitations and can happen at various levels. There can be a *one-to-one* encounter where two people meet and discuss issues involving the boundary relationships of relevance to them. Another type of encounter is an *immersion*.

Figure 1. A community of practice is embedded within an organization. Organisational learning involves a two-way exchange across the boundary between the community members and the rest of the organisation.

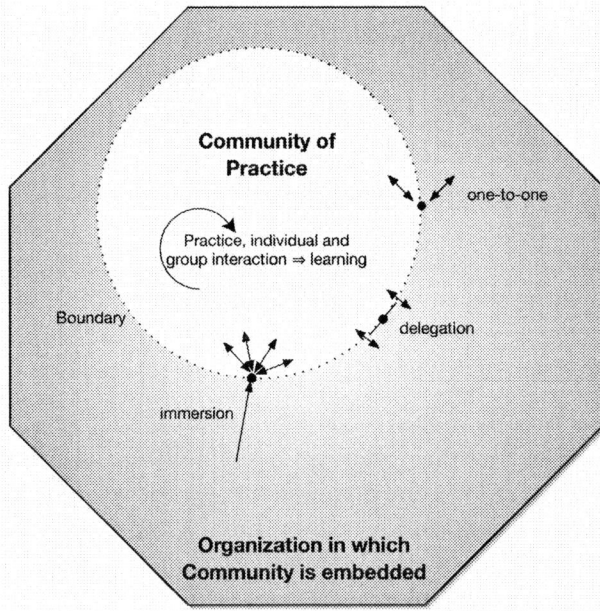

This can take the form of a visit to a practice. "This kind of immersion provides a broader exposure to the community of practice being visited and how its members engage with one another" (Wenger, 1998, p. 112). One disadvantage of this type of encounter is that the passage of information is essentially one way. The members of the visited community ascertain very little about the community the visitor belongs to. The final type of encounter is a *delegation* where multiple participants from each community meet for a mutual exchange of knowledge. In this type of exchange, meaning is negotiated between members of each community and across the boundary. An example of this type of boundary encounter would be a manager meeting with CoP members for an exchange of ideas, concerns, or insights into the CoP's expertise.

The links or boundary encounters that exist or need to be created between a CoP and an organisation are analogous to those that are needed or existing between different CoPs. This is true especially if the CoP under consideration is wholly embedded in the organisation.

FUTURE RESEARCH

In a recent article, Wenn and Burgess (in press) employ Wenger's ideas of boundary encounters to advance some ideas of how the links between academic IS research and real-world practice could be encouraged and maintained. The academics were one CoP while the practitioners were another. Among these was the need to ensure that researchers, students who will eventually become practitioners, and already existing practitioners are encouraged to adopt a more reflective attitude toward their work and consider how it may benefit members of their own communities and their own learning, as well as interested and concerned members of communities external to the one they are currently in.

CONCLUSION

By their very nature, communities of practice are involved in learning and constructing knowledge as a group. Employing Wenger's social theory of learning with its four components of learning as doing, experience, becoming, and belonging allows the individual and group nature of learning to be revealed. CoPs are by their very nature bounded entities, but this boundary is permeable, and knowledge can flow from the CoP to the organisation or CoP to CoP by several mechanisms. If an organisation is to survive and thrive, it is important that the organisation nurtures CoPs and that the CoP shares its knowledge by allowing these knowledge flows to continue.

REFERENCES

Foucault, M. (1986). Of other spaces. *Diacritics, 16*(1), 22-27.

Lave, J., & Wenger, E. C. (1991). *Situated learning: Legitimate peripheral participation*. Cambridge: Cambridge University Press.

Nidumolu, S. R., Subramani, M., & Aldrich, A. (2001). Situated learning and the situated knowledge web: Exploring the ground beneath knowledge management. *Journal of Management Information Systems, 18*(1), 115-150.

Saint-Onge, H., & Wallace, D. (2003). *Leveraging communities of practice for startegic advantage*. Amsterdam: Butterworth-Heinemann.

Star, S. L., & Griesemer, J. (1989). Institutional ecology, "translations" and boundary objects: Amateurs and professionals in Berkeley's Museum of Vertebrate Zoology, 1907-1939. *Social Studies of Science, 19*, 387-420.

Wenger, E. C. (1998). *Communities of iractice: Learning, meaning and identity*. Cambridge: Cambridge University Press.

Wenger, E. C., & Snyder, W. M. (2000). Communities of practice: The organizational frontier. *Harvard Business Review, 78*(1), 139-145.

Wenn, A., & Burgess, S. (in press). IS research, practical outcomes, real world practice: Making sustainable links between communities. *Journal of Information Technology Theory and Applications*.

KEY TERMS

Boundary Encounters: The ways in which different CoPs may meet to exchange knowledge.

Boundary Objects: Boundary objects describe actants that are able to bring a degree of commensurability to the knowledge practices of different communities for some shared purpose. They take knowledge from one community and present it to another in such a way that it makes sense to that community.

Knowledge Flows: The ways in which knowledge can move between a CoP and the larger organization or between members of the CoP. A free flow of knowledge may be facilitated by human or technological means or may result as a natural exchange between individuals, for example, a conversation.

Organizational Learning: Organizational learning is where the organization employs its skills to create, acquire, and disseminate new knowledge. A learning organization is able to modify its behavior in light of the new knowledge it acquires (Saint-Onge & Wallace, 2003).

Social Learning: Learning arising from social participation, involving four components: learning by doing, by participation, by becoming, and by experience.

Community of Practice and the Special Language "Ground"

Khurshid Ahmad
University of Surrey, UK

Rafif Al-Sayed
University of Surrey, UK

INTRODUCTION

The creation of new knowledge is essential for survival in a global economy, for providing better public services, and for maximising profits over long timeframes. People in an organisation create knowledge, and the organisation facilitates or hinders the creation. Knowledge is created through interaction, and this interaction often results in the emergence of a community. The community invariably develops a common social, philosophical, and cognitive ground amongst the members of an organisation, and helps members to share and learn knowledge of others. In an organisation people learn and share knowledge by watching each other, by talking to each other, by reading documents written by each other to gain a *common understanding*. Common understanding helps in creating a community. The community is a dynamical eco-system where new ideas are nurtured, existing ideas pruned, and some 'killed off'. The understanding supports quiescent changes and *paradigm shifts* as well.

A community is defined as a body of people "organized into a political, municipal, or social unity"—a body that shares values, beliefs, and aspirations and creates its own icons. All communities have an exchange system—rewards for good behaviour and opprobrium for bad. And language is amongst one of the important icons for communities as diverse as national and regional communities, and scientific and technical communities.

A specialist community uses the language of the populace and then starts to specialise the meaning of certain words within the existing stock of words of the parent language, creates its own words, and places similar restrictions on the grammar of the populace at large when used within the community.

This specialisation process results in the language of the community, and the language is called *language for special purposes* (LSP), *language for specific purposes*, or just *special language of X*, where X refers to a specific branch of human enterprise—language of physics, of business, of sports. There are further specialisations: LSP of nuclear physics, financial trading, and football.

Special languages can be differentiated from the language of everyday usage at the level of vocabulary; the differences are increasingly less discernible at the levels of grammar, syntax, and semantics. Special languages are in many ways a social phenomenon: consciously created to foster a sense of common purpose amongst a group of people and sometimes used to exclude. Special languages are key instruments of personal and group promotion. A specialist community, individually and collectively, weaves a fabric of facts and imagination (Goodman, 1978); in essence, the weave is a collection of specialist texts. We attempt to relate the development of an LSP to a specific community (of practice).

PRACTICE OF A COMMUNITY

Community of practice, or communities of practice (CoPs), is defined variously in organisational behaviour (Nonaka & Takeuchi, 1995), in human resources management (Lesser & Storck, 2001), in discourse analysis (Clark, 1996), and by military planners (Bennet & Bennet, 2003) and computing professionals (Seely-Brown, 1998). For these authors the term CoP helps to articulate how individuals, within the context of the formally created enterprise (*the organisation* or *the firm)* identify their beliefs (values and aspirations) with that of the enterprise over a period of time.

Common Ground and Community of Practice

Common ground (CG) is defined in terms of the interaction between two people: "The sum of their mutual, common, or joint knowledge, beliefs, and suppositions…a form of self-awareness" (Clark, 1996, pp. 93-94). The ground evolves in a principled manner.

In language-based communication, the common ground enables two language users to coordinate their actions. Clark divides the shared bases for the coordination into two types: *communal common ground*, defining cultural communities, and *personal common ground*, helping to distinguish between friends and strangers. The communal common ground is "a large mental encyclopaedia…divided into chapters by cultural communities" (Clark, 1996, p. 106). The communal CG could be divided into five content areas: human nature—"people in general"; cultural facts, norms, and procedures; and "ineffable experiences that others cannot understand unless they have them; grading of information; and communal lexicon." Cultural communities develop communal lexicon, and "many inferences are based narrowly on the language communities we know some one belongs to […] Word knowledge […] divides into […] sets of word conventions in individual communities." Such lexicons include dialects, technical terminology, 'academese', and 'medicalese' (Clark, 1996, p. 107): essentially the language of individual communities or Language for Special Purposes of the community.

The *personal common ground* (PCG) relies on 'joint personal experiences': strangers share no PCG, acquaintances have limited PCG, friends have extensive PCG, and intimates have extensive PCG and private information. The personal *lexicon* facilitates communication between acquaintances, friends, and intimates. This lexicon appears to be that of everyday language.

Community Practice and Knowledge Creation

For Nonaka and Takeuchi (1995), knowledge is created through a cyclical process of the tacitly held experience-based knowledge being converted into formalised, symbolic, and publicly available *explicit* knowledge, and vice versa. The word *tacit* is rooted in the Latin *tacitus* meaning 'silent', and its synonyms include *understood, implied, unexpressed,* and *silent*. 'Explicit' contrariwise means 'to unfold', and its synonyms include *categorical, definite, express,* or *specific*. There are two exemplar instances of how tacit knowledge is transferred from a practitioner to another within an organisation or across organisations. Heath and Luff (1996) describe how operators in the control room of an underground railway system learn from a (more experienced) colleague by watching the colleague carefully while he or she is involved in making critical decisions. Nonaka and Takeuchi (1995, p. 63) describe how an innovative team at Matsushita Electric Company (Japan), comprising production engineers and software developers, involved in the design of a home bakery, learnt the intricacies of bread making from a well-known master baker at an Osaka hotel, especially the final twist of the dough before the master placed the dough in the oven. The final twist was then engineered within Matsushita's home bakery system, which could then be used to make bread of about the same quality as that bought from a bakery. Both groups—the underground operators within an organisation, and the production engineers/software developers in one organisation and the master baker in another—learnt the knowledge, which is seldom articulated.

Explicit knowledge, formalised and symbolically coded, is grounded in one or more theories, it is usually independent of the context, and the explication is generally equated with rationality. The developments in science, engineering, and lately in biomedicine, demonstrate the triumph of rationalism. The individual is almost excluded in the explicit articulation of knowledge, parenthesised or rendered into numbers in a footnote. The use of mathematical and logic notation and systems makes the explicit knowledge independent of context (see, for instance, Kuhn, 1962, 1999).

Nonaka and Takeuchi (1995, pp. 62-73) discuss how knowledge can be created through a conversion process: the conversion of tacit knowledge into explicit and vice-versa through four modes: *socialisation* converting tacit into tacit, *externalisation* converting tacit into explicit, *combination*—explicit into explicit, and *internalisation*—explicit into tacit.

The use of language varies from a minimal use in the socialisation mode (the mode for sharing the 'ineffable background' (Clark, 1996)), through to a sophisticated use of the language as in the creation of metaphors and analogies in the externalisation mode. The combination mode relies extensively on both spoken and written language, and includes telephone and face-to-face conversation, written reports, and inter-office memos: this language-based communication relies on the existence of a communal lexicon. The internalisation mode involves the use of oral stories in addition to the texts used in the externalisation mode. Once internalised the knowledge is put into practice, and the socialisation mode starts up followed by the other three—knowledge spirals in the organisation. Nonaka and Takeuchi (1995, p. 14) emphasise the need for a 'common cognitive ground' among employees for disseminating tacit knowledge and spreading new explicit knowledge through the organisation.

It is the lexical choice of the key protagonists in the community, and the consensual acceptance or rejection of the chosen lexical items by the others in the community, which can be observed relatively easily. It was Thomas Kuhn who posited the lexicon centrally in knowledge evolution: "to possess a lexicon, a structured vocabulary, is to have access to the varied set of worlds which that lexicon can be used to describe" (Kuhn, 1999, p. 300).

SPECIAL LANGUAGE AND COMMUNITIES OF SPECIAL INTEREST

According to the *Shorter Oxford English Dictionary* (1973), the words in a language like English "are classifiable according to the sphere of their currency and usage." There are 'common words' in which literary and colloquial usage meet. 'Scientific', 'foreign', and 'archaic' words are the specially learned outposts of the literary language; 'technical' and 'dialectal' words blend with the common language. 'Slang' touches the terminology of trades and occupations; 'slang', dialect, and 'vulgar' speech "form a group of lower or less dignified status" (1973, p. x).

Scientists and technicians build the stock of words related to their science and technology over the years initially. This stock is built using three mechanisms. First, the specialists deploy a word that is in current literary/colloquial usage to tell each other about their experiments, observations, and theories. Second, the specialists borrow words from other languages to do the same. Third, and comparatively infrequently, the specialists invent a neologism or add a new word to the stock of their language. The emergence of a community of practice, comprising scientists, technicians, novices, supporters, and dissenters, depends upon specialising a select stock of their natural language for furthering their aims and aspirations.

The stock of words is used for communicating within a small community of a given specialism, for example as learned journal articles or personal correspondence. A selection of these words is then used to transmit the knowledge of the application, as in installation/repair manuals and technical notes for example, and for inducting novices into a given specialism, with the chosen words appear in popular science magazines and advanced textbooks. If the community's influence increases still further, then these words are used in texts for teaching the specialism in primary and secondary school curricula. Often the words and works of specialists attract the attention of the public at large and enter the realm of newspaper speak. Words are used to weave a variety of text types—some for the use of the few and others for the many.

The world of the specialist relies on language, and language appears to cope well with the various leaps of imagination: Louis Pastuer articulated the germ theory of disease; James Watson, Francis Crick, Maurice Wilkins, and the 'forgotten heroine' Rosalind Franklin (Maddox, 2003) argued that nucleic acids are the bases of life. Language is used to represent the objects of immediate interest to one or more specialists, and helps the specialist to describe real and imaginary worlds; words of everyday usage are incorporated into complex phrases and sentences. For instance, we have '*rules* of behaviour', '*laws* of nature', 'genetic *code*', and there is '*parallel distributed* cognition'. Occasionally, scientists add to the stock of words—there are *quarks* and *leptons* in English and other languages; the acronym *LASER* (*L*ight *A*mplification by *s*timulated *E*mission of *R*adiation) is a contribution of the specialists together with lasers, lasered substrates and logos, and lasering-in. Special language texts show how a natural language can be supplemented by logic and nomenclature and graphs and images and all subsumed in text. Scientists have to learn to be good at using language to express the unseen, the counter-intuitive, and the novel. Such accounts "seem tightly congruent with repeated experience and precisely predictive of future experience" (Bazerman, 1988, p. 292). The same language is used sometimes to produce golem science: inaccurate claims that nuclear fusion can occur at ordinary temperatures (Collins & Pinch, 1998). Recently, there was a case where a group of scientists in semiconductor physics was reprimanded for making inaccurate claims (Service, 2002; Beasley, Datta, Kogelnik, Kroemer & Monroe, 2002), and the scientists had to retract over 40 journal papers published in prestigious journals (Schön & Bao, 2003).

Special languages make use of the systematic nature of language. The productive use of morphology is rife in a special language where essentially a small vocabulary, ten or hundreds of very frequently used nouns, is used to write documents that comprise millions of words. A term is used on its own, suffixes (in English) added to make plurals, adjectives and other nouns used with the select class of frequently used headwords to make compound terms, and infrequently used verbs made into nouns and then frequently used terms. The inflection (singulars into plurals), compounding (adding adjectives and other nouns with nominal heads), and derivation (involving changing the grammatical category) are morphological processes used to great effect in science and technology.

The role of a community of specialists, and eventually the larger community that hosts the specialists, is quite crucial in accepting or rejecting keywords, writing styles, and rhetorical devices. The acceptance of the concept of *zero* illustrates the role of the specialists and the hosts. Zero started life 1,500 years ago with the Sanskrit/Hindi *sunya* and thence onto the Arabic *sifr*. Initially Europeans dismissed the concept as "Saracen magic" (Seife, 2000; Kaplan, 2000), and it was only in the 17th century that *zero* was introduced to European mathematics.

Next, the role of a special language, especially the communal lexicon, in creating a community will be outlined with the help of a case study.

CREATING A COMMUNAL LEXICON: A CASE STUDY

Healthcare has many stakeholders, and it is essential that the stakeholders understand each other as much as possible. Cancer care shows the importance of communication in a life-critical area: cancer is a disease of our times. Community action related to this disease ranges from international charities to village-based self-help groups; scientists, research funding agencies, and pharmaceutical companies; and the media are equally interested.

The advent of the Internet shows new forms of community action. The Web site of the American Cancer Society (ACS) and the U.S. National Cancer Institute (NCI) shows the action translated into an information service—the service is available to different types of users: researchers, professionals, patients, survivors, ACS supporters, and 'everyone.' The Web site has information about types of cancer, and the patients, researchers, and professionals are provided with 'tools' to select and be informed about various disease aetiologies. Information for researchers and professionals is written by and large by the members of the group itself. For other groups it appears that the primary authors are the researchers and the professionals, together with technical writers. The natural language of information in the ACS is English. The two text types we are interested in are the papers written by researchers and professionals for communicating amongst themselves, together with the third type, written for patients.

A measure of cohesion in the community will be the commonality of keywords and a common understanding of the same (the emergent communal lexicon of the society supported by an online glossary). Our hypothesis is that terms emerge and become established over time through a process of in-text negotiation, or the community's neglect kills them off. Researchers focus on new ideas and associated keywords; the professional summarily takes note of the researchers' fashion. Professionals prefer novel ideas where the risks are well understood and their use of terminology will reflect this. The information for patients has to be couched in terms that may be familiar to the patients and focused on tried and tested concepts.

Shared Lexicon and the Emergence of a Community: Borrowing from General Language

We use methods developed by (computational) lexicographers for deciding whether or not a word is eligible for entry in a standard reference dictionary. Such decisions are largely based on the intuition of the lexicographer concerned. During the 1970s, scholars led by Randolph Quirk (Quirk, Greenbaum, Leech & Svartvik, 1985) and John Sinclair (1987) challenged this orthodoxy and suggested that lexicographical decisions should be informed by evidence from language users. Quirk and Sinclair suggested that texts produced in a language can be systematically sampled, collected, and analysed for gathering evidence for *language in use*—such a collection is usually called a *corpus*. Herbert Clark appears to follow the same approach to language. Major dictionaries of English, and increasingly other languages,

use a standard reference corpus. Such a corpus is a systematically sampled collection of texts and speech excerpts drawn from a large population of texts and speech communities. For Quirk et al. and Sinclair, a text corpus is a starting point of linguistic description or a means of verifying hypotheses about a language. And for us, a corpus is a starting point for studying specialist communities.

The key to corpus-based analysis of linguistic output is that the frequency of usage of a linguistic unit—words, phrases, and grammatical and semantic patterns—correlates with its acceptability within a linguistic community. Frequency metrics are then augmented by other statistical considerations. This method has been used to construct terminology dictionaries, knowledge bases, and ontology systems (Ahmad, 2001) on the one hand and to conduct studies in the evolution of science and technology on the other (Ahmad, 2002).

We have created a corpus of texts for monitoring the emergence of the cancer-care community by examining the language used in the corpus. Essentially there are three sub-corpora: the first written by experts for experts, the second written by professionals and experts for professionals, and the third written by professionals/experts or copywriters for patients. The texts in each corpus were randomly selected from cancer-care Web sites including the ACS and NCI (see Table 1 for details). The texts in the corpora were published between 1980-2004.

The analysis of these three corpora was compared and contrasted with a 'representative' sample of British English, between 1960-1990, created by

Table 1. The typology, composition, and sources of the three corpora

Corpus	Number of Texts	Total Number of Tokens	Source
Expert	300	114,394	Cancer research journals – mainly titles & abstracts.
Professional	1,000	226,464	Web sites: U.S. National Cancer Institute, National Library of Medicine – mainly full-text articles. Journal of the American Medical Association – mainly titles & abstracts.
Patient	800	464,000	Web sites: ACS, NCI, Cancer Research UK, Alliance of Breast Cancer Organisations, and Bay Area Tumor Institute (California) – mainly full texts.

academics and lexicographers and funded by the UK government, called the British National Corpus (BNC). The BNC comprises over 100 million words in more than 4,000 texts drawn from 10 different text types, including fiction, news reportage, scientific, and business texts (Aston & Burnard, 1998). The BNC represents the language of everyday usage, and the three corpora represent various sub-communities in the larger 'cancer' community.

The 'Weirdness' of Special Languages

The general language genre is replete with closed class words, words like *the, and, if* belonging to grammatical categories whose stock is not renewed regularly and includes determiners, pronouns, conjunctions, and prepositions. These are also called grammatical or function words. The stock of open class words, nouns, adjectives, some verbs, and adverbs is renewed regularly. There are only two open class words, *time* and *person*, amongst the 100 most frequent word tokens in the BNC, and the rest are closed class words. These 100 make up just under half of the 100 million words, and the first 10 comprise a quarter of the BNC. The three corpora—Patient, Professional, and Expert—show the dominance of closed class words as well. As in the BNC, the first 10 most frequent word tokens account for 25% of each of the corpora and the first 100 account for just under 40% of all the texts. But there is a clear ingress of the open class words. A comparison of the first 50 most frequent words clearly shows that the two tokens *breast* and *cancer* are amongst the first 10 most frequent in all three cancer corpora; the experts show a penchant for abbreviations and use the newly found BRCA1 (first reported in 1993/1994) amongst the 10 most frequent (see Table 2 for details).

The next 10 most frequent tokens comprise one open class word in the Patient corpus (*women*), three in the Professional corpus (*women, risk,* and *patients*), and four in the Expert's corpus (*BRCA2, families, risk, mutation,* and *mutations*) (see Table 3 for details).

The next 30 most frequent words show that there are 12 and 14 open class words in the Professional and Expert corpus; the Patient corpus has only 7. Experts have a greater tendency of using plurals (*families, mutations, cells*). Note the very low frequency of personal pronouns (e.g., *I, you, your,*

Table 2. First 10 most frequent words in four corpora in rank order

RANK	BNC	PATIENT	PROFESSIONAL	EXPERT
1	the	the	a	Of
2	of	of	of	The
3	and	to	the	In
4	to	and	and	and
5	a	a	in	cancer
6	in	breast	cancer	To
7	that	cancer	to	A
8	it	is	breast	breast
9	is	in	with	BRCA1
10	was	or	for	with

Table 3. The 11th-20th most frequent words in the four corpora

RANK	BNC	PATIENT	PROFESSIONAL	EXPERT
11	I	for	women	for
12	for	are	was	that
13	s	you	were	BRCA2
14	on	that	risk	were
15	you	be	patients	is
16	he	your	or	was
17	be	with	is	by
18	with	have	that	families
19	as	women	at	risk
20	by	it	on	mutation

they, their) in the Professional and Expert corpora as compared to the BNC and the Patient corpus; the use of this category shows a conscious attempt towards inclusiveness.

On the basis of the first 50 most frequent terms, it appears that frequently used terms in the Professional corpus (*tamoxifen, chemotherapy, estrogen*) are used with lesser frequency in the Expert corpora. Professionals are a kind of a halfway house for a term initially used very frequently by experts. We have noted that *BRCA1* and *BRCA2* are making

Table 4. Weirdness ration of frequent 'open' class word in the four corpora—Inf stands for infinity (i.e., a number); frequency in the specialist corpus is divided by zero in the BNC

Expert (N=114,394)	f_{Exp}	f_{Exp}/f_{BNC}	Professional (N=226,464)	f_{Prof}	f_{Prof}/f_{BNC}	Patient (N=464,000)	f_{Pat}	f_{Pat}/f_{BNC}
cancer	1.87%	443	cancer	1.41%	320	breast	2.19%	769
breast	1.39%	831	breast	1.25%	430	cancer	2.18%	465
BRCA1	1.37%	Inf	women	0.64%	11	women	0.96%	15
BRCA2	0.71%	Inf	risk	0.56%	43	treatment	0.61%	47
mutation	0.49%	1014	patient	0.53%	24	risk	0.47%	33
families	0.53%	63	treatment	0.27%	22	therapy	0.32%	153
risk	0.50%	41	therapy	0.23%	116	surgery	0.28%	100
ovarian	0.39%	7893	tamoxifen	0.21%	7,149	chemotherapy	0.26%	969
gene	0.33%	148	chemotherapy	0.20%	757	cells	0.30%	23
carriers	0.33%	512	estrogen	0.20%	Inf	lymph	0.29%	1316
women	0.23%	7	disease	0.20%	19	radiation	0.20%	108
DNA	0.23%	68	BRCA1 & BRCA2	0.20%	Inf	biopsy	0.18%	177
protein	0.22%	76	ovarian	0.19%	3,687	mastectomy	0.16%	5360
tamoxifen	0.21%	7242	family	0.13%	4	tamoxifen	0.15%	5265

inroads in the Professional corpus with commensurate ranks.

The ratio of relative frequency of a single word in a specialist corpus with that of the same word in a general language corpus may perhaps reveal the extent to which the particular word is used as term. If the ratio is close to unity, then the word is generally a closed class word or a noun of everyday usage. But if the ratio is much greater than unity, then the word usually belongs to the open class category and possibly is a term. If the word is not found in the general language corpus, then this word is neologism or a spelling mistake. The ratio has been called *weirdness:* higher weirdness reflects the use of a word preferentially in one specialist domain as compared to its everyday usage. Table 4 is a computation of the weirdness ratio for the 10 open class words amongst the first 100 most frequent in our Expert's corpus as measured against the BNC. The results for the Professional and Patient's corpus are given for comparison as well.

Single words like *mutation, ovarian,* and *tamoxifen* are 'terms' as their weirdness is very high—over 1,000. The very frequent use of *breast* and *cancer*, weirdness of at least 300, suggest that these are 'terms' as well.

Morphological Productivity and Compound Terms

Domains are distinguished by the productive use of certain terms and, apart from inflectional and derivational use of these terms, much of the productivity manifests itself in the frequently used compound noun phrases that comprise one or more highly frequent single words that give the idiosyncratic lexical signature to a given specialist domain. Compound words often convey a semantic relationship between the constituents' words as well; semantic deals with the study of the meaning of words.

Consider a compound word $a+b$ where a and/or b are both highly frequent single-word terms. In our three corpora, $a=breast$ and $b=cancer$. It is not easy to contrast the frequency of specialist compound words across corpora, and indirect and usually illustrative statistics are used. We have used the mutual information (MI) measure. This measure explains the amount of information provided by occurrence of the term a about the occurrence of the term b in a compound $a+b$. If MI for a pair of single words is greater than zero, then the co-occurrence of the pair is not by chance and suggests that the pair may be a compound term. The occurrence of the term *breast cancer* in four corpora under consideration is related to the mutual information of the terms 'breast' and 'cancer' (see Table 5a).

Table 5a. Mutual Information (MI = log₂(f(a,b)/(f(a)×f(b))) computation for frequently occuring compounds

Corpus Name (& Number of Tokens)	BNC (100,106,008)	Patient (473,346)	Professional (226,464)	Expert (114,394)
Tokens	f_{BNC}	$f_{Patient}$	$f_{Professional}$	f_{Expert}
breast cancer	207	2559	859	376
breast	1615	9843	3200	1594
cancer	4204	9718	2840	2138
Mutual Information (MI)	11.58	3.66	3.66	5.90

Table 5b. Mutual Information amongst the eight most frequent single words (breast & cancer; hormone & therapy; estrogen & receptor; BRCA1 & mutations) in the three corpora

Corpus Name (& Number of Tokens)	BNC (100,106,008)	Patient (473,346)	Professional (226,464)	Expert (114,394)
Compound	MI_{BNC}	$MI_{Patient}$	$MI_{Professional}$	MI_{Expert}
breast cancer	11.58	3.66	3.66	5.90
hormone therapy	8.90	8.86	7.13	8.36
estrogen receptor	11.35	6.04	5.56	7.19
BRCA1 mutations	N/A	2.79	5.40	4.97

The MI for all the four corpora is greater than zero, indicating that *breast cancer* is a 'term'; this is perhaps related to the fact that the constituents of the compound occur very frequently. Candidate compound terms can be extracted and mutual information computed to determine for termhood, as shown in Table 5b.

The high value of MI in the three corpora suggests that these are indeed compound terms.

Shared Lexicon and the Emergence of a Community: Neologisms and Borrowing from Other Languages

Neologisms

The establishment of a new term within a communal lexicon can be visualised by looking at the development of the idea that breast cancer is a hereditary disease. The discovery that "human breast cancer is usually caused by genetic alterations of somatic cells of the breast, but occasionally, susceptibility to the disease is inherited" was made in 1990. By 1993, there was an agreement that "breast cancer is known to have an inherited component." The researchers did not yet have a name for this gene. Two

papers published in the same issue of the journal *Science* on October 7, 1994, named the candidate gene, or more accurately protein, as BReast CAncer gene 1 (BRCA1): "It is a paradoxical gene located on chromosome 17 that normally helps to restrain cell growth, however inheriting an altered version of BRCA1 predisposes an individual to breast, ovary, and prostate cancer." BRCA2 was named in 1995 and is located on chromosome 13. BRCA1 "confers higher risk of ovarian cancer" and BRCA2 that of "male breast cancer."

A diachronic analysis of the two terms shows how these concepts start life as 'candidate genes' and 10 years later these two become breast cancer 'predisposition genes'. The search engine Google™was given the phrases 'BRCA1' and 'BRCA2'. Most of the retrieved documents are held in a text archive, mainly of journal articles, created by the U.S. National Institute of Health. We selected 10 papers: five from the two half years (October 1994-June 1995) in which the terms were apparently coined, and the others in the period of consolidation (1996-2004). A keyword in context concordance shows this metamorphosis from concept to 'reality', and the two terms entered the communal lexicon of the cancer community: from a mere suggestion and 'strong candidate' in 1994, the

Table 5c. *An example of the diachronic analysis of the terms BRCA1 and BRCA2*

1994 October	familial tumours *suggest* that	**BRCA1**	a gene that confers susceptibility to ovarian and
1994 October	a *strong candidate* for the 17q-linked	**BRCA1**	gene, which influences susceptibility to breast... cancer
1994 December	mutations in the *coding region of the*	**BRCA1**	candidate gene
1995	*dominant* susceptibility genes	**BRCA1**	and BRCA2
1995		**BRCA1**	*confers* higher risk of ovarian cancer
1995	*confers* higher risk of ovarian cancer and	**BRCA2**	much higher risk of male breast cancer.
1996	*the breast* cancer susceptibility gene	**BRCA2**	in mammary epithelial cells
2000		**BRCA1**	**and BRCA2** *are* breast cancer susceptibility genes.
2001	*2 breast cancer* predisposition genes,	**BRCA1**	**and BRCA2** in the mid-1990s
2004	the	**BRCA1**	*protein* presents a paradox to the scientists

term within a year becomes 'dominant' and 'confers' risks. In 2000 the terms *BRCA1* and *BRCA2* are emphatically the genes (see Table 5c).

Borrowings

The terminology of medicine is largely based on terms that have roots in the two classical languages of Greek and Latin, and modern bio-medicine has terms from German as well. The specialist domain of breast cancer shows a mixture which borrows many of its frequently used terms from these languages as was confirmed by an analysis of our corpora. Additionally, pharmaceutical companies sometimes name drugs by using a combination of the letters of the chemical formula for the drugs. Cancer care drugs show this tendency as well (Table 6 shows some exemplar borrowings).

THE FUTURE: A KNOWLEDGE MANAGEMENT SPIDER

The information dissemination activities of U.S.-based cancer charities and research foundations appear related to similar efforts in knowledge management for monitoring and maintaining repositories of formal and informal documents generated and used by a variety of users. Siemens' *Merger* and *Acquisitions Knowledge Environment* (MAKE) was developed for managing the various stages of corporate mergers or acquisitions (Kalpers, Kastin, Petrikat, Scheon & Späth, 2002). *MAKE*, an information *spider*, is designed to capture the knowledge of mergers and acquisitions (M&A) experts together with documentation related to the instruments of M&A, and oral stories of successful and failed M&As. MAKE additionally comprises a

Table 6. *Etymology of some frequently used terms in our three specialist corpora*

Term	Etymology
biopsy	Greek
cancer	Greco/Latin
chemo-therapy	German
estrogen	Latin
gene	German
lymph	Greek
mastectomy	Greco/Latin
mutation	Latin
ovarian	Latin
protein	German
tamoxifen	Acronym of T(rans) AM(INE)+OXY+PHEN(OL)*
therapy	Latin

* *with alteration of y (for i) and ph (for f)*

glossary of terms. Lesser and Storck (2001) have stressed the need for a repository which will act as a virtual workspace for the community to share and learn.

An information spider is a document management system designed to facilitate the work of teams that coalesce into a community. This work is to engender a sense of common identity and purpose amongst the merging organisations, or amongst those who are taking over and those taken over. The system gives users access to a range of documents through indexing and cross-indexing programs. The glossary can be updated manually.

We have developed a prototype information spider for studying how to disseminate tacit and explicit knowledge in an emergent discipline, cancer care for example. This system can classify documents at different levels of linguistic description—keyword and factor analytic descriptions. The prototype system can: (a) extract (candidate) terms, (b) index and cross-index documents, (c) identify names of the original authors and cited authors together with their respective organisations, and (d) summarise documents. The architecture of the prototype is shown in Figure 1, where K-D represents a *knowledge document*.

The prototype currently comprises the three corpora mentioned above. The spider is currently powered by *System Quirk*—a text and terminology management system (Ahmad, 2001). The prototype is being developed in close collaboration with the University of Surrey's Centre for Healthcare Workforce Management, and the initial results are encouraging.

CONCLUSION

The emergence, establishment, and eventual extinction of a community of practice almost invariably leaves a trace—the linguistic output of the community. This output, a small yet significant record of the activities of the community, is written in a subset of the everyday language. The special language of the community has a small lexicon, which is used productively and in many ways quite creatively. For us, the key here is that the vocabulary, or terminology of a community, evolves through a consensual process. The coinage, currency, usage, and ultimate obsolescence of terms show how a community of practice creates a common ground. The consensus, occasionally subverted for personal gratification, is essential: (a) for reaching a common understanding of the beliefs and values the community seeks to promote, (b) for developing a shared understanding of complex systems of ideas that the community develops, and (c) for sharing knowledge within and across organisations. For some, the relationship between knowledge and language is a symbiotic one: knowledge (the 'message') is articulated through lan-

Figure 1. The architecture of the Surrey Health Care Spider

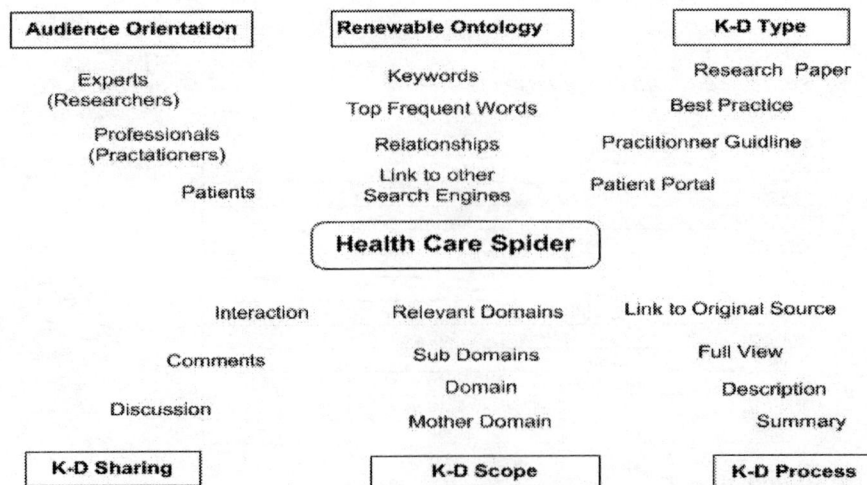

guage (the 'medium'), and language, in turn, is enriched by the community of knowledge workers through the re-interpretation of extant words, the occasional creation of a new word, and through new styles of writing (for example, the learned paper, the poster presentation) and speech (for instance, the plenary lecture, the public lecture, awareness campaigns). For others, language is just a symbol set, a weak placeholder for profound thoughts and complex artefacts; this one of the many symbol sets available to humans including mathematical symbolism, graphical symbolism, and so on. We conclude by noting that what is important is that after a community member departs from the community or the world itself, all we have is his or her writings (and now speeches)—a trace of endeavour, a trace of contribution to the establishment or extinction of a community.

ACKNOWLEDGMENTS

Khurshid Ahmad wishes to thank the UK Economics and Social Sciences Research Council's e-Science Programme (FinGrid, Grant Number: RES-149-25-0028) for its support. Rafif Al-Sayed wishes to thank the University of Surrey for a doctoral scholarship. Both the authors wish to thank Sophie Gautier O'Shea for her diligent proofreading.

REFERENCES

Ahmad, K. (2001). The role of specialist terminology in artificial intelligence and knowledge acquisition. In S.E. Wright & G. Budin (Eds.), *Handbook of terminology management* (vol. 2, pp. 809-844). Amsterdam; Philadelphia: John Benjamins.

Ahmad, K. (2002). Scientific texts and the evolution of knowledge—The need for a critical framework. In M. Koskela, C. Lauren, M. Nordman, & N. Pilke (Eds.), *Proceedings of the 13th International Conference on Language for Special Purposes, LSP,* August 2001 (pp. 123-135). Vaasa: University of Vassa Publications Unit.

Aston, G., & Burnard, L. (1998). *The BNC handbook.* Edinburgh: Edinburgh University Press.

Bazerman, C. (1988). *Shaping written knowledge: The genre and activity of the experimental article in science.* Madison; London: The University of Wisconsin Press.

Beasley, M.R., Datta S., Kogelnik H., Kroemer, H., & Monroe, D. (2002). *Report of the investigation committee on the possibility of scientific misconduct in the work of Hendrik Schön and co-authors.* Retrieved November 8, 2004, from *http://www.lucent.com/news_events/pdf/summary.pdf*

Bennet, A., & Bennet, R. (2003). The partnership between organizational learning and knowledge management. In C.W. Holsapple (Ed.), *Knowledge management handbook: Volume 1—knowledge matters* (pp. 439-455). Berlin: Springer-Verlag.

Clark, H. (1996). *Using language.* Cambridge: Cambridge University Press.

Collins, H M., & Pinch, T. (1998). *The golem: What you should know about science.* Cambridge; New York: Cambridge University Press.

Goodman, N. (1978). *Ways of worldmaking.* Indianapolis, IN: Hackett Publishing Company.

Heath, C., & Luff, P. (1996). Convergent activities: Line control and passenger information on the London underground. In Y. Engestrom & D. Middleton (Eds.), *Cognition and communication at work* (pp. 96-129). Cambridge: Cambridge University Press.

Kalpers, S., Kastin, K., Petrikat, K., Scheon, S., & Späth, J. (2002). How to manage company dynamics: An approach for mergers and acquisitions knowledge exchange. In T. Davenport & G. Probst (Eds.), *Knowledge management case book—best practises* (2nd ed., pp. 187-206). Munich: Publicis Corporate Publications and John Wiley & Sons.

Kaplan, R. (2000). *The nothing that is: A natural history of zero.* London: Allen Lane The Penguin Press.

Kuhn, T. (1962). *The structure of scientific revolutions.* Chicago: The University of Chicago Press.

Kuhn, T. (1999). *The road since structure.* Chicago: The University of Chicago Press.

Lesser, L., & Storck, J. (2001). Communities of practice and organizational performance. *Knowledge Management. IBM Systems Journal, 40*(4), 831-841.

Maddox, B. (2003). The double helix and the "wronged heroine." *Nature, 421,* 407-408.

Nonaka, I., & Takeuchi, H. (1995). *The knowledge-creating company.* New York; Oxford: Oxford University Press.

Quirk, R., Greenbaum, S., Leech, G., & Svartvik, J. (1985). *A comprehensive grammar of the English language.* London; New York: Longman Publishers.

Schon, J.H., & Bao, Z. (2003). Retractions: Nanoscale organic transistors based on self-assembled monolayers. *Applied Physics Letters, 82*(8), 1313-1314.

Seely-Brown, J. (1998). *Research that reinvents the corporation. Harvard business review on knowledge management* (pp. 153-180). Boston: Harvard Business School Press.

Seife, C. (2000). *Zero: The biography of a dangerous idea.* New York; London: Viking Penguin Books.

Service, R.F. (2002). Scientific misconduct: Bell Labs fires star physicist found guilty of forging data. *Science, 298,* 30-31.

Shorter Oxford English Dictionary. (1973). Oxford: Oxford University Press.

Sinclair, J. (Ed.). (1987). *Looking up: An account of the COBUILD project in Lexical computing.* London; Glasgow: Collins ELT.

KEY TERMS

Common Ground: A form of self-awareness in the interaction between two people; the sum of their mutual, common, or joint knowledge, beliefs, and suppositions.

Corpus: Any systematic collection of speech or writing in a language or variety of a language. A corpus is often large and diverse, and can be classified according to contexts or styles tagged and indexed for specific features.

Golem: "A creature of Jewish mythology...a humanoid made by man from clay and water, with incantations and spells...The idea of a golem takes on different connotations in different legends...in some the golem is terrifyingly evil [and in others it is a] metaphor of bumbling giant." (Collins & Pinch, 1994, pp. 1-2)

Lexicographer: A compiler or writer of a dictionary for practical use or for any other purposes.

Languages for Special Purposes (LSP): The languages used for particular and restricted types of communication (e.g., for medical reports, scientific writing, air-traffic control). They are semi-autonomous, complex semiotic systems based on and derived from general language; their use presupposes special education and is restricted to communication among specialists in the same or closely related field.

Mutual Information (MI): A measure that explains the amount of information provided by the occurrence of an event, say *a*, about the occurrence of another event, say *b*.

Mutual Information $(MI = \log_2(f(a,b)/(f(a)^2 \times f(b)))$

Spider System: A document management system which acts as spider—each leg of the spider represents a dimension or category that serves to classify documents.

Weirdness Ratio: The ratio that reflects preferential use of a word in a specialist domain as compared to its everyday usage.

A Comparison of the Features of some CoP Software

C

Elayne Coakes
University of Westminster, UK

INTRODUCTION

There are now significant numbers of software houses supplying services and solutions for community collaboration. In this article we briefly review the requirements for virtual support and the current offerings. This is not intended as a comprehensive survey, but rather an overview of what might be available.

BACKGROUND

In 2004 the Directorate of Science and Technology Policy (DSTP) in Canada produced a report reviewing portal technology. In particular, DSTP reviewed a specific subset or portals for community support. They looked at four specific program offerings, operating under portals, across eight areas of functionality. These eight areas were:

1. ongoing interactions,
2. work,
3. social structures,
4. conversation,
5. fleeting interactions,
6. instruction,
7. knowledge exchange, and
8. documents.

These program suites—Tomoye, communityZero, iCohere, and Communispace—were all strongly oriented towards Fleeting interactions and Instruction (apart from iCohere), but weakly supportive of social structures, knowledge exchange, and documents. In addition, all software suites contained taxonomy, a local search, an experts database, discussion, and an events notification facility. None provided audio- or video-supported meetings or webinars, and only Communispace provided a (limited) virtual meeting space. All, except for Tomoye, provided community governance and polls.

Other Software Offerings

Enable2 was not considered by DSTP. It is provided by Fount Solutions, who claims that it provides the essential capabilities required for CoP support. These, they say, would include: content management (to generate domain-specific content), discussion forums, document management, member profiles, and a search engine. As we see, the 'missing' capabilities of this software suite are also missing from the software reviewed by the DSTP—that is, support for audio and video meetings, webinars, and virtual meeting spaces. Fount also recommends the provision of weblogs so that users can publish specific content and a tool called Really Simple Syndication

Table 1. Core technology features

Relationships	Learning	Knowledge	Action
Member networking profiles; Member directory with 'relationship-focused' data fields; Subgroups that are defined by administrators or that allow members to self-join; Online meetings; Online discussions.	Recorded PowerPoint presentations; E-learning tools; Assessments; Web conferencing; Online meetings; Online discussions; Web site links.	Structured databases; 'Digital stories'; Idea banks; Web conferencing; Online meetings; Online discussions; Expert database and search tools; Announcements; Web site links.	Project management; Task management; Document collaboration; File version tracking; File check-in and check-out; Instant messaging; Web conferencing; Online meetings; Online discussions; Individual and group calendaring.

or RSS. RSS is used to enable users to subscribe to content sources that match their specific interests.

Livelink for Communities of Practice is relatively new software that was launched in 2004 by OpenText ™Corporation (www.contentmanager.net and www.opentext.com/solutions/platform/collboration/communities-of-practice). Livelink also provides weblogs, FAQs, webcasts, an experts database, forums with threaded discussions, and role-based permissions for community users so that they can perform specified tasks.

Sitescape (BillIves, 2004) also launched new CoP software in 2004. This software provides both synchronous and asynchronous communication facilities, document management, shared scheduling, and instant messaging, as well as a number of task- and process-based tools. Web meetings, white boards, videoconferencing, and voiceover IP are also supported.

iCohere in its CoP Design guide (available from www.icohere.com) states that there are four focal areas for CoPs—relationship building, learning and development, knowledge sharing and building, and project collaboration. The company also provides Table 1, which allocates core technical features to each focal area. Obviously, iCohere considers that its software offering provides these necessary features.

Member networking profiles;Member directory with 'relationship-focused' data fields;Subgroups that are defined by administrators or that allow members to self-join;Online meetings;Online discussions. Recorded PowerPoint presentations; E-learning tools; Assessments; Web conferencing; Online meetings; Online discussions; Web site links. Structured databases; 'Digital stories'; Idea banks; Web conferencing; Online meetings;Online discussions; Expert database and search tools;Announcements;Web site links. Project management; Task management; Document collaboration; File version tracking; File check-in and checkout; Instant messaging; Web conferencing; Online meetings; Online discussions; Individual and group calendaring.

FUTURE TRENDS

The software market for KM and IC management is competitive. Features and facilities are changing rapidly and developing in complexity. Increasingly the software for community support is being subsumed into the general KM management software, which is, in its turn, being incorporated into organisational portals. DSTP (2004) expects to see a rapid growth in the portal development market, with organisations integrating their applications to "facilitate creation, sharing and preservation, and intellectual capital management...This trend is eroding the benefits of specific community of practice tools" (p. 3).

CONCLUSION

Whilst this article was not intended as a comprehensive review of software for supporting CoPs, it has shown that many offerings lack some (apparent) essentials for a virtual community. Whilst many would be useful additions for a 'physical' community by providing shared documents and databases, few provide virtual meetings spaces and the possibilities for synchronous communication and the sharing of tacit knowledge—this latter being, of course, the prime driver behind the development of CoPs.

NOTE

The May/June issue of *KM Review* (Melcrum Publishing) also contains a useful comparison of CoP and collaboration software.

REFERENCES

BillIves. (2004). Retrieved from *http://billives.typepad.com/portals_and_km/2004/07/supporting_com.html*

Content Manager. (2004). Retrieved from *www.contentmanager.net/magazine/news_h9697_opentext-launches_*

DSTP. (2004). Retrieved from *http://pubs.drdc_rddc.gc.ca/inbasket/dguertin.040317-1030.p521012.pdf*

Fount Solutions. (n.d.). Retrieved from *www.enable2.com*

KEY TERMS

Portals: "Frameworks for integrating tools, applications, collaboration, and information that is shared across an organisation" (DSTP, 2004, Abstract, p. 3). Collects applications and Web sites together to provide a common look and feel.

Webinars: Seminars conducted 'on the Web' through the use of an intranet or the Internet.

C

The Concept of Communities of Practice

Elayne Coakes
University of Westminster, UK

Steve Clarke
University of Hull, UK

INTRODUCTION

This article looks at the concept of communities of practice (CoPs) in the workplace. The theories surrounding these types of communities are still very new and in the process of development. The practice and the importance of these communities for knowledge transfer is also still being explored as to best methods for establishing such communities and how to support and encourage them. Below we discuss the background and main threads of theory that are under development.

BACKGROUND

Communities of practice are becoming increasingly important in many organisations. As the APQC (2004) says:

CoPs are becoming the core knowledge strategy for global organizations. As groups of people who come together to share and learn from one another face-to-face and virtually, communities of practice are held together by a common interest in a body of knowledge and are driven by a desire and need to share problems, experiences, insights, templates, tools, and best practices.

To define a community of practice, it is worth considering the words of Etienne Wenger (2001), who is considered one of the foremost experts in this field. He says:

Communities of practice are a specific kind of community. They are focused on a domain of knowledge and over time accumulate expertise in this domain. They develop their shared practice by interacting around problems, solutions, and insights, and building a common store of knowledge.

The initial concept of communities of practice came out of work by Lave and Wenger (1991) relating to situated learning in the workplace and other communities with related interests. Thus such communities are an aggregation of people who are bound (in their specific context) to accomplish tasks or engage in sense-making activities (Brown & Duguid, 1991; Lave & Wenger, 1991). Learning, to Lave, was the transformation of practice in situated possibilities. Newcomers to a group learn from the old participants, bearing in mind that practices will change over time and place due to changes in circumstances. In addition, intergenerational relationships will affect the learning situation. There may well be a fear from the older group members in transferring knowledge to the younger—implying a loss of power and importance—or a fear from the new or younger group members of demonstrating ignorance. So the social process of knowledge acquisition affects the practice of knowledge sharing and the desire for knowledge sharing.

The context or domain for these communities is related to the subject matter around which they are formed. Within this domain, communities interact, learn, and build relationships in order that they may practice their skills through tools, frameworks, idea sharing, artefacts, or documents.

In this *Encyclopedia of Communities of Practice in Information and Knowledge Management*, a number of particular issues are covered in a multi-layered form. Here we see that such communities are governed by internal informal and unspoken rules dominated by specialised language development. We also see that there are issues in measuring the output and value of such communities for an organisation, that strategy needs to be developed uniquely for each community as well as for the organisation in general, and that how, or even whether, to reward participants is a matter of some debate. The psychology of participants and the difficulties with creating a shared meaning within a community can be explored through philosophy and psychology as well as organisational studies, and we find that many perspectives are available to understand communities and their actions. This being the case, many fields of study have a view on how and why communities work, and how and why people should or could participate in this work.

Focus on Communities

If we accept that the role of CoPs in the business environment is to share knowledge and improve the way the organisation does business, whether in the public or private sector; and that they are a community workplace where people can share ideas, mentor each other, and tap into interests (APQC 2002); each CoP can be a focus of learning and competence for the organisation. Much of the organisation's work can be facilitated or conversely frustrated through these communities, depending on how permissive or permitted they are. Organisational culture, it would seem, plays a great part in communities and how they operate. The members of a community need to trust the other members before they are willing to share their experience and understanding.

The bonds that tie communities together are both social and professional, and whilst they can be fostered and supported by organisations, they are not formed by them. Convincing people to participate in communities requires an ongoing commitment from the leaders within an organisation to permit communities to self-organise and collaborate, as they see fit, with suitable encouragement and support. Education plays a part in this encouragement, but so too does enthusiasm from amongst the community's members which will come from seeing the benefits to their own self-knowledge and development as well as a business value. Over-regulation or under-structuring can lead to a stale community or a community that fails to develop and thus eventually 'fails'. In addition, due to the voluntary nature of membership of such a community, some are affected when they become too prominent in an organisation and may disappear from view (Gongla & Rizzuto, 2004). This can happen in a number of ways. The community may apparently disappear whilst continuing to operate under the organisational surface, not wishing to become too obvious to the formal organisational structure or bound by its requirements. Other CoPs stop operating, merge with other communities, or re-define themselves. CoPs that become formal organisational structures because their work becomes necessary to organisational functioning lose much of what makes them a CoP and transform into project teams.

Vestal (2003) suggests that there are four main types of communities:

- **Innovation Communities:** that are cross-functional to work out new solutions utilising existing knowledge;
- **Helping Communities:** to solve problems;
- **Best-Practice Communities:** attaining, validating, and disseminating information;
- **Knowledge-Stewarding:** connecting people, and collecting and organising information and knowledge across the organisation.

Each of these community types will require different amounts, levels, and functionality of support. However, it is unwise for any business to rely on CoPs performing these tasks continuously or to a set standard, as their voluntary nature means that outside control should not, or cannot, be exercised directly or they may cease to comply with the tasks at hand.

Building a Community

Communities are easy to destroy but difficult to construct. Membership, and choice, of a community needs to be voluntary, otherwise members may not participate in the knowledge-sharing that is their 'raison-d'être'.

McDermott (1999) concludes that there are four challenges when building communities: (1) the design of the human and information systems to help the community members think together and interact; (2) to develop communities such that they will share their knowledge; (3) to create an organisational environment that values such knowledge; and (4) to each community member being open and willing to share.

CoPs differ from traditional team-working approaches in that they are most likely to be cross-functional and multi-skilled. They therefore align themselves closely to the sociotechnical ideals of inclusivity and fluid boundaries. CoP members will be drawn from those who wish to involve themselves, and who desire to share knowledge and learn from others about a specific topic, wherever in an organisation (and in some cases, outside the organisation too) they may be located. Functional position is irrelevant; topic knowledge or interest is all that is necessary to join a CoP. The diversity of a CoP's population may encourage creativity and problem solving, and linkages to external communities will also enhance their activities. CoPs are the legitimate place for learning through participation. They additionally provide an identity for the participator in terms of social position and knowledge attributes and ownership. CoPs will have a shared domain and domain language, and some members may become apprentices as they are acculturated into this domain and knowledge development. It is also important when establishing CoPs to think about the embedded habits, assumptions, and work practices or cultural norms that exist in the organisation. Communication and how, and where, as well when, people communicate are extremely important in relation to information sharing.

Communities (Brown argues in Ruggles & Holtshouse, 1999) are also the places that provide us with different perspectives and lenses through which to view the world. Successful communities maintain a clear purpose and active leadership (McDermott, 2004) and support innovation and staff creativity through collaboration and collective solutions. CoPs also provide members with the ability to self-start and search for information and support as required (Heald, 2004), including extended expertise—that is, expertise outside their immediate work environment.

FUTURE TRENDS

The evidence from the workplace is that ICT-supported strategies for CoP development are better than ICT-led strategies (Kling & Courtright, 2003) and that the sociotechnical approach is valid for CoP development. ICT has different roles to play as knowledge management systems are established and evolve in organisations—it moves from being the underlying infrastructure to the linking mechanism, to the support mechanism (Pan & Leidner, 2003). Yet without an understanding of the underlying work practices and organisational social and cultural aspects, the ICT support will not match the specific elements that make this organisational culture unique and thus will be ineffective. As Nick Milton of Knoco argues (*KMOnline,* 2004):

The best software to use is the one the community is most familiar with and is most prepared to use. Ideally one they are already using on a routine basis…why not let the community make the decision?…they can do much of their business through e-mail alone. Do they really need anything further?

In addition, in the same article (a collection of comments from an online community), Giles Grant of BNFI argues: "*IT should only be an enabler for sharing and collaboration. It isn't the community; the community is the people.*"

The future of CoPs, it would seem therefore, is an interesting one. There is increasing evidence that they are being formalised into organisational structures with budgets, resources, and tasks—thus becoming more like project teams with an aim and a strategy. As such, those who saw them as a means of social support and informal tacit knowledge sharing may choose go underground as discussed above, and the value of such groups to an organisation may be lost.

CONCLUSION

Thus we see from the discussion above some of the issues that surround CoPs and their establishment in the workplace. Too close to the formal structure and the community will transform into a project team and thus lose the learning and voluntary nature of participation that is so important. Too far from the formal structure and the community may not work towards an organisational goal. There is little agreement about how to support CoPs through technology or through organisational means. However, there is much evidence that communities are best left to self-organise and self-manage, and that any organisational outcomes are a benefit and not an expectation.

This article is but a brief summary of some of the more salient points relating to CoPs. It cannot cover all the issues and indeed is not intended to do so. It is instead intended to indicate to the reader the issues and potential areas of study that are related to current thinking.

REFERENCES

APQC. (2002). *Communities of practice.* Houston: APQC.

APQC. (2004). Retrieved February 14, 2004, from *http://www.apqc.org/portal/apqc/site/generic?path=/site/km/communities.jhtml*

Brown, J.S., & Duguid, P. (1991). Organisational learning and communities of practice: Towards a unified view of working, learning and organisation. *Organisation Science, 2*(1), S40-S57.

Gongla, P., & Rizzuto, C.R. (2004). Where did that community go? Communities of practice that disappear. In P. Hildreth & C. Kimble (Eds.), *Knowledge networks: Innovation through communities of practice* (pp. 295-307). Hershey, PA: Idea Group Publishing.

Heald, B. (2004). Convincing your staff on CoPs—what's in it for them? *KMOnline, 7*(2), 3.

Kling, R., & Courtright, C. (2003). Group behaviour and learning in electronic forums: A sociotechnical approach. *Information Society, 19*(3), 221-235.

KMOnline. (2004). In the know: Expert perspectives—what is the best software supplier for communities of practice? *KMOnline, 7*(2), 4.

Land, F.F. (2000). Evaluation in a socio-technical context. *LSE Working Papers.* London.

Lave, J., & Wenger, E. (1991). *Situated learning—legitimate peripheral participation.* Cambridge, MA: Cambridge University Press.

McDermott, R. (1999). Why information technology inspired but cannot deliver knowledge management. *California Management Review, 41*(4), 103-117.

McDermott, R. (2004). How to avoid a mid-life crisis in your CoPs. *KMOnline, 7*(2).

Pan, S.L., & Leidner, D.E. (2003). Bridging communities of practice with information technology in pursuit of global knowledge sharing. *Journal of Strategic Information Systems, 12,* 71-88.

Ruggles, R., & Holtshouse, D. (1999). *The knowledge advantage: 14 visionaries define marketplace success in the new economy.* Oxford: Capstone.

Vestal, W. (2003). Ten traits for a successful community of practice. *Knowledge Management Review, 5*(6), 6.

Wenger, E. (2001). Retrieved April 2004 from *http://www.ewenger.com/tech*

KEY TERMS

Community of Practice: A group of individuals which may be co-located or distributed, motivated by a common set of interests, and willing to develop and share tacit and explicit knowledge.

Domain: Scope or range of a subject or sphere of knowledge.

Domain Language: The language including specific technical terms, phrases, and shortcuts/abbreviations of speech that are unique and specific to the sphere of knowledge.

Sociotechnical: Derives from *Socius*—Latin for associate or companion—here meaning society and technical—that is, a solution produced by technological means, which derives from *Technologia*—Greek for systematic treatment.

Sociotechnical Thinking: A part of social theory and of philosophy. Its original emphasis was on organisational design and change management. The term "sociotechnical" means a task design approach that is intended to optimise both the application and development of technology, and the application and development of human knowledge and skill. The underlying philosophy of sociotechnical approaches is based essentially on two ideas focusing on the individual and the organisation. The first is the *humanistic welfare paradigm,* involving the redesign of work for autonomy, self-actualisation, the use of self-regulating teams, individual empowerment, and thus stress reduction. In this view the design of work systems is performed to improve the welfare of employees. The second (and perhaps contradictory philosophy) is the *managerial paradigm,* focusing on improving the performance of the organisation.

Context Based Approach to Applying Virtual Reality

Manumaya Uniyal
Loughborough University, UK

Ray Dawson
Loughborough University, UK

INTRODUCTION

It was Ivan Sutherland, nearly 30 years ago, who introduced the modern concept of VR in his thesis work (Sutherland, 1963). It has been 14 years since Jaron Lanier (1996) coined the term *virtual reality* to collectively present such ideas as formulated since Sutherland. Since then, VR has been offered as a one-stop solution for tackling issues as diverse as ranging from manufacturing and design to tourism. In fact, the liberal usage of the word *virtual*, often drawn from the term *VR*, is best summed up by Professor J. Vince when he says, "today we have virtual universities, virtual offices, virtual pets, virtual graveyards, virtual exhibitions, virtual wind tunnels, virtual actors, virtual studios, virtual museums, virtual doctors—and all because of VR" (Vince, 1998, p. 1). Unfortunately, even such worldwide media attention has been unable to help VR penetrate and broad-base itself across all market segments as had been predicted.

This article builds upon the above-mentioned issues while specifically focusing on:

a. Finding and understanding the reasons for an overall lack of enthusiasm for VR usage in the Cost Sensitive Organizations (CSO).
b. To develop and present a VR application methodology specifically for CSOs based on the findings of point "a".

BACKGROUND: A HISTORICAL OVERVIEW OF VR

One of the earliest examples of a VR-like representation can be traced back to the works of Mond.

Mond and Mackay, during 1914 to1916 (Mitchell, 1999), created a photographic archive of the interior of the Tomb of Menna. A rail-based camera rig was set up by Mond along the inside walls of the tomb. He shot 3,300 black and white photographs using 10.5cm by 8cm photographic plates. Photographs were placed so their edges would overlap. They were then touched up using a paint brush and rephotographed, thereby creating seamless panoramic shots. The panoramas were pasted onto cardboard walls, put up based on the floor plan of the tomb. As people walked along the cardboard walls, it created an impression as if they were walking inside the real Tomb of Menna.

Modern day VR can be said to start from the time of Ivan Sutherland who is aptly called the father of VR. It is he who, in his 1968 work, "The Ultimate Display", referred to computer rendered space as virtual worlds with the chief characteristic of the space being realism. Later in the year, he used a head mounted display for viewing his virtual worlds. In his paper "A Head-Mounted Three-Dimensional Display," he wrote, "Our objective in this project has been to surround the user with displayed three-dimensional information" (Sutherland, 1968).

Although numerous projects (mainly in America) were experimenting with VR, it was in 1989 that the term *virtual reality* was coined by Jaron Lanier. "VPL performed the first experiment in what I decided to call VR in the mid to late 1980's. VR combines the idea of virtual worlds with networking, placing multiple participants in a virtual space using head mounted displays" (Lanier, 1996).

It should be noted that realism in terms of virtual worlds and the hardware used to view virtual scenarios, that is, head-mounted displays, were considered important since the early days of VR—an image that persists today. Perhaps this was because, in its early days, aerospace sector, in general, and flight simulators, in particular, were one of the main areas of application where VR was experimented with.

VR AND ITS PERCEPTIONS

VR is predominantly considered a technology that falls in the realm of flight simulators, offshore oil rig simulation, construction industry, and so forth. Often, organizations (be they private corporations or government sectors like the Ministry of Defence) applying VR tend to have large budgets at their disposal. This popular view of VR regards the technology as a means of providing an alternative reality where the level of realism blurs the distinction between what is real and what is provided by the technology. Image of VR created in the films, sometimes termed *Hollywood Factor*, too, presents virtual worlds in the same light of high realism, as can be seen in films like *The Lawn Mower Man* and *The Matrix*.

Perception that high-end hardware, software, and a high level of reality are a prerequisite to VR is also confirmed in two surveys. The UKVR awareness survey (Bevan & Leston, 1999) looked at the UK business potential for VR in the small and medium enterprise (SME) sector confirming this expectation of VR among the business community. A further survey of 25 professionals in companies developing/applying VR applications or with a potential do so, conducted by the authors, also echoed findings similar to that of the UKVR awareness survey. It was found that SME's view of VR was often driven by the Hollywood ideal coupled with a perception of high costs. Another point that emerged was the fact that only few SMEs could find any corelation between VR and the real value it could add within their business process. UKVR awareness survey also had such responses as can be seen by the following statement, "VR has been well received and is clearly working, but it's difficult to pin down the value of a process improvement", according to a department head of an automotive industry.

The survey conducted by the authors also found an alternative perception of VR to exist that came from the games industry and the Internet, one which is usually considered to be gimmicky. Once again, business professionals failed to see any real value in this relatively less complicated use of VR other than entertainment. As a consequence, this level of VR is generally dismissed as not being of use for any serious application.

These perceptions seriously hinder the exploitation of the real potential of an intermediate level of VR, where serious business applications can be serviced by relatively inexpensive VR hardware/software that offer a suitable and not necessarily the highest level of realism. Perhaps VR needs to be redefined to bring about a change in its stereotypical image and perceptions.

VR REDEFINED

As has been highlighted in pervious sections and upon examining numerous definitions from Sutherland's time to late the 1990s, the key point that emerges is the increasing stress on the importance of realism within VR scenarios. This in turn means investing into high end technology to be able to attain such realistic environments even without knowing what returns that investment would give. As Professor Roy Kalawsky says, "VR is perhaps an unfortunate term....Press and media speculations about VR provided the platform for its world-wide exposure. However, these speculations were in the danger of 'over claiming' what could be delivered with existing technologies" (Kalawsky, 1993, p. 3).

VR, according to the authors of this article, is simply defined as a real-time interactive communication and visualization tool that uses computer-generated models to build any given scenario. Such models are drawn and viewed across all three axis—x, y, and z—thereby creating an illusion of a three dimensional visual representation of the given scenario (in a two dimensional space).

The above definition makes VR independent of any specific hardware, software, and levels of realism while focusing on the actual application nature of VR. Equipment used and the level of realism required become a function of the given project's context of application and budget.

CONTEXT-BASED FRAMEWORK: FOR WHO AND WHY?

Since controlling costs is equally important to all organizations, the level of cost consciousness is the same irrespective of the organization's size. However, not all organizations have the same level of cost sensitivity. This is because small organizations are more susceptible to market fluctuations as opposed to large organizations. Such organizations are termed *cost sensitive organizations* and cover areas such as small, medium, and micro enterprises, departments of large companies that function like complete entities in themselves and governmental/nongovernmental business in developing countries, and so forth. The Context Based Virtual Reality (CBVR) methodology is target predominantly at CSOs.

Need for a context-based approach comes from the fact that applying VR with a model that either is based on perceptions of VR, as discussed earlier, or following one that is used by large companies might not be cost beneficial for CSOs. For example, taking an analogy from the airline industry, the business and service model applied by a regular airline is different from that of a budget airline. While certain features are common due to their being mandatory for running an airline, for safety issues and so forth, the differences usually stop there. This is mainly due to difference in the client profiles (which often fall within the context of their expectations and the ability to spend to match those expectations) of a regular and budget airline. Often when it comes to VR and CSO organizations, the solutions offered are generic in nature, mainly serving the front end such as walkthroughs, displaying interactive models. Such VR applications often serve as no more than interactive advertisements that require relatively large budgets for development while offering a product that has a short lifespan and whose benefits are usually difficult to measure. This generic approach usually means overlooking the overall business context within which VR can be embedded and just applying VR at a superficial level.

Context-Based Framework: Main Features

The matrix shown in Figure 1 summarizes the CBVR methodology of applying virtual reality (CBVR). It starts with examining the application area (1) in as much detail as possible in order to derive a context (2) within which VR can be applied. Once the context has been defined, the required level of recognition/realism (3) is to be established. The next stage is to examine how much, if any, of the existing hardware (4) and software (5) within the given company can be used. From then on, VR is applied to develop a virtual scenario.

The green part of the matrix offers a system to calculate Return on Investment (ROI) when it comes to applying VR. This area is very important to address, especially as surveys conducted by the authors found that most CSOs could not find a clear system of measuring ROI. Most organizations have three main business objectives when they apply a new technology or tool as shown in boxes 6a, 6b, and 6c. These objectives themselves can serve as cost benefit indicators. Measuring them is also easier since they are quantifiable.

Recognition vs. Realism

One of the key features of CBVR is the way the issue of realism is tackled. The importance of realism within VR is such that, in conventional thinking, they can be considered synonymous. This need for realism can play a critical role when it comes to establishing the entry point in terms of investment in VR applications. Therefore, it is important to handle the issue of realism early on.

Figure 1.

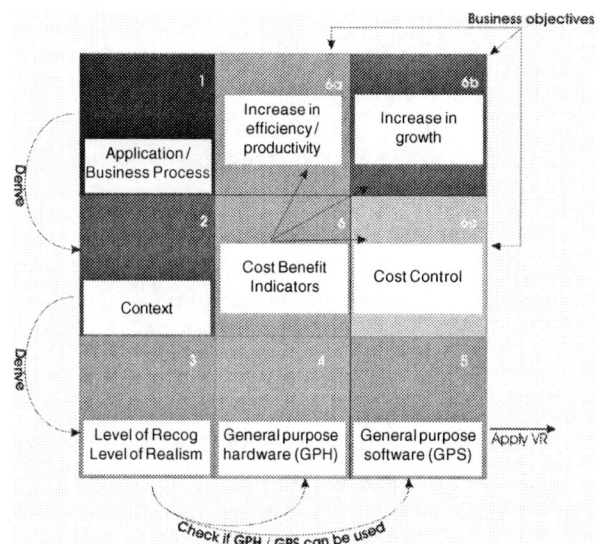

99

Figure 2a.

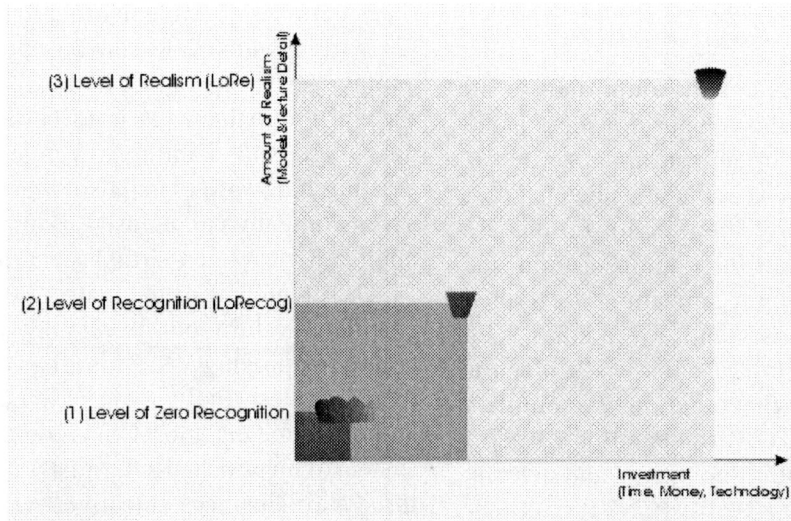

CBVR framework proposes using recognition as opposed of realism. The following analogy illustrates this point more clearly.

Take the example of a lump of clay, Figure 2a.

- **Stage 1:** Looking at Figure 2a at the Level of Zero Recognition, the potter starts with a lump of clay. Looking at the lump of clay, it is difficult to identify what will be the ultimate shape of the product or its quality other than the fact it is just a lump of clay.
- **Stage 2:** Here the lump of clay starts to look like a crude basic shape, yet it is visibly distinct from the lump of clay. Here the shape resembles a cup. Most people will recognize the shape as a cup and not as a lump of clay, even though there might be issues of its visual quality (quality of model in VR terms), colors (quality of textures in VR terms), and so on. It can therefore be said that, at this stage, the shape is recognized as a cup. This stage is termed the Level of Recognition and shows the entry requirements in terms of investments time, technology, and cost needed to develop a lump of clay into a recognizable shape of a cup.
- **Stage 3:** Any more work put beyond Stage 2 falls in the Level of Realism area. Any expense to move to increasing Levels of Realism in terms of a detailed modeling or texturing should be determined by examining the project criticality and client's budgetary constraints. The entry points vary depending on the nature of the project as can be seen in Figure 2b where the investments to be put in for making a recognizable model of a train carriage is on a different point as compared to the cup model.

General Purpose Hardware

When starting a VR project, it should be considered if any existing hardware present with the CSO can be used to accomplish the task rather than recommending investing in a high-end VR system. Even though specialist hardware might have a qualitative edge, a cost benefit analysis of investing into such system vs. the context of application is very important. Often the idea of making a big investment can be off putting to CSOs.

General Purpose Software

Once again, like hardware, using (as much as is possible) software that is already present will help to keep the costs on the lower side.

Cost Benefits Indicators

After investing into a VR application, it is important to have tools that aid measurement of value added by a given VR application. This is especially important when it comes to either reuse or widening the area of VR applications with a given CSO.

Figure 2b.

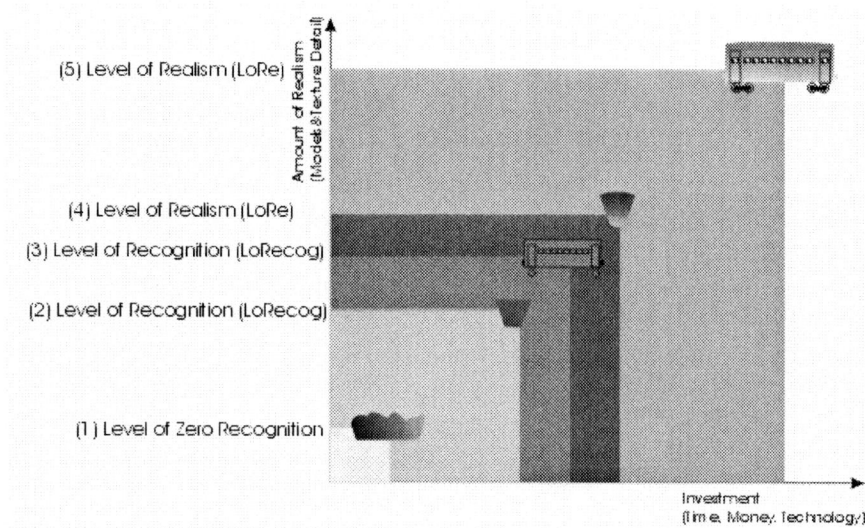

PROJECT LIBRARY

The following case study was conducted at the Loughborough University library to test some of the features proposed in the CBVR approach.

The library at Loughborough University was chosen because:

- In terms of its size and budget, it falls within the CSO category.
- It was possible to establish an application context for VR.
- There was a chance that it would positively affect:
  - Productivity (saving staff time by reducing the number of directional queries put to staff; 70% of inquiries made to the library staff were relating to direction).
  - Improvement of quality of service provided.
- It was easy to get a large sample base.

In informal talks with the library staff, it was found that they felt that a real-time interactive visualization system showing the layout and other features of the library would be a useful addition to the existing systems of navigation in the library. The existing systems of handling navigational queries were:

- Printed map.
- Physically answering direction queries (sometimes taking individuals to the exact spot where a certain resource could be found).

The CBVR framework required an understanding of what type of virtual environment was contextually appropriate to this VR project. Just a walkthrough showing a three dimensional library model was found not to be of much use. The system designed was not based on its visual realism and richness but based on points along which value addition could be seen and measured.

Visual realism was kept to a recognizable level without highly rich textures and excessive detail. It was also important to have a system that would build upon and complement the value presented by the existing printed map.

The two key points always kept in focus while designing the VR application were:

a. **User value:** would the users find the system helpful, and, if so, how? This would enhance the quality of service provided.
b. **Cost value:** would the number of directional inquiries be sufficiently reduced to make the development of the VR system cost beneficial?

Some other features that would add value not offered by existing systems were:

1. Internet access to the VR environment: this would allow library users to familiarize themselves to the library layout before arriving at the library leading to saving time.
2. It would provide information that is otherwise difficult to show in the real world (metaphorical).
3. Users could use it as an Intuitive Interactive Interface for browsing information.
4. Library staff could use the system internally for space management planning and retraining staff.
5. Audio feed in/feed out navigation system of the virtual world allowing an easier access to people with disabilities.
6. Guided audio tour of the virtual environment in multiple languages

Since the library already had personal computers using Windows, it was decided to use them as development platforms (features 3 and 4 of CBVR framework). Textures used were of a visual quality that would allow users to recognize the library rather than go for a visually realistic environment. The virtual environment was developed on a personal computer using

- AMD K6-2 350Mhz CPU and
- 32MB RAM.

Minimum requirements to run the environment were:

- Windows 98;
- 133Mhz Intel Pentium; and
- 16MB RAM.

The zipped size of the environment was 0.678kb. Since the environment was also available on the Web, the download times for file were

- At 56.6kbps modem, the VR scenario tool approx 1.5 to 2 minutes;
- If using DSL/Cable modem, the download speed was usually under 1 minute.

RESULTS

Eighty-six subjects used the library virtual environment and then filled out a detailed questionnaire.

These subjects were divided into three categories based upon their knowledge of the layout of the library.

1. **First Time Users:** Going to the library after using the virtual environment.
2. **Regular Users:** Going to the library up to five to seven times a week.
3. **Occasional Users:** Going to the library three to four times a month.

The last two categories of users (even though they had been some idea of the library layout) behaved very much like first time users when they had to deviate from the fixed path and section they most often visited.

Some basic facts that emerged from the questionnaire were:

- 72 users believed that it would save time for a first time user.
- 78 users said such a system would reduce directional queries.
- 72 users found that VR environment was more effective than the printed map for the same job.
- 81 users felt it was easy to recognize the library, and texturing was appropriate.

In comparative terms as to how close the virtual environment resembled the real library:

- six users thought that, even though the textures were not realistic, the two environments looked up to 80% similar.
- 47 users thought it to be 70% similar.
- 21 users thought it to be 60% similar.
- two users thought it to be 25% similar.

By making the virtual environment recognizable rather than totally realistic, it was possible to place signage in the virtual environment that would not be shown in the real world. As seen in Figure 3, on the left is the virtual image while on the right is the real image; whereas it is not possible to place a big sign in the real world pointing to the short loan section, it can be easily done in the virtual environment. This also illustrates that, at times, not being totally realistic can be advantageous when it comes to presenting an additional set of information.

Figure 3.

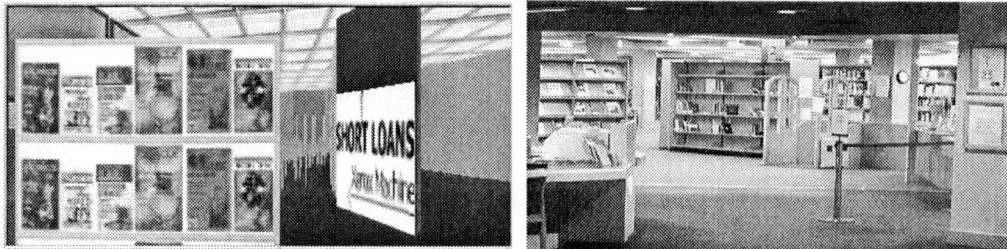

After examining the data generated by the questionnaire, it becomes evident that a high visual realism need not be a necessary prerequisite to all VR projects. In fact, by reducing the level of realism, not only the actual developmental time and costs can be brought down to a level that is acceptable to a CSO but also at times can help present more information than would be possible in realistic environments. This is especially important from CBVR point of view since the very definition of VR within CBVR framework is based on the view that VR is an information communication and visualization tool and not alternative to reality.

CONCLUSION AND FUTURE TRENDS

VR, even though an extremely powerful tool, has not been able to broad-base itself. There are almost two distinct application styles. One is used in large sector areas like aerospace and so forth where cost benefits and savings accruing due to VR are clearly visible and documented. The other is more of a superficial application of VR that has a short shelf life and is often difficult to measure in a quantifiable way. The other major difference is that often large scale sectors apply VR internally (for back end operations/business to business) where as most VR solutions sold to CSO sectors are external (for front end operations/business to customers) in their nature. Since the large scale sectors use VR internally, it becomes simple to layout contexts of application whereas the CSOs, due to the very style of VR application, are limited when it comes to defining their contexts. Further, this lack of contextual connection makes it difficult to calculate investments and cost benefits for a given VR application. It is therefore necessary to create a contextual framework before applying VR, especially if it has to become a tool that can used and applied across all market sectors as predicted by most VR enthusiasts.

REFERENCES

Bevan, M., & Leston, J. (1999). *VR market awareness survey.* Cydata Ltd, UK: VR Forum.

Kalawsky, R. (1993). *The science of virtual reality and virtual environments.* London: Addison-Wesley.

Lanier, J. (1996, September 15). Interview with Jaron Lanier. *Scientific American.* Retrieved from *http://www.sciam.com/article.cfm?chan ID=sa004&articleID=00070110-E9C1-1CD9-B4A8809EC588EEDF*

Mitchell, W. L. (1999). Moving the museum onto the Internet: The use of virtual environments on education about ancient Egypt. J. Vince & R. Earnshaw (Eds.), *Virtual worlds on the Internet* (p. 263). IEEE Computer Society Press.

Sutherland, I. E. (1963). *A man machine graphical communication system.* MIT, Department of Electricity Engineering.

Sutherland, I. E. (1965). The ultimate display. *Proceedings of the IFIP Congress 65* (Vol. 2, pp. 506-508).

Sutherland, I. E. (1968). A head-mounted three-dimensional display. Retrieved from *http://www.cise.ufl.edu/~lok/teaching/dcvef03/papers/sutherland-headmount.pdf*

Vince, J. (1998). *Essential virtual reality fast.* UK: Springer.

The Contribution of Communities of Practice to Project Management

Gillian Ragsdell
Loughborough University, UK

INTRODUCTION

More and more organisations are using projects as a means of managing their business; increasingly, 'new initiatives' are the focus of organisational life. Such initiatives could include cultural change programmes, organisation redesigns, or process improvements. Tackling the sociological and psychological aspects of the project is a great enough challenge, but there is often a requirement to develop a technological dimension too. Accelerating technical advancements brings an extra level of complexity to the projects so that, in general, projects have become more complex—not only do they tend to have a wider variety of customers to satisfy, but they also tend to utilise more sophisticated technology and have more far-reaching implications than ever before. It is not too surprising that some projects 'fail'; the increased complexity of projects brings an obvious rise in the associated risks. However, the increased complexity of projects also brings a rise in the opportunities for learning through the management of knowledge therein. These are opportunities that are not being fully exploited at present, as illustrated by the continuation of the 'failure-to-learn' and 'learning-to-fail' themes in the literature (e.g., Lyytinen & Robey, 1999; Cannon & Edmondson, 2004); a more active stance would consciously draw lessons from projects, from 'successes' and 'failures' alike.

Parallel to the growing emphasis on projects in organisational life and their changing nature, there is growing recognition of the interplay between the fields of project management (PM) and knowledge management (KM). Reference has already been made to the opportunities for more effectively managing knowledge within a project setting. This article operates at a finer level of detail and draws attention to the potential synergy between project teams and a much popularised social network derived from the KM arena—that of communities of practice (CoP). In doing so, the disciplines of PM and KM are explicitly bridged and, it is put forward, the prospect of breaking the 'learning-to-fail' and 'failing-to learn' loops is raised.

BACKGROUND

New Knowledge and a Commitment to Action

The following brief literature review is a platform from which to launch the main thrust of the article when CoPs are compared and contrasted with project teams. Inevitably the reference material is taken from the second-generation KM arena where human and social aspects are central. Most authors agree on the general characteristics of CoP; this agreement can be tracked chronologically. Of more interest and significance to this article is the changing emphasis on CoPs' intention to act and the distinction that is, at times implicitly, made about the possibility of CoPs generating new knowledge.

Seminal works on CoPs are those of Lave and Wenger (1991) and, later in that decade, Wenger (1998). The concept is now well known throughout the second-generation KM movement and used by various authors. Pór (1998) describes communities as "connecting islands of knowledge into self-organising, knowledge sharing networks." Skyrme (1999, p. 170) goes on to say:

While some communities focus on a particular profession or discipline, the most powerful communities are customer or problem focused. They transcend disciplines and bring in different perspectives. They exchange, develop and apply knowledge.

The indication from Skyrme is that CoPs share knowledge and in turn increase their knowledge base and their sphere of application. However, this is through the development of knowledge rather than through its creation.

When distinguishing between their concept 'enabling context' and CoPs, Von Krogh, Ichijo, and Nonaka (2000, pp. 179-180) assert:

While a community of practice is a place in which members learn knowledge that is embedded there, an enabling context helps create new knowledge. The boundary of a community of practice is firmly set by the task, culture, and history of that community, but an enabling context is determined by the participants and can be changed easily. Membership in a community of practice is fairly stable, and it takes new members time to become full participants. But the many organisational members who interact in an enabling context come and go. Instead of being constrained by history, an enabling context has a here-and-now quality—and it is this quality that can spark real innovations.

There are various angles from which Von Krogh et al.'s (2000) work could be challenged—aspects such as the stability of a group and notions of 'participation' and 'task' will be clarified in the next section. However, Wenger (2000, p. 206) confronts the aspect of whether CoPs generate new knowledge when he states:

What these groups have in common is that engaging with each other around issues of common interest, sharing insights and information, helping each other, or discussing new ideas together are all part of belonging to the group.

He goes on to be more specific when he states that CoP provide "the resources that members use to make sense of new situations and to create new knowledge" (Wenger, 2000, p. 209), and refers to good practice in World Bank and Daimler Chrysler. So the notions of new knowledge and of action are reinstated, and Von Krogh et al.'s (2000) interpretation of CoP is refuted.

In current times authors, such as Lehaney, Clarke, Coakes, and Jack (2004) retain Wenger's (2000) line, even though, initially, the foci on new knowledge and

on action are not obvious. Lehaney et al. (2004, p. 46) say that CoPs are "willing to develop and share tacit and explicit knowledge" and that CoPs have become important means for "sharing information within professions and between like-minded people" (p. 50). However, the balance is redressed when they also say that CoPs may encourage creativity and problem solving through the diversity of their population (p. 49).

This article goes forward with the understanding that KM is concerned with the generation, capture, storage, and sharing of knowledge with an intent to take action in order to increase an organisation's competitive advantage. Therefore, if CoPs are an effective tool for KM, their capacities to create new knowledge (as well as to develop 'old' knowledge) and to apply knowledge, are crucial. The same is true for the argument to embrace CoPs in PM. Otherwise, there could be a tendency to continue to work with the same knowledge with little intention to move into the practical arena, and the 'learning-to-fail' and 'failing-to-learn' cycles of current projects are supported.

COMPARISON OF COMMUNITIES OF PRACTICE WITH PROJECT TEAMS

Sapsed, Besant, Partington, Tranfield, and Young (2002) and Crawford and Cooke-Davies (2000) started to draw together the notions of KM and teams, and KM and PM respectively. I move specifically to project teams and communities of practice, at a different level of focus from these authors. CoPs and project teams probably implicitly co-exist in practice. This section makes their co-existence explicit by identifying their differences and their areas of overlap. In doing so, it is anticipated that the areas of overlap and their associated untapped synergy will be better managed.

A holistic approach is taken and attention is given to features such as purpose, culture, composition, structure, and accountability. Melcrum (2000) was a useful starting point for a comparison, and the table below builds on some of his key themes.

Project teams and communities of practice are not one and the same. One is accepting that project teams are formed for the purpose of completing a project within agreed time, budgetary, and quality constraints, and that CoP develop as a result of a common interest in a field or problem area. However, the table shows that there is some common ground—for instance, their common intent to share and apply knowledge, and their potential interdisciplinary nature and possible cultural overlap.

FUTURE TRENDS

At first sight, CoPs and project teams may appear to be at opposite ends of the spectrum. Nevertheless, CoPs are increasingly being recognised by companies as an effective vehicle for increasing the sharing of learning about projects and, in turn, for improving the success rate of their projects. It is noted, however, that the 'CoP' may be referred to as a 'network' or a 'special interest group', and that this lack of clarity over use of language can cause confusion. With the growing emphasis on projects as a means of managing organisations, it is anticipated that this trend will continue.

There is much work to be done in this area, and I would suggest that future research aim for a deeper discussion about the underpinning ideology of project teams and CoPs. This in turn will enable the features in Table 1 to be expanded upon. Of a more practical nature, empirical work needs to be undertaken to discover how to manage the aforementioned similarities more effectively and how to minimise the potential conflict that the differences between project

teams and CoPs may bring. Indeed, in "Project Teams and CoPs in the Construction Industry" Remington and Ragsdell explore some specific challenges for CoPs in traditional project environments such as construction.

CONCLUSION

This article has continued a theme that is evident in recent literature—that of building bridges between the disciplines of project management and knowledge management. This has been achieved by advancing discussion of the two groupings known as project teams and CoPs. In doing so, it has drawn attention to their complementary and contradictory aspects, and raised the possibility of tapping the synergy between the complementary aspects.

REFERENCES

Cannon, M.D., & Edmondson, A.C. (2004). Failing to learn and learning to fail (intelligently): How great organisations put failure to work to improve and to innovate. *Journal of Long Range Planning*.

Crawford, L., & Cooke-Davies, T. (2000). Managing projects: Managing knowledge: Sharing a journey towards performance improvement. *Proceedings of the World Project Management Week Conference*, Eventcorp, Brisbane, Australia.

Table 1. Comparison of communities of practice and project teams

	Communities of Practice	Project Teams
Purpose		
Overall	To increase capacity for improved action	To successfully complete project
In relation to knowledge	To generate, capture, store, and share knowledge	To share and apply knowledge within project boundaries
Composition		
Membership	Voluntary	Contractual
Profile of membership	Can be multi-disciplinary	Tend to be multi-disciplinary
Life span	'Natural' lifetime	Project lifecycle
Accountability		
Reporting structure	Informal	Formal
Rigidity of structure	Fluid	Relatively fixed
Measures of effectiveness	Intangible	Tangible
Culture		
Based on	Trust	'Esprit de corps'
Leadership style	Empowering	Variable

Lave, J., & Wenger, E. (1991). *Situated learning: Legitimate peripheral participation.* Cambridge, MA: Cambridge University Press.

Lehaney, B., Clarke, S., Coakes, E., & Jack, G. (2004). *Beyond knowledge management.* Hershey, PA: Idea Group Publishing.

Lyytinen, K., & Robey, D. (1999). Learning failure in information systems development. *Information Systems Journal, 9*(2), 85-101.

Melcrum. (2000). Teams versus communities of practice. Retrieved from *www.melcrum.com*

O'Donnell, D., Porter, G., McGuire, D., Garavan, T.N., Heffernan, M., & Cleary, P. (2003). Creating intellectual capital: A Habermasian community of practice (CoP) introduction. *Journal of European Industrial Training, 27*(2/3/4), 80-87.

Pór, G. (1998). Knowledge ecology and communities of practice: Emergent twin trends of creating true wealth. *Proceedings of the Knowledge Summit '98*, London (cited in Skyrme, 1999).

Sapsed, J., Besant, J., Partington, D., Tranfield, D., & Young, M. (2002). Teamworking and knowledge management: A review of converging themes. *International Journal of Management Reviews, 4*(1), 71- 85 (cited in Lehaney et al., 2004).

Skyrme, D.J. (1999). *Knowledge networking: Creating the collaborative enterprise.* Oxford: Butterworth-Heinemann.

Von Krogh, G., Ichijo, K., & Nonaka, I. (2000). *Enabling knowledge creation: How to unlock the mystery of tacit knowledge and release the power of innovation.* Oxford University Press.

Wenger, E. (1998). *Communities of practice: Learning, meaning & identity.* Cambridge, MA: Cambridge University Press.

Wenger, E. (2000). Communities of practice: The structure of knowledge stewarding. In C. Despres & D. Chauvel (Eds.), *Knowledge horizons: The present and the promise of knowledge management* (pp. 205-224). Oxford: Butterworth-Heinemann.

Wenger, E.C., McDermott, R., & Snyder, W.M. (2002). *Cultivating communities of practice.* Boston: Harvard Business School Press.

KEY TERMS

Project: A unique set of coordinated, goal-oriented activities that brings about change. A project has definite starting and finishing points, and is undertaken by an individual or team within defined parameters of time, cost, and quality.

Project Team: A group of individuals who have been brought together to form a cohesive whole in order to successfully complete a project. The team is mutually accountable for meeting predefined project objectives, and the team members may be characterised by their diverse backgrounds, skills, and personalities.

Second-Generation Knowledge Management Movement: Understood to have taken the emphasis away from discussion of technological issues related to knowledge management and to have placed it onto discussion of human and social factors with particular interest in the creation of new knowledge.

Creating the Entrepreneurial University

Lorraine Warren
University of Southampton, UK

INTRODUCTION

This article discusses how the concept of the Community of Practice (CoP) can be useful in developing more entrepreneurial universities. Following a brief introduction, the argument is developed through the exploration of a research-led university in the UK.

BACKGROUND

While the term *entrepreneurial university* is open to a wide range of interpretations, it has been associated with knowledge transfer through the formation of spin-out companies and the exploitation of intellectual property rights by faculty and students of universities. This encourages the transfer of science and technology innovation to the business sector, contributing to economic development at the regional and national levels. The idea of the entrepreneurial university has become increasingly prominent on the UK government's agenda in recent years, as a third mission for higher education, alongside teaching and research (Ost.gov.uk, 2004). For example, as part of the government's desire to foster a new entrepreneurial climate, the Science Enterprise Challenge (SEC) was launched in 1999. This endeavor is based in part on the understanding that it is possible to design entrepreneurship and innovation courses that provide graduates with entrepreneurial skills to enable them to exploit innovative ideas generated during their involvement with their university. In other words, the university becomes a more entrepreneurial environment through:

- teaching enterprise and entrepreneurship to science and technology students;
- making ideas and know-how available to business to support competitiveness and wealth creation; and
- encouraging the growth of new businesses by supporting start-ups, including spin-out companies based on innovative ideas developed by students and faculty within the universities.

Yet, achieving the aims listed above is not straightforward. There are extensive debates around many aspects of this kind of provision, particularly where pedagogy is concerned. Of course, students may set up businesses without ever encountering entrepreneurship courses; or if they do, they may only draw on them to a limited extent in their business activities. However, inherent within entrepreneurship education is the assumption that some students may make significant changes in their career activities at some point, as a result of being influenced by their learning experience. Thus, there is a more overt developmental agenda present in an entrepreneurship course in a typical knowledge-based science/technology module. This presents a challenge to educators that the CoP can help resolve.

AN ACADEMIC FOCUS FOR CoPs

Raising awareness of entrepreneurship for students is relatively straightforward, conforming to academic norms through teaching examples, case studies, exposure to external speakers, and the supervision of student-centered projects. Beyond this, however, a number of authors note the importance of experiential learning to entrepreneurial learning (Deakins & Freel, 1998; Garavan & O'Cinneide, 1994; Gibb, 1987; Gorman, Hanlon & King, 1997; Jack & Anderson, 1999; Rae & Carswell, 2000). Rae and Carswell (2000) note that while this seems to be a reasonable conclusion, there is as yet little research on how successful entrepreneurs have turned their experience into learning, essential knowledge if effective education and training programs are to be developed. Jack and Anderson (1999) argue that there are pedagogic difficulties in teaching the practice of enterprise, in part due to variability within enterprises, in part because entrepreneurship is about process rather than stasis. Gibb (1996) also argues that the academic focus on understanding and critical analysis contrasts with the reality of the entrepreneur operating with incomplete information under time pressures.

It is not obvious how the experiential aspect of entrepreneurial learning can be built into modular programs of short duration, such as those typically available to technology students for the study of enterprise. Clearly, there is a limit to the amount of experiential learning that could be built into a given program: internships or work experience where a student could participate in day-to-day decision making in an entrepreneurial setting are not an option in this setting. An understanding of the dynamics of CoP can help overcome this difficulty.

Brown and Duguid (1991), Lave and Wenger (1991), and Wenger (1998) have argued for a community-based analysis of learning, which seeks to support a unified view of learning and innovation. They argue that learning is best achieved through supporting access to and membership of the target community of practice, not by explicating abstractions of individual practice. Such communities are reflexive, socially constructed, and emergent, existing outside formal organizational structures. For learners, a position on the periphery of practice is important with access to formal and informal meetings, picking up know-how—information, manner, and technique—from being on the periphery of competent practitioners going about their business.

In a university, a CoP can be identified around the spin-out domain. There is an informal grouping of actors centered on universities generally (though not necessarily members of the university) who are engaged in the spin-out activity, either directly as part of a company or as part of a broad range of supporting activities. This group includes:

- Faculty (academic/academic-related/contract research staff) engaged in spin-out activity;
- Students engaged in spin-out activity (or other small business endeavors designed to supplement funding in the short term);
- Interested academic staff: potential spin-outs, plus business school staff;
- University support staff: technology transfer staff, incubation center staff;
- Senior management of the university;

- Representatives of local support agencies: Business Link in the UK, for example;
- Local technology companies: from SMEs to global corporations;
- Local professional services: financiers, lawyers, business development consultants; and
- Successful entrepreneurs in the local community with strong university linkages.

Given the above, the challenge is to enable students to interact with the innovation community in a meaningful way, given that:

- modular degree courses focus on discrete blocks of understanding, clashing with the interconnected nature of the innovation system overall; and
- the potential for experiential learning through placement, or similar, is very small.

At one level, this could be addressed through the development of networking skills with a particular emphasis in the local context without drawing on the notion of the CoP. Research on the networks, in which entrepreneurs participate, has emerged as an important field of inquiry within entrepreneurship over the last 15 to 20 years (Hoang & Antoncic, 2003) with key contributions from Aldrich and Zimmer (1986); Birley (1985); Chell and Baines (2000); Dyer and Singh (1998); Granovetter (1973); Johannisson (1987); Larson and Starr (1993); Uzzi (1997a, b). A number of authors argue that such interconnectedness means entrepreneurship needs to be understood as a social process, not as an isolated individual activity (Bygrave, 1989; Bygrave & Hofer, 1991; Gartner, 1985).

One criticism of the network literature, however, is that much of it tends to focus on the morphology of aggregation, rather than how such aggregation takes place—there appears to be little discussion of how new (or potentially new) members of the community are assimilated and contribute to emergent properties of the network overall. Although the network approach overall is useful, the CoP approach adds value by shedding light on the learning processes of new community members as they begin to participate in new settings. Although they do not refer specifically to entrepreneurship, Lave and Wenger (1991) and, later, Wenger (1998) argue that developing a practice of any kind requires the formation of a community, however loosely defined whose members can engage with one another, thus acknowledge and legitimize each other as participants. In Lave and Wenger's conceptualization, students engaging with this CoP are undergoing a process of becoming part of this community, not just encountering it—in other words, they are undergoing a transformative social process.

Wenger (1998) highlights the importance of identity in this process of becoming in arguing that the formation of a CoP is also the negotiation (and renegotiation) of identities (p. 198). There is a huge literature on identity in the social sciences; much of it as a lens on exploring the mutual constitution between organizations and groups (Giddens, 1991; Strauss, 1959). There is also a more limited literature on entrepreneurial identity, notably Cohen and Musson (2000); du Gay (1996); du Gay and Salaman (1992); Ritchie (1991); and Warren (2004). This latter group of authors argues that there is an emphasis on how individuals are reflexively constituted by the discourse of enterprise. This resonates with Wenger's view of identity as a negotiated process.

Wenger (1998) expands his argument on identity and CoPs at length, drawing parallels between practice (as negotiation of meaning) and identity (as negotiated experience of self), in terms of participation and reification (p. 150). A central pillar of the argument, however, is the identification of three co-existing modes of belonging (and their associated mechanisms) as a source of identity (Table 1). Wenger argues that these modes co-exist to different extents at different times:

- **Engagement:** active involvement in mutual processes of negotiation of meaning;
- **Imagination:** creating images of the world and seeing connections through time and space by extrapolating from our own experience; and
- **Alignment:** coordinating energy and activities to fit with broader structures and contribute to broader enterprises.

In terms of entrepreneurship then, the CoP notion goes beyond teaching students better networking skills to putting the student at the center of a process of identity in transition. The ethos is to facilitate learning through developing participative routines for students where they might begin (or continue) to conceptualize themselves as practicing entrepreneurs through interacting with, and negotiating meaning through, a local CoP. Thus, the most involved students might undergo a period of becoming an entrepreneur, not just listing potential network contacts in the classroom, but learning the norms, values, and vocabulary of a social grouping, albeit more nebulous than the formal departmental structure for staff and students on campus. This approach should go some way to addressing the difficulties of experiential entrepreneurial learning—and is some way from a traditional modular approach.

Thus, looking to the classroom setting, students need to be treated as if they are learning *about* entrepreneurial practice as a basis for choosing, at some point in the future, whether to engage in the innovation domain in an entrepreneurial manner or not. Rather, the emphasis needs to be on character-

Table 1. Modes of belonging (adapted from Wenger, 1998)

Mode of belonging	Mechanisms	Nature
Engagement	• shared histories of learning • relationships • interaction • practices	• peripherality • bounded by physical limits of time and space
Imagination	• images of possibilities • images of the world • images of the past and future • images of ourselves	• unconstrained by time or space
Alignment	• discourses • coordinated enterprises • styles	• can span vast distances both socially and physically, but more focused than imagination

izing the student as *already being* a practitioner (entrepreneur) at the outset, albeit at the periphery of the innovation community through their possession of ideas and the potential to develop them.

Creating the Entrepreneurial University

The university's senior management had made a commitment to support the government's agenda toward developing entrepreneurial universities by first supporting business school teaching initiatives to develop suitable courses for science and engineering students. Second, organizational structures had been developed to support the spin-out process itself. An Enterprise Office (EO) dedicated to the management of intellectual property, and the establishment of spin-out companies had been enlarged and given a more strategic mission. The EO director now participates in middle management committees in the university. Connections to senior management are through an identified pro-vice chancellor. The EO is the first port of call for staff and students wishing to establish spin-outs or develop licensing deals, or any other intellectual property issue such as patents. The newly launched EO has established an impressive track record, claiming:

- 260 consultancy projects a year;
- £27.6M research income for 2002/2003;
- around 30 patents a year;
- highest start-up rate of companies in the UK per research pound; and
- 23 spin-outs nurtured to date.

The EO remit also includes the management of a new Innovation Centre on the main campus, providing a range of accommodation, advice, and mentoring services. In addition, it organizes the Enterprise Club (EC), a monthly club where entrepreneurs and business professionals who champion innovation meet to exchange ideas and support. Each month, a different, practically based workshop or presentation is held providing members with the opportunity to network informally. The EC organizes the Local Technology Initiative (LTI), a local networking and support organization for technology and knowledge-based organizations in the region. The LTI currently contacts 200 companies ranging from small high-technology start-ups to global corporations; members of the LTI are eligible to attend EC events.

Clearly, there are networking opportunities at the university that extend beyond campus boundaries. Yet observations at EC events suggest that the grouping fills the criteria of the CoP as depicted by Wenger previously.

1. EC events were generally well attended with a good mix of participants from the groupings identified previously. Good interaction was taking place between new members and stalwarts with a lively, informal atmosphere over refreshments.

2. Seminars on skill areas, such as marketing and finance employment law, were generating to-the-point discussions of individual problems or opportunities; there was no need for preamble, and there was an atmosphere of trust between old and new participants.

3. Regarding the above, there were open discussions, assessment of management tools and processes, and a developmental atmosphere between presenters and participants.

4. There were apparent shared understandings of how to access resources within and without the university.

5. There were evident shared understandings of the ways of dealing with the university policy and culture toward spin-outs.

6. There was a development of innovative ideas through contact with local industry experts.

7. In addition, there was a rapid-fire development of contacts toward problem solving (not just cursory exchanges of business cards).

8. There was an apparent shared perspective on the excitement, as well as the risks and hard work, of the spin-out process.

In summary, following Lave and Wenger's initial conceptualization, participants engaging with this CoP are undergoing a process of becoming part of this community, not just encountering it—in other words, they are undergoing a transformative social process.

This ethos of introducing students on enterprise courses to the local spin-out CoP was based on the experiences of a successful student entrepreneur from the university, discussed in depth in Warren and Stephens (2004). In short, design students from the university met a venture capitalist during his degree show (where he exhibited work from his final year project) and eventually set up a company around his design. Although that meeting took place in June 2000, the company was not established until February 2001. During that time, the student became a peripheral member of the university's innovation community, working to carry out patent searches, negotiating licensing deals with associated companies, performing due diligence, and developing a business plan as well as continuing to work on the product. During this time, the student learned about the potential of becoming a director of his own company, rather than working for a larger company. In other words, he became familiar with the norms, values, and vocabulary of the spin-out community in general, prior to finally going ahead with company formation.

This chance process of peripheral participation could be mirrored and enhanced through:

- Introductions to the university's innovation support staff, in particular, the Student Enterprise Manager who provides individual counseling and advice to potential student entrepreneurs, as well as organizing enterprise-related campus-wide initiatives and events.
- Information about local events and structures such as business planning competitions, innovation fairs, innovation fellowships, the university's Innovation Centre (which allows for student hot-desking, as well as catering for established spin-out businesses).
- Participation in the Students' Enterprise Forum, which meets at least twice a semester, where students can meet entrepreneurs from the university and the local community, as well as interested students from all faculties of the university, undergraduate and postgraduate alike.
- Introduction to ex-students (via the forum or through workshops) who run spin-out companies, or who work in related areas such as venture capital or business proposal assessment.
- Business planning exercises supervised by mentors with entrepreneurial background, drawn from the local business community.

- A general proactive approach where information was to be gleaned from the community, not just texts and tutors.

In other words, the students are introduced to the local innovation system as a community of scholars and practitioners where they were free to form and reform relationships and groups to suit their own needs and as opportunities arose. In this way, an expectation was developed that the knowledge students were gaining from a range of sources, including the teaching settings, would take on significance in the target community, albeit at the periphery. They could engage, imagine, and maybe align with the innovation community through the formation of a spin-out company. And this might take place quite soon on leaving the university or perhaps later in life. Thus, a two-way interaction, engendering more entrepreneurial students and a more entrepreneurial university, could take place.

FUTURE TRENDS

Though more research needs to be carried out on the theory and practice of this ethos, there is evidence that the approach was successful in terms of participation, activity generation, and personal development (Warren, 2003 a, b, 2004). Yet, for some, there are question marks over entrepreneurship education for science and technology students—that it detracts from academic achievement, is a distraction, or is some kind of low-level training activity. Yet, building on general understandings in the literature of entrepreneurship as a networked, processual activity, and developing the CoP concept can counter this criticism by providing intellectual underpinning to support participative activity.

CONCLUSION

Of course, while it is possible to justify the approach academically in this way, there are inevitably a few individuals in universities, as in all organizations, who find such cross-boundary interactions a little threatening. Our attempts to foster entrepreneurial universities must challenge these assumptions if we are to provide students with the richness of opportunity they deserve.

REFERENCES

Aldrich, H., & Zimmer, C. (1986). Entrepreneurship through social networks. In D. Sexton & R. Smiler (Eds.), *The art and science of entrepreneurship* (pp. 3-23). New York: Ballinger.

Birley, S. (1985). The role of networks in the entrepreneurial process. *Journal of Business Venturing, 1,* 107-118.

Brown, J. S., & Duguid, P. (1991). Organizational learning and communities-of-practice: Toward a unified view of working, learning and innovation. *Organization Science, 2*(1), 40-57.

Bygrave, W. D. (1989). The entrepreneurship paradigm: A philosophical look at its research methodologies. *Entrepreneurship Theory and Practice, 14*(1), 7-26.

Bygrave, W. D., & Hofer, C. W. (1991). Theorising about entrepreneurship. *Entrepreneurship Theory and Practice, 16*(2), 13-22.

Chell, E., & Baines, S. S. (2000). Networking, entrepreneurship and microbusiness behaviour. *Entrepreneurship and Regional Development, 12*(3), 195-205.

Cohen, L., & Musson, G. (2000). Entrepreneurial identities: Reflections from two case studies. *Organization, 7*(1), 31-48.

Deakins, D., & Freel, M. (1998). Entrepreneurial learning and the growth process in SMEs. *The Learning Organisation, 5*(3), 144-155.

du Gay, P. (1996). *Consumption and identity at work.* London: Sage.

du Gay, P., & Salaman, G. (1992). The cult(ure) of the consumer. *Journal of Management Studies, 29*(5), 615-633.

Dyer, J. H., & Singh, H. (1998). The relational view: Cooperative strategy and sources of interorganizational competitive advantage. *Academy of Management Review, 23*(4), 660-679.

Garavan, T. N., & O'Cinneide, B. (1994). Entrepreneurship education and training programmes: A review and evaluation – part 1. *Journal of European and Industrial Training, 18*(8), 3-12.

Gartner, W. B. (1985). A conceptual framework for describing the phenomenon of new venture creation. *Academy of Management Review, 10*(4), 696-706.

Gibb, A. A. (1987). *Enterprise culture: Its meaning and implications for education and training.* Bradford, Yorkshire, UK: MCB University Press.

Gibb, A. A. (1996). Entrepreneurship and small business management: Can we afford to neglect them in the 21st century business school? *British Academy of Management, 7,* 309-321.

Giddens, A. (1991) *Modernity and identity.* Cambridge: Polity Press.

Gorman, G., Hanlon, D., & King, W. (1997). Some research perspectives on entrepreneurship education, enterprise education and education for small business management: A ten year literature review. *International Small Business Journal, 15*(3), 56-77.

Granovetter, M. (1973). The strength of weak ties. *American Journal of Sociology, 78*(6), 1360-1380.

Hoang, H., & Antoncic, B. (2003). Network-based research in entrepreneurship, a critical review. *Journal of Business Venturing, 18,* 165-187.

Jack, S. L., & Anderson, A. R. (1999). Entrepreneurship education within the enterprise culture: Producing reflective practitioners. *International Journal of Entrepreneurial Behaviour and Research, 5*(3), 110-125.

Johannisson, B. (1987). Anarchists and organizers: Entrepreneurs in a network perspective. *International Studies of Management and Organization, 17*(1), 49-63.

Larson, A., & Starr, J. A. (1993). The network model of organisation formation. *Entrepreneurship Theory and Practice, 18,* 5-15.

Ost.gov.uk. (2004). Retrieved May 4, 2005, from *http://www.ost.gov.uk/enterprise/knowledge/sec.htm*

Rae, D., & Carswell, M. (2000). Using a life story approach in researching entrepreneurial learning: The development of a conceptual model and its implications in the design of learning experience. *Education and Training, 42*(4/5), 220-227.

Ritchie, J. (1991). Enterprise cultures: A frame analysis. In R. Burrows (Ed.), *Deciphering the enterprise culture*. London: Routledge.

Strauss, A. (1959). *Mirrors and masks: The search for identity*. Glencoe, IL: Free Press.

Uzzi, B. (1997a). Social structure and competition in interfirm networks. *Administrative Science Quarterly, 42*(2), 417-418.

Uzzi, B. (1997b). Social structure and competition in interfirm networks: The paradox of embeddedness. *Administrative Science Quarterly, 42*(1), 35-67.

Warren, L. (2003a, June). What are the educational challenges of the Science Enterprise Challenge? (CD-ROM). *Proceedings of the 48th International Council for Small Business Conference*, Belfast, Ireland.

Warren, L. (2003b, November). Meeting the Science Enterprise Challenge: Developing communities of practice. *Proceedings of 26th ISBA National Small Firms Policy and Research Conference SMEs in the Knowledge Economy*, Surrey, United Kingdom.

Warren, L. (2004, June 7-8). Developing communities of practice: The dynamics of the Science Enterprise Challenge. *Proceedings of the Entrepreneurship Research Workshop, University of Nottingham Institute for Enterprise and Innovation*, Jubilee Campus, UK.

Warren, L., & Stephens, J. (2004). From classroom to community of practice. *International Journal of Entrepreneurship Education, 2*(3), 329-350.

Wenger, E. (1998). *Communities of practice: Learning, meaning and identity*. New York: Cambridge University Press.

KEY TERMS

Entrepreneurial Identity: There is a view that individuals 'become' entrepreneurs as they are reflexively constituted by the discourse of enterprise. This is, they begin to characterise themselves as entrepreneurs as they adopt the norms practices and values of the entrepreneurial community with which they engage.

Entrepreneurial University: A wide-ranging term, used here to describe universities associated with high rates of knowledge transfer through the formation of spin-out companies and the exploitation of intellectual property rights by faculty and students of universities.

Experiential Learning: Learning achieved through everyday practice of an activity, rather than in formal classroom setting, or other directed training programmes; experience as a source of learning and development.

Intellectual Property: The concept of intellectual property (IP) allows individuals or organisations to own their creativity and innovation in the same way that they can own physical property. The owners of IP can control its use, and be rewarded for it; in principle, this encourages further innovation and creativity.

Spin-Out: A new company formed by an organisation, often a university, to develop and market a new technology or process invented within the host organisation, usually (but not always) involving the inventor in a management capacity.

Technology Transfer: The process by which new ideas and technologies diffuse from research settings, usually in universities or research institutes, into the wider community. Typically, this involves the commercialisation of new knowledge through the establishment of intellectual property rights.

Third Mission: The desire of the UK government to extend the remit of UK universities beyond their traditional research and teaching role to a broader set of commercially-oriented activities, including the establishment of spin-out companies.

Creating Value with Regional Communities of SMEs

Cecily Mason
Deakin University, Australia

Tanya Castleman
Deakin University, Australia

Craig Parker
Deakin University, Australia

INTRODUCTION

This article provides a conceptual argument that the knowledge management (KM) approach of communities of practice (CoPs), and their virtual equivalents (VCoPs), can create value for clusters of regional small and medium enterprises (SMEs). The article firstly shows that value creation in regional clusters occurs by encouraging collective learning and reciprocal knowledge exchange. The article then shows that CoPs, and VCoPs in particular, have been the most successful value creation mechanism in large organisations. We argue that VCoPs hold considerable potential for value creation in regional clusters of SMEs by promoting innovation, more effective knowledge sharing, and recognising the value of VCoPs as capital. The strategic integration of SMEs in regional clusters is analogous to large organisations' global operations. In this environment VCoPs combine industry-specific knowledge with firm specific knowledge and emerge as a new source of social capital.

BACKGROUND

Towards the end of the 20th century, a new global knowledge-based economy emerged as global knowledge became increasingly sophisticated and diversely located. Concurrently, developments in information and communication technologies (ICT) have significantly increased the ability to create, transfer, and maximise knowledge worldwide (Kulkki, 2002). Today, knowledge is the primary source of competi-

tive advantage and the key to success for organisations in the knowledge economy (Grant, 2002; MacKinnon, Cumbers & Chapman, 2002; Patriotta, 2003). To capitalise on the value of knowledge, organisations "need to know precisely what gives them competitive advantage, keep this knowledge on the cutting edge, deploy it, leverage it in operations and spread it across the organisation" (Wenger, McDermott & Snyder, 2002, p. 6). People, as the creators of innovation and renewal, are the sources of value in knowledge. This human capital has the potential to create value at all times by generating something that did not previously exist. In fact Chattel (1998) asserts that human capital is core for the design of the future. It is also the basis on which communities of practice are built (Stewart, 1997). With the development of ICTs, these communities of practice now have the ability to operate over large areas, their communication facilitated by the Internet. These virtual communities of practice have the potential to create value to an even greater extent if they are properly understood and nurtured.

Creating value through knowledge and the use of VCoPs is an issue for organisations and for other economic formations, including clusters or networks of individual businesses. This article presents a conceptual approach to the creation of value through VCoPs based on existing knowledge management (KM) and cluster research. We conclude that VCoPs should be an important value-creating mechanism for regional clusters of small and medium enterprises (SMEs). Considerable attention has been paid to how VCoPs can be established and promoted (e.g., Ardichvilli, Page & Wentling, 2003). What is needed is an understanding of how KM systems create value, and what features of these initiatives enable SMEs in regional clusters to be self-sustaining. It is our assertion that the *personalisation* technique inherent in CoPs, and used successfully in large organisations, has great potential to enable regional clusters of SMEs to create value. These concepts provide a foundation for future empirical research to identify the key elements for value creation through *personalisation*, specifically CoPs, and their associated VCoPs. This article will therefore establish the conceptual groundwork for future theoretical and empirical research which will investigate the practical mechanisms required to develop CoPs into regional clusters of SMEs.

Value Creation through Knowledge

Value creation through knowledge occurs when the organisation obtains value from its intellectual capital in the form of intangible assets. These intangible assets include *hard intangibles,* including such things as patents, royalties, copyrights, and databases, as well as *soft intangibles,* which are individual skills, expertise and capabilities, organisational culture, loyalties, and trust (Stewart, 1997). The sources of value in intellectual capital are:

- **Human capital:** people as the creators of innovation and renewal;
- **Structural capital:** organisational infrastructures including information systems (IS), procedures, and processes; and
- **customer or relationship capital:** relationships with external people (Stewart, 1997).

The underlying characteristic of value creation is that it is a mutually advantageous process of co-creation between the various organisational stakeholders (Prahlad, 2004; Prahalad & Ramaswamy, 2004; Rowley, 2004; Skoog, 2003). The participants clearly understand that this co-creation provides access to intellectual capital (human capital, structural capital, and customer/relationship capital) that would not otherwise be available.

REGIONAL CLUSTERS AS VALUE CREATORS

The importance of value creation in the context of regional clusters was first recognised by Marshall (1947). In this resource-based view, value was obtained by access to resources, labour, and technological improvements resulting from knowledge spillovers. In the 1980s a new form of technologically dynamic industrial district emerged where competing firms cooperated by adhering to norms of reciprocity (Lawson & Lorez, 1999). This enabled them to access collective goods and services such as education and research and development, and to reduce the risks involved in developing new products and processes. The most frequently cited successful examples of this type of regional cluster are the Emilia-Romagna region in Italy and the Baden-Wurttemberg region of Germany (e.g., Lawon & Lorenz, 1999; Hospers & Beugelsdijk, 2002; Humphrey & Schmitz, 2002). In both cases, geographically co-located SMEs provide specialised activities for a single stage of production in vertical value-added chains. This form of social embeddedness has the potential to lower transaction costs and in the process create value for all involved (Tallman, Jenkins, Henry & Pinch, 2004).

In the 1990s a new form of regional cluster emerged: the *innovative milieu* or *learning region.* The shift towards the knowledge economy changes the comparative advantage clusters obtained from physical resources to competitive advantage based on learning and knowledge (Mitra, 2000). In this innovative milieu, knowledge is the most important resource and learning is the most important process (MacKinnon et al., 2002). Innovative clusters typically have collective learning processes sustained by continuing exchange of knowledge and ideas; labour mobility; a high degree of openness based on geographically embedded social networks; and porous intra-firm, inter-firm, and intra-region boundaries (Saxenian, 1994; De Bernardy, 1999; Sternberg & Tamasy, 1999; MacKinnon et al., 2002). A paradoxical situation results in which firms have to cooperate in order to remain competitive. Success stories, such as the much publicised Silicon Valley and the Cambridgeshire Phenomenon, entail technological regional clusters where cooperation and competition co-exist with an innovating economy, and illustrate that firms are able to resolve this paradox (Saxenian, 1994; Lawson & Lorenz, 1999; Hospers & Beugelsdijk, 2002). Universities have played a key role in product development based on histories of transfer of tacit knowledge and know-how between employees, students, and staff in trusting relationships. The Grenoble region in France is indicative of the types of SMEs that typically emerge in high-technology clusters. Small start-ups or spin-offs arise from research centres which in turn mature into medium-sized, research-intensive, innovation-oriented enterprises. There are also customer-oriented subcontractors and high performance family SMEs (De Bernardy, 1999).

There has been a resurgence of government interest in regional clusters worldwide (e.g., DIIRD, 2003; Hwa, 2003; DTI, 2004). Many are promoting regional development in the belief that it will increase organisational productivity, the innovation will generate growth, and new businesses will expand the cluster and make it stronger (Porter, 1998). A study of 350 manufacturing firms in the Valencia region in Spain found a significant positive relationship between the firm's membership of the industry cluster and its ability to create value, measured in terms of the number of innovations produced (Molina-Morales & Martinez-Fernandez, 2003).

Merely clustering firms together in a geographic location will not necessarily create the innovative milieu conducive to value creation. Staber's research on a declining cluster warns regional planners to be aware of this and ensure that any efforts to establish regional clusters create an "institutionally conducive environment for collective learning and business success" (Staber, 2001, p. 339). However, as indicated above, collective learning is a cumulative process that takes place over time. It is based on trusting relationships that emerge out of reciprocal exchanges of knowledge and information, and are bound by strong social ties (MacKinnnon et al., 2002). For this reason, it makes sense to examine how this collective learning takes place in other contexts to see if there is potential for using similar approaches to support value creation in regional clusters of SMEs.

CoPs AS VALUE CREATORS IN LARGE ORGANISATIONS

The popularisation of Senge's (1990) concept of the *learning organisation*—with its promise that the five disciplines of systems thinking, personal mastery, mental models, shared vision, and team learning would enable organisations to innovate continuously—led to the organisations seeking a KM technique that would facilitate these. Wenger enunciated the communities of practice technique specifically for this purpose, proposing it as a means of addressing the five disciplines and in the process enabling firms to become learning organisations (refer to Wenger et al., 2002).

Early attempts by large organisations to tap into human capital (the source of intellectual capital) using KM techniques focussed on *codification* of knowledge by locating, capturing, and storing it in databases for later reuse by decision makers (Boisot, 2002). These initial KM efforts failed to deliver the expected benefits as they did not access the valuable tacit knowledge and know-how held by individuals. Attention was then directed to making the best use of knowledge by systematically supporting knowledge sharing (Yoo & Torrey, 2002). These *personalisation* approaches linked KM with business strategy. A knowledge sharing culture was developed, work processes were redesigned to incorporate knowledge flow, and an emphasis was placed on behavioural change (Davenport & Prusak, 2000).

CoPs have become the quintessential knowledge sharing and collaborative mechanism available to large organisations. The focus on CoPs as a coordination technique, rather than on organisational functional units, enables the transfer and integration of knowledge to occur across traditional organisational boundaries (Grant, 2002). Thus CoPs create value by addressing the two roles of economic exchange: (1) improving efficiency by continual reallocation of resources to more productive use, and (2) stimulating new productivity by combining resources in new ways (Huizing & Bouman, 2002).

Since 1997, large organisations have been actively adopting CoPs as a major element of their KM initiatives (Lee, Parslow & Julien, 2002;

Zboraiski, Gemuenden & Lettl, 2004). However, CoPs cannot be managed in the same way as other organisational initiatives. This is because they are not standardised, are hard to locate and define, lack a formal structure, and rely on voluntary participation and acceptance by other members of the community (Hackett, 2002). Consequently, organisations have created environments in which CoPs are identified and nurtured, enabling them to flourish and in the process provide organisational access to their knowledge potential. For example IBM has set in place a formal group to support its CoPs (Vorbek, Heisig, Martin & Schutt, 2003).

Many leading multinational companies acknowledge the contribution of CoPs to their success in the knowledge economy (Wenger et al., 2002). Perhaps the most frequently cited of these is the World Bank, which has successfully used CoPs as a means of tapping into the vast knowledge that already exists within its organisation, rather than trying to discover new knowledge (Stewart, 2001). It now has more than 100 CoPs/VCoPs throughout the world (King, 2002). Similarly Shell Oil uses its CoPs/VCoPs as a means of retaining technical excellence throughout its worldwide operations (Wenger et al., 2002; Burress & Wallace, 2003).

We can see that large organisations are successfully using CoPs and their associated VCoPs as a major technique for creating value by tapping the organisation's knowledge assets. By contrast, SMEs do not have access to the knowledge resources and capabilities of large organisations. This raises the question of how participation in CoPs and VCoPs by SMEs might enable them to tap into these resources and capabilities, and generate value for the regional cluster.

COPS CREATE VALUE FOR SMEs IN REGIONAL CLUSTERS

CoPs/VCoPs are the KM technique with the greatest potential for value creation by regionally based SMEs which are members of, or seek to establish, a regional cluster. SMEs are critical constituents of clusters as they "create a hub of learning through cooperation and competition among themselves" (Mitra, 2000, p. 232). We now consider what kinds of value CoPs can create for the region or cluster and how this value creation can be fostered. We

argue that value in this context is created by promoting innovation, using knowledge more effectively, and recognising the value of VCoPs as capital.

Promoting Innovation

Innovation is dependent on knowledge and is increasingly important for the success of regions. The collaboration and communication capacity of networks, or intraregional ties, provides the means of accessing the intangible assets of innovation (Fuchs, 2002).

The collective learning environment of the innovative milieu displays striking similarities to Senge's learning organisation, perhaps unsurprising as it is also known as the learning region. CoPs developed specifically for the learning organisation appear to be eminently suitable as the KM mechanism with greatest potential for SMEs in innovative clusters. Spence's (2004) research in Canadian high-technology SMEs illustrates how a process that maximises *personalisation* creates sustainable value. Close geographical proximity and close personal relationships are essential features in the early stages of partner development. In fact, long-term value cannot be created by efficiency elements of speed, reliability, and innovation alone. Instead it must be accompanied by an ongoing dialogue in transparent and customised relationships, based on the intangible elements of tacit knowledge, reputation, integrity, and technical competence, from which trust emerges. This example illustrates the value-creating potential CoPs, as a *personalisation* mechanism, have for high-technology SMEs in regional clusters.

CoPs are able to support innovation regardless of the industry base. Innovativeness is an attitude that can apply to low-tech activities and is particularly relevant for many regional areas. Albonies and Moso (2002) attest that there is more to innovation than the technology-based innovative milieu. Few regions have the capability and resources required to develop high-technology clusters. Instead there needs to be an emphasis on innovativeness, an evolutionary process of innovative behaviour based on daily operations. They describe this innovative process in the Basque Country region of Spain as a highly industrialised, as opposed to high-technology, cluster. *Working groups* are established around special interests and needs of firms. Group members are involved in cooperative ongoing knowledge exchange, where business management knowledge is collected and disseminated, and international business management learning and exchange occurs. This collaboration provides SMEs in the cluster the opportunity to learn new ways of operating. "Never has innovation been more related to discovery" (Albonies & Moso, 2002, p. 352). These working groups bear the hallmarks of CoPs and indicate their applicability as value creators for SMEs in regional clusters.

CoPs can create value through non-high-technical innovation. The traditional artisan jewellery industry from St. Petersburg in Russia is an example of this type of medium-to-low technology innovation cluster. Value is dependent creativity and technique based on design know-how that is transferred via tacit knowledge and achieved through practice (Forsman & Solitander, 2004).

More Effective Use of Knowledge

SMEs have deep professional, social, and business networks where *personalisation* is the preferred mechanism for knowledge transfer, and tacit and mutual trust is developed over long histories of interaction.

Sharing knowledge about the region and its resources can add value to individual SMEs and to the region. SMEs are good at knowledge creation, but are poor at retaining that knowledge (Levy, Loebbecke & Powell, 2003). This means that many SMEs fail to fully utilise the knowledge that enables them to grow and develop, namely in supporting customers and in managing the business. CoPs provide a mechanism where synergistic relationships can be developed by SMEs through local collaborations in their region. This mechanism, based on the social capability and prior experience, provides access to the absorption of knowledge (Almeida & Kogut, 1999). Thus SMEs are more easily able to recognise the value of knowledge and the value-adding potential of their regional knowledge exchanges. Capello (1999) describes this knowledge sharing where collective learning is developed as the *club good*.

Sharing knowledge resources among SMEs can help them overcome limitations of size. Large organisations have access to many resources such as expertise, infrastructure, and physical and intellectual resources within the confines of the firm.

SMEs do not often have access to such resources and facilities internally. SMEs that are involved in flexible and cooperative regional networking, or intraregional ties, have the potential to overcome these limitations (Fuchs, 2002). CoPs are an effective mechanism for SMEs to conduct these intraregional networking interactions. These intraregional ties enable SMEs to obtain competitive production value. Fuchs (2002) is concerned that these ties are extended to include global links which provide value through added know-how and access to international markets. Perhaps the greatest value of extending these ties is that it prevents the CoPs from becoming too inflexible. The discussion that follows indicates how some SMEs are accessing these global resources through the use of VCoPs.

FUTURE TRENDS

Virtual Communities of Practice (VCoPs) as Capital

VCoPs are able to create value by driving strategy, starting new lines of business, quickly solving business problems, transferring best practice, developing professional skills, and supporting the recruitment and retention of talented employees. VCoPs emerged in large organisations to address the needs of their globally dispersed operations. This is analogous to the composition of the regional cluster's agglomeration of SMEs. VCoPs appear to be a natural evolution of CoPs for value creation in regional cluster-based SMEs.

Ho, Au, and Newton (2003) describe how the successful use of VCoPs is contributing to value creation in the apparel industry cluster of Hong Kong. VCoPs have arisen out of virtual trading communities. These provide members with access to portals where vast amounts of relevant information are distributed daily to supply chain members. Trade and professional associations, and academic institutions are also providing non-profit portals with free access to information. These professional associations and educational institutions are playing an important role in establishing VCoPs. For example *apparelkey.com*, established by the Hong Kong Polytechnic University and Chinese University of Hong Kong, provides different channels for knowledge sharing including threaded discussion forums and a page where members can discuss problems with experts. VCoPs have changed the way intellectual capital is acquired and leveraged for product/process improvement and innovation. They are instrumental in finding the best sources of supply and demand. In fact VCoPs are a new source of capital that is obtained when industry-specific information from them is combined with internal firm-specific knowledge.

VCoPs necessitate new ways of managing knowledge. No longer is the firm able to confine the knowledge within its boundaries, as they have become permeable to external knowledge flows. SME managers therefore need a clear understanding of the internal and external knowledge that provides value to the organisation and ensures that appropriate channels are established to maximise this value.

Large organisations have found that it is not easy to manage CoPs, as they are fundamentally informal and self-organising, and are not amenable to organisational structures; thus, efforts to institutionalise them may well limit their potential. However, the very nature of the SME would indicate that these issues are unlikely to create problems for SMEs. Instead CoPs provide SMEs value-creating potential as it is known that SMEs firms investing in their external relationships are more likely to succeed.

CONCLUSION

With the move to the knowledge economy, large organisations are successfully creating and sustaining VCoPs to access the valuable knowledge that exists within their organisation. SMEs, without the organisational resources available to large organisations, can utilise the principles of such KM initiatives by linking with other SMEs. Regional clusters can create value from knowledge, and communities of practice are essential to this process. VCoPs are the most suitable mechanism because of their facility for rapid, inclusive communication, coupled with their ability to draw in knowledge resources from a variety of sources and to manage the clusters' knowledge most effectively. To achieve this, VCoPs must develop appropriate practices to foster trust, an ethos of innovation, and commitment to the regional area. These developments present not just an opportunity, but a mandate for action given the global nature of competitiveness. Further research on VCoPs in regional clusters will help us understand how to enhance the value of the collective knowledge of SMEs, and how regional areas can establish and sustain SME-based clusters and derive value from them.

REFERENCES

ACTETSME. (1998). SME profile Australia. Retrieved August 17, 2004, from *http://www.actetsme.org/aust/aus98.html*

Albonies, A.L., & Moso, M. (2002). Basque Country: The knowledge cluster. *Journal of Knowledge Management, 6*(4), 347-355.

Almeida, P., & Kogut, B. (1999). Localisation of knowledge and the mobility of engineers in regional networks. *Management Science, 45*(7), 905-917.

Ardichvilli, A., Page, V., & Wentling, T. (2003). Motivation and barriers to participation in virtual knowledge-sharing communities of practice. *Journal of Knowledge Management, 7*(1), 64-77.

Benner, C. (2003). Learning communities in a learning region: The soft infrastructure of cross-firm learning networks in Silicon Valley. *Environment and Planning A, 35*, 1809-1830.

Boisot, M. (2002). The creation and sharing of knowledge. In C.W. Choo & N. Bontis (Eds.), *The strategic management of intellectual capital and organisational knowledge* (pp. 65-77). Oxford, UK: Oxford University Press.

Burress, A., & Wallace, S. (2003). Brainstorming across boundaries at Shell. *KM Review, 6*(2), 20-23.

Capello, R. (1999). Spatial transfer of knowledge in high-technology milieu. *Regional Studies, 33*(4), 353-365.

Chattell, A. (1998). *Creating value in the digital era: Achieving success through insight, imagination and innovation*. London: Macmillan.

Davenport, T.H., & Prusak, L. (2000). Knowledge management: On to phase two. Retrieved February 13, 2003, from *http://bhsworkingknowledge. hbs.edu/item.jhtml*

De Bernardy, M. (1999). Reactive and proactive local territory: Cooperation and community in Grenoble. *Regional Studies, 33*(4), 343-352.

DIIRD. (2003). *Clusters: Victorian business working together in a global economy.* Victoria, Australia: Department of Innovation, Industry & Regional Development.

DTI. (2004). *A practical guide to cluster development.* London: Department of Trade and Industry.

Forsman, M., & Solitander, N. (2004, April). Knowledge transfer and industrial change in the jewellery industry: An ecologies-of-knowledge approach. *Proceedings of the 5th European Conference on Organisational Knowledge, Learning and Capabilities*, Innsbruck, Austria.

Fuchs, G. (2002). The multimedia industry: Networks and regional development in a globalised economy. *Economic and Industrial Democracy, 23*(3), 305-333.

Grant, R.M. (2002). The knowledge-based view of the firm. In C.W. Choo & N. Bontis (Eds.), *The strategic management of intellectual capital and organisational knowledge* (pp. 133-148). Oxford, UK: Oxford University Press.

Hackett, B. (2002). Beyond knowledge management: New ways to work. In C.W. Choo & N. Bontis (Eds.), *The strategic management of intellectual capital and organisational knowledge* (pp. 725-738). Oxford, UK: Oxford University Press.

Ho, D.C.K., Au, K.F., & Newton, E. (2003). The process and consequences of supply chain virtualisation. *Industrial Management & Data Systems, 103*(6), 423-433.

Hospers, G.-J., & Beugelsdijk, S. (2002). Regional cluster policies: Learning by comparing? *KYKLOS, 55*(3), 381-402.

Huizing, A., & Bouman, W. (2002). Knowledge and learning, markets and organisations. In C.W. Choo & N. Bontis (Eds.), *The strategic management of intellectual capital and organisational knowledge* (pp. 185-204). Oxford, UK: Oxford University Press.

Humphrey, J., & Schmitz, H. (2002). How does insertion in global value chains affect upgrading in industrial clusters? *Regional Studies, 36*(9), 1017-1027.

Hwa, K.K. (2003, February/March). Knowledge powers Singapore economy. *Information Age,* 30-33.

King, K. (2002). Banking on knowledge: The new knowledge projects of the World Bank compare. *A Journal of Comparative Education, 32*(3), 311-326.

Kulkki, S. (2002). Knowledge creation of global companies. In C.W. Choo & N. Bontis (Eds.), *The strategic management of intellectual capital and organisational knowledge* (pp. 501-519). Oxford, UK: Oxford University Press.

Lawson, C., & Lorenz, E. (1999). Collective learning, tacit knowledge and regional innovative capacity. *Regional Studies, 33*(4), 305-317.

Lee, L.L., Parslow, R., & Julien, G. (2002). *Leading network development practices in 2002: A literature review.* Melbourne: BHP Billiton.

Levy, M., Loebbecke, C., & Powell, P. (2003). SMEs, co-opetition and knowledge sharing: The role of information systems. *European Journal of Information Systems, 12*(1), 3-17.

Loecher, U. (2000). Small and medium-sized enterprises: Delimitation and the European definition in the area of industrial business Bradford. *European Business Review, 12*(5), 261.

MacKinnon, D., Cumbers, A., & Chapman, K. (2002). Learning, innovation and regional development: A critical appraisal of recent debates. *Progress in Human Geography, 26*(3), 293-311.

Marshall, A. (1947). *Principles of economics: An introductory volume.* London: Macmillan.

Mitra, J. (2000). Making connections: Innovations and collective learning in small businesses. *Education + Training, 42*(4/5), 228-236.

Molina-Morales, F.X., & Martinez-Fernandez, M.T. (2003). The impact of industrial district affiliation on firm value creation. *European Planning Studies, 11*(2), 155-170.

Patriotta, G. (2003). *Organisational knowledge in the making: How firms create, use, and institutionalise knowledge* (1st ed.). Oxford, UK: Oxford University Press.

Porter, M.J. (1998). Clusters and the new economics of competition. *Harvard Business Review,* (November/December), 77-90.

Prahalad, C.K. (2004). The blinders of dominant logic. *Long Range Planning, 37,* 171-179.

Prahalad, C.K., & Ramaswamy, V. (2004). Co-creating unique value with customers. *Strategy & Leadership, 32*(3), 4-9.

Rowley, J. (2004). Just another channel? Marketing communications in e-business. *Marketing Intelligence & Planning, 22*(1), 24-41.

Saxenian, A.L. (1994). Silicon Valley versus Route 128. *01628968, 16*(2).

Senge, P.M. (1990). *The fifth discipline: The art and practice of the learning organization.* New York: Doubleday/Currency.

Skoog, M. (2003). Visualising value creation through the management control of intangibles. *Journal of Intellectual Capital, 4*(4), 487-504.

Spence, M. (2004). Efficiency and personalisation as value creation in interorganisationalising high-technology SMEs. *Canadian Journal of Administrative Sciences, 21*(1), 65-78.

Staber, U. (2001). Spatial proximity and firm survival in a declining industrial district: The case of knitwear firms in Baden-Württemberg. *Regional Studies, 35*(4), 329-342.

Sternberg, R., & Tamasy, C. (1999). Munich as Germany's no. 1 high-technology region: Empirical evidence, theoretical explanations and the role of small firm/large firm relationships. *Regional Studies, 33*(4), 367-377.

Stewart, T.A. (1997). *Intellectual capital: The new wealth of organisations.* New York: Doubleday.

Stewart, T.A. (2001). *The wealth of knowledge: Intellectual capital and the twenty-first century organisation.* New York: Currency.

Tallman, S., Jenkins, M., Henry, N., & Pinch, S. (2004). Knowledge, clusters, and competitive advantage. *Academy of Management Review, 29*(2), 258-271.

Vorbek, J., Heisig, P., Martin, A., & Schutt, P. (2003). Knowledge management in a global company: IBM Global Services. In K. Mertins, P. Heisig & J. Vorbek (Eds.), *Knowledge management: Concepts and best practices* (2nd ed., pp. 292-304). Berlin: Springer-Verlag.

Wenger, E. (1999). *Communities of practice: Learning, meaning and identity.* Cambridge, UK: Cambridge University Press.

Wenger, E., McDermott, R., & Snyder, W.M. (2002). *Cultivating communities of practice: A guide to managing knowledge.* Boston: Harvard Business School Press.

Yoo, Y., & Torrey, B. (2002). National culture and knowledge management in a global learning organisation. In C.W. Choo & N. Bontis (Eds.), *The strategic management of intellectual capital and organisational knowledge: A case study* (pp. 421-434). Oxford, UK: Oxford University Press.

Zboraiski, K., Gemuenden, H.G., & Lettl, C. (2004, April). A members' perspective of the success of communities of practice: Preliminary empirical results. *Proceedings of the 5th European Conference on Organisational Knowledge, Learning and Capabilities,* Innsbruck, Austria.

KEY TERMS

Cluster: A group of organisations that are linked together around a particular industry.

Coopetition: A situation where organisations, usually SMEs, are cooperating with each other and at the same time they are also competing against each other

E-Clusters: Digitally enabled communities of organisations that come together on a needs basis, in varying formations of virtual organisations, to meet a temporary business opportunity.

Knowledge Economy: An economy where the resource of most value is knowledge.

Regional Clusters: Geographic concentrations of organisations, predominately SMEs, in the same or related industries that share resources and have access to other institutions important to competition, for example educational and training facilities. This close proximity creates a network of alliances that enables more productive operation, facilitates innovation, and lowers barriers to new entrants.

Small to Medium Enterprises (SMEs): Businesses defined by their small scale in contrast to large corporations. Criteria for defining small business vary from one context to another. Europe defines SMEs as organisations that employ fewer than 250 employees and have a maximum of 40 million Euro annual turnover, a maximum of 27 million Euro annual balance sheet total, a minimum of 75% owned by company management, where owners-managers/their families manage the company personally (Loecher, 2000). In countries with smaller total populations, the SME definition reflects this. An example is Australia, where SMEs are separated into two sectors: *Manufacturing*, where small enterprises employ fewer than 100 employees and medium enterprises 100 to 199 employees, and *services sectors,* where small enterprises have fewer than 20 employees and medium enterprises 20-199 employees (ACTETSME, 1998).

Value: In an organisational context, refers to anything that assists in achieving that organisation's objectives (e.g., Chattel, 1998).

Value Creation: In an information systems context, refers to the process of utilising intellectual capital (IC) to realise organisational value (e.g., Stewart, 1997).

Virtual Communities of Practice (VCoPs): CoPs enabled by online interactive technologies made possible by rapid ICT developments, often necessitated by the globalisation of operations. They are the most recent strategy for a personalised KM approach, and multinational corporations have made VCoPs the preferred KM technique (Ardichvilli et al., 2003). Traditionally, CoP members interact on a face-to-face basis, but online VCoPs enable disparate members' ongoing participation. An example of a VCoP is the Silicon Valley *Webgrrls*, established by female professionals to counter the masculine dominance of the IT profession in that region (Benner, 2003).

Critical Success Factors for the Successful Introduction of an Intellectual Capital Management System

Brenda Elshaw
IBM, UK

It has long been recognized that one of the most valuable assets an organization possesses is the knowledge and experience of its employees. Yet, month by month, many organizations allow a great part of this knowledge to walk out the door as their employees leave. In some cases, even while the employee is still there, their knowledge is not captured and reused, as its value to the organization is not recognized.

Several years ago, a well known car manufacturer was designing its next generation vehicle. Wishing to repeat a previous success, both in design and marketing, the company tried to identify what factors had contributed to this success. However, as the lessons learned from the previous exercise had not been documented, not to mention, no record existed of the team members who had worked on the original project, this valuable experience was ultimately lost. How different this could have been had they captured the intellectual capital resulting from the design and been able to build upon the best practices to help repeat their earlier success.

Organizations that put processes in place to capture their intellectual capital can substantially reduce costs due to time lost by employees reinventing the wheel and can often increase revenue by the reuse of selected assets.

Stewart, in his book *The Wealth of Knowledge: Intellectual Capital and the Twenty-First Century Organization* (2002) states:

It has become standard to say that a company's intellectual capital is the some of its human capital (talent), structural capital (intellectual property, methodologies, software, documents, and other knowledge artifacts, and customer capital (client relationships)." It is the *"knowledge that transforms raw materials and makes them more valuable.* (pp. 12-13)

Figure 1.

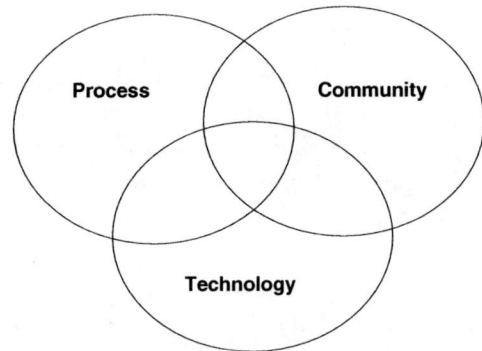

In most organizations, the majority of knowledge is held via various data storage mechanisms, usually computer based. However, its true value is realized only when it has context added to it by the application of the knowledge and skill of the practitioners involved in its creation and application.

For that reason, an intellectual capital management system (ICMS) has to be more than just an efficient data storage and retrieval system. An effective ICMS takes into account three components—technology, process, and community.

TECHNOLOGY

At the heart of any successful ICMS is an efficient data storage and retrieval system. The system has to be:

- Capable of holding large amounts of data in various formats;
- Able to be easily searched;

- Easily accessible by the user community with reasonable response times for downloading large data files;
- Secure enough to give the appropriate level of access to those who use it—readers, editors, submitters of content, and so forth; and
- Available during those hours when users need access.

In addition, the taxonomy used to categorize the content should be meaningful to the user population it serves. This may mean that data repositories could be categorized in different ways to suit different user communities; for example, a technical community may require content to be categorized according to different technologies, while a project-based community may require their content to be categorized to correspond to the stage or activity of the project. This may lead to separate repositories configured for different communities rather than a single, enterprise-wide one.

Each piece of intellectual capital held should have a meaningful summary that will enable users to validate the usefulness of the information contained without having to download large files unnecessarily.

Lastly, once a user has identified a suitable piece of intellectual capital, the content must be accurate and up to date. The success of an ICMS is dependent upon the users perceiving its content to have value for them so that they not only go to it as the primary source of reliable reusable content, but also contribute their own experiences to it, therefore adding to the value for others. Once users discover the content to be either inaccurate or out of date, they are unlikely to go back for a second try, and the system rapidly becomes dysfunctional.

PROCESSES

Once an ICMS system has been implemented, its usefulness and eventual success is dependent upon the amount of good quality content it holds. While it needs a critical mass of content before it is made available to the user community, that content has to be maintained and supplemented by an active user population. In order for that to happen, users should be encouraged to check:

1. for any relevant intellectual capital that can be reused at the start of any new initiative; and
2. if anything has been created that will have reuse value for others who will undertake similar activities and, if so, submit it for inclusion in the system.

This should happen as a natural part of the work environment and not be seen as an extra chore to add to the workload. To encourage this to happen seamlessly, the use of the system needs to be incorporated into accepted work processes, including:

- **Role Definitions and Responsibilities:** While it is encouraging to believe that all employees will naturally see the benefit of intellectual capital reuse and will generously donate time and expertise to the maintenance and updating of the content, in practice, this rarely happens. Already overworked employees will not volunteer to do activities they perceive as extra to their expected job responsibilities. In order to ensure they contribute to the system, its usage should be included in all appropriate job definitions; for example, a project manager should have the responsibility for ensuring that members of the team search for items that can be reused as part of the project task planning and for ensuring that all items that have reuse value are submitted to the ICMS as part of the project closure. This implies that time is allowed in the project plan for these activities to occur.
- **Goal and Reward Systems:** Many organizations link the usage of an ICMS into the individual business goals and commitments of their employees. While it is easy to set up targets for each employee to submit a certain number of items into the system, these are the least successful motivators for the creation of any real value and often fail for one of two main reasons:
  i. The emphasis should be on the quality of the content submitted rather than quantity. Numeric submission targets usually result in an overload of submission processes for no great gain as employees search their

filing systems for items to submit in order to meet their year-end targets.

ii. Many of the most valuable items are the culmination of teamwork rather than individual efforts; individual submission targets set team members against each other rather than encouraging teamwork.

Where targets are set or rewards given, these should be related to the amount of reuse of a particular element of intellectual capital, which is more likely to result in the collection of high quality content. It is far better to gather a few items of quality content than to have a content repository containing everything but with no indication of its usefulness.

- **Relevant Training Programs:** The checking of ICMS for relevant content and the submission of new items of intellectual capital resulting from work processes should be incorporated into all relevant training programs including induction training. Once it is seen as a natural process in multiple business-related activities, its use will become automatic, and the time and effort taken to do it will no longer be questioned.

- **Intellectual Capital Submission:** While many ICMS systems are set up to hold anything that is submitted to them, the most successful ones are those where the quality and relevance of the content to the user population can be guaranteed. This means that there has to be a process whereby content is submitted, evaluated, and either selected for inclusion and published, sent back to the submitter with suggestions for further enhancement, or rejected with appropriate feedback and encouragement for improvement. If this process is to be successful, users need to have trust in and respect for the panel sitting in judgment on the content. This happens most easily when the panel consists of a team of their peers whose subject matter expertise is beyond question.

- **Content Evaluation and Archiving:** In today's fast-moving world, most intellectual capital has a limited shelf-life. This means that content which has proved to be valuable last year may have been superseded by new content and may be valueless this year. For this reason, users of a successful ICMS must have the reassurance that the content not only had value when it was submitted, but is still current. Consequently, the content should be regularly re-evaluated to identify its timeliness and relevance. It is advisable that a regular process of re-evaluation and archiving should be set up and essential that an effective content management regime is implemented from day one.

COMMUNITY

For an ICMS to be effective and to be the first port of call for users looking for content, it has to be perceived to have relevance to them and the job they are doing. A successful ICMS is one whereby the tool used facilitates the knowledge sharing activities of an existing community, rather than being an end in itself. As content is specific to the different activities of user communities, this is best achieved by making the ICMS closely tie into the communities of practice.

If the content is collected and disseminated via a recognized community rather than by a series of individuals posting content into a database, it is more likely that the content resides in the corporate memory of several people rather than just an individual. That way, should the originator of the content leave the organization, there is more chance that the tacit knowledge surrounding the use of the content is retained by other community members.

Each community that could benefit from sharing knowledge and experience should build a "knowledge network", consisting of a business sponsor, a community of practice with information they would benefit from sharing, and a dedicated core team of people to drive the community activities. Roles should be established for a recognized and respected core team of people recruited from within the community to drive the collection of the content as part of a wider community-building role.

Activities to be undertaken by the core team would include:

- Development of the taxonomy for the community's intellectual capital that is meaningful in the context of their community and content types.

- Collection, assessment, publishing, and content management of the intellectual capital submitted by the community.
- Analysis of areas where no intellectual capital currently exists in the system and encouragement of community members with known areas of expertise to submit their content or work together to create content to fill the gaps.
- Coordination of community activities including seminars, facilitation of formal and informal knowledge sharing activities, and so forth.
- Creation and delivery of a communication plan to keep the community informed regarding community activities, new intellectual capital of interest, feedback on reuse, and so on.
- Provision of management statistics to the business sponsor and key stakeholders regarding the usage of the system.
- Acting as information brokers and links to the community to help members build their informal networks.

The system will only work if community members are prepared to share their expertise with each other. In order for that to happen, the following have to be in place:

- **Trust and a Willingness to Share:** People need to have faith that their content will not be misused and that they will get recognition as the owner and originator of the content. This is more likely when it is driven within a recognized and attributable community than when put into an anonymous data repository.
- **Respect for Intellectual Property Rights:** Guidelines need to be in place to ensure that information is used with care and that content of a confidential or business sensitive nature will not be abused. When setting up an ICMS, many organizations make acceptance of a set of usage guidelines a condition of initial access to the system.
- **Respect for Subject Matter Experts:** Those who are sitting in judgment of the relevance of content have to be respected as thought leaders in their community. For that reason, the core team members driving the collection of content should be recruited from those members of the community who have shown thought leadership,

although one or two junior members could be recruited as part of their career advancement.

If a culture can be generated whereby being part of the core team is seen as something to be desired, maybe even seen as a good career move, so that community members aspire to become part of the team, then the ICMS will have credibility and will more easily gain acceptance as a primary source of useful information.

CONCLUSION

A successful ICMS is more than just a collection of documents in a database. It is something that should be created, managed, and disseminated via a community. That way, the content should have relevance for the community members and will have a quality guarantee assured by respected community experts. If the content is well organized and managed, users of the system will have the confidence that the content they find there is of value and will also be more likely to submit their own valuable content for use by others.

For this to happen, equal prominence has to be given to the community-building aspects and the supporting-business processes in addition to the technical tools used to store the data. This is an ongoing commitment for the lifetime of the ICMS, not just during the early stages.

There is a considerable amount of investment required for the implementation of a successful ICMS in terms of time required:

a. by the provision of a dedicated team to drive the collection and maintenance of the intellectual capital as part of other community building activities.
b. by the community members to search for existing content for reuse and to submit their content when appropriate.

In order for the system to be successful, there has to be buy-in from the organization at all levels—from executive management who fund it, the operational management who ensure that people are allowed the time to use it, and the user population who freely submit their valuable content for reuse.

It is likely that, during the initial phases of setting up an ICMS, more time will be spent submitting content than downloading and reusing it, and the costs will outweigh the benefits. Once there is a critical mass of content and a set of supported and accepted processes for its use, the benefits will greatly outweigh the costs.

FURTHER READING

Mayo, A. (2001). *Valuing people as assets: Monitoring, measuring, managing.* Nicholas Brealey.

Conway, S., & Sligar, C. (2002). *Unlocking knowledge assets.* Microsoft Press.

Lesser, E., & Prusak, L. (Eds.). (2004). *Creating value with knowledge: Insights from the IBM Institute for Business Value.* Oxford: Oxford University Press.

Skyrme, D. J. (1999). *Knowledge networking: Creating the collaborative enterprise.* Butterworth-Heinemann.

REFERENCES

Stewart, T. A. (1998). *Intellectual capital: The new wealth of organizations.* Nicholas Brealey.

Stewart, T. A. (2002). *The wealth of knowledge: Intellectual capital and the twenty-first century organization.* Nicholas Brealey.

KEY TERMS

Community of Practice: A group of people who have work practices in common.

Intellectual Capital: Knowledge gathered by an organization and its employees that has value and would help the organization gain benefit when re-used.

Intellectual Capital Management System (ICMS): A combination of communities, processes and technology brought together to identify, value, categorize and capture intellectual capital for re-use.

Knowledge Network: A self-managing community of people who share mutual trust and respect and come together to share their knowledge.

Tacit Knowledge: Knowledge that is not written down but is held in the heads and minds of people.

Taxonomy: Classification into categories and sub-categories.

Discovering Communities of Practice through Social Network Analysis

Cindi Smatt
Florida State University, USA

Molly McLure Wasko
Florida State University, USA

INTRODUCTION

The concept of a community of practice is emerging as an essential building block of the knowledge economy. Brown and Duguid (2001) argue that organizations should be conceptualized as consisting of autonomous communities whose interactions can foster innovation within an organization and accelerate the introduction of innovative ideas. The key to competitive advantage depends on a firm's ability to coordinate across autonomous communities of practice internally and leverage the knowledge that flows into these communities from network connections (Brown & Duguid, 2001). But how does an organization do this? A key challenge for management is understanding how to balance strategies that capture knowledge without killing it (Brown & Duguid, 2000).

BACKGROUND

Typically, top-down business processes aimed at leveraging knowledge flows end up stifling creativity by institutionalizing structures promoting rigidity. In order to understand knowledge flows, managers need to change their focus away from a process view of knowledge creation to a practice-based view. When individuals have a common practice, knowledge more readily flows horizontally across that practice, creating informal social networks to support knowledge exchange (Brown & Duguid, 2001). Therefore, the key to understanding knowledge flows within organizations is to switch the conceptualization of work away from formal processes to that of emergent social networks.

Social network perspectives focus on the pattern of relationships that develop between members of a community of practice, suggesting that individuals and their actions are interdependent, rather than autonomous occurrences. In contrast to focusing on work tasks as the unit of analysis, a social network perspective of work focuses on how relational ties between individuals lead to outcomes, such as knowledge exchange and innovation. The ties that develop between community members are characterized by their content, direction, and relational strength, all of which influence the dynamics of individual interactions. The content of ties refers to the resource exchanged, such as information, money, advice, or kinship. The direction of ties indicates the giver of the resource and the receiver. The relational strength of ties pertains to the quality of the tie. For instance, the relational strength of ties indicates the amount of energy, emotional intensity, intimacy, commitment, and trust connecting the individuals.

When the resource being exchanged in the network is knowledge, prior research indicates value is derived from bridging "structural holes" or gaps (Burt, 1992). As a result, individuals who develop ties with disconnected communities of practice gain access to a broader array of ideas and opportunities than those who are restricted to a single community of practice. In addition, individuals who network with others from diverse demographic categories benefit because different people have different skills, information, and experience. Such ties bridge structural holes in the larger organization, and thereby enhance its capacity for creative action.

FUTURE TRENDS

Managers interested in understanding where the communities of practice are, and how these commu-

Figure 1. Sample of a social network graph

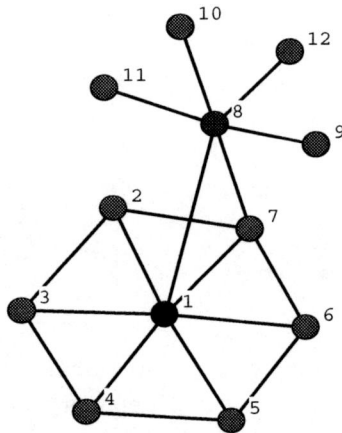

CONCLUSION

Social network analysis should be considered an essential tool for mapping actual knowledge flows. In contrast to focusing on business processes to formalize knowledge flows, taking a social network perspective allows management to redesign knowledge flows by adjusting network structures. Using social network analysis techniques to discover communities of practice within the organization allows managers to influence knowledge flows without killing innovation.

REFERENCES

Brown, J.S., & Duguid, P. (2000). Balancing act: How to capture knowledge without killing it. *Harvard Business Review,* (May-June), 73-80.

Brown, J.S., & Duguid, P. (2001). Knowledge and organization: A social-practice perspective. *Organization Science, 12*(2), 198-213.

Burt, R. (1992). *Structural holes: The social structure of competition.* Cambridge, MA: Harvard University Press.

KEY TERMS

Social Network: Interconnected people who directly or indirectly interact with or influence one another.

Social Network Analysis: Tools and techniques for identifying the patterns of connections among individuals in exchange relations.

Structural Hole: A gap or lack of a connection in a social network.

nities link to one another, could create a knowledge map of the organization using social network analysis. Social network analysis is a tool that can be used to depict the informal flows of knowledge both within and between communities of practice. By using social network analysis to examine the organization, managers are better able to understand what type of knowledge is being exchanged and the pattern of its exchange. This would facilitate not only the identification of barriers to knowledge exchange and areas of the organization that need better integration into the knowledge network, but would also indicate key personnel in the network through which knowledge is currently flowing.

A picture of a social network resulting from social network analysis is illustrated in Figure 1.

This knowledge map indicates two communities of practice that are bridged by individual 8. Individual 1 is central to the network, yet if this individual left the organization, it would have relatively little effect on knowledge flows. Although this would increase the distance between certain individuals, knowledge flows among all individuals remain possible as there is no fragmentation.

Discovering Implicit Knowledge from Data Warehouses

M. Mehdi Owrang O.
American University, USA

INTRODUCTION

Today, every corporation faces the problem of how to acquire, store, and share information. Knowledge management (KM) has been introduced to accomplish these tasks (Adams, 2004; Barquin, 2000; Frappaolo & Wilson, 2004). Fundamental to KM is the realization that knowledge exists in two basic forms: explicit and tacit (Adams, 2004; Barquin, 2000; Frappaolo & Wilson, 2004; Orr, 2004). Organizations have data, in the form of operational databases and/or data warehouses, which contain implicit knowledge. Some knowledge believed to be tacit (experiential and intuitive) can be transformed into explicit knowledge. Getting to implicit knowledge requires taking a look at tacit knowledge resources (i.e., domain experts or data warehouses) to determine whether that knowledge could be codified if it were subjected to some type of mining and translation process. Then, it requires implementing that mining/translation process. The majority of an organization's knowledge is presumed to be tacit. Yet, the majority of the KM applications seem to focus on the explicit knowledge base: working on existing corporate knowledge or making individuals more effective at sharing explicit knowledge (Frappaolo & Wilson, 2004). Efforts have been put in creating an organized explicit knowledge repository, called data warehousing (Bischoff & Alexander, 1997) that is continuously fed and leveraged. Knowledge management is not truly possible without data warehousing (Barquin, 2000). It is the real-time access to an enterprise's integrated data stores through data warehousing that complements an individual's tacit knowledge of how something is done.

Knowledge discovery is defined as the nontrivial extraction of implicit, previously unknown, and potentially useful information from data (Adriaans &

Zantinge, 1996; Agrawal, Imielinski & Swami, 1993; Brachman et al., 1996; Fayyad, 1996; Inmon, 1996). The automatic knowledge acquisition in a nondata warehouse environment has been on the operational databases which contain the most recent data about the organizations. Summary and historical data, which are essential for accurate and complete knowledge discovery, are generally absent in the operational databases. A data warehouse is an ideal environment for rule discovery since it contains the cleaned, integrated, detailed, summarized, historical, and metadata (Bischoff & Alexander, 1997; Inmon, 1996; Meredith & Khader, 1996; Parsaye, 1996).

In this article, we are looking at the discovery of implicit knowledge from the data warehouses. Most of the success of knowledge discovery resides in the ability of the system to elicit the right level of detail as well as accuracy from the data warehouse which has the implicit data. We look at the knowledge discovery process on detailed, summary, and historical data. Also, we show how the discovered knowledge from these data sources can complement and validate each other.

KNOWLEDGE DISCOVERY IN DATA WAREHOUSES

Knowledge discovery on operational relational databases could lead to inaccurate and incomplete discovered knowledge. Without first warehousing its data, an organization has lots of information that is not integrated and has little summary or history information. The effectiveness of knowledge discovery on such data is limited. A data warehouse environment integrates data from a variety of source databases into a target database that is optimally designed for decision support. A data warehouse includes integrated data, detailed and summary data, historical data, and metadata. Each of these elements enhances the knowledge discovery process (Adriaans & Zantinge, 1996; Barquin & Edelstein, 1997; Bischoff & Alexander, 1997; Meredith & Khader, 1996).

There are several benefits in rule discovery in a data warehouse environment. First, in a data warehouse environment, the validation of the data is done in a more rigorous and systematic manner. Using metadata, many data redundancies from different application areas are identified and removed. The cleansing process will remove duplication and reconcile differences between various styles of data collection. Second, data warehouses are not concerned with the update anomalies since update of data is not done. This means that at the physical level of design, we can take liberties to optimize the access of data, particularly in dealing with the issues of normalization and physical denormalization. Universal relations can be built in the data warehouse environment for the purposes of rule discovery, which could minimize the chance of undetecting hidden patterns.

Figure 1 shows a general framework for knowledge discovery in a data warehouse environment. External data, domain knowledge (data that is not explicitly stored in the database, that is, male patient cannot be pregnant), and domain expert are other essential components to be added in order to provide an effective knowledge discovery process in a data warehouse environment.

KNOWLEDGE DISCOVERY FROM DETAILED DATA

Most of the knowledge discovery has been done on the operational relational databases. An operational database stores the most recent and detailed data. In addition, the goal of the relational databases are to provide a platform for querying data about uniquely identified objects. However, such uniqueness constraints are not desirable in a knowledge discovery environment. In fact, they are harmful since from the data mining point of view, we are interested in the frequency with which objects occur (Adriaans & Zantinge, 1996). In the following, we discuss two main problems associated with the knowledge discovery in the operational relational databases, namely, the possibility of discovering incorrect and incomplete knowledge.

INCORRECT KNOWLEDGE DISCOVERY FROM RELATIONAL DATABASES

In general, summary data (aggregation) is never found in the operational environment. Without discovery process on summary data, we may discover incorrect knowledge from detailed operational data. Discovering rules based just on current detail data may not depict the actual trends on data. The problem is that statistical significance is usually used

Figure 1. A framework for knowledge discovery in a data warehouse environment

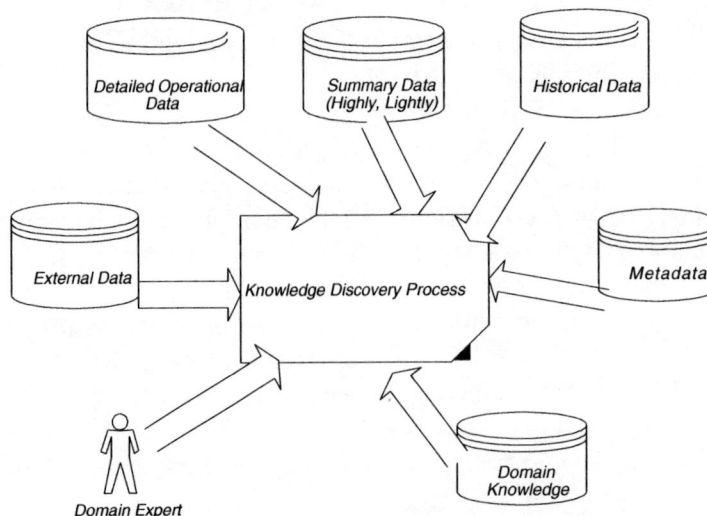

in determining the interestingness of a pattern (Giarrantanto & Riley, 1989). Statistical significance alone is often insufficient to determine a pattern's degree of interest. A "5% increase in sales of product X in the Western region", for example, could be more interesting than a "50% increase of product X in the Eastern region". In the former case, it could be that the Western region has a larger sales volume than the Eastern region; thus its increase translates into greater income growth.

The following example (Matheus, Chan & Piatetsky-Shapiro, 1993) shows that we could discover incorrect knowledge if we only look at the detailed data. Consider Table 1, where the goal of discovery is to see if product color or store size has any effect on the profits. The data are not large, but they show the points.

Assume we are looking for patterns that tell us when profits are positive or negative. We should be careful when we process this table using discovery methods such as simple rules or decision trees. These methods are based on probabilities that make them inadequate for dealing with influence within aggregation (summary data). A discovery scheme based on probability may discover the following rules from Table 1:

- **Rule 1:** IF Product Color = Blue Then Profitable = No   CF = 75%
- **Rule 2:** IF Product Color = Blue and Store Size > 5000 Then Profitable = Yes CF = 100%

The results indicate that blue products in larger stores are profitable; however, they do not tell us the amounts of the profits which can go one way or another. Now, consider the modified table, where the third row in Table 1 is changed for the Profit to be 100 instead of 7000. Rules 1 and 2 are also true in the

Table 1. Sample sales data

Product	Product Color	Product Price	Store	Store Size	Profit
Jacket	Blue	200	S1	1000	-200
Jacket	Blue	200	S2	5000	-100
Jacket	Blue	200	S3	9000	7000
Hat	Green	70	S1	1000	300
Hat	Green	70	S2	5000	-1000
Hat	Green	70	S3	9000	-100
Glove	Green	50	S1	1000	2000
Glove	Blue	50	S2	5000	-300
Glove	Green	50	S3	9000	-200

modified table. That is, from a probability point of view, Table 1 and the modified one produce the same results.

However, this is not true when we look at the summary tables (product color = Blue, Profit = 6400, based on Table 1) and (product color = Blue, Profit = -500, based on modified Table 1). The former summary table tells us that Blue color product is profitable, and the latter summary table tells us it is not. That is, in the summary tables, the probability behavior of these detailed tables begins to diverge and thus produce different results. We should be careful when we analyze the summary tables since we may get conflicting results when the discovered patterns from the summary tables are compared with the discovered patterns from detailed tables. In general, the probabilities are not enough when discovering knowledge from detailed data. We need summary data as well.

INCOMPLETE KNOWLEDGE DISCOVERY FROM RELATIONAL DATABASES

The traditional database design method is based on the notions of functional dependencies and lossless decomposition of relations into third normal forms. However, this decomposition of relations is not useful with respect to knowledge discovery because it hides dependencies among attributes that might be of some interest. To provide maximum guarantee that potentially interesting statistical dependencies are preserved, knowledge discovery process should use the universal relation (Parsaye et al., 1991) as opposed to normalized relations in order to reveal all the interesting patterns.

Consider the relations Sales (Client Number, Zip Code, Product Purchased) and Region (Zip Code, City, Average House Price) (Adriaans & Zantinge, 1996) which are in third normal form. The relation Sales-Region (Client Number, Zip Code, City, Average House Price, Product Purchased) shows the universal relation which is the join of the two tables, Sales and Region. From the universal relation, Sales-Region, we may discover that there is a relationship between the Average Price of the House and the type of Products Purchased by people. Such relationship is not that obvious on the normalized relations.

One possible scheme for validating the completeness/incompleteness of the discovered knowledge is to analyze the discovered rules (known as statistical dependencies) with the available functional dependencies (known as domain knowledge). If new dependencies are generated that are not in the set of discovered rules, then we have an incomplete knowledge discovery. For example, processing the Sales relation, we may discover that if Zip Code = 11111, then Product Purchased = Wine with some confidence. We call this a statistical dependency that indicates that there is a correlation (with some confidence) between the Zip Code and the Product Purchased by people. Now, consider the Region relation, where the given dependencies are Zip Code —> City and City —> Average House Price which gives the derived new functional dependency Zip Code —> Average House Price due to the transitive dependency. By looking at the discovered statistical dependency and the new derived (or a given dependency in general), one may deduce that there is a relationship between the Average House Price and the Product Purchased (with some confidence). If our discovery process does not generate such a relationship, then we have an incomplete knowledge discovery that is the consequence of working on normalized relations as opposed to universal relations.

KNOWLEDGE DISCOVERY FROM SUMMARY DATA

In knowledge discovery, it is critical to use summary tables to discover patterns that could not be otherwise discovered from operational detailed databases. Summary tables have hidden patterns that can be discovered. For example, a summary table (Product Color, Profit) based on Table 1 tells us that Blue products are profitable. Likewise, a summary table (Product, Profit) based on Table 1 tells us that Hat products are not profitable. Such discovered patterns can complement the discoveries from the detailed data (as part of the validation of the discovered knowledge).

Accurate knowledge, however, cannot be discovered just by processing the summary tables. The problem is that the summarization of the same data set with two summarization methods may produce the same or different results. Therefore, it is extremely important that the users be able to access metadata that tells them exactly how each type of summarized data was derived so that they understand which dimensions have been summarized and to what level. Otherwise, we may discover inaccurate patterns from different summarized tables. For example, based on summary tables from Table 1, it is the Green Hat in small stores (Store Size <= 1000) that makes profit, and it is the Green Hat product in large stores (Store Size > 1000) that loses money. This fact can only be discovered by looking at all different summary tables and knowing how they are created (i.e., using the metadata).

VALIDATING POSSIBLE INCORRECT RULES

It is possible to use the patterns discovered from the summary tables to validate the discovered knowledge from the detailed tables. The following cases are identified for validating possible incorrect/correct discovered rules.

- **Case 1:** If the discovered pattern from the summary tables completely supports the discovered knowledge from the detailed tables, then we have more confidence in the accuracy of the discovered knowledge.
- **Case 2:** The patterns discovered from the detailed and summary tables support each other, but they have different confidence factors. Since the discovered patterns on the summary tables are based on the actual values, they represent more reliable information compared to the discovered patterns from the detailed tables which are based on the occurrences of the records. In such cases, we cannot say that the discovered pattern is incorrect, but rather it is not detailed enough to be considered as an interesting pattern. Perhaps, the hypothesis for discovering the pattern has to be expanded to include other attributes (i.e., Product or Store Size or both) in addition to the Product Color.
- **Case 3:** The patterns discovered from the detailed and summary tables contradict each

other. The explanation is the same as the one provided for Case 2.

- **Case 4:** There are cases where the discovered knowledge from summary tables is based on statistical significance. If the discovered knowledge from detailed and summary tables support each other with a different confidence factor, then additional information from other sources (perhaps from domain expert, if possible) is needed to verify the accuracy of the discovered knowledge.

KNOWLEDGE DISCOVERY FROM HISTORICAL DATA

Knowledge discovery from operational/detailed or summary data alone may not reveal trends and long-term patterns in data. Historical data should be an essential part of any discovery system in order to discover patterns that are correct over data gathered for a number of years as well as the current data. For example, we may discover from current data a pattern indicating an increase in students' enrollment in the universities in the Washington, DC area (perhaps due to good Economy). Such pattern may not be true when we look at the last 5 years of data.

Using Historical Data for Knowledge Discovery

There are several schemes that could be identified in using historical data in order to detect undiscovered patterns from detailed and summary data and to validate the consistency/accuracy/completeness of the discovered patterns from the detailed/summary data.

1. Validate discovered knowledge from detailed/summary data against historical data

We can apply the discovered rules from detailed and/or summary data to the historical data to see if they hold. If the rules are strong enough, they should hold on the historical data. A discovered rule is inconsistent with the database if examples exist in the database that satisfy the condition part of the rule, but not the conclusion part (Giarranttanto & Riley, 1989). A knowledge base (i.e., set of discovered rules from detailed and summary data) is inconsistent with the database if there is an inconsistent rule in the knowledge base. A knowledge base is incomplete with respect to the database if examples exist in the database that do not satisfy the condition part of any consistent rule.

If there are inconsistent rules, that means we have some historical data that contradict the rules discovered from detailed/summary data. It means we may have anomalies in some of the historical data. This is the case where any knowledge from external data, domain expert, and/or domain knowledge could be used to verify the inconsistencies. Similarly, if we have incomplete knowledge base, then there are some historical data that could represent new patterns or some anomalies. Again, additional information (i.e., domain expert) is necessary to verify that.

2. Compare the rules discovered from detailed/summary data with the ones from historical data

We perform the knowledge discovery on the historical data and compare the rules discovered from the historical data (call it H_RuleSet) with the ones discovered from detailed/summary data (call it DS_RuleSet). There are several possibilities:

a. If H_RuleSet $\cap$ DS_RuleSet $= \varnothing$ Then, none of the rules discovered from detailed/summary data hold on the historical data.

b. If H_RuleSet $\cap$ DS_RuleSet $= X$ Then
    - If DS_RuleSet - X $= \varnothing$ Then, all of the rules discovered from detailed/summary data hold on the historical data.
    - If X $\subset$ DS_RuleSet Then, there are some rules discovered from detailed/summary data that do not hold on the historical data (i.e, N_RuleSet - X). We can find the data in the historical data that do not support the rules discovered from the detailed/summary data by finding the data that support the rules in N-RuleSet and subtract it from the entire historical data. This data can then be analyzed for anomalies.

c. If H_RuleSet - DS_RuleSet $!= \varnothing$ (or DS_RuleSet $\subset$ X) Then, there are some rules discovered from historical data that are not in the set of rules discovered from the detailed/summary data. This means we discovered some new patterns.

CONCLUSION AND FUTURE TRENDS

Current research in knowledge management involves the tools and techniques to acquire the tacit knowledge from the domain experts. We presented an approach for the automatic acquisition of some tacit knowledge from the implied knowledge that may be presented in the organizations' data warehouses. There are several issues/concerns that need to be addressed before we could have an effective knowledge discovery process. One major issue is the size of the data warehouses. The larger a warehouse, the richer its patterns would be. However, after a point, if we analyze too large a portion of a warehouse, patterns from different data segments begin to dilute each other, and the number of useful patterns begins to decrease (Parsaye, 1996). We could select segment(s) (i.e., a particular medication for a disease) from data that fits a particular discovery objective. Alternatively, data sampling can be used to foster data analysis. However, we lose information because we throw away data not knowing what we keep and what we ignore. Summarization may be used to reduce data sizes; although, it can cause problem too, as we noted.

In automatic discovering of implied knowledge from data warehouses, there is definitely some tacit knowledge that can be discovered and verified by experts. However, we may find some implied knowledge that may not be verifiable, as even the experts do not know the truth of the discovered knowledge. The significance and interestingness of such knowledge may become apparent in the future after the discovered knowledge is actually used in the organization.

REFERENCES

Adams, K. C. (2004). Information architecture translate KM theory into practice. KM World Magazine Archives. Retrieved June 10, 2005, from *http://www.kmworld.com/publications/maxine/index.cfm*

Adriaans, P., & Zantinge, D. (1996). *Data mining.* Reading, MA: Addison-Wesley.

Agrawal, R., Imielinski, T., & Swami, A. (1993). Database mining: A performance perspective. *IEEE Transactions on Knowledge and Data Engineering, 5*(6), 914-925.

Barquin, C. R. (2000). Knowledge management in the public sector. Performance & Results, Management Concepts, Inc. Retrieved June 10, 2005, from *http://www.barquin.com/documents/km-public-sector.pdf*

Barquin, C. R., & Edelstein, H. A. (1997). *Building, using, and managing the data warehouse.* Upper Saddle River, NJ: Prentice Hall PTR.

Bischoff, J., & Alexander, T. (1997). *Data warehouse: Practical advise from the expert.* Upper Saddle River, NJ: Prentice Hall.

Brachman, R. J., Khabaza, T., Kloesgen, W., Piatetsky-Shapiro, G., & Simoudis, E. (1996). Mining business databases. *Communications of the ACM, 39,* 42-28.

Fayyad, U. (1996). Data mining and knowledge discovery: Making sense out of data. *IEEE Expert, 11,* 20-25.

Frappaolo, C., & Wilson, L. T. (2004). After the gold rush: Harvesting corporate knowledge resources. Retrieved June 10, 2005, from *http://www.intelligentkm.com/feature/feat1.jhtml?-requestid=387284*

Giarrantanto, J., & Riley, G. (1989). *Expert systems: Principles and programming.* Boston, MA: PWS-Kent.

Inmon, W. H. (1996). The data warehouse and data mining. *CACM, 39,* 49-50.

Matheus, C. J., Chan, P. K., & Piatetsky-Shapiro, G. (1993). Systems for knowledge discovery in databases. *IEEE Transactions on Knowledge and Data Engineering, 5*(6), 903-913.

Meredith, M. E., & Khader, A. (1996). Designing large warehouses. *Database programming design, 9*(6), 26-30.

Orr, K. (2004). Ken Orr Institute. Retrieved June 10, 2005, from *http://www.kenorrinst.com*

Parsaye, K. (1996). Data mines for data warehouses. *Supplement to Database Programming & Design, 9,* S6-S11.

Parsaye, K., Chignell, M., Khoshafian, S., & Wong, H. (1991). Intelligent data base and automatic discovery. In B. Soucek (Ed.), *Neural and intelligent systems integration.* New York: John Wiley & Sons.

KEY TERMS

Discovery Tools: Programs that enable users to employ different discovery schemes, including classification, characteristics, association, and sequence for extracting knowledge from databases.

Data Quality: Most large databases have redundant and inconsistent data, missing data fields and/or values as well as data fields that are not logically related and that are stored in the same data relations.

External Data: Traditionally, most of the data in a warehouse have come from internal operational systems such as order entry, inventory, or human resource data. However, external sources (i.e., demographic, economic, point-of-sale, market feeds, Internet) are becoming more and more prevalent and will soon be providing more content to the data warehouse than the internal sources.

MetaData: Metadata are used to describe the content of the data (e.g., description of the data tables; fields; constraints; data transformation rules such as profit = income-cost; domain knowledge such as male patient cannot get ovarian cancer) as well as to define the context of the data.

Operational Data: Contains the most recent data about the organization and are organized for fast retrieval as well as avoiding update anomalies.

Optimization Process: This process is used to focus the search (or guide the search) for interesting patterns as well as to minimize the search efforts on data.

Pattern Interestingness: A pattern is interesting not only to the degree to which it is accurate but to the degree which it is also useful with respect to the end user's knowledge and objectives.

D

Distinguishing Work Groups, Virtual Teams, and Electronic Networks of Practice

Molly McLure Wasko
Florida State University, USA

Robin Teigland
Stockholm School of Economics, Sweden

INTRODUCTION

Communities of practice are promoted within organizations as sources of competitive advantage and facilitators of organizational learning. A community of practice is an emergent social collective where individuals working on similar problems self-organize to help each other and to share perspectives about their work practice, resulting in learning and innovation within the community (Brown & Duguid, 1991; Wenger, 1998). Recent advances in information and communication technologies have enabled the creation of computer-supported social networks similar to communities of practice, where individuals are able to discuss and debate issues electronically. Given the success of communities of practice for facilitating knowledge exchange, both electronically and in face-to-face settings, management has recently focused on how to formally duplicate these networks and gather their benefits in work groups and virtual teams. However, with the evolution of new technology-enabled organizational forms, theoretical development is needed to distinguish between these different types of organizational forms since there are significant differences in the dynamics of formal vs. informal membership groups and between electronic and face-to-face interactions (Hinds & Kiesler, 2002).

BACKGROUND

Recently, the concept of networks of practice (Brown & Duguid, 2000) has emerged as a means to describe informal, emergent social networks that facilitate learning and knowledge sharing between individuals conducting practice-related tasks. These authors propose that communities of practice are a localized and specialized subset of networks of practice, typically consisting of strong ties linking individuals engaged in a shared practice, typically face-to-face. They describe networks of practice as consisting of weak ties, where individuals may never get to know one another or meet face-to-face. In networks of practice, individuals generally coordinate through third-party organizations, such as professional associations, or by indirect means, such as newsletters, Web sites, or bulletin boards (Brown & Duguid, 2000).

In contrast to the use of formal controls to support knowledge exchange, such as contractual obligation, organizational hierarchies, monetary incentives, or mandated rules, networks of practice promote knowledge flows along lines of practice through informal social networks. Therefore, one way to distinguish between networks of practice and work groups created through formal organizational mandate is by *the nature of the control mechanisms*.

A second distinguishing property is the *primary media channel* used for communication between members, for example, face-to-face interactions, remote computer-mediated channels such as newsletters or discussion boards, or a combination of these techniques. The communication media is important for understanding networks of practice, for this is the channel through which the resource of knowledge is exchanged. In electronic networks of practice, the primary communication channel of asynchronous computer-mediated communication has a profound influence on how knowledge is actually shared.

Additionally, networks of practice and formal work groups vary in terms of their *size*, ranging from a few select individuals to very large, open electronic networks consisting of thousands of participants. These groups also vary in terms of *who can participate*. Work groups and virtual teams typically consist of members who are formally designated and assigned. In contrast, networks of practice consist of

Table 1. Macrostructural properties distinguishing formal work groups and networks of practice

Property	Work Groups	Virtual Teams	Communities of Practice	Electronic Networks of Practice
Control	Formal control, not voluntary	Formal control, not voluntary	No formal control, voluntary	No formal control, voluntary
Communication channel	Face-to-face	Text-based computer-mediated, e.g., e-mail, listservs	Face-to-face	Text-based computer-mediated, e.g., listservs, discussion boards
Network size	Small	Small	Small	Large
Access	Restricted, assigned by a formal control	Restricted, assigned by a formal control	Restricted, locally bounded, limited to collocation	Open, no limitations other than access to technology
Participation	Jointly determined, specific task outcomes	Jointly determined, specific task outcomes	Jointly determined	Individually determined

volunteers without formal restrictions placed on membership.

Finally, networks of practice and formal work groups vary in terms of *expectations about participation*. In formal work groups and virtual teams, participation is jointly determined, and members are expected to achieve a specific work task or goal. Participation in communities of practice is jointly determined, such that individuals generally approach specific others for help. In electronic networks of practice, participation is individually determined; knowledge seekers have no control over who responds to their questions or the quality of the responses. In turn, knowledge contributors have no assurances that seekers will understand the answer provided or be willing to reciprocate the favor. The properties and different organizational structures are summarized in Table 1.

FUTURE TRENDS AND CONCLUSION

Although there has been a significant increase in networked communication and a growing interest in virtual organizing, to date, researchers have yet to establish consistent terminology and have paid little attention to how specific characteristics of electronic communication or formal organizational structure influence social dynamics such as knowledge contribution within the various organizational forms. To address this gap, we have developed a table summarizing the different properties and how they are relevant for distinguishing between electronic and face-to-face as well as formal and informal structures. Identification of these key properties should help managers better understand how to create strategies to ensure the success of these different collectives and recognize that strategies that work in one area may not transfer across all collectives.

REFERENCES

Brown, J. S., & Duguid, P. (1991). Organizational learning and communities-of-practice: Toward a unified view of working, learning, and innovation. *Organization Science, 2*(1), 40-57.

Brown, J. S., & Duguid, P. (2000). *The social life of information.* Boston: Harvard Business School Press.

Hinds, P., & Kiesler, S. (2002). *Distributed work.* Cambridge, MA: MIT Press.

Wenger, E. (1998). *Communities of practice.* Cambridge, UK: Cambridge University Press.

KEY TERMS

Community of Practice: A relatively tightly knit, emergent social collective, in which individuals working on similar problems, self-organize to help each other and share perspectives about their work practice generally in face-to-face settings.

Electronic Network of Practice: A relatively large, emergent social collective, in which individuals working on similar problems, self-organize to help each other and share perspectives about their work practice through text-based computer-mediated means, for example, listservs, discussion boards, and so forth.

Virtual Team: A relatively small, formally designated, and formally controlled group that generally works together through text-based computer-mediated means, for example, e-mail, listservs, and so forth.

Work Group: A relatively small, formally designated, and formally controlled group that generally works together in face-to-face situations.

Document Management, Organizational Memory, and Mobile Environment

Sari Mäkinen
University of Tampere, Finland

INTRODUCTION

Wireless networks and new tools utilizing mobile information and communication technologies (ICTs) challenge the theories and practices of document management, in general, and records management, in particular. The impact of these new tools on document management as a part of organizational memory is as yet unexplored because the wireless and mobile working environment is a new concept. Recent studies of mobile environment have focused on mobile work itself or technologies used, and the aspect of document management, especially records management, has been ignored.

BACKGROUND

Records form one important part of the memory of an organization. From the organizational perspective, one method of managing intellectual resources is to augment the organization's memory. A standard connotation of organizational memory is a written record, although this is only one form of memory. Organizational memory has explicit and implicit forms and can be retained in several places like databases and filing systems, but also in organizational culture, processes, and structures (Ackerman, 1996; Walsh & Ungson, 1991). Megill (1997) specifies organizational memory to include all the active and historical information in an organization that is worth sharing, managing, and preserving for use. It is an important asset encompassing all types of documented and undocumented information that an organization requires to function effectively.

Digital documents and records can be found in every area of administration and business activities. Official records are produced in carrying out business or administrative processes, decision-making processes or procedures. These records are vital

and must be preserved for later use, as documentation and evidence and for cultural and historical reasons. Records are not preserved only for the use of the organization; they must be made accessible to individuals and customers (Young & Kampffmeyer, 2002). With a growing number of people using mobile tools, new kinds of problems are emerging. These problems arise because documents are created, processed, stored, managed, and shared through various mobile ICT tools and technologies. In a mobile working environment, it is essential that every piece of an organization's explicit memory is accessible, searchable, and preservable. This is vital, especially in the case of official and business records.

The literature on document management focuses mainly on the technologies used or the functionality of the document management systems created by practicing consultants. Academic research is rare (Bellotti & Bly, 1996; Eldridge et al., 2000; Luff, Heath & Greatbatch, 1992). Mobile working environment has been examined from the social-scientific and social interaction perspectives (Brown, Green & Harper, 2001; Katz & Aakhus, 2002). The mobile working environment in relation to the aspects of document management is an uninvestigated area and a new research topic.

ORGANIZATIONAL MEMORY

The concept of organizational memory is not new. Its roots go back to the organizational science and information-processing theories of the 1950s (Walsh & Ungson, 1991). Research on organizational memory increased especially in the 1990s in the field of information systems research. Understanding of the concept is limited, and the term is vague but commonly used. Mostly organizational memory is seen from the perspective of the organizational member.

It refers to the stored information on the organization's history that can be brought to bear on present decisions (Walsh & Ungson, 1991).

The perspectives of information systems scientists on organizational memory are pragmatic, more often concentrating on the development of databases and information systems supporting organizational memory, since examining the contents of the concept is the focus of organizational scientists. Walsh and Ungson's (1991) classic study, in turn, is completely conceptual. Bannon and Kuutti (1996) claim that the concept of organizational memory does not belong exclusively to any particular research area or discipline and that a variety of definitions is available in such different fields as administrative science, organizational theory, change management, psychology, sociology, design studies, concurrent engineering, and software engineering. The viewpoint taken in archival science (see, e.g., Hedstrom, 2002; Yates, 1990, 1993) is on the historical mission of organizational memory. The purpose of archives is to retain and store the historical memory of an organization. Organizational memory research has been criticized for perceiving organizational memory as only a problem of information technology. The problem of how databases serve users is not the most essential (Koistinen & Aaltio-Marjosola, 2001).

On the basis of a through concept analysis, the definition of organizational memory is the organized knowledge of an organization, a process which is individual and distributed and past preserving, which has an effect on organizational learning, competitiveness and decision-making, and which can be supported by information technology. (Mäkinen & Huotari, 2004).

The preservation and use of organizational memory refer strictly to working life and information used in work-related settings. The empirical case studies on organizational memory pertain particularly to carrying out a task (Mäkinen & Huotari, 2004).

Schwartz, Divitini, and Brasethvik (2000) note that organizational memory has become a close partner of knowledge management (KM), denoting the actual content that a knowledge management system purports to manage. They perceive knowledge as the key asset of the knowledge organization. They also argue that organizational memory amplifies this asset by capturing, organizing, disseminating, and reusing the knowledge. Generally, the purpose of KM is seen to make these resources available for use. This approach refers to knowledge as an object (Sveiby, 1996), and thus, brings KM close to the traditional role of information management.

Wilson (2002) argues that the information systems orientation dominates the approaches and implicit conceptions presented in the research papers, consulting practices and university curricula of KM. According to him, the theoretical foundation of this orientation is similar to that of information management research; that is, the term *knowledge* is in fact used to refer to information. Wilson argues that we cannot manage individual knowledge because it resides in human minds. Research on organizational memory information systems also supports this view by serving the needs of information retrieval and information seeking in the case of an explicit preserved form of organizational memory (Mäkinen & Huotari, 2004).

DOCUMENT MANAGEMENT IN MOBILE WORKING ENVIRONMENT

The issues of records management are not taken into account utilizing mobile tools for document management. The current need is to combine the perspectives of both document management and records management. For example, it has been suggested that about 12% of organizational knowledge is in its structured knowledge base and the majority (46%) lies scattered about organizations in the form of paper and electronic documents (Kikawada & Holtshouse, 2001). We can assume that the mobile working environment does not improve this situation.

Mobile devices can be defined in many ways. A mobile device can be described as an application of mobile technology—a technical device utilizing mobile technology and is designed to be mobile. Mobile devices, for example, include laptop computers, personal digital assistants (PDAs), mobile phones, and other handheld devices for data transfer and communication (Allen & Shoard, 2004; Weilenmann, 2003). Mobile technology is also about personal communication technologies (PCTs), which is a broader category and includes video cassette recorders, TVs, interactive voice response units (VRUs), beepers, and e-mail (Katz & Aakhus, 2002). The essential character of a mobile device is that it is mobile; it can be carried wherever you have to be and uses information and communication technology. The use of a mobile device is independent of time and space.

Even today, mobile professionals need to take paper documents with them when traveling. Paper is immediately viewable and is frequently used for ad hoc reading activities. This is still the case in spite of the amazing boom in mobile devices. The potential of combining, for example, mobile phone use with other kinds of information-related activities is being investigated in IT and telecommunications companies (O'Hara et al., 2002).

Mobile professionals have particular needs for technologies such as flexibility to accommodate their information needs in unpredictable circumstances. Mobile phone and paper documents respect this need and allow creative use while traveling (O'Hara et al., 2002).

For a mobile worker, the most important features of mobile document management are easy access, timely access, user interface, ubiquity, and compliance with security policies (Lamming et al., 2000). These features are also practical differences between document management using conventional ICT and mobile ICT. Current solutions in document management do not necessarily meet these requirements. The problems of access are probably the most familiar to mobile workers: how to unpack and plug in a laptop in an unfamiliar environment, how to access remote documents, how to transfer a file, how to print a file, and how to secure a confidential file.

Organizational memory should be understood in a novel manner when its content, that is, documents, is managed in a wireless and mobile operating environment. The utilization of documents produced in mobile devices in knowledge processes and the problems caused by mobile environment to the lifecycle of these documents require attention regarding their creation, transfer, storage, dissemination, sharing, use, and disposal (Mäkinen, 2004).

The challenges of mobile document management and organizational memory augmentation become even more evident among communities of practice. This concept was introduced by Lave and Wenger (1991). Communities of practice are about relations among people, activity, and world in relation to other tangential and overlapping communities of practice. A newcomer learns from old-timers, and newcomers see communities of practice as an intrinsic condition for the existence of knowledge. It is a flexible group of professionals having common interests and interacting through independent tasks and embodying common knowledge (Davenport & Hall, 2002; Kimble, Hildreth & Wright, 2001). In mobile working environment communities of practice share knowledge through technological tools, but it has been argued that some types of knowledge are unsuitable for electronic storage and retrieval (Davenport & Hall, 2002).

FUTURE TRENDS

In recent years, there has been an explosion in mobile computing and telecommunications technologies. A lot of work is done outside the office in different and unpredictable locations (Allen & Shoard, 2004; Weilenmann, 2001). Mobile working environment poses challenges on organizational document management and augmentation of organizational memory. How do mobile produced documents become a part of organizational memory, and what is the relation of these documents to the intellectual capital of an organization?

The future research challenge is to increase understanding of the current state of document management and records management in mobile environments in relation to the development of organizational knowledge and intellectual capital. The focus of future research could be on the role and utilization of mobile documents produced in the joint knowledge processes and the problems caused by wireless and mobile environments, the lifecycle of these documents.

Another important research topic is the idea of access: what problems do mobile professionals have in accessing information sources of their organizations? Problems which a user encounters when trying to connect organizational information systems with mobile devices need to be studied. Using mobile devices and digital records, we also need to be convinced of the integrity of data, that it has not been modified or manipulated. If a document has been created and disseminated utilizing, for example, a mobile phone, what happens to the data when it is transferred to another information system, like document management system?

Social factors have an impact on document management practices. Wireless and mobile tools are technical innovations, but there may also be social innovations in use in the organizations when these tools are used. Organizational changes (flexible working hours), new services (use of Web pages in marketing), and new social arrangements (telework at home) are examples of social innovations. This relates to the concepts of intellectual capital and social capital.

The idea of studying communities of practice and mobile working environments provides new perspectives on mobile computing and joint value creation. It has been stated that really important and useful information for improvement is too complex to put online. Workers might be afraid of job security and sabotage knowledge management systems (Davenport & Hall, 2002). Online communities of practice have the characteristics of material communities of practice, but they may be ephemeral, and the individuals involved may never have met.

CONCLUSION

The challenges of mobile devices and mobile working environment to document management and especially records management are varied and still largely unexplored. It is clear that the explosion of mobile computing will not improve or ease the augmentation of organizational memory, which is strictly connected to individuals.

The analysis of the concept of organizational memory suggests that its characteristics are contradictory, thereby reflecting the complex nature of the phenomenon. The explicit form of organizational memory is emphasized, but simultaneously, the individual and abstract nature of the concept are also underlined. Organizational memory, in recorded form, is concrete and palpable like paper records in an archive. However, organizational memory was also manifest implicitly and defined as invisible, mute, fuzzy, and easy to lose.

Understanding of the issues related to the management of an organizational memory is essential for enhancing the generative, productive, and represen-

tative knowledge processes in the joint value creation of different stakeholders. New knowledge is created in generative processes and with the new knowledge organization is able to provide new products and services. The new, generated knowledge is used in productive processes to provide the basis for products and services, and knowledge is transmitted to the customer as final products and services in representative processes (Huotari, 2000; Huotari & Chatman, 2001; Normann & Ramiréz, 1994).

The theoretical foundation of the organizational memory is more closely related to the multidisciplinary research area of KM and enhancement of knowledge construction based on organizational learning as a source of competitive capability than to information management. This indicates a shift from an individual organizational member's way of applying his/her own knowledge and use of information toward distributed knowledge, communication, and information and knowledge sharing, also through the use of information systems. This characteristic of the concept refers to the social nature of knowledge and information, implying that knowledge is socially constructed; that is, knowledge is a process, not an entity. The process perspective is rarely applied to studies on organizational memory, mostly in relation to an information system and its use (Ackerman & Halverson, 1998). The strategic perspective has gained more emphasis in economics (e.g., Hatami, Galliers & Huang, 2002).

REFERENCES

Ackerman, M. (1996). Organizational memory. Retrieved June 11, 2005, from *http://www.eecs.umich.edu/~ackerm/om.html*

Ackerman, M., & Halverson, C. (1998, November). Considering an organizational memory. *Proceedings of the Computer-Supported Cooperative Work (CSCW'98)*, Seattle, Washington. Retrieved June 11, 2005, from *http://www.eecs.umich.edu/~ackerm/pub/98b24/cscw98.om.pdf*

Allen, D. K., & Shoard, M. (2004). Spreading the load: Mobile information and communication technologies and their effect on information overload. *Proceedings of the ISIC Conference*, Dublin, Ireland.

Bannon, L. J., & Kuutti, K. (1996, January 3-6). Shifting perspectives on organizational memory: From storage to active remembering. *Proceedings of the 29th Hawaii International Conference on System Sciences (HICSS-29)* (pp. 156-167), Maui, Hawaii. Los Alamitos: IEEE Computer Press.

Bellotti, V., & Bly, S. (1996). Walking away from the desktop computer: Distributed collaboration and mobility in a product design team. *Computer Supported Cooperative Work '96,* Cambridge, MA (pp. 209-218).

Brown, B., Green, N., & Harper, R. (Eds.). (2001). *Wireless world: Social and interactional aspects of the mobile age.* London: Springer-Verlag.

Davenport, E., & Hall, H. (2002). Organizational knowledge and communities of practice. *Annual Review of Information Science and Techn ology, 36,* 171-227.

Eldridge, N., et al, (2000). Studies of mobile document work and their contributions to the satchel project. *Personal Technology, 4,* 102-112.

Hatami, A., Galliers, R. D., & Huang, J. (2002). Exploring the impacts of knowledge (re)use and organizational memory on the effectiveness of strategic decisions: A longitudinal case study. *Proceedings of the 36th HICSS.*

Hedstrom, M. (2002). Archives, memory and interfaces with the past. *Archival Science, 2,* 21-43.

Hofman, H. (1996, May 30-31). Lost in cyberspace – Where is the record? *Proceedings of the 2nd Stockholm Conference on Archival Science and the Concept of Record.*

Huotari, M.-L. (2000). Information behaviour in value constellation—An example from the context of higher education. *Swedish Library Research, 3/4,* 3-20.

Huotari, M.-L., & Chatman, E. (2001). Using everyday life information seeking to explain organizational behaviour. *Library and Information Science Research, 23*(4), 351-366.

Katz, J. E., & Aakhus, M. A. (2002). Conclusion: Making meaning of mobiles—a theory of Apparatgeist. In Katz & Aakhus (Eds.), *Perpetual contact: Mobile communication, private talk, public performance* (pp. 301-320). New York: Cambridge University Press.

Kikawada, K., & Holtshouse, D. (2001). The knowledge perspective in the Xerox Group. In I. Nonaka & D. J. Teece (Eds.), *Managing industrial knowledge: Creation, transfer and utilization* (pp. 283-314). London: Sage.

Kimble, C., Hildreth, P. & Wright, P. (2001). Communities of practice: Going virtual. In Y. Malhotra (Ed.), *Knowledge management and business model innovation* (pp. 216-230). Hershey, PA: Idea Group Publishing.

Koistinen, P., & Aaltio-Marjosola, I. (2001, July 5-7). Organizational memory in partnership. *Proceedings of the EGOS 2001 Conference.* Lyon, France.

Lamming, M., Eldridge, M., Flynn, M., Jones, C., & Pendlebury, D. (2000). Satchel: Providing access to any document, any time, anywhere. *ACM Transactions on Computer-Human Interaction, 7*(3), 322-352.

Lave, J., & Wenger, E. (1991). *Situated learning. Legitimate peripheral participation.* Cambridge: Cambridge University Press.

Luff, P., Heath, C., & Greatbatch, D. (1992). Tasks-in-interaction: Paper and screen based documentation in collaborative activity. CSCW'92. Retrieved June 11, 2005, from *http://portal.acm.org*

Luff, P., & Heath, C. (1998). *Mobility in collaboration. Proceedings of the CSCW'98.*

Mäkinen, S. (2004, May 23-26). The use of mobile ICT in organizational document management in the context of organizational memory. *Proceedings of the Information Resources Management Association International Conference IRMA2004,* New Orleans, Louisiana.

Mäkinen, S., & Huotari, M.-L. (2004, May 23-26). Organizational memory: Knowledge as a process or information as an entity. *Proceeedings of the Information Resources Management Association International Conference IRMA2004,* New Orleans, Louisiana.

Megill, K. (1997). *The corporate memory: Information management in the electronic age.* London: Bowker & Saur.

Megill, K. A., & Schantz, H. (1999). *Document management. New technologies for the information services manager.* London: Bowker & Saur.

Normann, R., & Ramiréz, R. (1994). *Designing an interactive strategy: From value chain to value constellation.* Chichester, UK: John Wiley & Sons.

O'Hara, K., Perry, M., Sellen, A., & Brown, B. (2001). *Exploring the relationship between mobile phone and document activity during business travel. Wireless World. Social and Interactional Aspects of the Mobile Age.* London: Springer-Verlag.

Schwartz, D.G., Divitini, M., & Brasethvik, T. (2000). On knowledge management in the Internet Age. In D. G. Schwartz, M. Divitini, & T. Brasethvik (Eds.), *Internet-based organizational memory and knowledge management* (pp. 1-23). Hershey, PA: Idea Group.

Sprague, R. H., Jr. (1995, March). Electronic document management: Challenges and opportunities for information systems managers. *MIS Quarterly.*

Sveiby, K.-E. (1996). What is knowledge management? *Quarterly, 19*(1), 29-49. Retrieved June 11, 2005, from *http://www.sveiby.com/articles/KnowledgeManagement .html*

Thomassen, T. (2001). A first introduction to archival science. *Archival Science, 1,* 373-385.

Walsh, J. P., & Ungson, G. R. (1991). Organizational memory. *Academy of Management Review, 16*(1), 57-91.

Weilenmann, A. (2001). Mobile methodologies: Experiences from studies of mobile technologies-in-use. *Proceedings of the 24ᵗʰ Information Systems Research Seminar in Scandinavia (IRIS 24).*

Weilenmann, A. (2003). Doing mobility: Towards a new perspective on mobility. *Proceedings of the 26ᵗʰ Information Systems Research Seminar in Scandinavia (IRIS 26).*

Wilson, T. D. (2002). The nonsense of "knowledge management". *Information Research, 8*(1), paper no. 144. Retrieved June 11, 2005, from *http://informationr.net/ir/8-1/paper144.html*

Yates, J. (1990). For the record: The embodiment of organizational memory, 1850-1920. *Business and Economic History, 2ⁿᵈ Series, 19,* 172-182.

Yates, J. (1993). *Control through communication: The rise of system in american management.* Baltimore: Johns Hopkins University Press.

Young, R., & Kampffmeyer, U. (2002). *Availability & preservation: Longterm availability & preservation of digital information* (AIIM Industry White Paper on Records, Document and Enterprise Content Management for the Public Sector). AIIM International Europe: Stephens & George Print Group.

KEY TERMS

Communities of Practice: A flexible group of professionals having common interests, interacting by independent tasks and embodying common knowledge (Davenport & Hall, 2002). Communities of practice are defined as a set of relations among people, activities, and the world (Lave & Wenger, 1991).

Document: Defined as a unit of recorded information structured for human consumption. Documents contain information in some structured way, and they are human creations. A document is created for a certain purpose (Megill & Schantz, 1999; Sprague, 1995).

Document Management: Covers the creation, modification, storage, and retrieval of documents required to meet users' needs and objectives (Megill & Schantz, 1999). Electronic Document Management (EDM) is the application of technology to save paper, speed up communications, and increase the productivity of business processes (Sprague, 1995).

Local Mobility: Refers to mobility within a certain space, as between rooms or floors.

Micro-Mobility: Refers to the way an artifact is mobilized and manipulated around a relatively circumscribed domain.

Mobile Device: Refers to an application of mobile technology, that is, to a technology which is designed to be mobile. Mobile devices, for example, include laptop computers, personal digital assistants (PDAs), mobile phones, and other handheld devices for data transfer and communication (Allen & Shoard, 2004; Weilenmann, 2003).

Mobility: Used here to signify the physical movement of nodes in a network or remote interaction between individuals who are far apart from each other using mobile technology. Mobility can be divided into micro mobility, local mobility, and remote mobility. (Luff & Heath, 1998; Weilenmann, 2001).

Organizational Memory: The organized knowledge of an organization, a process which is individual and distributed and past preserving, which has an effect on organizational learning, competitiveness, and decision making, and which can be supported by information technology (Mäkinen & Huotari, 2004).

Record: Regarded as process-bound information: a record is generated by work processes, structured and recorded by these work processes in order to be retrieved from the context of that work process (Thomassen, 2001). A record has four elements: recorded (physically), it contains information (content), it is an outcome of the process in which it was created (context), and it has a certain form or manifestation (structure) (Hofman, 1996). Contextual information is necessary for defining a document as a record. Unlike a document, a record needs to have contextual information. Records are also documentation of transactions, and they are preserved for evidential, historical, and cultural purposes.

Remote Mobility: Refers to remote users interacting with each other using technology.

D

Economic Issues of Online Professional Communities

Ettore Bolisani
University of Padova, Italy

Enrico Scarso
University of Padova, Italy

Matteo Di Biagi
University of Padova, Italy

INTRODUCTION AND BACKGROUND

The term *online professional community* (OPC) is employed here to identify groups of professionals sharing information and knowledge for business purposes by means of Internet-based technologies (see also Plant, 2004). It is not our aim to thoroughly discuss the concept of community, but since this term may be used with various shades of meaning, it is thus convenient to indicate more precisely where our definition can be placed in the broader picture (Figure 1). This will be made by discussing two important dimensions that are relevant here.

First, a distinction can be made between non-profit communities (e.g., groups of people sharing cultural interests, hobbies, and so on) and communities having a business purpose. A community of practice, that is, a group of people sharing insights,

experience, and competence on a particular domain in order to deepen their expertise on an ongoing basis (Brown & Duguid, 2001; Maier, 2002; Wenger, McDermott & Snyder, 2002), can generally be placed in the latter category.

The second relevant feature pertains to the infrastructure supporting relations among members. We can distinguish between traditional communities, where interactions are based on conventional means (i.e., face-to-face contacts, meetings, publications, etc.) and communities whose nature is closely intertwined with the use of network technologies (namely, the Internet). Such communities are generally called online or virtual communities. Even if there is a lack of consensus about the definition of online community (de Souza & Preece, 2004; Kardaras, Karakostas & Papathanassiou, 2003), the concept generally indicates a collective group of entities (individual or organizations) who interact though an electronic medium for a common purpose or interest (Plant, 2004; Preece, 2000). In other words, online communities develop around a shared idea or task, rather than a physical place. Thus, members can interact and share knowledge across organizational boundaries, geographical barriers, and time zones (Johnson, 2001).

By combining the two illustrated factors, we will focus on OPCs as a particular type of virtual community whose aim is to enhance the business potentials of its participants. Several examples of such communities can now be found in professional fields (e.g., legal practice, medical practice, fiscal consulting, engineering and product design, informatics), where they are used to combine and integrate com-

Figure 1. Online professional communities within the broader concept of community

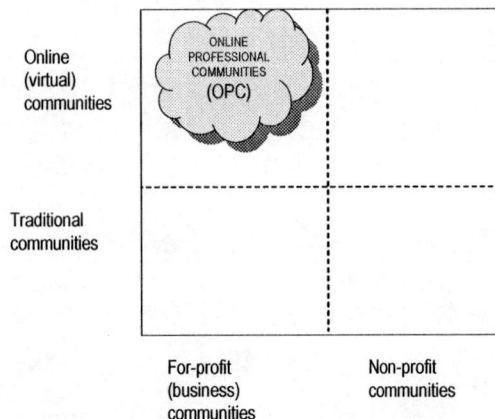

petencies of professionals specializing in different fields. Those OPCs also represent an emerging field of experimentation of new profitable business models (Plant, 2004).

As a result, the issue of creating and managing OPCs necessarily has multiple dimensions (de Souza & Preece, 2004; Dyer & Nobeoka, 2000; von Krogh, 2002): a social dimension (i.e., the nature and structure of relations among participants), a technological dimension (namely, the technical infrastructure used to communicate and exchange knowledge), and an economic dimension (i.e., the value and costs of participation). While the social and the sociotechnical perspective still dominate the studies of virtual communities (Koh & Kim, 2004), the economic side is less considered. The aim of this article is to explicitly analyze this element and its relationships with the others (Figure 2). In doing this, we argue that a knowledge management view can be particularly useful to understand (a) the economic mechanisms underpinning the social processes of knowledge transfers in an OPC and (b) the relations of such

economic mechanisms with the technological infrastructure used to perform those processes.

ONLINE PROFESSIONAL COMMUNITIES AND KM

Many scholars (Ardichvili, Page & Wentling, 2003; Pan & Leidner, 2003; Sharratt & Usoro, 2003) underline that the communities of practice represent a good example of social context where KM activities and strategies can be effectively implemented. In other terms, since a community is considered an effective tool for knowledge creation and sharing, its success can be seen and understood through KM lens. Again, when OPCs are considered, technologies (or better, Knowledge Management Systems—KMS) have to be included in the analysis.

With this purpose, it is appropriate to draw attention to some key elements that define KM activities. An essential starting point is the working definition of knowledge, commonly adopted in the KM literature: knowledge can be regarded as actionable information, that is, used to make business decisions or take actions. This notion emphasizes two key aspects that are relevant for our purpose: (a) knowledge differs from pure information or simple data, meaning that the "bits processed by computers" or transferred through the Internet do not (yet) represent knowledge; and (b) knowledge is built on data and information, since it is the active involvement of the individual that transforms them into knowledge, hence, decisions.

In KM studies, knowledge has also been frequently distinguished based on (a) different forms

Figure 2. Dimensions of analysis

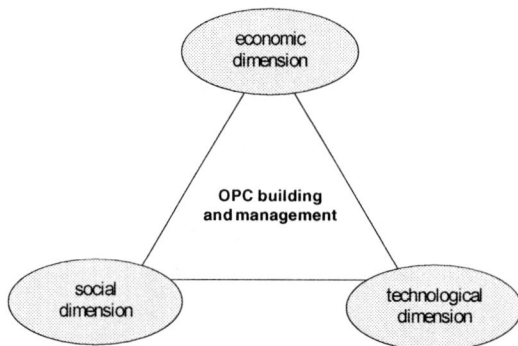

Table 1. Cultivating a community: A list of key issues from a KM view

- Understanding the main hurdles hindering the exchange of knowledge among members
- Making knowledge easy to use by organizing it according to the natural way in which members think about their practice
- Evaluating how the community brings benefits to its members by sharing knowledge
- Motivating people to share knowledge
- Developing trust by ensuring that members do not misuse the shared knowledge (e.g., by taking advantage of confidential information), that the community is a source of reliable information, etc.
- Establishing coordinating roles such as sponsors, champions, facilitators, practice leaders, and infomediaries
- Making the access easy through the development of the technological infrastructure for KM

149

(explicit—relatively easy to formalize, transfer, or store vs. implicit—pertaining to ideas, feelings, experience, and thus much more complex to share; Polanyi, 1967); (b) contents (e.g., know-about, know-who, know-how, etc.; Alavi & Leidner, 2001); and (c) owners, namely, individuals or organizations (Bhatt, 2001). This might allow singling out the different knowledge contents that are (or can be) exchanged in a particular OPC, as well as the mechanisms and the technologies employed for this.

In addition, KM scholars define the approaches to the systematic management of knowledge and connote the fundamental KM processes and activities (Alavi & Leidner, 2001; Maier, 2002): (a) creation or acquisition; (b) storage and retrieval; (c) transfer, distribution, and sharing; and (d) application or exploitation. Within a community (and thus within an OPC), such processes imply different capabilities, roles, mechanisms, and tools. An analysis of the specific role of each OPC member in the transfer processes is therefore essential.

In the case of OPCs, it is also important to mention that further implications derive from the inter-organizational nature that characterizes OPCs. As a matter of fact, OPCs can be seen as communities of professional firms, each one running its peculiar business. In other words, knowledge exchanges well extend beyond the walls of the single firm, and the principles and contents elaborated in traditional intra-organizational KM (Nonaka & Takeuchi, 1995) have to be reframed accordingly. An emerging theme in the literature concerns the building and management of knowledge networks, that is, inter-organizational arrangements to share knowledge (Jarvenpaa & Tanriverdi, 2002; Peña, 2002). Such extended approach raises additional problems, since each member of the network may have specific goals, languages, values, mental schemes, competencies, tools, and so forth. In addition, managing a knowledge network implies the subdivision of cognitive tasks and KM activities among the participants and can require new skills and innovative roles.

In summary, the concepts and practice of network KM represent the grounds for building and managing online communities in that they are, as previously said, networked entities aimed to improving knowledge sharing and generation. It must be remembered, in fact, that every community, even if spontaneous and informal, has to be cultivated according to what

is described in Table 1 (Ardichvili, Page & Wentling, 2003; McDermott, 2000; Smith & McKeen, 2003); a virtual one must be designed, too (Johnson, 2001), especially because its technological infrastructure has to be planned, implemented, and managed.

THE ECONOMIC SIDE OF OPC

The second step is to link the KM analysis of OPC with the economic factors that affect the effectiveness of the community in enabling useful knowledge transfers among its (business) members. This is the theme of the present section, where the issues raised by OPC management are handled according to a perspective that integrates KM and economic concepts.

With regard to this, it is worth noting that each participant can be seen as an independent economic player. Hence, to understand how such a community can profitably work, it is necessary to examine the economic and business mechanisms underpinning the decisions of the single players to "share what they know" to make profits. Based on this, one can easily develop the analogy with transactions of physical products (Figure 3).

Figure 3. Knowledge transactions within an OPC

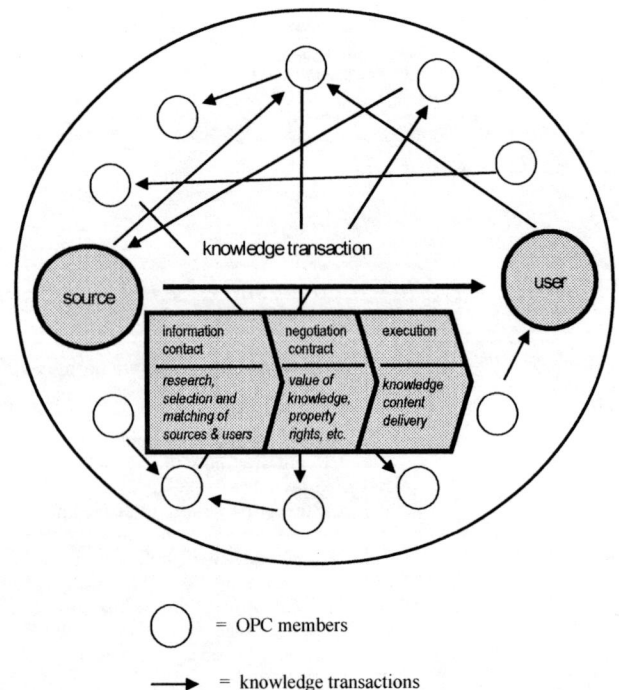

○ = OPC members

→ = knowledge transactions

E

Roles of Participants

While in a transaction of a physical good, there is a seller and a buyer; in the case considered here, there is a source of knowledge that transfers contents to a user. In our case, both sources and users are internal active members of the community. Differently from physical goods, the role of source and receiver can be interchangeable, even in the same transaction. Community members can also assume other roles (see also point G) and responsibilities, thus contributing in different ways to the success of an OPC. A question here is what form of economic incentives can motivate the participants to play an active role and avoid opportunistic behaviors.

Contents Exchanged

As an object of transaction, information/knowledge exhibits special features:

- While the characteristics of a physical product can be (partly) tested before buying it, information and knowledge are *experiential goods* (Choi, Stahl & Whinston, 1997), in that their value can be precisely appraised once the transaction has been completed;
- Knowledge expressed in *explicit form* can be easily reproduced and delivered at almost zero cost (Shapiro & Varian, 1998). On the contrary, tacit knowledge is highly localized, and its transfer—requiring the direct involvement of sources and users and the active processes of teaching and learning—cannot be repeated in the same way by other partners;
- Property rights are crucial in case of explicit knowledge. Consequently, policies for protecting property rights can be adopted to avoid illegal appropriation or replication. Conversely, the legislative protection of tacit knowledge contents may be useless and, in any case, much harder to apply;
- The exchange of knowledge contents can activate particular mechanisms for fixing prices and establishing versions (Shapiro & Varian, 1998). For instance, the same content can be transferred in different formats to distinct receivers.

Activities Performed

Each transaction can be subdivided into three stages (Figure 4; Bakos, 1998), that is, information/contact (parties search for each other, match information about supply and demand), negotiation/contract (parties define the terms of a transaction), and execution (delivery, etc.). In case of knowledge transactions, special issues arise, as illustrated in the KM literature. The information/contact stage implies the *selection* of sources and/or target users. However, knowledge needs are generally not clear or explicit in advance, and it is therefore difficult to exactly match potential sources and users. A complex interaction may be required, especially when multidimensional contents are considered. Similarly, the *negotiation/contract* stage implies agreements on aspects that are difficult to associate with the content to be exchanged (e.g., what the value/price of knowledge is, etc.); in addition, the problem of *property rights* and their legal management is also relevant (see above). Finally, the *execution* stage recalls the problem of *knowledge content delivery* (i.e., formats, supports, etc.), which is particularly challenging in the case of tacit contents.

Also, knowledge transactions within an OPC can require explicit arrangements that define or have an effect on the specific processes of knowledge transfer. Such arrangements include:

- duration and frequency of the transactional relation (spot, repetitive, or project-based);
- mechanisms used for the economic exchange (formats, technology supports, contractual specifications, etc.—see above);
- payments (membership fees, pay-per-use services, sponsorships, etc.).

Perceived Values and Costs of Participation

The active involvement in an online community depends on the *perceived value* ascribed to knowledge transactions by each participant (Wenger, McDermott & Snyder, 2002). Such value might largely vary, generally depending on the ability of individual members to exploit their participation:

- **Short-term vs. long-term value:** Short-term value pertains to immediate benefits (exploiting suggestions of colleagues, finding better solutions and making immediate decisions), while long-term value comes from continuous advancements in practice that may result in improved customer response and business value (members develop professionally, keep abreast of new advancements, etc.);
- **Tangible vs. intangible value:** Tangible results may come from the improved skills and reduced costs through faster access to information by the single member, the diffusion of good practices, and the standardization of service quality within the community. Less tangible results are associated to the building of a *sense of identity* within the community and an increased ability to innovate;
- **Old vs. new business strategies:** OPCs can be a way to realize previously defined strategies (cost reduction, market improvement, higher quality of services, improved professional competence) or to implement new ones (new services, novel ways to serve old customers, innovative outsourcing policies).

The expected valued should be compared with the *costs of participation*. *Tangible costs* include investments and operational costs for the single member (network access, Internet services, software, etc.), costs of KM services provided by the community, and costs of building it (central Web services, databases, knowledge management systems, operational structure, etc.). On the other hand, *intangible costs* consist of training and learning costs, business reengineering, organizational efforts, and such.

The *distribution* of costs between members should be fair and in line with the expected benefits. With regard to this, it should be noted that the costs for the single member might regard not only the single services used (or the knowledge contents received), but also the contribution to the management of the community.

Trust

A frequently used definition of trust is "the willingness of a party to be vulnerable to the actions of another party based on the expectations that the other will perform a particular action important to the trustor" (Mayer, Davis & Schoorman, 1995, p. 712). Such definition recalls both sociological and economic aspects. Also, it raises issues such as the evaluation of the risks associated to potentially opportunistic behaviors by partners and the selection of appropriate mechanisms for reducing such risks. Trust can build on different grounds (Ford, 2003), but for the purposes of our study, a basic distinction is between hard and soft trust mechanisms (Roberts, 2003). While the former depend on abstract and/or institutional systems (contracts, IPR regime, regulatory and legal mechanisms), the latter rely on social and cultural structures, reputation, and interpersonal relations. In environments that cannot count on established institutions (Adler, 2001), soft trust mechanisms can represent a vital element. This case appears to be particularly important here because legal mechanisms to protect knowledge transactions are very hard to establish in an OPC. In other words, even if economic or contractual solutions can still be arranged to limit opportunistic and unfair behaviors by the members, the mechanisms of soft trust appear to be much more significant. For instance, a careful selection of partners and the establishment of shared unwritten rules, on which each participant is expected to agree, may be essential ingredients for the stability of the OPC.

Intermediation

Some typical activities of intermediaries in traditional trade (e.g., identification of demand needs, information on products and suppliers, comparisons, customer targeting, demand orientation, balancing of conflicting interests) clearly imply *cognitive contents*, as often explicitly underlined in the economic literature (Choi, Stahl & Whinston, 1997; Pratt & Zeckhauser, 1985). In substance, a significant part of the value added by an intermediary consists of bridging over the *cognitive gap* between buyers and sellers, thereby facilitating the exchange of knowledge for settling transactions. In this sense, functions of *intermediation* appear essential in an OPC, where the object of transaction is knowledge itself. Specific players can act as an interface between the different members and help them to solve some of the problems of knowledge transfer and the associated economic issues. Even the KM literature draws the attention to the importance of mediators between knowledge sources and users (Bolisani, Di Biagi & Scarso, 2003; Markus, 2001). In addition, theoretical and empirical studies show that several kinds of new mediators (or *cybermediators*) are emerging in the Internet environments (Sarkar, Butler & Steinfield, 1995). Within a virtual community, the role of intermediary may be played by particular members, such as the founder(s) or sponsor, the manager, the Web master, and so forth. Intermediaries have to:

- **Manage community development:** membership promotion, incentives to active participation, community support, selection and management of technological infrastructure;
- **Set the rules:** selection of members; identification of (online) partners; role of referee in disputes; pricing services; distributing costs and access to services; technical help to members; interface with external environments and services;
- **Arrange knowledge transactions:** matching sources and receivers within the community; selection of knowledge contents to be transferred; implementation and maintenance of the KMS (e.g., Web portal, databases, communication systems between parties, etc.);
- **Support knowledge transactions:** codification/decodification of contents; format conversions; quality control of knowledge exchanged; management of payments;
- **Regulate competition** among partners potentially in conflict over the same business.

The organization of the intermediary functions may require a trade-off between conflicting goals. On one hand, the more the intermediary is perceived as *neutral* (unbiased), the more each member can trust its capacity of distributing benefits/costs and mediating disputes. On the other hand, since a virtual professional community should be intended as a community of businesses, the intermediary should have *business skills* (capability to select suitable information services, to negotiate with external service providers, etc.), but this may favor some users instead of others.

CONCLUSION: IMPLICATIONS FOR THE FUTURE OF OPC MANAGEMENT

While OPCs are emerging organizational solutions that have the potential to change the way businesses operate, how can a virtual community for businesses and professionals be sustained and developed? As a result of the analysis conducted above, we can articulate this question by considering two viewpoints. On one hand, a social perspective of OPCs highlights aspects such as *knowledge sharing* and *cooperation*, and the KM literature offers useful insights into the way these processes can be effectively done in operative terms. On the other hand, since professional communities are business-oriented, the economic aspects of knowledge transactions have to be integrated with the social and technical issues of KM. In other words, *collaboration* (i.e., knowledge transfers and sharing) should be combined and balanced with the *creation of economic value* (i.e., fair distribution of costs and benefits, protection and enhancement of the competitiveness of each single member, and of the community as a whole). In addition, considering the virtual nature of the community, this target should be reached by implementing the appropriate technological infrastructures and online services that favor effective knowledge exchanges and facilitate access and use by the various participants.

The development and the long-term sustainability of an OPC can thus be described as in Figure 4, where a mix of economic, organizational, and technological actions are essential to start up a *positive circle*. As the experience shows (Bolisani, Scarso & Di Biagi, 2003; Wenger, McDermott & Snyder, 2002), this virtuous circle can be vital for the reaching of a *critical mass* of users, thus making the participation in the community profitable. Conversely, the growth in the participating number raises the issues of *managing organizational complexity* of the community itself, as well as *potential competition* among participants. We also argued that a community *manager* or *mediator* plays central functions for this.

E

Figure 4. Development and sustainability of an OPC

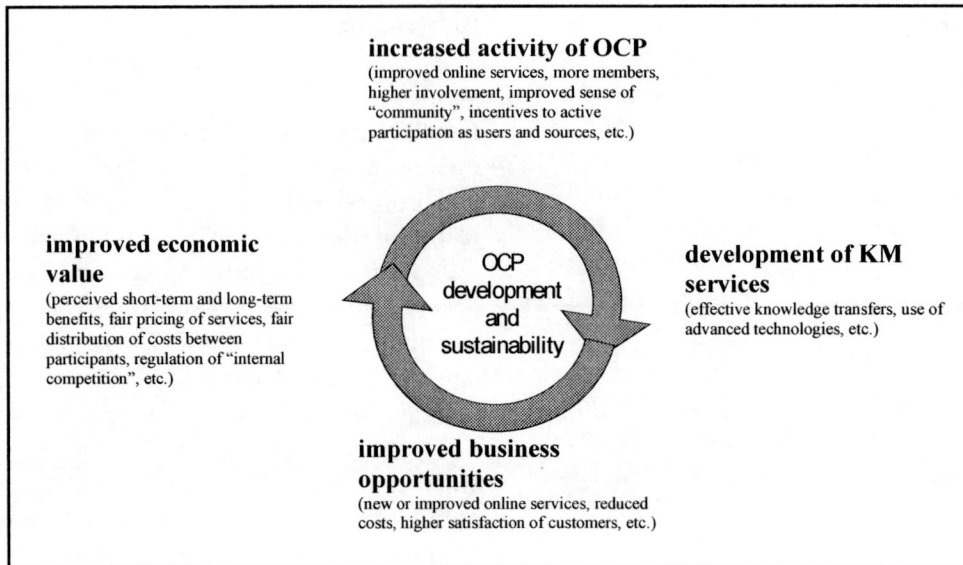

REFERENCES

Adler, P. S. (2001). Market, hierarchy, and trust: The knowledge economy and the future of capitalism. *Organization Science, 12*(2), 215-234.

Alavi, M., & Leidner, D. E. (2001). Knowledge management and knowledge management systems: Conceptual foundations and research issues. *MIS Quarterly, 25,* 107-136.

Ardichvili, A., Page, V., & Wentling, T. (2003). Motivation and barriers to participation in virtual knowledge-sharing communities of practice. *Journal of Knowledge Management, 7*(1), 64-77.

Bakos, Y. (1998). The emerging role of electronic marketplaces on the Internet. *Communications of the ACM, 41*(8), 35-42.

Bhatt, G. D. (2001). Knowledge management in organizations: Examining the interactions between technologies, techniques, and people. *Journal of Knowledge Management, 5*(1), 68-75.

Bolisani, E., Di Biagi, M., & Scarso, E. (2003). Knowledge intermediation: New business models in the digital economy. *Proceedings of the 16th Bled eCommerce Conference* (Vol. 1, pp. 987-999).

Bolisani, E., Scarso, E., & Di Biagi, M. (2003, September 18-19). Knowledge management between co-operation and business: A case-study of virtual community of practice (pp. 101-110). *Proceedings of the 4th European Conference on Knowledge Management,* Oriel College, Oxford, UK.

Brown, J. S., & Duguid, P. (2001). Knowledge and organization: A social-practice perspective. *Organization Science, 12*(2), 198-213.

Choi, S., Stahl, D. O., & Whinston, A. B. (1997). *The economics of electronic commerce.* Indianapolis: MacMillan.

de Souza, C. S., & Preece, J. (2004). A framework for analysing and understanding online communities. *Interacting with Computers, 16,* 579-610

Dyer, J. H., & Nobeoka, K. (2000). Creating and managing a high-performance knowledge-sharing network: The Toyota case. *Strategic Management Journal, 21,* 345-367.

Ford, D. P. (2003). Trust and knowledge management: The seeds of success. In C. W. Holsapple (Ed.), *Handbook on knowledge management* (pp. 553-575). Heilderberg: Springer.

Jarvenpaa, S. L., & Tanriverdi, H. (2002). Leading virtual knowledge networks. *Organizational Dynamics, 31*(4), 403-412.

Johnson, C. M. (2001). A survey of current research on online communities of practice. *Internet and Higher Education, 4*, 45-60.

Kardaras, D., Karakostas, B., & Papathanassiou, E. (2003). The potential of virtual communities in the insurance industry in the UK and Greece. *International Journal of Information Management, 23*, 41-53.

Koh, J., & Kim, Y.-G. (2004). Knowledge sharing in virtual communities: An e-business perspective. *Expert Systems with Applications, 26*(2), 155-166.

Maier, R. (2002). *Knowledge management systems. Information and communication technologies for knowledge management.* Berlin: Springer.

Markus, M. L. (2001). Toward a theory of knowledge reuse: Types of knowledge reuse situations and factors in reuse success. *Journal of Management Information Systems, 18*(1), 27-93.

Mayer, R., Davis, J., & Schoorman, F. (1995). An integrative model of organizational trust. *The Academy of Management Review, 20*(3), 709-734.

McDermott, R. (2000). Knowing in community: 10 critical success factors in building communities of practices. Retrieved May 6, 2005, from *http://www.co-i-l.com/coil/knowledge-garden/cop/knowing.shtml*

Nonaka, I., & Takeuchi, H. (1995). *The knowledge-creating company.* Oxford: Oxford University Press.

Pan, S. L., & Leidner, D. E. (2003). Bridging communities of practice with information technology in pursuit of global knowledge sharing. *Journal of Strategic Information Systems, 12*, 71-88.

Peña, I. (2002). Knowledge networks as part of an integrated knowledge management approach. *Journal of Knowledge Management, 6*(5), 469-478.

Plant, R. (2004). Online communities. *Technology in Society, 26*, 51-65.

Polanyi, M. (1967). *The Tacit dimension.* Garden City, NY: Doubleday Anchor.

Pratt, J. W., & Zeckhauser, R. J. (Eds.). (1985). *Principals and agents. The structure of business.* Boston: Harvard Business School Press.

Preece, J. (2000). *Online communities: Designing usability, supporting sociability.* Chichester, UK: Wiley.

Roberts, J. (2003). Trust and electronic knowledge transfer. *International Journal of Electronic Business, 1*(2), 168-186.

Sarkar, M. B., Butler, B., & Steinfield, C. (1995). Intermediaries and cybermediaries: A continuing role for mediating players in the electronic marketplace. *Journal of Computer Mediated Communication, 1*(3). Retrieved from *http://www.ascusc.org/jcmc/vol1/issue3/sarkar.html*

Shapiro, C., & Varian, H. R. (1998). *Information rules: A strategic guide to the network economy.* Boston: Harvard Business School Press.

Sharratt, M., & Usoro, A. (2003). Understanding knowledge-sharing in online communities of practice. *Electronic Journal on Knowledge Management, 1*(2), 187-196.

Smith, H. A., & McKeen, J. D. (2003). Creating and facilitating communities of practice. C. W. Holsapple (Ed.), *Handbook on knowledge management* (pp. 393-407). Heilderberg: Springer.

von Krogh, G. (2002). The communal resource and information systems. *Journal of Strategic Information Systems, 11*, 85-107.

Walsham, G. (2001). Knowledge management: The benefits and limitations of computer systems. *European Management Journal, 19*(6), 599-608.

Wenger, E., McDermott, R., & Snyder, W. (2002). *Cultivating communities of practices.* Boston: Harvard Business School Press.

KEY TERMS

Extended Knowledge Management: KM methods and practices applied to inter-firm knowledge management. The adjective *extended* means activities that span the organizational boundaries and involve suppliers, customers, vendors, business partners, and institutions.

155

Knowledge Management Systems: Information systems and technologies designed for knowledge management. They include both old software tools rearranged and new ones.

Knowledge Sharing: A process of mutual exchange of knowledge between two or more parties.

Knowledge Transaction: An economic exchange whose object is knowledge.

Knowledge Transfer: The process concerning the transfer of knowledge from a source to a recipient or user. It is the most studied process in the KM literature.

Online Community: A collective group of entities (individual or organizations) who come together through an electronic medium for a common purpose and who are governed by norms and policies.

Online Professional Community: Group of various business services firms that make a private use of the network to pursue an immediate business purpose, that is, to improve and extend the service provided to the final customer beyond the peculiar competence and practice of the single member.

Professional Services: Companies delivering business consulting services, including computer and related activities, research and development, and other business activities as legal, accounting, marketing, advertising, and engineering.

Encouraging Research through Communities of Practice

Sandra Corlett
Northumbria University, UK

Patricia Bryans
Northumbria University, UK

Sharon Mavin
Northumbria University, UK

INTRODUCTION

This article explores the role communities of practice (CoPs) can play in encouraging individual and organizational research capacity and capability. Through three case study CoPs with members from UK universities, we highlight this potential contribution. We argue that processes of social learning underpin the CoPs and use reflexive learning as the methodology to explore individual experiences. We focus particularly on the identity construction for new researchers and identity transformation for experienced researchers enabled by participation in the CoPs. Emerging knowledge domains that bind together individual members are also discussed.

While this account focuses on CoPs and the development of competencies and identities of researchers, we believe the process and benefits of collaboration discussed have application to other organizational contexts for the development of core competencies and organizational learning.

BACKGROUND: THE UK HIGHER EDUCATION CONTEXT

The pre-1992 UK universities have emerged grounded in research cultures, but in the ex- polytechnics, the move of focus from teaching to research is a relatively recent phenomenon.

As a result, the UK new universities may have achieved excellence in teaching, but performance in terms of research has been slower to develop. However, the necessity for developing research profiles within these new universities is increasingly placed on the strategic agenda, and there is an increasing expectation for academics to be research active.

Oliver (1997) argues that research is more a process than a product. From this perspective, research is an endless activity, and systematic inquiry becomes a way of life, encouraging us to become different people. Traditionally, research has been perceived as an individual, isolated endeavor with the "ivory tower" images of academic contexts (Bryans & Mavin, 2004). However, with the pace of change and the development of new knowledge, there are benefits to interdisciplinary collaboration. The task for new universities is to foster and develop research cultures which encourage collaboration between experienced and new researchers in order to leverage research capacity and capability.

CoPs

One way to foster this type of collaboration is through CoPs. Par (2002) argues CoPs are "communities that learn. Participants meet to learn from each other and share and benefit from each other's expertise." Wenger, McDermott, and Snyder (2002) argue that people who come together in CoPs do so within and across the boundaries of teams, departments, and organizations to create, share, deepen, and apply knowledge to their problems and passions about a topic. Far from being a new idea, CoPs are, in fact, our first knowledge-based social structure (Wenger et al., 2002).

Fox (2000) asserts that learning theory has developed from its roots in mainly psychological traditions through organizational learning to social constructionist influences resulting in an emphasis on social learning theory. He views Lave and Wenger's (1991) account of situated learning as a specific version of social learning theory, arguing that its principle element is the notion of the CoP in which members learn by participating in shared activity. We view social learning on two levels: first, that we learn with and from others in all our social relationships and, second, that the social context mediates and structures the sense making and meanings we experience and generate as we perform and interact within this context.

Wenger et al. (2002) claim that CoPs form naturally when people come together around common interests but argue that organizations need to become more proactive and systematic about developing and integrating them into their strategy. However, this may raise issues of alignment where individual and organization needs may not coincide. They observe that, in the development of a CoP, there are five stages: potential, coalescing, maturing, stewardship, and transformation, but these are not fixed linear points.

Taking a fluid approach, Brown and Duguid (1991) argue that communities are significantly emergent, "that is to say that their shape and membership emerges in the process of activity, as opposed to being created to carry out a task" (p. 49). They comment that much of the literature refers to the design or creation of new groups, but their interest is with the detection and support of emergent or existing communities.

IDENTITY

Participating in new activities and social groupings encourages us to become different people, impacting on our identities. The concept of individualized identity is problematic and has a long history of discussion in social studies. Identities are individualized through the narration of one's own story (Gherardi, 1996, p. 188), constructed within a repertoire accessible in a situated time and space (Meyer, 1986 and Czarniaswka–Joerges, 1994 as cited in Gherardi, 1996, p. 188) and part of a discourse of historically related sets of thoughts, expressions, and practices (Foucault, 1984). From this perspective, identity is a subjective concept which changes over time and in different situations (Mavin, 2001).

Occupations and, more particularly, professions have been increasingly important sources of identity for individuals. Not only has academic work provided the conditions for strong identities but also the building of individual identities that are nevertheless embedded in defined communities (Henkel, 2000, p. 13). Taylor (1989) as cited in Henkel (2000) emphasized the importance of a "defining community" in the formation of identity which provides the language through which we understand ourselves and interpret our world: "Your identity is essentially tied up with what you are committed to, what you overwhelmingly value and what you strive for" (p. 145). This embedded nature of identity is fluid and influenced by the institution's changing values and agendas.

CoP is a theory about learning as socialization, where increasing participation in CoP is the key to both how learning happens and identity formation (Fox, 2000, p. 859). "Participation is both a kind of action and a kind of belonging. Such participation shapes not only what we do but also who we are and how we interpret what we do" (Wenger, 1998, p. 4), and building an individual identity consists of negotiating the meanings of our experience of membership in social communities (Wenger, 1998, p. 145).

In terms of experienced and less experienced members of CoPs, Wenger (1998) notes that membership translates into an identity as a form of competence. From CoP theory, we know that masters of a practice show novices what to do: they act like obligatory points of passage enabling the novice to work/learn and move toward the center of the CoP gaining in legitimacy as they do (Fox, 2000, p. 861).

METHODOLOGY

We approached the research from a subjective stance, acknowledging that meaning comes into existence in and out of our engagement with the realities in our world, where meaning is not discovered, but socially constructed (Crotty, 1998). We developed a reflexive approach to three case studies whereby, as authors of this paper, we are both the researchers and the researched, as participants in the case study CoPs. We acknowledge individual's experience as a basis for research and the accountability of researchers to research participants and to a wider research community. We have taken a reflexive perspective on the research as part of a knowledge validation process (Griffin, 1995).

The boundaries of this article prevent us from exploring further the subjective perceptions of the research participants and the fluidity of forming and joining the CoPs discussed in this research. We draw upon narrative data from individual participants involved in three case study CoPs. Participants produced written self-reflective statements concerning their reasons for joining the CoP, the outcomes they wanted from the process, and the benefits they saw from membership. We selected particular voices to highlight participants' experiences and explore whether CoPs are a means of encouraging research and developing researcher identities.

THE CASE STUDIES

All three CoPs emerged informally rather than as a result of formal organizational strategy. The first case study COP was a research group of academic women from three different UK universities who, for three years, focused on supporting members to complete their doctoral study and begin publication. Members gained support and credibility for research approaches which were not from the positivist tradition and gained support as women, as they felt marginalized in their own universities. Having achieved its purposes of members' PhD completion and publication, the CoP progressed to the final life stage of transformation (Wenger et al., 2002). This transformed into the second case study CoP with a broader membership of new and experienced researchers from three different UK universities. This remains a community of women academics now focused upon a qualitative research knowledge domain and is reaching the maturing stage (Wenger et al., 2002). Here the CoP champions and legitimizes alternative approaches to positivism and focuses on qualitative methods.

The third case study community was created in 2003 with members, drawn from the same UK university, explicitly invited to join a CoP to facilitate the development of shared research interests and encourage publications in the areas of individual and organizational learning and knowledge management. However, this CoP agreed instead on a knowledge domain of exploring communities of practice and is at the coalescing life stage (Wenger et al., 2002), engaging an

interdisciplinary group of men and women academics who are new and experienced researchers.

ANALYSIS

Taking individual experiences from across the three case studies, we identified five themes: building relationships, building and sharing knowledge, building confidence, legitimizing research activity and approaches, and raising critical awareness and enhancing reflexivity.

Building Relationships

Participants across all three CoPs remarked that CoP participation builds relationships with colleagues. As an experienced researcher, Tina, a member of CoP 1 and 2, emphasized the particular benefit of the supportive environment:

[It's] very much a sharing of ideas. We can voice concerns, fears, whatever in a way that others understand. They don't see it as a weakness. ...I was nervous about my PhD viva...this network will try a rehearsal with me. I don't want to put onto others because it's a lot of work but they want to help. In return I am stopping them making mistakes I made, pointing out pitfalls. Over the last year, I've got to know the women in the network better. I felt I needed peer support.

The benefit of peer support is also highlighted by Philippa, an experienced researcher who participated in all three CoPs: "Commitment to CoPs makes me feel both supported and challenged."

Tina emphasizes the need for trust within what is not only a supportive but also challenging environment: "Trust is a key issue—someone has to bare their soul. People can feel very exposed but it's not just a touchy feeling thing."

The benefits of CoPs are not just social in nature. As Wenger et al. (2002) assert, "They help each other solve problems, discuss their situations, their aspirations and their needs. They ponder common issues, explore ideas and act as sounding boards" (pp. 4-5). The participants emphasized the importance of these aspects such as building of knowledge bases and problem solving.

Building and Sharing Knowledge

The issue of "like-mindedness" and knowledge sharing was a common motivation for membership of the three CoPs. Yin, an experienced researcher and member of CoP 3, states:

CoP is in essence a practitioners' community whereby each and every member could learn by sharing/sifting what others have got to put forward; it is also a community for practitioners who have something valuable to show to others. The practices of sharing and showing thus characterises and distinguishes the CoP from network or work team.

Laura, a new researcher and member of CoPs 2 and 3, comments:

I am still deciding how to make sense of what a "CoP" really means as I suspect there are nuances about it that take it beyond a meeting of like-minded people who share experience and knowledge in solving problems. ... For me, it feels like it should be something that enables people with common purpose/ aims to communicate with one another about their practice—so it is about us as researchers communicating about research.

Sarah, a new researcher and member of CoP 3, highlights the impact of membership of CoP on her own identity construction in "becoming a researcher" and learning from others, as well as developing her knowledge:

I am interested in participating in this CoP as it will provide me with the opportunity to deepen my knowledge of research, and my self perception and capabilities as a researcher, by interacting with others who are interested, actively involved and experienced in research. In that regard, my concern is that I have a lot to learn and therefore to gain from others but am not sure what balance I will be able to achieve between giving and taking.

Building Confidence

Wenger et al. (2002:4) note participation informally binds members:

by the value they find in learning together. This value is not merely instrumental for their work, it also accrues in the personal satisfaction of knowing colleagues who understand each other's perspective and of belonging to an interesting group of people. They develop personal relationships and established ways of interacting and may even develop a common sense of identity. (p. 4)

All three CoPs helped to build the confidence of inexperienced members and their identity as new researchers and this seemed to apply also to experienced researchers. Sarah reflects:

I wouldn't be here now (writing this paper) if it hadn't been for my participation in the CoP. I have noted a change in the way that I "see" myself as a researcher. Initially I introduced myself to the CoP as a non-active researcher and now I would describe myself as a developing one. Colleagues have commented on a change in my attitude towards, and confidence in, my capabilities as a researcher.

Paechter (2003:70) argues, "contributions become more complex and important...as they progress towards full participation. Through this they develop not just their expertise in the practice itself but their understanding of and embeddedness in the culture which surrounds it". Sarah notes:

I now look forward to researching and writing papers, particularly with others, as I appreciate how much I learn from them through the process and how outcomes from the collaboration enable me to "measure" my own progress.

Philippa comments:

Being in a CoP has given me the confidence to be prepared to contribute more formally to the development of new researchers. I've led staff training sessions...I wouldn't have been prepared to lead either of these sessions before my participation in CoPs. Through CoPs I know people I didn't know before and have developed a better sense of my existing expertise and the areas I want to continue to develop.

Geraldine, an experienced researcher and member of CoP 2, draws attention to the supportive environment of her CoP as a "nice comfort zone". Tina expresses the downside of this confidence building within a CoP. She emphasizes the "uncomfortable zone" depending on the individual's stage of development:

People can feel very exposed but it's not just a touchy feeling thing. It's also about keeping up conceptually. In this network you've got to be at the appropriate stage in your academic development or you will feel out of your depth.

Legitimizing Research Activity and Approaches

Belonging to a CoP appears to enable members to align their individual interests with organizational requirements and builds legitimacy and credibility. Philippa highlights this and comments:

Participating in CoPs really helps me to prioritise and legitimise my research activity. You gain a sense of commitment and purpose from working and learning with others. Discussions help me to work out what I know and what I think, and also challenge my thinking and broaden and deepen my knowledge base.

Raising Critical Awareness and Enhancing Reflexivity

CoP membership enables members to develop processes of reflexivity applied to themselves as researchers, as academics and in deepening their understanding of their knowledge domain. Sophie comments:

Belonging to the CoP has opened up new avenues for thinking, practice and research. The process of the CoP enabled reflexivity to emerge as a new knowledge domain for some of the members and this now shapes my approach to research and has impacted on my practice.

Daniel notes:

The more I work in this field, the more I'm convinced that what we're really trying to do is to help individuals and organisations to LEARN…learning is essentially an individual, socially-driven process that we can't control (and shouldn't try); but we can perhaps stimulate and nurture it, and learn(!) new ways to "light the blue touch paper" and watch our good ideas spread like wildfire. And all of that is what appeals to me about the ideas described as "Communities of Practice".

Barbara, a new researcher and member of CoP 3, uses the CoP for furthering her research on more critical approaches to management. At an early meeting, she raised the issues of management and academic fads and has consistently encouraged other members to consider the "dark side" of CoPs in light of a literature base which emphasizes benefits. James similarly points to the disadvantages of CoPs in practice: "Much has been written about the benefits of CoPs to organisations, but is it all plain sailing?…In terms of practical research [outcomes], our own CoP could form the basis for some of this work."

CONCLUSION AND FUTURE TRENDS

CoP participation has brought together academics from a range of backgrounds and disciplines whose paths might not otherwise have crossed. The informal nature of CoP meetings allows the building of relationships which can transfer themselves into other areas of organizational practice.

The sharing of knowledge from different disciplinary backgrounds in all three CoPs helps deepen and broaden individual knowledge and subject expertise. In turn, this helps build individual confidence, not only to contribute to the CoP, but also to extend the contribution to the wider organization in supporting and developing others' research interest and capabilities.

Sharing knowledge between the CoP members and appreciating conceptual connections supports individuals to demonstrate connectivity through their research and teaching activities.

Given that the primary university has not been a research-led institution, when members have met as a CoP, it has legitimized research as an activity and allowed participants to prioritize and progress with their research. Each CoP has given participants the impetus to generate research outputs and, for new researchers, the confidence to write for publication. The second case study CoP provides an example of how the CoP legitimized alternative research approaches and methodologies. Individuals are encouraged through the collaborative support of the CoP in their belief that their research approaches are acceptable and can be defended in the organization and the wider research community.

Members of all three case study CoPs commented on how their participation helped to build their confidence as researchers. Both new and experienced researchers experienced the benefits of giving and taking. In the case of new researchers, their legitimate peripheral participation (Lave & Wenger, 1991) contributes to their sense of themselves as research active and shapes their identity as researchers. For the experienced researchers, their participation allows them to appreciate their expertise and encourages them to give the benefit of their experiences to other CoP members and to the wider organizational community.

REFERENCES

Brown, J. S., & Duguid, P. (1991). Organisational learning and communities of practice: Toward a unified view of working, learning and innovation. *Organization Science, 2*(1), 40-57.

Bryans, P., & Mavin, S. (2004, August 30-September 1). Monks, hunters and white-water rafting: Images of research and researchers. *Proceedings of the British Academy of Management Conference*, St. Andrews, Scotland.

Crotty, M. (1998). *The foundations of social research: Meaning and perspective in the research process*. London: Sage.

Foucault, M. (1984). *l'usage des plaisirs,* Paris, S. Gherardi, S. (1996) Gendered Organisational Cultures: Narratives of women travellers in a male world, *Gender Work and Organisation, 3*(4), 187-201.

Fox, S. (2000). Communities of practice, Foucault and actor-network theory. *Journal of Management Studies, 37*(6), 853-868.

Gherardi, S. (1996). Gendered organisational cultures: Narratives of women travellers in a male world. *Gender Work and Organization, 3*(4), 187-201.

Griffin, C. (1995, March). Feminism, social psychology and qualitative research. *The Psychologist.*

Henkel, M. (2000). *Academic identities and policy change in higher education.* London: Jessica Kingsley.

Lave, J., & Wenger, E. (1991). *Situated learning: Legitimate peripheral participation.* New York: Cambridge University Press.

Mavin, S. (2001). *The gender culture kaleidoscope: Images of women's identity and place in organisation.* Unpublished PhD Thesis, University of Northumbria at Newcastle, Newcastle Business School.

Oliver, P. (1997). *Teach yourself research for business, marketing and education.* London: Hodder and Stroughton.

Paechter, C. (2003). Masculinities and femininities as communities of practice. *Women's Studies International Forum, 26*(1), 69-77.

Par. (2002). What is a community of practice: Awakening technology. Retrieved May 5, 2005, from *http://www.coil.com/coil/knowledge-garden/cop/definitions. html*

Wenger, E. (1998). *Communities of practice: Learning, meaning and identity.* Cambridge: Cambridge University Press.

Wenger, E., McDermott, R., & Snyder, W. M. (2002). *Cultivating communities of practice.* Boston: Harvard Business School Press.

Exploring the Role of Communities of Practice in Regional Innovation Systems

Robin Teigland
Stockholm School of Economics, Sweden

Andrew Schenkel
Stockholm School of Economics, Sweden

INTRODUCTION

In the past two decades, the related concepts of regional innovation systems and clusters have become widely circulated in both academic and policy circles. Both concepts depart from the idea that innovations predominantly occur as a result of interactions between various actors, rather than as a result of a solitary genius (Håkansson, 1987; von Hippel; 1988; Lundvall, 1992), and that innovation and industrial transformation result from interactions across sets of actors within a spatially defined territory (e.g., countries, regions). Researchers within this field posit that most innovations are based on some form of problem solving in which someone generally perceives a problem and turns to someone else for help and advice (Teigland, Lindqvist, Malmberg & Waxell, 2004), and that spatial proximity seems to enhance the processes of interactive learning and innovation (Malmberg & Maskell, 2002). These assumptions draw striking parallels to the traditional concept of communities of practice (Brown & Duguid, 1991; Orr, 1990; Wenger, 1998), which are emergent groups of people who know each other relatively intimately and who primarily work together directly in face-to-face situations since learning and knowledge are situated within a physical setting (Teigland, 2003). Thus, the purpose of this short article is to provide a brief discussion of clusters and regional innovation systems, and propose broad areas of future research in which the community of practice concept can contribute to our understanding of clusters and regional innovation systems.

BACKGROUND

While no one universal definition exists, one definition of a cluster is a *spatial agglomeration* of similar and related economic and knowledge-creating activities (Waxell, 2005). Regional innovation systems are networks of organizations, institutions, and individuals within which the creation, dissemination, and exploitation of new knowledge and innovations occurs (Cooke, Heidenreich & Braczyk, 2004). The link between clusters and regional innovation systems is that within these spatial systems, groups of similar and related firms (e.g., large and small firms, suppliers, service providers, customers, rivals, etc.) comprise the core of the cluster, while academic and research organizations, policy institutions, government authorities, financial actors, and various institutions for collaboration and networks make up the innovation system of which the cluster is a part (Teigland et al., 2004).

Cluster and regional innovation system researchers argue that interactive learning and innovation processes are not space-less or global, but on the contrary, geographical space plays an active role in these processes. Spatial proximity carries with it, among other things, the potential for intensified face-to-face interaction, short cognitive distance, common language, trustful relations between various actors, easy observations, and immediate comparisons (Malmberg & Maskell, 2002).

Two areas of primary investigation within regional innovation systems are: (1) Why are some regions more competitive than others? and (2) How can regional innovation systems be supported? Much of the extant literature on regional innovation systems and clusters tends to focus on formal interactions between actors. However, in one of the most well-known studies, Saxenian (1996) proposes that one of the primary reasons for the relative success of the Silicon Valley area over that of Route 128 in Boston is that knowledge is easily shared through informal relationships similar to those of communities of practice between individuals belonging to competing firms as well as other organizations in the Silicon Valley region. This is in direct contrast to the Route 128 area in Boston where informal interorganizational fraternization was discouraged.

REGIONAL INNOVATION SYSTEMS AND COMMUNITIES OF PRACTICE: FUTURE AREAS OF RESEARCH

The concept of a community of practice is emerging as an essential building block of the knowledge economy. These communities develop through the mutual engagement of individuals as they participate in shared work practices, supporting the exchange of ideas and knowledge between people, which results in learning and innovation within the community (Brown & Duguid, 2000). Communities of practice may also extend across a firm's legal boundaries, and relationships between such a community's members facilitate the flow of knowledge between organizations (Brown & Duguid, 2000; Teigland, 2003), and as indicated above in Saxenian's research, they may play a vital role in regional innovation systems.

Community of practice research may be divided into two broad areas: (1) the cognitive and relational aspects and (2) the structural aspects. Prior research in the first area has emphasized the importance of shared identity, language, values, and norms, as well as relations built on mutual trust and reciprocity for knowledge exchange and learning (e.g., Wenger, 1998). Thus, one area of future research could investigate the cognitive and relational aspects of communities of practice that span organizational boundaries within a regional innovation system to better understand the dynamics of knowledge flows and innovation. A second area of future research within regional innovation systems could focus on understanding the structural aspects of these interorganizational communities of practice (e.g.,

Schenkel, Teigland & Borgatti, 2001)—for example, the relationship between community structure and knowledge sharing, how structures change over time, and how community structure influences the cognitive aspects of shared language, values, and goals.

CONCLUSION

The cluster and regional innovation system concepts provide a means to describe the systemic nature of an economy, that is, how various types of industrial activity are related. Using the communities of practice perspective as a lens provides an additional scope for analyzing the interactions and interdependencies between firms and industries across a wide spectrum of economic activity.

REFERENCES

Brown, J.S., & Duguid, P. (1991). Organizational learning and communities of practice. *Organization Science, 2*(1), 40-57.

Brown, J.S., & Duguid, P. (2000). *The social life of information.* Boston: Harvard Business School Press.

Cooke, P., Heidenreich, M., & Braczyk, H.-J. (Eds.). (2004). *Regional innovation systems. The role of governances in a globalised world* (2nd ed.). London: Routledge.

Håkansson, H. (1987). *Corporate technological behaviour: Co-operation and networks.* London: Routledge.

Lundvall, B.-Å. (Ed.). (1992). *National systems of innovation: Towards a theory of innovation and interactive learning.* London: Pinter.

Malmberg, A., & Maskell, P. (2002). The elusive concept of localization economies—towards a knowledge-based theory of spatial clustering. *Environment and Planning, A,* 34.

Orr, J. (1990). *Talking about machines: An ethnography of a modern job.* Unpublished Doctoral Dissertation, Cornell University, USA.

Saxenian, A. (1996). *Regional advantage: Culture and competition in Silicon Valley and Route 128.* Boston: Harvard University Press.

Schenkel, A., Teigland, R., & Borgatti, S.P. (2001). Theorizing structural properties of communities of practice: A social network approach. *Proceedings of the Academy of Management Conference.*

Teigland, R. (2003). *Knowledge networking: Structure and performance in networks of practice.* Published Doctoral Dissertation, Stockholm School of Economics, Sweden.

Teigland, R., Lindqvist, G., Malmberg, A., & Waxell, A. (2004). Investigating the Uppsala biotech cluster: Baseline results from the 2004 Uppsala biotech cluster survey. *CIND Research Report,* 1. Retrieved from *www.cind.se*

von Hippel, E. (1988). *The sources of innovation.* Oxford: Oxford University Press.

Waxell, A. (2005). Uppsalas biotekniska industriella system: En ekonomisk-geografisk studie av interaktion, kunskapsspridning och rörlighet på arbetsmarknaden (*Uppsalas biotech industrial system: An economic-geographic study of interaction, knowledge dissemination, and workforce mobility*). Published Doctoral Dissertation, Uppsala University.

Wenger, E. (1998). *Communities of practice: Learning, meaning, and identity.* Cambridge: Cambridge University Press.

KEY TERMS

Cluster: A spatial agglomeration of similar and related economic and knowledge-creating activities (Waxell, 2005).

Community of Practice: Generally tightly knit, emergent groups of people who know each other relatively intimately and who primarily work together directly in face-to-face situations, since learning and knowledge are situated within a physical setting (Teigland, 2003).

Regional Innovation System: A network of organizations, institutions, and individuals within which the creation, dissemination, and exploitation of new knowledge and innovations occur (Cooke et al., 2004).

Exploring the Selection of Technology for Enabling Communities

Keith Patrick
University of Westminster, UK

Andrew Cox
Loughborough University, UK

Rahman Abdullah
South Bank University, UK

INTRODUCTION

Communities, whilst represented and apparent in their members, are most evident in the technological entity—the technology and tools that support the common and communal activities. Technology acts as the *enabler* linking a group of individuals who are most likely dispersed, in terms of time and place, and *facilitates* their interaction.

So you have decided or been asked to create or facilitate the activities of a community, but how do you select an appropriate solution for the requirements of a particular community and its members? Do you follow a traditional systems and software acquisition route: establish the requirement, develop the system, approach a consultant, call the IS department to see how you can adapt existing technologies? Or do you adopt an approach that can also be viewed as part of the community development process through the generation of involvement, engagement leading to ownership of the community and therein its future activities?

The bottom line is that technology—either a system or a tool—is still required; the following exploration is based on the original research of the Knowledge Library project (Cox, Patrick & Abdullah, 2003). The K-Library project, facilitated by Research Development funding from London South Bank University, started from the premise that it would be useful to assemble a group of librarians from across the library sectors to share their understanding of the concept of knowledge management (KM), and to look at how some of the ideas drawn from the KM literature could be applied to library practice. It also reflects the considerations of a project team with its participants for the selection of a suitable system to support this community of interest. This project and subsequent work provided the basis for this exploration of tools and technologies that support communal activity; be that a community of intent, interest, purpose, or practice, which seeks to identify the user requirements for the technology or systems to support the building of an online community. The aim is to review the type of technologies and varying features, and to explore how they can assist a community through its initiation and maintenance phases without intruding or hampering the intention and activities of that community. This has enabled the creation of a list of features supported by explanation of these functions and options, and the creation of a checklist (see Appendix). This checklist can be used to help potential users identify their expected needs by distinguishing essential, useful, and non-essential features. The resulting checklist can be found at the end of the exploration and explanation of functions and functionality that could be adopted in the technology to support a community. From this exercise and other user requirements, it is possible to evaluate different systems for their compliance with user requirements. In the K-Library project, this enabled the selection of a suitable system, replicating the majority of the features the community considered to be appropriate for their activities. This exploration is augmented with observation by the authors and the reflections of the communities arising from the selection and use of a system/technology and the building of the community.

BACKGROUND

On commencing the K-Library project, we found that there was little to directly lead us toward making a choice of supporting technology. A difficulty that lay in the wide and expanding range of technologies available for supporting communities, extending from simple bulletin boards or e-mail archiving systems through to sophisticated suites of integrated software with document repositories and content/records management, calendars, task management, and workflow. Also it was found that at the time of initiating the project, the literature on communities was focused toward the conceptual and practical aspects of running communities. This literature was focussed on selecting technologies rather than more generally on systems and software development, and texts were geared to specific software—groupware or intranets. These texts, whilst relevant, did not offer the fuller picture we initially perceived and then revealed as relevant in the discussions with the K-Library Communities membership.

Additionally, the speed of technological development has continued apace, making much material dated with the arrival of newer technologies and the convergence and integration of these technologies with other broader ranging technologies and standards, like instant messaging, Weblogs, and RSS feed. These are significantly changing how information can be collated and distributed, with increasing opportunity for customisation and personalisation. This is coupled with systems becoming more pervasive with enterprise information portals, integrated content management systems, wireless networks, and mobile commerce, and these devices becoming smaller and more mobile themselves. An important impact can be seen in the increasing standardisation occurring as useful technologies and practice evolve, particularly the Web-based metaphor for the look and feel of systems for familiarity and ease of use, with many of these technologies or practices being viewed as standards in the field. Although in terms of a true defining of a standard reflecting the technology domain, these may be transitional, emerging, de-facto, or proprietary standards like XML, XHTML, Java, or DHTML. The essential aspect being that this semblance of a common standard is facilitating the development of more consistent systems, and the genuine use and reuse of documents and artefacts

for a range of the devices without the need for re-creation, providing the ability to use or adapt them with relative ease, and enabling a group of individuals to form and evolve into a genuine and mutually beneficial community.

MAIN FOCUS

Identifying the Necessary Features

The question therefore is what functions or functionality does your community require? A simple bulletin board or mailing list (e.g., listserv, jiscmail) based on specific software, or a Web-based group with discussion threading and the ability to push e-mails (e.g., e-group or Yahoo group) may be enough. Is real time functionality—like chat or instant messenger (e.g., AOL, MSN, ICQ) or Internet telephony (e.g., Skype)—required? What about reminders for outstanding action 'to-do' or task list activity areas/places (e.g., Groove, Lotus Notes/Domino, First Class)? Are options required relating to the creation and dissemination of work via content management systems, record management, workflow, sign-offs, and portal-based enterprise systems (e.g., Opentext Livelink, or Hyperwave or Microsoft's Sharepoint)? Wenger (2001), in analysing tools for the suitability to run communities of practice, argues that "the ideal system [for this purpose] at the right price does not yet exist." He points out that most available systems were developed primarily with other applications in mind; specifically he identifies tools developed such as: knowledge portals, project type software, Web site communities, discussion groups, online meeting spaces, e-learning spaces, knowledge exchange, and knowledge repositories.

It is very easy in choosing a system to overlook the lack of important features that can be critical as to whether members are encouraged or discouraged from actively and consistently engaging with a community. The stress in functionality varies from system to system; for example, project management tools will stress document management and task listings much more than a Web site community tool. Also the specific terminology used may be appropriate to a particular application—that is, 'task lists' would more appropriately be called 'activity lists', say, in a Web site community as opposed to a project management system. Clearly it makes sense in selecting software for one type of community as far as possible to look for a system that was developed with that type of community in mind.

Conversation Space

This represents the core of any system or tool, the appropriate communication tools being vital to the success of a potential community. The more sophisticated tools will often provide several different spaces for shared and individual activities, and synchronous and asynchronous communication.

Asynchronous Communication

Asynchronous communication comes in two main forms: (1) e-mail-based systems where e-mails to a central account are forwarded to members, with a Web archive existing for reference; and (2) bulletin board systems, where messages are posted on the Web and users have to log in to read messages.

An issue that impacts on discussions lies in how messages are delivered or read via e-mail and reply can be made via e-mail or the Web site. Impacted can be the number of e-mails received in a single day according to energy of the debate and length, when in digest form, characteristically containing messages plus the replies, creates long and difficult-to-comprehend e-mails/documents. It is possible to reply via the individual message and reflect in the discussion as the thread has evolved. How previous messages are shown is significant if they are differentiated from the new message through the use of quotes or indentation, and these can be quite critical to readability. Some systems encourage the use of emoticons to partially overcome the limitations of communication based just on text; or an alternative allows users to add formatting to text, sometimes with HTML coding.

An addition to this is the potential for taking activities offline for contribution, that is, editing or contributing to collaborative documents; however, some form of synchronisation is beneficial for managing version control and ensuring the correct document is being improved.

It may be useful to offer the option to make anonymous contributions, to give users the opportunity to experiment with points of view that are contro-

versial. This would need to address in the communities the development of operating protocols as to how it is used to ensure that it is not used negatively or to the determent of free-flowing discussions.

Synchronous Communication

For highly spontaneous interactions many community building systems offer real-time chat. Ideally there should be some means of archiving these discussions, though the most effective way may be through human reporter. Transcripts are difficult to read, and to respond to, particularly if there are multiple participants, and can quickly disappear off the screen with the lack of threading.

Often the chat facility has the ability to allow a user to check 'who else' is logged on—this is apparently a popular feature in some community building systems, even where communication is actually continued using asynchronous channels. This can also include shared activity spaces that allow co-viewing and co-editing.

Announcements

This is different than alerts, which are explained below. It is possible to simply make announcements through the asynchronous communication channel, but it is useful to have another channel through which to issue important community-wide news. Announcements are often posted at the top of the homepage. In some systems different types of announcements are distinguished and colour coded.

What's New

There are two main approaches to defining what is new: either this can be determined by when the user last logged on, or more simply by elapsed time. The system we selected used the latter, and it was obviously not as satisfactory if one had been away a week; it also alerted one to one's own postings and uploads.

Ideally the administrator should be able to control what types of items are monitored for changes and be given some options about how new items are marked, such as where they appear on the homepage, colours, and so forth.

Alerts

An alternative approach to making people aware of changes is the option to alert all, or some, users of changes when they are made. This has the merit of putting the choice in the hands of the community member as to who to notify of a particular change. With the system adopted by K-Library, the alerts only contained links to the site, not a direct link to the source of the alert. This additional step to using the system was viewed as detrimental and a hindrance by the members.

Polling

Polls are a powerful tool for helping communities to foster active participation and even to make collective decisions. While it is possible to canvas opinion simply through asynchronous communication channels, this is haphazard and difficult to control. The ability to automatically close a poll at a specific time will also encourage members to cast their votes within the allotted period.

A poll posted on the front page of the site allows members to follow its progress from the start to when the poll is finished and is especially important in circumstances where members are allowed to change their mind while the poll is still open. An additional option to allow anonymous votes can also be beneficial in encouraging members to vote freely.

Ideally there should be the flexibility for the member to add his or her own answer, if the option he or she wants to choose is not listed. Some systems provide automatic e-mail notifications when polls are running, closed, and when results are posted on the site.

Membership Directory

The importance of a membership directory lies in its ability to encourage users to share information about themselves, such as their areas of expertise, interests, and work experience, and so connect people. It might be useful to build a link from postings to membership entry or vice versa.

Ideally the administrator should be able to define suitable fields for entries in the directory, and select what level of information is required to be provided.

Users should be able to have latitude to express themselves and reflect their interest and personality. The richer the record they can generate, the more other members will be able to relate to that person. It is useful to allow users to add a photograph of themselves—or links or documents; equally, users must be able to withhold information if they wish. This an area for the community to decide what the criteria is to be; a deadline for this information is also useful, as whilst members may agree, they may still abstain from publishing the agreed information.

Contacts List

Users can be encouraged to share contacts through a dedicated contacts directory; this can allow access or publication of a restricted set of information to that of the membership directory. This also overcomes any potential reluctance to enter full information within the Membership Directory, arising from concern over who will be able to see, use or pass on that information. Some members will find the real value of the community is to be able to access expertise through the back-channel, although the group experts may not take the same viewpoint.

Document Repository

The document repository can be one of the core features of a communications system. It is possible simply to e-mail files to a discussion list, but there are advantages in having a specific document store. Documents in a repository are easier to locate than those that are somewhere in a long-threaded discussion. Some systems offer something approaching a full document management system with check-in/check-out of documents and version control, and an archive of previous versions. This is essential where multiple authors are expected to be working on the same document, as in project work. In other contexts this may feel unnecessarily cumbersome and inhibit usage of the files. We felt it would be useful to build associated discussion areas around a document or to link to other parts of the system.

A large number of documents being shared can rapidly eat into storage space, possibly affecting system performance—or incurring extra expense if an Application Service Provider (ASP) is hosting the service. It is essential that commonly used document formats are supported with the ability to add new ones. If it is a hosted service, the evaluation should include looking for evidence that they will support new file types as they emerge. In a small community it should be possible and fair to negotiate standards about what authoring software to use. In projects it is common to standardise on a version of Word, for example. Selective alerting of major document changes to named individuals or groups is useful.

Link Store

The ability to add links to the site encourages members of the community to share online resources. Systems that have the ability to create folders and sub-folders with clear headings, ideally with additional details such as who added it, a short description, and the dates and times it was created, can make it even easier for members to access the information. Some systems provide additional features that allow for the creation of personal links, which are only visible to the individual member. This can be useful for members to personalise the site, and can help encourage them to visit and use the site regularly, but at the same time can also create an insular environment and can cause barriers to sharing.

Calendar

The calendar can be used as a record of events and to alert users of community events or relevant events in the outside world.

Task Lists

This lists the tasks and activities of the community, typically chronologically ordered, with additional attributes for start/finish, individuals taking part, and can be subdivided. Some systems provide integration with Calendar and Alerts features.

Help

The most common form of help facility is online documentation, which users can access by clicking the help button and going through a series of links. Ideally, the help button should be prominently situated on the front page so that users can see it clearly, and information should be obtained in no more than three clicks. There should also be context-sensitive help.

It would be useful if the administrator could customise the help screens, to use the local version of the language. In our experience we found that the use of American English proved to have a negative impact on the users in attempting to use help, and therein use the tool and participate in the community.

An addition found was the availability of personalised help through an online 'Live Person' support through a chat-like system. Off-line telephone communications are similar forms of help facilities that some systems offer, usually during normal office hours. The latter should also be considered in light of global time zones and where the provider may be based, as the K-Library community found there was only a short period where any real-time help was available. Finally, whilst useful, it needs due consideration if the service is not free.

Searching

The internal search facility is important once the size of data archived in the community has reached any size. The search facility should ideally be able to search all internal documents, links, and messages, with an option to search by type of file. Some systems we looked at only searched messages. Full-text searching of deposited documents would be ideal, failing that users should be required to attach an abstract at the point of adding the document or link; otherwise the search will only be on document/link titles.

Many systems offered a 'search the Web' option; this only seemed to be of added value if the target search engine could be chosen or if it worked to expand a search on the internal archive.

User Tracking and Statistics

Typically the user tracking functions on most of the systems examined during the study were minimal at best, improving with regard to the size and cost of the system/tool. This feature was more prevalent on the larger and high-end systems/technologies, particularly those with workflow, sign offs, or record management. We wanted to know who was using the site, how often they logged in at what time of day, whether they responded to announcements or alerts by logging in, or primarily used the site habitually at one time of day. It would have been interesting to see whether members were logging in work hours or not. Such tracking is essential, especially as such a high proportion of community members are 'lurkers'.

Clearly there are issues arising from the data protection and privacy perspectives, with this data at an individual level, but pseudonymous data would be invaluable to judge how well the community was working.

Usability and Terminology

It may well be critical to a community how usable the tool or system is perceived to be regardless of whether it is or is not. We found that simple things such as speed of response times may be a key to use of the system. This is obviously a function of many factors: the host system, but also user network connections and the speed of their own machines. For users with a modem connection, a simple e-mail system may be preferred.

Customisation

This is the ability to reorder or personalise the view presented on opening a tool to suit the priorities and interest of the individual user. Typically this is by choosing a theme for the layout or colouring, or the selection via on or off buttons of a series of features available. These methods are not untypical of common applications and grow with the sophistication of the product, especially portal-based solutions.

Security

In a private community a level of security is required, depending on how sensitive the discussions or documents posted are. The option to save one's password as a cookie on one's own machine can be useful.

In the system we used, there was a guest access facility, with very limited rights to view some areas of the site. Ideally there should be scope for the administrator of the community to define precisely what guests can and cannot do at a fine granular level.

Integration and Synchronisation

It most cases the community being created will not be core to participants' work, so it is essential that mechanisms exist to exchange data with other systems, and to maintain some form of version control.

Additional Factors

Clearly in choosing a system or tool, functionality is not the only criteria that is significant or impacts on the selection process.

Cost

There are a number of free services in the marketplace, but there is a price to be paid in using them in terms of intrusive advertising or the risk of the service being withdrawn probably on short notice.

The alternative is to licence software and install it, or to find a third-party ASP who for a fee will run the site for you. The total cost of running a server securely and reliably, along with the support for self-hosting, is likely to be significant.

Third-party services are generally cheaper to set up, but there are likely to be drawbacks. We found limitations on the level of customisation of interfaces and functions, and of usage statistics. Some of these limitations would probably also have existed if we had installed a system ourselves, at least from an out-of-the-box installation. There were limits on the total size of files we could save. We also found that one service (synchronous communication channel) was withdrawn without notice. When we finished using the service, there was no way to archive the site.

Moderation/Facilitation

Is the community to be self-maintaining or does it require a moderator? It should be noted that some systems like e-groups/Yahoo groups require at least one individual to be designated as the administrator/moderator.

Skill Levels

This relates to the technical skills of the community levels of competence and ease with technology, which introduces issues of usability and accessibility, and how adaptable the selected technology is or needs to be. There is also the aspect of the competence and technical skills of those facilitating the community; obviously in-house IS department will have skill levels, if not time. There is an increasing amount of Open Source options, these are packages based on PHP and MYSQL (e.g., Metadot, TYPO3, Invision Power Board). These offer an application-based, typical Web-based front end for the design and administration of a community without the need for detailed knowledge of either PHP or MYSQL, although the ability to customise and differentiate your communities from another design with such a tool is enhanced.

FUTURE ISSUES

As we indicated in the introduction, technological developments continue a pace, with new roles determined for older technologies, and new or recent technologies coming of age as the infrastructure also evolves to cope with their requirements. Changes are reflected in functionality and capability regarding the software, hardware, and the devices we use to access technologies, such as Web-logs—commonly known as blogs—for individual journal/diary-like software publishing via the Internet, which has evolved with a group-based authoring version. RSS feeds are becoming increasingly available, more ubiquitous means of linking and collating disparate information sources and data streams. Many of the 'coffee-and-book' shopping chains offer Internet connections via wireless networks. Link this to the expansion of the 3G mobile telephony networks and the success of the Blackberry handheld device. The technology does not need to be new, but recognition of an existing technology that is underused or could be used differently may be appropriate. The most successful (financially) technology in 2G mobiles was the initially overlooked 'texting' (SMS) facility. Do 3G mobiles have a community-orientated facility yet to be discovered or acknowledged?

CONCLUSION

As with any exercise in gathering users' conscious requirements prior to experience of a specific application, the results can be misleading as a guide to how people will actually use the system that is chosen and value particular functions. The K-Library found this to be an issue with several requested features being turned off, particularly the alerts, although this was followed by comments about not realising anything was happening on the communities online entity! Other analysis is required to identify the full user requirements, but forms a basis for the community, and encouraging involvement and engagement—two prime attributes for creating a successful community.

The checklist tool and descriptions presented here could also be applied to gather users' evaluations of software they have used in a community building exercise, and to triangulate with the results of actual user behaviour captured through site statistics or direct observation of behaviour.

Selecting the most appropriate technology in the form of a system or a tool is a significant part of creating and facilitating a community and can be part of the process of forming the community. It does not have to be sophisticated or heavily featured; it should be related to the community objectives and the community members, its formality or informality, its location outside an organisation or inside an organisation, or bridging organisations or interest groups.

To finish, one should remember Wenger's (2001) consideration that in analysing tools for their suitability to run communities of practice, he argues that "the ideal system [for this purpose] at the right price does not yet exist." It is our hope that our experiences can help you to consider what features your community requires prior to selecting a suitable system/tool for your community.

REFERENCES

Cox, A., Patrick, K., & Abdullah, R. (2003). Seeding a community of interest: The experience of the Knowledge Library project. *Aslib Proceedings, 55*(4), 243-252.

K-Library. (2002). The Knowledge Library project archive. Retrieved from *http://litc.sbu.ac.uk/klibrary/* [note: this Web site is no longer available online].

Lloyd, P., & Boyle, P. (1998). *Web-weaving: Intranets, extranets and strategic alliances.* Boston: Butterworth-Heinemann.

Ruggles, R.L. (1997). *Knowledge management tools.* Boston: Butterworth-Heinemann.

Webopedia. (2004). Definitions Retrieved November 14, 2004, from *http://www.webopedia.com*

Wenger, E. (2001). Supporting communities of practice—A survey of community-oriented technologies. Retrieved November 14, 2004, from *http://www.ewenger.com/tech/executive_summary.htm*

Wenger, E., McDermott, R., & Snyder, W.M. (2002). *Cultivating of practice: A guide to managing knowledge.* Boston: Harvard Business School Press.

Further materials have been taken from the relevant product Web sites, demonstrations, and personal use of versions of the products described, including:

Opentext Livelink—*www.opentext.com*

Groove Networks—*www.groove.net*

Lotus Notes/Domino—*www.lotus.com/notes/*

First Class—*www.firstclass.com*

Hyperwave—*www.hyperwave.com*

Microsoft Sharepoint—*www.microsoft.com/sharepoint/*

Skype—*www.skype.com*

Metadot—*www.metadot.com*

TYPO3—*www.typo3.com*

Invision Power Board—*www.invisionboard.com*

[All accessed May/June 2002 and/or June/November 2004]

Additionally, a number of reviews from a range of computing and Internet magazines have been drawn upon:

.Net, publisher Future Publishing—*www.netmag.co.uk*

Internet Magazine, publisher Emap—www.internet-magazine.com

Internet Works, publisher Future Publishing—*www.iwks.com*

Internet World, publisher Penton Media Inc.—*www.internetworld.com*

Mac User, publisher Dennis Publishing—*www.macuser.co.uk*

Mac Format, publisher Future Publishing—*www.macformat.co.uk*

KEY TERMS

Back-Channel: The standard use of e-mail, person to person, without routing through the community's available channels.

Blog: A Web page that serves as a publicly accessible personal journal for an individual. Typically updated daily, blogs often reflect the personality of the author (Webopedia, 2004).

DHTML: Short for Dynamic Hypertext Markup Language; presents the same Web page differently (dynamically) each time it is viewed, based on a range of possible parameters (Webopedia, 2005).

Emoticon: A small icon composed of keyboard characters and used in e-mails and instant messaging that indicates the mood and/or emotion of the writer. This is typically a representation of facial expressions, such as a smiley :-) (Webopedia, 2004).

Internet Telephony: A category of hardware and software that enables people to use the Internet as the transmission medium for telephone calls. For users who have free or fixed-price Internet access. Voice Over the Internet (VOI) or Voice Over IP (VOIP) products (Webopedia, 2004).

Java: A high-level object-orientated programming language devised by Sun Microsystems, whilst a propriety language has become accepted as a de-facto standard (Webopedia, 2004).

Lurker: A participant who reads posts and monitors the communities' activities without directly or explicitly participating. Often and increasingly viewed as a negative due to the lack of contribution; however, the individual may utilise learning gained from monitoring the activities.

MYSQL: An open source relational database management system (RDBMS) that relies on SQL for processing the data in the database (Webopedia, 2004).

PHP: Short for Hypertext Preprocessor, an open source, server-side, HTML-embedded scripting language used to create dynamic Web pages (Webopedia 2004).

RSS: Short for RDF Site Summary or Rich Site Summary, an XML-based format for syndicating Web content (Webopedia, 2004).

SGML: Short for Standard Generalized Markup Language; a system for organizing and tagging elements of a document developed and standardized by the International Organization for Standards (ISO) in 1986 (ISO8879). SGML itself does not specify any particular formatting; rather, it specifies the rules for tagging elements. These tags can then be interpreted to format elements in different ways (Webopedia, 2004).

SMS: Short for Short Message Service; similar to paging, SMS is a service for sending short text messages to mobile phones.

SQL: Short for Structured Query Language, a database query language that was adopted as an industry standard in 1986 (Webopedia, 2004).

Thread: A series of messages that have been posted as replies to each other (Webopedia, 2004).

XHTML: Short for Extensible Hypertext Markup Language, a hybrid between HTML and XML specifically designed for Net device displays. XHTML is a mark-up language written in XML; therefore, it is an XML application (Webopedia, 2004).

XML: Short for Extensible Markup Language (XML), a specification developed by the W3C. XML is a pared-down version of SGML, designed especially for Web documents. It allows designers to create their own customized tags, enabling the definition, transmission, validation, and interpretation of data between applications and between organizations (Webopedia, 2004).

APPENDIX: A COMMUNITY AND FEATURES CHECKLIST

Function	Attributes	Essential	Desirable	Neutral
1. Asynchronous Channel				
Pushed E-Mail	Individual			
	Daily Digest			
Bulletin Board				
	Threading			
	Emoticons			
	Formatting			
Anonymous Posting Option				
2. Synchronous Channel				
Whiteboard				
Instant Messaging				
Chat				
	Automatic Archiving			
Who-Is-Logged-On-Now Check				
Shared Surfing				
Conferencing				
	Video			
	Audio			
3. Announcements				
Posted To Homepage?				
4. What's New				
Since Last Log In				
In the Last n Days				
Customisation of Items Monitored				
5. Alerts				
For New Documents				
For New Polls				
Tasks				
Messages				
6. Polling				
Can Set Open Other Answer				
User Can Only Vote Once				
User Can Change Vote				
Pre-Set Closing Time				
7. Membership Directory				
Customisation of Fields in Entries				
Customisation of What Are Required Fields				
Photo Can Be Added				
Links or Documents				
8. Contacts List				
9. Document Repository				
Metadata Required				
PDF, Word Supported				
Check In/Version Control				
Discussion Can Be Directly Associated with Document				
Directory Structure				

APPENDIX: A COMMUNITY AND FEATURES CHECKLIST, CONT.

16. Usability	Server Response Time <1 Second		
17. Customisation			
18. Security	Encrypted Sessions		
	Cookie-Based Password Save		
19. Integration and Synchronisation			
20. Technical Requirements	Support Netscape Pre-Version 6		
	Administrators Only		
21. Cost	(For Self-Hosted)	Licence Cost	
		Cost of Support	
	(For Third-Party Service)		
		Cost Per User	
		One Off Costs	
16. User Tracking	Total Usage Per Individual Over Time		
	Analysis of Usage by Time of Day		
	Paths Taken Through Site		
	Most Popular/Least Popular Pages Analysis		
	Data Protection Compliance		

10. Link Store	Metadata Required		
	Directory Structure		
11. Calendar			
12. Task Lists	Individual		
	Activity Based		
13. Help	System		
	Community		
	Context Sensitive		
	Local Customisation by Administrator		
	Online Help		
14. Internal Searching	Includes Full Text of Messages and Metadata on Documents/Links		
	Includes Full Text of All Files		
	Full Boolean		
	Search Can Be Limited by Type of File, e.g., E-Mails or Documents		
15. External Search	Choice of Search Engine		
	Automatic Expansion of Internal Search		

Facilitating and Improving Organisational Community Life

Angela Lacerda Nobre
ESCE-IPS, Portugal

INTRODUCTION: A CHANGE OF PARADIGM

The growth in importance of communities within organisational settings is a sign of a change in paradigm. When management and organisational theory introduce the critical notion of communities, in parallel to the concepts of collaborative work and of knowledge sharing, there is an internal revolution going on. Therefore, communities of practice theory (Lave & Wenger, 1991; Wenger, 1999; Wenger, McDermott & Snyder, 2002; Brown & Duguid, 1991) has a critical role to play in today's development of management and organisation theory.

At a broader level, there is an ongoing metamorphosis that is highly visible through the vertiginous development of technology, the globalisation of markets, and the acceleration of the increase in complexity. Equally important are the less visible, and thus harder to acknowledge, changes in the way we think, reason, communicate, and construct our image of ourselves and of the world.

The changes brought by the knowledge society of the information age (Kearmally, 1999) triggered the development of theoretical approaches to management. Among these, *knowledge management* and *organisational learning* have developed. These theories have acknowledged the importance of information and communication technology within organisations, and have explored alternative insights into mainstream management approaches. The knowledge management and organisational learning sub-disciplines represent an innovation effort that affect areas of organisational life which had been marginalised or ignored under traditional management theory. Communities of practice is the single most important example. Therefore, communities of practice represent a critical aspect of the present understanding of the complexity of organisational life.

Within the broad and varied development of organisational theories, *semiotic learning* emerges as a particular approach to organisational learning. Semiotic learning may be described as a dynamic practice. It incorporates theoretical contributions from social philosophy and adapts them to a specific approach to facilitate learning at the organisational level. It is a learning and development tool for action at the organisational level. The central aspect of the semiotic learning approach is the focus on the quality of community life at the organisational level.

Through a semiotic learning approach to organisational learning and development, it is possible to intensify and to unleash the true potential of current challenges at personal, organisational, and societal levels. By focusing on the social practices, structures, and processes which underlay human interaction, and by calling attention to the way we construct ourselves and our image of the world through those interactions, it enables the development of a rationale that supports collaborative as well as transformative forms of work and learning.

BACKGROUND: SUBTLE AND HIDDEN NEEDS

The contribution of the semiotic learning approach to the fields of organisational learning, knowledge management, and communities of practice is that it offers an integrative theoretical approach. The organisational learning, knowledge management, and communities of practice theories deal with issues which they cannot themselves explain, or rather with issues which they cannot *yet* explain. The call for a greater depth and breadth in terms of theoretical grounding may be achieved by the semiotic learning perspective. This approach consists of an organisational design and organisational development instrument to be implemented in parallel with other existing initiatives. The rationale is simple and direct: it aims at recovering the balance between the necessary functionalistic efforts, and the subtle and hidden needs of the organisation's community life. It is the projection of powerful insights arising from philosophy mediated, translated, and adapted to organisational reality. It is based on both theory and practice, as it is through its application that it becomes reified. The theory, the description, the narrative is just a means to an end, an end which is an action-based and action-led organisation.

The semiotic learning approach rests on three broad groups of theories: social semiotics, critical realism, and action theory. The semiotic learning approach works as a cascade, so that within social semiotics, it includes Bakhtin Circle's social theory of discourse; within critical realism, it takes a pragmatic perspective; and within action theory, it questions current epistemological positions and recovers an ontological and hermeneutic perspective. The breadth of theory is directed at informing and illustrating how rich and diverse the universe of options is in terms of approaches that directly answer to the 'subtle and hidden' needs of organisational life. This diversity could even be extended, never reduced. It is this diversity which expands our thought horizons so that it becomes a kind of didactic or pedagogic process, thus the word "learning" in semiotic learning. This corresponds to the "thought-possibilities" and "action-possibilities" that Jaspers explored.

Heidegger's disciple, Karl Jaspers, explored thought-possibilities and action-possibilities following the guide of Max Weber's approach as an historian:

...in order to grasp reality, we must see the possibilities...Weber employs the category of 'objective possibility' in his historical appraisal of past situations. The historian considers a situation. His knowledge enables him to construct the possibilities of the day. By these constructions he first measures the possibilities of which protagonists were aware. And then, by the possibilities, he measures what really happened, in order to ask: for what specific reason did a particular possibility among several materialise? (Young-Bruehl, 1981)

Theoretical approaches such as critical realism (Archer, Bhaskar, Collier, Lawson & Norrie, 1998), complex systems theory (Chekland, 1999), social semiotics (Halliday, 1978; Lemke, 1984, 1995), or hermeneutics (Ricoeur, 1998) did not develop as an answer to today's virtual, fast-paced, and often chaotic communication forms. However, their insights represent a powerful tool in order to understand, cope with, and enable one to profit from the opportunities that are being opened by the knowledge society. Similar to the implicit unity between the individual and the social, there is a close connection between theory and practice. All these theoretical approaches have an intrinsic pragmatic nature and therefore do not separate the individual from the social or the theory from the practice.

It is within communities that organisations continually create and redefine meaning, and it is this meaning-making capacity that conditions the organisational identity, degree of cohesion, and potential to innovate. Meaning-making is part of the action- and thought-possibilities, part of the horizon of possibilities that is open through the practices and processes of organisational community life. Semiotic theory is unavoidably related to meaning—and therefore the importance of the term "semiotic" in semiotic learning.

MAIN FOCUS: MEANING-MAKING AND SOCIAL SEMIOTICS

Social semiotics developed out of the work of sociologists interested in language issues and of linguistics interested in the social influences within language use. Under this perspective, human development is as much the development of individuals as that of the social communities to which they belong, and language is the working tool and enabler of this process. Semiotics is commonly related to language, though it covers all forms of communication or rather 'characterisation' of a practice so that dressing, teenage gear, wrestling, or cooking have a semiotic content. Barthes (1996) developed this approach, including studies of advertising, media, and cinema. From another perspective and according to Umberto Eco, the implicit domain of semiotics is the whole history. Sebeok (1994) and other authors study semiotics in all life forms, because the ability to manufacture and recognise signs is a basic survival strategy. So they study biosemiotics, zoosemiotics, semiochemistry, and phytosemiosis. Semiotics was already present in ancient Greece through the works of Plato and Aristotle. In the Middle Ages, it continued to develop through the works of Augustine in the fourth and fifth centuries, and Ockham in the fourteenth century. Locke, in the seventeenth century, also focused on signifying processes.

Semiotics as a discipline is simply the analysis of signs or the study of sign systems. The idea that sign systems are of great consequence is easy enough to grasp. Yet the recognition of the need to study sign systems belongs to the modern age. A full-blown semiotic awareness arose at the turn of the century, and in the early 20th century, through the works of Saussure in Europe and Peirce in North America. Different schools of thought emerged from these roots, giving rise to diverse currents that deeply influenced both the linguistic turn and the context turn throughout the twentieth century. From Saussure's work, structuralism developed, as well as different branches, including one that would give rise to social semiotics. From Peirce's work, pragmatism developed, which was one of Peirce's creation, later followed by James, Dewey, Popper, Morris, Sellars, Putman, and others (Delanty & Strydom, 2003).

Jay Lemke is a contemporary social semiotitian who started as a physicist. According to Lemke (1995), Bourdieu's and Bernstein's work as sociologists, between 1970 and 1990, and Halliday's and Gunther Kress' work as a linguists, during more or less the same period, gave rise to that which is known today as social semiotics.

*Halliday's social theory of discourse suggests that our uses of language are inseparable from the social functions, the social contexts of actions and relationships in which language plays its part...This is what is meant by seeing language as a **social semiotic**, a resource to be deployed for social purposes.* (Lemke, 1995, p. 27; emphasis used in original text)

Lemke also relates the origins of social semiotics with the work on discursive formations by Foucault and, most importantly, to the early works of the Russian Bakhtin's Circle starting in 1918 (Lemke, 1984, 1995).

The works of Bakhtin and his colleagues introduced highly rich and complex terms and concepts which capture meaning, or cultural and semiotic related issues within texts, which can be literary texts, or else any social reality that may be read and interpreted as a text (Brandist, 2002). Bakhtin's work became known internationally in the 1970s; in the 1980s it spread within the areas of literary and cultural theory, and since then it has continued to disseminate, now influencing social theory, philosophy, and psychology. Bakhtin's four essays (1984), edited by Holquist, with 12 prints until 2000, were published in Moscow in 1975, after his death, and were written in the late 1920s and in the 1930s. Concepts such as heteroglossia, intertextuality, chronotope, dialogue, and dialogic relationships are some of the key concepts of Bakhtin's group of thinkers. Lemke (1995) refers to five major theories of discourse which have emphasised its social dimension: M. Bakhtin was the first one, then M. Foucault, M. Halliday, B. Bernstein, and finally P. Bourdieu, though he was not a discourse theorist.

Although semiotics is a wide area, enabling the coexistence of radically different approaches, including reminiscences of past positivist trends, social semiotics represents the cultural, historical, political, and social readings, interpretations, and meaning-making processes present in all symbolic systems. It is this richness as well as this complexity which makes social semiotics an attractive and powerful theoretical framework for organisational learning initiatives.

FUTURE: THE MEDIATION ROLE OF SEMIOTIC LEARNING

Social semiotics, together with hermeneutics and critical realism, acknowledge the importance of social structures in determining social practices and vice-versa. The notions of self and agency are intrinsically connected with how individuals are determined, and how they themselves determine the social contexts to which they belong. Pragmatism stresses the importance of integrating the individual and social dimensions into a single whole.

The semiotic learning approach takes into account these contributions and mediates their insights by adapting them to an organisational context.

Critical realism is a multidisciplinary movement in philosophy and the human sciences which started with Roy Bhaskar's publication in 1975 of *A Realist Theory of Science* (Archer et al., 1998). The term 'critical' suggests affinities with Kant's idealism and rationalism, though the term 'realism' indicates that there are fundamental differences. Bhaskar's philosophy is reflexive as transformatively practical, or presents a transformational model of social activity. According to this theory, social life has a recursive character, as agents reproduce and transform the social structures they use and are constrained by in their substantive activities.

...actors' accounts are both corrigible and limited by the existence of unacknowledged conditions, unintended consequences, tacit skills and unconscious motivations...but, in opposition to the positivist view, actors' accounts form the indispensable starting point of social enquiry. (Archer et al., 1998, p. xvi)

Critical realism, as a theory of science, presents three distinctive features:

(i) it recognises science as a social practice, and scientific knowledge as a social product; (ii) it recognises the independent existence of the objects of scientific knowledge; (iii) it has an account of scientific experiment and discovery as simultaneously material and social practices, in virtue of which both (i) and (ii) are sustained. (Benton & Craib, 2001, p. 130)

The fundamental links between individual and social arenas is a central issue to an organisational learning initiative. On one hand, individual action and thought—and the social structures and practices that contributed in the shaping and moulding of that action and that thought—and, on the other hand, the human agency capacity to strike back—to reconfigure or sustain those social structures and practices—may be acknowledged and recognised through the use and application of social philosophy theory.

The semiotic learning approach focuses on four philosophical categories: action, language, knowledge, and meaning. It claims that knowledge and meaning are already prefigured by action and language. Communities of practice theory implicitly acknowledges the social embeddeness of individual cognitive processes. In order to further develop the importance of the relations between social and individual processes, the semiotic learning approach uses six working concepts that are derived from the works of six social philosophers. These are Bakhtin's concept of dialogism (1981), Halliday's notion of grammar, Wittgenstein's idea of language-games (1958), Foucault's discursive formations (1972), Heidegger's instance of being-in-the-world (1996), and White's master tropes (1978). These philosophical contributions are introduced into four learning steps that represent the practical and applied nature of the semiotic learning approach. These are: *icebreak*—raising key issues; *experiencing*—confronting reality; *action horizons*—transformative learning; and *innovative practice*—open dynamism. There is a mediation between central notions of philosophical inquiry and key organisational issues. These key issues are: appreciative inquiry, open complex systems, socio-technical systems, collaborative work and learning, knowledge creation and sharing, reflexive practice and double-loop learning, and trust and social capital.

CONCLUSION

To finalise, two comments: Probably the best image to describe the subtle and hidden issues that the semiotic learning approach claims to tackle in order to facilitate organisational learning is that of the gap between what an organisation is supposed to be, its formal and public discourses, and the way it is felt and lived by the people actually involved, both directly and indirectly, from within and from without, from above and from below. The second issue is that the semiotic learning perspective reflects and projects a continual questioning and a work in progress—not in the sense of an unfinished work but in terms of its internal dynamic; its furthest aim is the broadening of horizons and the revolution of mentalities.

The subtle and hidden thought revolution starts within communities in general, and within communities of practice in particular. It is through communities that it reaches people at an individual level. If knowledge is in people's heads, communities of practice tell us how it got there.

REFERENCES

Archer, M., Bhaskar, R., Collier, A., Lawson, T., & Norrie, A. (1998). *Critical realism*. London: Routledge.

Bakhtin, M. (1981). *The dialogic imagination*. University of Texas Press.

Barthes, R. (1996). *Mythologies*. London: Vintage.

Benton, T., & Craib, I. (2001). *Philosophy of social science*. UK: Palgrave.

Brandist, C. (2002). *The Bakhtin Circle: Philosophy, culture and politics*. London: Pluto.

Brown, J., & Duguid, P. (2000). *The social life of information*. Boston: Harvard Business School Press.

Brown, J., & Duguid, P. (2001). Knowledge and organisation: A social-practice perspective. *Organisational Science, 12*(2), 198-213.

Chekland, P. (1999). *Soft systems methodology: A 30-year perspective*. West Sussex, UK: John Wiley & Sons.

Delanty, G. (2003). Michel Foucault. In A. Elliot & L. Ray (Eds.), *Key contemporary social theorists*. Oxford, UK: Blackwell.

Delanty, G., & Strydom, P. (2003). *Philosophies of social science*. Berkshire, UK: McGraw-Hill.

Foucault, M. (1970). *The order of things: An archaeology of the human sciences*. London: Tavistock.

Foucault, M. (1972). *The archaeology of knowledge*. London: Tavistok.

Halliday, M. (1978). *Language as social semiotic*. Victoria, Australia: Open University.

Heidegger, M. (1996). *Being and time*. Albany: State University of New York. Original publication [1927] Max Niemeyer Verlag, Tübingen, Germany.

Kearmally, S. (1999). *When economics means business*. London: Financial Times Management.

Lave, J., & Wenger, E. (1991). *Situated learning: Legitimate and peripheral participation*. Cambridge: Cambridge University Press.

Lemke, J. (1984). *Semiotics and education*. Toronto: Victoria College/Toronto Semiotic Circle Monographs.

Lemke, J. (1995). *Textual politics, discourse and social dynamics*. London: Taylor & Francis.

McHoul, A., & Grace, W. (1993). *A Foucault primer: Discourse, power and the subject*. New York: New York University Press.

Ricoeur, P. (1998). *Hermeneutics and the human sciences*. Cambridge, UK: Cambridge University Press.

Sebeok, T. (1994). *Signs: An introduction to semiotics*. Toronto, Canada: University of Toronto Press.

Wenger, E. (1999). *Communities of practice: Learning, meaning and identity*. Cambridge, MA: Cambridge University Press.

Wenger, E., McDermott, M., & Snyder, W. (2002). *Cultivating communities of practice*. Boston: Harvard Business School Press.

White, H. (1978). *Tropics of discourse: Essays in cultural criticism*. Baltimore: Johns Hopkins University Press.

Wittgenstein, L. (1958 [1953]). *The blue and brown books. Preliminary studies for the "philosophical investigations."* G. Anscombe (trans.). London: Basil Blackwell.

Young-Bruehl, E. (1981). *Freedom and Karl Jaspers philosophy*. Alpine.

KEY TERMS

Abduction, Induction, and Deduction: Abduction is the process by which a new concept is formed on the basis of an existing concept that is perceived as having something in common with it. Therefore abduction focuses on associations. Induction is the process of deriving a concept from particular facts or instances. Inductive knowledge is empirical knowledge, of facts and information. Deductive knowledge is formal rationalism, mathematical knowledge, and logical reasoning.

Critical Realism: One branch of social science that developed in the 1970s and which poses fertile intellectual challenges to current understanding of communities, organisations, and other social structures. It argues that without the concept of a social structure, we cannot make sense of persons as any predicate which applies to individuals, apart from a direct physical description, and presupposes a social structure behind it. Though we need the notion of a social structure, the only way to acknowledge it is through the social practices that it incarnates and reifies, which in turn are embedded in the actions of its members. A social structure is not visible or witnessable, only its social practices are. Though implicit and invisible, structures are enabling or constraining, as they open up or else severely restrict the actions of its members. However, structures are not simply a medium for social practices, as these practices also change and influence the structures themselves. This implies that structures are both a medium and a product of its practices. Social structures are reproduced and transformed by the practices of its members. Thus individuals have an agency capacity to interfere back, and thus promote social change—not isolated individuals, however, but units and collectivities of individuals. Individuals are persons, and their acts are situated in a world constituted by past and present human activity, thus a humanised natural and social world. Because social structures are incarnate in the practices of its members, this means that they do not exist independently of the conceptions of the persons whose activities constitute, and thus reproduce or transform them. It is because persons have beliefs, interests, goals, and practical and tacit knowledge, not necessarily cognitively available, acquired in their early stages as members of a society, that they do what they do and thus sustain, or transform, the structures to which they belong. Critical realism thus proposes a transformational model of social activity. It states that reality exists independently of our knowledge about it, and it takes science as a social practice, and scientific knowledge as a social prod-

uct. Communities of practice theory implicitly incorporates a critical realism perspective.

Hermeneutics: A branch of philosophy which studies interpretation processes. Methodology is the choice of the adequate method for performing research or any scientific activity. Hermeneutics can be understood as a parallel and complementary process to a method. The method identifies, organises, and orders the necessary steps to accomplish a certain activity and a fixed objective. And hermeneutics calls attention to the intrinsic necessity of constantly interpreting the laws and adapting the rules, norms, and indications given by the method to the idiosyncrasy of a concrete situation, the context, the situatedness, and the horizons of the interpreting community. Hermeneutics developed out of the interpretation of sacred texts. It started from relatively rigid and formalised procedures, and then developed to highly dynamic and flexible approaches. The idea behind modern hermeneutics, developed from the 1950s onwards, though with roots from the late nineteenth century, is that all reality is a text analogue so that it may be read and interpreted as a text. This interpretation process, again, is not rigid and static, but rather it has an ontological and epistemological dimension. Through interpretation, reality manifests itself ontologically, and through this process the resulting knowledge is organised in an epistemic way, that is, creating meaningful structures and conceptualised hierarchies. This approach to hermeneutics proposes a highly creative, constructive, and transformative process of dealing with human interpretation of reality, thus its potential for fertile change and development at the community and organisational levels.

Pragmatism: Derives from the Greek word *pragma* which means action. Emphasises the concept of human beings as agents, and focuses on their practical relation to the world. The principle according to which experience forms the basis of all knowledge is shared by pragmatists and empiricists, these later including positivists and neo-positivists. Pragmatists contrast with positivists by the former focusing on creative inquiring and the latter on passive observation. At a deeper level, the contrast is even greater because of pragmatism's origin in Peirce's critique of Descartes, and the overcoming of precisely those Cartesian dualisms which are presupposed by modern western philosophy, including positivism. Examples of these dualisms are subject and object, body and mind, perception and conceptualisation, theory and fact, fact and value, deduction and induction, reality and copy, nature and culture, individual and society, sign and signified, and so forth. To overcome these dualisms, pragmatism rejects some of the basic guiding ideas which inform not only positivism, but also interpretative and structuralist traditions. Among these are the notions such as the subject of knowledge as an individual, observation as presuppositionless activity, truth as a picture or representation corresponding with reality, knowledge as being built up of observation and logical inference, social science as being exclusively concerned with culture (and hence the interpretation and understanding of symbolic meaning), knowledge as involving an arbitrary or conventional twofold sign relation, and so on. Pragmatism, by contrast, stresses the anchorage of knowledge in real collective problems, and knowledge as being dependent on the mediation of signs, which means that it regards knowledge as being social by nature. It focuses on the development of knowledge which it sees as taking place in different ways and in a variety of contexts. Pragmatism in centred on abduction, not induction nor deduction, and not only on the individual creativity, but rather on the cooperative search for truth within a community through interpretation, discussion, and argumentation—that is, through the creative collective overcoming of action problems.

Semiotic Learning: A dynamic-practice approach to facilitate organisational learning. It focuses on improving and developing the quality of organisational community life. The semiotic learning approach is directed to recovering the balance between the necessary mechanistic, utilitarian, transactional, and functionalistic procedures present in all kinds of organisations, and the less visible, implicit, subtle, yet powerful issues related to the meaning-creation process which constitutively sustains any organisation and forms the bedrock of its community. The semiotic learning approach is supported and informed by social semiotics, critical realism, and action theory. It mediates philosophical forms of inquiry and adapts them to organisational contexts. Organisational learning promotes environments and establishes an organisational design that is condu-

cive to knowledge creation and sharing, as well as to collaborative forms of work and learning. Semiotic learning goes beyond traditional organisational learning and knowledge management approaches by capturing the dynamism, innovation, and creativity spontaneously present in every organisational context, and turning this potential into effective and meaningful action at the organisational level. Therefore this approach is action based and action driven.

Semiotics: The science of signs. It covers the analysis of signs and the study of sign systems. A sign is something that stands for something else. So semiotics can also be understood as the study of meaning-making or semiosis. A culture is the system of daily living that is held together by a signifying order, including signs, codes, texts, and connective forms. Semiotics is not just a theory, but it is a common and unavoidable practice present in all forms of communication. Thus the study of semiotics in human communication may be a form of studying cultural anthropology. Semiotics can have a positivist application when focusing solely on the formal, explicit, and visible structures of language or other sign systems, or else it may take a more interpretative approach when focusing on the dynamic and transformative nature of the same sign systems.

Social Semiotics: Explicitly takes a non-positivist approach as it focuses on the contexts, prerequisites, and conditions of possibility for a meaning creation process to occur. All meanings are made within a community. The analysis of sign systems and of sense-making processes cannot be separated from the social, historical, cultural, and political dimensions of these communities. Social semiotics also takes a non-cognitivist approach: instead of referring solely to meaning-making as the result of minds and of brain processes, it points to the role of social practices within communities. Communities are thus interpreted not as a collection of interacting individuals, but as a system of interdependent social practices. Social semiotics may be understood as a discourse on meaning-making where the aim is to examine the functions and the effects of the meanings we make in everyday life, within communities, organisations, and society.

Textuality, Intertextuality, and Contextuality: In philosophical and ontological terms, it is within a text that a discourse is woven, thus realising, by the nature of its own movement, the topography of the world and the maximum meaning of reality. Philosophy not only is linguistic in nature, but it is also characteristically literary, which means that it has to become involved in the fundamental conditions of the functioning of language. Philosophy, or rather the philosophical discourse, could be said to lead language to achieve its best expression, in the sense that it realises its fundamental aim, which is to manifest the broadest sense possible. Textualisation is part of the global action of emergence of meaning; humans enter this action as agents and as a result of the process itself. Where the act of textualisation is more visible and apprehensible, taking the perspective of its comprehension, and where we can measure the extent and reach of the intertextualisation, is at the fundamental moment of contextualisation. Intertextuality exists when there are referents present in one text which allude to referents in other texts. Intertextualisation is the intertwining of different texts, the relationship between texts. Each text meaning or action is generated by a previous text meaning and action. If authors are texts, action has stronger reason to be a text, as it is the distension of meaning from where other meaning may emerge. Contextualisation takes the work of the manifestation of meaning—the textualisation—and integrates it within the continuity of the meaning previously given. Contextualisation is the organisation of a text: it deepens the world of the interpreter; it captures the vectors of different texts in their fusion; and it produces a new text, which reflects the increase in meaning. Contextualisation shows the dynamic nature of the manifestation of meaning: the intentionality that transverses it; the active participation of the interpreter; and the historical, temporal, differentiated, and simultaneously global nature of that manifestation.

Facilitating Technology Transfer Among Engineering Community Members

Barry A. Cumbie
Auburn University, USA

Chetan S. Sankar
Auburn University, USA

P.K. Raju
Auburn University, USA

INTRODUCTION

A report, entitled "Benchmarking University-Industry Technology Transfer in the South and the EPSCoR States," by Waugaman and Tornatzky (1998) found that universities in Alabama have a poor technology-transfer record when compared to their counterparts in other states. Not a single university was listed among the "best-in-class" universities that produced significant economic impact among the 72 universities that were analyzed in this report. This report notes that those universities that emphasize technology development as an essential extension of basic research will generally also be more competitive in research. This problem of a poor technology-transfer seems strange given the enormous effort put forth by many individuals and organizations at state and local levels. The problem seems to arise not from any one group's deficiencies, but rather from the fragmentation of effort across a wide spectrum. Better communication among non-affiliated groups could increase the effectiveness of technology transfer, thus potentially creating a significant positive economic impact in Alabama.

To change this trend and position, the Auburn University College of Engineering, as a leader in

Figure 1. ISIRP organization chart

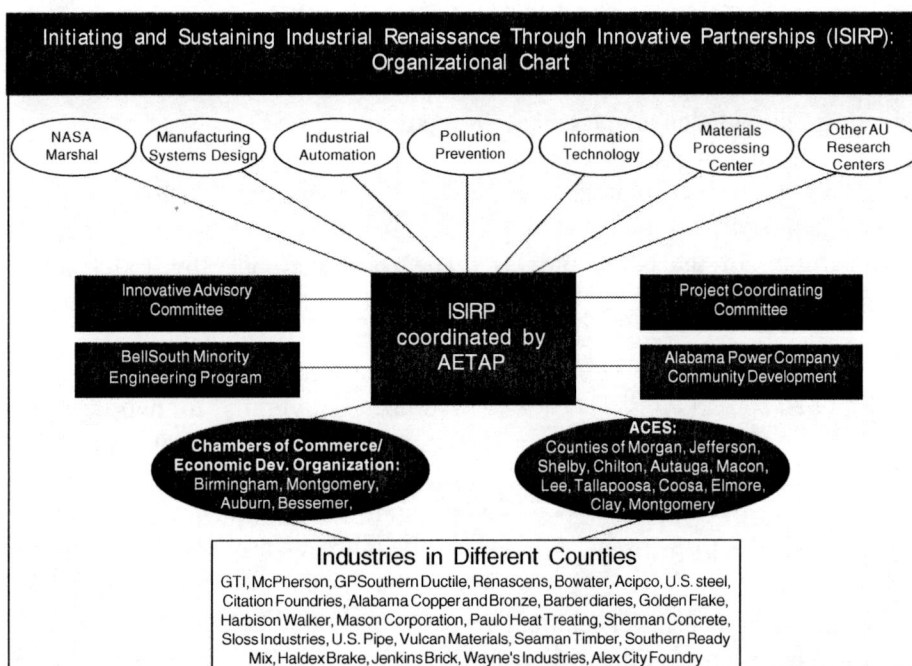

technology transfer, formed a partnership among Auburn University, NASA Marshall Space Flight Center, state agencies, local governments, and owners of industries in targeted counties (Figure 1). Such a partnership has become yet more critical as Alabama has lost approximately 48,000 jobs from over 273 plant layoffs and closings since 1998 (State Dislocated Worker Unit, 2002; Vision of Alabama, 1994; Whittington, 2002). In addition, the partnership has the ability to address the strong need to incorporate technology development and transfer into engineering education (Wyckoff & Tornatzky, 1998; National Science Board, 2000; Fox, 1984; Tornatzky, Lovelace, Denis, George & Eliezer, 1999).

By establishing partnerships across the groups of the community, all members will be aware of what the others are doing so that efforts are not duplicated and a greater impact is achieved. The infrastructure of this community will consist of people—county agents of the Alabama Cooperative Extension System (ACES), students, and researchers—and an information system component described as a knowledge sharing system (KSS). Auburn's Engineering Technical Assistance Program (AETAP) has taken the leadership role in forging community ties and developing the KSS as the centerpiece of the community.

The next section will describe the organizations that comprise the engineering community, followed by a profile of the community members. Subsequent sections will describe opportunities for improvement within the community, and how the KSS addresses these opportunities by creating a supportive virtual community and enacting a knowledge management (KM) strategy. The next section then discusses the relationship between communities of practice and the KSS. The last section reports the current status and looks toward the future of the KSS, and is followed by definitions of key terms.

BACKGROUND

Engineering Community Organizations

AETAP (www.eng.auburn.edu/aetap)

During 1997, the Auburn Engineering Technical Assistance Program, originally a part of the Auburn

Industrial Extension System (AIES) and later designated as AETAP in 2000, was conceived as a unit of the College of Engineering to help increase Alabama's industrial competitiveness in the rapidly changing world of international commerce. The infrastructure used by AETAP for technology transfer consists of identifying industry needs and fulfilling them through technical assistance and technology-transfer projects performed by researchers. The unit also maintains a Web site that provides information on the services offered by this unit. Since its inception, this unit has undertaken 716 projects providing technical assistance to industries. More than 40 researchers participated in these projects. During 1997-2003, a total of $1.2 million was generated to support this program through a combination of company-specific projects and funding from NASA, NIST, and other sponsoring agencies. AETAP was also involved in managing NASA technology-transfer programs to industries. At present, most of AETAP's efforts are focused on mobilizing the College of Engineering's resources to form new partnerships to provide technical assistance to industries.

During 2003, AETAP received a grant from the National Science Foundation and other funding sources totaling $1.3 million to improve Alabama's economic outlook by improving technology transfer, and to create and sustain innovative partnerships among members of the region's engineering community. The title of this project is Initiating and Sustaining Industrial Renaissance through Innovative Partnerships (ISIRP). AETAP manages and supports this project and the community by implementing the KSS.

ACES (www.aces.edu)

The Alabama Cooperative Extension System (ACES) is a statewide continuing education and technical assistance network that links the land-grant university campuses at Alabama A&M University and Auburn University to people and communities throughout Alabama. Alabama A&M University is a historically black college, and Auburn University is the state's 1862 land grant university. ACES serves as the outreach arm for both universities. Throughout the system of county offices, ACES personnel are accustomed to working with their local community and business leaders. This relationship places

ACES personnel in an advantageous position with respect to serving as contact agents for facilitating the technology-transfer function.

Engineering Community Members

University Faculty

These are primarily professors working with AETAP who interact with students and industry in several ways. They meet with students in a typical classroom setting, but also guide senior-level capstone projects that aim to solve real industry problems from a participating company. Graduate students participate in more in-depth projects that involve working closely with faculty and company personnel. To gain partners from the industry, faculty contact companies in an informal process to inquire about potential capstone projects. Faculty may also interact with people in industry by providing their expertise and resources in a consultant role.

University Students

Undergraduate students interact with their professors most often within the classroom while the students learn concepts of a specific engineering field. In their senior year, students participate in a senior capstone project with a group of their peers supervised by a faculty member, to design a prototype that will meet the specifications delivered from a company with a specific problem. Graduate students also interact with industry, often working onsite, to apply their skills in a real-world setting, thus providing an enhanced level of education.

County Agents

County agents employed by ACES are located in many counties of Alabama to serve as outreach liaisons for land-grant universities in Alabama. Individual agents create and sustain close relationships with industries located within their county. They stay abreast of research coming from the universities in their areas of concentration, such as Community and Economic Development, and create programs to economically grow their communities. Agents stay in touch with university faculty mainly on an ad hoc

basis—that is, placing a phone call in response to an industry inquiry.

Industrial Contacts

Industry contacts mainly refer to representatives of small and medium-sized manufacturers. Industry contacts hold relationships primarily with county agents but may also have direct contact with faculty.

Community Shortcomings

The current interactions of the community have many positive characteristics, including the indirect connectivity of all members and the specialized role that members serve. The engineering community members interact without established formal communications, which in turn creates communication silos and restricts access among members who are not part of the informal networks. The role of informal communication within a community is certainly important. Within the engineering community, faculty, industry, and county agents often collaborate on an informal basis with favorable results. Despite past successes, opportunities for improvement were discovered from extensive interactions with the members of the engineering community. A research team participated in weekly AETAP meetings for over one year, presented ideas at two ACES workshops, conducted one-on-one interviews with AETAP faculty and ACES county agents, as well as interacted with industry partners. From this close interaction with the engineering community, three major deficiencies were identified: the lack of accountability, the lack of manageability, and the lack of transferability:

- **Create Accountability:** A county agent who interacts with an industry contact in need of assistance may phone a faculty member. After the initial phone call, the county agent is placed out of the communication loop, and does not know if and to what extent the faculty member followed up with the company. If no follow-up occurs, the county agent's relationship with the company may become strained.

- **Allow Manageability:** From a project management standpoint, if no record of communication or action exists, then managing the project is very difficult. The manageability aspect is very important, not only in properly allocating resources, but also in the justification of supporting a community and creating sustainability.
- **Establish Transferability:** Within all communities people are mobile, entering and leaving as circumstances dictate. The entry or departure of an individual within a community can create confusion if the primary means of communication is informal. Recently the ACES organization underwent a restructuring in which many long-term employees retired. The faculty who often worked with retiring agents feared that their contacts might be lost, as no formal record had been made.

In reaction to the three identified opportunities for improvement, a process to formalize the communication within the community was proposed. Formalizing communication allows the county agent to remain in the loop and:

- creates accountability along the communication chain;
- generates records as to when, in what manner, what content was discussed, and what was the result of the communication to lead to better manageability; and
- allows for a smooth transition as the community membership changes.

The proposed solution included the development of an information system, the KSS, based upon the latest information technology.

Establishing the Community

Building a Non-Virtual Community

The members of this engineering community potentially stand to benefit greatly by progressing beyond informal connection to a formal community as supported by AETAP. A formal community with formal means of communication, the KSS will lead to manageability and a thoughtful allocation of re-

sources. Even so, the intention of the KSS is not to create a virtual community that will wholly supplant the interactions of the members: rather it will support the community and improve the communication channels as they currently exist. AETAP will conduct a variety of activities such as workshops, industry visits, student projects, technical assistance projects, and short courses to effectively transfer knowledge to industry. These activities will establish trust and commitment, the two basic requirements of a community according to Hagel and Armstrong (1997).

By participating in face-to-face interactions, AETAP and ACES members are able to build trusting relationships and show that they are committed to helping Alabama's industry. A show of commitment involves more than simply offering a Web address to industries in need of technical assistance. The engineering community, as established by interpersonal interactions, can be effective as evidenced by AETAP's 1997-2003 success of 716 projects involving 40 researchers to raise $1.2 million. However, this traditional community could be enhanced by the inclusion of a virtual element to increase the effectiveness of all technology-transfer efforts.

Building a Supportive Virtual Community

By supporting the existing community-building efforts with a virtual component, the expected benefits as outlined by Hagel and Armstrong (1997) include: (1) reducing the costs for both companies to find assistance from research institutions and vice versa, (2) reducing a company's perceived risk of undertaking a university-sponsored project by providing information about past projects, (3) enhancing the ability to focus resources where they will have the greatest impact, and (4) re-focusing existing services to add greater value.

Another potential benefit is reaching new markets, not only by virtue of presence on the Web, but also by "using other people's knowledge instead of re-creating it" (Kisielnicki, 2002). The KSS was developed with the vision that the content (i.e., knowledge) captured within the system will serve not only as a means to measure AETAP's progress (recording the number of projects and their economic impact for reporting needs), but also as a rich

Table 1. The KSS as a virtual community according to the elements defined by Hagel and Armstrong (1997)

Virtual Community Defining Elements	KSS Characteristics
Distinctive focus	The KSS user base focuses on small to medium-sized manufacturers within a 200-mile radius of Auburn, AL.
Capacity to integrate content and communication	The KSS includes an engineering knowledge repository, complete with project and research documentation with contact information of the document authors and domain experts.
Appreciation of member-generated content	University faculty and students as members will generate the majority of content, with county agents and industry partners contributing industry problems, project requests, and feedback of project results.
Access to competing publishers and vendors	The KSS is being designed with the intention to be integrated with more research institutions providing technology-transfer services.
Commercial orientation (generating profit)	As a non-commercial sponsored system, the KSS will not seek to generate profit per se, but will facilitate ROI on faculty and student research, attract partner contributions (cash and in-kind), and improve economic standings of partner companies.

electronic document repository that can be quickly searched to provide usefulness beyond the project's typical impact area. Thus, a project completed today, with its association to research documentation and experts, can be used tomorrow to either solve a new, similar problem within the community or at least provide a method to communicate to the experts in the appropriate discipline. Providing content and communication in this manner is but one of the defining elements of a virtual community as listed in Table 1.

Managing Knowledge for the Community

Knowledge Management Architecture

Kisielnicki (2002) continues to explain that by entering the virtual realm, the task of managing knowledge becomes increasingly difficult. KM is a concept that includes: (1) an organizational context, (2) integration of knowledge with business processes, (3) interpersonal connections, (4) document repositories, and (5) involvement of the practitioners of the knowledge (Wenger, 2004). The organizational context of KM in this case is AETAP, which includes interpersonal relationships among partners. To complete the concept of KM, the KSS must have an integration of knowledge with business processes,

include a document repository, and involve the practitioners of the knowledge. Michael Zack presents a KM architecture theory consisting of a repository, refinery, and roles in his 1999 article "Managing Codified Knowledge." This literature is helpful in describing the KM strategy for the KSS developed by Auburn University.

Repository

In discussing a repository Zack (1999) expands on the concepts of structure and content as being important components of knowledge; the content of the repository is interpreted within the context of the structure. The structure lends itself to "a high degree of viewing flexibility" in which users may "dynamically and interactively combine views to more easily apply the knowledge to new contexts and circumstances" (Zack, 1999). Individual knowledge units are described by Zack (1999) as "a formally defined, atomic packet of knowledge content that can be labeled, indexed, stored, retrieved, and manipulated" independently of their particular "format, size, and content." The knowledge unit being formally defined can serve as the basis for formal communication to be applied throughout the whole community, and the KSS revolves around the projects (Figure 2) performed by AETAP.

AETAP projects focus on a problem for a specific company, connect students and faculty with

Figure 2. Knowledge unit of the KSS

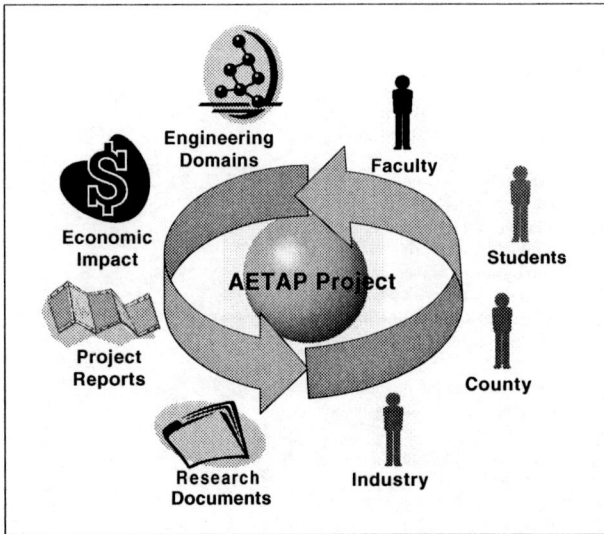

industry contacts via county agents, produce project documentation necessary to measure the effectiveness of the community and justify its sustainability, and produce research documentation that is classified by several dimensions: relevant engineering domain, category, type and format, author/expert, and associated project. By connecting AETAP projects, a member of the community can search for a term such as 'hazardous waste plan' and be able to retrieve: (1) relevant documentation under the folder Chemical Engineering and the sub-folder

Pollution Prevention, (2) information about the success of relevant past projects in terms of economic impact, (3) information about duration and costs of past projects, and (4) contact information and communication channels to reach persons (experts) involved with the project (primarily county agents and faculty, but also students).

Refinery
Knowledge repositories at a high level can be described to capture, store, and retrieve knowledge. Zack offers five processes that describe these core processes more in-depth: acquisition, refinement, storage and retrieval, distribution, and presentation (1999). As documentation is generated, it is uploaded (acquired) into the KSS. The refinement process involves defining the structure and associations of the knowledge content. Because of a rigidly defined knowledge unit, the KSS is capable of reducing the refining process to simply associating a document to a project and an engineering domain such as Chemical Engineering. The capture process (acquire and refine) of the KSS is shown in Figure 3. Barriers to capturing knowledge are countered by tightly binding the capture of knowledge with AETAP's business process of conducting projects. The KSS then indexes every word within a document regardless of format so that subsequent searches will return documents containing the search term.

Figure 3. Capture process of the KSS

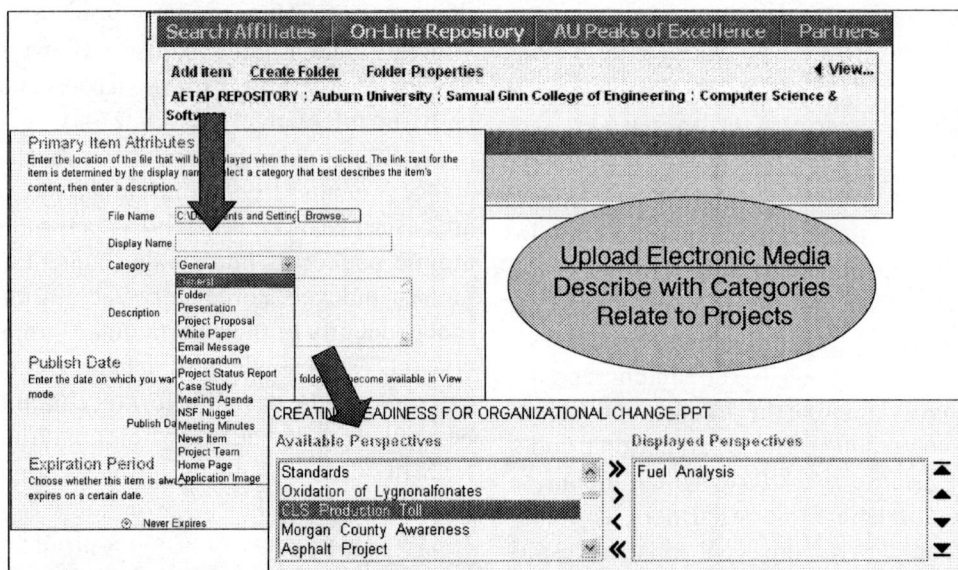

The problem of mislabeling or misrepresenting a document to create amiable search results is precluded. This is similar to Internet search engines, which are not based upon user-defined search keywords, but the occurrence of keywords within the Web page. The storage of documents is done according to the prescribed engineering domain in a robust relational database, and retrieval can be accomplished by searching any part of the document. Distribution is provided over the Web on an ad hoc basis and presented to the users in a variety of views so that they can navigate to any document, project, or person relevant to their search. A second layer of refinement occurs when users retrieve knowledge units and find redundancies or inaccuracies within the system. They may either report their findings, or some users will have the power to remove redundant records. Responsibility for the integrity of content within the system is shared across the community.

Roles

Zack discusses roles that refer to how information technology (IT) acts as an enabler in KM systems (1999). The KSS uses IT to provide applications over the Web that allow users to easily contribute content and a powerful database system to store, retrieve, and index content so that it may then be distributed back over the Web and presented in a myriad of views.

The Community as a Community of Practice (CoP)

Often times when knowledge is discussed, CoPs are not far behind. A CoP comprises like-minded individuals who not only share the same interest, but also have passion to advance the knowledge of their community (Wenger, 2004) and are unified in action to benefit the community at large (Swan, Scarborough & Robertson, 2002). The engineering community served by the KSS is a cross-organizational community that advances knowledge in their domain to achieve benefits beyond their own community by enacting technology transfer and can thereby be considered a CoP.

The actionable arm of the engineering community is the technology transferred from research institutions to the industrial community. The KM

component of the KSS, the knowledge repository, captures, refines, and disseminates the knowledge units of the engineering community. It is important to note that the KSS is still in early stages of development even while AETAP members conduct technology-transfer activities. Thus, the CoP is not dependent on the KSS; however, the fact remains that many manufacturing jobs are moving out of Alabama. The KSS is meant to support the CoP, not supplant it, making the CoP more effective in the activity of technology transfer. As the KSS supports the CoP, it will be the members of the community who request, create, modify, and consume the knowledge contained within the KSS.

The repository, refinery, and roles of the KSS describe the integration of knowledge with business processes and the inclusion of a document repository. Integration is achieved by the knowledge unit, which tightly binds knowledge dimensions around the core process of the AETAP project. The final component of KM as proposed by Wenger is involving the practitioners of the knowledge in the KM process. In AETAP's KM strategy, the concept of a community of practice explains the involvement of the practitioners of the community's knowledge.

FUTURE TRENDS

The Community: Today and Tomorrow

By effectively managing the knowledge content that will be used by the engineering community, the engineering community at Auburn University will have a powerful tool to perform technology-transfer activities. The virtual community will allow community interactions to be formalized across the KSS and be less linear; members of the community will be able to communicate with each other via the KSS. Industry partners when posting a problem will be able to reach all county agents and faculty members in one action. If the problem results in a project, the status will be transparent to all users, thereby ensuring that the problem is being thoroughly addressed.

Using IT as a community enabler, the KSS redefines how community members interact by providing a virtual medium complete with a document repository. Although not designed to completely replace the physical community, the KSS is a forum from which all community members can interact directly with, resulting in the once linear communication channels redrawn into a star pattern. The repository will act as a KM tool to capture knowledge units centered on AETAP projects, and make them readily available to community members so that captured knowledge can be leveraged to solve new problems and in turn create more knowledge.

To date, the KSS is successfully capturing project documentation generated by faculty and students. The next component, the repository, was designed and released in the fall of 2004. The base of users is and will continue to increase gradually to include more faculty, county agents, and industry partners. As the user base grows, the system will iterate through versions to ensure the needs of each community member is met.

A prototype KSS was reviewed by county agents during a Short Course Training on January 10, 2003. They rated the system as very valuable and expected it to greatly assist them in achieving their industrial liaison tasks. Specifically, agents noted the system's capabilities as essential in ensuring accountability and timely compliance of deadlines. The system will also provide a bird's-eye view of each project by allowing industry partners to monitor the progress of their request online at any time.

CONCLUSION

As the community grows to include its full range of users (faculty, students, agents, and industry partners) and the KSS becomes stable, the next level for the community will be to expand its member base and broaden its focus. By including other members with similar needs, the community is free to grow and reap the benefits of a virtual presence. The potential economic impact of the project will be to improve the efficiency of existing manufacturing operations, bring new manufacturing facilities into Alabama, provide a better-trained workforce, disseminate successful practices to industry, and produce better-educated university graduates, with higher potential for employment. The societal impact of the KSS will provide a new, more "hands-on" approach to the education of undergraduate students, will enable students to act as technology ambassadors, will provide county agents with a tool

to be more effective industry contact liaisons, will provide rapid feedback on technology transfer ideas, and will enable replication of successful results of this project to other counties in Alabama and, later, to other states. By growing, supporting, and reinforcing the technology-transfer community, all members can attain benefits, greater than the sum of their individual efforts, by working together.

ACKNOWLEDGMENTS

Special recognition for their work in this area is extended to all members of the project, past and present, including Dr. Larry Benefield, Dr. David Beale, Dr. Robert Cooper, Dr. Evelyn Crayton, Dr. Warren McCord, Dr. Tony Overfelt, Dr. Jeff Smith, Dr. Harry Strawn, and Dr. Art Tarrer.

Special thanks to Dr. John C. Hurt, Program Director (PFI), National Science Foundation, Grant EEC #0332594.

The project reported in this article is partially funded by the National Science Foundation, Grant EEC #0332594. Any opinions, findings, and conclusions or recommendations expressed in this article are those of the authors and do not necessarily reflect the views of the National Science Foundation.

REFERENCES

Fox, J.L. (1984, June 29). NSF Studies Cooperating R&D. *Science, 224*(June 29), 1411.

Hagel, J., & Armstrong, A. (1997). *Net gain: Expanding markets through virtual communities.* Boston: Harvard Business School Press.

Kisielnicki, J. (2002). *Modern organizations in virtual communities.* Hershey, PA: IRM Press.

National Science Board. (2000, June). *Science and engineering indicators 2000.* Washington: National Science Foundation.

State Dislocated Worker Unit. (2002). Listing of plant closing and employee layoffs. *Private Communications,* Alabama Department of Economic and Community Affairs, Montgomery, AL.

Swan, J., Scarbrough, H., & Robertson, M. (2002). The construction of 'communities of practice' in the management of innovation. *Management Learning, 33*(4), 477-497.

Tornatzky, L., Lovelace, K., Denis, G.O., George, W.S., & Eliezer, G. (1999). Promoting the success of U.S. industry/university research centers: The role of leadership. *Industry and Higher Education, 13*(2), 101-111.

Vision of Alabama: Economic Development Partnership of Alabama. (1994, February 14). *Vision Alabama: A plan for quality growth.* Montgomery, AL.

Wenger, E. (2004). Knowledge management as a doughnut: Shaping your knowledge strategy through communities of practice. *Ivey Business Journal,* (January/February), 1-8.

Whittington, R. (2002, May 2). Economically distressed counties: Ranking in Alabama. *Private Communication, Alabama Development Office,* Montgomery, AL.

Wyckoff, A.W., & Tornatzky, L.G. (1998). State-level efforts to transfer manufacturing technology: A survey of programs and practices. *Management Science, 34*(4), 469-481.

Zack, M.H. (1999). Managing codified knowledge. MIT *Sloan Management Review, 40*(4), 45-58.

KEY TERMS

Community of Practice: Like-minded individuals who not only share the same interest, but also have passion to advance the knowledge of their community (Wenger, 2004) and are unified in action to benefit the community at large (Swan et al., 2002).

Knowledge Management: A concept that requires: (1) an organizational context, (2) integration of knowledge with business processes, (3) interpersonal connections, (4) document repositories, and (5) involvement of the practitioners of the knowledge (Wenger, 2004).

Knowledge Repository: The conceptual representation of a knowledge management platform consisting of a repository (structure and content), refinery (capture, storage, and retrieval), and roles (IT as an enabler) (Zack, 2000).

Knowledge Unit: "A formally defined, atomic packet of knowledge content that can be labeled, indexed, stored, retrieved, and manipulated" independently of their particular "format, size, and content" (Zack, 2000).

Technology Transfer: The process of enacting research innovations by properly channeling research to actionable outlets.

Virtual Community: An electronic forum including the following attributes: distinctive focus, capacity to integrate content and communication, appreciation of member-generated content, access to competing publishers and vendors, and commercial orientation (generating profit) (Hagel & Armstrong, 1997).

F

Formal Work Groups and Communities of Practice

Jie Yan
Grenoble Ecole de Management, France

Dimitris Assimakopoulos
Grenoble Ecole de Management, France

INTRODUCTION AND BACKGROUND

Knowledge and innovation management scholars (see, for example, Leonard-Barton, 1995; Nonaka and Takeuchi, 1995) have recently emphasized the role of formal work groups, in particular, project teams, in organizational innovation and learning. In the late 1990s, the concept of communities of practice (CoPs) has also become a key concept in understanding the creation, sharing, and application of knowledge within and across organizational boundaries. This article discusses the relationship between CoPs and formal work groups, such as project teams, based on the authors' recent empirical research in software development work in China (Yan & Assimakopoulos, 2003a, b). Moreover, this article discusses how the division of formal work groups in project teams influences the knowledge activities in CoPs.

HOW CoPs DIFFER FROM FORMAL WORK GROUPS

As can be seen from Table 1, CoPs are informal social structures that are neither defined by authority, nor reflect formal organizational structures as described in organizational charts (Brown & Duguid, 2001; Wenger, 1998). CoPs come into being and grow organically during people's daily-based informal work related interactions. Generally, these interactions are ongoing without any formal hierarchy or management structure overseeing their activities. People are involved in a community of practice as a result of their interest in a specific knowledge domain and need to develop and share related knowledge to this particular area of common interest. The growth and sustenance of a community rely on the existence of a common practice, a need to share knowledge, and passion of its members for continuing membership. According to Wenger, McDermott, and Snyder (2002), CoPs may fade away or transform themselves to other informal or formal social structures, following environmental changes, such as the emergence of new markets or new organizational structures underpinned by technological changes such as the Internet and its ability to foster virtual communities of common interests.

By contrast, formal work groups are always defined and organized by management authority to fulfill a certain business function or achieve a specific work task such as serving a specific market segment or developing a new product. Formal work groups often have the responsibility and power to allocate resources, manage business processes, assign working roles, and be held responsible and accountable for business outcomes. The information flows, reporting relationships, and affiliation structures are also highly formalized in formal work groups with clearly described job specifications and organizational charts depicting in an unambiguous manner who is in and who is out of a group and what role(s) play each and every member. The life cycle of formal work groups also depends on organizational authority or, in the case of a project team, on the accomplishment of a particular project.

In general, however, CoPs do not overlap with the boundaries of formal work units (Wenger, 1998). CoPs may exist within a business function or department, stretch across divisional boundary, or even go farther beyond the boundary of an organization. Based on common practice and the need to share knowledge, people from different work groups may interact with each other and constitute a CoP. The boundary of a CoP is often fuzzy because its membership is based on personal voluntary participation and may have different levels of involvement depending on the members' specific positions in the under-

Table 1. Distinctions between CoPs and formal work groups

	Communities of Practice	**Formal Work Groups/Project Teams**
Objectives	Create and share knowledge based on common interest and practice	Achieve a specific work related task defined by business function(s) (e.g., R&D)
Organization	Informal, self-organizing, lacks clear hierarchy	Formal, affiliation and reporting relations are defined by organizational authority
Activities	Daily-based work practices, in which knowledge is collectively created and shared among peers	Organizational functions and routines define operations, goals, and specific work tasks shared among group members
Boundary	Fuzzy	Clear as defined by formal group membership
Membership	Voluntary participation	Defined by organizational function(s)
Hierarchy	Absent or established after LPP	Leadership and reporting relations are defined and managed by organizational authority
Power	Based on LPP process, a member's power is gradually recognized and accepted by peers	Mainly based on formal organizational and reporting arrangements
Cohesion	Stemming from personal passion and commitment to a common practice	Stemming from work requirements, shared responsibility, and common goals defined by organizational authority
Ending	Depending on a continuing need and personal passion of members	Depending on an organizational decision or accomplishment of a specific project

lying Legitimate Peripheral Participation (LPP) process (Lave & Wenger, 1991; Wenger, 1998). Sometimes, a project team or a business office forms a CoP as the result of long-term knowledge related interactions. In this case, the boundary of the CoP overlaps with the formal work group. In some other cases, there are several project teams existing within a CoP or several CoPs within a function or department. In addition, CoPs do not have any internal hierarchy, though power relationships exist as a manifestation of the LPP process and the core periphery pattern of ongoing interactions over a sustained period of time. The different levels of participation determine power relationship among CoP members. For example, people standing at the core are often more powerful in knowledge activities of the community than newcomers starting at peripheral positions.

SPONSORED CoPs

Up to the late 1990s, it was thought that CoPs start enthusiastically without much intervention from an organizational authority. People spontaneously come together because they need to support each other and share knowledge and experience. In recent years with the increasing recognition of the positive role CoPs play in organizational learning and innovation, an increasing number of organizations make systematic efforts to intentionally foster the birth and development of CoPs

(Wenger et al., 2002). CoPs may get moral and financial support from organization and management (especially, knowledge managers) and therefore have more resources to enable its activities such as having real (or virtual) meeting places, holding social events, developing Web sites, or printing publications.

Today CoPs getting organizational support and managerial attention are generally expected to produce measurable results benefiting the sponsoring organization(s). More and more CoPs and the organizations involved make efforts to formalize CoP-related activities. For example, members of a CoP may set formal agendas for knowledge creation and sharing, have regular meetings and publish newsletters. In most cases, however, whether a CoP accepts support from an organization or a degree of institutionalization does not mean a higher level of formality. The work of organizational supporting "is not to formalize them by making them follow procedures or meet efficiency goals, but rather to strengthen them as informal entities" (Wenger et al., 2002, p. 217). In this sense, CoPs can still keep their internal drive and informal workings and do not have any commitment to formal authority and management structures.

PROJECT TEAMS AND KNOWLEDGE ACTIVITIES IN COPS

Project teams have long been emphasized as flexible working units to achieve some specific organizational objectives. A project team is basically a group of people, "drawn from within and/or outside an organization to undertake a specific project. When the project ends, the team disbands and members are reabsorbed into the organization and into new projects" (Keegan & Turner, 2001, p. 78). Often, a project team is cross-functional in nature and is regarded as the unit of analysis and location where innovation occurs (Leonard-Barton, 1995). A project team is often where individual knowledge is shared, codified, and integrated as a basis for developing new organizational knowledge. Knowledge is therefore created in project teams through a series of conversions between tacit and explicit knowledge. These conversions generally take place at project team level (Nonaka & Takuechi, 1995). Generally, a team has a small size and membership to facilitate good communication and coordination.

A critical research question is how the division of formal work groups in project teams affects knowledge activities of CoPs. Yan and Assimakopoulos (2003a, b) conducted empirical research looking into this question within the broader context of software development in China. Their investigation focused on six small entrepreneurial companies where software engineers work in a single development room and form a CoP, divided in project teams of three to five engineers. Computerized statistical and network analysis reveals that software engineers' informal advice seeking behavior within the project team is significantly more intensive than across team boundaries. The advice seeking ties within team are more than double compared with the ties across team boundaries. Nearly three quarters of the strong ties exist within team, while only a quarter cross team boundaries.

When software engineers were asked to corroborate this finding, most of them insisted that they do not care about who is a team colleague and who is not when they have a problem begging for discussion with colleagues. What they care about most is who may be able to help and has experience for providing useful information. However, engineers did agree that they unconsciously chose team colleagues in most cases. Common understanding of the tacit and context specific knowledge of a particular project is the key factor contributing to the engineers' choice of team col-

leagues as advice seeking targets. A discussion with team colleagues usually is more efficient and effective, compared with a discussion with somebody who knows little about the internal organization of a particular project. This finding is also consistent with the software engineering literature, emphasizing the shared understanding of software development at team level (Lennartsson & Cederling, 1997; Microsoft, 1999; SEI, 2002).

Our research, therefore, highlights the discontinuous context of CoPs as a result of the division of project teams in CoPs. Each project team has its particular subcontext and specific properties in terms of tacit knowledge and techniques involved, such as those associated with different software programming languages used for developing different projects. Members' daily knowledge activities are shaped by these subcontexts, and these subcontexts unavoidably shape knowledge creation and sharing at CoP level. Software engineers intensively interact with colleagues within their project team and, to a much lesser extent, interact with colleagues in other teams. This finding suggests the existence of subgroups in CoPs where knowledge may be asymmetrically distributed according to different subgroups, rather than symmetrically spread in a CoP as a whole.

When discussing the structure of CoPs, Wenger (1998) suggests a core-periphery structure based on the LPP process in which newcomers learn to participate into the community. People in CoPs have different positions according to their levels of participation, that is, full participation (insider), legitimate peripherality, marginality, and full non-participation (outsider). As far as the literature reviewed, this is the only set of positions defining the underlying structure of a CoP. One can argue that our findings above suggest a new, more refined model of internal knowledge structure of CoP, in which, participants have dual or multiple membership as CoP members. Software engineers are positioned according to their work related specific subcontexts, thus maintain basic membership of a CoP, but also their knowledge activities are lodged in the practice of project teams or subgroups, enabling new knowledge creation and sharing according to a discontinuous shared practice.

CONCLUSION AND FUTURE TRENDS

Formal work groups in organizations stem from formal divisions of function, task, or labor. Typical formal work groups include functional departments, project teams, service units, and so on. As an informal structure, a CoP has significant differences compared to formal work groups in terms of objectives, organization, membership, and so forth (see Table 1). At the core, the distinction is that formal work groups are organized by management authority, while CoPs are generally self-organized and governed. However, with the increasing recognition of CoP and its role in organizational learning and innovation, some organizations have made systematic efforts to intentionally facilitate the birth and development of CoPs. In most cases, these efforts have not changed CoPs' informal nature, though, as our research suggests, the division of formal work groups by project teams makes the context of CoP discontinuous and suggests a multigroup structure of CoPs. More research is therefore needed in shedding light on how knowledge is asymmetrically produced and distributed in these subgroups, rather than spread equally in a community of practice as a whole.

REFERENCES

Brown, J. S., & Duguid, P. (2001). Structure and spontaneity: Knowledge and organization. In I. Nonaka & D. J. Teece (Eds.), *Managing industrial knowledge: Creation, transfer and utilisation*. London: Sage.

Keegan, A., & Turner, J. R. (2001). Quantity versus quality in project-based learning practices. *Management Learning, 32*(1), 77-98.

Lave, J., & Wenger, E. (1991). *Situated learning: Legitimate peripheral participation*. Cambridge, UK: Cambridge University Press.

Lennartsson, B., & Cederling, U. (1997). Team understanding: A successor of the process improvement paradigm? Retrieved May 5, 2005, from *http://www.ida.liu.se/~bin/www-bln-sw-architecture/hawaii-98-3.pdf*

Leonard-Barton, D. L. (1995). *Wellsprings of knowledge: Building and sustaining the sources of innovation*. Boston: Harvard Business School Press.

Microsoft. (1999). Microsoft solutions framework: Overview white paper. Retrieved May 5, 2005, from *http://www.microsoft.com/business/services/mcsmsf.asp*

Nonaka, I., & Takeuchi, H. (1995). *The knowledge creating company*. Oxford: Oxford University Press.

SEI. (2002). The team software process: An overview and preliminary results of using disciplined practices. Retrieved May 5, 2005, from *http://www.sei.cmu.edu/publications/documents/00.reports/00tr015.html*

Wenger, E. (1998). *Communities of practice: Learning, meaning and identity*. Cambridge: Cambridge University Press.

Wenger, E., McDermott, R., & Snyder, W. M. (2002). *Cultivating communities of practice*. Boston: Harvard Business School Press.

Yan, J., & Assimakopoulos, D. (2003a). Knowledge sharing and advice seeking in a software engineering community. In L. M. Camarinha-Matos & H. A. Afsarmanesh (Eds.), *Processes and foundations for virtual organizations* (pp. 341-350). Dordrecht: Kluwer Academic.

Yan, J., & Assimakopoulos, D. (2003b, September). The influence of project team to the knowledge activities of community of practice. *Proceeding of the British Academy of Management Annual Conference*, Harrogate, UK.

F

The Globalization Paradigm and Latin America's Digital Gap

Heberto Ochoa-Morales
University of New Mexico, USA

INTRODUCTION

The Andean Community of Nations (CAN) and others countries in Latin America (LA), as any less developed countries (LDCs), are located by inception on the wrong side of the "digital gap". Therefore, these countries confront an enormous challenge from the network revolution that is unfolding. Globalization represents a new paradigm composed of integrated and interdependent economies. The Globalization Index (GI) determines the rank of the countries within the model. This index is composed of several variables in which economic integration and technology, among others, play a very important role in country classifications. Currently, a diminishing trend of FDIs is preponderant in the region, and this affects the knowledge-based society and also the efforts to make these countries members of the new globalization paradigm.

Dessler (2004) stated that globalization is the tendency of firms to augment their sales, ownership, and manufacturing facilities to new markets located abroad. The research literature is consistent with the definition of globalization. Hill (2003), among others, agreed that the term *globalization* refers to a new paradigm in which the world economy is more integrated and interdependent. Therefore, this integration demands new methodologies and mechanisms to allow countries to perform their new roles within this emerging framework. A preponderant element in this new array is the convergence of computer-based power and telecommunications. These parameters are interrelated to computing infrastructure, new communication technology, and governmental policies that will make the old telecommunication model, a monopoly, obsolete; therefore, a new paradigm will evolve that makes this technology accessible to everyone through a new system that promotes and encourages competition within the private sector (Ochoa-Morales, 2003c).

Also, convergence that is taking place with computing and telecommunication demonstrates the importance of the development of this sector and the socioeconomic impact on the economic perspective and to the stimulus of economic growth (Ochoa-Morales, 2003a).

Kearney (2003) classified countries using a Globalization Index (GI), which determines the rank of the country as a more global country. Sixty-two countries that represent 85% of the world's population compose the sample used. The index is epitomized by 13 variables grouped in four baskets: (1) economic integration, (2) personal contact, (3) technology, and (4) political engagement. Economic integration is represented by trade, foreign direct investments (FDIs) and portfolio capital flows, and income payments and receipts. Personal contact consists of international travel and tourism, international telephone traffic, and cross-border transfers. Technology is characterized by number of Internet users, Internet hosts, and secure servers; and political engagement is characterized by number of memberships in international organizations, UN Security Council missions in which each country participates, and the quantity of foreign embassies hosted by the countries. The ranking for the year 2003 shows Ireland as number one, Switzerland number two, and the United States as eleventh. Ireland has large investments in high-tech and information technology. Its Internet infrastructure is still growing, and the number of secure servers has increased 32.6% from 337 to 500 in 2002. Also, it has been the most talkative country in the world, included heavy domestic and international traffic. The above is unequivocal proof of the high correlation that exists between technology, a parameter of the new paradigm, and access to new markets that will be the cornerstone of globalization.

According to Kearney (2003), one variable is economic integration in Latin America (LA), and the

Caribbean economic integration is extant. Numerous regional and multilateral agreements are present such as the Andean Community of Nations (CAN), composed of Bolivia, Ecuador, Colombia, Peru, and Venezuela; MERCOSUR, composed of Brazil, Paraguay, Uruguay, and Argentina; The Group of Three (3), composed of Colombia, Mexico, and Venezuela; and the CARICOM, composed of English speaking countries (Islands) within the Caribbean Basin (Secretaria, 1998). Ochoa-Morales (2001) stated that, from an economic perspective, the outcome is trade and therefore stimulus to economic growth. Foreign direct investments (FDIs) can greatly contribute to a host country's economy providing the required factors of production are present, making the countries more competitive within the globalization framework. Schuler and Brown (1999) emphasized that the most important occurrence in the location of the FDIs is the support or impediment exercised by the institutions in the host country.

Another important factor within the GI is technology characterized among other parameters by Internet users and Internet hosts. In LA, the growth rate of the Internet has been the highest in the world, and the number of users has increased 14-fold within the 1995 to 1999 period (UIT, 2000). The literature defines teledensity as the number of main telephone lines for every 100 inhabitants, excluding wireless access. This term is also used as a parameter to measure the level of telecommunication infrastructure of any country. A review of the literature also shows the existence of a high correlation between teledensity and economic development, and a negative one between teledensity and population size has been found (Mbarika, Byrd & Raymond, 2002).

PURPOSE

Globalization is a new paradigm within the world's economies. The less developed countries (LDCs) such as in LA, the Andean Community of Nations (CAN), are marginal to such a model as a consequence of the digital gap. The quest of the article is to demonstrate some of the variables that compose the GI that would negatively affect this geographical region to acquire the necessary rank within the paradigm framework to be considered a global country.

BACKGROUND

Based on Kearney's (2003) GI, the four main realms will be analyzed: economic integration, personal contact, technology, and political engagement and, within them, the variables that compose them. Technology and FDIs represent the primary factors because both are interrelated and have a high correlation with computer-based power and telecommunication convergence. Economic integration is extant in LA and the Caribbean Basin, not only under the model of regional agreements, but also as multilateral ones. By the year 2000, LA's regional agreements, CAN and MERCOSUR without considering other regional pacts with Chile, have a potential market of 310 million consumers (UN-CEPAL, 1999). Chile's contribution alone is 15.2 million potential customers. The latter represents a very large concentration of population very well suited to be penetrated and to be converted in a market expansion. Ochoa-Morales (2003b) stated that, under the scheme of regional integration, theoretically, an unrestricted no trade tariff or barriers to a high flow of goods, services, and investment among the countries will be originated. From an economic perspective, the outcome is greater comparative advantage to the countries and, consequently, stimulus to economic growth. Also, a high correlation exists between economic growth and the demand for capital, technology, and management resources. In the telecommunications domain, the development of new technologies has performed a critical role in the process. Cellular phones and cable television are, among others, new technologies imbedded in the new paradigm. As a result, more people are interconnected and better informed. In Venezuela and Paraguay, there are more cell phones than conventional ones (UIT, 2000). The privatization of the communication industry within the telephony subsector in LA has increased the parameters that reflect teledensity. Therefore, the domestic and international telephone traffic has grown accordingly. Ochoa-Morales (2002a) stated that the privatization and deregulation of the communication sector act, as an incentive to bring to the LDCs foreign direct investments, not only provide the financing required to develop the industry, but also provide the know-how embedded therein. It is critical to accentuate the fact that, to attract these investments, a well-defined legal and political framework must be in place. The only way these countries located on the wrong side of the digital gap could be evolved within the technology environment rests on foreign sources of funding. At the same time, the developed countries (DCs) could augment and/or expand their markets, investing their financial resources and technology in LA (Ochoa-Morales, 2002a).

The flow of information has been present as an integral part of activities related to production, trade, and investments, among others. Therefore, historically, a strong correlation exists among economic and networking development. Also, the latter plays a very important role in the development of modern social and institutional structures (World Bank, 2000).

The amount of investments in projects to be developed by multinational enterprises (MNEs) during the next five years in LA and the Caribbean that was announced between January 2001 and April 2002 is $31.896 billion. The services sector takes 80% of the total, the oil and gas 15.7%, and the manufacturing sector 4.3% (UN-CEPAL, 2002b). Regarding the European Union (EU), the amount of FDIs toward LA has increased from $1.6 billion in the quinquennial 1990-1995 to $19.5 billion in 1996-1999. Also, LA captured 13.5% FDIs outside of the EU and 6.5% of the ones generated in the EU (UN-CEPAL, 2002b).

Another variable in the GI is political engagement; the countries that composed the region are members of all international organizations and maintain political and commercial relationships with all the countries of the globe. It is necessary to emphasize the fact that, recently, presidential elections have brought to the governments a cadre of leadership that is inclined to the left and considers privatization and liberalization primordial components of a new sociopolitical and economical model called neo-liberalism. The leadership thinks that the model will only benefit the DCs, the same way that globalization does. These political issues have generated a diatribe of opinions and an unstable political climate not conducive to the attraction of foreign investments. During the year 2001, the flow of FDIs toward CAN has diminished considerably, and there are no signs of improvement for the year 2005 (Table 1, Charts 1 and 2). This is not only applicable to new investments but also to mergers and acquisitions. The receding FDIs

are attached to the end of economic reforms, especially the privatization of state enterprises in the realm of energy and basic services, and there is also the relevance of China's attraction to FDIs as a powerful incentive to redirectioning them (UN-CEPAL, 2002b).

DISCUSSION

The year 2001 shows almost no privatization; acquisitions and mergers were reduced to a minimum due to the fact that only a few large entities remain within the state ownership, especially in the hydrocarbon sector in CAN's country members such as Colombia and Venezuela. Within the telecommunication sector, the most important one concerning this discussion is the new concessions in the wireless subsector of Venezuela. Telecom Italia Mobile (TIM) with presence in Bolivia, Venezuela, and Peru, among other LA countries, will invest another $200 million in Peru to reach a half a billion dollars in the wireless telecommunication sector (Cadena Global, 2002). Also, the economic growth in the LA countries is almost nil due to a regional financial crisis, the poor performance of the LA country currency, and the political instability generated by the "lefties" and populist governments of the region. As proof, the second quarterly reports of the Spaniards banks, Santander Central Hispano and Banco Bilbao Vizcaya Argentia (BBVA), for the year 2002 will be affected greatly. LA provides between one-fifth and one-half of the income of these banks (El Nacional, 2002).

In 1998, only 1% of the population of LA and the Caribbean were connected to the Internet. It is necessary to emphasize the fact that the region has shown the most rapid growth in the world. Today, due to FDIs and the implementation of policies that attract them, 84% of the telecommunication infrastructure is digital and completely automatic. Within the wireless sector in the first quarter of 2001, 70 million subscribers existed. E-commerce usage in the region is less than one-fifth of subscribers of the Internet that is still at incipient levels (UN-CEPAL, 2002a). There is also a high correlation between Internet connectivity and the Gross Domestic Product (GDP) of countries, but governments of LDCs should be responsible for the utilization of their

Figure 1.

	A	B	D	K	M	N
Ireland	1	1*	16	1*	24	16
Switzerland	2	5*	7*	2*	11	10*
United States	11	50	1*	16	4*	4*
Colombia	55	48	44	28	45	41
Peru	59	55	31	30	37	44
*Top 10 in the category						

Table 1. FDIs in the CAN (million $) (CEPAL)

	93	94	95	96	97	98	99	2000	2001	2002
BOLIVIA	125	147	391	472	728	952	983	693	647	721
COLOMBIA	719	1298	712	2784	4753	2032	1336	1905	2386	1864
ECUADOR	474	576	452	500	724	870	648	720	1330	1335
PERU	687	3108	2048	3242	1697	1880	1969	662	1063	1943
VENEZUELA	-514	455	894	1676	5036	4262	2789	4357	2684	1200
TOTAL FDIs	1491	5584	4497	8674	12938	9996	7725	8337	8110	7063
(million $)										
GROWTH RATE		2.75	-0.19	0.93	0.49	-0.23	-0.23	0.08	-0.03	-0.13

political power to create the necessary mechanisms, so the mass population will be able to have access to the benefits provided by technology and, consequently, be constituents of the knowledge-based society (Ochoa-Morales, 2002b).

The flow of FDIs has contributed in an almost incommensurate manner to the economic growth and development of the member countries of the CAN, not only in the realm in which they were invested, but also as a sequel in the regional economy as a whole. The GDP index is affected due to the fact that the positive effects of FDIs are present in diverse elements of production. Emphasis should be placed on the high correlation between teledensity and GDP indexes stated in the research literature (Ochoa-Morales, 2003d). The caveat resides in some factors that should be present for the FDIs to work: political stability, improvement in education and developing of the human resources with the managerial skills necessary to perform its function, macroeconomic stability, liberalized trade regimes, and the political and legal framework required to attract foreign investments (Ochoa-Morales, 2002b).

Table 1 depicts the growth rate of FDIs into the CAN. Year 1997 shows the highest amount of FDIs, $12,938 million (Chart 1), with a rate of growth equal to 0.49; after this peak, the trend becomes negative. The latter depicts large disinvestments caused by,

among other factors, the ones described above as the caveat. The consequence of this trend is a diminishing of the GDP and teledensity that will affect interconnectivity, especially in a region of scarce economical resources. Another negative factor is the lack of critical mass necessary to accelerate the process of access to the Internet. All this plays an important role against convergence, which is a relevant parameter in the knowledge-based society (Ochoa-Morales, 2003d).

During the year 1995, all of the country members were at a low level of connectivity. By the year 2000, only two countries, Colombia and Venezuela, moved to the median position; the others, Bolivia, Ecuador, and Peru, remain at a low level in comparison to the world countries. Only Venezuela moved from below WBM to expected value based in GDP income per capita. At year 2000, the growth rate of FDIs is still positive; after that, the data declines to a negative position (Table 1). The lack of external source of funding will affect economic development and growth. As a corollary, the knowledge-based society will come to a phase of stagnation (Ochoa-Morales, 2003d).

The consequences of the stagnation regarding the knowledge-based society in LCD countries, such as the CAN, will widen the digital gap and prevent the latter to conform to the variables that compose the Global Index. Therefore, LDCs will not be participants in the new globalization paradigm in which information and communication technology (ICT) convergence plays the most important role.

CONCLUSION AND FUTURE TRENDS

The GI (Kearney, 2003) ranks only three of the five countries that composed the CAN: Colombia is ranked 55, followed by Peru at 59 and Venezuela at 60. Ecuador and Bolivia do not have any rank. It is necessary to mention that the first ranking position belongs to Ireland, the last one, and 62, to Iran. Switzerland is ranked number 2, and the United States is ranked 11. To establish a comparison among the countries mentioned above, some of the variables such as rank (A), economic integration (B), technology (D), telephone (K), Internet users (M), and Internet hosts (N) will be used.

Venezuela ranked last of all the LA countries; it has dropped from fifty-seventh to sixtieth place. The country's decline in 2001 was due to the temporary drop in oil prices, causing FDIs to plummet. One positive aspect is the fact that in LA the number of wireless telecommunication subscribers grew 33% (more than 86 million people) in 2001, or double the world growth rate (Kearney, 2003).

The aforementioned demonstrate the difficulties for countries located at the wrong side of the digital gap such as the country members of CAN to be participants in the new paradigm of globalization. Unless the DC implement, in conjunction with international organizations, that is, the United Nations (UN), supranational policies to create the environment necessary for the LDCs to be in compliance with the parameters that would qualify them as global countries, the LDCs will continue to be unable to participate fully and develop their economies to the fullest extent. The research literature states that the negative effects caused by not being global will exponentially condemn these countries to continue to exist suboptimally.

REFERENCES

Cadena Global. (2002, June 27). Telecom invertira US$200 millones adicionales en Peru.

Dessler, G. (2004). *Management: Principles and practices for tomorrow's leaders* (3rd ed.). Upper Saddle River, NJ: Pearson Prentice Hall.

El Nacional. (2002, July 24). America Latina restara brillo a resultados de bancos espanoles. Retrieved June 7, 2005, from *http://www.el-nacional.com/L&F*

Hill, C. W. (2003). *International business: Competing in the global market place* (4th ed.). New York: Irwin McGraw-Hill.

Kearney, A. T. (2003, January-February). Measuring globalization: Who's up, who's down. *Foreign Policy,* p. 134.

Mbarika, V. W., Byrd, T. A., & Raymond, J. (2002, April-June). Growth of teledensity in least developed countries: Need for a mitigated euphoria. *Journal of Global Management, 10*(2), 16-17.

Ochoa-Morales, H. (2001). The digital gap between the industrialized countries and the less developed (LDC) ones: The transition toward a knowledgeable society in Latin America. *Journal of Issues in Information Systems, 2,* 337-342.

Ochoa-Morales, H. (2002a). The dynamic changes in the telecommunication sector in Latin America and its effects on the knowledgeable society. *Communications of the IIMA, 2*(1), 84-93.

Ochoa-Morales, H. (2002b). The impact of reforms in the telecommunication sector and its effects on Latin America. *Journal of Issues in Information Systems, 3,* 483-489.

Ochoa-Morales, H. (2003a). Social responsibility and the transition toward a knowledgeable society in Latin America. In R. Azari (Ed.), *Current security management and ethical issues of information technology.* Hershey, PA: IRM Press.

Ochoa-Morales, H. (2003b). The tendency of the European Union Foreign Direct Investments, into the Andean Community of Nations (CAN). *Trust, responsibility and business 2003* (pp. 653-658). SAM Publishing.

Ochoa-Morales, H. (2003c). Teledensity, privatization, and the Andean Community of Nations (CAN): The Peruvian case. *Information technology and organizations: Trends, issues, challenges and solutions 2003* (pp. 1136-1138). Hershey, PA: Idea Group.

Ochoa-Morales, H. (2003d). Diminishing foreign direct investments (FDIs) in the Andean Community of Nations (CAN) and the knowledge-based society. *Journal of Issues in Information Systems, 4*(2), 639-644.

Schuler, D. A., & Brown D. S. (1999). Democracy, regional integration, and foreign direct investments. *Business and Society, 38*(4), 451.

Secretaria de la Comunidad Andina (SCA), Cooperacion Francesa y CEPAL. (1998, May 1-2). *Multilaterismo y regionalismo.* Seminario effectuado, Santa Fe, Bogota.

UN-CEPAL (Comision Economica Para America Latina y El Caribe). (1999a). America Latina: Poblacion total, urbana y rural y porcentaje urbano por paises. Cuadro 11. *Boletin Demografico, 63*, 1-6. Retrieved May 7, 2005, from *http://www.eclac.cl/publicationes/Poblacion/2/LCG2052/BD63.11html*

UN-CEPAL (Comision Economica Para America Latina y El Caribe). (2002a, April). *Globalization y desarrollo.* Informe LC/G.2157 (SES.29/3).

UN-CEPAL (Comision Economica Para America Latina y El Caribe). (2002b, May). *La inversion extranjera en America Latina y El Caribe.* Informe LC/G.2178-P.

UIT. (2000, April). Indicadores de telecomunicaciones de las Americas 2000. *Resumen Ejecutivo,* pp. 1-22

World Bank. (2000, August 17). The network revolution and the developing world. *Analysis Report,* p. 216.

G

Hybrid Knowledge Networks Supporting the Collaborative Multidisciplinary Research

Stanislav Ranguelov
University of the Basque Country, Spain

Arturo Rodríguez
University of the Basque Country, Spain

INTRODUCTION

Virtual networks are becoming increasingly important instruments for knowledge and collaboration management. In addition, research, development, and innovation performances are among the most important activities in modern organisations. These two issues deal with complex problems that companies, universities, and other organisations can only face with multidisciplinary, geographically widespread teams.

This article describes the setup of a model of a hybrid knowledge network that can group and connect together universities and researchers and enable them to collaborate. The proposed model for the virtual network is based on the conjunction of the personal and organisational aspects of collaboration. Due to this union within the organisational structure, two main levels of collaboration have been envisaged, namely the institutional one and the individual one.

BACKGROUND

Nowadays, it is claimed that the main source of sustainable competitive advantage is based on the possession of valuable information and the capacity to exploit, produce, and obtain new knowledge.

Networks in general, and virtual networks in particular, have gradually become more and more important instruments for knowledge management. Early references in this field can be found in the research made by Drucker (1989), Savage (1990), Keen (1991), Donaht (1998), and Koch, and Wörndl (2001). A huge number of definitions have been identified to characterise collaborative networks and the organisation of communities of practice (Koschtzky, 2001), but it has been quite difficult to define a clear border between the different types of knowledge networks that exist.

In the research and development (R&D) process, there is no doubt that communities of practice as specific forms of such networks have become the most important tools to implement knowledge management and to accelerate the transference of innovation. They bring together people with common goals and interests who are physically remote and are working in different types of organisations. Using new technologies they can join together and work as a team towards the objectives set.

This article deals with a new form of knowledge network which, on one side, groups together elements from the traditional virtual community of interest and from the more sophisticated communities of practice, and on the other side, promotes collaborative multidisciplinary research that produces high-quality research results and stimulates their transfer.

MAIN FOCUS—KNOWLEDGE NETWORKS: CONCEPT AND APPROACHES

The main concept treated in the networks economy theory is the cooperation between organisations based on the mutual trust, without hierarchical structures, and that considers knowledge networks as an intermediary stage between the free market and the rigid organisation. In their theoretical approach to the concept of knowledge networks, Seufert, Krogh, and Bach (1999, p.182) define them as structures established between individuals, groups, and organisations in which not only bilateral relations, but also all activities carried out by the knowledge network are important.

From the socio-economic viewpoint, networks are interpreted as a specific set of linkages between a defined set of actors, with the additional property that the characteristics of these linkages as a whole may be used to interpret the social behaviour of the involved actors (paraphrased from Mitchell, 1969, p. 2, as cited by Alba, 1982, p. 40). Therefore, the term "network" covers strong social relationship and includes players who may be individuals, groups, or even whole companies. From this viewpoint, networks can be structured formally or informally. The relationships that can be identified within them are interpreted as long-term connections, which may be personal/organisational or technological/organisational.

Knowledge networks usually share a series of characteristics among which, most important according to Seufert et al. (1999) and Real Communities Inc. (2000), are the following:

- Networks exist to *create and disseminate new knowledge*.
- They are structured and operate to *increase the rate of creation* of new knowledge.
- They provide *clear, recognisable benefits* to all participants.
- Network *membership is by invitation, based on merit or prior review* of the purposes of the project.
- Networks are *usually inter-disciplinary, and cross over the frontiers* between sectors of activity and areas of knowledge.
- Through networking, a *transfer between the tacit knowledge of individuals and the explicit knowledge* held at organisations takes place.

COMMUNITIES OF PRACTICE

The term "communities of practice" was presented first by Lave and Wenger (1991). These communities are groups of people who share an interest in a domain of human endeavour and engage in a process of collective learning that creates bonds between them. A basic characteristic of the community of practice is the specific way in which learning takes place, through a process of *"legitimate peripheral participation"* (LPP). Therefore it is not exclusively based on practical teaching, but also comprises a process of development of knowledge based on experience. The

elements of legitimacy, participation, and peripheral define specific characteristics of communities of this type (Wenger, 1999).

Communities of practice therefore have the role of integrating specific knowledge¾that is, turning individual knowledge into collective knowledge according to the capabilities of the team involved. In short, the creation of "team knowledge" is the result of interaction between individual and collective knowledge, and between tacit and explicit knowledge, according to the development of the "knowledge spiral" (Nonaka & Takeuchi, 1995).

There are certain features that distinguish communities of practice from formally configured teams (see Table 1).

Some Key Factors for Success of the Communities of Practice

- **The knowledge network should be focused on the needs of its members.** In line with this, management should seek to study the profiles of members: what knowledge they possess and what they need.
- **The knowledge network must invest in content.** Much of the effort put in must go into the generation of new content and new contributions, as this is the only way to increase the knowledge of the organisations.
- **Adopt the assumption that the community cannot operate on its own.** This means that members must identify who can act as informal moderator and lead the remaining members towards the problems to be dealt with and provide basic working methods.
- **External factors like the organisational culture are extremely important.** One of the main aims of group moderators must be to achieve a common culture so that knowledge sharing is a natural activity, not a special effort.
- **The understanding that the activities of knowledge network are not limited to discussion groups.** Communities are much more complicated than that.

As a concluding remark for this section, online communities can be said to hold clear advantages for research groups.

Table 1. Differences between communities of practice and formal teams (Lesser & Storck, 2001)

Communities of Practice	Formal Teams
Relationships are formed around one practice	Relationships are formed when a team is assigned
Authority relationships emerge through interaction around expertise	Authority in teams are organisationally determined
Communities are responsible to their members	Teams have goals which are established by people not involved in the team
Communities develop their own procedures	Teams have processes that are organisationally defined
High flexibility provided by the absence of any hierarchical dependences	Rigid structure with less possibilities to reorganise the work
Bidirectional relations between the members of the community	Unidirectional relations with the organisational structure
Less bureaucracy defined by the absence of contractual arrangements between the members	Strong bureaucracy processes defined by the hierarchical structure

HYBRID KNOWLEDGE NETWORKS

The process of creating a virtual network supporting collaborative multidisciplinary research has its roots in the project "Knowledge Management at a Public University: The Process of Research, Development, and Transfer of Scientific and Technical Knowledge," which was set up in 2000 at the University of the Basque Country in Spain with a three-year timeframe. The project brought together a multidisciplinary group of 22 researchers with the purpose of drawing up a model for knowledge management that could cover the whole process from pure research to the transfer of scientific and technical knowledge from the university to businesses and institutions.[1]

The idea of organising virtual research centres has come a step closer to reality with major advances in technology and the need for researchers to respond to rapid changes in the socio-economic environment and the increasing uncertainty that they bring. Several attempts are currently under way to set up non-physical centres, but doubts remain as to the management method best suited to creating and structuring them so that they are capable of providing an open climate for cooperation. Some schemes in Japan and the U.S. have shown the benefits of cooperation by large groups of researchers from different organisations (Echeverri-Carroll, 1999; Jin, 1999). Many of these knowledge networks were set

up with short-term aims, but others are structured as long-term alliances.

The development of these networks has also been fostered by the *arrival of the age of cooperation for innovation*. The cost of innovation and the speed of change create a major need for cooperation to achieve progress. The promotion of research and pre-competitive cooperation between organisations of different types is an essential factor in reducing R&D&I costs and bringing new products, processes, and services to the market as quickly as possible.

Another aspect is the break-up of the old research process. The frontiers between basic and applied research and between research and innovation are disappearing. The time lapse between obtaining basic research results, developing applications in a specific sector, and marketing those applications is becoming ever shorter, making it harder to distinguish the different phases of the process.

FUTURE WORK

Based on these elements, the idea of developing a new type of network of universities for knowledge management is radically different from the way in which the process is currently organised. First of all, a change is introduced in the purpose of applied research: it no longer has to be contained within a specific framework of action, but can instead be carried on in various areas hitherto untouched, but related to particular topics. Furthermore, this network must bring together people of very different types from different branches of science, public administration, and interested companies.

In view of the foregoing, a virtual research network should set itself the following immediate goals:

- To organise the resources needed for research at the lowest possible cost and the greatest possible speed.
- To attract renowned researchers by offering them broad cooperation, flexibility, and new concepts of organisational culture, along with a policy of openness.
- To promote the principle of blending team spirit with individual creativity to increase competitiveness and create the conditions for a substantial increase in technological development.

The model that we propose for a virtual network of universities for knowledge management is fully in line with this approach. The research is broad in its purpose and can take place in different areas concerned with management of the transfer of knowledge from universities to companies and to society. Furthermore, the scope and variety of the group of researchers and universities who make up the network provides a multidisciplinary outlook on problems with multiple possibilities for cooperation and the appearance of innovative ideas.

We propose to establish two distinct levels for the collaboration, with different names: a *"networked virtual centre for universities for knowledge management"* and a *"virtual network of university personnel for knowledge management."*

Membership of the higher level (i.e., the "networked virtual centre") would be restricted to research centres committed to carrying out knowledge management projects concerned with the process of research, development, and transfer of scientific and technical knowledge. Their basic purpose would be to exchange and share information, knowledge, and experience in regard to knowledge management at universities through an online structure, with particu-

lar reference to the managing of the transfer of scientific and technical knowledge to businesses and institutions. On this basis, forums, seminars, joint projects, and so forth on the relevant topics can be arranged.

The second level is the virtual network of university researchers for knowledge management (though the possibility of including non-university researchers might be considered, depending on their level of involvement) interested in going deeper into the field of knowledge management in general, and knowledge management at universities in particular, with special emphasis on managing the transfer of scientific and technical knowledge from universities to companies and institutions. This network would work mainly as a forum for discussion and for the exchange of information, ideas, and knowledge online, without excluding the possibility of joint projects and face-to-face meetings.

Figure 1 shows the structure of the two-level model proposed.

As can be observed, our proposal to some extent contains elements from both types of knowledge network considered.

Like communities of practice, they have distinct levels of participation, but the distinction in levels is not based, as it is in those communities, on the degree of involvement in tasks, but on such characteristics as the personality of members (individual or institutional), their degree of commitment (drawing up of a knowledge management project or mere commitment to confidentiality), and the type of tasks to be performed (exchange of ideas, knowledge, and experi-

Figure 1. Model of a virtual network of universities for knowledge management (Rodríguez, Araujo & Ranguelov, 2001)

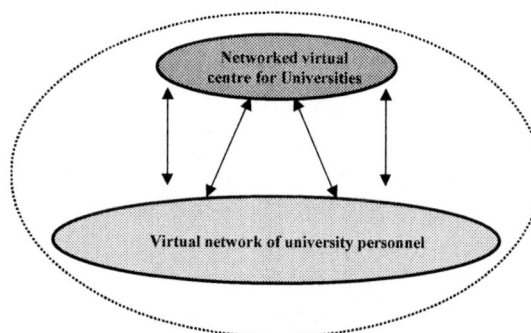

ence on projects and even joint projects, or merely a forum for debate and exchange on topics).

Our proposal could therefore be said to envisage two levels of network. As in communities of practice, there may be some process of learning and gradual integration, as researchers from a university may first come into contact with a topic in the virtual network and then undertake a project under the auspices of their university which raises them to the higher level of the network (i.e., the virtual centre).

CONCLUSION

This article introduced a project that develops a new model of virtual knowledge management network, which groups together universities and research teams, and enables them to exchange information and knowledge, and share their experience in the R&D area.

The proposed model for an innovative virtual network presents two levels of integration, namely a "networked virtual centre for universities for knowledge management" and a "virtual network of university researchers for knowledge management." This hybrid knowledge network includes characteristics of both types of virtual communities considered in the article. The two levels are not based on the degree of involvement in the activities, but on the personality of members, their degree of commitment, and the type of tasks to be performed. Our proposal could therefore be said to envisage two different but connected virtual communities linked by a common theme, though a learning process is possible which would enable movement from one to the other.

On that basis, the presented model introduces four important innovation elements in the field of R&D and network communities. The first element is the creation of mutual trust and adaptation of members' activities and to promote the interdisciplinary research. The second element is the development of an information network, along with virtual and interactive work methods between universities and individual university researchers. The third point considers the development of an exchanging mechanism of knowledge, good practices, and experiences in the field of managing knowledge in the research, development, and transference process. And the fourth element is about the joint management of the knowledge produced inside the knowledge network.

REFERENCES

Adams, E.C., & Freeman, C. (2000), Communities of practice: Bridging technology and knowledge assessment. *Journal of Knowledge Management, 4*(1), 38-44.

Alba, R.D. (1982). Taking stock of network analysis: A decade's results. In S.B. Bacharach (Ed.), *Research in sociology of organizations* (pp. 39-74). Greenwich: JAI Press.

Debakere, K., Clarisse, B., & Rappa, M.A. (1996). Dismantling the ivory tower: The influence of networking on innovative performance in emerging fields of technology. *Technological Forecasting and Social Change, 53*(2), 139-154.

Donath, J.S. (1998). Identity and deception in the virtual community. In P. Kollock & M. Smith (Eds.), *Communities in cyberspace.* London: Routledge.

Drucker, P.F. (1989). *The new realities.* New York: Harper Business.

Echeverri-Carroll, E.L. (1999). Knowledge flows in innovation networks: A comparative analysis of Japanese and U.S. high-technology firms. *Journal of Knowledge Management, 3*(4), 296-303.

Jin, Z. (1999). Organizational innovation and virtual institutes. *Journal of Knowledge Management, 3*(1), 75-83.

Koch, M., & Wörndl, W. (2001, September). Community support and identity management. *Proceedings of the European Conference on Computer-Supported Cooperative Work* (ECSCW 2001), Bonn, Germany.

Koschtzky, K. (2001). *Räumliche Aspekte im Innovationsprozess. Ein Beitrag zur Neunen Wirtschaftsgeographie aus Sicht Der Regionalen Innovarionsforschung.* Münster: Lit-Verlag.

Lave, J., & Wenger, E. (1991). *Situated learning. legitimate peripheral participation.* Cambridge: Cambridge University Press.

Lesser, E.L., & Storck, J. (2001). Communities of practice and organizational performance. *IBM Systems Journal, 40*(4), 831-841.

Lodge, G.C., & Walton, R.E. (1989). The American corporation and its new relationships. *California Management Journal, 31*, 9-24.

Mitchell, J.C. (1969). The concept and use of social networks. In J.C. Mitchell (Ed.), *Social networks in urban situations* (pp. 1-12). Manchester: Manchester University Press.

Nonaka, I., & Takeuchi, I. (1995). *The knowledge creating company. How Japanese companies create the dynamics of innovation.* Oxford: Oxford University Press.

Real Communities Inc. (2000). Shared knowledge and a common purpose: Using 12 principles of civilization to build Web communities. White Paper retrieved from *http://www.realcommunities.com/*

Rodríguez, A., Araujo, A., & Ranguelov, S. (2001). Redes virtuales de universidades para la gestión del conocimiento. Una propuesta. In F.C. Morabito & P. Laguna (Eds.), *Nuova Economia, Vecchi Problemi. Best Papers Proceedings 2001, X International Conference AEDEM* (pp. 891-901), European Association of Management and Business Economics, Reggio Calabria, Italy.

Savage, C. (1990). *Fifth generation management. Integrating enterprises through human networking.* Boston: Butterworth-Heinemann.

Seufert, A., Krogh, G., & Bach, A. (1999). Towards knowledge networking. *Journal of Knowledge Management, 3*(3), 180-190.

Wenger, E. (1999). *Communities of practice: Learning, meaning and identity.* Cambridge, UK: Cambridge University Press.

KEY TERMS

Communities of Practice (CoPs) Membership: In these communities, new members are included in tasks concerned with the practices of the group and they acquire knowledge from more expert members. Participation in communities of practice therefore involves movement from the periphery towards full participation in the group.

Hybrid Knowledge Network: On one side, it groups together elements from the traditional virtual community of interest and from the more sophisticated communities of practice, and on the other side, promotes collaborative multidisciplinary research that produces high-quality research results and stimulates their transfer.

Knowledge Network(s): These networks are viewed like a reflection of the growing dependence of the companies in front of external sources of knowledge. Therefore, they are not only sensitive to the social contacts, but rather they are also important for the mobility of the knowledge and, in consequence, at the space distance among the components of the network.

Legitimacy in CoPs: The element that defines the force and authority in relations within the group, but that legitimacy need not be formal.

Levels in the Hybrid Network: The proposed model of network includes distinct levels of participation, but the distinction in levels is not based on the degree of involvement in tasks, but on the personality of members, their degree of commitment, and the type of tasks to be performed.

Participation in CoPs: The key to understanding these organisations, as it is the most important factor if they are to be managed satisfactorily. With the active participation of all members, the working language can be unified, the goals pursued clarified, and the achievement of those goals brought closer.

Peripheral in CoPs: Does not define a physical measurement of remoteness from a hypothetical "centre," since the concepts "remaining on the periphery" and "full integration" indicate first and foremost degrees of commitment to the community.

ENDNOTE

[1] More detailed information about this project can be seen in Rodríguez et al. (2001). The project received funding in 2000 and 2001 from the Spanish Ministry of Science and Technology in 2000 and 2001, and from the Programme for fostering the Basque Technological network of the Basque Government in 2001-2002.

Identifying Knowledge Flows in Communities of Practice

Oscar M. Rodríguez-Elias
CICESE, Mexico

Ana I. Martínez-García
CICESE, Mexico

Aurora Vizcaíno
University of Castilla-La Mancha, Spain

Jesús Favela
CICESE, Mexico

Mario Piattini
University of Castilla-La Mancha, Spain

INTRODUCTION

Knowledge sharing is a collective process where the people involved collaborate with others in order to learn from them (Huysman & de Wit, 2000). This kind of collaboration creates groups of people with common interest called communities of practice where each member contributes knowledge about a common domain (Wenger, 1998).

Communities of practice enable its members to benefit from the knowledge of each other (Fontaine & Millen, 2004). To achieve this, different techniques and technologies can be used, such as shared documentation, groupware tools, lessons learned systems, and so forth. Therefore, to increase and improve knowledge sharing in communities of practice, it is important to study the mechanisms used by a particular community and understand how the knowledge flows through its members (Guizzardi, Perini & Dignum, 2003).

This article presents a qualitative approach for studying and understanding how knowledge flows in communities of practice within organizations. The goal is to provide a methodological guide for obtaining useful information for the development of knowledge management tools for supporting *knowledge flows* in these communities.

The content of the article is organized as follows. First the importance of supporting *knowledge flows* in communities of practice is highlighted. Then, a quali-

tative methodology for identifying *knowledge flows* in communities of practice is described, followed by some examples from a study conducted in the field of software maintenance. Finally, we present our conclusions of this work and future research.

MAIN BODY: KNOWLEDGE FLOWS IN COMMUNITIES OF PRACTICE

In a knowledge-intensive organization, employees constantly have to deal with a changing environment where knowledge is crucial to make decisions and adapt to these changes. To obtain the required knowledge for making those decisions, employees generate communities where each member collaborates with the others sharing knowledge about a common domain. On the other hand, to facilitate their adaptation, the organization's processes must become dynamic, that is, they must be designed to change based on the knowledge involved and on the activities performed by the members of the organization. Knowledge management (KM) can help address this issue, since it provides methods, techniques, and tools for facilitating organizations to become adaptable to these changing environments (Davenport & Prusak, 2000; Tiwana, 2000).

One of the main objectives of KM is to make available the appropriate knowledge, in the right place, at the right moment, to whoever needs it; therefore the flow of knowledge is very important for managing the knowledge of an organization (Nissen & Levitt, 2002). In fact, it has been considered the central component of a KM system (Borghoff & Pareschi, 1998). Communities of practice stimulate this flow of knowledge through organizations, since knowledge flows easily in these communities because they enable face-to-face interaction between their members (Brown, 2002; Fontaine & Millen, 2004). Even though direct interaction between members of the community is very important for sharing their tacit knowledge, other kinds of knowledge transfer must be considered such as documents sharing. Hence, provision of mechanisms that facilitate, increment, and improve the transfers of both tacit and explicit knowledge into communities of practice it is required. Therefore, *knowledge flow* must be one of the most important issues for supporting KM in these communities, since the goal is that the knowledge of each member can be used by the others (Borghoff & Pareschi, 1998; Guizzardi et al., 2003).

To provide support to the *knowledge flow* of a community, it is important to identify specific issues of the dynamics of *knowledge flows* in the processes and activities performed by the members of that community, as well as the social, cultural, and technological aspects which can affect those flows, in order to provide useful insights for the definition of requirements for designing KM systems that support the flow of knowledge in the community (Rodríguez, Martínez, Favela, Vizcaíno & Piattini, 2004a). A process modeling approach, as used in business processes reengineering (Curtis, Kellner & Over, 1992), can be appropriate for this purpose, since it provides techniques for analyzing technological and social aspects in organizations, as well as for modeling the dynamics of their processes. Once identified and understood how the knowledge flows through the community and which are the main elements that affect that flow, other approaches can be used for implementing the support systems—for example, an agent-oriented approach such as the proposed by Guizzardi et al. (2003, 2004).

In the following section we present a qualitative methodology for identifying *knowledge flows* in communities of practice; this is a methodology that we have defined and followed to obtain requirements for the design of a KM support system for a software maintenance group.

KOFI: A METHODOLOGY FOR KNOWLEDGE FLOWS IDENTIFICATION

To design and develop support systems, such as for KM, for communities of practice, it is important to consider the contextual issues of the customers or those who will use the system (Beyer & Holtzblatt, 1998). We think *knowledge flow* must be a central aspect for supporting communities of practice; therefore, to understand the context of those communities, it is important to understand which kinds of knowledge are important for the community, which knowledge sources they share and how to obtain that knowledge, which mechanisms they use to consult the sources, and how all of these interact in the processes and activities performed by the members of the community—in general, how the knowledge flows through the community (Rodríguez et al., 2004a). To obtain answers for these questions, we have defined a qualitative methodology to guide the process of identifying how knowledge flows in a community of practice, and how to provide support to facilitate, increment, and improve the flow of knowledge in the community by identifying the problems that affect that flow.

THE METHODOLOGY

The methodology is composed of four stages, as shown in Figure 1. In stage one the main sources of knowledge and information are identified and classified (documents and people); then, in stage two, the knowledge contained in those sources is also defined and classified; in the third stage the main processes and activities performed by the members of the community are modeled to identify the people involved, how they collaborate to complete their tasks, and how the knowledge and sources interact in those activities; finally, in stage four the main problems that can affect the flow of knowledge are highlighted through the definition of scenarios. The process proposed to carry out the above stages is iterative, since each stage could generate information to complement the others. For example, if we identify a new kind of knowledge source while we are modeling flows of knowledge, we can add the source

Figure 1. Stages of the methodology for identifying knowledge flows

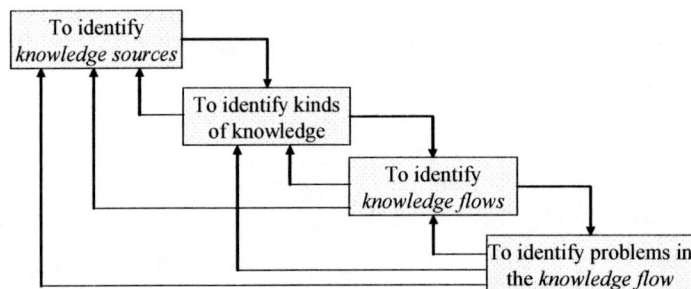

kind to the ontology and then identify the kinds of knowledge that can be obtained from it.

In the following subsections we describe more details about each stage and present some examples about how they can be carried out. These examples have been obtained from a case study in a software maintenance organization, where a multi-agent knowledge management system was designed with requirements obtained from the results of the study (Rodríguez et al., 2004a; Rodríguez, Vizcaino, Martínez, Piattini & Favela, 2004b).

IDENTIFYING AND CLASSIFYING KNOWLEDGE AND KNOWLEDGE SOURCES

The first step starts by identifying the main documents and people involved in the community. Then, in stage two, the documents are analyzed in order to define the kinds of knowledge that can be obtained from those, together with the kinds of knowledge that the people involved can have or require for their activities. Taxonomies can be defined to classify the knowledge sources found and the kinds of knowledge these sources have; also an ontology can be designed to help define the relations between the sources and the kinds of knowledge.

Ontologies are conceptual models for specifying meanings or knowledge about a common domain; they can be used to provide a framework for sharing these meanings or knowledge (Gruber, 1995; Maedche, Motik, Stojanovic, Studer & Volz, 2003). Therefore, ontologies can be used for specifying information sources and the knowledge they can have, as well as the connections between them, in order to develop a conceptual framework of these relations.

Figure 2 presents a general ontology used for classifying knowledge and its sources in the case

Figure 2. A generic ontology of knowledge sources and knowledge topics

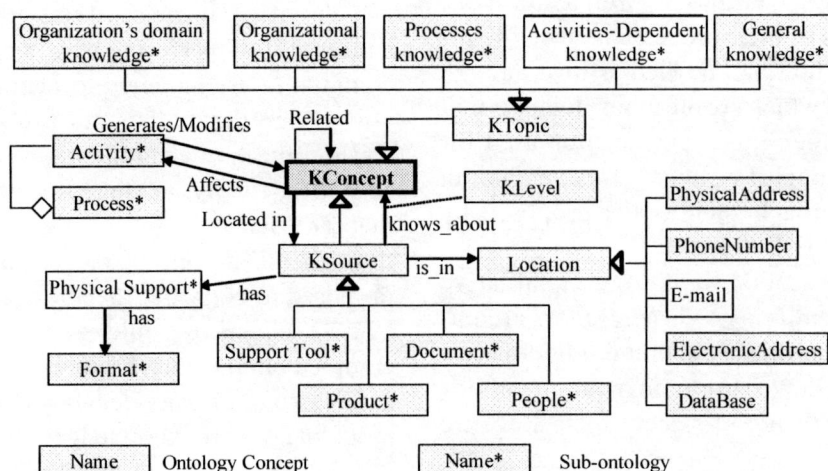

study carried out. This ontology is used for identifying *knowledge concepts* (KConcept) which can be both *knowledge sources* (KSource) or *knowledge topics* (KTopic). The *knowledge concepts* involved in an activity can affect that activity in some way; for example, in order to perform an activity, some *knowledge topics* can be necessary or some *knowledge sources* can be required; moreover, an activity can generate or modify some topics or sources of knowledge. Some elements of the ontology have been defined as sub-ontologies, and their structure must be specified for the particular needs of the studied organization or community.

Knowledge sources can be people, documents, support tools (such as organizational memories, experience repositories, etc.), and the products developed or built by the organization. For example, in a software organization, the systems developed (source code and executable program) can be a very useful source of knowledge. Each *knowledge source* can have a specified physical support (such as paper, electronic file, audiotape, videotape, etc.) and a format (such as Word document, PowerPoint presentation, etc.); they can also have one or more locations which define how they can be consulted; and finally, the sources have a level of knowledge about *knowledge topics* or other *knowledge sources*.

Knowledge topics have been classified in five main groups:

1. those related to the organization's domain knowledge, for example, if the organization develops software for telephonic services management, it must know about the call fees of the different kinds of calls of each telephone company;
2. knowledge about the structure of the organization, its norms, its culture, and so forth;
3. knowledge about the processes of the organization, for example, the activities, the people involved, and so forth;
4. knowledge dependent of specific activities, for example the procedures or support tools used for performing the activity, and so forth; and
5. other kinds of knowledge that can be important, for example, it can be useful to know which employees speak foreign languages or have other skills that are not used in their daily work.

This ontology can be used for defining and classifying the kinds and sources of knowledge and how all these are related. This information can be later used for defining the structure of a knowledge base, for example, by specifying the most important knowledge topics for the organization, the sources of knowledge available and the kinds of knowledge that can be obtained from those sources.

In the third stage of the methodology, we have followed a process modeling approach to identify the flows of knowledge by modeling the activities performed by the community, the knowledge required and generated in the activities, the people in charge of them, and the sources of knowledge used, modified, or generated during the activities. This approach is presented in the following section.

KNOWLEDGE FLOWS MODELING

A process modeling (Curtis et al., 1992) approach can be very useful to identify how the knowledge and sources of information are involved in the activities performed by the community. To do this, the main activities of the processes carried out by the community must be identified, as well as the decisions that the people involved must make while they perform those activities. A graphical modeling technique, such as rich picture (Monk & Howard, 1998), can be used to model these activities. Rich pictures are cartoon-like representations that identify actors, roles, their concerns, and some of the structure underlying the work context. Thus, these kinds of representations can be useful to model the people and roles involved in some activities, the knowledge required by them to perform the activities, and the sources they consult or those that could have information to help them to complete their activities. These models can be later used to analyze how the knowledge flows through the group while its members perform their activities.

Figure 3 illustrates an example of a graphical model, which shows the main activities performed in the definition of the modification plan carried out by the group studied. The model shows the people involved in those activities, the knowledge they have together with their relevance to the activities modeled, and the main sources used, created, or modified in the activities.

Table 1. Schema used to identify knowledge in decision making

Role	Project leader	
Activity	To define modification plan	
Decision	To define required resources	
	To define main tasks to perform	
	To assign tasks to the participants of the project	
	To estimate the time the project would consume	
Knowledge	Previous projects' experiences	
	Requirements and restrictions of the project	
	Abilities and experience of each of the possible participants of the project	
Sources of information		
Name	Information	Consulted at
Chief of the department	Available resources; time and cost restrictions	Telephone, Physical address, Email
Software engineers	Experience with the system that will be modified; time that could consume the modifications; time availability	Telephone, Physical address, Email
Previous projects' documentation	Resources required by previous projects	Documents' files, modifications' logbook

Once the activities have been modeled, the next step is to define the decisions that must be made by the people involved. To do that, we used the schema shown in Table 1. This schema helps to identify the knowledge that the people in charge of the activities must have to make the decisions required, and the sources they consult to obtain information that helps them to make those decisions. At this step, it is important to identify the mechanisms that people can use to consult the sources, as well as those used to share the knowledge generated in the activities—for example, the documentation of the modifications' plan in Figure 3.

The analysis of the activities performed by the members of the community, using the graphical model and the information from the tables, are later used to understand how the knowledge flows through the community, and what techniques they use to share and obtain that knowledge. Finally this analysis can help to identify the problems that are affecting that flow. We next describe how scenarios can be used for this purpose.

SCENARIOS FOR IDENTIFYING FAULTS IN THE KNOWLEDGE FLOWS

In the fourth stage of the methodology, the models generated in the previous phase are analyzed to find the problems that could be affecting the flow of

Figure 3. An example of a model of activities performed by members of a maintenance group

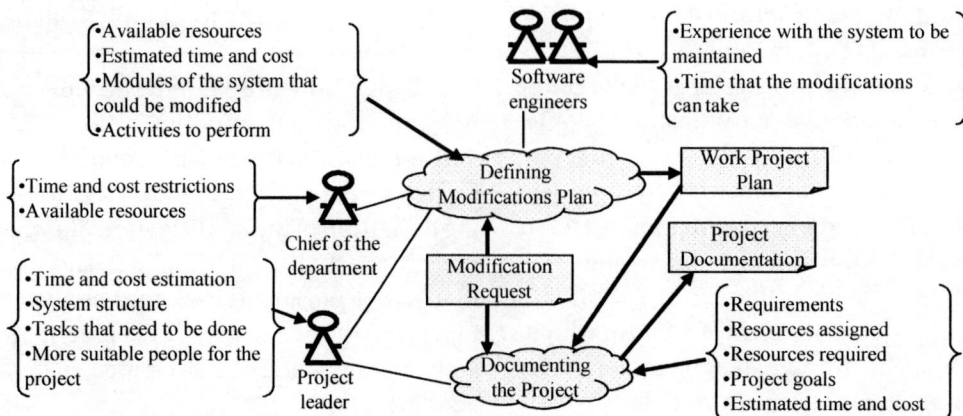

Table 2. An example of a problem description scenario and an alternative scenario

Kind:	Expert finding (knowledge sources management)
Problem description:	
Mary is a software engineer that must make some changes in the finances system. Since her knowledge in the domain of finances is not good enough, the changes to the system are taking more than a week of the estimated time. At the end of the week, Susan, the chief of the department, while she was checking the advances of the project, detects the delay and asks Mary the reasons of that delay. Mary tells Susan the problem and since Susan has experience with finances, she tells Mary how the problem could be solved. Finally, Mary solves the problem the same day.	
Alternative:	
When Mary decides to solve the problem of the finances system, the tool where Mary manages her tasks detects this action. This tool knows about Mary's knowledge, and identifies the kind of knowledge that Mary needs to make the changes in the finances system, so the tool identifies that Mary probably will need to consult some sources of knowledge and decides to search for those sources to help Mary do her Job. The tool founds some sources that can be relevant to the task Mary will perform, thus the tool informs Mary about it. Then, Mary decides to see the kind of knowledge those sources can have, and based on that, decides to consult Susan who is one of the sources found by the tool.	

knowledge—for example, if the information generated from the activities is not captured, or if there are sources that could help in performing some activities, but they are not consulted by the people in charge. In this stage, *problem scenarios* can help identify how the problems detected affect the *knowledge flow*, and how these could be addressed. These *problem scenarios* could be later used to obtain design requirements to the development of tools to address these problems, since scenarios enable the identification of design requirements for software systems and make feasible the participation of users during the requirements specification stage (Chin, Rosson & Carroll, 1997).

A scenario is a textual description of the activities that people might engage in while pursuing a particular concern (Carroll & Rosson, 1992). Hence, the *problem scenarios* can be structured as a story of particular problems detected from the analysis of the information obtained in the previous stages. Then these scenarios can be studied in order to discuss how those problems can be tackled. Table 2 presents an example of the description of a *problem scenario* obtained from the group studied and an alternative scenario where the knowledge sources are provided by a system. These kinds of descriptions can provide insights, which can later be used for defining requirements for developing support tools focused on addressing the problems identified.

As we mentioned before, the methodology has been applied in a case study in the software maintenance field (Rodríguez et al., 2004a). The first two phases of the methodology helped us to identify the main knowledge sources available for the members of

the maintenance groups, as well as the kinds of knowledge these sources have. This information was useful for developing a knowledge base to help find knowledge sources for maintainers to do their jobs. Then, the third phase guided us in identifying the activities where these sources are involved, the kinds of knowledge required or generated in those activities, and the mechanisms maintainers used to consult those sources or to obtain the required knowledge; that is, this last phase helped us to identify how the knowledge was flowing in the maintenance community. Finally, the scenarios defined in the fourth phase were used to obtain design requirements to develop a knowledge management system for helping maintainers to reduce the loss and waste of knowledge by facilitating the search of knowledge sources related to the activities they perform (Rodríguez et al., 2004a, 2004b).

CONCLUSION AND FUTURE POSSIBILITIES

The flow of knowledge is a very important factor for communities of practice, since one of the goals of these communities is to provide an environment where their members could share knowledge with others in order to learn together. Thus, for providing support to these communities, we think that the flow of knowledge through their members must be considered a central aspect of the design of the support tools. To address these issues, in this article we presented a qualitative methodology for studying how the knowledge flows through communities of practice in organizations, and how to identify the problems that can be affecting that flow, in order to use all this information to provide tools to support the flow of knowledge between the members of a community. The proposed methodology has been applied in a case study in a software maintenance group, where an appropriate knowledge management system according to the results obtained in this study has been designed.

We think it is important to consider the particular aspects of each community to provide better support for its particular needs. Thus, it is important to identify the knowledge needed by the members of the community, the sources they use to obtain that knowledge, the particular processes and activities carried out by them, as well as the main decisions they must make. All these aspects are considered by the proposed methodology.

Nevertheless, in order for the methodology to be more useful, we consider it necessary to provide tools for managing the information obtained by applying it—for instance, tools for defining the structure of the ontology of knowledge and knowledge sources, and for capturing information about the specific knowledge topics and sources in a knowledge base that could be later used by the tools developed to support the community. At the moment, we are working on providing this kind of support for the methodology.

REFERENCES

Beyer, H., & Holtzblatt, K. (1998). *Contextual design*. San Francisco: Morgan Kaufmann.

Borghoff, U.M., & Pareschi, R. (1998). *Information technology for knowledge management*. Berlin: Springer-Verlag.

Brown, J.S. (2002). How does your knowledge flow? An interview with John Seely Brown. *CSC World*, (Spring/Summer), 24-25.

Carroll, J.M., & Rosson, M.B. (1992). Getting around the task-artifact cycle: How to make claims and design by scenario. *ACM Transactions on Information Systems, 10*(2), 181-212.

Chin, G.J., Rosson, M.B., & Carroll, J.M. (1997). Participatory analysis: Shared development of re-

quirements from scenarios. *Proceedings of the Conference on Human Factors in Computing Systems* (CHI97), Atlanta, GA.

Curtis, B., Kellner, M.I., & Over, J. (1992). Process modeling. *Communications of the ACM, 35*(4), 75-90.

Davenport, T.H., & Prusak, L. (2000). *Working knowledge: How organizations manage what they know*. Boston: Harvard Business School Press.

Fontaine, M.A., & Millen, D.R. (2004). Understanding the benefits and impact of communities of practice. In P. Hildreth & C. Kimble (Eds.), *Knowledge networks: Innovation through communities of practice* (pp.1-13). Hershey, PA: Idea Group Publishing.

Guizzardi, R.S.S., Perini, A., & Dignum, V. (2003). *Using intentional analysis to model knowledge management requirements in communities of practice* (pp. 3-53). CTIT Technical Report, University of Twente, The Netherlands.

Guizzardi, R.S.S., Perini, A., & Dignum, V. (2004). Providing knowledge management support to communities of practice through agent-oriented analysis. *Proceedings of the 4th International Conference on Knowledge Management* (I-KNOW'04), Granz, Austria.

Gruber, T.R. (1995). Towards principles for the design of ontologies used for knowledge sharing. *International Journal on Human-Computer Studies, 43*(5/6), 907-928.

Huysman, M., & de Wit, D. (2000). *Knowledge sharing in practice* (vol. 4). Dordrecht: Kluwer.

Maedche, A., Motik, B., Stojanovic, L., Studer, R., & Volz, R. (2003). Ontologies for enterprise knowledge management. *IEEE Intelligent Systems, 18*(2), 26-33.

Monk, A., & Howard, S. (1998). The rich picture: A tool for reasoning about work context. *Interactions, 5*(2), 21-30.

Nissen, M.E., & Levitt, R.E. (2002, November). *Dynamic models of knowledge-flow dynamics*. Working Paper #76, Center for Integrated Facility Engineering, Stanford University, USA.

Rodríguez, O.M., Martínez, A.I., Favela, J., Vizcaíno, A., & Piattini, M. (2004a). Understanding and supporting knowledge flows in a community of software developers. In de Vreede et al. (Eds.), *Groupware: Design, implementation, and use* (pp. 52-66). Berlin: Springer-Verlag (LNCS 3198).

Rodríguez, O.M., Vizcaino, A., Martínez, A. I., Piattini, M., & Favela, J. (2004b). How to manage knowledge in the software maintenance process. In G. Melnik & H. Holz (Eds.), *Advances in learning software organization* (pp. 78-87). Berlin: Springer-Verlag (LNCS 3096).

Tiwana, A. (2000). *The knowledge management toolkit: Practical techniques for building a knowledge management system.* Englewood Cliffs, NJ: Prentice-Hall.

Wenger, E. (1998). *Communities of practice: Learning, meaning, and identity.* Cambridge, UK: Cambridge University Press.

KEY TERMS

Graphical Process Modeling Technique: A technique for representing models of processes with a graphical notation.

Knowledge Concept: A concept that is part of an ontology used for defining and describing the knowledge related with a community, such as the kinds of knowledge or the sources of knowledge.

Knowledge Flow: Defines how the knowledge flows through the activities performed by a community according to the kinds of knowledge and knowledge sources involved in the activities, the mechanisms used by the people involved in the activity to obtain or share that knowledge, and so forth.

Knowledge Source: A source of information which can be useful to obtain knowledge for practical application such as know-how, know-what, know-where, and so forth—for example, lessons learned or members of the community.

Knowledge Topic: Definition of a particular area of knowledge useful for a person or the members of a community.

Ontology: An explicit and formal representation of a shared conceptualization. Ontologies are conceptual models for specifying meanings or knowledge about a common domain.

Problem Scenario: A textual description of a problem observed in a community studied, which has the form of a story that illustrates the problem and possible solution alternatives.

Process Modeling: Collection of techniques used to model systems' behavior. These models help in analyzing the current state of organizations as facilitators of organizational learning.

The Impact of Communities of Practice

Katja Zboralski
Berlin University of Technology, Germany

Hans Georg Gemünden
Berlin University of Technology, Germany

INTRODUCTION

In today's knowledge-based and networking economy, an organization's ability to acquire, develop, and strategically leverage knowledge has become a crucial factor for global competitiveness (Drucker, 1993; Kogut & Zander, 1992; Leonard-Barton, 1995; Nonaka & Takeuchi, 1995). Consequently, a growing number of companies have introduced knowledge management systems into their organizations. The purpose of these efforts is to use the resource knowledge more effectively and efficiently and thereby gain strategic advantages (Davenport & Prusak, 1998; Probst, Raub & Romhardt, 1999). In this context, the concept of communities of practice (CoPs) has gained considerable attention as one of the central means of implementing knowledge management.

For more than a decade, the term *community of practice* (CoP) has been the subject of various discussions in theory and practice alike. The origin of CoPs lay in Lave and Wenger's (1991) seminal research toward a social theory of learning. By investigating learning in groups, the researchers called a community of practice an active system about which members share their understanding of what they do and which are united in action and in the meaning this action has. The increasing popularity of the concept in the scientific discourse and managerial practice brought about various interpretations of the term. Therefore, no universal definition of the term exists. The same applies for the name of this organizational phenomena. Nevertheless, while different organizations use different names, they share the underlying idea. Existing CoP definitions commonly stress the *activities* of these learning communities: to work together; exchange information, knowledge, and experiences, and thereby, learn and generate new knowledge and common practices

(Lesser & Storck, 2001; McDermott, 1999; Stewart, 1996; Wenger, 1998a). CoPs were initially understood as self-emerging and self-organizing organic networks in which everyone can participate (Wenger, 1998b). Current practice, however, shows that organizations strategically support existing networks and deliberately establish communities with managed memberships (Storck & Hill, 2000).

In the following, CoPs are defined as a group of people in an organization who interact with each other across organizational units or even across organization boundaries due to a common interest or field of application. Their objective is to learn and support one another in order to create, spread, retain, and use knowledge relevant to the organization.

STRATEGIC IMPACT OF COMMUNITIES OF PRACTICE

Communities of practice are particularly used by multinational companies in knowledge-intensive industries (APQC, 2000; Hildreth, Kimble & Wright, 2000). Related to a specific business topic, these networks are fostered and established in order to build strategic capabilities within the organization by leveraging learning and knowledge sharing (Lesser & Prusak, 1999; Saint-Onge & Wallace, 2003). But do communities of practice really meet managers' high expectations? Which concrete value do communities of practice deliver? In answering these questions, two levels are examined: the individual and the organizational.

Individuals as community members profit directly from their participation in the community. Although personal goals and individual motivation influence their perception of community benefits, the following general outputs can be distinguished: By communicating frequently, the community members develop a common language and a collective knowledge base. The shared context, increased *networking* between members, as well as emerging interpersonal relationships support not only access to new knowledge sources, but also the development of social capital (Lesser & Storck, 2001; Nahapiet & Ghoshal, 1998). Members' knowledge is reused and modified, and thereby transformed into new knowledge (Wenger, 1998b). Hence, the *personal knowledge* of the community members is increased, and new competences are gained which are beneficial for *improved performance* (McDermott, 2002). Due to advanced competences, community members are regarded as *experts* in a specific field which in turn leads to a higher reputation within the organization. This has a positive impact on their professional development and, as a consequence, on their *work satisfaction* (Schoen, 2001).

Strategic advantages for an organization result, above all, from community impacts on the *organizational level*. As emphasized by several authors, CoPs are forums for shared learning and action (Brown & Duguid, 1991; Smith & McKeen, 2003) and thereby, tools to increase *organizational learning capabilities* (Brown & Duguid, 2001; Hedberg & Holmquist, 2001). The underlying mechanism in this context is that community activities support the *externalization of knowledge*. Particularly, close and intense communication among community members foster the transfer of hitherto tacit knowledge which has been identified as a central mode of knowledge creation and a source of competitive advantage (Leonard & Sensiper, 1998; Nonaka, 1994). A common knowledge base is not only created at the individual level, but also at the organizational level. Existing know-how is improved, and new *organizational competences* are developed (Tsai & Goshal, 1998). Communities of practice exhibit a climate which may stimulate creativity through an open communication, the exchange of interdisciplinary knowledge and, thereby, the development of mutual trust (Storck & Hill, 2000). As members are encouraged to articulate new ideas and "think outside of the box" truly creative activities are fostered. Hence, communities enhance the *creative capacity* and, by this, the *innovative capability* of the organization (Brown & Duguid, 1991).

Besides these relatively intangible benefits, the impact of CoPs can also become apparent by hard facts. *Resource savings* result because CoPs may not only promote better solutions for problems and easier and faster access to knowledge, but also decrease training periods for new employees as well as help to avoid double work. Shared experiences, communicated, for example, as best practices and lessons learned, lead to decreased learning curves. Optimized and *accelerated processes* together with the developed knowledge base will potentially lead to higher customer satisfaction, as customer needs can be addressed in a more flexible manner (Lesser & Storck, 2001; Wenger & Snyder, 2000).

Last but not least, communities of practice can change the existing *organizational culture* in a favorable way. On one hand, the development of collective sense-making, a common language as well as the emergence of networks among members affect the culture. On the other hand, people's attitudes toward knowledge sharing change as it is actively approved and rewarded.

CRITICAL ISSUES

Although communities of practice have been applied in several organizations, there are ongoing discussions in research and practice on the concrete value CoPs create. Particularly, the question of how to measure these benefits is addressed (McDermott, 2002; Wolf, 2003). The reasons for this are twofold. First, evaluating community outcomes in terms of financial ratios is rather problematic (Schoen, 2001). Effects cannot always be directly linked to activities of the CoP but could also result from other contextual factors. Moreover, effects may only become apparent after a certain time lag. Besides, most of the community outcomes are intangible assets, therefore, difficult to measure (Adler & Kwon, 2002; Bontis, 2001; Carmeli, 2004). Second, assessing the exact costs of a community is challenging. They consist not only of technology investment, but of costs for participation in the community (opportunity costs, salaries, incentives), costs directly related to meetings, costs for maintaining the technical infrastructure, and costs for content publishing, promotional material, and so forth (Millen, Fontaine &

Muller, 2002). Consequently, the assignment of community benefits to costs is difficult. However, it should be stressed that the assessment of benefits and cost of *alternative means* of knowledge generation and diffusion is also very difficult.

As the objective of CoPs is to improve organizational performance, it is important to link CoPs and their value assessment to the organization's business strategy. In accordance with this strategy, a knowledge strategy should be developed taking into account existing knowledge-based resources and capabilities as well as knowledge required for products and processes (Zack, 1999). *Strategic goals of knowledge management*, particularly the desired level of knowledge depth (expertise), of knowledge diffusion in an organization (degree of externalization), and of knowledge innovation, should be mapped. Therefore, valuable knowledge fields have to be identified in a moderated dialogue between management and experts (Hofer-Alfeis & van der Spek, 2002).

Another critical aspect is that communities of practice do not, per se, solve the problem of sharing knowledge in an organization (von Krogh, 2002). They are rather a complement to traditional work structures (Smith & McKeen, 2003). Communities of practice cannot simply be launched into action; rather, they need to *emerge in an organic way*. As literature on knowledge management emphasizes (Davenport, De Long & Beers, 1998; Mertins, Heisig & Vorbek, 2003), *management support* and sponsorship are of high importance for successful knowledge management initiatives such as CoPs. An appropriate technical infrastructure provides the foundation of these learning networks but does not guarantee success (Barret et al., 2004; Jarvenpaa & Staples, 2000; Pan & Leidner, 2003). Above all, management needs to foster a *climate* which encourages people to participate in communities (De Long & Fahey, 2000; Wasko & Faraj, 2000).

Finally, potential threats of CoPs have not been considered in detail yet. In this context, negative group effects, for example, group-thinking, that have been observed in teams (Högl, 1998) might also appear in communities. Furthermore, power conflicts between the CoP and the formal organization due to knowledge edges and improved control, particularly in the context of measuring benefits, can also hinder

the exchange of knowledge (Fox, 2000; Swan, Scarbrough & Robertson, 2002).

FUTURE RESEARCH AND TRENDS

Identified *success factors for communities of practice* are mainly based on qualitative, case-study oriented research (Dyer & Nobeoka, 2000; Gongla & Rizzuto, 2001; Storck & Hill, 2000; Swan et al., 2002). Thus, further empirical research is needed to explore whether these findings can be generalized. Particularly, the *organizational context* in which CoPs need to be embedded, the effectiveness of various *information and communication technologies* with respect to different *types of knowledge* as well as the role of a *community broker* should be addressed by further studies. To broaden the perspectives of CoPs, the *potential of interacting with other partners* should be studied in more detail. On one hand, the interactions between different communities within an organization are of interest. On the other hand, the possible interplay of an organization's *CoP* with external partners, such as customers, universities, and research institutions, is a highly relevant research area.

From a research design perspective, a *longitudinal study* could take the *evolutionary aspect* of communities into account and could also address the question of measuring the impact of CoPs. Furthermore, a study in different organizations would make it possible to study effects of different contextual variables.

CONCLUSION

Taking into consideration all different benefits of community activities, one can say that communities of practice have the potential to improve organizational performance and, thereby, contribute to the reinforcement of an organization's long-term strategic advantages. Communities can be of paramount importance for the organization. Hence, management should actively facilitate community work by providing required resources and by establishing necessary prerequisites in the organization. Notwithstanding the potential of communities, CoPs should not be considered as "miracle weapons" in a company's pursuit of competitiveness. Organizations have to incorporate the organizational context, integrate CoPs in the processes and routines, and get people "to live" knowledge management. In the long term, organizations will only be able to survive in the knowledge-based economy if they manage the organizational and cultural change necessary.

REFERENCES

Adler, S. P., & Kwon, S.-W. (2002). Social capital: Prospects for a new concept. *Academy of Management Review, 27*(1), 17-40.

APQC. (2000). *Building and sustaining communities of practice: Final report*. Houston: American Productivity & Quality Center.

Barret, M., Cappleman, S., Shoib, G., & Walsham, G. (2004). Learning in knowledge communities: Managing technology and context. *European Management Journal, 22*(1), 1-11.

Bontis, N. (2001). Assessing knowledge assets: A review of the models used to measure intellectual capital. *International Journal of Management Reviews, 3*(1), 41-60.

Brown, J. S., & Duguid, P. (1991). Organizational learning and communities-of-practice: Toward a unified view of working, learning, and innovation. *Organization Science, 2*(1), 40-57.

Brown, J. S., & Duguid, P. (2001). Knowledge and organization: A social-practice perspective. *Organization Science, 12*(2), 198-213.

Carmeli, A. (2004). Assessing core intangible resources. *European Management Journal, 22*(1), 110-122.

Davenport, T. H., De Long, D. W., & Beers, M. C. (1998). Successful knowledge management projects. *Sloan Management Review, 39*(2), 43-57.

Davenport, T. H., & Prusak, L. (1998). *Working knowledge: How organizations manage what they know*. Boston: Harvard Business School Press.

De Long, D. W., & Fahey, L. (2000). Diagnosing cultural barriers to knowledge management. *Academy of Management Executive, 14*(4), 113-127.

Drucker, P. F. (1993). *Post-capitalist society*. New York: Harper Business.

Dyer, J. H., & Nobeoka, K. (2000). Creating and managing a high-performance knowledge-sharing network: The toyota case. *Strategic Management Journal, 21*(3), 345-367.

Fox, S. (2000). Communities of practice, Foucault and actor-network theory. *Journal of Management Studies, 37*(6), 853-867.

Gongla, P., & Rizzuto, C. R. (2001). Evolving communities of practice: IBM Global Services experience. *IBM Systems Journal, 40*(4), 842-862.

Hedberg, B., & Holmquist, M. (2001). Learning in imaginary organizations. In M. Dierkes, A. Berthoin Antal, J. Child, & I. Nonaka (Eds.), *Handbook of organizational learning & knowledge* (pp. 733-752). New York: Oxford University Press.

Hildreth, P., Kimble, C., & Wright, P. (2000). Communities of practice in the distributed international environment. *Journal of Knowledge Management, 4*(1), 27-38.

Hofer-Alfeis, J., & van der Spek, R. (2002). The knowledge strategy process: An instrument for business owners. In T. H. Davenport & G. J. B. Probst (Eds.), *Knowledge management case book: Siemens best practices* (pp. 24-39). Erlangen: Publicis Corporate Publishing and John Wiley & Sons.

Högl, M. (1998). *Teamarbeit in innovativen projekten: Einflußgrößen und Wirkungen*. Wiesbaden, Germany: Deutscher Universitätsverlag

Jarvenpaa, S. L., & Staples, D. (2000). The use of electronic media for information sharing: An exploratory study. *Journal of Strategic Information Systems, 9*(2/3), 129-154.

Kogut, B., & Zander, U. (1992). Knowledge of the firm, combinative capabilities and the replication of technology. *Organization Science, 3*(3), 383-397.

Lave, J., & Wenger, E. (1991). *Situated learning: Legitimate peripheral participation*. Cambridge, UK: Cambridge University Press.

Leonard, D., & Sensiper, S. (1998). The role of tacit knowledge on group innovation. *California Management Review, 40*(4), 112-132.

Leonard-Barton, D. (1995). *Wellsprings of knowledge: Building and sustaining the sources of innovation.* Boston: Harvard Business School Press.

Lesser, E. L., & Prusak, L. (1999). *Communities of practice, social capital and organizational knowledge* (White Paper). IBM Institute for Knowledge Management.

Lesser, E. L., & Storck, J. (2001). Communities of practice and organizational performance. *IBM SystemsJournal, 40*(4), 831-841.

McDermott, R. (1999). Learning across teams: The role of communities of practice in team organizations. *Knowledge Management Review, 2*(8), 32-36.

McDermott, R. (2002). Measuring the impact of communities. *Knowledge Management Review, 5*(2), 26-29.

Mertins, K., Heisig, P., & Vorbeck, J. (2003). *Knowledge management: Concepts and best practices* (2nd ed.). Berlin: Springer.

Millen, D. R., Fontaine, M. A., & Muller, M. J. (2002). Understanding the benefit and costs of communities of practice. *Communications of the ACM, Special Issue: Supporting Community and Building Social Capital, 45*(4), 69-73.

Nahapiet, J., & Ghoshal, S. (1998). Social capital, intellectual capital and the organizational advantage. *Academy of Management Review, 23*(2), 242-266.

Nonaka, I. (1994). A dynamic theory of organizational knowledge creation. *Organization Science, 5*(1), 14-37.

Nonaka, I., & Takeuchi, H. (1995). *The knowledge-creating company: How Japanese companies create the dynamics of innovation.* Oxford: Oxford University Press.

Pan, S. L., & Leidner, D. E. (2003). Bridging communities of practice with information technology in pursuit of global knowledge sharing. *Journal of Strategic Information Systems, 12*, 71-88.

Probst, G. J. B., Raub, S., & Romhardt, K. (1999). *Wissen managen: Wie Unternehmen ihre wertvollste Ressource optimal nutzen* (3rd ed.). Frankfurt am Main, Wiesbaden: Frankfurter Allg. Zeitung für Deutschland; Gabler.

Saint-Onge, H., & Wallace, D. (2003). *Leveraging communities of practice for strategic advantage.* Amsterdam: Butterworth-Heinemann.

Schoen, S. (2001). *Gestaltung und unterstützung von communities of practice.* München: Herbert Utz Verlag.

Smith, H. A., & McKeen, J. D. (2003). Creating and facilitating communities of practice. In C. W. Holsapple (Ed.), *Handbook on knowledge management: Knowledge matters* (Vol. 1, pp. 393-407). Berlin; Heidelberg; New York: Springer.

Stewart, T. A. (1996, August 5). The invisible key to success: Shadowy groups called community of practice. *Fortune Magazine,* 173-176.

Storck, J., & Hill, P. A. (2000). Knowledge diffusion through "strategic communities". *Sloan Management Review, 41*(2), 63-74.

Swan, J., Scarbrough, H., & Robertson, M. (2002). The construction of "communities of practice" in the management of innovation. *Management Learning, 33*(4), 477-496.

Tsai, W., & Goshal, S. (1998). Social capital and value creation: The role of intrafirm networks. *Academy of Management Journal, 41*, 464-476.

von Krogh, G. (2002). The communal resource and information systems. *Journal of Strategic Information Systems, 11*, 85-107.

Wasko, M. M., & Faraj, S. (2000). "It is what one does": Why people participate and help others in electronic communities of practice. *Journal of Strategic Information Systems, 9*, 155-173.

Wenger, E. C. (1998a). Communities of practice: Learning as a social system. *The Systems Thinker, 9*(5), 1-5.

Wenger, E. C. (1998b). *Communities of practice: Learning, meaning and identity.* Cambridge, UK: Cambridge University Press.

Wenger, E. C., & Snyder, W. M. (2000). Communities of practice: The organizational frontier. *Harvard Business Review, 78*(1), 139-145.

Wolf, P. (2003). *Erfolgsmessung der Einführung von Wissensmanagement. Eine Evaluationsstudie im Projekt "Knowledge Management" Der Mercedes- Benz Pkw- Entwicklung der DaimlerChrysler AG*. Münster: Monsenstein und Vannerdat.

Zack, M. H. (1999). Developing a knowledge strategy. *California Management Review, 41*(3), 125-145.

KEY TERMS

Community Broker: In this context, a community broker is the leader of a community who provides the overall guidance and management to establish and/or maintain a community. He/she supports community activities, promotes the community within the organizations and acts as the contact person for both community members and people interested in (joining) the community.

Knowledge: Knowledge refers to gained and preserved results of experiences, perception, learning, and reasoning. It can be considered information associated with intentionality; it allows the interpretation of situations and the generation of activities.

Knowledge Management: Knowledge management is the intentional influencing of an organization's knowledge resources and capabilities. Its activities aim to capture, codify, transfer, share, and create knowledge relevant for the organization.

Knowledge Management Systems: A knowledge management system is an information and communication technology platform combining and integrating functions for handling knowledge.

Organizational Learning: Organizational learning describes an organization's ability to gain insight and understanding from experience based on experimentation, observation, analysis, and an existing willingness to examine and learn from both successes and failures.

Social Capital: An organization's social capital can be defined as the collective value of the network of relationships and the benefits that arise from these networks.

Information Security as a Community of Practice

Paul Drake
University of Hull, UK

INTRODUCTION AND BACKGROUND

This entry concerns a live application in which the principles of communities of practice have been used to supplement the delivery of a critical business process. The company concerned is a multinational pharmaceutical organization with annual sales in the order of £20bn and a workforce of 100,000 employees worldwide.

One of the more critical IT services provided is that which defends the organisation's computer systems against attack by malicious software (commonly called computer viruses). This service draws significant direct and indirect resources to provide an acceptable level of defence for the organization. The service manages the provision of this defence from the gathering of intelligence concerning latest threats through deployment of protective measures to reporting of metrics showing service performance and adequacy of defences.

SERVICE DELIVERY THROUGH A COMMUNITY OF PRACTICE

The service is delivered through a retained team who provides the core service management capabilities. This is supplemented by a number of nonretained people who provide some aspect of the wider service. These can be categorized as follows:

1. **Local Representation:** Provide service capabilities locally such as training, reporting, remediation, and incident management.
2. **Other Services:** Make a contribution to the Malicious Software Service. For example, server operations provide infrastructure that hosts deployment and reporting; personal computer operations provide deployment of updates driven by the malicious code service.
3. **Business Representation:** Provide conduit between business and service.

The core and extended service fora are summarized in Table 1.

Table 1. Core and extended service fora

Forum	Purpose	Membership	Size and Structure
Core Team	Provide core functions of the service	Employees dedicated to fulfilling core service functions	Formal, 8 people
Extended Service Group	Provide elements that make up end to end service		Disparate, approx. 500 people
Governance Board	Provide direct resources Approve strategy	Line management	Formal, 15 people
Steering Group/ Customer Board			Community of Practice, 15 people
Project Teams	Deliver a specified service improvement within agreed time scales and cost		Formal, varies but typically 5 to 10 people
Users	Derive benefit from the service Ultimately sustain resource for the service		Informal, approx. 100,000 people

The Steering Group/Customer board comprises representation from all three categories above and also from users and representation from the body corporate. The group establishes the broad direction for the service including new technologies and significant updates to existing technologies. Many of the defences managed by the service are highly labor intensive. The steering group provides a forum through which experiences are shared and expertise continually improved and developed. As would be expected in a multinational corporation, the group is not sited in a single country. Therefore, extensive use is made of collaborative IT tools and telephone conference capability. Despite the geographical diversity, no location dominates or is considered the core location of the group. This is a good indication of the maturity of the group and a community of practice.

CONCLUSION AND FUTURE TRENDS

The group provides a rapid flow of information which has proven to be very effective. A similar membership forms the core of the organisation's incident response team which has become a highly competent command and control capability. The group provides a fast and effective route to innovative solutions through brainstorming and similar but more informal exchange of ideas. In most cases, issues are resolved quickly and effectively. This has materialized through a shared and common understanding of the aims and objectives of the service.

Members of the group have developed a means of communicating, which continues to mature through language shortcuts such as acronyms and jargon, providing a rapid means of framing problems and communicating solutions.

One of the strongest characteristics of the group is a very wide acceptance of the competencies and specializations of the various members. As the group has matured, the roles of individuals have become firmly embedded in its operation. This broad competency base very effectively overcomes many deficiencies that are evident in the formal service but are difficult to adequately formalize.

Already, other CoPs have been formed to address critical issues affecting the business which have proved intractable when approached by traditional means. The learning derived from this process has enabled us to consider CoPs as part of an overall strategy, as complementary to other approaches to managing the business.

KEY TERMS

Critical Business Process: An operation, or group of operations, within an organization that is key to its effectiveness.

Information: Knowledge acquired through experience or study.

Malicious Software: Generic term for a variety of well-known and less well-known means of disrupting or damaging computer systems and users.

Integrating Knowledge, Performance, and Learning Systems

Scott P. Schaffer
Purdue University, USA

Ian Douglas
Florida State University, USA

INTRODUCTION AND BACKGROUND

Considerable effort has been devoted recently to development of systems or platforms that manage the learning, performance, or knowledge delivered to students and employees. These systems are generically labeled learning management systems (LMS), learning and content management systems (LCMS), performance support systems, and knowledge management systems (Rockley, 2002). Organizations increasingly use content management systems to deliver content objects to employees on a just-in-time basis to support knowledge and performance requirements (Rosenberg, 1999).

While systems are developed that efficiently manage learning, knowledge, or performance, it seems desirable to consider how integration of each of these areas into a single system would benefit organizations. A major challenge to developing such systems has been the degree to which they are interoperable and the components within each are reusable. Reuse of data or information for learning or performance solution development is considered the primary driving force behind the movement toward object-based architectures for such systems (Douglas & Schaffer, 2002; Schaffer & Douglas, 2004).

Ideas for integrating different sources of support for individuals and making its construction more cost effective have begun to take shape. Some efforts have focused on reusable and interchangeable (between different delivery systems) content objects, such as the U.S. Department of Defense Advanced Distributed Learning initiative (http://www.adlnet.org). A big challenge in development of support is the lack of a pedagogical model that takes advantage of object-based architectures while promoting collaboration and knowledge capture and sharing. A

significant move in this direction has been outlined by Collis and Strjker (2003) who view the learner as a contributor of knowledge that may be captured and stored for reuse by future learners or course designers. An expansion of this idea, focused on in this article, is the reuse of the contributions of various members of a design and development team. This includes artifacts, decisions, and rationales related to activities such as the analysis of needs, identification of metrics, and identification of causes and possible solutions to workplace problems. This approach essentially attempts to link the analysis and design processes related to initial development of solutions with the ongoing adaptation and evaluation of the solutions in practice.

MOVING FROM E-LEARNING TO E-PERFORMANCE DEVELOPMENT

Advances in technology have made integration of various types of information for the purpose of just-in-time learning and performance development more viable (Greenberg & Dickelman, 2000). The Internet and World Wide Web, along with various authoring tools, have facilitated development of digital materials that are easily accessible by learners and performers. The technology that has lagged is the pedagogy and design thinking and strategies required to make all of this digital information reusable and targeted toward adding value (Clark & Meyer, 2002). Structured training or learning experiences do not always translate into better performance, and, given the fast changing nature of modern organizations, workers need to access critical and specific knowledge and performance support exactly when they need it. The traditional training approach relies on acquisition of knowledge in the hope that it will be useful and be remembered when needed. Unfortunately, much of this knowledge acquisition is explicit and context-specific and does not often transfer well to problem-solving situations (Smith, 2002).

Software development has for a number of years progressed toward embedding knowledge acquisition in context rather than rely up-front on training courses. This is evident through context-sensitive help, task-oriented help, task automation, and task wizards. For example, an LMS will often support a particular task such as entering a new course or adding new students to a course. Furthermore, content management systems are becoming object-based and will allow learners and designers to actively "pull" learning content on an as-needed basis. The development of tools to support the selection of content and to guide this kind of designing "on-the-fly" is also on the rise, as the new wave of user support tools are designed with an object-oriented architecture in mind (Spector, 2001).

Integrating knowledge, performance, and learning within a single system requires thinking of both the whole and the parts. The learners and performers who use the system will interact with an interface that is integrative and allows them to filter and select information most important to them (Gery, 1991). The kinds of information made more readily available to a particular user should be determined by their job role, function, performance objective, and organizational goal. Visual modeling tools are proposed as one way to aid in such integration during problem analysis. Such tools may allow collaborators to construct system models that identify key requirements and subsystems. The veracity of the models is tested as collaborators with multiple perspectives on the system provide feedback and revisions to the model. Subsequent KPL solutions developed from these models would thus more accurately reflect actual workplace situations, constraints, resources, and interactions.

An integrated KPL system would support learners and performers as they (1) access and construct knowledge; (2) perform a specific task; and (3) learn about a topic or objective. Such a system may take many forms. A knowledge management system may essentially be a digital library of artifacts such as manuals, guides, and company records that are stored in a database for retrieval on an as-needed basis. More recently, such systems support collabo-

ration that builds and promotes sharing of knowledge across learners, roles, or organizations through the use of tools such as discussion forums and online white boards (Greenberg & Dickelman, 2000; Shadbolt & Wielenga, 1990).

Performance support systems are typically role or job related and guide performers as they perform specific tasks. An example of performance support could be an electronic job aid with procedures for calibrating a monitoring device in a chemical facility. These kinds of systems purport to offer users a greater level of simplicity and efficiency as they seek to manage courseware, knowledge, and performers. Blended solutions incorporating online knowledge building and learning activities, workplace performance support, and face-to-face classroom learning experiences are powerful examples of how knowledge, performance, and learning integration can be accomplished in a collaborative manner. Collis, Waring, and Nicholson (2004) describe a project at Shell in which workers collaborate online in preparation for classroom activities. Collaboration is supported by a LCMS where contributions are stored in a repository and may be accessed by other learners or by facilitators for classroom use. Participants learn at their own workplace and are able to improve individual and organizational performance as a result of online participation in discussions with other Shell employees across the world.

OBJECTS AND THE CONTENT REPOSITORY

Object thinking, dividing knowledge into discrete granular chunks, represents the next step in the progression toward increasing reuse potential within a KPL system. Object thinking should not be constrained to end products, for example, learning objects used in courses; it also applies to analysis and design knowledge (Due, 2002). By integrating object thinking into analysis, a higher level of reusability as well as adaptability, interoperability, and durability may be achieved (Schaffer & Douglas, 2004). An object approach with a results focus (i.e., each object relates to a specific result required on a job) applied throughout the development process can make it easier to obtain, develop, and implement the solutions to organizational problems or opportunities.

What Is an Object?

Gibbons, Nelson, and Richards (2002) refer to a learning object, educational object, knowledge object, intelligent object, or data object as an instructional object. However, since the focus of this framework is problem solving, any learning, performance, knowledge, or instructional object is referred to as a sharable content object (SCO), taken directly from SCORM (n.d.) version 1.2.

The SCORM defines an SCO as "a set of representations of media, text, images, sounds, web pages, assessments objects, or other pieces of data that can be delivered to a Web client" (SCORM, n.d.). A single representation, according to SCORM 1.2, is called an asset. A single asset is unusable in an educational/performance setting, but by conjoining these assets, a shareable content object is created. A set of shareable content objects is referred to as content aggregation. Content aggregation consists of "a map (content structure) that can be used to aggregate learning resources into a cohesive unit of instruction, to apply structure, and to associate learning taxonomies" (SCORM, n.d.).

For an object to be SCORM compliant, it must meet specific criteria. Any object developed for performance/instructional purposes must be accessible, interoperable, durable, and reusable. "Without them [the criteria], anyone with a significant investment in either content or a learning system is locked in to that particular content or system" (Robson, 2001).

To ensure that the criteria exist within an object, metadata is tagged to each asset, SCO, and/or content aggregate. Metadata is tagged to an asset, SCO, and content aggregate to ensure that during the process of content creation, the information within each is reusable as well as discoverable. By integrating

The basic asset level is combined into SCOs, the SCOs into packages, and the packages into digital support solutions. Currently, the thinking around SCOs and packages is for them to be delivered in traditional training courses accessed through an LMS or documents accessed through a content or knowledge management system. In a KPL system, the idea is to have all digital support blended together and organized by the particular performance goals people have to achieve in their employment.

Figure 1 illustrates the manner in which SCO's may be packaged to create customized performance support solution packages. In the example shown, a personal digital assistant-based (PDA) support system is developed and slotted together with other solutions available to support a particular role. Examples of other support solutions shown in Figure 1 include a mentor network, which provides a collaborative community for support and knowl-

Figure1. Example of performance support solution blend

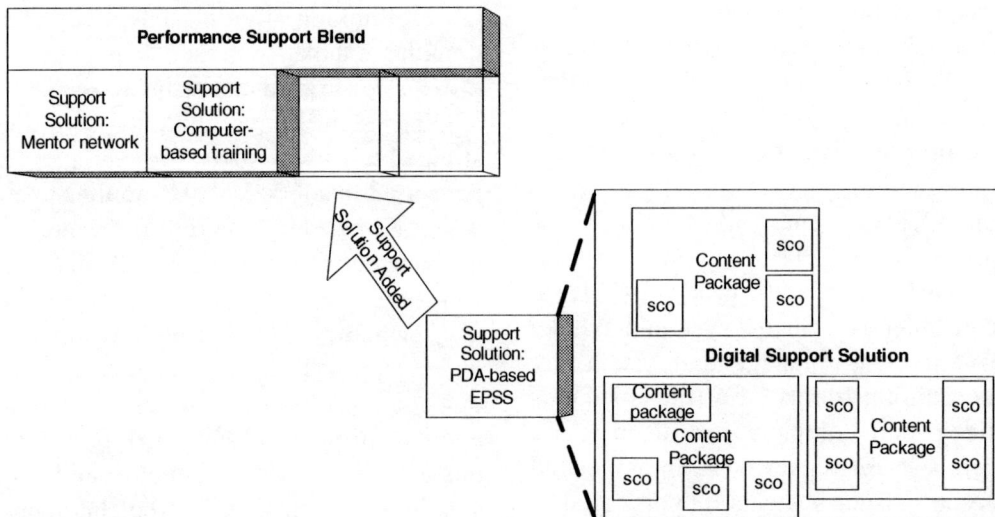

edge development in a particular performance role, and computer-based training which is specific, structured instructional activities. These are just some of the many possible forms of support that could be developed to support a particular performance role.

Why are Important Objects Within the Framework?

The reasons for using objects are simple: they enhance the resulting solution package by defining the separate components of problem-solving knowledge, provide methods for standardization, and offer potential economic advantages through reuse.

There has been a growing emphasis on objects within the fields of instructional design and performance technology. Peters (1995) states:

objects enabled by [an] emergent artifact of digital libraries will be much more like 'experiences' than they will be like 'things,' much more like 'programs' than 'documents,' and readers will have unique experiences with these objects in an even more profound way than is already the case with books, periodicals, etc.

This statement leads Gibbons, Nelson, and Richards (2002) to suggest the need for "model components that can be brought together in various combinations to create the environments and systems" to represent a variety of problems. A contribution-oriented approach to using artifacts and objects to represent problems supported by visual modeling tools goes a long way toward experiencing problems as opposed to simply categorizing and storing them. Furthermore, a comprehensive framework combining an analysis, design, and object orientation in a sequential process would allow such problem representation. Collaborative approaches to the development of objects enable the users/learners to continually reflect upon and evaluate the usefulness of objects. Repositories of analysis and design knowledge provide analysis and design teams with support throughout their respective processes. A content object repository has the potential to provide learner-designers with solution packages that match the recommended solutions identified during analysis. The representation of problems as a

result of collaborative and systematic analysis ensures that resulting objects created through design and development processes may be evaluated following use.

Repositories

The purpose of repositories is to support problem solvers, designers, or learners by providing a centralized location for the storage and reuse of standard artifacts and objects. An artifact generally refers to any template, documentation, data, visual model, or component of a visual model that can be accessed and used during any phase of an analysis and design process, for example. We envision that interlinked artifacts will exist for various levels of performance (organizational, process, and individual). For example, an object could contain specifications for the support requirements of a specific task, which will enable early identification of content objects that may be useful in the construction of customized solutions related to performance of that task.

The purpose of a repository is to allow creation of an easier, adaptable, and reusable analysis, design, and development process. This process would also support the collaborative development of organizational problem-solving capacity and ultimately link to the identification of solutions. Creating a common standard for artifacts and objects will enable the sharing of information about common performance problems or opportunities across different organizations. Tools such as groupware and visual modelers support teams as they create shareable objects. As contributions to a repository are reused, they may also be evaluated, revised, or replaced depending on utility and perceived value by members of the community.

SYSTEM DEVELOPMENT

In system and object development, there are two parallel tasks. Initially, at the systems level for each solution selected, design teams must decompose the system into subsystems (e.g., a course unit in the case of a training system), distribute the objects to the subsystems, and create the packaging and se-

Figure 2. A model for the services provided by a performance support portal

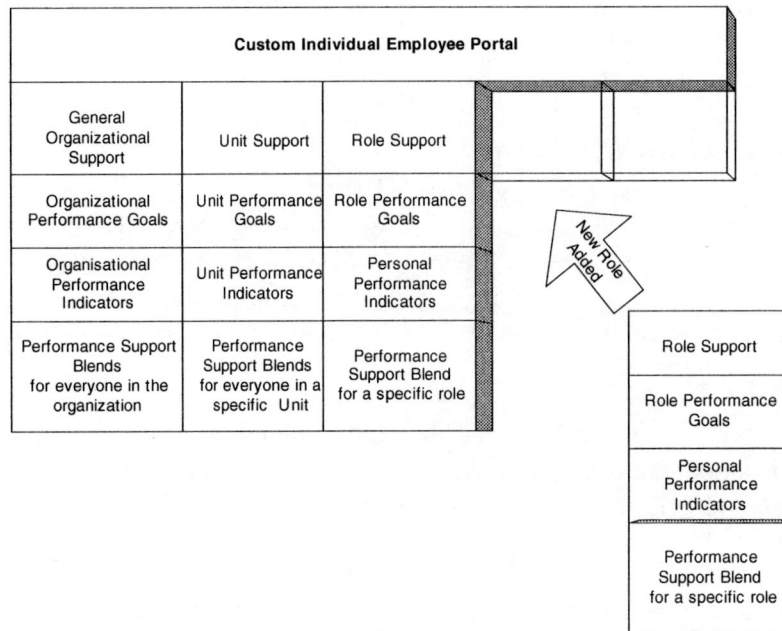

quencing structure to bind the objects together. Second, for those analysis objects not matched against pre-existing objects, teams must design and develop content, package it into a SCORM compliant object, and submit a copy to the repository. Figure 2 illustrates how these packages and sequencing structures might be bound to create support solutions.

Decomposing the system into subsystems is critical if aligning the results of KPL systems across individual, functional, and organizational levels is desired. For example, in Figure 2, a custom individual employee (student) portal is modeled that illustrates the emphasis on roles that a problem solver in an organization would play rather than using the traditional job title to identify required performance. Roles often cut across traditional job title designations and may be performed by many different employees. Consider the many roles related to a formal title such as manager. Roles often include budgeting, staffing, developing people, proposing projects, and so on. These roles may be performed by many other employees in various job titles across an organization in much the same way.

Organizations that are continuously developing their capacity to solve problems, that is, learning organizations, are able to store and share relevant knowledge, performance, and learning support related to each of these roles. Figure 2 illustrates how goals, indicators of success in achieving goals, and related support are aligned across the organization, unit (department, function), and individual levels. A new role is shown as it is added to a particular employee's portal. This employee has likely assumed this role as part of changing job requirements or as a member of a problem-solving team.

INTEGRATING THE SILOS

There is still a training-oriented bias within the standard setting community in that objects are conceived as learning objects. The main solution considered is computer-based training delivered through a learning management system. As noted in the introduction, the trend is toward thinking in terms of integrated solutions rather than being fixated on the training solution that assumes a knowledge or skill gap for the performer. This not only requires research into how problems are analyzed and solutions are selected, it also requires a reconsideration of how solutions are delivered and managed.

Figure 3. A model for a comprehensive organizational support system incorporating reuse

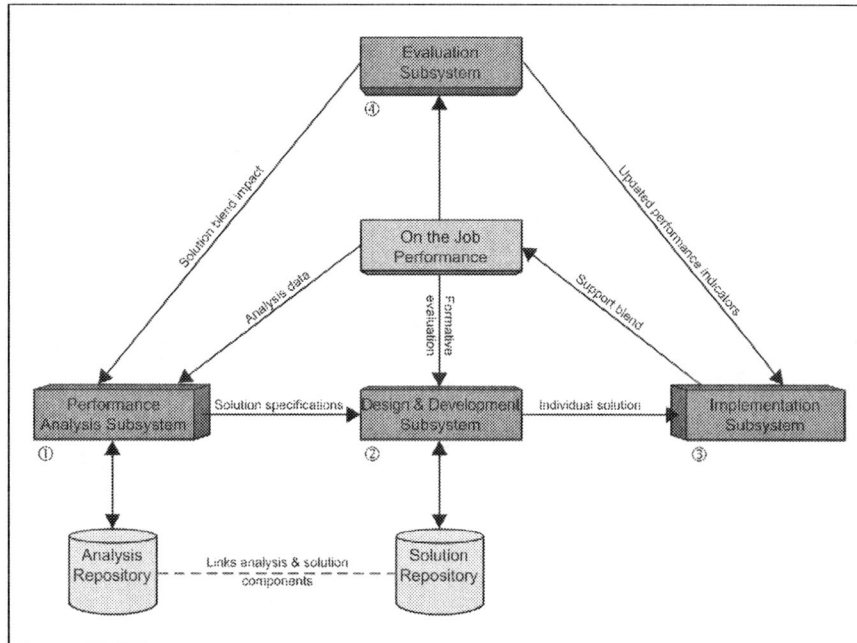

If we look at the way support solutions are currently delivered, we see a lack of integration with systems often developed and delivered independently in silos. Such silos often require that learners and performers discover, integrate, and synthesize the resources that are available to support them. This can lead to usability problems, as users have to contend with a variety of different systems with different interface designs. There may also be reduced utility in some of the systems due to the redundant, irrelevant, inadequate information. Learning management systems solve some of these problems for online learning, but they do not solve the problem for performance support in general.

A KPL is conceptualized as a dynamic performer/learner-defined system that links to a database of packaged KPL support systems. Performers using personal digital assistants, wearable computers, or desktop PCs can access available system and subsystem packages. Performance managers or instructors can create a customized performance support environment for a particular individual based on the roles they will perform or tasks to be completed. We would envisage the possibility for a certain amount of resequencing and packaging of systems within this environment. In addition to pro-

viding customized access to available performance support systems, the management system should act as a collection point for evaluation information concerning the systems use.

Figure 3 presents an initial model for how a KPL organizational support system might work. The model features two types of repositories. At the top is the reusable analysis knowledge repository which supports problem solvers by providing access to previous problem cases. These cases are linked to objects that may be useful in solving the current problem. Another repository, the reusable solution repository, supports solution developers by linking them to potentially useful objects that are related to the problem identified in the analysis.

The core of the model is on-the-job performance or, in the case of a learning environment, learning goals or problems. Analysis of performance roles or specific goals results in solution recommendations to close gaps between desired and actual role performance. The performance support development system locates reusable solutions if any exist or supports the design and development of SCOs to be packaged into a performance support system. This system is then made available to the performer via the performance support portal. This portal would automatically be made available to any performer with the responsibility of performing that role.

A key element of this model is the connection between on-the-job performance, performance evaluation, and the performance support portal. Indicators of successful performance as related to the performance support for a given role is constantly fed back to the performer. The evaluation subsystem is a key to continuously improving the quality and fidelity of the objects created within this model. Data from actual performance is relayed through the evaluation system to the performer and to the analysis team. Evaluation data is a key ingredient to successful integration of KPL systems since it allows for determination of the effect of a particular type of support system on individual and organizational effectiveness. Over time, patterns of particularly successful solutions may be quickly identified and made accessible to performers automatically. An automated system could monitor patterns of access and use, and automatically generate and administer questionnaires to gather qualitative data from performers when certain patterns are detected.

CONCLUSION AND FUTURE TRENDS

An outline of a new model for the IT systems support for the collaborative development and delivery of KPL management systems has been presented. The unifying strands of the framework are that it be performance, learning, and object oriented. The role of the KPL system is currently taken by learning management systems (LMS) or content management systems (CMS), and many of these systems are facilitating the technical aspect of learning objects; however, they are still rooted in the thinking that formal courses are the main solution to learning and performance problems and that objects are created by content experts. A reusable object and performance orientation should run through an entire support system from its initial conception to its delivery to the end users and the evaluation of its impact of the organization. Within this orientation is a contribution focus that supports the development of reusable objects by analyzers, designers, develop-

ers, and performers within the system. These contributors are supported in their efforts by tools such as discussion forums, other groupware, and visual modeling software that are integrated with object repositories.

REFERENCES

Brennan, M., Funke, S., & Anderson, C. (n.d.). The learning content management system: A new elearning market segment emerges. Retrieved May 6, 2005, from *http://www.trainingfoundation.com/tfimages/ftp/IDCLCMSWhitePaper.pdf*

Clark, R., & Mayer, R. (2002). *E-learning and the science of instruction*. San Francisco: Jossey-Bass.

Collis, B., Waring, B., & Nicholson, G. (2004). Learning from experience, for experienced staff. In M. R. Simonson (Ed.), *Proceedings of Selected Research and Development Paper Presentations AECT, 1*.

Collis, B., & Strjker (2001). New pedagogies and reusable learning objects: Toward a new economy in education. *Journal of Educational Technology Systems, 30*(2), 137-157.

Douglas, I., & Schaffer, S. P. (2002). Object-oriented performance improvement. *Performance Improvement Quarterly, 15*(3), 81-93.

Due, R. T. (2002). *Mentoring object technology projects*. Upper Saddle River, NJ: Prentice Hall PTR.

Gery, G. (1991). *Electronic performance support systems*. Boston: Weingarten.

Gibbons, A. S., Nelson, J., & Richards, R. (2002). The nature and origin of instructional objects. In D. A. Wiley (Ed.), *The instructional use of learning objects* (pp. 25-58). Bloomington, IN: AIT/AECT.

Greenberg, J., & Dickelman, G. (2000, July). Distributed cognition: A foundation for performance support. *Performance Improvement*, 18-24.

Peters, P. E. (1995, July-August). Digital libraries are much more than digitized collections. *Educom*

Review, 30(4) Retrieved May 6, 2005, from *http://www.educause.edu/pub/er/review/reviewArticles/30411.html*

Robinson, D., & Robinson, J. (1995). *Performance consulting: Moving beyond training*. San Francisco: Berrett-Kohler.

Robson, R. (2001). All about learning objects. Retrieved May 6, 2005, from *http://www.eduworks.com/LOTT/tutorial/learningobjects.html*

Rockley, A. (2002). *Managing enterprise content: A unified content strategy*. New Riders.

Rosenberg, M. (2000). *Strategies for delivering knowledge in the digital age*. New York: McGraw-Hill.

Schaffer, S. P., & Douglas, I. (2004). Performance support for performance analysis. *TechTrends, 3*(2), 34-39.

SCORM. Version 1.2. Retrieved May 6, 2005, from *http://www.adlnet.org/*

Shadbolt, N., & Wielinga, B. (1990). Knowledge based knowledge acquisition: The next generation of support tools. In B. Wielinga, J. Boose, B. Gaines, G. Schreiber, & M. Van Somerau (Eds.), *Current trends in knowledge acquisition*. Washington, DC: IOS Press.

Spector, M. (2002). Knowledge management tools for instructional design. *Educational Technology Research and Development, 50*(4), 37-46.

Smith, D. (2002). Real-world learning in the virtual classroom: Computer-mediated learning the corporate world. In E. Rudestam & J. Schoenholtz-Read (Eds.), *Handbook of online learning: Innovations in higher education and corporate training*. Thousand Oaks, CA: Sage.

International Knowledge Transfer as a Challenge for Communities of Practice

Parissa Haghirian
Kyushu Sangyo University, Japan

INTRODUCTION AND BACKGROUND: TRANSFERRING KNOWLEDGE WITHIN COMMUNITIES OF PRACTICE

Knowledge is widely recognized as a primary resource of organizations (Drucker, 1992). Some authors propose that knowledge is a company's only enduring source of advantage in an increasingly competitive world (Birkinshaw, 2001). The problem and challenge companies encounter is managing it in an effective way to increase their competitive advantages. Knowledge management is therefore concerned with various aspects of creating, examining, distributing, and implementing knowledge.

But knowledge management theory often leaves us with the impression that knowledge can be as easily managed like products and commodities (Shariq, 1999). This Cognitive Model of Knowledge Management (p. 82) is founded on the belief that knowledge is an asset that needs to be managed, but is strongly contrasted by the Communities in Practice Model of Knowledge Management (p. 83), which looks at knowledge managment and transfer from a sociological perspective (Kakabadse, Kakabadse & Kouzmin, 2003).

In fact, the transfer of knowlege happens between individuals; it is a mainly human-to-human process (Shariq, 1999). Knowledge has no universal foundation; it is only based on the agreement and the consensus of communities (Barabas, 1990), which make people and communities the main players in the knowledge transfer process. They can share or conceal knowledge; they may want to know more and want to learn. For knowledge transfer on an individual as well as on a corporate level, there "has to be a voluntary action on behalf of the individual" (Dougherty, 1999, p. 264). Knowledge transfer happens for individuals and is conducted by individuals. The base of knowledge transfer is therefore a simple communication process transferring information from one individual to another. Two components of the communication are essential: The source (or sender) that sends the message and the receiver to receive the message. Person A (sender) intends to send information to person B (receiver). Person A codifies the information into a suitable form and starts the process of sending the information or knowledge to B. This can take place via talking or writing. The channel which transmits the information might influence the flow of the message and its reception. Receiver B receives the information and decodes it. After this, B tries to understand the information received in his/her context and implements the knowledge in the surrounding environment. The communication model also includes the feedback of the receiver. B starts the whole process again and codifies and sends information back to A. A receives, decodes, and interprets the information or knowledge received.

A prerequisite for effective knowledge transfer is a high level of trust among the individuals and work groups and a strong and pervasive culture of cooperation and collaboration. This trust is developed through work practices that encourage and allow individuals to work together on projects and problems (Goh, 2002). Knowledge transfer is thus performed by communities of practice, which are described as groups of professionals informally bound to one another through exposure to a common class of problems, common pursuit of solutions, and thereby embodying a store of knowledge (Manville & Foote, 1996). Their members show a collectively developed understanding of what their community is about. They interact with each other, establishing norms and relationships of mutuality that reflect these interactions. Communities of practice generally produce a shared repertoire of communal resources, for example, language, routines, sensibilities, artifacts, tools, stories, and so forth. Members need to understand the community well enough to be able to contribute to it. They furthermore need to engage with the community and need to be trusted as a partner. Finally, they need to have access to the shared communal resources and use them appropriately (Wenger, 2000).

Communities of practice develop strong routines for problem solving via communication and knowledge exchange. If knowledge is transferred within communities of practice, both sender and receiver have a common understanding about the context, the way knowledge is transmitted, its relevance, and integration into the knowledge base of the corporation. Accordingly, communities of practice are generally agreed on to have a positive influence on knowledge transfer processes. Members of a community of practice are informally bound by the gains they find when learning from each other and by efficient problem-solving activities via communication (Wagner, 2000).

INTERNATIONAL KNOWLEDGE TRANSFER

Internationalization of business is also making company operations more geographically distributed (Hildreth, Kimble & Wright, 2000). Consequently, knowledge transfer does not only happen between members of one single community of practice but, on an international scale, between members located in culturally and geographically dispersed company units.

International knowledge transfer refers to the transfer of knowledge between two distant units of an MNC or between two different functional units at the headquarters, between a vendor and a customer, or even between countries. The use of *transfer* implies an image of flow: knowledge flows from its primary holder to the secondary holder (Doz & Santos, 1997). Knowledge transmitted is either expertise or external market information of global relevance, but not internal administrative information (Gupta & Govindarajan, 1991).

An organization can create competitive advantages by the usage of internal or external knowledge (Stock, 2000). Gupta and Govindarajan (1993) state that knowledge can be transferred more effectively

and efficiently through internal organizational, rather than external market, mechanisms. External knowledge transfer involves knowledge that is received from sources outside the company and so connects the corporation to external partners and their knowledge. These partners can be competitors or partner corporations, universities, R&D organizations or consultants (Leonard-Barton, 1995). Internal knowledge transfer takes place among people, groups or communities, departments, company units, and subunits (von Krogh & Köhne, 1998) and is the transfer of either expertise (e.g., skills and capabilities) or external market data of strategic value within the organization. The knowledge transferred refers to input processes (e.g., purchasing skills), throughout processes (e.g., product designs, process designs, and packaging designs), or output processes (e.g., marketing know-how, distribution expertise) (Gupta & Govindarajan, 1991). The importance of internal knowledge transfer is generally higher than external knowledge transfer.

Knowledge flows and knowledge transfers are strategically important to organizations for several reasons. They transmit localized know-how, which is generated in one subunit to other locations in the organization. Knowledge transfers also facilitate the coordination of work flows linking multiple, geographically dispersed subunits. Furthermore, they can enable organizations to capitalize on business opportunities requiring the collaboration of several subunits. Knowledge flows are also crucial to the orchestrated execution of unified strategic responses to moves of competitors, customers, and suppliers. Finally, knowledge flows enable the recognition and exploitation of economies of scale and scope (Schulz & Jobe, 2001). For multinational corporations, the effectiveness of these processes are of major importance. The differences between local markets and the home market require adaptation of products and operations to local conditions. The capability of multinational corporations (MNCs) to efficiently combine knowledge from different locations around the world is becoming increasingly important as a determinant of their competitive success (Doz et al., 1997). This implies overcoming geographical, cultural, and lingual barriers to transfer knowledge effectively. The company has to be constantly aware that the knowledge accumulated in various parts of the organizations needs to be localized and examined because it can be reused at another location within the organization (Quinn, 1992).

Since the communication model described above only deals with human communication, it has to be extended to meet the needs of intra-organizational knowledge transfer. International knowledge transfer does not only take place via oral communication but by many other means. The transfer of knowledge can happen in the form of data, information, blueprints, parts, subassemblies, machines, or other means to represent knowledge. It can also happen via persons, individual or teams (Doz & Santos, 1997). In any event, sender A has to think about the way knowledge is to be codified and starts the sending process. Receiver B receives the knowledge in its codified form, has to decodify it, and implement it into his working environment. Codifying knowledge is influenced by various contextual factors as well as transmission of knowledge and reception by receiver B. Even the simplest attempt to transfer knowledge in a codified form requires that the recipient is located in a context where it is possible to interpret and reconstruct the knowledge received sufficiently close to the structure and meaning intended by the sender (Shariq, 1999). These are important points which can easily affect the effectiveness of the whole process.

FUTURE TRENDS: CHALLENGES FOR COMMUNITIES OF PRACTICE

Iverson and McPhee (2002) state that knowledge within communities of practice needs to be cultivated to be spread throughout the organization and not only among members who are close to each other. Communities of practice usually show a strong understanding of how to share information and knowledge and how to implement it. In doing so, they produce a high level of context-related know-how about knowledge transfer processes. Members of the communities need to understand routines and communal resources to be competent in transferring knowledge (Wenger, 2000). However, these routines and common understandings are challenged when knowledge is transferred internationally. Know-how of routines and resources may differ in communities of practice in overseas company units. As shown before, the

context in which the transfer of knowledge takes place is of major importance for its success. Considering this, the strengths that communities of practice develop in conveying knowledge between their members may become a weakness when trying to share knowledge with members of foreign communities they have not interacted with before.

Interacting with members of overseas communities of practice naturally means experiencing boundaries, exposed to a foreign competence, which can lead to potential difficulties, for example, the communities of practice can become "hostage of their history, become insular, defensive, closed in and oriented to their own focus" (Wenger, 2000, p. 233). Facing international knowledge transfer, communities of practice do not only experience social boundaries, but also physical and temporal boundaries, which will have certain connotations for the notion of participation (Hildreth et al., 2000). Interacting with international company units and their members often enhances these difficulties as cultural distance can increase the reluctance to implement and use knowledge received from a foreign subsidiary (Haghirian & Kikima, 2004).

In fact, communities of practice gain their strength in the overall participation of their members and their being informed about about routines and showing a common understanding. Participation of community members is central to the efficiency of a community and to the relationships which help develop the sense of trust and identity defining the community (Hildreth et al., 2000). How can they now continue to operate their practices when exposed to geographically distributed company operations? How can they develop new strategies to overcome these barriers and to form communities of practice across cultures?

The first step to establish communities of practice across cultures will be the development of routines in interacting with communities of practice located at overseas units or to communicate with single members of these communities. Exchanging managers within subsidiaries and company units allows access to various communities located in different cultures. They can act as knowledge brokers and introduce elements of one practice into another located in another country or company unit (Wenger, 2002). Contact and communication can furthermore be enhanced by visits, discussions, and sabbaticals, activities which are meant to create trust and a personal relationship between members.

Since the foundations of efficient knowledge transfer practices within communities are mutual trust and a high degree of participation, difficulties will be found in attempts to integrate all members of an international community of practice. Modern technology can support intensive communication, but maintaining face-to-face contact is still very important. Only having established an intensive relationship with employees located in overseas company units, communities of practice across cultures can develop. Knowing each other can give people a great sense of unity and a common purpose and thus supports a constituted team to form into a community of practice overcoming geographical and national boundaries (Hildreth et al., 2000).

CONCLUSION

The Community of Practice Model of Knowledge Transfer implies that knowledge is not simply a company asset but embedded in a sociological and historical surrounding. Communities of practice are therefore important carriers of informal knowledge and developers of knowledge transfer routines.

In general knowledge, transfer takes place in a certain environmental context. This context influences knowledge sending and receiving decisions. Communities of practice develop intensive know-how on these contexts and so promote effective knowledge transfer and increase knowledge sharing.

In case knowledge transfer happens in an international setting, communities of practice are facing new challenges because their expertise is mainly based on participation and trust between their members within the community. When transferring knowledge internationally, their competencies are naturally decreasing since interactors are geographically seperated from each other.

Communities of practice therefore need to increase their interactions with members of international communities or with communities of practice located at an overseas company unit. The overall goal will be the establishment of communities of practice across cultures to increase the effectiveness of international knowledge transfers.

REFERENCES

Barabas, C. (1990). *Technical writing in corporate culture*. Norwood, NJ: Ablex.

Birkinshaw, J. (2001). Why is knowledge management so difficult? *Business Strategy Review, 12*(1), 11-18.

Dougherty, V. (1999). Knowledge is about people, not databases. *Industrial and Commercial Training, 31*(7), 262-266.

Doz, Y., Asakawa, K., Santos, J., & Williamson, P. (1997). *The metanational corporation*. Fontainebleau, France: INSEAD Working Papers Series.

Doz, Y., & Santos, J. F. P. (1997). *On the management of knowledge: From the transparency of collocation and co-setting to the quandary of dispersion and differentiation*. Fontainebleau, France: INSEAD Working Papers Series.

Drucker, P. F. (1992). The new society of organizations. *Harvard Business Review, 70*(5), 95-104.

Goh, S. C. (2002). Managing effective knowledge transfer: An integrative framework and some practical implications. *Journal of Knowledge Management, 6*(1), 23-30.

Gupta, A. K., & Govindarajan, V. (1991). Knowledge flows and the structure of control within multinational corporations. *Academy of Management Review, 16*(4), 768-792.

Gupta, A. K., & Govindarajan, V. (1993). Coalignment between knowledge flow patterns and strategiy systems and processes within MNCs. In P. Lorange (Ed.), *Implementing strategic processes* (pp. 329-346). Oxford, UK: Basil Blackwell.

Haghirian, P., & Kikima, O. (2004). Intercultural knowledge transfer between headquarters and subsidiaries: The case of Japanese multinational corporations. In Oliver H. M. Yau, C. S. Tseng, & F. S. L. Chueng (Eds.), *Harmony vs. conflict:Euro-Asian management in a turbulent era. Proceedings of the 21st Annual Conference of the Euro-Asia Management Studies Association,* Hong Kong.

Hildreth, P., Kimble, C., & Wright, P. (2000). Communities of practice in the distributed international environment. *Journal of Knowledge Management, 4*(1), 27-38.

Iverson, J. O., & McPhee, R. D. (2002). Knowledge management in communities of practice. *Management Communication Quarterly, 16*(2), 259-266.

Kakabadse, N. K., Kakabadse, A., & Kouzmin, A. (2003). Reviewing the knowledge management literature: Towards a taxonomy. *Journal of Knowledge Management, 7*(4), 75-91.

Leonard-Barton, D. (1995). *Wellsprings of knowledge: Building and sustaining the sources of innovation.* Boston: Harvard Business School Press.

Manville, B., & Foote, N. (1996, July 7). Harvest your workers' knowledge. *Datamation, 42*(78-80).

Quinn, J. B. (1992). *The intelligent enterprise.* New York: Free Press.

Schulz, M., & Jobe, L. A. (2001). Codification and tacitness as knowledge management strategies: An empirical exploration. *Journal of High Technology Management Research, 12,* 139-165.

Shariq, S. Z. (1999). How does knowledge transform as it is transferred? Speculations on the possibility of a cognitive theory of knowledgescapes. *Journal of Knowledge Management, 3*(4), 243-251.

Stock, W. G. (2000). *Informationswirtschaft.* München, Wien: Oldenbourg.

von Krogh, G., & Köhne, M. (1998). Der Wissenstransfer in Unternehmen: Phasen des Wissenstransfers und wichtige Einflussfaktoren. *Die Unternehmung, 5*(6), 235-252.

Wagner, E. (2000). Communities of practice: The structure of knowledge stewarding. In C. Despress & D. Chauvel (Eds.), *The present and the promise of knowledge management* (pp. 225-246). Boston: Butterworth-Heinemann.

Wenger, E. (2000). Communities of practice and social learning systems. *Organization, 7*(2), 225-246.

KEY TERMS

Data: Data are carriers of knowledge and information. They consist mostly of signs and are the raw material to be further processed. Data represent observations or facts out of context that are not directly meaningful. Both information and knowledge are communicated through data.

Information: The term information refers to details about an event or situation in the past or simply a scientific fact. Information can be regarded as a piece of knowledge of an objective kind. It results from placing data within some meaningful context, often in the form of a message. It is purely descriptive and explicit, does not enable decisions or actions, nor does it trigger new questions.

Knowledge: Knowledge is created and organized by the very flow of information, anchored on the commitment and beliefs of its holder. Knowledge provides links between different information. That includes an interpretation and sense-making process.

Know-How: Know-how is knowledge embedded in the individual. Examples of know-how are skills, capabilities, and expertise.

Knowledge Codification: The process of packaging knowledge into formats that enable the organization to transmit it to other parts and thus facilitate knowledge transfer. Codification of knowledge is therefore a means of representing it and enables all members of the organization to access relevant knowledge.

Inter–Organisational Knowledge Transfer Process Model

Shizhong Chen
University of Luton, UK

Yanqing Duan
University of Luton, UK

John S. Edwards
Aston Business School, UK

INTRODUCTION

Knowledge management (KM) is an emerging discipline (Ives, Torrey & Gordon, 1997) and characterised by four processes: generation, codification, transfer, and application (Alavi & Leidner, 2001). Completing the loop, knowledge transfer is regarded as a precursor to knowledge creation (Nonaka & Takeuchi, 1995) and thus forms an essential part of the knowledge management process. The understanding of how knowledge is transferred is very important for explaining the evolution and change in institutions, organisations, technology, and economy. However, knowledge transfer is often found to be laborious, time consuming, complicated, and difficult to understand (Huber, 2001; Szulanski, 2000). It has received negligible systematic attention (Huber, 2001; Szulanski, 2000), thus we know little about it (Huber, 2001). However, some literature, such as Davenport and Prusak (1998) and Shariq (1999), has attempted to address knowledge transfer within an organisation, but studies on inter-organisational knowledge transfer are still much neglected.

An emergent view is that it may be beneficial for organisations if more research can be done to help them understand and, thus, to improve their inter-organisational knowledge transfer process. Therefore, this article aims to provide an overview of the inter-organisational knowledge transfer and its related literature and present a proposed inter-organisational knowledge transfer process model based on theoretical and empirical studies.

BACKGROUND: AN OVERVIEW OF KNOWLEDGE TRANSFER AND RELATED LITERATURE

Knowledge Transfer within an Organisation

Knowledge transfer implies that knowledge is transferred from the sender(s) (person, group, team, or organisation) to the recipient(s) (person, group, team, or organisation) (Albino, Garavelli & Schiuma, 1999; Lind & Persborn, 2000). It may happen within an organisation or between organisations. Szulanski (2000) argues that knowledge transfer is a process in which difficulty should be seen as its characteristic feature. This process view may help organisations identify difficulties in the knowledge transfer. He further proposes a process model for intra-organisational knowledge transfer as shown in Figure 1, which contains four stages: initiation, implementation, ramp-up, and integration.

In the *initiation* stage, the effort aims to find an opportunity to transfer and to decide whether to pursue it. An opportunity to transfer exists as soon as the seed for that transfer is formed, that is, as soon as a gap is found within the organisation, and the knowledge to address the gap is thought to be available. In the *implementation* stage, following the decision to transfer knowledge, attention shifts to the exchange of information and resources between the source and the recipient, that is, "learning before doing" for the recipient. In the *ramp-up*

Figure 1. The process for knowledge transfer within an organisation (Szulanski, 2000)

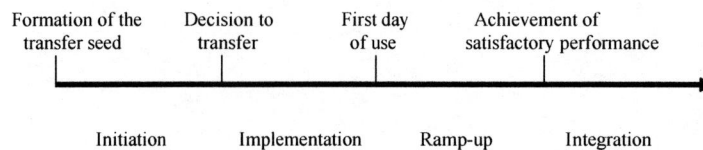

Formation of the transfer seed	Decision to transfer	First day of use	Achievement of satisfactory performance

Initiation   Implementation   Ramp-up   Integration

stage, the recipient begins using acquired knowledge, and tries to ramp-up to satisfactory performance, that is, "learning by doing" for the recipient. In the *integration* stage, the recipient takes subsequent follow-through and evaluation efforts to integrate the practice with its other practices (Szulanski, 2000).

The process model demonstrates that knowledge transfer within an organisation is complex and difficult. However, knowledge transfer between organisations is even harder and more complicated. When knowledge is transferred within an organisation, the organisation should try to expand the amount of shared knowledge among its employees to an appropriate level (or to the highest level possible) (Lind & Seigerroth, 2000) so as to develop (or preserve) its competitive advantage. When transferring knowledge between organisations, the organisations have to face "the boundary paradox" (Quintas, Lefrere & Jones, 1997), which involves more complicated factors impinging on the transaction. It also requires the negotiation between participating parties, strict governance mechanisms to regulate the transfer content, and higher loyalty by relevant employees.

Inter-Organisational Knowledge Transfer

Inter-organisational knowledge transfer may have different types. For instance, von Hippel (1987) classifies know-how trading between firms into two types: informal and formal. He defines informal know-how trading as the extensive exchange of proprietary know-how by informal networks in rival (and nonrival) firms. Here is an example, when a firm's engineer who is responsible for obtaining or developing the know-how his/her firm needs finds that the required know-how is not available in-house or in public sources; the engineer may, through his/her private relationships, seek the needed information from professional counterparts in rival (and

nonrival) firms. Formal know-how trading is referred to as official knowledge exchange agreements between firms such as agreements to perform R&D cooperatively or agreements to license or sell proprietary technical knowledge (von Hippel, 1987). von Hippel further argues that the main differences between the informal and formal trading are (1) the decisions to trade or not trade proprietary know-how in the former are made by individual, knowledgeable engineers; no elaborate evaluations of relative rents or seeking of approvals from firm bureaucracies are involved; however, the decisions for the latter are made by firm bureaucracies; (2) the value of a particular traded module in the former is too small to justify an explicit negotiated agreement to sell, license, or exchange, but the traded module in the latter is of considerable value. In fact, the fundamental difference between the so-called informal and formal inter-organisational knowledge transfer is that the former is carried out through employees' private relationships without the direct involvement of their corporate management, but the latter has direct involvement of their corporate management.

This article is mainly concerned with the formal knowledge transfer process between organisations.

Inter-Organisational Learning

From an organisational learning perspective, inter-organisational knowledge transfer is actually the process of organisations learning from each other, that is, inter-organisational learning.

Organisational learning may occur when the organisation acquires information (knowledge, understanding, know-how, techniques, or practices) of any kind and by whatever means (Argyris & Schon, 1996). It is individuals that make up an organisation; thus each organisational learning activity actually begins from individual learning. Individual learning is a necessary condition for organisational learning which is institutionally embedded (Beeby & Booth,

2000). However, individual learning is not sufficient. It is generally accepted that the acquisition of knowledge by individuals does not represent organisational learning (Beeby & Booth, 2000; Nonaka & Takeuchi, 1995). To achieve the necessary cross-level effects, that is, successful organisational learning, individual learning should be on the organisation's behalf (Argyris & Schon, 1996) and must be shared through communication which is supported by institutional processes for transferring what is learned by individuals to the organisation as well as for storing and accessing that which is learned (Beeby & Booth, 2000).

Literature review shows that study on organisational learning mainly focuses on learning within an organisation, that is, on how to convert individual learning into organisational learning once the individuals have acquired the needed knowledge. Issues related to how and from where the individuals acquire the needed knowledge are more or less ignored. When organisations learn from each other, it is normally some individuals who learn on their organisation's behalf from other individuals on another organisation's behalf. Then the learner's individual learning will be further converted into organisational learning. Therefore, inter-organisational knowledge transfer process, as a kind of inter-organisational learning, can be divided into two subprocesses: (1) inter-employee learning between employees from different organisations and (2) organisational learning within the receiving organisation by converting individual learning to organisational learning through the organisation's internal mechanisms (Chen, Duan & Edwards, 2002).

Social Networks

Social relationships play an important role in social networks. Granovetter (1985) points out that all activities are embedded in complex networks of social relations which include family, state, educational and professional background, religion, gender, and ethnicity.

From the social network perspective, inter-organisational knowledge transfer activities can be regarded as activities within social networks. Assuming the influence from a third party is ignored, the network may have four actors: receiving organisation and receiving employee, giving organisation and

Figure 2. The relationship mechanism for inter-organisational knowledge transfer

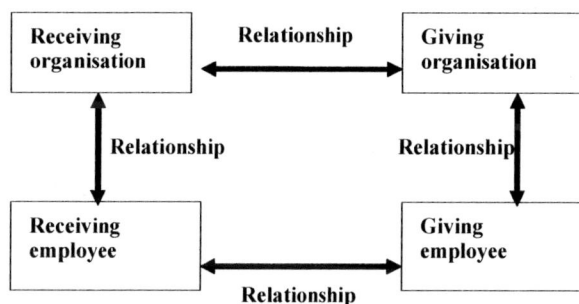

giving employee. The actors' behaviours will be influenced by their relationships. In the first subprocess (i.e., inter-employee learning between employees from different organisations), when the receiving organisation requests knowledge from the giving organisation, they will establish their own knowledge transfer strategies based on the relationship between two organisations. Then the organisations may use their relationships with their own employees to influence and guide the employees' learning behaviours to conform to their knowledge transfer strategies. The personal relationship between the receiving and giving employees will also influence their individual learning effectiveness. In the second subprocess, the relevant actors will be the receiving organisation and receiving employee. The key point for the receiving organisation is to establish its internal mechanisms to promote the conversion from the receiving employee's individual learning into organisational learning. The internal mechanisms may be considered as being embedded in the relationship between the receiving organisation and receiving employee. Therefore, there is a relationship mechanism, as depicted in Figure 2. This mechanism coordinates and influences the relevant actors' behaviours for inter-organisational knowledge transfer.

AN INTER-ORGANISATIONAL KNOWLEDGE TRANSFER PROCESS MODEL

Through the above review, it is known that inter-organisational knowledge transfer process can be divided into two subprocesses. Drawing on

Szulanski's (2000) process model in Figure 1, the first subprocess can be further divided into three stages: initiation, selection, and interaction; the second subprocess may be called conversion. So, a similar four-stage model for inter-organisational knowledge transfer is offered in Figure 3.

In the *initiation* stage, two organisations try to find an opportunity to transfer and to decide whether to pursue it through negotiation. In the *selection* stage, the receiving and giving organisations select an employee as a receiving and giving employee respectively (more than one employee may be involved, of course, in either organisation). In the *interaction* stage, the giving employee transfers his/her knowledge to the receiving employee. In the *conversion* stage, the receiving employee transfers his/her acquired knowledge to his/her employer— the receiving organisation. The conversion stage is only related to the receiving organisation and receiving employee.

The relationship between the process model in Figure 3 and Szulanski's (2000) process model may be seen as follows: (1) The initiation and interaction stages of the former are similar to the initiation and implementation stages of the latter. (2) In the conversion stage of the former, the receiving employee plays two roles: first, he/she, as a recipient, will apply his/her acquired knowledge to his/her work and have to experience the ramp-up and integration stages; second, he/she is also a source for his organisation as his/her colleagues may learn from him/her. So, the conversion stage contains the ramp-up and integration stages, as well as the whole transfer process within an organisation.

Based on Figures 2 and 3, annd in addition to suggestions from some empirical evaluation with company managers (e.g., the initiation stage should be further divided into two stages: identification and negotiation to highlight their importance), a process model can be proposed for the inter-organisational knowledge transfer and is illustrated in Figure 4. The

following explanation is provided for the five stages, although there may be no clear-cut division between them.

1. **Identification:** In this stage, the receiving company internally finds its knowledge gap, identifies its needs for acquiring external knowledge and the external knowledge source.
2. **Negotiation:** In this stage, the receiving company negotiates (or discusses) with the giving company on the knowledge transaction, or any problems happening in the transfer process, to reach an agreement or oral commitment.
3. **Selection:** It is a stage in which a giving (or receiving) employee is selected by the giving (or receiving) organisation to specifically carry out the agreed transfer task.
4. **Interaction:** It is a stage in which both the giving and receiving employees iteratively contact each other to transfer the agreed knowledge.
5. **Conversion:** It occurs when the receiving employee contributes his/her acquired knowledge to the employer (i.e., the receiving organisation), the individual learning will be converted into organisational learning to successfully improve the receiving organisation's business.

The proposed process model not only identifies the important stages in the inter-organisational knowledge transfer process, but also shows the dynamic interactions between the organisations involved. More importantly, the model emphasises the repetitive nature of the process among stages and demonstrates the necessity of iterative loops between some stages. The transfer process may, sometimes not simply progress in the stage sequence but in iterative loops, as it may be necessary to go back to the previous stage. For example, once the receiving organisation initially identifies its needs for acquiring

Figure 3. The inter-organisational knowledge transfer process

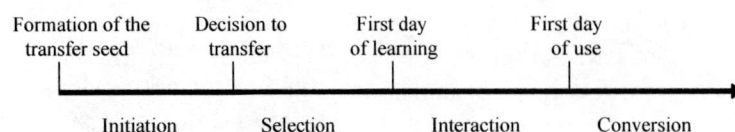

Figure 4. The inter-organisational knowledge transfer process model

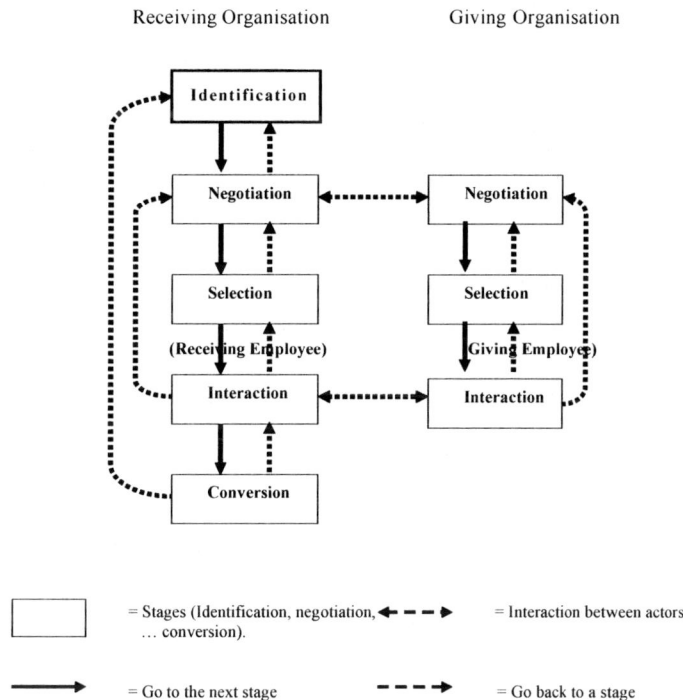

external knowledge and the external knowledge source (i.e., the giving organisation), the former will negotiate or discuss with the latter to further clarify what the former exactly wants. Sometimes, the needs initially identified by the receiving organisation may be found to be inaccurate; thus it is necessary for the receiving company to go back to the identification stage to further clarify its needs. Then it will negotiate or discuss with the giving organisation again. This process may carry on until the true needs for the receiving organisation are correctly identified. Although the selection of a receiving employee is the receiving organisation's internal affair, sometimes the receiving organisation may inform or consult the giving organisation about its arrangements for the receiving employee. So, there is a feedback loop that goes from the selection stage to the negotiation stage until the receiving employee is finally selected. Further, the transfer process in the receiving organisation may also have iterative loops during its interaction with the giving organisation. Similar things may happen in the giving organisation as well.

In the conversion stage, the receiving employee will apply the acquired knowledge into the receiving

organisation's business. The receiving employee may still need the giving employee's help because he/she may not fully understand the acquired knowledge or not fully absorb the knowledge needed for the application. This will initiate a feedback loop from the conversion stage to the interaction stage, then back to the conversion stage again. Furthermore, different organisations have different environments. The application of the knowledge in the new environment may trigger some new problems, which may cause the receiving organisation to identify its new needs for knowledge acquisition. Some of them may be internally met in the conversion stage. Some of them may cause the receiving organisation to seek a new external knowledge source and begin a new round of inter-organisational knowledge transfer. So, there is a backward loop from the conversion stage to the identification stage.

CONCLUSION

Through a review of the relevant literature on knowledge transfer, organisational learning and social networks, an inter-organisational knowledge transfer process model is developed. As shown in the model, inter-organisational knowledge transfer is a complex process and difficult to understand. As a result, the success of the transfer can be affected by many factors and pose serious challenges to organisations. Some empirical research has been carried out to test the model, and the preliminary findings suggest that managers feel that the process model is a sound attempt to reflect companies' knowledge transfer practices and can help the companies to better understand the nature, the mechanism, and the process of the knowledge transfer.

FUTURE TRENDS

Future research needs to be undertaken to identify the important factors in each stage. For instance, in the interaction stage, the receiving employee will learn from the giving employee, the former's absorptive capacity and prior experience, the latter's openness, prior experience and expressiveness, as well as the trust between both of them (Cohen & Levinthal, 1990; Wathne, Roos & von Krogh, 1996; Chen, Duan, & Edwards, 2002) could be identified as the important factors for the stage. Furthermore, inter-organisational knowledge transfer strategies for both receiving and giving organisations can be developed to help them to address the "boundary paradox" (Quintas, Lefrere, & Jones, 1997) more effectively and maximise the potential benefits of knowledge sharing for both organisations involved.

REFERENCES

Alavi, M., & Leidner, D. (2001). Knowledge management and knowledge management systems: Conceptual foundations and research issues. *MISQ Quarterly, 25*(1), 107-136.

Albino, V., Garavelli, A. C., & Schiuma, G. (1999). Knowledge transfer and inter-firm relationships in industrial districts: The role of the leader firm. *Technovation, 19*, 53-63.

Argyris, C., & Schon, D. A. (1996). *Organisational learning II: Theory, method, and practice.* London: Addison-Wesley.

Beeby, M., & Booth, C. (2000). Networks and inter-organisational learning: A critical review. *The Learning Organisation, 7*(2), 75-88.

Chen, S., Duan, Y., & Edwards, J. S. (2002, September 24-25). Towards an inter-organisational knowledge transfer framework for SMEs. *Proceedings of the 3rd European Conference on Knowledge Management,* Dublin, Ireland.

Cohen, W. M., & Levinthal, D. A. (1990). Absorptive capacity: A new perspective on learning and innovation. *Administrative Science Quarterly, 35,* 128-152.

Davenport, T. H., & Prusak, L. (1998). *Working knowledge: How organisations manage what they know.* Boston: Harvard Business School Press.

Granovetter, M. (1985). Economic action and social structure: The problem of embeddedness. *American Journal of Sociology, 91*(3), 481-510.

Huber, G. (2001). Transfer of knowledge in knowledge management systems: Unexplored issues and suggested studies. *European Journal of Information Systems, 10,* 72-79.

Ives, W., Torrey, B., & Gordon, C. (1997). Knowledge management: An emerging discipline with a long history. *Journal of Knowledge Management,* 1(4), 269-274.

Lind, M., & Persborn, M. (2000, September 12-14). Possibilities and risks with a knowledge broker in the knowledge transfer process. *Proceedings of the 42nd Annual Conference of the Operational Research Society,* University of Wales, Swansea.

Lind, M., & Seigerroth, U. (2000, September 12-14). Development of organisational ability through team-based reconstruction: Going from personal to shared contextual knowledge. *Proceedings of the 42nd Annual Conference of the Operational Research Society,* University of Wales, Swansea.

Nonaka, I., & Takeuchi, H. (1995). *The knowledge-creating company.* Oxford: Oxford University Press.

Quintas, P., Lefrere, P., & Jones, G. (1997). Knowledge management: A strategic agenda. *Long Range Planning, 30*(3), 385-391.

Shariq, S. Z. (1999). How does knowledge transform as it is transferred? Speculations on the possibility of a cognitive theory of knowledgescapes. *Journal of Knowledge Management, 3*(4), 243-251.

Szulanski, G. (2000, May). The process of knowledge transfer: A diachronic analysis of stickiness. *Organisational Behaviour and Human Decision Process, 82*(1), 9-27.

von Hippel, E. (1987). Cooperation between rivals: Informal know-how trading. *Research Policy, 16,* 291-302.

Wathne, K., Roos, J., & von Krogh, G. (1996). Towards a theory of knowledge transfer in a cooperative context. In G. von Krogh & J. Roos (Eds.), *Managing knowledge: Perspectives on co-operation and competition.* London: Sage.

KEY TERMS

Absorptive Capacity: Reflects the receiving employee's ability to absorb the knowledge sent by the giving employee.

Boundary Paradox: In the knowledge transfer process, the giving and receiving organisations' borders must be open to flows of information and knowledge from the networks and markets in which they operate, but, at the same time, the organisation must protect and nurture its own knowledge base and intellectual capital.

Expressiveness: The ability of giving employees to use oral or facial expression and body language to clearly express what they know.

Knowledge Transfer: Knowledge is transferred from the sender(s) (person, group, team, or organisation) to the recipient(s) (person, group, team, or organisation).

Openness: The giving employees' willingness to transfer their knowledge in a collaborative interaction. This stresses the attitude of giving employees involved in the knowledge transfer of not hiding their knowledge so that potential learning is facilitated.

Social Network: Refers to a set of social entities (or persons) and social relationships which connect them.

Trust: A social actor extrapolates that another actor will behave as expected. Trust is a risky engagement.

IS Design for Community of Practice's Knowledge Challenge

Kam Hou Vat
University of Macau, Macau

INTRODUCTION

The last decade of the 20th century saw explosive growth in discussions about knowledge—knowledge work, knowledge management, knowledge-based organizations, and the knowledge economy (Cortada & Woods, 2000). Against this backdrop, enterprises including educational institutes are challenged to do things faster, better, and more cost-effectively in order to remain competitive in an increasingly global environment (Stalk, Evans & Shulman, 1992). There is a strong need to share knowledge in a way that makes it easier for individuals, teams, and enterprises to work together to effectively contribute to an organization's success.

This idea of knowledge sharing has well been exemplified in the notion of a learning organization (LO) (Senge, 1990; Garvin, 1993; King, 1996; Levine, 2001). Essentially, a learning organization could be considered as an organization that focuses on developing and using its information and knowledge capabilities in order to create higher-value information and knowledge, to modify behaviors to reflect new knowledge and insights, and to improve bottom-line results. Consequently, there are many possible instances of information system (IS) design and realization that could be incorporated into a learning organization. The acronym "LOIS" (Learning Organization Information System) (Williamson & Lliopoulos, 2001) as applied to an organization is often used as a collective term representing the conglomeration of various information systems, each of which, being a functionally defined subsystem of the enterprise LOIS, is distinguished through the services it renders. For example, if a LOIS could support structured and unstructured dialogue and negotiation among the organizational members, then the LOIS subsystems might need to support reflection and creative synthesis of information and knowl-edge, and thus integrate working and learning. Also, if each member of an organization is believed to possess his or her own knowledge space, which is subject to some level of description, and thus may be integrated into an organization's communal knowledge space (Wiig, 1993; Davenport & Prusak, 1998; Levine, 2001), the LOIS subsystems should help document information and knowledge as it builds up, say, by electronic journals. Or, they have to make recorded information and knowledge retrievable, and individuals with information and knowledge accessible. Collectively, a LOIS can be considered as a scheme to improve the organization's chances for success and survival by continuously adapting to the external environment. That way, we stand a better chance of increasing social participation and shared understanding within the enterprise, and thus foster better learning. More importantly, the philosophy underlying the LOIS design should recognize that our knowledge is the amassed thought and experience of innumerable minds, and LOIS helps capture and reuse those experiences and insights in the enterprise. Indeed, the cultivation of an organization's communal knowledge space—one that develops new forms of knowledge from that which exists among its members, based on seeing knowledge as a social phenomenon, and not merely as a 'thing'—is fundamental to enterprises that intend to establish, grow, and nurture a learning organization, be it physical or digital (Hackbarth & Groven, 1999), where individuals grow intellectually and expand their knowledge by unlearning inaccurate information and relearning new information.

The theme of this article is to examine the knowledge processes required of the learning organization viewed from the community of practice viewpoint, to develop and sustain the communal knowledge space through the elaboration of suitable LOIS support so as to expand an organization's capacity to adapt to future challenges.

THE BACKGROUND OF COMMUNITIES OF PRACTICE

According to Wenger, McDermott, and Snyder (2002, p. 4), communities of practice are groups of people who share a concern, a set of problems, or a passion about a topic, and who deepen their knowledge and expertise by interacting on an ongoing basis. As they spend time together, they typically share information, insight, and advice. They help one another solve problems; they ponder common issues, explore ideas, and accumulate knowledge. Oftentimes, they become informally bound by the value that they find in learning together. This value is not merely instrumental for their work. It also accrues in the personal satisfaction of knowing colleagues who understand each other's perspectives and of belonging to an interesting group of people. Over time, they develop a unique perspective on their topic, as well as a body of common knowledge, practices, and approaches. They also develop personal relationships, a common sense of identity, and established ways of interacting.

Indeed, communities of practice are not a new idea (Wenger, 1998). They were our first knowledge-based social structures, back when we lived in caves and gathered around the fire to discuss strategies for cornering prey, the shape of arrowheads, or which roots were edible. They have captured our focus today because organizations have come to realize that knowledge has become the key to success (OECD, 1996), and their competitive edge is mostly the intellectual capital of their employees (Stewart, 1997), and they need to be more intentional and systematic about managing knowledge through harnessing their human resources in order to stay ahead of the pack. Undeniably, in today's knowledge-intensive economy, organizations are increasingly expecting their employees to continually improvise and invent new methods to deal with unexpected difficulties and to solve immediate problems, and to share these innovations with other employees through some effective channels.

In this regard, the idea of the community of practice has inspired many an organization to initiate their collective learning based not so much on delineated learning paths, but rather on experience sharing, the identification of best practices, and reciprocal support for tackling day-to-day problems in the workplace. Cultivating communities of practice in strategic areas is considered as a practical way to manage knowledge in terms of critical knowledge domains; organizations need to identify the people and the specific knowledge needed for their work, and explore how they connect them into suitable communities of practice so that together they could steward the necessary knowledge. From this viewpoint, the cultivation of an organization's communal knowledge space is literally the cultivation of the various communities of practice throughout the organization.

UNDERSTANDING THE KNOWLEDGE CHALLENGE FOR LEARNING ORGANIZATIONS

Nowadays, enterprises need to understand precisely what knowledge will give them a competitive advantage. They then need to keep this knowledge on the cutting edge, deploy it, leverage it in operations, and spread it across the organization. However, many an organization still has no explicit, consolidated knowledge strategy to steward the required knowledge. Instead, many attempts at knowledge management have simply counted on new information technologies to capture all the possible knowledge of an organization into databases that would make it easily accessible to all employees (King, 1999; Levine, 2001). This philosophy of regarding knowledge as a "thing" that can be managed like other physical assets has not been quite successful for several obvious reasons. One is the apparent difficulty concerned with knowledge capture and the issue of tacit-to-explicit transformation. Another is the question of intellectual asset management. Third is the myopic interpretation of knowledge management in terms of information management, which involves breaking information into smaller chunks that can be detected throughout the organization, stored for later use, manipulated by being combined with other chunks, and transferred where they are needed. The ultimate goal of such knowledge management efforts is to get the right information to the right people at the right place with the right information technologies. It is believed that a knowledge strategy must be based on understanding what the knowledge challenge is. The essence of this challenge comes down to a few key points about the nature of knowing (Nonaka & Takeuchi, 1995; O'Leary, 1998; Wenger, 1998, 2002).

- **Knowledge lives in the human act of knowing:** In many instances of our daily living, our knowledge can hardly be reduced to an object that can be packaged for storage and retrieval. Our knowledge is often an accumulation of experience—a kind of residue of our actions, thinking, and conversations—that remains a dynamic part of our ongoing experience. This type of knowledge is much more a living process than a static body of information.

- **Knowledge is tacit as well as explicit:** Not everything we know can be codified as documents or tools. Sharing tacit knowledge requires interaction and informal learning processes such as storytelling, conversation, coaching, and apprenticeship. The tacit aspects of knowledge often consist of embodied expertise—a deep understanding of complex, interdependent elements that enables dynamic responses to context-specific problems. This type of knowledge is very difficult to replicate. This is not to say that it is not useful to document such knowledge in whatever manner serves the needs of practitioners. But even explicit knowledge is dependent on tacit knowledge to be applied.

- **Knowledge is dynamic, social, as well as individual:** It is important to accept that though our experience of knowing is individual, knowledge is not. Much of what we know derives from centuries of understanding and practice developed by long-standing communities. Appreciating the collective nature of knowledge is especially important in an age when almost every field changes too much, too fast for individuals to master. Today's complex problem solving requires multiple perspectives. We need others to complement and develop our own expertise. In fact, our collective knowledge of any field is changing at an accelerating rate. What was true yesterday must be adapted to accommodate new factors, new data, new inventions, and new problems.

In short, what makes managing knowledge a challenge is that it is not an object that can be stored, owned, and moved around like a piece of equipment or a document. It resides in the skills, understanding, and relationships of its members, as well as in the tools, documents, and processes that embody aspects of this knowledge. In response to such knowledge challenge in a learning organization, it is interesting to observe some of the interpretations from the standpoint of the communities of practice (CoPs).

Firstly, it is not a CoP's practice to reduce knowledge to an object. They often make it an integral part of their activities and interactions, and they serve as a living repository for that knowledge. Secondly, a CoP is in the best position to codify knowledge since their members can combine its tacit and explicit aspects. They also can produce useful documentation, tools, and procedures because they understand the needs of practitioners. Such CoP products are often not considered as just objects by themselves, but are part of the life of the community. Thirdly, what counts as collective knowledge is often produced through a process of communal involvement, including all the possible controversies, so as to develop the specific body of knowledge. This collective character of knowledge creation does not mean that individuals do not count. In fact, the best communities welcome strong personalities and encourage disagreements and debates. Besides, that knowledge is not static does not mean that a domain of knowledge lacks a stable core. One of the primary tasks of a community of practice is to establish a common baseline of knowledge and standardize what is well understood so that people can focus their creative energies on the more advanced issues.

CONCEIVING KNOWLEDGE PROCESSES FOR COMMUNITIES OF PRACTICE

In order to create the communal knowledge space through cultivating various communities of practice for the entire organization, it is important to have a vision that orients the entire organization to the kind of knowledge it must acquire, and wins spontaneous commitment by the individuals and groups involved in knowledge creation (Dierkes, Marz & Teele, 2001; Kim, 1993; Stopford, 2001). It is top management's role to articulate this knowledge vision and communicate it throughout the organization. A knowledge vision should define what kind of knowledge the organization should create in what domains. It helps determine how an organization and its knowledge base will evolve in the long run (Leonard-Barton, 1995; Nonaka & Takeuchi, 1995). On the other hand, the central requirement for organizational knowledge synthesis is to provide the organization with a strategic ability to acquire, create, exploit, and accumulate new knowledge continuously and repeatedly. To meet this requirement, we need an actionable framework, which could facilitate the development of this strategic ability through the communities of practice. It is likely that there are at least three major processes constituting this synthesis framework of a learning organization, including the personal process, the social process, and the organizational process. What follows is our appreciation of these three important knowledge processes considered as indispensable in the daily operations of the learning organization. Of particular interest here is the idea of appreciative settings, which according to Vickers (1972, p. 98) refer to the body of linked connotations of personal interest, discrimination, and valuation which we bring to the exercise of judgment and which tacitly determine what we shall notice, how we shall discriminate situations from the general confusion of ongoing events, and how we shall regard them. The word "settings" is used because such categories and criteria are usually mutually related; a change in one is likely to affect others.

• **The Personal Process:** Consider a human being as an individual conscious of the world outside his or her physical boundary. This consciousness means that we can think about the world in different ways, relate these concepts to our experience of the world, and so form judgments that can affect our intentions and, ultimately, our actions. This line of thought suggests a basic model for the active human agent in the world. In this model we are able to perceive parts of the world, attribute meanings to what we perceive, make judgments about our perceptions, form intentions to take particular actions, and carry out those actions. These change the perceived world, however

slightly, so that the process begins again, becoming a cycle. In fact, this simple model requires some elaborations. First, we always selectively perceive parts of the world, as a result of our interests and previous history. Secondly, the act of attributing meaning and making judgments implies the existence of standards against which comparisons can be made. Thirdly, the source of standards, for which there is normally no ultimate authority, can only be the previous history of the very process we are describing, and the standards will themselves often change over time as new experience accumulates. This is the process model for the active human agents in the world of individual learning, through their individual appreciative settings. This model has to allow for the visions and actions, which ultimately belong to an autonomous individual, even though there may be great pressure to conform to the perceptions, meaning attributions, and judgments which belong to the social environment, which, in our discussion, is the community of practice.

• **The Social Process:** Although each human being retains at least the potential selectively to perceive and interpret the world in his or her own unique way, the norm for a social being is that our perceptions of the world, our meaning attributions, and our judgments of it will all be strongly conditioned by our exchanges with others. The most obvious characteristic of group life is the never-ending dialogue, discussion, debate, and discourse in which we all try to affect one another's perceptions, judgments, intentions, and actions. This means that we can assume that while the personal process model continues to apply to the individual, the social situation will be that much of the process will be carried out inter-subjectively in discourse among individuals, the purpose of which is to affect the thinking and actions of at least one other party. As a result of the ensuing discourse, accommodations may be reached which lead to action being taken. Consequently, this model of the social process which leads to purposeful or intentional action, then, is one in which appreciative settings lead to particular features of situations, as well as the situations themselves, being observed and interpreted in specific ways by standards built up from previous experience. Meanwhile, the standards by which judgments are made may well be changed through time as our personal and social history unfolds. There is no permanent social reality except at the broadest possible level, immune from the events and ideas, which, in the normal social process, continually change it.

- **The Organizational Process:** Our personal appreciative settings may well be unique since we all have a unique experience of the world, but oftentimes these settings will overlap with those of people with whom we are closely associated or who have had similar experiences. Tellingly, appreciative settings may be attributed to a group of people, including members of a community, or the larger organization as a whole, even though we must remember that there will hardly be complete congruence between the individual and the group settings. It would also be naïve to assume that all members of an organization share the same settings, those that lead them unambiguously to collaborate together in pursuit of collective goals. The reality is that though the idea of the attributed appreciative settings of an organization as a whole is a usable concept, the content of those settings, whatever attributions are made, will never be completely static. Changes both internal and external to the organization will change individual and group perceptions and judgments, leading to new accommodations related to evolving intentions and purposes. Subsequently, the organizational process will be one in which the data-rich world outside is perceived selectively by individuals and by groups of individuals. The selectivity will be the result of our predispositions to "select, amplify, reject, attenuate, or distort" (Land, 1985, p. 212) because of previous experience, and individuals will interact with the world not only as individuals but also through their simultaneous membership of multiple groups, some being formally organized and others informally. Perceptions will be exchanged, shared, challenged, and argued over, in a discourse that will consist of the inter-subjective creation of selected data and meanings. Those meanings will create information and knowledge which will lead to

accommodations being made, intentions being formed, and purposeful action undertaken. Both the thinking and the action will change the perceived world, and may change the appreciative settings that filter our perceptions. This organizational process is a cyclic one and a process of continuous learning; it should be richer if more people take part in it. And it should fit into the context of a learning organization.

AN ORGANIZATION SCENARIO OF KNOWLEDGE SYNTHESIS FOR COMMUNITIES OF PRACTICE

From the discussion built up so far, we can understand that knowledge synthesis is a social as well as an individual process. Sharing tacit knowledge requires individuals to share their personal beliefs about a situation with others (Nonaka, 2002). At that point of sharing, justification becomes public. Each individual is faced with the tremendous challenge of justifying his or her beliefs in front of others—and it is this need for justification, explanation, persuasion, and human connection that makes knowledge synthesis a highly dynamic process (Markova & Foppa, 1990; Vat, 2003).

To bring personal knowledge into a social context, within which it can be amplified or further synthesized, it is necessary to have a field that provides a place in which individual perspectives are articulated, and conflicts are resolved in the formation of higher-level concepts. In the organizational context of our investigation, this field for interaction is provided in the form of a community of practice, made of members perhaps from different functional units.

It is a critical matter for an organization to decide when and how to establish such a community of interaction in which individuals can meet and interact. This community triggers organization knowledge synthesis mainly through several stages. First, it facilitates the building of mutual trust among members, and accelerates creation of an implicit perspective shared by members as tacit knowledge. Second, the shared implicit perspective is conceptualized through continuous dialogue among members. Tacit field-specific perspectives are converted into explicit concepts that can be shared beyond the boundary of the community. It is a process in which one builds concepts in cooperation with others. It provides the opportunity for one's hypothesis or assumption to be tested. As Markova and Foppa (1990) argue, social intercourse is one of the most powerful media for verifying one's own ideas. Next comes the step of justification, which determines the extent to which the knowledge created within the community is truly worthwhile for the organization. Typically, an individual justifies the truthfulness of his or her beliefs based on observations of the situation; these observations, in turn, depend on a unique viewpoint, personal sensibility, and individual experience. Accordingly, when someone creates knowledge, he or she makes sense out of a new situation by holding justified beliefs and committing to them. Indeed, the creation of knowledge, from this angle, is not simply a compilation of facts, but a uniquely human process that cannot be reduced or easily replicated. It can involve feelings and belief systems of which we may not even be conscious. Nevertheless, justification must involve the evaluation standards for judging truthfulness. There might also be value premises that transcend factual or pragmatic considerations. Finally, we arrive at the stage of cross-leveling knowledge (Nonaka, 2002). During this stage, the concept that has been created and justified is integrated into the knowledge base of the organization, which comprises a whole network of organizational knowledge.

CRITICAL CHALLENGES OF ARCHITECTING IS SUPPORT FOR COMMUNITIES OF PRACTICE

Undeniably, setting up an organizational IS support for various communities of practice is a social act in itself, requiring some kind of concerted action by many different people (Vat, 2004a); and the operation of any LOIS subsystem entails such human phenomena as attributing meaning to manipulated data and making judgments about what constitutes a relevant category (Vat, 2004b). Subsequently, an organization is often seen at core as a conversational process in which the world is interpreted in a particular way which legitimates shared actions and establishes shared norms and standards. There is no single body of work which underlies this soft approach to IS, but the works of Sir Geoffrey Vickers (1965) provide quite an interesting reference. For Vickers, organizational members set standards or norms rather than goals, and the traditional focus on goals is replaced by one on managing relationships according to standards generated by previous history of the organization. Furthermore, the discussion/debate, which leads to action, is one in which social action is based upon personal and collective sense making (Weick, 1995). Thereby, organizations are also regarded as networks of conversation or communicative exchanges in which commitments are generated (Ciborra, 1987; Winograd & Flores, 1986). And LOIS support should be thought of as making such exchanges easier—the exchange support systems.

Consequently, a strategy for IS support needs to be thought of, through which desirable change and organizational learning are often considered as the aims. Its stages of development could be characterized as follows with plausible iterations in stages 3, 4, and 5 (Wilson, 2002, pp. 6-10):

1. define the situation that has provoked concerns;
2. express the situation with different sets of concerns;
3. select concepts that may be relevant;
4. assemble concepts into an intellectual structure;
5. use this structure to explore the situation;
6. define changes to the situation as the challenges to be explored; and
7. implement the change processes.

Given the great variety of organizational design problems for CoP-based LOIS support, considerable flexibility must exist in the concepts and structures available to the analysts. It is believed that unless the particular methodology is assembled as a conscious part of the analysis, it is very unlikely that the changes and solutions identified will represent an effective output of the analysis. More importantly, the specific methodology needs to be explicit in order to provide a defensible audit trail from recommendations back to initial assumptions and judgments.

Thereby, thinking about how to think in designing LOIS support is about planning the intellectual process to follow up with the design itself. And there are numerous challenges (Carroll, 1995, 2000) in the underlying process. First, there is often an incomplete description of the problem to be addressed, but it is always necessary to identify the relevant description of the current situation that is to be altered by the design work. Secondly, the problem space of allowable and possible moves is often not determined beforehand. In fact, there is often no guidance on possible design moves in reasoning from a description of the current situation toward an improved version of the situation. Thirdly, design problems themselves characteristically involve many trade-offs; any move creates side effects, such as impacts on human activities. Accordingly, it is by no means a routine process in the IS design for organizational communities of practice.

FUTURE TRENDS OF IS DESIGN FOR COP-BASED KNOWLEDGE SYNTHESIS

According to Checkland and Holwell (1995), the main role of an information system is that of a support function helping people in their purposeful actions. Many of today's information systems are difficult to learn and awkward to use; they often change our activities in ways that we do not need or want. The problem lies in the IS development process. Oftentimes, IS designers have to face convoluted networks of trade-off and inter-dependence, the need to coordinate and integrate the contributions of many kinds of experts, and the potential of unintended impacts on people and their social institutions. It has been observed that traditional textbook approaches to IS development (Checkland & Holwell, 1998) seek to control the complexity and fluidity of design using techniques which filter the information considered, and weakly decompose the problems to be solved. In contrast, the scenario-based design approach (Vat, 2004a, 2004b; Carroll, 1995, 2000) belongs to a complementary tradition that seeks to exploit the complexity and fluidity of design by trying to learn more about the concrete elements of the problem situation. Thereby, John Carroll characterizes scenarios as concrete stories about use through which IS architects could envision and facilitate new ways of doing things and new

things to do. Specifically, scenarios provide a vocabulary for coordinating the central tasks of systems development—understanding people's needs, envisioning new activities and technologies, designing effective systems and software, and drawing general lessons from systems as they are developed and used. Namely, scenarios help IS designers analyze the various possibilities by focusing first on the human activities that need to be supported and allowing descriptions of those activities to drive the quest for correct problem requirements. It is expected that through maintaining a continuous focus on situations of and consequences for human work and activities, IS designers could become more informed of the problem domains, seeing usage situations from different perspectives, and managing trade-offs to reach usable and effective design outcomes (Carroll, 1994, 1995).

Consequently, through the appropriate use of design scenarios, the problems of designing CoP-based LOIS support for knowledge work should never be thought of as something to be defined once and for all, and then implemented. Instead, it must be based on the observation that all real-world organizational problem situations contain people interested in trying to take purposeful action (Checkland, 1999). Pragmatically, the idea of a set of activities linked together so that the whole, as an entity called the human activity system (HAS) from the viewpoint of soft systems methodology (SSM) (Checkland & Holwell, 1998; Checkland & Scholes, 1999) could pursue a purpose, could indeed be considered as a representative organizational scenario for architecting LOIS support, which is never fixed once and for all. In practice, given a handful of the HAS models, namely, models of concepts of purposeful activity built from a declared point of view, we could create a coherent structure to debate about the problem situation and what might improve it (Checkland, Forbes & Martin, 1990; Checkland, 1981, 1983).

Subsequently, from the IS architect's point of view, while conceiving the necessary IS support to serve the specific organizational knowledge requirements, the fundamental ideas could be integrated as follows: Always start from a careful account of the purposeful activity to be served by the system. From that, work out what informational support is required (by people) to carry out the activity. Treat the creation of that support as a collaborative effort between technical experts and those who truly understand the purposeful action served. Meanwhile, ensure that both system creation and system development and use are treated as opportunities for continuous learning. In this way, models of purposeful human activities can be used as scenarios to initiate and structure sensible discussion about LOIS support for the people undertaking the real-world problem situations. Thereby, the process of IS development needs to start not with attention quickly focused on data and technology, but with a focus on the actions served by the intended organizational system. Once the actions to be supported have been decided and described, which can usefully be done using activity models, we can proceed to decide what kind of support should be provided. The key point is that in order to create the necessary IS support which serves the intended organizational scenario, it is first necessary to conceptualize the organizational system (different communities of practices) that is to be served, since this order of thinking should inform what relevant services would indeed be needed in the IS support.

CONCLUSION

This article describes an initiative to develop an actionable framework of knowledge processes, which are aimed to facilitate the creation and sustenance of communities of practice in the context of a learning organization. Our discussion has paid particular attention to the design issues in support of participatory knowledge construction, which is essential for the growth of any CoP in the organizational workplace. In particular, we have elaborated the design issues of three important knowledge processes (the individual, the social, and the organizational), which have tremendous implications for the design of suitable IS support (Vat, 2004b) to help structure and facilitate knowledge creation in the specific organizational setting, where a community of people can conceptualize their world and hence the purposeful action they wish to undertake. This renders a perspective of a knowledge context in a learning organization in which social reality is continually defined and re-defined in both the talk and action of the various communities within the organization. The

article concludes by reiterating the challenge of designing LOIS support so that the purposeful actions of the CoPs can be accommodated. It is important that the examination of meanings and purposes should be broadly based, and its richness will be greater the larger the number of people who take part in it. This consequently provides the basis for ascertaining the development of an organization's communal knowledge space: namely, what IS support is needed by those undertaking their actions, and how modern information technologies can help to provide that support to the various communities of practice.

REFERENCES

Carroll, J.M. (1994). Making use a design representation. *Communications of the ACM, 37*(12), 29-35.

Carroll, J.M. (2000). *Making use: Scenario-based design of human-computer interactions.* Cambridge, MA: MIT Press.

Carroll, J.M. (Ed.). (1995). *Scenario-based design: Envisioning work and technology in system development.* New York: John Wiley & Sons.

Checkland, P. (1981). *Systems thinking, systems practice.* Chichester, UK: John Wiley & Sons.

Checkland, P. (1983). Information systems and systems thinking: Time to unite? *International Journal of Information Management, 8,* 230-248.

Checkland, P. (1999). Systems thinking. In W.L. Currie & B. Galliers (Eds.), *Rethinking management information systems.* Oxford University Press.

Checkland, P., Forbes, P., & Martin, S. (1990). Techniques in soft systems practice, part 3: Monitoring and control in conceptual models and in evaluation studies. *Journal of Applied Systems Analysis, 17,* 29-37.

Checkland, P., & Holwell, S. (1995). Information systems: What's the big idea? *Systemist, 17*(1), 7-13.

Checkland, P., & Holwell, S. (1998). *Information, systems, and information systems: Making sense of the field.* Chichester, UK: John Wiley & Sons.

Checkland, P., & Scholes, J. (1999). *Soft systems methodology in action.* Chichester, UK: John Wiley & Sons.

Ciborra, C.U. (1987). Research agenda for a transaction costs approach to information systems. In Boland and Hirschheim (Eds.), *Critical issues in information systems research.* Chichester, UK: John Wiley & Sons.

Cortada, J.W., & Woods, J.A. (Eds.). (2000). *The knowledge management yearbook 2000-2001.* Butterworth-Heinemann.

Davenport, T.H., & Prusak, L. (1998). *Working knowledge: How organizations manage what they know.* Boston: Harvard Business School Press.

Dierkes, M., Marz, L., & Teele, C. (2001). Technological visions, technological development, and organizational learning. In M. Dierkes, A.B. Antal et al. (Eds.), *Handbook of organizational learning and knowledge* (pp. 282-304). Oxford University Press.

Garvin, D.A. (1993). Building a learning organization. *Harvard Business Review, 71*(4), 78-91.

Hackbarth, G., & Grover, V. (1999). The knowledge repository: Organization memory information systems. *Information Systems Management, 16*(3), 21-30.

Kim, D. (1993). The link between individual and organizational learning. *Sloan Management Review,* (Fall), 37-50.

King, W.R. (1996). IS and the learning organization. *Information Systems Management, 13*(3), 78-80.

King, W.R. (1999). Integrating knowledge management into IS strategy. *Information Systems Management, 16*(4), 70-72.

Land, F. (1985). Is an information theory enough? *The Computer Journal, 28*(3), 211-215.

Leonard-Barton, D. (1995). *Wellsprings of knowledge: Building and sustaining the sources of innovation.* Boston: Harvard Business School Press.

Levine, L. (2001). Integrating knowledge and processes in a learning organization. *Information Systems Management,* (Winter), 21-32.

Markova, I., & Foppa, K. (Eds.). (1990). *The dynamic of dialogue.* New York: Harvester Wheatsheaf.

Nonaka, I. (2002). A dynamic theory of organizational knowledge creation. In C.W. Choo & N. Bontis (Eds.), *The strategic management of intellectual capital and organizational knowledge* (pp. 437-462). Oxford University Press.

Nonaka, I., & Takeuchi, H. (1995). *The knowledge creating company: How Japanese companies create the dynamics of innovation.* Oxford University Press.

OECD. (1996). *The knowledge-based economy.* Organization for Economic Co-operation and Development, OCDE/GD(96)102, Paris.

O'Leary, D.E. (1998). Enterprise knowledge management. *IEEE Computer, 31*(3), 54-61.

Senge, P. (1990). *The fifth discipline: The art and practice of the learning organization.* London: Currency Doubleday.

Stalk, G. Jr., Evans, E., & Shulman, L.E. (1992). Competing on capabilities: The new rules of corporate strategy. *Harvard Business Review,* (March-April).

Stewart, T.A. (1997). *Intellectual capital: The new wealth of organizations.* New York: Doubleday.

Stopford, J.M. (2001). Organizational learning as guided responses to market signals. In M. Dierkes, A.B. Antal et al. (Eds.), *Handbook of organizational learning and knowledge* (pp. 264-281). Oxford University Press.

Vat, K.H. (2004a, July 21-25). Conceiving scenario-based IS support for knowledge synthesis: The organization architect's design challenge in systems thinking. *Proceedings of the 10th International Conference on Information Systems Analysis and Synthesis* (ISAS2004), (pp. 101-106), Orlando, FL.

Vat, K.H. (2004b, November 4-7). Systems architecting of IS support for learning organizations: The scenario-based design challenge in human activity systems. *CD-Proceedings of the 2004 In-*

formation Systems Education Conference (ISECON2004), Newport, RI.

Vat, K.H. (2003, June 24-27). Toward an actionable framework of knowledge synthesis in the pursuit of learning organization. *CD-Proceedings of the 2003 Informing Science + IT Education Conference* (IsITE2003), Pori, Finland.

Vickers, G. (1965). *The art of judgment.* London: Chapman and Hall.

Vickers, G. (1972). Communication and appreciation. In Adams et al. (Eds.), *Policymaking, communication and social learning: Essays of Sir Geoffrey Vickers.* New Brunswick, NJ: Transaction Books.

Weick, K.E. (1995). *Sense-making in organizations.* Thousand Oaks, CA: Sage Publications.

Wenger, E. (1998). *Communities of practice: Learning, meaning, and identity.* Cambridge University Press.

Wenger, E., McDermott, R., & Snyder, W.M. (2002). *Cultivating communities of practice: A guide to managing knowledge.* Boston: Harvard Business School Press.

Wiig, K.M. (1993). *Knowledge management: The central management focus for intelligent-acting organizations.* Arlington, TX: Schema Press.

Williamson, A., & Lliopoulos, C. (2001). The learning organization information system (LOIS): Looking for the next generation. *Information Systems Journal, 11*(1), 23-41.

Winograd, T., & Flores, F. (1986). *Understanding computers and cognition: A new foundation for design.* Reading, MA: Addison-Wesley.

Wilson, B. (2001). *Soft systems methodology: Conceptual model building and its contribution.* New York: John Wiley & Sons.

KEY TERMS

Appreciative Settings: A body of linked connotations of personal or collective interest, discrimination and, valuation which we bring to the exercise of judgment and which tacitly determine what we shall notice, how we shall discriminate situations of concern from the general confusion of ongoing events, and how we shall regard them.

IS Support: An information systems (IS) function supporting people taking purposeful action. This is often done by indicating that the purposeful action can itself be expressed via activity models, through a fundamental re-thinking of what is entailed in providing informational support to purposeful action. The idea is that in order to conceptualize and so create an IS support which serves, it is first necessary to conceptualize that which is served, since the way the latter is thought of will dictate what would be necessary to serve or support it.

Knowledge Processes: These are processes to leverage the collective individual learning of an organization such as a group of people, to produce a higher-level organization-wide intellectual asset. This is supposed to be a continuous process of creating, acquiring, and transferring knowledge accompanied by a possible modification of behavior to reflect new knowledge and insight, and to produce a higher-level intellectual content.

Knowledge Synthesis: The broad process of creating, locating, organizing, transferring, and using the information and expertise within the organization, typically by using advanced information technologies.

Knowledge Vision: A root definition of what knowledge will give the organization a competitive edge in the knowledge-based economy.

Learning Organization: An organization that helps transfer learning from individuals to a group, provide for organizational renewal, keep an open attitude to the outside world, and support a commitment to knowledge. It is also considered as the organization that focuses on developing and using its information and knowledge capabilities in order to create higher-value information and knowledge, to modify behaviors to reflect new knowledge and insights, and to improve bottom-line results.

Meaning Attribution: An intellectual activity involving one's body of linked connotations of personal or collective interest, discrimination, and valuation which we bring to the exercise of judgment and which tacitly determine what we shall notice, how we shall discriminate situations of concern from the general confusion of ongoing event, and how we shall regard them.

Soft Systems Methodology: A methodology that aims to bring about improvement in areas of social concern by activating in the people involved in the situation a learning cycle which is ideally never-ending. The learning takes place through the itera- tive process of using systems concepts to reflect upon and debate perceptions of the real world, taking action in the real world, and again reflecting on the happenings using systems concepts. The reflection and debate is structured by a number of systemic models of purposeful activities. These are con- ceived as holistic ideal types of certain aspects of the problem situation rather than as accounts of it. It is also taken as given that no objective and complete account of a problem situation can be provided.

An IT Perspective on Supporting Communities of Practice

Fefie Dotsika
University of Westminster, UK

INTRODUCTION AND BACKGROUND

An increasing number of organisations have come to recognise the fact that encouraging and maintaining communities of professionals with common interests, aims, and objectives can reduce costs and increase profits. From enhancing customer responsiveness to increasing innovation and preventing reinvention, communities of practice (CoPs) are seen as an important vehicle to the improvement of organisational performance.

Even as the role of CoPs has been gaining momentum, the information technology (IT) community has become aware of the evolving opportunities and is consequently involved in attempting to provide the relevant software tools. This article investigates the requirements for the efficient IT support of CoPs, explores the advantages and pitfalls of supporting 'computerised' versions of these communities, reviews a number of existing software tools, and looks into emerging technologies considering their role and appropriateness.

CoPs AND IT

CoPs are often viewed as a catalyst to the success of a particular organisation's KM system. Their mission is the capturing and sharing of knowledge among practitioners: a task that has traditionally relied upon communicating organisational knowledge via personal interaction and sharing of experiences, problems, and best practices.

One might question whether the deployment of IT in supporting CoPs is justifiable, and whether it would offer a clear return on investment. Those who are for IT support argue that providing easy access to critical market intelligence through, say, a portal, is always good for business. Those who are against tend to overemphasise the problems that electronic systems have created over the years for managers and users alike.

But in spite of such problems, bad press, and disaster cases that come under the umbrella of system failure scenarios, it is an undeniable fact that an ever-increasing amount of vital business informa-

Figure 1. Flow of activities

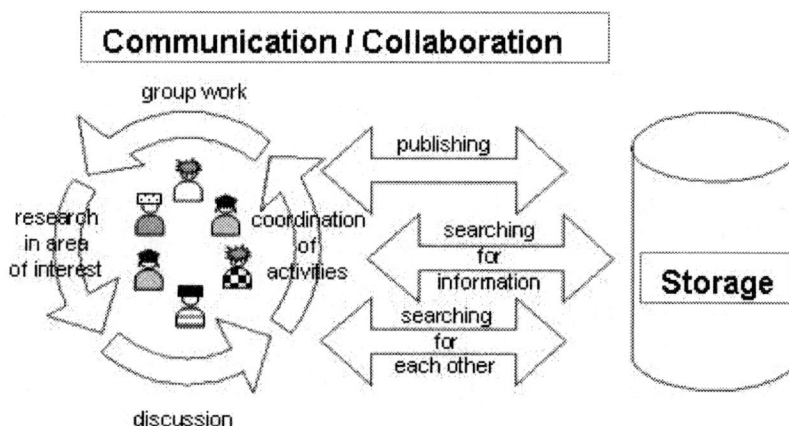

tion spends its whole lifecycle in digital format. This fact alone challenges the nature of old-fashioned communication/collaboration between the members of a group and adds to the need of consolidating the way information is handled.

While many communities are supported by Web sites providing knowledge sharing by means of online libraries, knowledge centres, specialist databases, information repositories, and white pages, only a few of them get the full necessary support. Terms like *online* and *virtual CoPs* are becoming common-place, reflecting thus the increasing tendency to form expanded and even globalised versions of the traditional groups of people who come together to share their knowledge. Despite the spatial difference between traditional groups and their online counterparts, the actual requirements remain the same. Figure 1 summarises these requirements, and depicts the main flow of basic activities within a CoP.

Looking into the role that IT plays in assisting communities of practice, we can distinguish three (often overlapping) categories: supporting the social actions inherent in CoPs, supporting the different stages of CoPs' lifecycles, and adaptive use of collaborative computing technologies generally servicing knowledge management issues and requirements.

Supporting Social Actions

Following the framework proposed by Ngwenyama and Lyytinen (1997), the four types of social actions can be seen at work here: instrumental, communicative, discursive, and strategic. According to this division we can now look at the four action clusters from an IT viewpoint, identifying what type of software/groupware would be appropriate for carrying out their respective tasks.

1. **Instrumental actions:** This category is supported by the so-called *research tools*. These are tools that provide the person executing the instrumental action with the relevant resources—that is, the relevant knowledge. Databases, data warehouses, data marts, electronic document management systems (EDMs), knowledge bases, and knowledge servers all play the role of *knowledge repositories* under

this category. The research tools that extract knowledge from these repositories come in all shapes and guises, from database query languages and search engine facilities to data mining and intelligent agents.

2. **Communicative actions:** Traditionally, this is the earliest and possibly most efficiently supported category. Use of e-mail, list servers, Internet, corporate intranets, and even remote login facilities, file transfer, and electronic messaging are examples of communication tools.

3. **Discursive actions:** Apart from the possible overlap with the previous category—such as the use of e-mail and listserv facilities—there are dedicated groupware packages that assist the setting up, customisation, and configuration of online discussion groups. Chat rooms and e-conferencing are also popular applications. In general, collaboration services come under two categories: *synchronous* and *asynchronous*. Instant messaging facilities, e-conferencing, and all sorts of audio and/or video streaming belong to the former category, whereas discussion forums, calendar postings, and e-mail belong to the latter.

4. **Strategic actions:** These form the last category, the only one with no evident IT support. Although closely related to instrumental actions to the extent that they both strive to achieve rational objectives, the two categories differ in their view of the opponent: the person executing the instrumental action treats the adversary as an organisational resource and not as a person capable of intelligent counter-action (Ngwenyama & Lee, 1997), which is the case in the strategic action. This "quirkiness" alone makes things hard as it predefines a requirement difficult to resolve with conventional IT tools. The solution is likely to come from the artificial intelligence community, with the use of intelligent agents. These are adaptive computer programs capable of reasoning and learning, and are collectively known as *bots*. There are many types of agents, each performing specific, specialised tasks (search bots, chatter bots, shopping bots, etc.). Their potential to support strategic actions derives from the fact that they are sociable—they can

Table 1. Stages, functions, and technologies of the Wenger communities evolution model

STAGES	MAIN FUNCTIONS	IT ENABLING TECHNOLOGIES
1	connect, plan, commit	e-mail, e-conferencing, listservers, online forums, Internet, corporate intranets
2	form framework, create context	as above, plus remote login facilities, file transfer, information repositories
3	operate, collaborate, grow, improve, mature	as above, plus online directories, analytical and decision-making tools, intelligent agents, e-surveying, and feedback facilities
4	sustain, renew, maintain, wind-down	also portals (see following section)
5	shut down	knowledge repositories may remain for use by future communities

interact and communicate with humans and other bots.

Supporting Different Stages of a CoP's Lifecycle

According to McDermott's *communities evolution model* (McDermott, 2000), we distinguish five stages of development in a CoP's lifecycle: planning, starting-up, growing, sustaining/renewing, and closing. This model emphasises the tensions and challenges that stimulate the community to develop and renew itself, and ends with the death of the community. Wenger (1998) proposes a very similar model with equal numbers of stages, which he calls potential, coalescing, active, dispersed, and memorable. At the end stage of this model, the community disappears, but parts of its knowledge remain in memories, stories, and artefacts. Table 1 maps each of the five stages into their main functions. The third column identifies the possible relevant enabling technologies.

Tailoring KM Support to Service CoPs

Apart from the above, there is a number of collaborative computing technologies used in the support of knowledge management that can also be put into use with CoPs. These tools can usually service the above action categories and the different stages of a CoP's lifecycle to varying degrees.

- **Knowledge management suites:** Provide solutions for creating centralized repositories for storing and sharing knowledge, allow for communication between the members of the group, and support group work. They thus integrate the storage, communications, and collaboration services into a single environment.

- **Portals:** Also known as *super-sites* or *enterprise knowledge portals*, they are an electronic doorway providing a comprehensive array of resources and services. Portals typically contain newsletters, e-mail services, search engines, online shopping, chat rooms, discussion boards, and personalised links to other sites. While portals attract a large number of visitors offering a wide range of contents, vortals (*vertical portals*, also known as *online communities*) are narrower in focus and address a specific industry, theme, or particular interest, a feature that has made them more appropriate for the support of CoPs.

- **Collaboration tools, often referred to as groupware:** A difficult to define class due to the diversity in the functions offered. Most packages comprise an information repository that can be accessed by team members who can collaborate working on common documents and can hold electronic discussions. Some groupware packages integrate calendars, group schedulers, and e-mail. Others offer e-conferencing facilities or other real-time meeting support.

EXISTING SOFTWARE PLATFORMS

We divide the software platforms into two distinct categories: software that offers IT support aimed especially at communities of practice and software designed to assist knowledge management in general, but also meets the requirements for the support of CoPs. Generally speaking, both KM and CoP support requirements are similar, though different emphasis is given to certain components. For instance, KMware demands broader content management techniques leading to more rigorous system interoperability requirements, whereas CoP support relies heavily on the communication layer. With a large and constantly increasing number of available platforms in each category, the list of products presented below is only representative of the range of services available, but is by no means exhaustive.

CoP-Dedicated IT Support

iCohere (2004) provides Web collaboration software tools for online communities, project teams, and distributed organizations. Specific applications include extranets, workgroup and virtual team collaboration, and online learning. Their technology and supporting processes enable engaging member communication, networking, knowledge sharing, collaboration, and learning. The groupware is available either as a hosted application on the iCohere servers or for use in the customers' own servers as a site licence. It supports a backend MS SQL server and Web-based dynamic DBMS access. Whether hosted or licensed, the software claims advanced security considerations. It supports https option, configurable password formats, and login timeouts. iCohere partners include universities, education-focused professional societies, corporate business, government agencies, healthcare associations, and non-profit organisations.

Tomoye's CoP platform *Tomoye Simplify 4.0* (2004) offers a similar set of resources. In efforts to meet the increasing customer demand for integrated services, Tomoye in March 2004 became a Microsoft-Certified Partner. On its Web site, the company demonstrates the different functionalities of the Simplify platform through two case studies: (a) oneFish, a CoP at the UN in Rome; and (b) Global Knowledge Partnership at the World Bank.

Tomoye built oneFish to enable 15,000 fisheries researchers from around the world to pool their knowledge, identity experts, and collaborate in online conversations. oneFish features 10,000+ records, cross referenced across 1,700 topics. It allows for easy navigation and provides threaded discussion forums, e-mail lists and digests, FAQs, content ratings, and a search engine over an XML database that includes multimedia content.

The second case study is the Global Knowledge Partnership at the World Bank, an organisation that comprises 65+ partner organisations, dedicated to the sharing of knowledge and best practices for sustainable development. Knowledge is modelled as knowledge objects, and each object (including people) can have its own discussions and FAQs. Users can further subscribe to a subject of interest and receive regular e-mail updates, digests, and links to new related objects. The online environment provides login facilities and membership privileges, customisation, navigation via bookmarks, search for knowledge and experts, discussion forums, and instant messaging.

Not all software houses providing CoP support software are big. *KnowNet* is a small company that was founded in 2000 to research and develop new architectures, ideas, and Internet software for collaborative knowledge development and learning (KnowNet, 2004). The company supports virtual online communities through integrated portals, collaborative content management interactive XML document repositories, structured discussion groupware, collaborative resource sharing, and metadata management. Its customers include the European Commission (Leonardo da Vinci Vocational Training, CEDEFOP, and STRATA programmes) and the British Library.

General KM Support

Open Text (2004) is one of the biggest players in groupware services, especially since it acquired fellow enterprise content management software firm IXOS in 2003. Both Open Text—better known for its collaboration and document management software—and IXOS—known for archiving and content management—had made several acquisitions in the months prior to the takeover, with the result of ending up with a surplus of software packages that needed sorting and integrating. The company offers Livelink, a KM software environment that manages corporate knowledge assets. Marketed as a "scalable and modular platform for the acquisition, creation, aggregation, management, and delivery of content," the Livelink interface brings together vari-

Table 2.

product	communication collaboration	group-work	coordination	discussion	extras	further support (KM)	pricing
iCohere	✓	✓✓	✓	✓	hosting or licence		£-££
Simplify	✓✓	✓✓	✓	✓			££
KnowNet	✓✓	✓	✓	✓			£
Open Text	✓✓	✓	✓	✓		✓	£££
HummingBird	✓✓	✓	✓	✓	powerful integration tools	✓	£££
iLevel	✓✓	✓	✓	✓	document/e-mail /content/management	✓	£££

ous collaborative applications supported on the Open Text platform and can be successfully used for CoP support. By leveraging best practices and lessons learned across different communities, Livelink connects and organises knowledge entities into knowledge-sharing networks and delivers an integrated system for collaborative work to globally distributed teams.

Another major KM software player is *Hummingbird* (2004), a global provider of enterprise software solutions. This player's integrated platform, Hummingbird Enterprise 2004, offers a comprehensive number of capabilities: content, document and record management, e-mail management, enterprise workflow, collaboration platform, wireless mobility, query and reporting facilities, and data integration. The *portal framework* integrates all components of Hummingbird Enterprise 2004 to deliver personalised content, applications, and collaboration capabilities within dynamic views or virtual workspaces, based on the role of the user in the business process. Hummingbird's customers cover a vast cross-section of industries: aerospace and defence, government, chemical, oil and gas, energy and utilities, automotive, telecommunications, financial services, life sciences and healthcare, education, manufacturing, retail, and so forth. Although the software is mainly marketed as enterprise content management, wherever available, the platform has enough attributes that make it an efficient CoP support tool.

iLevel Software (2004) provides solutions that enable teams to collaboratively manage the entire lifecycle of business content using a unified, tightly integrated platform and repository. The iLevel envi-

ronment offers extensive XML content management, Web-based document management, Web content management, and intranet/extranet access to business information, but also a number of services that improve knowledge exchange and retrieval, such as enterprise search, categorisation facilities, alerts, and collaborative capabilities.

Table 2 provides a relative comparison of the various products mentioned above.

THE SEMANTIC WEB AND THE USE OF ONTOLOGIES

The unprecedented expansion of the World Wide Web has triggered a significant increase in the expectations for Web-based information retrieval, knowledge sharing, and collaborative working, all of which work well within a tight frame of reference but become problematic when this frame expands. With the appearance of the Semantic Web (Berners-Lee, Hendler & Lassila, 2001), the rapidly developing form of Web content that is readable by computers, Web-based knowledge representation relies on languages that express information in a machine 'process-able' form.

The "conventional" Web relies on encoding schemes based on technologies such as HTML and XML (eXtensible Markup Language) (Decker et al., 2000). However, information that adheres to this encoding lacks explicit semantics. To this extent, the Semantic Web deploys two further enabling technologies: RDF (Resource Description Framework) (Brickley, 1999) and ontologies (Fensel, 2001). If we think of HTML as a mark-up language for displaying data, and XML as another for describing it, then RDF provides the semantic mark-up, and ontology languages supply a shared common understanding of a domain.

More specifically, RDF models knowledge as directed graphs, represented as triples. The semantic structure of these triples is the assertion that *subjects* are associated with *objects* by means of *predicates*, hence the subject-predicate-object relationship. Each of these terms can be represented by a URI (Universal Resource Identifier).

With the semantic mark-up in place, ontologies provide the formal specification of a knowledge domain, often along with an inference engine. A particular knowledge domain consists of classes, their instances, and the relationships between them. This domain specification can then be communicated between heterogeneous application systems, enhancing knowledge sharing and retrieval (Davies, Duke & Stonkus, 2002). Consequently ontologies are particularly useful for: (a) sharing a common understanding of a domain among the members of the community, (b) analysing and/or reusing domain knowledge, and (c) making explicit any domain assumptions.

The deployment of semantic mark-up, together with ontologies, revolutionises Web information retrieval and sharing, a fact that is of particular interest to CoPs (Domingue et al., 2001; Motta, Buckingham-Shum & Domingue, 2000), some of which are already working towards common encoding standards. Among them, the linguistic community is developing GOLD (General Ontology for Linguistic Description) (Farrar & Langendoen, 2003).

Nevertheless, the Semantic Web is not the only use of ontologies related to CoPs. Another use focuses on systems used to identify CoPs within an organisation, a process presently done by means of structured interviews. ONTOCOPI (Ontology-based Community of Practice Identifier) (Harith, O'Hara & Shadbolt, 2002) is such a system, capable of identifying CoPs by examining the connectivity of instances in a knowledge base with regard to their type, weight, and density.

CONCLUSION AND FUTURE TRENDS

There is a number of software platforms that are designed to assist communities of practice. Some of them provide dedicated support, whereas others are general KM environments able to offer CoPs the required IT facilities. But while online communities benefit from technology, and face-to-face member interaction can be substituted by virtual contact to various degrees, knowledge manipulation still poses a significant and often decisive obstacle to the flow of knowledge inside these communities. The emergence of the Semantic Web seems to tackle a number of these problems, though the process of migration is currently rather cumbersome and requires specialist knowledge of the technologies involved. However, software for the computerised adding of semantics to Web information is being developed, while the design and development of tools for the automated capturing, sharing, and retrieval of knowledge are underway.

REFERENCES

Berners-Lee, T., Hendler, J., & Lassila, O. (2001). The Semantic Web. *Scientific American,* (May).

Brickley, D. (1999). Semantic Web history: Nodes and arcs 1989-1999. The WWW proposal and RDF. Retrieved from *http://www.w3c.org/1999/11/11-WWWProposal/*

Davies, J., Duke, A., & Stonkus, A. (2002). OntoShare: Using ontologies for knowledge sharing. *Proceedings of the WWW2002 Semantic Web Workshop, 11th International WWW Conference,* Hawaii.

Decker, S. et al. (2000). The Semantic Web: The roles of XML and RDF. *IEEE Internet Computing,* (September-October), 63-75.

Domingue, J. et al. (2001). Supporting ontology-driven document enrichment within communities of practice. *Proceedings of the 1st International Conference on Knowledge Capture (K-Cap 2001),* Victoria, British Columbia, Canada.

Farrar, S., & Langendoen, T. (2003). A linguistic ontology for the Semantic Web. *Glot International, 7*(3), 97-100.

Fensel, D. (2001). *Silver bullet for knowledge management and electronic commerce.* Berlin: Springer-Verlag.

Harith, A., O'Hara, K., & Shadbolt, N. (2002). ONTOCOPI: Methods and tools for identifying communities of practice. *Proceedings of Intelligent Information Processing 2002*, Montreal, Canada.

Hummingbird Enterprise 2004. (2004, June). Enterprise content management platform. Retrieved from *http://www.hummingbird.com/products/enterprise/index.html*

iCohere. (2004, June). Creating collaborative communities. Retrieved from *http://www.icohere.com*

iLevel Software. (2004, June). Retrieved from *http://www.iLevelSoftware.com/*

KnowNet. (2004, June). Retrieved from *http://www.theknownet.com/*

McDermott, R. (2000). Community development as a natural step: Five stages of community development. *KM Review, 3*(5).

Motta, E., Buckingham-Shum, S., & Domingue, J. (2000). Ontology-driven document enrichment: Principles, tools and applications. *International Journal of Human-Computer Studies, 52*, 1071-1109.

Ngwenyama, O.K., & Lee, A.S. (1997). Communication richness in electronic mail: Critical social theory and the contextuality of meaning. *MIS Quarterly, 21*(2), 145-167.

Ngwenyama, O.K., & Lyytinen, K. (1997). Groupware environments as action constitutive resources: A social action framework for analyzing groupware technologies. *Computer Supported Cooperative Work: The Journal of Collaborative Computing, 6*(1), 71-93.

Open Text and IXOS. (2004, June). Retrieved from *http://www.opentext.com*

Tomoye Simplify 4.0. (2004, June). Retrieved from *http://www.tomoye.com/*

Wenger, E.C. (1998). Communities of practice: Learning as a social system. *Systems Thinker, 9*(5), 2-3.

KEY TERMS

Knowledge Management Suites: Software packages that provide solutions for creating centralised repositories for storing and sharing knowledge, support content management, allow for communication between the members of the group, and assist group-work.

Ontology: Originally, a branch of metaphysics that studied the essence of beings, or first principles. In IT, it is the working model of entities and interactions in some particular domain of knowledge or practices. In AI, an ontology is the specification of a conceptualisation.

Portal: An electronic doorway providing a comprehensive array of resources and services. Portals typically contain newsletters, e-mail services, search engines, online shopping, chat rooms, discussion boards, and personalised links to other sites.

RDF (Resource Description Framework): A recommendation from the World Wide Web Consortium for creating metadata structures that define data on the Web. It is designed to provide a method for classification of data on Web sites in order to improve searching and navigation.

Semantic Web: A collaboration of the World Wide Web Consortium and others to provide a standard for defining data structures on the Web (www.w3.org/2001/sw).

Vortal: A vertical portal. A vertical industry, or market, or specific group portal on the Internet.

XML (eXtensible Markup Language): A subset of SGML (Standard Generalised Markup Language), designed to describe data. It incorporates features of extensibility, structure, and validation, and is currently playing an increasingly important role in the exchange of a wide variety of data on the Web and elsewhere.

Knowledge Communities, Communities of Practice, and Knowledge Networks

Tobias Müeller-Prothmann
Free University Berlin, Germany

INTRODUCTION

In the last few years, the social perspective has emerged as the dominant paradigm in information and knowledge management studies. First-generation knowledge management, characterised by a technical and technological process view, has given way to new approaches that examine social dimensions of knowledge creation, transfer, and management. This shift of focus takes into account the perspective that the majority of individual knowledge transfer does not follow formal hierarchies or processes, but is instead driven by personal and informal communications. Such a social constructionist view of knowledge exchange considers not only single individuals, but also social aggregates and their structural patterns. Even so, despite a growing literature on the socially derived related concepts of knowledge communities (see, e.g., Botkin, 1999; Erickson & Kellog, 1999; 2001; Lesser, Slusher & Fontaine, 2000; Schmidt, 2000), communities of practice (see, e.g., Brown & Duguid, 1991; Lesser, 2001; Wenger, 1999), and knowledge networks (see, e.g., Collinson & Gregson, 2003; Liyanage, Greenfied & Don, 1999; Nohria & Eccles, 1992; Powell, 1998; Seufert, von Krogh & Bach, 1999), there is confusion over their conceptual and applied distinctiveness. Could it be, for example, that they are just different labels for the same phenomenon? Or are there justifiable and valid differences that demand a more careful and reflective use of terminology? This article provides basic steps to the exploration of similarities and differences between the concepts of knowledge communities, communities of practice, and knowledge networks.

BACKGROUND

Despite the existence of concise theoretical constructs that enable us to identify the unique concepts of communities of practice, knowledge communities, and knowledge networks, there remains a great deal of definitional misinterpretation and misapplication in both the literature and in practice. Below the three concepts are introduced before examining some of the most common misconceptions and practical inconsistencies.

Knowledge communities, also called communities of knowing (Boland & Tenkasi, 1995), are "groups of people with a common passion to create, share, and use new knowledge for tangible business purposes" (Botkin, 1999, p. 30). After Botkin, they are characterised through shared values and a common commitment that create a sense of belonging, trust, and openness amongst their members. Thus, knowledge communities provide a context for the sharing of knowledge. Moreover, "they are based primarily on the sharing of knowledge rather than practice" (Scarbrough & Swan, 2001, p. 13). Indeed, while Scarbrough and Swan's analysis of knowledge communities in innovation management distinguishes between IT-based and community-based approaches, most authors focus on knowledge communities as communities based on or at least supported by IT systems, often known as virtual knowledge communities (see, e.g., Diemers 2001; Erickson & Kellog, 1999, 2001; Schmidt, 2000).

Communities of practice are commonly constituted through shared work practice over a period of time (see Brown & Gray, 1998). Often, they are compared to an apprenticeship model where soft knowledge is transferred through the situated learning that takes place in apprenticeship environments. But the central communities of practice concept of "legitimate peripheral participation" is not restricted to apprenticeships alone. Rather, communities of practice "imply participation in an activity about which all participants have a common understanding...The community and the degree of participation in it are in some senses inseparable from the practice" (Hildreth, Kimble & Wright, 2000, p. 29). From this perspective, communities of practice are a social context for "learning as legitimate peripheral participation" (Hildreth et al., 2000, p. 28).

According to Lave and Wenger (1991), communities of practice may be oriented towards hierarchy or collegiality. Hierarchical communities of practice allow for socialisation of novices through expert masters into local understandings of the meaning of the work through opportunities for "legitimate peripheral participation" (Lave & Wenger, 1991). In collegially based communities of practice, "informed dialogue among members is central to the on-going co-evolution of meaning and capabilities" (Liedtka, 1999, p. 7).

One of the defining characteristics of a community is its bounded nature: it has a boundary in terms of social interaction and membership. This applies for a knowledge community as well as for a community of practice. Networks, including of course knowledge networks, are not characterised through clearly defined boundaries. Rather, the analysis of networks aims at tracing social relationships wherever they may go (on the boundary specification problem in network analysis, see Laumann, Marsden & Prensky, 1989). Discussions of network structures in management literature were influenced for example by Drucker (1989) and Savage (1990). Networks can be seen as a third form of organisation (Powell, 1990) or as a hybrid form of organisation between market and hierarchy (Thorelli, 1986). All these discussions state the increasing importance of networks.

Networks can be distinguished according to their level as between individuals, groups, communities, organisational units, organisations, collectives of organisations, or even between societies. Network research in knowledge management tends to stress the importance of informal networks have a long history of study. Informal social relations in organisations have been subject to research since at least the 1930s with the classical Hawthorne studies (Roethlisberger & Dickson, 1947).

Often too, networks are viewed in the context of knowledge management as an activity, that of "networking." Seufert et al. (1999) "use the term 'knowledge networking' to signify a number of people, resources, and relationships among them, who are assembled in order to accumulate and use knowledge primarily by means of knowledge creation and transfer processes, for the purpose of creating value." They also distinguish between emergent and intentional knowledge networks. "Intentional knowledge networks are seen as networks that are built up from scratch, whereas emergent knowledge networks already exist but have to be cultivated in order to become high performing" (p. 184).

Although the unique dimensions of the three distinctive concepts of knowledge communities, communities of practice, and knowledge networks can be clearly discerned, such distinctions are rarely found in the literature. Following Botkin (1999), the difference between communities of practice and knowledge communities is that communities of practice "are informal groups, shaped by circumstances, visible mainly to social anthropologists," whereas knowledge communities "are purposely formed…and their purpose is to shape future circumstances. They are highly visible to every business person in the organization" (p. 31).

Scarbrough and Swan (2001) try to distinguish knowledge communities from communities of practice in that they are based primarily on the sharing of knowledge rather than practice; however, they are able "to interface" with existing communities of practice (Scarbrough & Swan, p. 13). Moreover, bringing into play a network perspective, Swan, Newell, Scarbrough, and Hislop (1999) develop a "networking community" perspective on knowledge management: "Networking as a social communication process, which encourages the sharing of knowledge among communities" (p. 263).

AN EXPERT VIEW

From August 2003 until January 2004, the author undertook explorative study[1] of the views and interpretations that expert knowledge management academics and practitioners have of the three distinct concepts of knowledge communities, communities of practice, and knowledge networks. In the study, the experts were asked to define the three concepts and to outline the differences between each other.

The criteria used to distinguish the three concepts were initially derived from a study of the literature on communities and social networks. They were then tested for validity using a qualitative survey of expert knowledge management academics and practitioners. Using this approach, the central criteria that distinguish the different concepts were identified as follows:

- goal orientation,
- organisation,
- shared practice,
- size,
- identity,
- cohesion.

Goal orientation focuses on whether the network/ community is oriented towards a defined target or whether the common intention is more diffuse. Organisation of the network/community can be formalised or informal. The members of the network/ community may share practice or not, that is they work together or do not necessarily work together. The size of the network/community is defined by the number of members. Identification of the network/ community members as a group and their sense of community is a matter of strong or weak identity. The network/group can be densely knit or loosely coupled depending on intensity, frequency, and type of the members' contacts and the continuity of the network/community. This is expressed by cohesion.

The next section uses the derived criteria and a qualitative analysis of the expert survey to summarise the similarities and differences between the concepts of knowledge communities, communities of practice, and knowledge networks.

SIMILARITIES AND DIFFERENCES

First of all, the results demonstrate that the concepts of communities are distinctive from knowledge networks. According to the experts' views, knowledge communities as well as communities of practice:

- try to achieve a common purpose, that is with specific tangible focused goals (relatively high goal orientation);
- are more formal (than networks) and can be recognised as such (relatively formal organisation);

- are active and exchange driven, the members share practice (strong shared practice);
- consist of a relatively small number of members, and membership is relatively clearly defined (small size);
- members know that they belong to the community, they share a stronger sense of identity, which at its best can be broadcasted by a clear name, logo, or organisation (strong identity);
- are densely connected, show a high rate of interactions and personal affiliation in form of (at least partial) face-to-face communication, and develop mutual commitment and trust (high degree of cohesion).

Knowledge networks on the other hand are characterised by the experts in the survey as:

- emergent structures of organising knowledge across the organisation, with participants who contact each other in current cases, driven by "finding the right expert" (diffuse goal orientation);
- informally organised (since they are an emergent structure of organising knowledge) and thus cannot be recognized as such; in addition, they build a structure that might surround and interlink a number of communities (relatively informal organisation);
- passive, without continuous participation of its members (without or only with little shared practice);
- with a relatively high number of participants, open membership (the network border is not clear) (large size);
- without (or with little) sense of belonging and identification of the participants with the network (weak identity);
- characterised by low rate of interactions and low continuity, sparsely connected, loosely coupled ("I know someone who…") (low degree of cohesion).

Using the criteria of goal orientation, organisation, shared practice, size, identity, and cohesion, communities and networks can be described quite clearly as different social entities. Knowledge networks can be found as the informal networks of knowledge exchange within a certain domain of knowl-

Figure 1. Communities vs. knowledge networks

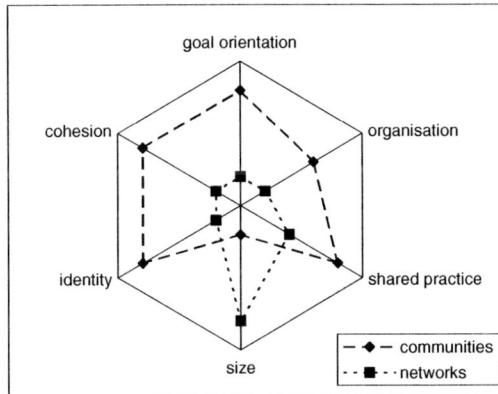

edge. These include not only networks within organisations, but also those networks across organisational boundaries, connecting people of multiple specialities and knowledge disciplines to achieve their individual objectives, like advice and support networks for example. Knowledge networks appear within organisations, institutions, cultures, and between them through dynamic interactions of their members. New networks emerge "by creating new meanings, new linguistic routines, and new knowledge" (Boland & Tenkasi, 1995, p. 352). And, of course, the different networks of the individual members differ. One participant of the survey illustrated the relative distinctiveness of communities and networks with regard to the appropriate Internet tool for communication: the tool for a community is the discussion forum, while a mailing list is the tool for a network. Examples of communities will be given below to highlight the different characteristics of knowledge communities and communities of practice.

In Figure 1, communities and knowledge networks are illustrated with regard to their orientation towards goal orientation, (formal) organisation, shared practice, (large) size, (strong) identity, and (high degree of) cohesion. These criteria build the vertices of a hexagon that represent the (extreme) poles for each mentioned criterion.

Knowledge communities and communities of practice are much more difficult to distinguish. While only very few participants do not distinguish between the different concepts of knowledge communities, communities of practice, and knowledge networks

works (especially practitioners who even ask about the usefulness of a precise definition of the different concepts), nearly half of the participants do not see any differences between knowledge communities and communities of practice. Instead, they consider a knowledge community as a kind of community of practice (or vice versa). Nonetheless, analysis of the answers still identifies important differences between knowledge communities and communities of practice:

- **Goal Orientation:** While communities of practice are focused on a specific topic, like for example on the development of a concrete product (that is, they are linked to a specific business process with a relatively clearly defined target), the domain of knowledge communities is more general.

- **Organisation:** In many cases, communities of practice are formally established in organisations, or at least they are officially supported, with a specific target. Knowledge communities are in most cases self-organised, more general, with a more altruistic motivation of the participants.

- **Shared Practice:** Communities of practice are organised for the purpose of practical implementation of knowledge derived from experience; knowledge communities are organised for research, development, and innovation, that is, for the generation of new knowledge. Members of a community of practice work together; knowledge community members do not necessarily have to work together. Focusing on practice means "how to" (communities of practice), while knowledge is more general (knowledge communities).

- **Size:** Size is not a distinguishing criterion between knowledge communities and communities of practice (some experts mentioned that the latter is smaller in size).

- **Identity:** While a strong sense of community develops through shared work practice in the case of communities of practice, identification with the community in case of knowledge communities is based on the awareness of the collective knowledge and on keeping it.

- **Cohesion:** Cohesion is viewed as being identical for communities of practice and knowledge communities by nearly all survey participants; only few viewed communities of practice as being based on more personalised relationships than knowledge communities.

FUTURE TRENDS

The study shows that there are some qualitative differences between communities of practice and knowledge communities. One practitioner expert put their distinctiveness in the context of organisational levels. In his view, communities of practice are a form of organisation on a meta level above the baseline organisation, while knowledge communities are a form of organisation on a meta-meta level.

Communities of practice may include engineers engaged in deep-water drilling within a large oil company or frontline managers in charge of check processing at a large commercial bank, for example (examples adopted with modifications from Wenger & Snyder, 2000). They all show a strong focus on a specific task or topic, could be formally organised within large enterprises, share practice, and develop a sense of community through common experience.

Examples of knowledge communities can be found in all kinds of distributed cognition within and between organizations, especially in research and development of knowledge-intensive firms. Also in science, knowledge communities are widely recognised in the various disciplines and, increasingly, between different disciplines with regard to highly specialised subjects of research. They focus on a certain domain of knowledge, mostly concerning the generation of innovations, are rather self-organised rather than formally established, show altruistic motivations of the participants, and identification arises due to the awareness of the collective knowledge.

In Figure 2, communities of practice and knowledge communities are illustrated in the hexagon with regard to their orientation towards the (extreme) poles of each criterion.

CONCLUSION

The results of the study show that knowledge networks are viewed as relationships of a large number

Figure 2. Communities of practice vs. knowledge communities

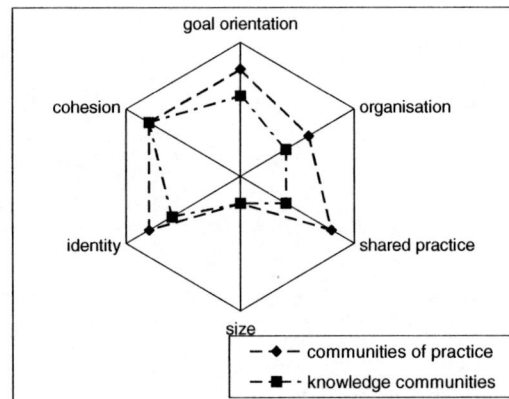

of loosely coupled participants with a diffuse common domain of knowledge and without clearly defined boundaries. Communities of practice are defined as relatively small groups of people who are strongly bound together; the groups are founded on core concepts of trust, shared work practice, and a common goal. Knowledge communities on the other hand are defined as relationships of trust between people within a wider domain of knowledge, but are difficult to distinguish precisely from communities of practice.

This article suggests that a greater theoretical foundation is necessary to facilitate the development of a common language and greater understanding of the popular concepts of knowledge communities, communities of practice, and knowledge networks. From a theoretical perspective, a comparative focus on the different concepts and the relationships between them is necessary. One theoretical approach is to conceptualise communities as a special form of social networks (for example as outlined by Poplin, 1979, pp. 14-18, or Wellman, 1997; on social network analysis as a method for leveraging communities, networks, and expertise, see Müeller-Prothmann & Finke, 2004 or Müeller-Prothmann, Siegberg, & Fink, 2005). Moreover, future attention is required to the "divide" that presently exists between well-founded theoretical conceptualisations on the one hand, and the use and interpretation of these concepts in the knowledge management practitioner community on the other. To resolve these inconsistencies, new forms of research that integrate theoretical distinctions, empirical studies, and practical relevance are needed. The expert study described in this article is a first step in this direction, and no more than an exploration of this subject and a basis for further research and discussions. Further research into theoretical foundations, empirical studies, and practical relevance could lead to a revised set of conceptualisations in theory and applications in practice of knowledge communities, communities of practice, and knowledge networks.

ACKNOWLEDGMENTS

I am very grateful to Chris Lawer, Cranfield School of Management and The OMC Group, England, for critical and fruitful comments and his valuable contribution in reviewing and correcting a draft version of this article.

REFERENCES

Boland, R.J., Jr., & Tenkasi, R.V. (1995). Perspective making and perspective taking in communities of knowing. *Organization Science, 6*(4), 350-372.

Botkin, J. (1999). *Smart business. How knowledge communities can revolutionize your company.* New York: The Free Press.

Brown, J.S., & Duguid, P. (1991). Organizational learning and communities of practice: Toward a unified view of working, learning, and innovation. *Organization Science, 2*(1), 40-57.

Brown, J.S., & Gray, E.S. (1995). The people are the company. *FastCompany Magazine.* Retrieved February 2, 2004, from *http://pf.fastcompany.com/magazine/01/people.html*

Collinson, S., & Gregson, G. (2003). Knowledge networks for new technology-based firms: An international comparison of local entrepreneurship promotion. *R&D Management, 33*(2), 189-208.

Diemers, D. (2001). *Virtual Knowledge Communities. Erfolgreicher Umgang mit Wissen im Digitalen Zeitalter.* Dissertation, Universität St. Gallen, Hochschule für Wirtschafts-, Rechts- und Sozialwissenschaften (HSG). Bamberg: Difo-Druck.

Drucker, P.F. (1989). *The new realities.* New York: Harper and Row.

Erickson, T., & Kellogg, W.A. (1999). *Towards an infrastructure for knowledge communities. Proceedings of the ECSCW '99 Workshop: Beyond Knowledge Management: Managing Expertise.* Unpublished manuscript.

Erickson, T., & Kellogg, W.A. (2001). *Knowledge communities: Online environments for supporting knowledge management and its social context.* Pre-print for a chapter in Ackerman, M., Pipek, V., Wulf, V. (Eds.), *Beyond knowledge management: Sharing expertise.* Cambridge, MA: MIT Press.

Hildreth, P., Kimble, C., & Wright, P. (2000). Communities of practice in the distributed international environment. *Journal of Knowledge Management, 4*(1), 27-38.

Laumann, E.O., Marsden, P.V., & Prensky, D. (1989). The boundary specification problem in network analysis. In L.C. Freeman, D.R. White & A.K. Romney (Eds.), *Research methods in social network analysis* (pp. 18-34). Fairfax, VA: George Mason University Press.

Lave, J., & Wenger, E. (1991). *Situated learning. Legitimate peripheral participation.* Cambridge, UK: Cambridge University Press.

Lesser, E.L. (2001). Communities of practice and organizational performance. *IBM Systems Journal, 40*(4), 831-841.

Lesser, E.L. Slusher, J., & Fontaine, M. (Eds.). (2000). *Knowledge and communities.* Boston: Butterworth-Heinemann.

Liedtka, J. (1999). Linking competitive advantage with communities of practice. *Journal of Management Inquiry, 8*(1), 5-16.

Liyanage, S., Greenfied, P.F., & Don, R. (1999). Towards a fourth generation R&D management model—research networks in knowledge management. *International Journal of Technology Management, 18*(3/4), 372-394.

Müeller-Prothmann, T., Siegberg, A., & Fink, I. (2005). Inter-organizational knowledge community building—Sustaining or overcoming organizational boundaries? *Journal of Universal Knowledge Management, 0*(1), 39-49.

Müeller-Prothmann, T., & Finke, I. (2004). SELaKT—social network analysis as a method for expert localisation and sustainable knowledge transfer. *Journal of Universal Computer Science, 10*(6), 691-701.

Nohria, N., & Eccles, R.G. (1992). Face-to-face: Making network organizations work. In N. Nohria & R.G. Eccles (Eds.), *Network and organizations* (pp. 288-308). Boston: Harvard Business School Press.

Poplin, D.E. (1979 (1972)). *Communities. A survey of theories and methods of research* (2nd ed.). New York: Macmillan.

Powell, W.W. (1990). Neither market nor hierarchy: Network forms of organization. In B.M. Staw & L.L. Cummings (Eds.), *Research in organizational behaviour* (vol. 12, pp. 295-336). Greenwich, CT: JAI Press.

Powell, W.W. (1998). Learning from collaboration: Knowledge and networks in the biotechnology and pharmaceutical industries. *California Management Review, 40*(3), 228-240.

Roethlisberger, F.J., & Dickson, W.J. (1947). *Management and the worker. An account of a research program conducted by the Western Electric Company, Hawthorne Works, Chicago*. Cambridge, MA: Harvard University Press.

Savage, C.M. (1990). *Fifth generation management. Integrating enterprises through human networking*. Boston: Butterworth-Heinemann (Digital Press).

Scarbrough, H., & Swan, J. (2001). Knowledge communities and innovation. In M. Huysman & P. Van Baalen (Eds.), *Trends in communication; Special issue on communities of practice* (pp. 7-20). Amsterdam: Boom.

Schmidt, M.P. (2000). *Knowledge Communities. Mit Virtuellen Wissensmärkten das Wissen im Unternehmen Effektiv Nutzen*. München: Addison-Wesley.

Seufert, A., von Krogh, G., & Bach, A. (1999). Towards knowledge networking. *Journal of Knowledge Management, 3*(3), 180-190.

Swan, J., Newell, S., Scarbrough, H., & Hislop, D. (1999). Knowledge management and innovation: Networks and networking. *Journal of Knowledge Management, 3*(4), 262-275.

Thorelli, H.B. (1986). Between markets and hierarchies. *Strategic Management Journal, 7*, 37-51.

Wellman, B. (1997). An electronic group is virtually a social network. In S. Kiesler (Ed.), *Culture of the Internet* (pp. 179-205). Hillsdale, NJ: Lawrence Erlbaum.

Wenger, E.C. (1999). *Communities of practice: Learning, meaning, and identity*. Cambridge, UK: Cambridge University Press.

Wenger, E.C., & Snyder, W.M. (2000). Communities of practice: The organizational frontier. *Harvard Business Review 78*(1), 139-145.

KEY TERMS

Cohesion: The strength of connection between the members of a community or network, and thus a community or network can be densely knit or only loosely coupled; cohesion is a result of intensity, frequency, and type of the members' contacts.

Communities of Practice: Relatively small groups of people who focus on a specific topic and are strongly bound together by trust, shared work practice, and a common goal.

Goal Orientation: The degree of orientation towards a defined target; low goal orientation means the common intention of a community or network is diffuse.

Identity: The degree of identification of the individual members with the community or network, and thus also an indicator for the identification of a community or network as such; strong or weak identity show the identification of the members as a group and their sense of community.

Knowledge Communities: Self-organised, altruistic relationships of trust between people for research, development, and innovation-oriented knowledge exchange within a wide domain of knowledge.

Knowledge Network: Relationships of a large number of loosely coupled participants with a diffuse common domain of knowledge and without clearly defined boundaries.

Organisation: The degree of formal organisation of social relationships and of a community or network as a whole; low organisation means informal self-organisation.

Shared Practice: The importance of sharing practice between the members of a community or a network; low shared practice means that the members do not (or rarely) work together.

Size: Indicated by counting the members of a community or a network; size is always used as a relative measure to compare different social entities and not as an absolute one.

ENDNOTE

[1] The "KM Expert Survey on Knowledge Management, Innovations, and Knowledge Communities" was conducted by the author from August 2003 until January 2004. Fifty-two knowledge management experts participated; by countries: 50% from Germany and 50% from other European countries; by business activity: 27% university/college/business school, 38% private companies (from which 50% consultancies), 25% research and service organisations, and 10% others.

Knowledge Exchange in Networks of Practice

Robin Teigland
Stockholm School of Economics, Sweden

Molly McLure Wasko
Florida State University, USA

INTRODUCTION

The concept of a community of practice is emerging as an essential building block of the knowledge economy. A community of practice consists of a relatively tightly knit group of members who know each other, work together face to face, and continually negotiate, communicate, and coordinate with each other directly. The demands of direct communication and coordination limit the size of the community, enhance the formation of strong interpersonal ties, and create strong norms of direct reciprocity between members (Brown & Duguid, 2000). These communities develop through the mutual engagement of individuals as they participate in shared work practices, supporting the exchange of ideas between people, which results in learning and innovation within the community (Brown & Duguid, 2000). However, typically not all of an organization's relevant knowledge resides within its formal boundaries or within its communities of practice. To remain competitive, organizations need to ensure that new knowledge found in the external environment is integrated with knowledge that is found within the firm (Cohen & Levinthal, 1990). Organizations must rely on linkages to outside organizations and individuals to acquire knowledge, especially in dynamic fields where innovation results from inter-organizational knowledge exchange and learning (Cohen & Levinthal, 1990).

BACKGROUND

Current research has focused on the role of communities of practice for encouraging knowledge exchange and innovation *within* organizations; however, we know much less about the role that mem-bers of communities of practice play in creating linkages to external knowledge sources. Previous research has found that organizational members may simultaneously be members of a community of practice as well as members of broader occupational communities (Van Maanen & Barley, 1984). These individuals perform the dual roles of generating local knowledge within an organizational community of practice while providing linkages to knowledge and innovations outside of the organization. These inter-organizational networks have been referred to as networks of practice. Networks of practice are social structures linking similar individuals across organizations who are engaged in a shared practice, but who do not necessarily know one another (Brown & Duguid, 2000). Although individuals connected through a network of practice may never meet one another face-to-face, they are capable of sharing a great deal of knowledge and may play a vital role in a firm's ability to acquire new knowledge.

While the participation of individuals in networks of practice is not a new phenomenon, the ability to access these networks has increased due to recent advances in information and communication technologies. Previous efforts to interact with others outside an organization's legal boundaries were often fruitless since they could be time-consuming or cumbersome, and individuals may not even have known whom to contact or how to find a relevant person. Furthermore, if management did not provide the resources to attend external conferences or other events, finding other like-minded individuals with whom to discuss work-related problems often proved difficult. However, communication technologies, such as cell phones, e-mail, IRC, chat rooms, bulletin boards, and so forth, have reduced the costs of informal inter-organizational communication. As a result, individuals may now easily access and discuss their work tasks with others outside their organization. These informal interactions are valued and sustained over time because the sharing of knowledge is an important aspect of being a member of a technological community or network of practice (Bouty, 2000).

Sharing knowledge across external organizational boundaries poses significant challenges to organizations attempting to manage their knowledge resources (Pickering & King, 1995). Through external sources, individuals gain access to knowledge not available locally and can interact informally, free from the constraints of hierarchy and local rules. However, accessing knowledge from external sources usually involves a high degree of knowledge trading and reciprocity. In order to receive help, individuals must be willing to give advice and know-how as well, some of which company management may consider proprietary (von Hippel, 1987). Of special interest to management is that individuals generally participate in networks of practice based on their own personal biases and preferences for others as opposed to what the formal organization dictates, and as a result, they may be exchanging knowledge with others who are working for direct competitors (Schrader, 1991). This makes the study of networks of practice of prime interest for researchers and practitioners.

PREVIOUS RESEARCH RELATED TO NETWORKS OF PRACTICE

Networks of practice are not a new phenomenon. They have existed for hundreds of years and have played an important role in the diffusion of knowledge through society. For example, the well-known term, invisible colleges, dates back to the 1640s when a group of 10 men who were well-educated within one field would meet informally in the taverns of London. These meetings later developed in 1660 into the Royal Society, the oldest scientific society in Great Britain (Price, 1963; Tuire & Erno, 2001). While there is considerable previous research on inter-organizational informal networks under a variety of names—such as scientific communities (Knorr-Cetina, 1981; Polanyi, 1962), co-citation networks (Usdiken & Pasadeos, 1995), invisible colleges

(Crane, 1972), epistemic communities (Haas, 1992; Holzner & Marx, 1979), thought-collectives (Fleck, 1935), paradigms (Kuhn, 1962), and occupational communities (Van Maanen & Barley, 1984)—a review of this literature reveals that research that explicitly focuses on knowledge sharing is quite limited. Below we present the relevant research and empirical studies that we found in our review. This research can be divided into two categories: (1) studies from the perspective of scientific communities and (2) studies from the perspective of high-technology firms.

Scientific Communities

Research on scientific communities suggests that knowledge sharing occurs between members as they engage in debate and discussion of each other's ideas and results, and through collaboration on joint research projects (Crane, 1972). Due to the universal nature of knowledge, shared language, and values within the scientific community, individuals can communicate relatively easily with one another (Tushman & Katz, 1980; Van Maanen & Barley, 1984). Thus, knowledge and innovations spread quickly across organizational, national, and cultural boundaries through these informal relationships. In many cases, these informal networks are more valuable for sharing knowledge than more formal channels, such as publications, since the results of failed experiments are rarely published, and learning about these can prevent their duplication.

In scientific communities, the central goals and values of the members are generally developed and spread throughout the network (Hagstrom, 1965). Strong norms that are well defined and socially imposed, such as reciprocity in knowledge sharing, respect for individuals' intellectual property rights, and honesty in research, facilitate knowledge exchange (Bouty, 2000; Liebeskind, Oliver, Zucker & Brewer, 1996). Trustworthy behavior and norms are enforced since the level of participation in the community is jointly determined by the community's members. Individuals who fail to follow the norms and implicit code of conduct can be excluded from participating in valuable exchanges with others (e.g., participation in research teams with leading researchers, access to the latest research findings, etc.). This exclusion can then negatively impact their career success (Tuire & Erno, 2001). As a result, the production and sharing of valuable knowledge is facilitated, allowing the frontier of knowledge to progress rapidly and at minimal cost

Structural studies of research-based communities of academic scientists have shown that these networks are generally characterized by a center and a periphery (Schott, 1988). The most important, visible, or active members are generally found in the center, and these individuals influence the direction of the development of the community's knowledge. The activities of the individuals in the core determine the community's dominating theoretical concepts, methods, and chosen research problems, which are then mediated through the community's links to individuals in the periphery (Schott, 1988). Through a process known as social contagion (Marsden, 1988), new members are socialized into the community and as such transform their personal identities, adapting their attitudes, behaviors, and values to those of the community (Holzner & Marx, 1979). Additionally, power is an integral part of scientific communities, with individuals often using knowledge strategies as components of power strategies (Holzner & Marx, 1979). Thus, the center of a scientific community is not only a realm of activity, but it also is a realm of identity and cultural values of the community (Schott, 1988; Tuire & Erno, 2001).

High-Technology Firms

Researchers have also taken the firm's perspective and focused on inter-organizational boundary spanning activity in high-technology firms. A major stream of this research began in the 1960s with an investigation into the communication patterns of scientists and engineers in R&D laboratories (see Flap, Bulder & Volker, 1998, for a review). One area within this research is why individuals communicate informally with others outside the organization. For example, von Hippel (1987) found that when specialist engineers could not find the required know-how in-house or in publications, they went outside their organization to their professional networks developed at conferences and other events. Further research has found that quite often professionals communicate with others in their professional networks in order to maintain contact with a reference group and to keep

abreast of technological changes (Aiken & Hage, 1968). Allen (1970) has also found that low-performing individuals choose to go outside for help. He argues that this choice is a way to avoid paying a psychological price of loss of face that occurs when an individual asks a colleague who is not a friend for advice.

A second area of investigation has focused on the informal flow of knowledge across a firm's boundaries in a limited number of settings, such as semiconductor, specialty steel and mini-mill industry, and R&D operations (Carter, 1989; Schrader, 1991; von Hippel, 1987). This research provides evidence that participation by individuals in inter-organizational networks leads to knowledge sharing across a firm's legal boundaries that is generally not governed by firm contracts or other market mechanisms (Liebeskind et al., 1996). In many instances, this knowledge sharing may even include the exchange of confidential organizational knowledge, even with others who might even be working in rival firms (Schrader, 1991; von Hippel, 1987). Thus, it is argued that knowledge "leaks" across the firm's legal boundaries (Mansfield, 1985; von Hippel, 1988). Bouty's research (2000) raises a very interesting point though—confidentiality is socially constructed, and as one of her interviewees noted, there are "open secrets." Research by Jarvenpaa and Staples (2001) further touches on this aspect of socially constructed confidentiality since they find that the more individuals view their knowledge as personal expertise, the more individuals regard such knowledge as their own property and not that of the organization.

However, this research suggests that individuals do not just give the knowledge away to others outside their firm. Rather they consciously exchange knowledge with other carefully chosen individuals with whom they often have a long-term relation built on mutual trust and understanding (Bouty, 2000; Schrader, 1991). Research conducted by Schrader (1991) finds that individuals often expect that their chances of receiving valuable knowledge in return are likely to increase after they provide knowledge. Thus, participation in inter-organizational emergent networks results in reciprocity and dyadic exchange of knowledge (von Hippel, 1987), with knowledge sharing viewed as an 'admission ticket' to the ongoing 'back room' discussions within professional networks (Appleyard, 1996).

As a result, participation in inter-organizational networks leads to *knowledge leaking in at the same time as it leaks out* of the firm (Brown & Duguid, 2000). Research on the relationship between this knowledge exchange and performance at any level, however, is scant. One of the primary reasons is that it is very difficult for firms to manage and evaluate the benefits since it occurs "off the books" (Carter, 1989). Secondly, data regarding the sharing of potentially firm proprietary knowledge are difficult to collect due to their sensitive nature. However, there is some initial evidence of a positive relationship between knowledge trading and firm performance (Allen, Hyman & Pinckney, 1983; Schrader, 1991), between knowledge trading and project performance (Allen, 1977), and between knowledge trading and individual performance (Teigland, 2003; Teigland & Wasko, 2003).

AREAS FOR FUTURE RESEARCH

Networks of practice are proposed to be a valuable *complement* to intra-organizational face-to-face communities of practice. The implication is that in order to be competitive, organizations should focus on sponsoring participation in *both* traditional communities of practice and networks of practice, as well as stimulating the *interaction* between the two. This leads to several interesting areas in need of further research. One area that deserves attention addresses the question of why individuals participate and access knowledge in networks of practice. While the research within high-technology firms provides some initial suggestions—for example, the inability to find the required knowledge in-house, the desire to maintain contact with a professional reference group or long-term relations with close ties, to keep abreast of technological changes, and even to avoid loss of face—the ability to access knowledge through weak tie connections basically requires depending on the kindness of strangers (Constant, Sproull & Kiesler, 1996).

Prior research has emphasized the importance of shared language, values, and goals, as well as long-term relations built on mutual trust for knowledge exchange. Thus, another area of research should

investigate the factors that lead to the creation of these facilitators within networks of practice, especially networks sustained through electronic communication. Future research should also investigate the relationship between network structure and knowledge sharing, how network structures change over time, and how network structure influences the cognitive aspects of shared language, values, and goals.

The studies reviewed above have also provided evidence that individuals in many instances participate in the exchange of confidential organizational knowledge, often making their own decisions to share knowledge without management's consensus or even awareness. As a result, knowledge "leaks" across an organization's boundaries, indicating additional areas for future research. For example, future research could investigate the factors leading to this leakage such as "open secrets," social construction of confidentiality, expectations of reciprocity, and so forth. Another factor to be investigated is that of commitment. Just as individuals have a certain degree of commitment to their organizations, they also have a degree of commitment to their profession or occupation (Van Maanen & Barley, 1984). In some professions, the degree of commitment to the profession can be so strong that the norms of the profession transcend the norms of the employing organizations. Finally, research on the relationship between knowledge leakage and performance at all levels is scant and is in need of significant research, especially due to management's concerns relating to the "leakage" of firm proprietary knowledge.

CONCLUSION

In conclusion, the purpose of this article was to direct our attention to networks of practice since current community of practice research has focused primarily on their role for encouraging knowledge exchange and innovation *within* organizations. While networks of practice are not a new phenomenon, a review of previous, related research reveals that the studies that explicitly focus on knowledge sharing are quite limited. As a result, we know much less about the role that members of communities of practice play in creating linkages to external knowledge sources and how participation in networks of practice influences performance at the firm, project, or individual level. Our review of the literature has also provided us with several areas for future research, and we hope that these suggestions, along with our review, will inspire researchers to further investigate networks of practice.

REFERENCES

Aiken, M., & Hage, J. (1968). Organizational interdependence and intra-organizational structure. *American Sociological Review, 33*(6), 912-930.

Allen, T.J. (1970). Communication networks in R&D laboratories. *R&D Management, 1*, 14-21.

Allen, T.J. (1977). *Managing the flow of technology*. Cambridge, MA: MIT Press.

Allen, T.J., Hyman, D.B., & Pinckney, D.L. (1983). Transferring technology to the small manufacturing firm: A study of technology transfer in three countries. *Research Policy, 12*, 199-211.

Appleyard, M.M. (1996). How does knowledge flow? Interfirm patterns in the semiconductor industry. *Strategic Management Journal, 17*(Special Issue), 137-154.

Bouty, I. (2000). Interpersonal and interaction influences on informal resource exchanges between R&D researchers across organizational boundaries. *Academy of Management Journal, 43*(1), 50-66.

Brown, J.S., & Duguid, P. (2000). *The social life of information*. Boston: Harvard Business School Press.

Carter, A.P. (1989). Know-how trading as economic exchange. *Research Policy, 18*, 155-163.

Cohen, W.M., & Levinthal, D.A. (1990). Absorptive capacity: A new perspective on learning and innovation. *Administrative Science Quarterly, 35*, 128-152.

Constant, D., Sproull, L., & Kiesler, S. (1996). The kindness of strangers: The usefulness of electronic weak ties for technical advice. *Organization Science, 7*(2), 119-135.

Crane, D. (1972). *Invisible colleges: Diffusion of knowledge in scientific communities*. Chicago: University of Chicago Press.

Flap, H., Bulder, B., & Volker, B. (1998). Intra-organizational networks and performance: A review. *Computational & Mathematical Organization Theory, 4*(2), 109-147.

Fleck, L. (1935). *Genesis and development of scientific fact*. Chicago: The University of Chicago Press.

Haas, P. (1992). Introduction: Epistemic communities and international policy coordination. *International Organization, 46*(1), 1-37.

Hagstrom, W.O. (1965). *The scientific community*. New York: Basic Books.

Holzner, B., & Marx, J.H. (1979). *Knowledge application: The knowledge system in society*. Boston: Allyn & Bacon.

Jarvenpaa, S.L., & Staples, D.S. (2001). Exploring perceptions of organizational ownership of information and expertise. *Journal of Management Information Systems, 18*(1), 151-183.

Knorr-Cetina, K.D. (1962). *The manufacture of knowledge: An essay on the constructivist and contextual nature of science*. Oxford: Pergamon Press.

Kuhn, T. (1962). *The structure of scientific revolution*. Chicago: University of Chicago Press.

Liebeskind, J.P., Oliver, A.L., Zucker, L., & Brewer, M. (1996). Social networks, learning and flexibility: Sourcing scientific knowledge in new biotechnology firms. *Organization Science, 7*(4), 429-443.

Mansfield, E. (1985). How rapidly does new industrial technology leak out? *The Journal of Industrial Economics, 34*(2), 217-224.

Marsden, P. (1988). Memetics and social contagion: Two sides of the same coin. *Journal of Memetics—Evolutionary Models of Information Transmission, 2*, 68-85.

Pickering, J.M., & King, J.L. (1995). Hardwiring weak ties: Interorganizational computer-mediated communication, occupational communities, and organizational change. *Organization Science, 6,* 479-486.

Polanyi, M. (1962). The republic of science. *Minerva, 2,* 54-73.

Price, D.J.D.S. (1963). *Little science, big science.* New York: Columbia University Press.

Schott, T. (1988). International influence in science: Beyond center and periphery. *Social Science Research, 17,* 219-238.

Schrader, S. (1991). Informal technology transfer between firms: Cooperation through information trading. *Research Policy, 20,* 153-170.

Teigland, R. (2003). *Knowledge networking: Structure and performance in networks of practice.* Published Doctoral Dissertation, Stockholm School of Economics, Sweden.

Teigland, R., & Wasko, M. (2003). Integrating knowledge through information trading: Examining the relationship between boundary spanning communication and individual performance. *Decision Sciences, 34*(2), 261-286.

Tuire, P., & Erno, L. (2001). Exploring invisible scientific communities: Studying networking relations within an educational research community. A Finnish case. *Higher Education, 42,* 493-513.

Tushman, M., & Katz, R. (1980). External communication and project performance: An investigation into the role of gatekeepers. *Management Science, 26*(11), 1071-1085.

Usdiken, B., & Pasadeos, Y. (1995). Organizational analysis in North American and Europe: A comparison of co-citation networks. *Organization Studies, 6*(3), 503-526.

Van Maanen, J., & Barley, S.R. (1984). Occupational communities: Culture and control in organizations. In B.M. Staw & L.L. Cummings (Eds.), *Research in organizational behavior.* Greenwich, CT: JAI Press.

von Hippel, E. (1987). Cooperation between rivals: Informal know-how trading. *Research Policy, 16,* 291-302.

von Hippel, E. (1988). *The sources of innovation.* New York: Oxford University Press.

K

KEY TERMS

Invisible Colleges: Groups of researchers within the same branch of science who have personal relationships with one another.

Knowledge Integration: The effectiveness of an organization to integrate the specialized knowledge of its members along three dimensions: efficiency, scope, and flexibility.

Knowledge Leakage: The flow of company proprietary knowledge across firm boundaries.

Network of Practice: An emergent social network linking similar individuals across organizations who are engaged in a shared practice but who do not necessarily know one another.

Social Contagion: The process through which new community members are socialized and as such transform their personal identities, adapting their attitudes, behaviors, and values to those of the community.

Knowledge Extraction and Sharing in External Communities of Practice

Ajumobi Udechukwu
University of Calgary, Canada

Ken Barker
University of Calgary, Canada

Reda Alhajj
University of Calgary, Canada

INTRODUCTION AND BACKGROUND

Communities of practice (CoPs) may be described as groups whose members regularly engage in sharing and learning, based on common interests (Lesser & Storck, 2001). Traditional communities of practice exist within organizations and are centered on work functions. These CoPs may be self-organizing or corporately sponsored. They exist to encourage learning and interaction, create new knowledge, and identify and share best practices for the organization's processes (Wenger, 1998). The members of a community of practice may be collocated (within an office) or spatially dispersed (e.g., a group may interact via electronic chat). There may also be communities of practice that are not centered on work functions. For example, several online groups exist for enthusiasts of new technology, politics, environment, and so forth. These groups qualify as bona fide CoPs. We classify the CoPs discussed so far as *active* communities of practice because the members actively seek to learn and share from each other. In this work, however, we examine *passive* communities of practice in which the members do not actively interact with each other. This class of CoPs shares the core characteristics of traditional communities of practice—the members can learn from each other, and the organization can gain useful knowledge capital and best practices. Our discussions will be based on user communities using cable-TV viewers as a case in point. In contrast to work-centered CoPs whose members share knowledge and learn how to perform their work tasks better, members of user-centered CoPs learn how to maximize the utility from the product/service of interest. In both cases, a learning organization can extract useful knowledge capital and best practices to improve its processes and products/services. In this work, we use the case of cable-TV viewers to show how useful knowledge can be learned and shared in passive user-centered communities of practice. Our techniques will be based on data mining and knowledge discovery, which are introduced in the subsection that follows.

DATA MINING AND NAVIGATIONAL PATTERN DISCOVERY

The widespread use of computers and the increased abilities to collect and store massive amounts of data have led to phenomenal growth in the popularity and use of data mining techniques. Data mining is the analysis of data with the goal of uncovering hidden patterns. Historically, technological advances that improve the collection of data have led to new domains for data mining. For example, advances in bar code technology and the ability to collect and store transaction data logs led to the growth of association rule mining (Agrawal, Inielinski & Swami, 1993) and its many variants (Fayyad, 1998). More recently, the widespread use of the World Wide Web and the ability to collect and store Web logs of user sessions have driven research interest in Web usage mining (Cooley, Mobasher & Srivatava, 1997; Srivastava et al., 2000). An interesting problem in Web usage mining that has attracted the attention of several researchers is the discovery of traversal patterns of Web users (Chen, Park & Yu, 1998; Nanopoulos & Manolopoulos, 2000). Mining path traversal patterns involves identifying how users access information of interest to them and travel from one object to another using the navigational facilities provided. Tracking user-browsing habits provides useful information for service providers and businesses, and ultimately should help to improve the effectiveness of the service provided. Previous works on identifying path-traversal patterns have been directed at traversals between relatively static objects (e.g., Web pages). By static, we mean information that can be regenerated by the user as required. Thus, dynamic Web pages fall under our definition of static objects because the user may regenerate the dynamic Web pages on each visit.

In this research, our focus is on navigational patterns in environments where the objects are continuously changing in time (i.e., streaming content). An example of such a system is cable-TV where the program sequence is continuously changing. The viewers of cable-TV are able to navigate from one object (channel/station) to another. However, if viewers navigate away from a station/channel and later return to that channel, the content displayed may have changed. Thus, there is a strong temporal component in the systems studied in this research. The temporal component in our framework motivates new techniques to capture navigational patterns, as existing techniques in the literature do not take temporal semantic information into consideration. Our framework can be applied to identifying navigational patterns in any environment with streaming content. However, the discussions in this article will be motivated by cable-TV viewing patterns. The choice of cable-TV viewing patterns is due to recent technological innovations that enable the collection of anonymous logs of viewing data through digital video recorders attached to cable-TV receivers. The logs are kept anonymous to protect the privacy of the viewers. This is similar to the ethical standards that have long been adopted in analyzing Web and transaction logs. In the past, there has been very limited ways to collect data on the viewing patterns of cable subscribers. The advent of digital video recorders and the ability to track and report logs on the channels viewed by subscribers (on a second-by-second basis) opens up several interesting areas for data mining. Digital video re-

corders are growing in popularity (with Tivo Inc. reporting over 700,000 subscribers in the US in 2003 and a projection of over a million subscribers by the year-end), and a massive deployment is expected in the near future (Tivo Inc., 2003). Digital video recorders keep track of the channels viewed through the cable receiver. The view logs are uploaded to the service provider daily. The challenge is to extract interesting patterns from all the view logs submitted to the service provider.

There are several interesting questions that can be addressed by analyzing the view logs. For example, an advertiser may be interested in knowing if more viewers stay tuned during the commercial breaks of prime-time programs than for regular programming. It may also be of interest to know the advertising slot that is most effective; that is, is it more likely for an advert to be viewed if it is the first ad during the commercial break or if it has the last slot, middle slot, and so on. It may also be interesting to discover the percentage of viewers that return to a program once they tune off during a commercial break. Several other interesting patterns may also be discovered. In our framework, we propose a novel technique that categorizes the dynamic content of sites into distinct event sequences and then explores the navigational patterns of users relative to the distinct event sequences. The behavioral/navigational patterns discovered may be used to improve the program sequencing for future broadcasts. The analysis may also be given a spatiotemporal dimension so that appropriate programming is directed at users based on their locations and times of broadcast. Viewers may be grouped or profiled based on common navigational behavior. In interactive TV environments, this would enable personalized programming and program recommendations tailored to particular viewer groups or individual viewers.

RELATED WORK

Several authors have studied communities of practice (CoPs) in organizations (Brown & Duguid, 1991; Hildreth, Kimble & Wright, 2000; Lesser & Prusak, 1999; Lesser & Storck, 2001; Wenger, 1998; Wenger & Snyder, 2000). These works are centered on work-related CoPs and differ from user-centered CoPs discussed in this article.

There are strong similarities between the behavioral pattern discovery techniques discussed in this work and Web usage mining. Web usage mining is the application of data mining techniques to discover usage patterns from Web data (Srivastava et al., 2000). The objects in our framework (e.g., channels) may be viewed as Web pages. Also, a viewer can jump to any object/channel just like a Web user may navigate to any URL. However, mining, viewing patterns in our framework, has a stricter temporal component. It is not sufficient to know the order in which the objects are viewed. There is a need to know the information content of the objects at the periods the viewer navigates to, or away from, the object. The work by Yang, Wang, and Zhang (2002) proposes an event prediction algorithm for Web usage mining. Their approach is aimed at predicting when Web accesses would occur. This is an extension of earlier works that only identify the order in which Web accesses would occur. The problem they address is different from the problem addressed in this article since we are interested in the information content of the objects at the times they are visited. Furthermore, the objects we study have streaming information content. Several other researchers have proposed techniques for identifying frequent path traversal patterns (Chen et al., 1998; Borges & Levene, 2000; Heer & Chi, 2002; Nanopoulos & Manolopoulos, 2000; Pei et al., 2000). However, these approaches do not incorporate the temporal semantics we introduce in our framework. Tivo Inc. (2003) has developed audience measurement tools that are able to report viewing statistics. However, their tools (just like tools for measuring Web hits) do not explore navigational patterns of users.

FRAMEWORK

The general framework of the class of information systems covered in our work consists of independent sites with links connecting all sites. Unlike Web pages that are grouped together into Web sites with internal navigational ordering, our framework is made up of stand-alone sites that are interlinked to each other. Using our example of a cable-TV system, each channel/station represents a site in our framework. A viewer is able to navigate from one channel to another either by following the ordering of the cable channels or by specifying the desired channel.

We define a *user session* as the complete set of activities by a user from the time the system is entered until departure. In our example, a *user session* starts when the user turns on the cable-TV and ends when the system is switched off. The system consists of sites with streaming content that can be divided up into categorical episodes. An episode is an event sequence that makes sense in its domain of application. In our example, we may identify three broad categories for the episodes: programming, commercials, and shutdown. The programming category can be further divided into specific types of programs (e.g., movies, sitcom, sports, news, etc.), and the commercials can be further divided into slots (i.e., first commercial slot, second, etc.). The categories may be abstracted further so that individual programs and commercials are identified. The choice of abstraction is determined by the data-mining analyst.

The information displayed by each site in the system can be broken into event sequences that fall into one of the episode categories defined. Thus, for each site, its streaming content (for 24 hours a day) can be categorized into definite episodes with the associated start and end times for each episode. Further, for each user of the system, the viewing patterns must be categorized for every user session during a given day.

IDENTIFYING NAVIGATIONAL PATTERNS

The first step in our framework is data preprocessing. The content/program information for each of the sites has to be preprocessed into a format suitable for mining. Similarly, the user logs have to be preprocessed. Each user session is counted independently; that is, one subscriber may have multiple user sessions in a day, and each of the sessions would be independently considered in the framework. For example, given that time is represented in the 24-hour format, *hh:mm:ss*, and that the numbers 4 to 62 represent channels/stations available to the user, a typical user log would specify the channels/stations the user viewed from

Table 1. An example of a user log

Time of Day	Channel Viewed
00:00:00	-
00:00:01	-
12:15:00	10
13:05:15	10
13:05:16	32
13:30:00	32
13:30:01	-
18:05:05	31
18:30:00	31
18:30:01	-
23:59:59	-

the start of a session to its end. Table 1 gives an example of a typical user log for one subscriber for a given day.

The logs record the viewing activity for each second of the day. The broken lines in Table 1 represent periods when there is no change in the channel viewed. From Table 1, we can identify two user sessions: the first starting at 12:15:00 and ending at 13:30:00, while the second session starts at 18:05:05 and ends at 18:30:00. Preprocessing the user log involves identifying all the user sessions and breaking each session into time brackets for the channels/stations viewed. The result of preprocessing the user log in Table 1 is shown below:

User session 1:
    Channel 10: 12:15:00 - 13:05:15
    Channel 32: 13:05:16 - 13:30:00
User session 2:
    Channel 31: 18:05:05 - 18:30:00

The program content for each site (channel/station) in the system is also preprocessed. The analyst specifies categories for each program. For example, given that a station airs its programs between 08:00:00 and 18:30:00 and, also, given the complete program schedule of the station. If the categories specified are as follows: N—news; S—sitcom; C—commercials; and M—movies, a pro-

gram episode can be represented by its category and an identifier. For example, the first news episode can be represented as N1, the second N2, and so on. Similarly, the first sitcom may be represented as S1 and the second S2, and so forth. The identifiers are necessary if it is of interest to keep track of complete program episodes since a program episode may be interleaved with another episode (e.g., several commercial episodes may interleave a program episode). It may also be of interest to separate the program content into slots. (For example, the first commercial in a commercial break takes slot one—C1, the next commercial takes slot two—C2, and so forth. Commercials that are not embedded within other programs may also be separated into a category, e.g., CO in Table 2.) Further, the analyst may choose to capture different segments of a program into separate categories. For example, it may be interesting to differentiate how users respond to the first segment of a program from how they respond to other segments, especially if they did not view the earlier segments. The salient point here is that categories may be defined for every program grouping of interest. Finally, the preprocessed program content for our example will be in a format similar to the one shown in Table 2. If the channel (or site) preprocessed in Table 2 is Channel 10 (from Table 1), then it is easy to extract the categories viewed during user session 1 (from Table 1).

Once the usage logs have been transformed into user sessions and the program schedules have been

Table 2. A partial listing of a sample program categorization for a channel/site

Program	Time Slot
N1	12:00:00 – 12:15:00
C1	12:15:01 – 12:17:00
C2	12:17:01 – 12:18:00
C3	12:18:01 – 12:20:00
N1	12:20:01 – 12:35:00
S1	12:35:01 – 12:55:00
C1	12:55:01 – 12:57:00
C2	12:57:01 – 12:58:00
C3	12:58:01 – 13:00:00
S1	13:00:01 – 13:30:00
CO	13:30:01 – 13:33:00

transformed into event categories, data mining procedures may then be performed on the processed data. The mining problem addressed in this work is formulated as follows:

• How do the users of the system navigate between sites in response to the contents displayed by the sites?

Details of the proposed techniques for discovering these behavioral and navigational patterns are discussed in the next section.

DISCOVERING EVENT-RELATED NAVIGATIONAL PATTERNS

This section examines the problem of identifying frequent navigational patterns of users relative to specific event categories (or collections of categories). For example, it may be of interest to know if viewers navigate away from a channel/station immediately after a news event begins, if they stay briefly before changing sites, or if they stay through the news event. It may also be of interest to discover if viewers navigate away at the commencement of a commercial break and whether they return to the same program once they navigate away. Figure 1 shows an example user navigation between two sites (channels). From Figure 1, it is easy to see that the user navigates away from program content that belongs to Category B, irrespective of the site/channel viewed. The data mining problem here is thus to determine all behavioral navigation patterns relative to program content that are frequently exhibited by various users of the system. Given that the

Figure 1. An example of user navigation relative to event categories

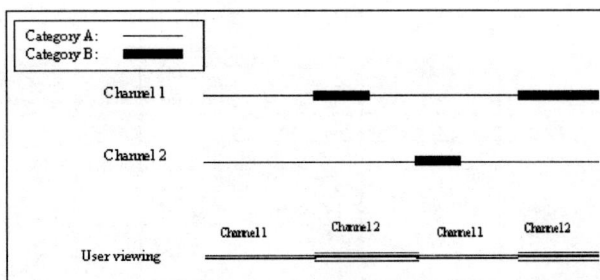

user logs and program schedules (content) have been preprocessed into user sessions and categories, respectively, the next step in the framework is to define the behavioral predicates that would capture the users' navigational patterns in response to dynamic content categories.

The behavioral predicates chosen may include the following:

• Navigate away
• Stay through
• Stay briefly
• Return to the same program content (i.e., after navigating away)

The set of behavioral predicates considered in the framework depends on the interests of the analyst. Further, the quantitative time units attached to some of the predicates (e.g., the definition of *briefly*) are set by the analyst. Given a threshold confidence (e.g., 0.75), it is then possible to discover rules of the form: *users of the system navigate away from program content in category B with x confidence (where x is a user defined threshold)*. It is also possible to capture navigational patterns of users in response to new program content in relation to the previous content they were viewing. An example of such a rule is: *users of the system navigate away from content in category B given that they were previously viewing content in category C with x confidence (where x is a user defined threshold)*. The details of the data mining process are given in the paragraphs that follow.

Recall that all the user logs are preprocessed into independent user sessions. Each user session details the channels viewed and the viewing times. By examining each user session against the program categories airing at the sites/channels viewed, it is possible to extract the program categories viewed during each user session and the behavioral navigational patterns of the user during the session examined. Given that X is the set of categories over which a rule R is defined, then the set of active user sessions with respect to rule R, A, is made up of user sessions with events in each of the categories in X. For example, given the rule: *users of the system navigate away from content in category B given that they were previously viewing content in category C with x confidence (where x is a user defined threshold)*, only user sessions with content in both categories B and C are active with respect to this rule (i.e., $X = \{B, C\}$). The confidence of a rule is calculated as the ratio of the support count of user sessions that satisfy the rule to the number of active user sessions with respect to that rule. The contribution of a user session to the support count of a rule is weighted and may range from 0 to 1. For example, if a user session encounters three instances of program content in category B and if in two of the three instances the user navigated away from the program content, then the contribution of this user session to the rule: *users of the system navigate away from program content in category B with x confidence (where x is a user defined threshold)* will be 0.67 (i.e., 2/3). The support count for the rule is then obtained by summing the support contributions of each user session for that rule. The outline of the algorithm is presented below.

INPUT:

- A set *U* of user sessions (obtained from preprocessing all the user logs)
- A set *P* of categorized program schedules for all the sites in the system
- A set *B* of behavioral predicates of interest
- An empty set *R* of all rules defined on the categories in *P*

OUTPUT:

Set of rules with associated confidence levels

PROCESSING:

FOR all $u \in U$ DO
Associate *u* with content categories by comparing its contents with relevant elements of *P*
Identify the rule set present in *u* with respect to *B*. If any rule found in *u* is not in *R*, add the rule to *R*.
Increment the count of active user sessions or all rules on program categories found in *u*.

FOR all rules $r \in R$ on categories found in *u* DO

Calculate the support contribution of *u* to *r*
Add the support contribution to the total support for rule *r*
END FOR
END FOR

FOR all $r \in R$ DO
Confidence of *r* = total support of *r* / number of active user sessions for *r*
END FOR

Given that *U* is the number of user sessions identified in the logs and $\bar{r}$ is the average number of rules defined on program categories in the user sessions, then the algorithm has a time complexity of $O(U\bar{r})$. However, $\bar{r} << U$, thus the algorithm runs in $O(U)$ time. (Note that rules are associated to user sessions if the behavioral predicates in the rule are present in the user session. This is analogous to constrained rule discovery using meta-rules.) The knowledge capital gained from the discovered patterns can be used by cable-TV companies and advertisers to improve their processes and produce better programming and scheduling. In interactive TV environments, the patterns can also be used to personalize programming and advertisements for individual viewers or viewer communities. The patterns may also be used to recommend programs of stations to viewers; thus individual viewers benefit from the collective knowledge of the viewing community.

CONCLUSION AND FUTURE TRENDS

This article abstracts external user communities as *passive*, user-centered communities of practice. Using a case of cable-TV viewers, we show that valuable knowledge capital can be learned from user-centered CoPs. The learned knowledge can be used to improve an organizations products and services and can also be filtered back to the user community to improve the overall user experience.

To achieve the learning of behavioral patterns from users, this article motivates a new domain for data mining that involves discovering user navigational patterns in information systems that disseminate dynamically changing (or streaming) content. The approach proposed in this work can be extended in several ways. For example, it may be of interest to separate ad-hoc and non-ad-hoc user sessions (i.e., some viewers may target certain programs while others may not). It may also be of interest to study the navigational behavior of users relative to the time of day the viewing occurred or navigational patterns relative to outlying content (e.g., a movie aired in a music channel). Several extensions to the framework are also possible.

REFERENCES

Agrawal, R., Imielinski, T., & Swami, A. (1993). Mining association rules between sets of items in large database. *Proceedings of the ACM SIGMOD 1993 International Conference on Management of Data* (pp. 207-216). Washington, DC.

Borges, J., & Levene, M., (2000). A heuristic to capture longer user Web navigation patterns. *Proceedings of the 1ˢᵗ International Conference on Electronic Commerce and Web Technologies (EC-Web, LNCS 1875* (pp. 155-164.)

Brown, J. S., & Duguid, P. (1991). Organizational learning and communities-of-practice: Toward a unified view of working, learning and innovation. *Organization Science, 2*(1), 40-57.

Chen, M.-S., Park, J. S., & Yu, P. S. (1998). Efficient data mining for path traversal patterns. *IEEE Transactions on Knowledge and Data Engineering, 10*(2), 209-221.

Cooley, R., Mobasher, B., & Srivastava, J. (1997). *Web mining: Information and pattern discovery on the World Wide Web. IEEE International Conference on Tools with Artificial Intelligence* (pp. 558-567), Newport Beach, CA

Fayyad, U. (1998). Mining databases: Towards algorithms for knowledge discovery. *IEEE Data Engineering Bulletin, 21*(1), 39-48.

Heer, J., & Chi, E. H. (2002). Mining the structure of user activity using cluster stability. *Proceedings of the Workshop on Web Analytics, SIAM Conference on Data Mining.*

Hildreth, P., Kimble, C., & Wright, P. (2000). Communities of practice in the distributed international environment. *Journal of Knowledge Management, 4*(1), 27-38.

Lesser, E. L., & Prusak, L. (1999). Communities of practice, social capital and organizational knowledge. *Information Systems Review, 1*(1), 3-9.

Lesser, E. L., & Storck, J. (2001). Communities of practice and organizational performance. *IBM Systems Journal, 40*(4), 831-841.

Nanopoulos, A., & Manolopoulos, Y. (2000). Finding generalized path patterns for Web log data mining. In J. Stuller et al. (Eds.), *ADBIS-DASFAA 2000, LNCS 1884* (pp. 215-228.) Berlin/Heidelberg: Springer-Verlag.

Pei, J., Han, J., Mortazavi-asl, B., & Zhu, H. (2000). Mining access patterns efficiently from Web logs. *Proceedings of the Pacific-Asia Conference on Knowledge Discovery and Data Mining (PAKDD'00), LNCS 1805* (pp. 396-407).

Srivastava, J., Cooley, R., Deshpande, M., & Tan, P.-N., (2000, January). Web usage mining: Discovery and applications of usage patterns from Web data. SIGKDD Explorations, 1(2), 12-23.

Tivo Inc. (2003). Retrieved July 1, 2003, from *http://www.tivo.com*

Wenger, E. (1999, December). *Communities of practice: Learning, meaning and identity* (New ed.). In R. Pea, J. S. Brown, C. Heath (Eds.). Cambridge: Cambridge University Press.

Wenger, E., & Snyder, B. (2000). Communities of practice: The organizational frontier. *Harvard Business Review, 78*(1), 139-145.

Yang, Q., Wang, H., & Zhang, W. (2002, December). Web-log mining for quantitative temporal-event prediction. *IEEE Computational Intelligence Bulletin, 1*(1), 10-18.

KEY TERMS

Data Mining: In today's increasingly complex database systems, data mining is the process of retrieving data from the system where there are no programs or utilities provided for the purpose.

Knowledge Capital: The sum or "worth" of the knowledge held within an organization. Often held to be of primary importance but difficult to quantify.

K

Knowledge Management in Civil Infrastructure Systems

Hyung Seok Jeong
Oklahoma State University, USA

Dolphy M. Abraham
Loyola Marymount University, USA

Dulcy M. Abraham
Purdue University, USA

INTRODUCTION

This article reviews current research and practice of knowledge management (KM) in the management of Civil infrastructure systems. Civil infrastructure systems, such as energy systems (electric power, oil, gas), telecommunications, and water supply, are critical to our modern society. The economic prosperity and social well being of a country is jeopardized when these systems are damaged, disrupted, or unable to function at adequate capacity. The management of these infrastructure systems has to take into account critical management issues such as (Lemer, Chong & Tumay, 1995):

- the need to deal with multiple, often conflicting objectives;
- the need to accommodate the interests of diverse stakeholders;
- the reliance of decision making on uncertain economic and social issues;
- the constraints in data availability; and
- the limitations posed by institutional structure.

BACKGROUND

KM approaches can play a central role in facilitating the effective management of these infrastructures. While well-designed information systems can get the right information to the decision maker at the right time, the age of the components of the infrastructure and a lack of available and usable records leads to utility managers frequent inability to take proactive measures to prevent system failures. Further, these infrastructures are interdependent, and managers at the various utilities and agencies need to work together to mitigate the risk of such threats and vulnerabilities. Analyzing each individual infrastructure system and the knowledge derived from managing each individual infrastructure becomes insufficient when managers have to make decisions at the intersection of multiple disciplines in a multihazard context. Sharing of information and ideas become critical to help detect and mitigate hazards and plan the recovery and response strategy.

Traditionally, utilities (especially the water utility) have been rich in "raw data but poor in the aggregated information derived from these data" (Rosen et al., 2003). Transforming the data into knowledge necessitates an understanding of the quality of the data and the aggregation measures used. KM approaches provide the basis for the development of relationships between different data structures and decision makers and by developing a higher level understanding of how information and process knowledge relate to one another.

Perez (2003) identified four common trends in the utility industry: the diminishing workforce, growing competition within the public sector, deterioration of employee loyalty, and increasing public involvement in government. Due to the concern about the potential negative impact of these trends on the ability to retain and share the institutional knowledge they currently possess, utilities have sought to find a method to efficiently maintain and improve the knowledge level of utility management.

Rosen et al. (2003) also point out that utilities lack a mechanism to aggregate, analyze, and restructure information in order to create knowledge. In general, many potential data users within a utility are not aware of a significant amount of the available data. Besides, in most cases, data are stored at multiple areas for the needs of the users. An organized directory of the entire data rarely exists. This creates redundancy of data and inefficiency of data retrieval.

Utilities have been recognizing the benefits of adopting KM strategies in their organizations. Foremost among these include the reduction in lost knowledge from downsizing and restructuring. Improving efficiencies of operations and workflow and improving customer satisfaction are also cited as reasons for moving toward a KM environment. Privatization of public and municipal utilities and increased regulation requires utilities to maintain a better handle of these physical and intellectual assets and liabilities.

However, there are several barriers impacting the access and use of information within a utility. These include a lack of awareness of what information (both internal and external) is available; difficulty in obtaining data access; lack of appropriate software for accessing, analyzing, and interpreting data; and the lack of complete historical data about the utility infrastructure and GIS base maps. In addition, the traditional "paper centric" nature of many utilities and lack of a
central repository of information make it harder to access information that is available within a utility. Further, "a large array of critical information for the utility is maintained in the heads of a few critical people" (Rosen et al., 2003).

These problems may be compounded in the future with new security requirements that are likely to restrict the flow of information. While the absence of complete historical data is a problem that is not easily fixed, information stored on paper can have implications that are both positive and negative. It is likely to be more secure than data stored electronically while the cost of use and maintenance is likely to be higher. It is necessary to find a solution that makes information available to utility managers so that they can do their job more effectively while also controlling access to the information more effectively.

EXAMPLES OF KM APPLICATIONS IN UTILITIES

The American Water Works Association (AWWA) Research Foundation completed a study in 2003 that investigated the feasibility of application of KM in the utility industry, specifically in drinking water utilities (Rosen et al., 2003). In this study, a review of 20 case studies related to utility management literature indicated that the primary areas of interest in upgrading information and knowledge management are:

- Supervisory Control and Data Acquisition (SCADA);
- Laboratory Information Management Systems;
- Geographic Information Systems (GIS);
- Maintenance Management Systems;
- Customer Relationship Management (CRM) and Tracking Systems; and
- Enterprise-wide Information Systems to tie other systems together.

The first five areas are operational systems focused on automation of specific department's current workflow and needs. The last area emphasizes synergy in the use of data. Currently, applications of KM in utilities are very limited, but the following case studies briefly demonstrate how KM approaches can be successfully applied to utilities.

- Columbus Water Works serves drinking water and treats wastewater for almost 200,000 residents in the Columbus, Georgia region. A data warehouse was implemented to act as a central source of all information from all of their applications. The star schema database adopted in this case consists of six primary "stars", including customers, employees, inventory, measurements, work orders, and accounting. A star schema allows for each sector data to be independent but also allows for integration of data through the central data warehouse. The utility reported that the main benefit is cost control, and it allows them to collect and distribute information between applications, better manage expenses, monitor data trends over time, and explore any unexpected variations.

- The OTAY Water District provides water and/or sewer facilities to about 100,000 people in San Diego County. They are using a new information management system, Myriad viewing software, allowing employees direct access to maps on their laptop, eliminating the need for the GIS department to reply to requests for drawings from employees in different departments.
- The Canadian electricity company, Ontario Hydro, is responsible for providing access to the 16,000 Material Safety Data Sheets (MSDSs) on the hazardous materials used at the utility to all employees. These sheets originated at the 2,100 companies that produced the hazardous chemicals and were then sent to Ontario Hydro. Since there has been a concern about the employees getting the most accurate information about the data due to losing MSDSs or spilling oil on them, Spicer's Imagination imaging soft water was implemented and linked to their existing database allowing for images to be scanned or faxed, which facilitates viewing the MSDSs for the employees.
- Michigan Consolidated Gas Company distributes natural gas to approximately 1.2 million residential, commercial, and industrial customers. IBM developed and implemented a system, Interactive Voice Response (IVR), which integrated the utility's voice and data systems. The IVR provides information to customers, including balance and payment histories, budget, billing, enrollment, payment agreements, service requests, and repair status. This automated service increased productivity in customer relations allowing the personnel more time to handle other issues.
- In 2003, the North Miami Beach Public Services Department was in the process of implementing a knowledge base for its water utility (Perez, 2003). In its early development stage, knowledge management focused on explicit knowledge sharing between internal subunits. The water utility operation was divided into three different segments including water production, water quality, and water distribution. All explicit knowledge in each segment, such as monthly operating reports, facility permits, record drawings, consultant studies and reports, and equipment data management sheet, was created in electronic format so that captured knowledge did not disappear. By placing the database for each segment within a secure network, critical information can be shared by the staff, improving efficiency of work by significantly reducing the amount of time spent on searching and retrieving a document.

The aforementioned cases illustrate that a variety of information technologies and systems are employed to support KM initiatives. What is important to note is that KM is not viewed as a technology

Figure 1. Infrastructure interdependency (Heller, 2002)

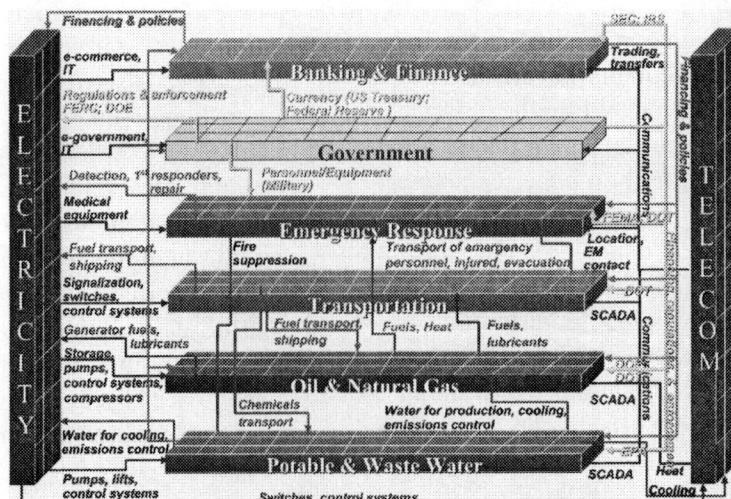

solution; rather, it is planned as a set of management processes that utilize appropriate information technologies in an effective manner.

IMPACT OF THE INTERDEPENDENCIES BETWEEN UTILITIES

The August 2003 electrical blackout in the Northeast USA emphasized that the functioning of any utility is dependent on the effective functioning of other utilities. For instance, after the blackout, residents of New York's many high-rise buildings lost water, since electric pumps are required to get water to the upper floors. The blackout also compromised the water supply in Detroit, Michigan and Cleveland, Ohio in at least two ways: first, by decreasing the pressure in water pumps, allowing bacteria to build up in municipal water systems and, second, by effectively shutting down sewage treatment facilities (Johnson & Lefebvre, 2003). Some communication utilities, including telephone service and Internet communications could not operate properly as they had an inadequate level of electricity back-up systems.

KM ACROSS CIVIL INFRASTRUCTURES

Figure 1 illustrates how the utilities are interwoven with each other for their proper operation. Even with the growing interdependency between these utilities, there are few communication routes between utilities. More importantly, information sharing or knowledge management paradigm across utilities is rare. Developing KM initiatives to support utility managers without considering the interdependencies may lead to decisions that emphasize local efficiency instead of global efficiency of the utility. Knowledge management processes across utilities will help the managers to respond rapidly to reduce the adverse consequences due to malfunction of dependent utilities, or even to plan ahead before incidents in order to eliminate the cascading effects. Moreover, once knowledge is shared between utilities, a solution for the global optimum for profit maximization and/or

service reliability can be sought by strategically managing their resources instead of developing locally optimal solutions solely for each utility's profit.

The issue of interdependence can be demonstrated using the case of Colorado Springs Utilities, an innovative water utility in the western US. This case study demonstrates how information sharing between utilities can simultaneously improve efficiency of operation and reduce costs of operation (Heller, 2001; Jentgen, 2004). The energy and water quality management system (EWQMS) of Colorado Springs Utilities is conceptually an extension of electric utilities' energy management systems (EMSs), which include power generation control and real-time power systems analysis.

Some aspects of the EWQMS are substantially more complicated than EMS. For example, in an EWQMS where hydropower is an option, decisions about pumped storage are coupled with the selection of electricity sources to exploit time-of-day electricity pricing. Alternatively, if spot market prices are exorbitant, hydropower might best be used to generate electricity for sale. Whereas EMS's power generation control has a short-term load-forecasting component, the EWQMS has two sets of demands to predict and satisfy: one for electricity and one for water. In addition, scheduling decisions also consider the following aspects:

- what quantity of raw water from which source is subject to water rights, quantity, and quality constraints, given variable pumping costs;
- what quantity of water is to be treated at which plant, given variable treatment costs; and
- which pumps are to be used for distribution, collection, and wastewater treatment and which ones should be taken off-line for maintenance.

The Colorado Springs Utility estimated that the cost reduction using the aforementioned approach would be worth more than $500,000 per year, not including windfalls from higher electricity prices. Heller, Von Sacken, and Gerstberger (1999) also discuss the concept of shared resources as a means of achieving regional eco-efficiency. In this context, "information system boundaries are extended to coordinate the shared production, consumption, treatment, or reuse of electricity, water, and wastewater resources among regional utilities and manufacturing facilities" (Heller, 2001).

BARRIERS TO THE PROPER APPLICATION OF KM IN UTILITIES

Many utility organizations are moving toward a KM approach to take advantage of the usefulness of information. However, implementing KM processes requires a significant change in a utilities organizational culture to be successful. For instance, the perception that data belong to one part of the organization to the exclusion of others must be changed. This cultural change is often more difficult than the investment and challenge of gathering and organizing all of the required information.

There is a general consensus that no single system will result in an effective KM initiative. Introducing KM processes will require careful thought and reorganization of workflow. Because management of a utility consists of various different business practices from laboratory operations to customer relations, careful studies regarding information structures and potential data transfer protocols and available information tools must be implemented before a KM initiative is launched.

A very low level of recognition regarding the need for information sharing and knowledge management framework across utilities is one of the significant barriers. Although the need to analyze the interdependencies between utilities were emphasized through the President's Commission Study (PDD 63, 1998), the mechanisms of these interdependencies is still not clearly known and is currently under investigation by researchers, including those at the Sandia National Laboratories, Argonne National Laboratory, and several universities.

The September 11, 2001 terrorist attacks and the blackout in the northeast US in August 2003 have propelled interest in further investigating these interdependencies. A report after the August 2003 blackout indicates that almost half of the companies that participated in a survey regarding the impact of the August 14th power failure were willing to invest their money to reduce the risk of their business to utility service outage in order to continue their business (Stoup, Slavik & Schnoke, 2004).

While it is possible to invest in information technology and systems to help reduce the risk of service outage, it is not possible to engineer a system that takes care of all possible eventualities. Because of interdependencies between utilities and causes for service outage are likely to be specific to a single utility, as well as pairs or groups of utilities, it is necessary for managers at these utilities to analyze interdependencies and to identify vulnerabilities. One approach to facilitate this analysis is the use of focus groups or brainstorming sessions with utility managers. Consecutive meetings of this group may help identify what data must be shared between utilities and what KM system can be structured in order to reduce risk to service outage due to outages in interdependent utilities.

Another concern is security of information. Utilities tend to hesitate sharing their internal information with others especially since September 11, 2001. Only minimal information is made available on the Web. The structure or topology of the utility network and information about operations is not disclosed to the public. While this trend may hinder the development of well-structured information systems to solve problems among utilities, it is necessary to find ways in which information and process knowledge can be shared without creating sources of vulnerability. This could be through predetermined and secure channels of communication.

KNOWLEDGE MANAGEMENT CONCEPTS RELEVANT TO MANAGEMENT OF CIVIL INFRASTRUCTURE

In managing critical infrastructures, the ability to capture and reuse tacit knowledge (Polanyi, 1967) is vital given the changing nature of threats, especially those that are intentional and human induced. Standard operating procedures and documented methods may not provide an appropriate guide under novel situations. In such a scenario, the decision maker needs support in the form of access to information about prior incidents and how they were handled. Further, given the interdependence between critical infrastructures, access to information about events in related utilities becomes important.

K

Each organization within an infrastructure works in a relatively unique manner based on its organizational culture and traditions. When comparing organizations in different infrastructures, the situation becomes more complex given disparate engineering set-ups and regulatory regimes. Information sharing about interdependencies is a starting point and is likely to be acceptable if the system links are designed with appropriate levels of security. In addition, given such a dynamic environment, the ability to "capture" discussions to glean ideas becomes extremely valuable.

The role that KM can play is one of accumulating knowledge from employees to make the organization work better. In the context of managing critical infrastructures, the purpose is to learn from each incident across infrastructures to enable a decision maker to handle a current or future unique incident for which there is not a standard operating procedure available.

Knowledge management is therefore about changing the way work is done. When threats are constantly changing, mere understanding of current threats and methods to mitigate them are not sufficient. It is necessary to have a process in place where the organization learns from each incident and develops strategies to improve the way in which vulnerabilities are assessed, disasters mitigated, and failures managed. KM processes create a structure that enables reflection of events, analysis of data, and procedures to facilitate learning and making more effective the decision-making process of utility managers. However, as Nidumolu, Subramani, and Aldrich (2001) point out "to be successful", KM initiatives "need to be sensitive to features of the context of generation, location and application of knowledge". The operating procedures, technology and information systems, and organizational culture will collectively impact the knowledge management efforts in infrastructure management.

CONCLUSION

The purpose of KM processes in civil infrastructure management is to assist owners to provide services economically and reliably. Utilities have to operate in an environment constrained by customer service requirements, regulations, shareholder expectations,

and an aging infrastructure. Civil Infrastructure managers need the capabilities provided by KM approaches to help them manage more effectively.

However, implementing KM processes goes beyond deploying specific types of information technologies and systems. Systems that support the sharing of information and process knowledge within civil infrastructures must be coupled with changes to operating and managerial procedures. These changes have to take into account the interdependencies between related infrastructures as well as the security risks imposed by release of critical information. It is necessary for the owners and top-level managers at critical infrastructures to take a much closer look at this issue so that they can allocate the appropriate resources to implement effective knowledge management processes.

REFERENCES

Heller, M. (2001). Interdependencies in civil infrastructure systems. *The Bridge, 31*(4), 9-15.

Heller, M. (2002, February 14). Critical infrastructure interdependencies: A system approachs to research needs. *Infrastructure and Information Systems, National Science Foundation, Presentation to National Academy Committee on Science and Technology for Countering Terrorism: Panel on Systems Analysis and Systems Engineering.*

Heller, M., Von Sacken, E.W., & Gerstberger, R. L. (1999). Water utilities as integrated businesses. *Journal of the American Waterworks Association, 91*(11), 72-83.

Jentgen, L. (2004). *Implementation prototype energy and water quality management system.* London: IWA.

Johnson, J., & Lefebvre, A. (2003). Impact of Northeast Blackout continues to emerge. Retrieved May 9, 2005, from *http://www.wsws.org/articles/2003/aug2003/blck-a20.shtml*

Lemer, A. C., Chong, K. P., & Tumay, M. T. (1995) Research as a means for improving infrastructure. *Journal of Infrastructure Systems, 1*(1), March, 6-15.

Nidumolu, S. R., Subramani, M., & Aldrich, A. (2001). Situated learning and the situated knowledge web: Exploring the ground beneath knowledge management. *Journal of Management Information Systems, Summer, 18*(1), 115-150.

PDD 63 (Presidential Decision Directive 63). (1998). An analysis of the consequences of the August 14th 2003 power outage and its potential impact on business strategy and local public policy. Retrieved May 9, 2005, from *http://www.terrorism.com/homeland/pdd63.htm*

Perez, A. L. (2003, June 15-19). Knowledge management for public utilities: A strategic approach for the retention of intellectual capital. *Proceedings of the 2003 American Water Works Association Annual Conference, Anaheim Convention Center,* Anaheim, CA

Polanyi, M. (1967). *The tacit dimension.* Garden City, NY: Doubleday.

Rinaldi, S. M., Peerenboon, J. P., & Kelly, T. (2001, December). Identifying, understanding, and analyzing critical infrastructure interdependencies. *IEEE Control Systems Magazine*, 11-25.

Rosen, J. S., Miller, D. B, Stevens, K. B., Ergul, A., Sobrinho, J. A. H., Frey, M. M., & Pinkstaff, L. (2003). *Application of knowledge management to utilities.* Denver: American Water Works Associations Research Foundation (AWWARF).

Stoup, G. M., Slavik, M. R., & Schnoke, M. S. (2004). An analysis of the consequences of the August 14th 2003 power outage and its potential impact on business strategy and local public policy. Retrieved May 9, 2005, from *http://www.acp-international.com/southtx/docs/ne2003.pdf*

KEY TERMS

AwwaRF: American Water Works Association Research Foundation.

Civil Infrastructure Systems: Refers to the physical infrastructure that enables basic services such as transportations (railroads, roads, airports), water supply, sewage disposal, electric power generation and supply, telecommunications services, ans so forth.

Data Warehouse: A place where managed data are situated after they pass through the operational systems and outside the operational systems.

Energy Management Systems (in electric utilities): The systems used to efficiently manage power network operation, coordinate optimized power distribution, and manage costs of electricity production and distribution.

Geographic Information Systems (GIS): Spatial data management systems that allow the user to deal with (save, retrieve, extract, manipulate, and visualize) the information of physical entities.

Hazard Communication Standard (HCS): OSHA's (Occupational Safety and Health Act) regulation on the Hazard Communication Standard states that "chemical manufacturers and importers must research the chemicals they produce and import. If a substance presents any of the physical and health hazards specified in the HCS, then the manufacturer or importer must communicate the hazards and cautions to their employees as well as to 'downstream' employers who purchase the hazardous chemicals."

Infrastructure Interdependency: Denotes a bidirectional relationship between two or more infrastructures, through which the state of each infrastructure influences or is correlated to the state of the other infrastructure.

Material Safety Data Sheets (MSDS): The source of the data to effectively communicate the hazard potential of different materials to users. They are required by The Hazard Communication Standard.

Supervisory Control and Data Acquisition (SCADA): Systems used to control the operations of utilities. These are typically computer-based applications that enable managers of utilities to operate the equipment and deliver services to customers.

Knowledge Management in Supply Chain Networks

Dolphy M. Abraham
Loyola Marymount University, USA

Linda Leon
Loyola Marymount University, USA

INTRODUCTION

This article reviews current research and practice of knowledge management (KM) and inter-organizational learning in supply chain networks. Knowledge management is the organizational process for acquiring, organizing, and communicating the knowledge of individual employees so that the work of the organization becomes more effective (Alavi & Leidner, 1999). Knowledge management is an increasingly important process in business organizations because "managing human intellect—and converting it into useful products and services—is fast becoming the critical executive skill of the age" (Quinn, Anderson & Finkelstein, 1998). Grover and Davenport (2001) state that KM becomes "an integral business function" when organizations "realize that competitiveness hinges on effective management of intellectual resources." Grover and Davenport also argue that knowledge management works best when it is carried out by all the employees of the organization and not just KM specialists.

Business organizations frequently partner with other firms to complement their core competencies. To collaborate effectively, partner firms have to communicate with each other information about business processes as well as share ideas of how to design or improve business processes. This phenomenon of knowledge sharing across organizational boundaries is called inter-organizational learning (Argote, 1999). Knowledge management, we posit, is necessary to facilitate inter-organizational learning and direct it in a way that supports the organization's overall objectives.

Supply chain systems are an example of business networks. Supply chains involve not only multiple corporate entities but also organizational units within a single organization. We present practices used in business organizations and networks of businesses to manage the information and knowledge sharing processes using the context of supply chain systems.

BACKGROUND

A supply chain consists of all parties involved, directly or indirectly, in fulfilling the end consumer's request. Its primary purpose is to satisfy customer needs while maximizing the overall profitability of the chain. A typical supply chain involves a variety of stages that may include customers; a distribution network of retailers, wholesalers, and distributors; manufacturing enterprises; and tiers of suppliers (Figure 1). Information, knowledge, funds, products. and services flow along both directions of the chain, where more than one player is often involved at each stage. The structure of supply chain systems can be described as a business network where inter-organizational learning and information sharing are critical factors in determining the chain's competitiveness.

The performance of a supply chain depends upon how well its processes are managed for the type of product that is associated with the chain. Fisher (1997) classifies products on the basis of their demand patterns, claiming that a product falls into one of two categories, either primarily functional or primarily innovative. Functional products satisfy basic needs and include the staples that people buy in a wide range of retail outlets such as grocery stores and gas stations. These products have stable, predictable demand, and long life cycles. Due to well-developed competition, low profit margins occur, requiring the chain to focus almost exclusively on

Figure 1. The integrated supply chain

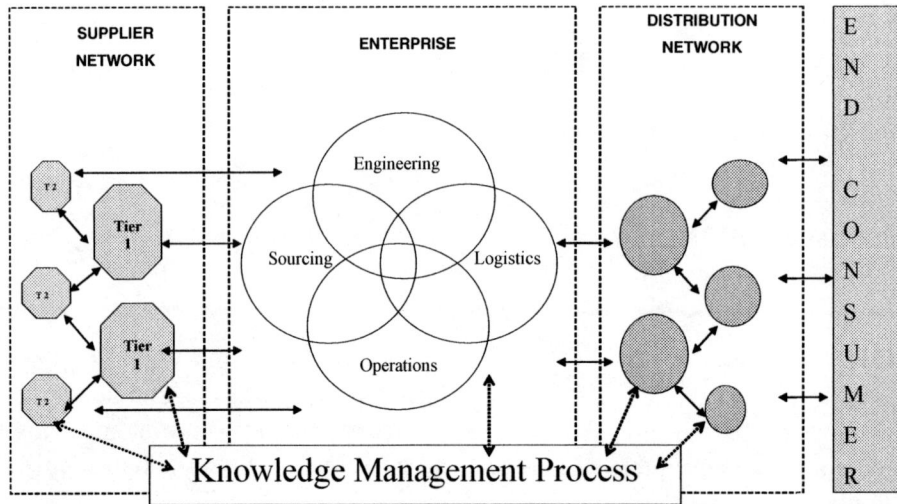

minimizing physical costs. Companies need to coordinate the ordering, production, and delivery of supplies in order to minimize inventory and maximize production efficiency in order to meet predictable demand at the lowest cost.

Innovative products, such as fashion apparel and technology products including plasma TVs, cellphones, and iPods, are differentiated from competition by their designer options and new features and capabilities. The novelty of these products allow higher profit margins, but also result in more demand uncertainty as it is difficult to predict how the market will respond to the newest design features and options. The life cycle for innovative products is short as ensuing competition forces companies to introduce newer innovations in order to maintain the higher profit margins. The short life cycles and the great variety typical of these products further increase demand unpredictability. The demand uncertainty from the market environment increases the risk and costs of shortages and excess supplies throughout the chain. To mitigate this risk, it is crucial that information flows not only within the chain but also from the marketplace to the chain. Fisher (1997) describes managers' primary focus in supply chains for innovative products as market mediation, the need to read early sales numbers or other market signals and have the chain respond quickly. The critical decisions to be made are not about minimizing costs but about where in the chain

to position inventory and available production capacity in order to hedge against the uncertain demand and be responsive. Supply chain systems reduce the external environmental uncertainty by introducing formal information and knowledge transfer mechanisms between supply chain partners. In supply chains with innovative products, suppliers are evaluated based on their reliability, speed, flexibility, and product development skills as well as for their cost.

Due to the emphasis on market mediation, supply chains for innovative products require more collaboration about product design and improvement of business processes than supply chains for functional products. While information sharing can improve the performance of functional product supply chains, inter-organizational learning is essential to support the overall objectives of the innovative product supply chain.

KNOWLEDGE MANAGEMENT IN SUPPLY CHAINS

As the goal for functional products is to minimize the physical costs associated with the production and delivery of the product, many supply chains have improved their coordination efforts by sharing information. Efficient Consumer Response (ECR) and Quick Response (QR) initiatives are efforts that certain industries have implemented to reengineer the supply channel to make it more responsive to customer demand. Controlling the flow of information and product between different stages within the chain is a major focus of these initiatives as it helps decrease the costs of inventory and shortages. Chains have used technologies such as electronic data interchange (EDI), the Internet, and satellite systems for transmitting point-of-sale data to provide real-time information. Improved coordination is achieved with this information even when the decision-making responsibilities remain decentralized. A $14 billion savings in the food service industry (Troyer, 1996) and $30 billion savings in the groceries industry (Kurt Salmon Associates, 1993) have been documented as a result of implementing these initiatives. Current MS/OR research (e.g., Lee, So & Tang, 2000) studies the value of different types of information that can be shared given the decentralized decision-making framework. For example, Wal-Mart's Retail Link program provides an online summary of point-of-sale data for Johnson & Johnson and Lever Brothers as well as direct satellite transmitted point-of-sale data. Cachon and Fisher (1997) describe the cost savings that Campbell Soup's continuous replenishment program generated for the grocery supply chain. In this instance, retailers transmitted daily inventory information via EDI to Campbell Soup, and the manufacturer assumed responsibility for managing retailer inventories, a process commonly referred to as vendor managed inventory. This particular continuous replenishment program reduced inventories in retailer distribution centers by 50%, while increasing service levels from 98.7% to 99.5%.

Some chains have extended their collaborative efforts to include information about processes as well as more centralized decision making. Aviv (2001) and Raghunathan (1999) describe Collaborative Forecasting and Replenishment (CFAR) as a new inter-organizational system that enables retailers and manufacturers to forecast demand and schedule production jointly. CFAR allows the exchange of complex decision support models and manufacturer/retailer strategies so that the two supply chain parties can reduce demand uncertainty and coordinate their decisions. Wal-Mart and Warner-Lambert embarked on the first successful CFAR pilot, involving Listerine products in 1996. Since then several major manufacturers of functional products, such as Procter and Gamble, have undertaken CFAR projects.

Due to the demand uncertainty and short life cycle, supply chains for innovative products need to develop strategies that will create flexibility and responsiveness within the chain. The exchange of knowledge about processes, innovations, and market interest are vital to the members of the chain as it works to design and distribute the newest product to the market quickly. The Knowledge and Learning in Advance Supply Systems (KLASS) pilot project (Rhodes & Carter, 2003) seeks to develop collaborative learning in networks of suppliers in the automotive and aerospace sectors. Focusing on the tiered supplier network, as illustrated on the left side of Figure 1, KLASS utilized an inter-company, computer-mediated learning network that focused on both immediate performance improvements and longer term objectives. It developed learning and knowledge to advance collaborative functioning and improved performance between the tiered companies linked in the supplier network. Similarly, to manage knowledge across the supplier network and enterprise boundaries, as shown in Figure 1, Chrysler developed a successful supplier-suggestion process to reduce costs and build collaborative relationships with its suppliers (Hartley, Greer & Park, 2002). Both of these are examples of inter-organizational systems.

In another example of an inter-organizational network, Mak and Ramaprasad (2003) introduce the idea of knowledge supply networks, which they define as "integrated sets of manufacturing and distribution competence, engineering and technology deployment competence, and marketing and customer service competence that work together to market, design and deliver end products and services to markets." They outline the nature of the business processes associated with designing and delivering innovative products and describe the need to effectively coordinate the knowledge in the market, design, and supply distribution chains. As costs for product development increase and faster time-to-market is expected, more and more original equipment manufacturers (OEM), such as Motorola and Nokia, are refocusing their competence on marketing, research and design, and critical high level design, and outsourcing everything else to contract manufacturers. This changes the chain structure as outlined in Figure 1 and requires the OEM to create a knowledge management network that will allow them to leverage the supplier and contract manufacturer knowledge, yet still preserve their own knowledge and control.

In order to be competitive, supply chains for innovative products must have processes in place to exchange product and market knowledge. Unlike the chains associated with functional products, innovative chains will not incur much competitive advantage as the result of demand and inventory information sharing. Due to high future demand uncertainty and potentially high profit margins, there is often distrust between the stages of the chain. This distrust can result in parties making uncoordinated decisions that are in their best individual interest (local optimization) and not in the best interest of the supply chain.

For instance, in the PC market, manufacturers suspect their distributors of inflating orders to ensure availability of the product. Dell Corporation removed the distributor stage in its supply chain so that it could improve its market mediation and receive end-consumer information directly. In the automotive and aerospace industry, manufacturers provide suppliers with forecasts that are often wrong, resulting in extreme shortages or excess capacity with no return on investment for the suppliers. Suppliers often make locally optimal decisions as a result. When a supplier could not provide an adequate supply of ashtrays and glove compartment doors, GM lost nearly two months of production of the Buick Roadmaster (Suris & Templin, 1993). In 1999, GM canceled two new models, leaving their suppliers with newly developed capacity and no return for their investment (Pryweller, 1999). In 1997, Boeing could not fill their plane orders largely due to a shortage of 500 different parts from 3,000 part suppliers who did not have enough capacity, resulting in a $1.6 billion charge against Boeing's third-quarter earnings (Cole, 1997).

A survey done by Lee and Whang (2000) identifies that firms are also concerned about the confidentiality of shared information when competition exists, and that it is one of the major hurdles that information sharing in a supply chain must overcome. Besides intentional information leaks, Li (2002) defines the leakage effect as the indirect effect of vertical information sharing that occurs when the shared information becomes known to competition as the result of observing the behavior of the party

that receives the information. Li illustrates how competing firms can react to the observed behavior and how this reaction can change the strategic interaction, causing additional gains or losses to the parties between which the information was directly exchanged. Powell, Koput, and Smith-Doerr (1996) point out that firms will continue to work with their partners, once the risks are managed at a "tolerable level."

In addition to trust, several other factors impact the ability of a firm to share information and knowledge productively with other parties in a supply chain. These include the technology infrastructure, application software used to manage the supply chain operations, and the culture of knowledge sharing within a firm as well as within the supply chain.

Scott (2000) studied the process of and reasons for information technology (IT) support for inter-organizational learning. Studying the disk drive industry, Scott identified the need for inter-organizational learning to help "cope with the complexity of new products and the capital intensity" in the industry. She noted that the industry had consolidated with several firms working very closely in a "vertically integrated virtual organization." IT helps the organizations streamline the information flow between them, making it easier to provide feedback between partners and facilitate learning across organizational boundaries. The model developed by Scott helps to explain the role of IT in lower and higher levels of inter-organizational learning. In lower level learning, an organization makes changes to its operations based on feedback from a partner. In higher order learning, partners change operating assumptions and procedures based on a new understanding typically resulting from collaborative work. An important finding from her study was that inter-organizational learning strategies "with customers and suppliers in the disk drive industry were facilitated by IT and constrained by lack of trust."

Kent and Mentzer (2003) conducted a study of the impact of inter-organizational information technology (IOIT) in a retail supply chain. They found that when suppliers invest in IOIT, retailers perceive a commitment to the relationship from the suppliers and are willing to reciprocate. The commitment to the relationship leads to the partners working together to improve logistics efficiency. Therefore, the payoff from the investment in IOIT comes from the reduction of costs in the supply chain as a result of improvements in logistics efficiency. Their findings also support Scott's (2000) observation about the role of trust between partners in a supply chain. In addition to the investment in IOIT, they show that firms must also "demonstrate characteristics of trust such as integrity and faithfulness." Only then will partners perceive the commitment to the relationship. As Kent and Metzer state, "relationship trust and investment in IOIT can lead to relationship commitment that can lead to logistics efficiency."

The above examples pertain to inter-organizational systems; however, KM plays an important role even within a single organization. Intra-organizational systems and initiatives are needed to facilitate knowledge sharing across organizational units. Edwards and Kidd (2003) treat knowledge management "as a process rather than as an organizational system" or a piece of technology. In order for this process to work across organizational boundaries, they identify trust, organizational culture, and the "relationship between top down strategy and bottom up organizational learning" as enabling factors. Trust needs to exist between individuals as well as between organizations. There needs to be compatibility between the cultures of organizations for knowledge management processes to work. Aligning the top down strategy with bottom up learning requires the organization to make its strategy for KM clear and create and maintain an atmosphere that supports organization learning. Trust between individuals and organizations can be enhanced by setting up exchange programs and by facilitating voluntary exchange of knowledge.

Another set of factors that Bessant and Kaplinsky (2003) recognize as necessary for learning in supply chain is the "accumulation and development of a core knowledge base," as well as the "long-term development of a capability for learning and continuous improvement across the whole organization." In order for each organization in a supply chain to manage how learning takes place, they need to have formal mechanisms and a clear understanding of the value of learning. Only then can long-term efforts be sustained.

This set of factors is based on the concept of *absorptive capacity* (Cohen & Levinthal, 1990). Cohen and Levinthal describe this as the capacity of

an organization to "recognize the value of new, external information, assimilate it, and apply it to commercial ends." Lane and Lubatkin (1998) suggest that the absorptive capacity of a firm is relative to its partner firms and is dependent on "the similarity of both firms' knowledge bases, organizational structures and compensation policies," and one firm's familiarity with the other firm's "set of organizational problems." Szulanski (1996) shows that a unit's lack of absorptive capacity, a distant relationship with other units and lack of a clear understanding of cause and effect relationships can all become impediments to intra-organizational learning.

In addition to efficiency in logistics is the issue of effectiveness of sourcing in enterprises that have multiple personnel handling the purchasing task. Rozemeijer, Van Weele, and Weggeman (2003) identify three constructs to help corporate purchasing officers create coordinating mechanisms that facilitate purchasing synergies within the corporation. The constructs include purchasing maturity, corporate coherence, and business context. The mechanisms they identified include:

- formal organizational mechanisms such as corporate steering boards or commodity teams;
- informal networking mechanisms such as annual purchasing conferences and job rotation;
- enterprise-wide information and communication systems; and
- advanced management and control systems.

These four mechanisms or systems represent different options for intra-organizational KM. These mechanisms were illustrated for a supply chain with functional products. Rozemeijer et al. (2003) also point out however that purchasing performance depends on increased coordination between multiple purchasing officers and requires constant monitoring to ensure that the purchasing function is aligned with the business context and corporate strategy, which is associated with the type of product and supply chain objectives involved.

CONCLUSION

As the studies cited above show, information technology that supports information sharing is a necessary element for knowledge management processes to work in supply chains. In supply chains for functional products, decentralized decision making can result in good decisions when IT support is effective. However, in innovative supply chains, IT by itself is not sufficient. The other factors are needed as well. Without the other factors, such as trust within and across organizations or an organizational culture that supports learning, the supply chain does not benefit. Even with good IT and information sharing, each company is likely to make decisions that are locally focused, resulting in suboptimal supply chain performance.

Practitioners interested in creating and managing knowledge management efforts in supply chains should consider the following issues carefully:

- The information flow between the tiers in the supply chain and how the systems and organization support information sharing.
- The utilization of information within an organization, especially to support decision making. Decentralized decision making can be beneficial for the supply chain as long as the firms trust each other and want to optimize the performance of the chain. Incentives need to be in place to ensure that each party is motivated to maximize the performance of the chain.
- The organizational culture with respect to learning within each firm and across the tiers in the supply chain. The culture must support learning and facilitate the process of learning.
- Trust between organizations in the supply chain is critical. Unless the firms develop this trust, they will make decisions solely based on their self interest, and that may be detrimental to supply chain performance.
- How the network assimilates the information and how the network changes its actions based on what it has learned is the effect of the knowledge management effort. In the case of functional products, this can be in the form of firms modifying processes to increase efficiency. In the case of innovative products, suppliers may be elevated to status of true partners, taking on more responsibility and risk and sharing in decision making as decisions are made in a more centralized manner.

REFERENCES

Alavi, M., & Leidner, D. E. (1999). Knowledge management systems: Issues, challenges and benefits. *Communications of the AIS, 1,* 1-37.

Argote, L. (1999). *Organizational learning: Creating, retaining and transferring knowledge.* Norwell, MA: Kluwer.

Aviv, Y. (2001). The effect of collaborative forecasting on supply chain performance. *Management Science, 47*(10), 1326-1343.

Bessant, J., & Kaplinsky, R. (2003). Putting supply chain learning into practice. *International Journal of Operations and Production Management, 23*(2), 167-184.

Cachon, G., & Fisher, M. (1997). Campbell Soup's continuous replenishment program: Evaluation and enhanced inventory decision rules. *Production and Operations Management, 6*(3), 266-276.

Cohen, W. M., & Levinthal, D. A. (1990, March). Absorptive capacity: A new perspective on learning and innovation. *Administrative Science Quarterly, 35*(1), 128-152.

Cole, J. (1997, June 26). Boeing, pushing for record production, finds parts shortages, delivery delays. *Wall Street Journal,* p. A4.

Edwards, J. S., & Kidd, J. B. (2003). Knowledge management sans frontiers. *Journal of the Operational Research Society, 54*(2), 130-139.

Fisher, M. (1997). What is the right supply chain for your product? *Harvard Business Review, 75*(2), 105-117.

Grover, V., & Davenport, T. H. (2001). General perspectives on knowledge management: Fostering a research agenda. *Journal of Management Information Systems, 18*(1), 5-21.

Hartley, J., Greer, B., & Park, S. (2002). Chrysler leverages its suppliers' improvement suggestions. *Interfaces, 32*(4), 20-27.

Kent, J., & Mentzer, J. (2003). The effect of investment in interorganizational information technology in a retail supply chain. *Journal of Business Logistics, 24*(2), 155-175.

Kurt Salmon Associates Inc. (1993). *Efficient consumer response: Enhancing consumer value in the grocery industry.* Washington, DC: Food Marketing Institute.

Lane, P. J., & Lubatkin, M. (1998, May). Relative absorptive capacity and interorganizational learning. *Strategic Management Journal, 19*(5), 461-477.

Lee, H., So, K., & Tang, C. (2000). The value of information sharing in a two-level supply chain. *Management Science, 46*(5), 626-643.

Lee, H., & Whang, S. (2000). Information sharing in a supply chain. *International Journal of Technology Management, 20*, 373-387.

Li, L. (2002). Information sharing in a supply chain with horizontal competition. *Management Science, 48*(9), 1196-1212.

Mak, K-T., & Ramaprasad, A. (2003). Knowledge supply network. *Journal of the Operational Research Society, 54*(2), 175-183.

Powell, W. W., Koput, K. W., & Smith-Doerr, L. (1996, March). Interorganizational collaboration and the locus of innovation: Networks of learning in biotechnology. *Administrative Science Quarterly, 31*(1), 116-145.

Pryweller, J. (1999, June 21). Delays jolt GM suppliers: Some part makers invest millions. *Automotive News.* Retrieved July 6, 2005, from *http://www.autonews.com/article.cms?articleId=20471*

Quinn, J. B., Anderson, P., & Finkelstein, S. (1998). Managing professional intellect: Making the most of the best. *Harvard Business Review of Knowledge Management,* 181-205.

Raghunathan, S. (1999). Interorganizational collaborative forecasting and replenishment systems and supply chain implications. *Decision Sciences, 30*(4), 1053-1071.

Rhodes, E., & Carter, R. (2003). Collaborative learning in advanced supply systems: The KLASS Pilot Project. *Journal of Workplace Learning, 15*(6), 271-279.

Rozemeijer, F., Van Weele, A., & Weggeman, M. (2003). Creating corporate advantage through purchasing: Toward a contingency model. *Journal of Supply Chain Management, 39*(1), 4-13.

Scott, J. E. (2000). Facilitating interorganizational learning with information technology. *Journal of Management Information Systems, 17*(2), 81-113.

Suris, O., & Templin, N. (1993, October 5). GM production problems hurt '94 model sales. *Wall Street Journal*, pp. A4.

Szulanski, G. (1996). Exploring internal stickiness: Impediments to the transfer of best practice within the firm. *Strategic Management Journal, 17*, 27-43.

Troyer, C. (1996, May 21-23). EFR: Efficient food service response. *Proceedings of the Conference on Logistics,* Palm Springs, California.

KEY TERMS

Collaborative Forecasting and Replenishment: An inter-organizational system that enables retailers and manufacturers to forecast demand and schedule production jointly by exchanging complex decision support models and manufacturer/retailer strategies so that the two supply chain parties can reduce demand uncertainty and coordinate their decisions.

Coordination within a Supply Chain: Occurs when the decisions made at different stages of the chain maximize the total supply chain's profitability. When a party makes a decision that maximizes its own local profitability, a lack of coordination can occur in the supply chain, as that decision may not be in the best interest of the entire chain.

Decentralized Decision Making: In supply chains, it involves decisions where each entity (member of the supply chain) has control over decisions at their stage. However, the decisions not only have a local impact but also impact the whole supply chain.

Functional Products: Have stable and predictable demand, long life cycles, and well-developed competition that results in low profit margins.

Innovative Products: Have uncertain demand, high profit margins, and short life cycles due to ensuing competition that forces companies to introduce newer innovations.

Interorganizational Information Technology (IOIT): Consists of networking and software applications that enable business networks to share data and information with each other. Networks (public or private) are used to provide connectivity between the software applications located within each organization. The software applications are designed to share data and information without the need for much human intervention. Examples of IOIT include Electronic Data Interchange (EDI) systems.

Interorganizational Learning: The sharing of information and process knowledge across organizational boundaries. The information and knowledge pertain to tasks or processes that are carried out by the various organizations. By making use of the information and process knowledge, these organizations can change the way they carry out these tasks and processes to improve performance.

Knowledge Supply Networks: A concept defined by Mak and Ramaprasad (2003) as "integrated sets of manufacturing and distribution competence, engineering and technology deployment competence, and marketing and customer service competence that work together to market, design and deliver end products and services to markets."

Supply Chain Performance: Often measured by the supply chain's profitability, which is the difference between the revenue generated from the customer and the total cost incurred across all stages of the supply chain. Supply chains that perform competitively provide a high level of product availability to the customer while keeping costs low.

Supply Chain Systems: One type of network of business organizations who work together to produce, distribute, and sell products. They typically consist of suppliers of raw materials as well as parts or components, manufacturers and assemblers, distribution companies, original equipment manufacturers, wholesalers, and retailers.

Knowledge Sharing through Communities of Practice in the Voluntary Sector

Lizzie Bellarby
Leeds Metropolitan University, UK

Graham Orange
Leeds Metropolitan University, UK

INTRODUCTION

Over the past few years, the concepts of knowledge management and knowledge sharing have been recognised as cognate areas of study and research. To date research has focussed mainly upon the commercial sector. However, this article looks at knowledge management and sharing through communities of practice within the voluntary sector. The work is based upon research carried out within a UK national voluntary counselling and advisory service. For reasons of privacy and confidentiality, the organisation shall remain anonymous and will be referred to as the 'organisation'.

This article considers the background to the study in terms of knowledge management and communities of practice. It then discusses the study's methodology and findings. It concludes that knowing and sharing are active processes, and that the natural disposition of the actors was found to be important in how knowledge sharing and learning was undertaken.

BACKGROUND

The term *knowledge management* (KM) has been around for some time, and has primarily been concerned with the capability of various software products to handle 'knowledge'. However during the 1990s there was a shift away from these technological issues and towards people. The research project upon which this article is based is people focussed, employing the Grounded Theory Method (discussed below). Contemporary theories of knowledge concentrate on the human dynamic, emphasising that this is more important than the tools designed to store and network information (Foskett, 1990; Streatfield & Wilson, 1999; Orlikowski, 2000; Prusak, 2001).

Whilst KM has been treated as a scientific discipline by those who have focussed mostly upon the exploitation of IT, there have been some qualitative studies that have explored the human side of knowledge management, and in particular have studied communities of people in the context of knowledge. It is important to study people and groups of people because "most practitioners…have begun to study networks and communities as the most productive units of analysis for doing knowledge work" (Prusak, 2001). Accordingly, links to sociology and anthropology have been explored by Snowden (2000), Delanty (2001), and Scharmer (2001), amongst others.

Knowledge Management and Communities of Practice

Wenger's (2000) analysis of Prusak and Davenport's concept of 'communities of practice' is an important one for this article, as the investigation of the voluntary organisation and the teams within it highlighted a strong resemblance to the 'communities of practice' described by Wenger.

"Communities of practice are the basic building blocks of a social learning system because they are the social 'containers' of the competencies that make up such a system" (Wenger, 2000, p. 229). According to Wenger, if organisations can encourage communities of practice, they can create a "social learning system," which will increase their knowledge generation, and therefore, hopefully, their economic success. Communities include teams and organisations, as well as virtual communities.

What are communities of practice? In brief, they're groups of people informally bound together by shared expertise and passion for a joint enterprise. (Wenger, 2000, p. 139)

Organisations need to "design themselves as social learning systems" (Wenger, 2000, p. 225) if they are to beat their competitors. Communities of practice are a component of these systems. Wenger "proposes a social definition of learning" (2000, p. 225), as opposed to an individualistic approach. This means that learning by individuals can take place effectively in a social context. By being identified with a group, individuals can learn within the boundaries of this dynamic setting, and the knowledge created is shared within the group. Wenger says that "knowing...is a matter of displaying competences defined in social communities [and] socially defined competence is always in interplay with our experience. It is in this interplay that learning takes place" (2000, 226).

An important facet of the human aspect of knowledge management is undoubtedly that of communication. Senge's work places strong emphasis upon articulation of knowledge and linguistics, and Snowden (2000) uses an anecdotal, storytelling approach to KM. Increasing peoples' personal and socially constructed knowledge becomes of more tangible value to firms when it is communicated to others. Perhaps this is why some IT consultants draw the conclusion that communicated knowledge should be captured and disseminated throughout the firm in order to gain maximum benefit from it. However, the capture and dissemination of knowledge may dilute it and even stifle knowledge creation. Indeed it could be argued that once knowledge is externalised (from the knower) and stored, it loses context, and ceases to be knowledge and becomes information (Orange, Burke & Boam, 2000; Wilson, 2002).

An area of importance is how people are motivated to share what they know, as it is often the issue that knowledge management programmes are the most difficult to address. Rewarding people to use knowledge and impart their own knowledge has been discussed in the literature in a very loose way (Koenig, 1999; Nonaka, 2000; Lubit, 2001), but there are few organisational solutions to motivate employees to create and share knowledge. The research explores motivation to create and share know-how where there are no financially bound incentives. For this reason, the research was conducted at a voluntary organisation.

Methodology

This research used the Grounded Theory Method (GTM). The method was founded by Glaser and Strauss in 1967 as an attempt to bring qualitative research methods in equal standing with quantitative, scientific research methods. They wanted to establish a method that would have as much authority as quantitative methods in terms of establishing 'truth'. GTM is a "particular style of qualitative analysis of data...for generating and testing theory" (Strauss, 1987). It is so called due to "its emphasis on the generation of theory and the data in which that theory is grounded" (Glaser, 1978).

We are offering more than a set of procedures. We are offering a way of thinking about and of viewing the world that can enrich the research of those who choose to use this methodology. (Strauss & Corbin, 1998, p. 4)

Strauss and Corbin refer to the procedures as being like a smorgasbord—that is, the researchers can pick and choose methods to suit their investigation. In this research it was decided to use GTM as per Strauss and Corbin's interpretation, that is to use the parts that are useful and filter out those which did not fit the purpose of the investigation, and to acknowledge the impact of the researcher on the results.

An important part of GTM is theoretical sampling, which is where data is collected in order to generate theory. The researcher collects data, in this case through the use of interviews, which is immediately coded and then analysed, before collecting further data. In this way, data collection is steered by the emerging theory.

Following interviewing, the researcher then looks at documents, including interview transcripts, memos, and field notes, for indicators of categories in events and behaviour. "This triad (of data collection/coding and memoing)...serves as a genuinely explicit control over the researcher's biases" (Strauss, 1987). This form of 'open coding' hopes to identify 'concepts'. These concepts or codes are then compared to each other to find consistencies and differences. Consistencies between codes (similar meanings or pointing to a basic idea) reveal categories.

Findings

It emerged that 'motivation' was the single most important category emerging from all of the data. Throughout the study it became apparent that there was a clear link between knowledge management and sharing/motivation. At the outset it was established that the interviewees tended to prefer the use of the terms 'learning' and 'know-how' rather than the term 'knowledge'. The research was conducted with this in mind when carrying out the interviews and reporting the findings.

The motivation of people links strongly with how willing they are to share what they know and continue their learning. When analysing the transcripts it was found that where the variable of motivation was considered to be high, there was a strong willingness to share and learn exhibited in the transcripts. That is, the volunteers were motivated, confident, talked of sharing know-how in teams, and were willing to learn.

According to Osbourne (1998), motivation within voluntary organisations is based on the service provided rather than financial incentives. The data gathered corroborates this, as most interviewees referred to the service element of what they do. Helping people appears to be a strong motivation for undertaking and continuing with their voluntary work. The National Council for Voluntary Organisations (NCVO, 1996) claims of the voluntary sector that "its ability to motivate staff is the envy of other sectors."

Through reading and re-reading the transcripts, it became clear that the interviewees felt that one had to be the 'right person', with a natural disposition and an open-minded attitude, in order to learn the requisite skills needed to do the voluntary work.

It was found that there was a link between natural disposition and learning, and also between natural disposition and attitude towards the organisation. From this it may be inferred that people are predisposed to learning and that it is not something that can be taught or forced. Typical comments from the transcripts which reinforce this are:

- "There has to be some sort of screening of people."
- "Sensitive, sensitivity to the team as well as to the caller—we look for these in selection, then training develops those qualities."
- "I must be doing something right to have been judged suitable."

A sense of self-confidence came through from the interviews. Many of them felt that they were already 'halfway there' to learning relevant skills before they even began their training. This was in part, they felt, due to their 'nature' or disposition, and also due to their life experiences. The data demonstrated a high degree of self-awareness amongst those interviewed, no matter what their age, length of service, or level of responsibility held. People with self-awareness are more likely to be selected to be volunteers at the organisation, which has a bearing on peoples' motivations to learn, and share what they know with others. There was a link between self-confidence, self-awareness, and a willingness to learn and share know-how with others. If organisations choose employees with high self-awareness, then this may contribute to a successful 'learning organisation' (Senge, 1992) or a 'knowledge creating company' (Nonaka, 1995).

Nonaka (1995) refers to the firm (Nonaka uses the term firm, but it can relate to a wider organisational context) as 'a knowledge-creating entity', which does not portray the full picture according to the findings of this research. The success of the voluntary organisation in terms of its motivation to learn appeared to be initiated by the careful selection of volunteers who have confidence in their ability to do the work, and who bring a broad range of life experiences to the organisation. So, the emphasis should perhaps then be upon the people chosen to belong to the organisation, rather than the organisation itself, as the creator of knowledge.

Communication was also viewed as highly important by the interviewees, as part of the learning process. This is emphasised in the transcripts, for example:

- "When people come off the phone, [they] share what they've had from the phone; and

that's an opportunity really, for additional learning."

- "If people come off the phone and say 'oh I
don't think I've managed that very well', folk
get a chance to talk about it, and to learn from
it."
- "If we don't communicate with each other we
don't learn."
- "So on every level, right from the day one of
training, you're encouraged to share any problems."
- "I'll talk with my team if, especially if, I think
I've screwed up."
- "There's certainly ample opportunity to seek
advice and share your experiences."

This aspect of communication fits in with the
'engagement' part of the communities of practice
model (Wenger, 2000). In Wenger's model, *Competence + Experience + Engagement = a Community of Practice.* Communication forms an aspect of
belonging to a social learning system that includes
doing things together and talking. Teams telling each
other stories or anecdotes are a way of belonging to
a community of practice, according to Wenger.
Snowden (1999) writes of storytelling as a deliberate
way by which some cultural groups share knowledge and expertise, and how this has been replicated
in his research projects as a way of coercing employees to follow the company vision. In addition,
Senge (1992) talks of the 'dialogue' that takes place
between employees as a way of sharing knowledge
within learning organisations. Wenger refers to
'shared repertoire' within communities of practice,
which includes language and stories, and forms one
of the building blocks upon which communities of
practice are based. The research data reinforces
this assertion. Sharing, talking, and discussing are
highlighted as being key components to volunteers'
learning.

The personal life experience of the volunteers
interviewed was identified as a significant part of the
learning progression. This compares favourably with
Wenger's (2000) idea that learning/knowledge is
made up of competence and personal experience.
Adding engagement to this creates the building
blocks for communities of practice. The teams at the
organisation displayed all three of these. Competence is exhibited by the volunteers' self-confidence

and positive attitude towards their skills and abilities
to perform the job; engagement is demonstrated by
their stress upon communication and active training
as a way of learning; and personal experience is
seen as important by the volunteers.

The main thread that emerged from the data was
that almost everything of consequence that the
interviewees talked about related to motivation.
That is, the motivation to learn and share knowledge
and understanding with others, and in a wider sense.

FUTURE TRENDS

These results reinforce the results of other studies
(Breu & Hemingway, 2002; Orlikowski, 2000;
Osborne, 1998; Davenport, 2001). Using GTM, the
findings provide an interesting verification of what
others have discovered in relation to learning and
sharing knowledge, but in another context that is in
a truly voluntary setting.

The findings of the research reinforce other
studies that have investigated knowledge sharing
practices amongst groups, in both the public and
private sector. The main results of the research
were that people (in this case, volunteers) are motivated to learn, communicate what they know with
others, and have a positive attitude towards their
organisation. There does not seem to be a huge
difference between how the volunteers operate and
communities of practice. The common link between
volunteers and employees both being motivated to
learn and share may be human nature or being part
of a community, rather than what sector they are in.

Being of service, affiliated to the organisation,
and self-awareness became key motivators for the
volunteers, which corresponds to research done in
1999 by Scott, Cox, and Dinham about English
school teachers. Scott et al. found that a sample of
English teachers were motivated by altruism, affiliation, and personal growth.

In addition, Breu and Hemingway (2002) found
that people are motivated to make social contributions, for example sharing knowledge. They did an
interpretative study of a community of practice in a
utilities company and found that the group was
bonded by choice, not forced by the organisation,
and that the members willingly shared what they
knew with others. This indicates that it may not be

the sector that is important, but the being part of a group and/or the freedom to choose whether to belong to that group. This reinforces Drucker's argument that in the knowledge economy we need volunteers, not conscripts, which is where this research project began (Snowden, 2000).

CONCLUSION

One of the main findings was that if someone with a natural disposition for learning is selected initially and given support from the organisation and their colleagues, then they will learn, remain motivated, and stay with the organisation. They will also then, voluntarily, provide support for others and continue to learn via this process. This feeds back into others in the organisation, who in turn support new volunteers, and so it perpetuates. This is a dynamic process, which fits in with Orlikowski's findings in 2002. Orlikowski found that "knowing is not a static embedded capability or stable disposition of actors, but rather an ongoing social accomplishment, constituted and reconstituted as actors engage the world in practice" (2000, p. 249). In other words, knowing and sharing are active processes. These results established that knowing, learning, and sharing are dynamic procedures; for example, most of those interviewed preferred 'hands-on' training, rather than more passive forms of learning, however the natural disposition of the actors was found to be as important.

If the right people are chosen in the first instance, if structures are in place to benefit communication between colleagues, and if there exists a culture of trust, then learning will take place and the knowledge base of the organisation will evolve.

REFERENCES

Breu, K., & Hemingway C. (2002). Collaborative processes and knowledge creation in communities-of-practice. *Creativity & Innovation Management 11*(3), 147-153.

Davenport, E. (2001). Knowledge management issues for online organisations: 'Communities of practice' as an exploratory framework. *Journal of Documentation, 57*(1), 61-75.

Davenport, T.H., & Prusak, L. (1998). *Working knowledge.* Boston: Harvard Business School Press.

Delanty, G. (2001). *Challenging knowledge: The university in the knowledge society.* Buckingham/Philadelphia: SRHE and Open University Press.

Drucker, P. (1969). *The age of discontinuity: Guidelines to our changing society.* London: Heinemann.

Foskett, D.J. (1990). *The communication chain. The information environment: A world view* (pp. 177-182). Amsterdam: Elsevier Science Publishers.

Glaser, B.G. (1978). *Theoretical sensitivity: Advances in the methodology of grounded theory.* Mill Valley, CA: Sociology Press.

Koenig, M. (1999). Education for knowledge management. *Information Services & Use, 19*(1), 17.

Lubit, R. (2001). Tacit knowledge and knowledge management: The keys to sustainable competitive advantage. *Organizational Dynamics, 29*(3), 164-179.

NCVO. (1996). *Meeting the challenge of change: Voluntary action into the 21st century. The report of the Commission on the Future of the Voluntary Sector.* London: Commission on the Future of the Voluntary Sector.

Nonaka, I.T. (1995). *The knowledge-creating company: How Japanese companies create the dynamics of innovation.* New York; Oxford: Oxford University Press.

Nonaka, I.T., & Nagata, A. (2000). A firm as a knowledge-creating entity: A new perspective on the theory of the firm. *Industrial & Corporate Change, 9*(1), 1.

Orange, G., Burke, A., & Boam, J. (2000). Organisational learning in the construction industry: A knowledge management approach. *Proceedings of ECIS 2000*, Vienna, Austria.

Orlikowski, W. (2000). Using technology and constituting structures: A practice lens for studying technology in organizations. *Organization Science, 11*(4), 404-427.

K

Orlikowski, W.J. (2002). Knowing in practice: Enacting a collective capability in distributed organizing. *Organization Science, 13*(3), 249-273.

Osbourne, S. (1998). *Managing in the voluntary sector: A handbook for managers in charitable and non-profit organisations*. London: Thomson Business Press.

Prusak, L. (2001). Where did knowledge management come from? *IBM Systems Journal, 40*(4), 1002-1007.

Scharmer, C.O. (2001). Self-transcending knowledge: Sensing and organising around emerging opportunities. *Journal of Knowledge Management, 5*(2), 137-150.

Scott, C., Cox, S., et al. (1999). The occupational motivation, satisfaction and health of English school teachers. *Educational Psychology, 19*(3), 287-308.

Senge, P.M. (1992). *The fifth discipline: The art and practice of the learning organization*. London: Century Business.

Snowden, D. (1999). The paradox of story: Simplicity and complexity in strategy. *Scenario and Strategy Planning, 1*(5), 16-20.

Snowden, D.J. (2000). Organic knowledge management: Part two—knowledge elicitation: Indirect knowledge discovery. *Inside Knowledge, 3*(9). Retrieved from *http://www.kmmagazine.com/*

Strauss, A. (1987). *Qualitative analysis for social scientists* Cambridge: Cambridge University Press.

Strauss, A.L., & Corbin, J.M. (1998). *Basics of qualitative research: Techniques and procedures for developing Grounded Theory*. Thousand Oaks, CA: Sage Publications.

Streatfield, D., & Wilson, T.D. (1999). Deconstructing 'knowledge management'. *Aslib Proceedings, 51*(3), 67-71.

Wenger, E.C. (2000). Communities of practice and social learning systems. *Organization, 7*(2), 225-246.

Wilson, T.D. (2002). The nonsense of 'knowledge management'. *Information Research, 8*(1). Retrieved from *http://informationr.net/ir/8-1/infres81.html*

KEY TERMS

Communities of Practice: Groups of people bound to one another through exposure to common problems, pursuit of common goals, employment of common practices. As a result they hold similar beliefs and values, and in themselves embody a common store of knowledge.

Knowledge: "A fluid mix of framed experience, values, contextual information, and expert insight that provides a framework for evaluating and incorporating new experiences and information. It originates and is applied in the minds of knowers. In organisations it often becomes embedded not only in documents or repositories, but also in organisational routines, processes, practices and norms" (Davenport & Prusak, 1998).

Knowledge Management: A combination of management awareness, attitudes, processes, and practices for creating, acquiring, capturing, sharing, and using knowledge to enhance learning and performance in organisations.

Voluntary Organisation: As defined by the Scottish Council for Voluntary Organisations, non-profit driven, non-statutory, autonomous, and run by individuals who do not get paid for running the organisation.

Volunteers: A person who performs or offers to perform a service on behalf of an organisation, cause, benefit, and so forth, out of his or her own free will, often without payment.

K

Knowledge Transfer within Interorganizational Networks

Jennifer Lewis Priestley
Kennesaw State University, USA

INTRODUCTION

Increasingly, knowledge is recognized as a critical asset, where a firm or an individual's competitive advantage flows from a unique knowledge base. The subsequent degree to which knowledge is then recognized and valued as a resource has been the theme of many papers on competitive advantage (Barney, 1991; D'Aveni, 1994; Nonaka & Teece, 2001; Prahalad & Hamel, 1990; Spender, 1996; Teece & Pisano, 1994). As a result, the ability to value and leverage external knowledge has become recognized as the basis of competitive advantage.

Gulati and Gargiulo (1999) suggest that membership in a networked community satisfies the need for knowledge as a way to help cope with environmental uncertainty. Consequently, inter-organizational networks or communities of practice represent a significant conduit for knowledge transfer to help manage this environmental uncertainty (Madhavan, Koka & Prescott, 1998).[1] Researchers in organizational learning have effectively concluded that organizations participating in a networked community will realize superior economic gains from their increased access to knowledge relative to independent or non-aligned firms (Argote, 1999; Baum & Ingram, 1998; Carlsson, 2002; Darr, Argote & Epple, 1995).

Although the implications of membership in a network having any structure versus no membership (and therefore no structure) are generally accepted, the implications of the different structural types that these networks can assume are less understood. Networks can accommodate, for example, different levels of competition, different degrees of centralization, and different operational objectives.

Knowledge may or may not transfer within different types of networked communities, raising an important question: Given that network membership is accepted as preferable for knowledge transfer relative to non-membership, does the specific network type in question have an effect on the degree to which knowledge will or will not transfer? This is the guiding research question of this article.

Prior to an exploration of this question, it should be noted that a multi-entity network (or community of practice) is very different from a dyad, and therefore represents unique challenges with respect to research. Unlike a dyadic relationship, networked communities can take on a life of their own that supersedes the presence of any individual member. Simmel (1950), who studied social relationships, found that social triads (and relationships involving more than three entities) had fundamentally different characteristics than did dyads. First there is no majority in a dyadic relationship—there is no peer pressure to conform. In any group of three or more people, an individual can be pressured by the others to suppress their individual interests for the interests of the larger group. Second, individuals have more bargaining power in a dyad. This is not only true because of percentages, but if one member withdraws from a dyad, the dyad disappears—this is not true in a networked community. Finally, third parties represent alternative and moderating perspectives when disagreements arise. As a result of these differences, multi-entity networks are more complex to study and less understood than dyads.

MAIN FOCUS: FACTORS OF KNOWLEDGE TRANSFER

A foundational concept from the Knowledge-Based View of the Firm is that within the context of knowledge management, knowledge is viewed as moving unencumbered by and transferring without cost within and among organizations (Grant, 1997; Kogut & Zander, 1992, 1996); although knowledge is recognized as an asset, unlike other assets its transferability has no associated costs. As von Hippel (1994) described, this may not be the case.

Knowledge has been described as a "sticky" asset that is costly to acquire and difficult to transfer between locations, even within the boundaries of a single firm. This stickiness is caused by, among other factors, the form of knowledge being transferred (Is the knowledge in question explicit? Or is it tacit?), as well as different attributes of the source(s) and the recipient(s), such as their situational absorptive capacity, their respective levels of causal ambiguity, and the degree of trust or motivation shared between the source and the recipient (Szulanski, 1996; von Hippel, 1994). I will refer to this last attribute as the source-recipient relationship. In this section, these three established factors of knowledge transfer—absorptive capacity, causal ambiguity, and the source-recipient relationship—are examined in terms of their effects on inter-organizational knowledge transfer.

Absorptive Capacity

Organizations must possess some degree of absorptive capacity to first recognize and then realize any value from the external knowledge to which it is exposed as a member of a network. The concept of absorptive capacity has received a significant amount of research attention since Cohen and Levinthal's seminal work on the topic (1990). Their definition of the concept is the most widely cited,

...the ability of a firm to recognize the value of new, external information, assimilate it and apply it to commercial ends is critical to its innovative capabilities. We label this capability as a firm's absorptive capacity. (p. 128)

In a networked context, the absorptive capacity of the recipient organization is integral to the success of the knowledge transfer process. In his work examining the effectiveness of inter-organizational alliances, Walker argues that firms that emphasize their relationships with other firms will be more successful, in large part because of their ability to recognize and apply new knowledge (1995). The ability to "sense" new external knowledge and have the processes in place to then bring it internal to the organization quickly becomes a competitive advantage when translated into economic rents. This "sensemaking" is a

critical function, which enables an organization to more effectively connect with its operating environment and allocate resources efficiently (Teece, 1998). Cohen and Levinthal (1990) and others (e.g., Lane & Lubatkin, 1998) suggested four types of commonalities, which represent the contributors to a recipient's overall level of absorptive capacity. These commonalities include language, knowledge base, process, and problem solving. If these commonalities are not present, absorptive capacity is considered to be low and knowledge transfer is less likely to occur.

Causal Ambiguity

The concept of causal ambiguity centers around "knowability" (the extent to which something *can* be known) and "knowness" (the extent to which something *is* known) of two sets of elements—(i) the organizational inputs and (ii) the causal factors that are used in combination to generate outcomes. Organizational inputs can be understood, for instance, as the raw materials used to manufacture a product, and the causal factors can be viewed as the processes used. When an organization does not know what combination of inputs and process factors cause the final outcome, its knowledge is, at best, causally ambiguous. For example, in the 1890s, Procter and Gamble had been manufacturing Ivory Soap (outcome) utilizing the same ingredients (inputs) and the same processes (causal factors). When an employee had inadvertently left one of the soap-making machines on during his lunch break, he returned to a frothy mixture unlike any soap mixture ever seen. Because none of the inputs had changed, P&G elected to package and distribute the soap as normal. Several months later, they were inundated with orders for the "floating soap". At this point, they were operating under causal ambiguity: having forgotten about the frothy accident several months before, they were unclear as to what input or causal factor could have generated the outcome of floating soap. Eventually the connection to the extra air in the soap-making process was discovered, and "It Floats" became an advertising slogan for Ivory Soap.

Causal ambiguity has been recognized as a factor in knowledge transfer difficulty across much of the research in organizational learning and knowledge management. For example, Wilcox-King and Zeithaml (2001) examined, in part, the tacitness of the knowledge in question as an indicator of causal ambiguity. Mosakowski (1997) developed a useful typology through which to examine the effects of causal ambiguity on the transfer of knowledge and decision making. Extending the work of Lippman and Rumelt (1982), Mosakowski determined that although increased causal ambiguity has the potential to increase competitive advantage by increasing the difficulties associated with imitation by competitors, increased causal ambiguity has the impact of decreasing knowledge transferability, and by association its application.

In the intraorganizational context, Szulanski (1996) found causal ambiguity to be an important contributor to knowledge transfer difficulty. Specifically, Szulanski identified "fundamentally irreducible" (or high) causal ambiguity as a factor in knowledge transfer failure.

Extending the logic of these findings, it can be stated that as causal ambiguity increases, the difficulty associated with knowledge transfer is also expected to increase.

Source-Recipient Relationship

There is uncertainty that exists within the context of the relationship between a knowledge source and a knowledge recipient. The basic premise here is that the knowledge recipient can put the received knowledge to more than one use. That is, the recipient can choose from multiple possible actions to follow once the knowledge has been received. Where the knowledge source can effectively reduce the potential action set of the recipient, uncertainty can be reduced. This reduction in uncertainty may occur as a result of contractual or legal obligations or through a mutually beneficial outcome. The alternative is also true—if the actions of the knowledge recipient cannot be known either because of lack of experience or threat of opportunistic behavior, the recipient action set is considered to be unbounded. In addition, the issue of network size is paradoxical—as the size of the network increases, the potential base of accessible knowledge increases. However, the decision to share knowledge becomes more complex because the knowledge source must consider multiple recipient action sets, translating into greater uncertainty and greater knowledge transfer difficulty. However, inter-organizational networks can mitigate the uncertainties related to initially

unbounded action sets through governance policies and controls. As will be seen below, some network forms can effectuate trust or eliminate the need of trust, and thus can help reduce uncertainties and reduce knowledge transfer difficulty. The greater the uncertainty associated within this relationship, the greater the negative impact to inter-organizational knowledge transfer (Szulanski, 1996).

To briefly summarize this section, three established factors of knowledge transfer have been explored: absorptive capacity is considered to have a negative relationship with knowledge transfer difficulty, while causal ambiguity and the uncertainty embedded within the source-recipient relationship are considered to have a positive relationship with knowledge transfer difficulty.

TYPES OF INTER-ORGANIZATIONAL NETWORKS

The basic concept of an inter-organizational network is generally understood and has received significant research attention. A simple network can be defined as "*nothing more (or less) than a system…consisting of objects and connections*" (Casti, 1995)—generally referred to as 'nodes' and 'linkages' in Social Network Theory. When addressing the concepts of inter-organizational communities of practice, a fundamental distinction should be made between durable, permanent "networks" and temporary "networks" (Westlund, 1999). My concern is with the former case, and conceptually I am more aligned with the characterism put forth by Johnson (1995) that networks are a static form of infrastructure, which support and constrain dynamic activity.

Researchers have studied inter-organizational networked communities from different vantage points. Thorelli (1986) and Almeida, Song, and Grant (2002) studied membership in a network from the perspective of Transaction Cost Economics as a strategy occupying the space between complete organizational self-sufficiency with no inter-organizational transactions and a complete outsourcing strategy with exclusively market-based transactions. Allee (2002) proposed the concept of a "value network" as the basis for understanding the activities related to the creation of intangibles such as knowledge. Carlsson developed a generalized framework of networks organized for the purposes of strategic knowledge management (2002).

Although these studies developed and defined networks, they did not examine the of knowledge transfer within different types of networked communities. Therefore, the question regarding *how* knowledge transfer difficulty, and its associated factors, varies with inter-organizational network type remains unanswered.

I approached the study of inter-organizational network communities from a slightly different perspective. Using the established foundational theories of Transaction Cost Economics, the Knowledge-Based View of the Firm, and Social Network Theory, I differentiate network types using three established characteristics—centrality of authority, scope of operations, and intensity of competition. I use these characteristics to create an abstract model of inter-organizational networks and then explain different inter-organizational networks that exist in practice.

Centrality of Authority

In an intra-organizational (and inter-personal) context, the two basic structures of centralization and decentralization have been studied extensively (e.g., Adler, 2001; Galbraith & Merrill, 1991; Van den Bosch, Volberda & de Boer, 1991; Volberda, 1998). Researchers have then looked to inter-organizational networks as an organizing principle residing between pure market transactions and complete organizational self-sufficiency. However, once *within* the community, the question of centralization remains, specifically to what extent governance and decision-making authority is centralized or decentralized.

Williamson (1973, 1975) describes a centralized structure as providing the authority to address issues related to opportunistic behavior, information impactedness, and bounded rationality. A (formal or informal) centralized authority would also have the ability to mandate standardization of operations, language, policies, and so forth. Alternatively, a decentralized structure is described as one of peer group associations, without subordination, involving collective and usually cooperative activities. This type of structure is deficient in its ability to address opportunism and "free-rider" abuses. However, recent researchers have found a decentralized struc-

ture to be particularly well adapted to facilitate innovation and new knowledge creation, without the encumbrance of the weight of a formal centralized hierarchy. Alternatively, the former structure has been found to better facilitate the diffusion and implementation of existing knowledge (Adler, 2001; Galbraith & Merrill, 1991; Van den Bosch et al., 1991; Volberda, 1998).

Scope of Operations

In this article, "scope" is defined as the degree of operational differentiation among the members of a community. Members of "narrow scope networks" engage in similar processes and exhibit some commonalities of knowledge base, process, language, problem solving, and so forth. On the other end of the continuum, the participating organizations within "broad scope networks" generally engage in very different types of business processes and often have different knowledge bases, use very different descriptive languages, and experience different types of problem-solving environments.

Intensity of Competition

Within the context of Social Network Theory, an important component of network structure that has been found to have significant impact on how well knowledge does or does not transfer is the ties or linkages among participants (Dacin, Ventresca & Beal, 1999; Granovetter, 1985; Uzzi & Lancaster, 2003). The linkages that exist among participants have been described as being either 'embedded' (integrated) or at 'arm's length' (Dacin et al., 1999). Integrated ties:

...are considered to create behavioral expectations that...shift the logic of opportunism to a logic of trustful cooperative behavior in a way that creates a...basis for knowledge transfer. (Uzzi & Lancaster, 2003, p. 384)

By contrast, linkages at 'arm's length' are:

...cool, impersonal, atomistic...motivated by instrumental profit seeking. (Uzzi & Lancaster, 2003, p. 384)

Although it may initially appear counterintuitive that organizations voluntarily join networks while maintaining 'arm's length' ties, consider VISA. Individual banks are fierce competitors, yet collectively benefit from the functionality of global payment card acceptance afforded by VISA—their relationships are "cool and impersonal," with linkages created for the purposes of decreased transaction costs. In addition, Powell, Koput, and Smith-Doerr (1996) found that as the technological sophistication of an industry increases, the intensity and number of competitive alliances also increases; although relationships are again "cool and impersonal," they come together to reduce the costs associated with R&D:

When there is a regime of rapid technological development, research breakthroughs are so broadly distributed that no single firm has all the internal capabilities necessary for success...Firms thus turn to collaboration to acquire resources and skills they cannot produce internally, when the hazards of cooperation can be held to a tolerable level. (1996, p. 117)

Using a more commonly accepted description of these integrated and arm's length linkages, I will refer to this characteristic as "intensity of competition" among the participants, with low intensity of competition equating to integrated linkages and high intensity of competition equating to arm's length linkages.

A graphical representation of these three network characteristics and the unique space occupied by four types of networks can be seen in Figure 1. These four types are explored.[2]

Three of the types identified in Figure 1 have already received some degree of research attention. These types include the *franchise network* (Argote, 1999; Darr et al., 1995; Thorelli, 1986), the *value chain network* (Dyer, 1997; Li, 2002; Thorelli, 1986), and the *innovation network* (Harris, Coles & Dickson, 2000; Harrison & Laberge, 2002). The value chain network type has been studied in three configurations, including dyadic (e.g., the manufacturer/supplier relationship), one-to-many relationships (e.g., the manufacturer with multiple suppliers), and N-to-N relationships (e.g., multiple manufacturers and multiple suppliers). A fourth type of inter-organizational network, the *co-opetive network*, has had the least amount of formal treatment in the literature. The term "co-

Figure 1. Types of inter-organizational networks

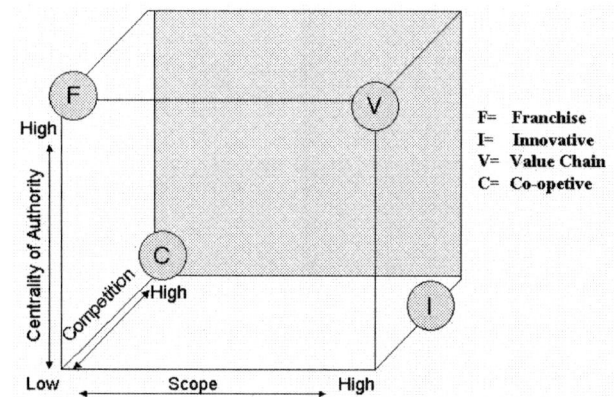

opetive" has been used to describe a situation where traditional competitors have agreed to cooperate to achieve a common objective (Brandenburger & Nalebuff, 1996; Loebecke, van Fenema & Powell, 1999; Shapiro & Varian, 1999). Using this accepted notion of "co-opetive," I extend this concept to define a co-opetive network as some formalized arrangement of N competitors collaborating to achieve some common objective.

A FRAMEWORK FOR CONSIDERATION

Based upon the discussion of knowledge transfer factors and the different types of networked communities above, I now put forth a framework to address the research question posed in the beginning of this article.

The model depicted in Figure 2 represents a framework to investigate how each of the four network types (franchise, innovation, value-chain, and co-opetive) affects the three factors of knowledge transfer difficulty (absorptive capacity, causal ambiguity, source-recipient relationship). As explored above, the respective relationships of each factor and knowledge transfer difficulty (the right side of the model) are relatively well established: an increase in absorptive capacity would be expected to lead to a decrease in knowledge transfer difficulty, an increase in causal ambiguity would be expected to lead to an increase in knowledge transfer difficulty, and more uncertainty in the source-recipient relationship would

Figure 2. Types of inter-organizational networks and factors of knowledge transfer

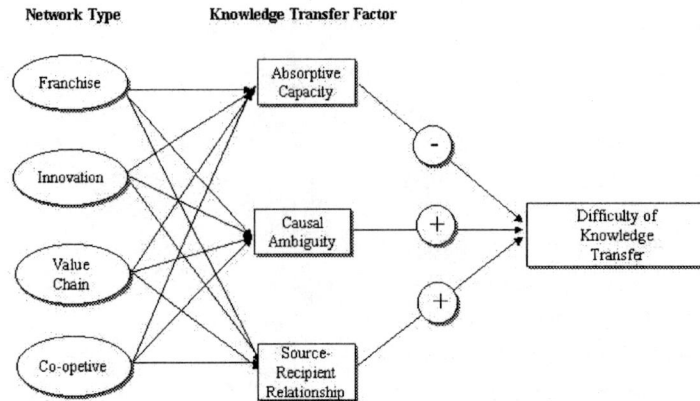

be expected to lead to an increase in knowledge transfer difficulty. It is the left side of the model or the relationships among the different network types and the factors of knowledge transfer that represent the greatest opportunity for further exploration.

Given the differences in the attributes of the networks—centralization, scope of operations, and intensity of competition—it would be reasonable to expect that each type would, in fact, experience each factor of knowledge transfer differently. For example, two of the characteristics of the franchise network are narrow operational scope and a hierarchical centralized structure. It could be argued that narrow scope provides a fertile environment for the four commonalities identified as necessary for absorptive capacity—language, knowledge base, process, and problem solving—while a strong hierarchical centralized structure has the ability to mandate standards governing, for example, service and quality. These standards would logically lead to the commonalities of process, knowledge, and language identified above—and therefore a high state of absorptive capacity. Similarly, the common processes that exist in a franchise community of practice would be expected to support a common knowledge of inputs and causal factors, both before and after outcomes associated with their use are known, thereby creating a low state of causal ambiguity. A related characteristic of causal ambiguity identified by Mosakowski (1997) is task complexity: the more complex tasks become, the more difficult it becomes to identify the specific cause and effect that each input or factor has on related outcomes. Where this complexity can be mitigated, causal ambiguity is

reduced. Simon (1962) determined that a strong, centralized/hierarchical structure can mitigate task complexity through specialization of labor and standardization. Given the expected hierarchical centralized structure of the franchise community of practice, complexity of task is expected to be low, and again thereby contribute to low causal ambiguity. Finally, the franchise network is characterized by a low intensity of competition amongst the network members. Franchisees are generally stakeholders within a larger entity—they are economically interdependent. Adler (2001) and Kogut and Zander (1996) refer to this interdependence as "shared destiny." Shared destiny would help to mitigate actions related to opportunistic behavior and contribute to a bounded recipient action set. Another characteristic of a franchise community of practice is limited organizational scope, evidenced in part by a commonality of operational processes, again decreasing the uncertainty related to the knowledge in question. In addition, the franchise community of practice is generally considered to have strong central governance. A hierarchical centralized structure would include an authority for punishment associated with opportunistic behavior amongst the franchises. Assuming this threat of punishment is severe enough to prevent defection, trust (or at least trust-like behaviors) could be mandated. As a result, the recipient action set would again be considered to be bounded.

Using this logic, the franchise network type would be expected to demonstrate a high level of absorptive capacity, a low level of causal ambiguity, and limited uncertainty, leading to a bounded recipient

Table 1. Proposed network type relationships

Factor/Network	Absorptive Capacity	Causal Ambiguity	Source-Recipient Relationship Uncertainty	Knowledge Transfer Difficulty
Franchise	High	Low	Low	Low
Innovation	Low	High	High	High
Value Chain	Low	Low	Low	Low-Med
Co-Opetive	High	High	High	Med-High

action set. Given the expected relationships between each of these factors and knowledge transfer difficulty, the franchise network would be expected to exhibit limited knowledge transfer difficulty. Using similar logic, the relationships between each network type and each factor of knowledge transfer difficulty could be reasoned. These proposed relationships are shown in Table 1.[3]

FUTURE IMPLICATIONS

Researchers have determined that the primary factors that affect knowledge transfer within and among firms include absorptive capacity, causal ambiguity, and the relationship between the source and the recipient (e.g., Cohen & Levinthal, 1990; Mosakowski, 1997; Szulanski, 1996). In addition, membership in some type of structured network has been generally accepted as superior to non-membership for the purposes of knowledge transfer (e.g., Argote, 1999; Darr et al., 1995; Powell et al., 1996). However, prior to this article, no researcher has attempted to frame or address the issues regarding how different network types affect these factors of knowledge transfer—and ultimately the transfer of knowledge itself—differently.

The theory developed in this article provides a framework through which to consider the differences that might exist among different network types regarding the transfer of knowledge. Understanding these differences, if they exist, would have implications for both practitioners as well as for researchers.

For practitioners currently operating within a networked community, this research study has several implications, including both descriptions of the phenomena as well as prescriptions for improvement. For example, if the states of the established

factors of knowledge transfer provided in Table 1 are proven to be true, then the strengths and weaknesses of different network types could be considered and addressed in terms of their ability to facilitate the transfer of knowledge. Those individuals or organizations currently operating within a particular network type would be better prepared to anticipate potential challenges to the transfer of knowledge among network members, and proactively allocate resources appropriately.

CONCLUSION

For those engaged in knowledge management research, this article provides two potential contributions. First, although each of the three factors of knowledge transfer identified in Figure 2 are well established as unilateral contributors to the transfer of knowledge, this article provides a framework through which to investigate the *relative* effects of all three factors on knowledge transfer difficulty simultaneously. Through this framework, it may be determined that causal ambiguity strongly influences the transfer of knowledge within one network type, but is dominated by absorptive capacity within the context of a second network type.

Second, Grant (1997) explains the role of strategic alliances in the Knowledge-Based View of the Firm, from the perspective of resource (knowledge) acquisition and utilization efficiencies within the boundaries of the firm versus outside of the firm—analogous to the foundations of Transaction Cost Economics. However, the Knowledge-Based View of the Firm is currently void of any specificity regarding the general forms that these alliances assume and how these forms then affect knowledge transfer. This work may provide a basis to frame this specificity, thereby contributing to, and possibly extending, this foundational theory of knowledge management.

REFERENCES

Adler, P. (2001). Market, hierarchy and trust: The knowledge economy and the future of capitalism. *Organization Science, 12*(2), 215-234.

Allee, V. (2002). Value networks and evolving business models for the knowledge economy. In C.W. Holsapple (Ed.), *Handbook on knowledge management* (pp. 605-621). Berlin: Springer-Verlag.

Almeida, P., Song, J., & Grant, R. (2002). Are firms superior to alliances and markets? An empirical test of cross-border knowledge building. *Organization Science, 13*(2), 147-162.

Argote, L. (1999). *Organizational learning. Creating, retaining and transferring knowledge.* Norwell, MA: Kluwer Academic Publishers.

Barney, J.B. (1991). Firm resources and sustained competitive advantage. *Journal of Management, 17*(1), 99-120.

Baum, J.A.C., & Ingram, P. (1998). Survival-enhancing learning in the Manhattan hotel industry 1898-1980. *Management Science, 44*(7), 996-1016.

Brandenburger, A.M., & Nalebuff, B.J. (1996). *Co-opetition.* New York: Doubleday.

Carlsson, S.A. (2002). Strategic knowledge managing within the context of networks. In C.W. Holsapple (Ed.), *Handbook on knowledge management* (pp. 623-650). Berlin: Springer-Verlag.

Casti, J. (1995). The theory of networks. In D. Batten, J. Casti & R. Thord (Eds.), *Networks in action. Communication, economics and human knowledge.* Berlin: Springer-Verlag.

Cohen, W.M., & Levinthal, D.A. (1990). Absorptive capacity: A new perspective on learning and innovation. *Administrative Science Quarterly, 35*(1), 128-153.

D'Aveni, R.A. (1994). *Hypercompetition.* New York: The Free Press.

Dacin, M., Ventresca, M., & Beal, B. (1999). The embeddedness of organizations: Dialogue and directions. *Journal of Management, 25*(3), 317-356.

Darr, E.P., Argote, L., & Epple, D. (1995). The acquisition, transfer and depreciation of knowledge in service organizations: Productivity in franchises. *Management Science, 41*(11), 1750-1762.

Dyer, J.H. (1997). Effective interfirm collaboration: How firms minimize transaction costs and maximize transaction value. *Strategic Management Journal, 18*(7), 535-556.

Galbraith, C.S., & Merrill, G.B. (1991). The effect of compensation program and structure on SBU competitive strategy: A study of technology-intensive firms. *Strategic Management Journal, 12*(5), 353-370.

Granovetter, M. (1985). Economic action and social structure: The problem of embeddedness. *The American Journal of Psychology, 91*(3), 481-510.

Grant, R.M. (1997). The knowledge-based view of the firm: Implications for management practice. *Long Range Planning, 30*(3), 450-454.

Gulati, R., & Gargiulo, M. (1999). Where do interorganizational networks come from? *American Journal of Sociology, 104*(5), 1439-1493.

Hamel, G., Doz, Y.L., & Prahalad, C.K. (1989). Collaborate with your competitors—and win. *Harvard Business Review, 67*(1), 133-139.

Harris, L., Coles, A.M., & Dickson, K. (2000). Building innovation networks: Issues of strategy and expertise. *Technology Analysis and Strategic Management, 12*(2), 229-238.

Harrison, D., & Laberge, M. (2002). Innovation, identities, and resistance: The social construction of an innovation network. *Journal of Management Studies, 39*(4), 497-518.

Ivory.com. Retrieved January 29, 2003, from *http://www.ivory.com/history.htm*

Johnson, J. (1995). The multidimensional networks of complex systems. In D. Batten, J. Casti & R. Thord (Eds.), *Networks in action. Communication, economics and human knowledge.* Berlin: Springer-Verlag.

K

Kogut, B., & Zander, U. (1992). Knowledge of the firm, combinative capabilities and the replication of technology. *Organization Science, 3*(3), 383-398.

Kogut, B., & Zander, U. (1996). What firms do? Coordination, identity and learning. *Organization Science, 7*(5), 502-519.

Lane, P.J., & Lubatkin, M. (1998). Relative absorptive capacity and inter-organizational learning. *Strategic Management Journal, 19*(5), 461-477.

Li, L. (2002). Information sharing in a supply chain with horizontal competition. *Management Science, 48*(9), 1196-1212.

Lippman, S.A., & Rumelt, R.P. (1982). Uncertain imitability: An analysis of interfirm differences in efficiency under competition. *Bell Journal of Economics, 13*(2), 418-439.

Loebecke, C., van Fenema, P.C., & Powell, P. (1999). Co-opetition and knowledge transfer. *The DATABASE for Advances in Information Systems, 30*(2), 14-25.

Madhavan, R., Koka, B.R., & Prescott, J.E. (1998). Networks in transition: How industry events (re)shape interfirm relationships. *Strategic Management Journal, 19*(5), 439-459.

Mosakowski, E. (1997). Strategy making under causal ambiguity: Conceptual issues and empirical evidence. *Organization Science, 8*(4), 414-442.

Powell, W., Koput, K., & Smith-Doerr, L. (1996). Inter-organizational collaboration and the locus of innovation: Networks of learning in biotechnology. *Administrative Science Quarterly, 41*(1), 116-146.

Prahalad, C.K., & Hamel, G. (1990). The core competence of the corporation. *Harvard Business Review, 68*(3), 79-92.

Priestley, J.L. (2004). *Inter-organizational knowledge transfer difficulty: The influence of organizational network type, absorptive capacity, causal ambiguity and outcome ambiguity.* Unpublished doctoral dissertation, Georgia State University, USA.

Shapiro, C., & Varian, H.R. (1999). *Information rules: A strategic guide to the network economy.* Boston: HBS Press.

Simmel, G. (1950). Individual and society. In K.H. Wolff (Ed.), *The sociology of Georg Simmel.* New York: The Free Press.

Simon, H. (1962). New developments in the theory of the firm. *American Economic Review, 52*(2), 1-16.

Spender, J.C. (1996). Making knowledge the basis of a dynamic theory of the firm. *Strategic Management Journal, 17*(Winter Special Issue), 45-62.

Stuart, T. (2000). Interorganizational alliances and the performance of firms: A study of growth and innovation rates in a high-technology industry. *Strategic Management Journal, 21*(8), 791-812.

Szulanski, G. (1996). Exploring internal stickiness: Impediments to the transfer of best practice within the firm. *Strategic Management Journal, 17*(Winter Special Issue), 27-43.

Teece, D.J., & Pisano, G. (1994). The dynamic capabilities of firms: An introduction. *Industrial and Corporate Change, 3*(3), 537-556.

Teecc, D.J. (1998). Capturing value from knowledge assets: The new economy, markets for know-how and intangible assets. *California Management Review, 40*(3), 55-79.

Teece, D.J. (2001). Strategies for managing knowledge assets: The role of firm structure and industrial context. In I. Nonaka & D.J. Teece (Eds.), *Managing industrial knowledge: Creation, transfer and utilization* (pp. 125-144). London: Sage Publications.

Thompson, J. (1967). *Organizations in action.* New York: McGraw-Hill.

Thorelli, H.B. (1986). Networks: Between markets and hierarchies. *Strategic Management Journal, 7*(1), 37-52.

Uzzi, B. (1996). The sources and consequences of embeddedness for the economic performance of organizations: The network effect. *American Sociological Review, 61*(4), 674-699.

Uzzi, B., & Lancaster, R. (2003). Relational embeddedness and learning: The case of bank loan managers and their clients. *Management Science, 49*(4), 383-400.

Van den Bosch, F., Volberda, H.W., & de Boer, M. (1991). Coevolution of firm absorptive capacity and knowledge environment: Organizational forms and combinative capabilities. *Organization Science, 10*(5), 551-568.

Volberda, H.W. (1998). *Building the flexible firm: How to remain competitive.* Oxford, UK: Oxford University Press.

von Hippel, E. (1994). "Sticky information" and the locus of problem solving: Implications for innovation. *Management Science, 40*(4), 429-438.

Walker, W. (1995). *Technological innovation, corporate R&D alliances and organizational learning.* Unpublished doctoral dissertation, RAND, Santa Monica, CA.

Westlund, H. (1999). An interaction-cost perspective on networks and territory. *The Annals of Regional Science, 33,* 93-121.

Wilcox-King, A., & Zeithaml, C. (2001). Competencies and firm performance: Examining the causal ambiguity paradox. *Strategic Management Journal, 22(1),* 75-99.

Williamson, O.E. (1973). Markets and hierarchies: Some elementary considerations. *American Economic Association, 63*(2), 316-325.

Williamson, O.E. (1975). *Markets and hierarchies: Analysis and antitrust implications.* New York: The Free Press.

KEY TERMS

Absorptive Capacity: The ability of a firm to recognize the value of new, external information, assimilate it, and apply it to commercial ends.

Causal Ambiguity: The "knowability" (the extent to which something *can* be known) and "knowness" (the extent to which something *is* known) of two sets of elements: (i) the organizational inputs, and (ii) the causal factors that are used in combination to generate outcomes.

Co-Opetive Network: A structured network of N organizations that are in simultaneous competition and cooperation (e.g., the VISA network). This network type is characterized by a decentral-

ized structure, high competition, and a common scope of operations among members.

Franchise Network: A structured network of N organizations sharing a common brand or public identity (e.g., Holiday Inns hotels). This network type is characterized by a centralized structure, low competition, and a common scope of operations among members.

Innovation Network: A structured network of N organizations sharing common goals related to research and/or development of new products/technologies (e.g., The Human Genome Project). This network type is characterized by a decentralized structure, low–medium competition and uncommon scope of operations among members.

Network: A community of practice of N organizations, where N is more than 2. In this article, network types are defined through three primary characteristics—the degree of centralization of authority, competition, and commonality of operations. An inter-organizational network is considered to be analogous to an inter-organizational community of practice.

Value Chain Network: A structured network of N organizations engaged in the manufacture/distribution/retail sales of a product (e.g., GM and its suppliers). This network type is characterized by a centralized structure, limited (vertical) competition, and uncommon scope of operations among members.

ENDNOTES

[1] For the remainder of this article, the term "network" will be used to designate an inter-organizational community of practice consisting of more than two organizations. Most of the logic used to develop arguments in this article are common to both inter-organizational as well as inter-personal communities of practice. Where specific logic or research applies to one or the other, these differences will be highlighted.

[2] The author acknowledges that while other configurations of inter-organizational networks may exist, these networks are common in practice and three configurations have already received significant research attention.

[3] A complete development of these propositions is included in Priestley (2004).

Leadership Issues in Communities of Practice

Iwan von Wartburg
University of Hamburg, Germany

Thorsten Teichert
University of Hamburg, Germany

INTRODUCTION

In another contribution in this encyclopedia, we presented the construct social structure as the context in which interactions between CoP members take place. Social structure has been defined along several dimensions, for example, group (CoP) longevity, norms for conflict resolution, and coordination of daily exchange. It has been disputed whether social structure can be deliberately influenced by management, that is, whether CoPs represent a social collective that is manageable to a degree as in, for example, a formal project team.

In this article, we argue that the social structure of CoPs can be influenced by using a certain leadership style as an influence tactic. We believe that for influencing the kind of social structure proposed for CoPs, transformational leadership is most suitable.

BACKGROUND

A prominent camp in leadership research differentiates between transformational and transactional leadership (Antonakis, Avorio & Sivasubramaniam, 2003; Avolio, Bass & Jung, 1999; Bass, 1985; Bass et al., 2003).

Transactional leadership denotes the situation where followers are rewarded contingent on the quality of carrying out their roles and assignments (Podsakoff, Todor & Skov, 1982). Transformational leadership, rooted in the notions of transforming leadership (Burns, 1978) and charismatic leadership (House, 1977), entails several dimensions derived by factor analyses, for example, inspirational motivation, individualized consideration, intellectual stimulation, and idealized influence. Roughly, it describes a situation where followers are intrinsically motivated to fulfill their role and tasks because they admire their leader for his or her personality or expert knowledge and share a strong identification with the proposed vision and mission.

We propose that transactional leadership is not suited to bring about knowledge creation in CoPs. The more active form of transactional leadership calls for closely monitoring deviances, mistakes, and errors in order to be able to take timely corrective action. However, this implies that explicit goals are formulated and a tight rule set for coordination is given. Without these explicit governing artifacts, a close monitoring is not possible. For CoPs, this is foreclosed by definition. A passive-avoidant form of transactional leadership called *laisser faire* is not beneficial for enabling learning processes in CoPs either. A laisser faire management *avoids* specifying any guideposts, that is, goals, expectations, and standards for interaction at all (Bass et al., 2003). This will leave CoP members without orientation and security about the support for and longevity of the learning collective.

Hence, we believe that transformational leadership (Antonakis et al., 2003; Avolio et al., 1999; Bass, 1985; Bass et al., 2003) fits most closely the characteristics of the proposed social structure in CoPs. First of all, transformational leadership ensures idealized influence, that is, being respected and trusted. Leading members of CoPs therefore adhere to norms and shared values in the CoP and thereby shape social structure. Second, transformational leadership brings about intellectual stimulation. Establishing CoPs that do not touch the very interests of potential contributors concerning intellectual stimulation will not lead to the emergence of social structure for relational social capital. Furthermore, since formal authority only coincidentally collapses with the kind of authority deployed for conflict resolution in CoPs, managers may create intellectual stimulation by explicitly and repeatedly claiming that conflicts have to be negotiated on the basis of rational arguments rooted in deep expertise. Third, transformational leaders pay attention to individual growth by acting as coach or mentor. This is called individualized consideration. The concept of LPP also mirrors this learning and growth motive. Fourth, by espousing enthusiasm and optimism about the content of innovative practice in a CoP, inspirational motivation can be fostered (Bass, 1985; Bass et al., 2003).

FUTURE DEVELOPMENT AND CONCLUSION

Managers seeking to implement CoPs for difficult and knowledge intense work processes may gain some advice from this discussion. They might find that establishing and enforcing some basic and ballparking rules may enable situated learning in actual practice more than devising precise decision cut-off criteria, standardized conflict resolutions processes based on a given distribution of authority, or predetermined internal transfer prices. Stated differently, managers who exert transactional leadership (Antonakis et al., 2003; Avolio et al., 1999; Bass et al., 2003) most likely will not stimulate knowledge creation processes in CoPs.

Although more research touching on this subject of leadership for successful learning dynamics in CoPs is needed, we make the claim that transformational leadership especially enables learning in actual innovative practice whereas both the active and passive-avoidant forms of transactional leadership will restrict the potential learning benefits during innovative practice offered by CoPs. By choosing a suitable leadership style, one will be able to meet objections stating that CoPs are emerging social phenomena which can by no means be designed or implemented in a deliberate manner.

REFERENCES

Antonakis, J., Avorio, B. J., & Sivasubramaniam, N. (2003). Context and leadership: An examination of the nine-factor full-range leadership theory using the multifactor leadership questionnaire. *The Leadership Quarterly*, *14*, 261-295.

Avolio, B. J., Bass, B. M., & Jung, D. I. (1999). Reexamining the components of transformational and transactional leadership using the multifactor leadership questionnaire. *Journal of Occupational and Organizational Psychology*, *7*, 441-462.

Bass, B. M. (1985). *Leadership and performance beyond expectations*. New York: Free Press.

Bass, B. M., Avolio, B. J., Jung, D. I., & Berson, Y. (2003). Predicting unit performance by assessing transformational and transactional leadership. *Journal of Applied Psychology*, *88*(2), 207-218.

Burns, J. M. (1978). *Leadership*. New York: Harper & Row.

House, R. J. (1977). A 1976 theory of leadership effectiveness. In J. G. Hunt & L. L. Larson (Eds.), *Leadership: The cutting edge* (pp. 189-207). Carbondale, IL: Southern Illinois Press.

Podsakoff, P. M., Todor, W. D., & Skov, R. (1982). Effect of leader contingent and non-contingent reward and punishment behaviors on subordinate performance and satisfaction. *Academy of Management Journal*, *25*, 810-821.

KEY TERMS

Leadership: Leadership denotes first a constellation of a person who is called a leader and other individuals who are called followers. Main research areas are first traits of the leading individual, context factors of the situation where leadership takes place, and third, different influence tactics employed by the leading person.

Transactional leadership: An influence tactic that involves (a) contingent reward for performance, (b) management by exception, that is, leaders monitor their followers' behavior and performance and take corrective action only if deviations from targets occur, and (c) *laisser-faire* leadership, that is, non-leadership behavior failing to fulfill leadership responsibilities like taking a stance in important questions and needs for assistance and feedback (Bass, 1985; Bass et al., 2003).

Transformational leadership: An influence tactic that involves (a) idealized influence (e.g., leaders

L

display conviction; emphasize trust; present their most important values), (b) inspirational motivation (e.g., leaders articulate an appealing vision of the future, talk optimistically and with enthusiasm), (c) intellectual stimulation (e.g., stimulate in others new perspectives and ways of doing things), and (d) individualized consideration (e.g., leaders deal with others as individuals and consider their individual needs) (Bass, 1985; Bass et al., 2003).

Leadership Issues within a Community of Practice

Barbara J. Cargill
Swinburne University of Technology, Australia

INTRODUCTION

Communities of practice are, and must be, fundamentally voluntary membership groups since they are about sharing of knowledge and expertise, something which cannot be effectively forced. Accordingly, no single person has positional leadership in a community of practice, as there is no formal structure in place to create such hierarchy.

However, wherever groups of people exist with any kind of shared task, there is leadership present, and leadership issues are repeatedly emerging aspects of the informal dynamics of that group or community of practice. It has often been noted that there actually is no such thing as a leaderless group. Informal leadership behaviours will come to the fore at certain points in any group's life, whether consciously evoked or not (Tyson, 1998).

Given this scenario, leadership exists in a community of practice (CoP) by informal agreement and negotiation. CoPs usually find it necessary to designate a leader for purposes of coordination and clarification, and possibly for direction of communications and to help structure the group interactions. Leaders are therefore created by the 'followers', and have only as much authority as the CoP group is willing to invest in the leadership role. Much research has been done into the psycho-dynamics of group relations, and it is often said that we place a little too much emphasis in our investigations and our speech on the phenomenon of leadership, when we also know that leaders, especially of voluntary groups, cannot function without followers who 'permit' the leader to act on their behalf. Leadership and 'followership' are thus flipsides of the same coin, and one cannot be understood without the other. Perhaps we should therefore focus on the needs of the followers to see what kind of leader will help the CoP serve its purposes (Long, 1992; Hirschhorn, 1991, 1997).

In the CoP, leaders will aid the workings of the community and therefore be granted limited authority by the group on the basis of:

- charismatic personality,
- superior expert knowledge,
- outstanding breadth of knowledge (not necessarily a specialist or expert, but holding some knowledge of a wide domain of interest to the CoP),
- high professional standing and reputation,
- high capacity to organise and mobilise the CoP (i.e., facilitation skills), or
- some combination of some or all of these aspects. (Tyson, 1998)

As with all voluntary groups, leaders with limited authority rely heavily on their capacity for positively influencing the work of the CoP and the interactions of its members. This influence takes the form of a number of leadership behaviours that are most likely to sustain the followers and keep the issue of leadership as a constructively assigned informal role. It is worth noting that leadership in this sense is a series of *functions*, and can be shared by more than one person. However, in order to avoid confused communications and expectations, a designated leader is normally more able to productively assign and direct other contributions of a leadership nature, setting in place a negotiated sharing of the role and functions.

The leadership functions of highest value to the CoP will be:

- balancing of members' interests and articulation of agenda items for the CoP, including identifying priority rankings on certain issues;
- attending to inclusiveness of the CoP, actively working at drawing in contributions from all members;

- facilitating interactions of the CoP, chairing, clarifying, encouraging, summarising, challenging at times, and articulating the issues the CoP may be struggling to articulate;
- especially encouraging a culture of egalitarianism and respect for 'junior partners', so that neither experience nor strong personality dominates the CoP.

These leadership behaviours relate to both the *task* elements of the CoP and the *relationship* elements (Hersey, Blanchard & Johnson, 2000), as they are described in the Situational Leadership literature. Task elements are more to do with getting the work of the CoP underway and done, and often require a degree of control and directiveness. Relationship elements are more to do with drawing the group together and maintaining its working relationships and sense of connectedness (to each other and to the CoP's tasks). Noticeably, the actions for leaders of CoPs are more to do with the relationship axis, except where some particular crisis might arise which demands urgent action and interaction from the community, in which rare case a more directive leadership style will work best for a short time.

Overall, leaders of communities of practice need to be prepared to renegotiate their leadership status frequently ("If you would prefer that I not operate as leader, that's OK with me, and we should decide to choose someone else now"), always recognising that they have no power other than that which the CoP members voluntarily surrender upwards to them. This kind of power base does not suit individuals who like clear positional authority, but favours those with high tolerance for ambiguity and strong 'people' skills. These are most likely to be able to draw a CoP together and hold it so for a sufficient period of time for it to operate as a sharing community.

REFERENCES

Hersey, P., Blanchard, K.H., & Johnson, D.E. (2000). *Management of organizational behaviour: Leading human resources* (8th ed.). Upper Saddle River, NJ: Prentice-Hall.

Hirschhorn, L. (1991). *Managing in the new team environment: Skills, tools and methods.* Reading, MA: Addison-Wesley.

Hirschhorn, L. (1997). *Reworking authority; Leading and following in the post-modern organization.* Cambridge, MA: MIT Press.

Long, S. (1992). *A structural analysis of small groups.* London: Routledge.

Tyson, T. (1998). *Working with groups* (2nd ed.). South Yarra, Australia: Macmillan.

KEY TERMS

Facilitation: The leadership contributions of structuring, enabling and encouraging good group interaction and process, normally with low reliance on positional authority and content expertise, but high reliance on communication and interpersonal skills and presence.

Formal Leader: One who is formally appointed or authorized by either the group or an external party to hold an officially designated leadership role for the group or team. Such a leader may or may not display strong leadership behaviour.

Informal Leader: One who demonstrates substantial leadership behaviour even though not officially appointed or designated to hold such authority. Informal leaders often emerge tacitly from within a group.

Leadership: The dynamic phenomenon in group life where an individual or individuals consciously or unconsciously influence and direct the activities and interactions of that group. Leadership may be formally constructed and embodied within an organisational role, or emerge informally from the interactions of the group by tacit processes.

Leadership Behaviours: The various actions that leaders contribute to group and team life that may be generally categorised as *task related* or *relationship oriented*. *Task behaviours* aim to structure, direct and progress the work of the group. *Relationship behaviours* aim to address the way the group members feel about and interact with each other, with the leader, and with the task they are tackling. The holding and communicating of a vision for the group is also generally acknowledged as a leadership behaviour that is both task related and relationship oriented.

Psychodynamics: The highly dynamic unconscious forces within individuals and between individuals that form and motivate outward behavioural manifestations in group life. The group-as-a-whole may develop a collective psychology that mirrors the psychological processes of an individual in many respects, but not that of any one of its members.

Life Cycle of Communities of Practice

Deepa Ray
Oklahoma State University, USA

INTRODUCTION

Today more and more companies are realizing the value of knowledge, and the benefits of capturing and leveraging it among all employees. Knowledge in the form of data, symbols, facts, and figures has been captured, but knowledge that is tacit (implicit) still continues to pose a challenge. How does such knowledge exist and where exactly does it thrive? The answer to that is "people"—the soft aspect of a company's asset. Thus intellectual capital (IC), which represents the human intelligence asset of a company, is where both tacit and explicit knowledge reside.

With technology advancing to connect different people across the world, support groups, developer forums, and message lists are probably the most immediate resources that professionals look to for knowledge or solutions to issues at work. This could very well be the first step a person takes to be a part of a community of practice (CoP). Interactions, discussions, exchange of ideas, and solving each other's problems is in itself a source of knowledge, and although no attempt is made to hold onto such knowledge or guard it as a secret, the wealth of information remains privy to the community that shares it.

Thus we can see that the continuum of informal discussions to a structured process of knowledge sharing can be represented by different stages in a lifecycle of a CoP. This article is an attempt to look at the lifecycle of a CoP, not just in terms of knowledge creation at each of its stages, but also as an example of how social networks are born and how they thrive. Understanding the lifecycle of CoPs will give greater insight into the knowledge sharing process resulting in more companies recognizing the importance of CoPs.

BACKGROUND

Communities of Practice: A Definition

A community of practice is formed when individuals with common interest (shared goals) come together on a mutual basis. Wenger and Snyder (2000) define CoPs as a group of people informally bound together by shared expertise and passion for a joint enterprise. Thus when people find ways to relate to each other by participating in a knowledge flow process, they form a CoP. CoPs can be formed across functional units, organizations, and even nations. In effect, a CoP succeeds in eliminating the creation of knowledge silos formed due to the "protect your own information" attitude of many organizations.

THE LIFECYCLE OF A COMMUNITY OF PRACTICE

As in any other entity, a CoP also goes through a lifecycle process. Wenger (1998) described the following stages in the lifecycle of a CoP:

1. **Potential:** At this stage, people face similar situations but have not yet formed a shared practice.
2. **Coalescing:** At this point, members have interacted and found one common emerging point and its potential.
3. **Maturing:** CoP sets standards, defines agenda, and develops relationships.
4. **Active:** At this stage, the community formed is most productive. Members develop shared practices.
5. **Dispersed:** CoP is no longer active, functions more as a repository of knowledge.

Since a CoP is formed when people with common interests come together, we can look at a CoP as an evolving social network. The stages that mark the lifecycle of CoP when it forms as a social network are:

1. **Scattered:** Individuals that are grappling with similar problems, in search of similar information.

2. **Informal Group:** One has informal contacts that can help with the current problem at hand. Interactions result in informal support groups that one turns to with questions.

3. **Community:** The focus of the informal groups becomes clearer as members come together with similar pursuits to define the core of the community.

4. **Decline:** Maturity of knowledge results in expansion of the core up to a maximum limit. Knowledge still exists, but the focus has fully developed and no more refinement is possible.

5. **Death:** Focus is no longer important or relevant to its members, resulting in a steady decrease in interactions between members. One must understand that death of a community of practice does not signify end of knowledge. Knowledge still exists, however the community that developed it has moved on.

According to Gongula and Rizzuto's (2001) community evolution model, the following stages form the lifecycle of CoP:

1. **Potential:** Where a community is forming.
2. **Building:** Where the community defines itself.
3. **Engaged:** Where community executes and improves its processes.
4. **Active:** Where the community understands and demonstrates benefits from knowledge management.
5. **Adaptive:** Community uses knowledge for competitive advantage.

Thus we can see that CoPs basically start off with the mere process of asking someone for help. As more and more people face similar problems, the need to codify the knowledge emerges. Groups of people with similar issues (for example, it could be developers using a particular platform or people with similar

research interests) start to come together to form what is then called communities of practice. One must however keep in mind that knowledge cannot cease to exist. Thus CoPs, which are basically containers of knowledge, cannot actually die out. What could happen though would be that the relevance of CoPs or the topic at the core of a CoP could face a decline.

Lifecycle of CoP and Knowledge Creation

Most organizations get involved in the process of knowledge management to document or codify ideas, information, or data to facilitate sharing and availability of information to all members of the organization. Thus knowledge management is basically about organizations trying to leverage their tangible and intangible assets.

CoPs act as containers of tacit as well as explicit knowledge. What distinguishes CoPs from other knowledge management techniques is the fact that CoPs are "live" containers of knowledge for that particular domain. It is "live," meaning that CoPs keep refining the main core of the domain. Thus knowledge that exists is not stagnant or just a repository, but is being constantly created by means of community members' interaction.

During the initial stages of formation of a CoP, much that is shared by the members is random and not cohesive in nature. Here the knowledge workers are searching for solutions to their individual problems. Thus "knowledge" is shared as needed by individuals. However once a CoP is formed and its core is defined, members engage in defining the practices and shared understanding of the domain of CoP. This in terms of the lifecycle of a CoP as a social network marks the formation of a knowledge network. It is at this stage that an attempt is made to formalize sharing of the knowledge by means of discussions and exchange of ideas between CoP members. Thus tacit to explicit knowledge conversion, as well as greater knowledge creation, marks the active stages of the lifecycle of a CoP.

Also, CoP members are not exclusive to their particular community. They might be members of other kinds of groups within the organization (for example: project teams, functional group, etc.). In fact, one of the most emphasized issues in the CoP literature is the existence of double-knit organizations, where people belong to more than one group at a given time.

As mentioned by Smits and Moor (2004):

The knowledge creation process is continuous and expanding: as the community matures, it accumulates and applies knowledge, resulting in an internal learning process. (p. 2)

This is why CoPs are examples of self-organized knowledge management practices in an organization.

FUTURE TRENDS: ISSUES IMPORTANT TO THE LIFECYCLE OF A CoP

The mere existence of a CoP does not ensure knowledge management or sharing. Companies need to understand the importance of CoPs that exist in their organization and make efforts to help foster CoPs to gain competitive advantage by using the knowledge wealth in a CoP. The following are some issues raised concerning the lifecycle of a CoP:

*   Lee and Valderrama (2003) talk about a leader that supports the CoP formation through a knowledge management plan. To create value for the organization, they mention three aspects of this plan. First, the formation of the CoP needs to align with strategic and tactical goals of the organization. Second, the important members in the domain need to support and advocate the CoP and its value. Finally, the long-term vision of the CoP should be documented such that it mentions the value proposition of the CoP.
*   In their article, LaContora and Mendonca (2003) state that the four critical success factors for sustaining a CoP are social structure, purpose, learning environment, and knowledge sharing.
*   In order to gain competitive advantage, a CoP needs to focus on learning. In his article, Schon (1983) talks about how, in the dynamic marketplace of today, it is important to "reflect in action" what one knows or has learned. Thus the emphasis falls on the learning that takes place in a CoP.

CONCLUSION

CoPs thus act as a mechanism of exchanging information and learning which in turn makes the organization members more competent and gives the organization a competitive advantage. Tacit knowledge is more easily shared and more importantly is not dormant. Discussion of the lifecycle of a CoP gives us insight into this knowledge conversion as well as sharing process. CoPs use this knowledge as a fuel for the learning cycle that generates more knowledge through medium of exchange and discussion.

CoPs can become good support systems for any organization for two reasons. First, belonging to a CoP would give every member a support group to turn to in case of difficulty, and second, a community instills a feeling of trust, thus enabling better flow of knowledge resulting in community benefit. Many organizations are actively pursuing different knowledge management techniques to gain competitive advantage. Understanding the different stages of the lifecycle of a CoP can help these organizations recognize the needs of a CoP and thus provide the necessary support resulting in thriving communities.

REFERENCES

Gongula, P. & Rizzuto, C.R. (2001). Evolving communities of practice: IBM Global Services experience. *IBM Systems Journal, 40*(4), 842-862.

LaContora, J.M. & Mendonca, D.J. (2003, August 11-13). Communities of practice as learning and performance support systems. *Proceedings of the International Conference on Information Technology: Research and Education.*

Lee, S.J. & Valderrama, K. (2003). Building successful communities of practice. *Information Outlook, 7,* 28.

Schon, D.A. (1983). *The reflective practitioner.* New York: Basic Books.

Smits, M. & Moor, A.D. (2004). Measuring knowledge management effectiveness in communities of practice. *Proceedings of the 37th Hawaii International Conference on System Sciences.*

Wenger, E. (1998, June). Communities of practice learning as a social system. *Systems Thinker,* 1-10.

Wenger, E., & Snyder, W. M. (2000). Communities of Practice: The Organizational Frontier. *Harvard Business Review, 78*(1), 139-145.

KEY TERMS

Core: The main domain (focus) of a particular community pf practice. It is the common interest that is the driving force behind the formation of a CoP.

Double-Knit Organizations: Organizations where people work in teams for projects, but importantly also belong to a much more enduring and lasting community of practice in order to keep their skills sharp.

Explicit Knowledge: Also known as information, it is that knowledge that is adequately and properly represented by facts, figures, symbols, and data.

Intellectual Capital: The human intelligence asset that belongs to a company. It consists of people, the skills, values, learning, and knowledge that they bring to the organization.

Knowledge Networks: Common name for communities of practice, basically referring to an informal network of professionals belonging to the different domains of interest for the communities of practice.

Knowledge Workers: People who look to different repositories or wealth of knowledge to find new solutions to challenges faced in a dynamic and changing work environment.

Tacit Knowledge: Knowledge that has not been explicitly codified. It resides in people's minds as an understood thing. Mostly ingrained in experience and learning of a person.

The Limits of Communities of Practice

Chris Kimble
University of York, UK

Paul Hildreth
K-Now International Ltd., UK

INTRODUCTION

When knowledge management (KM) began to emerge in the late 1980s, it was seen as an innovative solution to the problems of managing knowledge in a competitive and increasingly internationalized business environment. At that time, the term was often used in conjunction with so-called expert systems that dealt with hard[1], structured knowledge (Hildreth, Wright & Kimble, 1999). During this period, knowledge was seen as something that had an independent existence; it could be captured from an expert, codified in a series of rules, and stored in a computer. However, many authors have argued that, in practice, KM was often little more than information management systems rebadged (Wilson, 2002).

More recently, there has begun to be recognition of the importance of softer, less structured types of knowledge (Hildreth, Wright & Kimble, 1999). There has been a growing awareness that knowledge is not found in rules, frames, cases, predicate logic, or document repositories but that other factors were at work. This inevitably raises questions about what these other factors are and how this new softer form of knowledge might be managed.

Communities of practice (CoPs) were identified by many as a means by which this softer type of knowledge could be created, shared, and sustained. From this, it was a small step to arguing that CoPs were in fact a new approach to KM that offered the solution to many of the shortcomings of the earlier, systems based attempts at KM. However, the concept of a CoP is built around a very different set of principles to those put forward by the proponents of KM, and it is not always clear that this argument will hold.

Much of what is now called KM has developed in a formal organization setting. In this setting, groups are often seen simply as collections of people who are brought together to complete a specific task; once the task has been completed, the group can be dissolved. These groups are often created in a top down fashion, and the structure of the group usually reflects the existing organizational hierarchy. The successful completion of the task (or repeated series of tasks) is usually the basis for financial or other reward. In contrast, CoPs tend to be self-perpetuating and self-directed. The focus of a CoP is not on a narrowly bounded task but on a living and dynamic practice; the rewards are intrinsic rather than financial. Authority and legitimacy are not a function of formal rank or hierarchy but of an informal status in the group. In summary, the members of a CoP have more in common with a troop of altruistic volunteers than a band of paid employees.

This contrast between the nature of CoPs and the demands of a high tech, global commercial enterprise raises two important questions that we will return to in the Communities of Practice Today section. First, do CoPs really offer a way to *manage* the softer aspects of knowledge? That is to say, can they be initiated and directed by management, or will the outcome always be the product of the emergent properties of a self-directed and self-organized group? Following on from this, the second question is: if they do offer ways to manage the softer aspects of knowledge, will they work in today's high tech and increasingly internationalized virtual world?

BACKGROUND: COMMUNITIES OF PRACTICE – A HISTORICAL PERSPECTIVE

When the term *communities of practice* was first used, it was used in relation to situated learning rather than knowledge management. The term was coined in 1991 when Jean Lave and Etienne Wenger (1991) used it in their exploration of the activities of groups of non-drinking alcoholics, quartermasters, butchers, tailors in Goa, and midwives in the Yucatan. What linked these diverse groups was a mode of learning based on what might broadly be termed an apprenticeship model, although the concept of CoPs is not restricted to this form of learning.

Lave and Wenger (1991) saw the acquisition of knowledge as a social process in which people participated in communal learning at different levels depending on their authority in a group, that is, whether they were a newcomer to the group or had been an active participant for some time. The process by which a newcomer learns from the rest of the group was central to their notion of a CoP; they termed this process Legitimate Peripheral Participation (LPP). However, LPP is more than simply learning situated in a practice; it is learning as an integral part of a practice that give meaning to the world: learning as "generative social practice in the lived in world" (Lave & Wenger, 1991, p. 35).

LPP is both complex and composite; legitimation, peripherality, and participation each play a part in defining the other. Legitimation is concerned with power and authority relations in the community but is not necessarily formalized. Peripherality is not a physical concept or a measure of acquired knowledge, but concerned with the degree of engagement with the community. Participation is engagement in an activity where the participants have a shared understanding of what it means in their lives. Taken separately, each has no meaning, but taken together, they form the central thread of a CoP activity.

For Lave and Wenger (1991), the community and participation in it were inseparable from the practice. Being a member of a CoP implied participation in an activity where participants have a common understanding about what was being done, what it means for their lives, and what it means for the community. Thus, it would appear that CoPs with their concentration on situated learning and shared understanding might be well suited to the management of the softer aspects of knowledge, but can this idea be applied to the business world?

COMMUNITIES OF PRACTICE TODAY

Interest in CoPs continued to grow throughout the 1990s, and several attempts were made to redefine Lave and Wenger's (1991) original model. In particular, several attempts were made to redefine CoPs in a way that was more relevant to the commercial environment (e.g., Brown & Duguid, 1991, 1996). One of the most widely cited, business related definitions of a CoP was offered by John Seely Brown and Estee Solomon Gray in their 1995 article called "The People Are the Company":

At the simplest level, they are a small group of people...who've worked together over a period of time. Not a team not a task force not necessarily an authorised or identified group...they are peers in the execution of "real work". What holds them together is a common sense of purpose and a real need to know what each other knows.

The main surge in interest in CoPs and business came in 1998, when Wenger (1998) published the results of a ground breaking ethnographic study of a claims processing unit of a large insurance company. In this study, he argued that CoPs were formed through mutual engagement in a joint enterprise and that these CoPs exploited a shared repertoire of common resources (e.g., routines, procedures, artifacts, vocabulary). His argument was that the CoPs he studied (1) arose out of the need to accomplish particular tasks in the organization and (2) provided learning avenues within, between, and outside that organization. Thus, his view of the business was not of a single monolithic community, but a constellation of interrelated CoPs that can even spread beyond the borders of the host organization.

The original description of CoPs as isolated groups based on LPP was now replaced by a different view. According to Wenger (1998), a CoP could now be defined in terms of three constructs.

What it is About

The focus of the CoP is a particular area of activity or body of knowledge around which it has organized itself. This is a joint enterprise in as much as it is based on a common or shared understanding that is continually renegotiated by its members.

How it Functions

People become members of a CoP through shared practices, and they are linked to each other through their involvement in common activities. It is this mutual engagement that binds the members of a CoP together in a single social entity.

What it Produces

The members of a CoP build up a shared repertoire of communal resources over time. Written files are a tangible example of this, although less tangible examples such as procedures, policies, rituals, and idioms may also form part of the repertoire.

The next step of linking CoPs to KM and the world of business came from the way in which Wenger describes the underlying processes that are at work in a CoP.

Linking Communities of Practice and Knowledge Management

In an earlier paper (Hildreth, Wright & Kimble, 1999), we argued that the various different approaches to KM often viewed knowledge in terms of mutually exclusive opposites. We used the terms hard and soft knowledge to describe these two opposites and argued that too often KM emphasized hard knowledge over soft. Our intention was not to add to the plethora of terms already used to describe knowledge but to attempt to bundle together a range of views so that the issues could be discussed without becoming too tied to a particular, pre-existing viewpoint.

We described hard knowledge as being unambiguous and unequivocal; it is something that can be clearly and fully expressed; it can be formalized, structured, and owned without being used. Hard knowledge is both abstract and static: it is about the world, but not in it. In contrast, soft knowledge is implicit and

unstructured. It is the sort of knowledge that cannot be easily articulated, although it can be understood even if it is not openly expressed. It is often knowledge that is associated with action; it can not be possessed; it is about what we do and can only be acquired through experience.

More recently, we argued (Hildreth & Kimble, 2002) that the underlying problem of KM was not simply that it privileged one form of knowledge over another; it was that KM had failed to recognize that knowledge itself was a duality consisting simultaneously of both hard and soft knowledge. Drawing on the Chinese concepts of Yin and Yang—a perspective of balance and continual change—we argued that hard and soft knowledge were not mutually exclusive but mutually dependant; one could not exist without the other.

Knowledge is not made up of opposites; regarding knowledge in these terms is a false dichotomy. Rather than seeing knowledge as opposites, perhaps we should think of it as consisting of two complementary facets: a duality consisting simultaneously and inextricably of both [hard and soft] knowledge.

The use of the device of a duality is not new (see, for example, Orlikowski, 1992); however, viewing knowledge in this way does allow us to make a conceptual link between KM and CoPs.

In his work with CoPs at the insurance company, Wenger (1998) identified two key processes that formed a duality: participation and reification. He described participation as "the social experience of living in the world in terms of membership in social communities and active involvement in social enterprises" (p. 55) and reification as "the process of giving form to our experience by producing objects that congeal this experience into thingness" (p. 58).

Wenger emphasizes that, like LPP, participation and reification are analytically separable but are inseparable in reality. Participation is the process through which people become active participants in the practice of a community, and reification gives concrete form to the community's experience by producing artifacts.

With these concepts in place, CoPs can now be seen as a way to manage knowledge. In their day-to-day work, people can both negotiate meaning through participation in shared activities and project that meaning onto the external world through the production of artifacts. Wenger's (1998) work with CoPs claimed to show that not only could CoPs exist in a business setting, but that the concept of a CoP could be applied to the management of knowledge in such settings.

Since then, several other authors have taken this idea and sought to identify specific quantifiable business benefits that can be associated with CoPs (e.g., Fontaine & Millen, 2004; Lesser & Storck, 2001). However, one problem remains: almost all of the previous work on CoPs has been based on collocated CoPs. With the increasing globalization of business and the heavy reliance on information and communication technology (ICT), the question of whether CoPs become virtual remains unanswered.

COMMUNITIES OF PRACTICE: GOING ONE STEP TOO FAR?

Having now examined the background to the use of CoPs to manage knowledge in a commercial setting, we will now, as indicated in the introduction, address two main questions: Are CoPs really applicable to a business environment? Can a CoP ever be truly virtual? To answer these questions, we will mostly draw on material from a series of studies in a recently published book (Hildreth & Kimble, 2004).

Given that much of the work quoted in the previous section seems to be related to CoPs in a business setting, this first question might seem rather strange. However, while there is little doubt that CoPs exist in industry and even some evidence that CoPs can add value to a business, this is not the same as asking: Are CoPs really suitable for use in a business setting? The aim of this article is to offer a critical view of CoPs; it is our belief that until now too much emphasis has been placed upon identifying the real and potential business benefits of CoPs and too little on identifying the potential costs and disadvantages. This is not to say that we believe that CoPs cannot be of benefit to a businesses but simply that without an understanding of the limitations of CoPs, their true value to the world of business and commerce will not be fully understood.

CoPs IN THE BUSINESS ENVIRONMENT

In the introduction of this article, we briefly outlined the tension between the way in which most business organizations view a team or a task group and the way in which CoPs view themselves. Most formal organizations view groups as project teams or task groups: a group of people that can be brought together and controlled by the larger organization, a group that exists solely for the benefit of the organization. CoPs, on the other hand, are self-directed and self-motivated entities; the engine that drives a CoP is the shared interests of its members, which may not be the same as the interest of the wider organization.

In their study of communities of practice that disappear, Gongla and Rizzuto (2004) provide several interesting examples of how this tension resolves itself in IBM. They identify four common patterns of disappearance—CoPs that drift into non-existence, CoPs that redefine themselves, CoPs that merge with others, and CoPs that become formal organizational units. For example, they note that if an organization spotlights a CoP and tries to manage too much of what it is and what it does, "the community may remove itself completely from the organisational radar screen [the members] may remove it from the organisational spotlight by pretending to disperse, but in reality continuing to function outside of the organisations purview" (p. 299).

If the organization has become reliant on the work of the CoP, this could be a serious problem. If this is the case, then frequently the last of the list of reasons given above for CoPs disappearing will come into play, and the CoP will be taken over and become a formal organizational unit, an outcome that results in the loss of many of the supposed advantages of a CoP.

Much play is also made of Wenger's (1998) view of a business being a collection of interrelated CoPs that provide avenues for learning both within and beyond the boundaries of the organization. Again, while it is undoubtedly true that CoPs can allow the sharing of knowledge between different groups, the capricious nature of CoPs means that this particular outcome cannot be guaranteed. Hislop (2004) examined three case studies of CoPs in large European organizations and concluded that only one was successful in sharing knowledge between communities; the other two failed to do so because of a lack of shared identity and a lack of consensual knowledge. He argues that because of a strong internal sense of identity, CoPs can actually lead to less knowledge sharing between communities rather than more. Similarly, Vaast (2004), in her four case studies of public and private sector organizations in France, noted in one case the strengthening of the internal sense of identity within a CoP resulted in a group of employees outside the CoP becoming marginalized.

The conclusion from these studies seems to be that CoPs as self-managing and self-directed entities may be of value to a business organization, but precisely because they are self-managing and self-directed, their contribution to the organization will always be uncertain. In this sense, the role that CoPs can play in core business activities must always remain peripheral.

CoPs IN THE VIRTUAL ENVIRONMENT

Internet-based networking technologies, which can provide a convenient single platform for groups or networks of groups to form within larger organizations, have led to a proliferation of various forms of virtual groups and communities. Subsequently, there has been much discussion about whether these virtual groups are CoPs or some other form of group.

Lueg (2000) draws a distinction between virtual and distributed CoPs based on what he claims are two salient features of a CoP: where the learning takes place and where the action takes place. He concludes that CoPs are deeply rooted in the lived in world and that moving CoPs to the virtual world raises some significant conceptual problems.

Rather than attempting to deal with virtual CoPs, Brown and Duguid (2000) coined the phrase "Networks of Practice" (NoPs) to describe groups of people who are geographically separate and may never get to know each other personally but share similar work or interests. Thus, NoPs share many of the features of CoPs but are organized at a more individual level than CoPs and are based on personal rather than communal social networks.

In a study of job seeking activity, Granovetter (1973) introduced the notion of strong and weak social ties. In terms of the above description, CoPs are characterized by strong social ties whereas NoPs are characterized by weak social ties. In this network view of virtual communities, CoPs are seen as providing a collocated hub for the wider network: providing a tightly knit subnetwork that serves as knowledge generating node for the larger NoP. CoPs can also act as bridges or brokers, drawing together different groups and combining knowledge in new ways. Finally, they can provide the access points for individuals to engage with the wider network and to establish a local identity within the larger organization.

Previous research has shown that the most common distributed form of a so called virtual CoP has a collocated active core (Hildreth, Kimble & Wright, 1998), which tends to support the networked view of distributed working. A more recent example of this was provided by Lundkvist's (2004) study of customer networks as sources of innovation. This case study was based on a long-term study of the Cisco Systems newsgroup, which identified user networks as peripheral and yet vital sites of innovation. In this case, the collocated core of the network was provided by a group of university technicians.

If wholly virtual, CoPs pose significant problems. What of the applicability of geographically distributed CoPs to the problems of knowledge management? In particular, how might the balance between reification and participation be maintained in virtual working? Hildreth (2003) describes how a geographically distributed CoP managed both hard (reified) and soft (social) knowledge. In this situation, it might have been expected that sustaining participation would be more difficult, and therefore, reification would play a greater role. However, the findings of the case study showed that this was not necessarily the case.

While the group was able to sustain itself using e-media, it was still dependent on the development of relationships in the physical environment through face-to-face meetings. A shared artifact, such as a planning document, did play an important role in virtual working, but the importance of social relationships remained paramount. Here the planning document stimulated discussion and problem solving, but through the process of working on it, it also acted as a focus for further participation.

A similar account can be found in Bradshaw, Powell, and Terrell (2004) that describes how a team of remote workers gradually developed into a CoP. They describe not only how the group deployed a variety of technologies to maintain contact but also the efforts that went into building commitment, ownership, engagement, and focus in the group. In this case, the members of the group were all engaged in collaborative research. Writing about their work and presenting papers for peer review was seen as a key factor in maintaining cohesion and developing the community's shared understanding of goals, development of knowledge, and sense of belonging.

CONCLUSION AND FUTURE TRENDS

CoPs began life as a way of describing the process of informal situated learning that took place in certain types of group. From here, the concept has been extended first into the formalized, hierarchical, and task centered world of business and commerce and later into the rather more esoteric world of knowledge management. The aim of this article is not to dismiss the work that has been done in this area. The authors do not wish to argue that CoPs do not exist in business, that CoPs are of no value to business, or that CoPs have no place in knowledge management. It is simply that we believe that in much of the current literature in this area, too much stress is placed upon the supposed business benefits of CoPs and too little on the problems of CoPs in a business setting.

Perhaps the most obvious area where this is the case is the singular failure to examine the consequences of having significant business activities built around self-directed, semi-autonomous groups such as CoPs. Gongla and Rizzuto's (2004) study is almost unique in examining this aspect of CoPs. We believe that too many authors focus exclusively on the creating and sustaining of CoPs without sufficient concern for the other end of the life cycle. We would argue that without a "warts and all" understanding of the reason for having and not having CoPs, their full potential will never be realized. Similarly, there is a paucity of studies which show CoPs either failing to deliver benefits (e.g., Hislop, 2004) or even acting in a way that could be seen as counterproductive to the wider business goals (e.g., Vaast, 2004). Again, our point here is not to try to show that CoPs fail but to try to gain a more balanced understanding of the strengths and weaknesses of CoPs as a solution to business problems.

Similarly, there seems to be an often unquestioned assumption that CoPs will seamlessly translate from the collocated physical world to the geographically distributed virtual world. Few would argue that the shift to the virtual is not a real feature of today's world, but few seem to have thought through the consequences for CoPs. Instead of inhabiting a world of fixed roles with easy access to collocated resources, today's workers are increasingly based in an individualistic world of weak ties where resources are frequently obtained through personal networks and individual relationships. Rather than being embraced by a collective CoP, workers often find themselves functioning as isolated individuals and building up networks, one contact at a time. Again, paradoxically, as social networks such as NoPs become more important to organizations, the fundamental unit for many examples of virtual working is not the group but the individual. This is not to say that collective groups such as CoPs and teams have ceased to be relevant but simply that the difficulty of building, and maintaining the strong social ties needed to build a sense of community in a virtual environment should not be underestimated.

In conclusion, we would like to urge both academics and practitioners who work in this area to take a moment to reflect on the current surge of interest in CoPs. There is a natural tendency among those who are enthusiastic and passionate about a topic to ignore, or simply not see, the downside. We also believe that CoPs have the potential to make a significant contribution to certain areas of the commercial world; however, we also believe that if CoPs are to reach their full potential, a more balanced view is needed.

REFERENCES

Bradshaw, P., Powell, S., & Terrell, I. (2004). Building a community of practice: Technological and social implications for a distributed team. In P. Hildreth & C. Kimble (Eds.), *Knowledge networks: Innovation through communities of practice* (pp. 184-201). Hershey, PA: Idea Group Publishing.

Brown, J. S., & Duguid, P. (1991). Organizational learning and communities of practice. *Organization Science, 2*(1), 40-57.

Brown, J. S., & Duguid, P. (1996). Universities in the Digital Age. *Change,* 11-19.

Brown, J. S., & Duguid, P. (2000). *The social life of information.* Boston: Harvard Business School Press.

Brown, J. S., & Gray, S. E. (1995). The people are the company. Retrieved May 5, 2005, from *http://www.fastcompany.com/online/01/people.html*

Fontaine, M. A., & Millen, D. R. (2004). Understanding the benefits and impact of communities of practice. In Knowledge Networks: In P. Hildreth & C. Kimble (Eds.), *Knowledge networks: Innovation through communities of practice* (pp. 1-13). Hershey, PA: Idea Group Publishing.

Gongla, P., & Rizzuto, C. R. (2004). Where did that community go? Communities of practice that disappear. In P. Hildreth & C. Kimble (Eds.), *Knowledge networks: Innovation through communities of practice* (pp. 295-307). Hershey, PA: Idea Group Publishing

Granovetter, M. (1973). The strength of weak ties. *American Journal of Sociology, 78*(6), 1360-1380.

Hildreth, P. (2003). *Going virtual: Distributed communities of practice.* Hershey, PA: Idea Group Publishing.

Hildreth, P., Kimble, C., & Wright, P. (1998, March). Computer mediated communications and communities of practice. *Proceedings of Ethicomp'98, Erasmus University,* The Netherlands, pp. 275–286. Retrieved May 5, 2005, from *http://www.cs.york.ac.uk/~kimble/research/EthiComp98.pdf*

Hildreth, P., Wright, P., & Kimble, C. (1999). Knowledge management: Are we missing something? In L. Brooks & C. Kimble (Eds.), Information systems: The next generation (pp. 347-356). *Proceedings of the 4ᵗʰ UKAIS Conference.* Retrieved May 5, 2005, from *http://www.cs.york.ac.uk/mis/docs/kmmissing.pdf*

Hildreth, P., & Kimble, C. (2002). The duality of knowledge. *Information Research, 8*(1), paper no. 142. Retrieved May 5, 2005, from *http://InformationR.net/ir/8-1/paper142.html*

Hildreth, P., & Kimble, C. (Eds.). (2004). *Knowledge networks: Innovation through communities of practice.* Hershey, PA: Idea Group Publishing.

Hislop, D. (2004). The paradox of communities of practice: Knowledge sharing between communities. In P. Hildreth & C. Kimble (Eds.), *Knowledge networks: Innovation through communities of practice* (pp. 38-46). Hershey, PA: Idea Group Publishing.

Kimble, C., Hildreth, P., & Wright, P. (2000). Communities of practice: Going virtual. In Y. Malhotra (Ed.), *Knowledge management and business model innovation* (pp. 220-234). Hershey, PA: Idea Group Publishing.

Lave, J., & Wenger, E. (1991). *Situated learning: Legitimate peripheral participation.* Cambridge, UK: Cambridge University Press.

Lesser, E. L., & Storck, J. (2001). Communities of practice and organizational performance. *IBM Systems Journal, 40*(4), 831-841. Retrieved May 5, 2005, from *http://researchweb.watson.ibm.com/journal/sj/404/lesser.pdf*

Lueg, C. (2000). Where is the action in virtual communities of practice? *Proceedings of the Workshop Communication and Cooperation in Knowledge Communities at the D-CSCW 2000.* Retrieved May 5, 2005, from *http://www.staff.it.uts.edu.au/~lueg/papers/commdcscw00.pdf*

Lundkvist, A. (2004). User networks as sources of innovation. In P. Hildreth & C. Kimble (Eds.), *Knowledge networks: Innovation through communities of practice* (pp. 96-105). Hershey, PA: Idea Group Publishing.

Orlikowski, W. J. (1992). The duality of technology: Rethinking the concept of technology in organizations. *Organization Science, 3*(3), 398-116.

Vaast, E. (2004). The use of intranets: The missing link between communities of practice and networks of practice? In P. Hildreth & C. Kimble (Eds.), *Knowl-*

edge networks: Innovation through communities of practice (pp. 216-228). Hershey, PA: Idea Group Publishing.

Wenger, E. (1998). *Communities of practice: Learning, meaning and identity.* Cambridge, UK: Cambridge University Press.

Wilson, T. D. (2002). The nonsense of "knowledge management". *Information Research, 8*(1), paper no. 144. Retrieved May 5, 2005, from *http://InformationR.net/ir/8-1/paper144.html*

KEY TERMS

Emergent Properties: A systems concept from which it is proposed that a whole system contains properties which are not seen within any of its components or subsystems. It gives rise to the idea that a system is more than the sum of its parts.

Expert Systems: Information systems which contain and help disseminate expert knowledge.

Information Management: The management of all aspects of information in an organization, generally seen to encompass technical, human, and organizational.

ENDNOTE

[1]   The terms hard and soft knowledge are dealt with in §2.1

Linking Communities of Practice and Project Teams in the Construction Industry

Gillian Ragsdell
Loughborough University, UK

Kaye Remington
University of Technology, Sydney, Australia

INTRODUCTION AND BACKGROUND

Continuing from Ragsdell's (Article, *The Contribution of Communities of Practice to Project Management*) discussion highlighting potential synergy between project teams and CoPs, Remington and Ragsdell move into the practical arena. This article interrogates the usefulness of CoP in the construction industry and the challenges they pose for project management practice therein. Emphasis is on the role of CoPs in addressing the problem of project knowledge transfer within and between project teams.

The concept of community of practice (CoP), defined initially by Lave and Wenger (1991) and later developed by Wenger (1998) and others (Barab & Duffy, 2000; Gherardi & Nicolini, 2002; Skyrme, 1999), has only recently received attention in the project management professional literature (Galarneau & Rose, 2002; Love, Huang, Edards, & Irani, 2005; Morris, 2002). This is perhaps a little surprising since, on initial analysis, the project environment would appear to comply with all three of Wenger's three dimensions of a CoP (Wenger, 1998, pp. 72-85). There exists a level of *mutual engagement* in practice between parties who are involved in projects. A project is certainly a *joint enterprise* between a number of individuals who might come from a variety of organizations and backgrounds to achieve agreed goals. At the wider level, within the project management professional community, there is a shared interest in improving practice. At least among project management professionals, there is a *shared repertoire* of language, routines, stories, and cultural artifacts. However, the peculiar nature of the range of initiatives that are referred to under the generic name *projects* suggests a number of struc-

tural and organizational barriers to free exchange and development of knowledge within the CoP model described by Wenger and others. This article goes on to explore a number of characteristics, peculiar to construction projects, which might influence effective application of the CoP concept.

THE PROJECT TEAM AS A SPECIAL CASE OF CoP

Discussion of the project team as a special case of CoP was undertaken in Ragsdell's article. Building on that discussion, the extended project team might be seen as a special case of CoP with a lifespan limited to the duration of the project. Although Huang and Newell (2003) concluded that "only limited strong ties can be developed purely by the project team members" (p. 173), they also note that through a process of referral (Burt, 1992), the strong ties were extended, allowing teams to expand their social networks to a broader network. Such boundary spanning into knowledge networks is critical (Ancona & Caldwell, 1990), particularly as small project teams often cannot include all the expertise needed for a particular project. Similarly, individuals with a certain technical expertise may often need to serve various projects simultaneously, thus prohibiting organizations from assigning them full-time to a single project. Furthermore, research into the groupthink phenomenon has shown that integrating external knowledge and experience is an important component of effective decision making, particularly in teams with complex and innovative tasks (Janis, 1995 as cited in Hoegl et al., 2003; Neck & Moorhead, 1995).

Distributed Teams as CoP

Project teams may or may not be collocated. In large construction and engineering projects, it is now common practice to have a centrally located project room, as a meeting room and repository for all documents, greatly assisting with version control, especially when projects are fast-tracked. However, with multinational projects and ever-increasing globalization of design and construction, teams are having to work across time and space, assisted by computer-mediated communications. As designers often work in isolation from team members, some researchers are suggesting that less formal social practices found in CoP facilitate the sharing of experience and knowledge more effectively than conventional teams (Pemberton-Billing et al., 2003). Nevertheless, for online CoPs in other disciplines, such as education, trust has been shown to be an issue (Kling & Courtright, 2003).

THE PECULIAR NATURE OF CONSTRUCTION PROJECTS

In contrast to the current fashion for describing any cooperative venture as a project, projects from traditional project disciplines, such as construction and engineering, tend to be conducted within highly formalized contractual conditions. In government and community sectors, more and more large-scale projects are also subject to increasingly high levels of public scrutiny. Most research into the efficacy of the CoP model has been done with projects in professional communities for which the risk of litigation is very low such as education (Hirst et al., 2004). It follows therefore that the terms of operation of a CoP for a multinational, politically sensitive construction or engineering project would need to be radically different from those governing projects which are not so constrained.

Size

Research about KM has tended to focus on large organizations; however, in the construction industry, the vast majority of organizations employ less than five people (Constructing Excellence, 2004). Inter-views conducted within the construction sector in the UK (Cushman et al., 2002) revealed differences between organizations in their ability to create and use knowledge, large firms demonstrating commitment to R&D, small firms seeing themselves as consumers of knowledge. SMEs reported feeling isolated from knowledge networks, and although they reported the need for knowledge networks, these networks had not been established. Recently, government initiatives have endeavored to improve access to industry knowledge through cross-industry knowledge portals (see, for example, Constructing Excellence, 2004, funded by the UK Department of Trade and Industry). However, as the following discussion suggests, knowledge transfer using the CoP model might be inhibited by the peculiar characteristics of the construction industry.

Trust and Security of Information

Trust (or, rather, lack of trust) has been recognized as a potential barrier to effective project delivery in the construction industry (Zahgloul & Hartman, 2002, 2003). Examples exist of alliances and partnerships which have successfully bridged the trust issue, parties working collaboratively to deliver the project (Cushman et al., 2002; Pitsis et al., 2003), but these are rare in an industry which is highly risk averse and contract driven. Project management contracts that force transfer of risk also inhibit free exchange of knowledge, and there exists an adversarial, rather than a problem solving, relationship between stakeholders. Additionally, networking that necessarily involves information transfer may result in critical breaches of confidentiality (Bouty, 2000). In these circumstances, project team members would be unable to participate in fully honest discussions in a CoP, particularly before the project has been completed and handed over. Ongoing litigation can even prevent free exchange of information until many years after project completion.

Another consequence of lack of trust is a reluctance to talk about project failures. Successes are communicated reasonably as effectively as "war stories", but failures are underreported except where public investigations demand. As research in progress is demonstrating, the "conspiracy of silence" becomes compounded within the higher echelons of project governance (Remington & Helm, 2004).

Kelleher and Levene (2001) found that KM was significantly affected by reluctance to admit ignorance with employees believing they are paid to solve problems. Consequent feelings of vulnerability meant they were unlikely to seek advice. Successful implementation of a CoP is linked with both corporate and project culture and prevailing attitudes to sharing of lessons learned.

RECOGNIZING THE NEED FOR LEARNING

Project Scale and Complexity

A CoP, by definition, is a voluntary association of individuals who wish to share knowledge and experiences. Without a perceived need for learning, it is unlikely that participants will freely engage in learning activities. Teams in the construction industry generally comprise people who like solving technical problems. Therefore, if learning takes place, preference will be for technical issues, and the harder, formal side of management such as contract law. Learning about the softer people-oriented side of management is likely to be a low priority. If the project is simple and relatively unchallenging, little perceived need for sharing learning is likely. Lack of perceived need for learning can be an important barrier to engagement with a CoP. As projects increase in scale and public accountability increases, teams are likely to be more diverse. The complexities of dealing with multiple stakeholders in addition to technical challenges provide opportunities for sharing both explicit and tacit forms of knowledge, as defined by Nonaka and Takeuchi (1995). In these situation, the propensity for recognizing the need for learning softer, people-centered forms of knowledge has the potential to increase.

Time

Time can be another barrier to perceived need for learning. Development and maintenance of networks is time consuming and may divert attention from apparently more productive activities (Ancona & Caldwell, 1990). Usually, a project is driven by milestones defined early in the planning phase; therefore, the nature of the learning tends to be restricted to what is perceived as essential for the delivery of each particular project. Research suggests that the time constraints experienced by project team members significantly influence the kind of learning that takes place during a project. Attendance at CoP meetings or involvement with a community online may not be seen as a direct cost to the project. Project personnel often work under conditions of considerable pressure: 12-hour days, 5.5 days per week the norm. The results of a study by Kelleher and Levene (2001) highlighted time as the most frequently cited barrier to KM. The strong emphasis on prioritization of tasks means that KM activity is seen as desirable but not essential.

TEAM TO TEAM KNOWLEDGE TRANSFER

Between Project Phases

It is common practice to change the project team at each phase of a project for the simple reason that different types of expertise are needed at different phases of the project. For instance, in a building project, design development, feasibility studies, and detailed documentation might be performed by one or more architectural teams, whereas implementation of the project would be the responsibility of the construction team and management of the final building by a property management team. Effective and efficient communication of knowledge about the project at each of these phase changes is fundamental to the successful delivery of the project. The ever-present fear of litigation in the construction industry has resulted in well-documented, formal knowledge transfer procedures. However, there is a propensity for breakdown of these procedures, particularly in large cross-functional projects involving nontraditional project sectors such, as community infrastructure projects. A CoP might assist in sharing peripheral knowledge that is normally excluded from official minutes and other records; however, the litigious nature of many project environments would limit the amount of "rich" information that could be exchanged online.

Project Team to Project Team

"Reinventing the wheel" is symptomatic of many project environments. An important potential application of the CoP is being trialed within some large organizations to assist in transfer of learning between projects and teams. The characteristic of uniqueness, which is often used to distinguish functional processes from projects, frequently results in lack of transfer of knowledge from one project to another or from one team to another working on concurrent projects. Once the project team has disbanded, members tend to move to other locations within the organization or even outside the organization, taking project knowledge with them. Hence, there is no guarantee that knowledge thus gained will be transferred effectively to other project situations. This phenomenon is poorly researched but widely recognized as highly problematic (Galarneau & Rose, 2002; Morris, 2002). Variations of CoP currently being explored include discipline specific interest groups, project executive groups, and in some organizations, professional groups for women in construction. Most of these CoP tend to be located within individual organizations. However, there are also examples of informal cross-organizational groups which have been established by practitioners dissatisfied with the quality of knowledge exchange within professional bodies. These groups should be distinguished from professional benchmarking organizations, which charge membership fees. They exhibit many of the characteristics of a CoP. Participation is voluntary; organization is fluid; trust is high; and there are no membership fees. However, like exclusive clubs, they are differentiated by the fact that membership tends to be by invitation and restricted to senior executives. Hence, the potential for master-apprentice communication at an informal level, which is ideally a byproduct of the CoP, is nonexistent.

CONCLUSION AND FUTURE TRENDS

While at face value, the CoP model offers several possibilities for more effective transfer of learning within and between project environments in the construction industry, applications of the CoP as a learning strategy may be subject to a range of constraints, depending upon the type and complexity of project. Issues of trust and security, especially in contractually bound or politically sensitive project environments, the diverse nature and inherent instability of project teams, special needs of SMEs, and distributed teams, all challenge accepted models of CoP. Bounded CoPs, with restricted membership confined along organizational or discipline lines, might provide part of the answer, but the rules of engagement would need to be carefully defined for optimal exchange of information and learning to take place. However, the bounded CoP is less likely to produce the richer forms of learning that can occur when access is entirely free. Transfer of learning from one project context to another remains a challenge for organizations. Further research is needed to explore how learning through the "spillovers" or "overflows" (Callon, 1998) might be more effective for KM than planned knowledge transfer events, particularly within the complex environments in which construction projects are managed today.

REFERENCES

Ancona, D. G., & Caldwell, D. F. (1990, March-April). Improving the performance of new product teams. *Research Technology Management, 33*(2), 25-29.

Barab, S. A., & Duffy, T. (2000). From practice fields to communities of practice. In D. Johanssnn & S. M. Land (Eds.), *Theoretical foundations of learning environments* (pp. 25-56). Mahwah, NJ: Lawrence Erlbaum Associates.

Bouty, I. (2000). Interpersonal and interaction influences on informal resource exchanges between R&D researchers across organizational boundaries. *Academy of Management Journal, 43*, 50-65.

Burt, R. (1992). *Structural holes: The social structure of competition.* London: Harvard University Press.

Callon, M. (1998). *The laws of the market.* Oxford: Blackwell.

Constructing Excellence. (2004). Retrieved May 7, 2005, from *http://www.constructingexcellence.org.uk*

Construction 2020. (2004). A vision for Australia's property and construction industry. Retrieved May 7, 2005, from *http://www.construction-innovation*

Cushman, M., Venters, W., Cornford, T., & Mitev, N. (2002, September 9-11). Understanding sustainability as knowledge practice. *Proceedings of the British Academy of Management Conference: Fast-Tracking Performance Through Partnerships*. London.

Galarneau, J., & Rose, K. H. (2002, September). Cultivating communities of practice (Book review). *Project Management Journal, 33*(3), 68.

Gherardi, S., & Nicolini, D. (2002). Learning in a constellation of interconnected practices: Canon or dissonance? *Journal of Management Studies, 39*(4), 419-436.

Hirst, E., Henderson, R., Allan, M., Bode, J., & Kokatepe, M. (2004). Repositioning academic literacy: Charting the emergence of a community of practice. *Australian Journal of Language and Literacy, 27*(1), 66-80.

Hoegl, M., Praveen, K., Parboteeah, K., & Munson, C. L. (2003). Team-level antecedents of individuals' knowledge networks. *Decision Sciences,* 34(4), 741-770.

Huang, J. C., & Newell, S. (2003). Knowledge integration processes and dynamics within the context of cross-functional projects. *International Journal of Project Management, 21*, 167-176.

Kelleher, D., & Levene, S. (2001). *KM: A guide to good practice.* London: British Standards Institute.

Lave, J., & Wenger, E. (1991). *Situated learning: Legitimate peripheral participation.* New York: Cambridge University Press.

Love, P., Huang, J., Edwards, D., & Irani, Z. (2005). Building a learning organisation in a project environ-ment. In P. Love, S. Fong, & Z. Irani (Eds.), *Management of knowledge in projects.* Oxford, UK: Elsevier.

Morris, P. W. G. (2002). Managing project knowledge for organizational effectiveness. *Proceedings of the PMI Research Conference 2002, Project Management Institute* (pp. 77-87).

Nonaka, I., & Takeuchi, H. (1995). *The knowledge creating company: How Japanese companies create the dynamics of innovation.* New York: Oxford University Press.

Pemberton-Billing, J., Cooper, R., Wootton, A., & North, A. (2003, April). Distributed design teams as communities of practice. *Proceedings of the 5th European Academy of Design Conference.*

Pitsis, T., Clegg, S., Marosszeky, M., & Rura-Polley, T. (2003). Constructing the Olympic dream: A future perfect strategy of project management. *Organization Science, 14*(5), 574-590.

Remington, K., & Helm, J. (2005). Effective sponsorship. *The evaluation of the role of executive sponsor in complex infrastructure projects by senior project managers, PMJ Special Issue.*

Skyrme, D. J. (1999). *Knowledge networking: Creating the collaborative enterprise.* Oxford, UK: Butterworth Heinemann.

Wenger, E. (1998). *Communities of practice: Learning, meaning and identity.* Cambridge: Cambridge University Press.

Zaghloul, R., & Hartman, F. (2002). Reducing contract costs: The trust issue, *AACE International Transactions,* CD161-164.

Zaghloul, R., & Hartman, F. (2003, August). Construction contracts: The cost of mistrust. *International Journal of Project Management, 6*, 419.

Linking Organisational Culture and Communities of Practice

Anzela Huq
Royal Holloway University of London, UK

Jawwad Z. Raja
Royal Holloway University of London, UK

Duska Rosenberg
Royal Holloway University of London, UK

INTRODUCTION AND BACKGROUND

The purpose of this article is to identify a link between organisational culture and communities of practice. We propose that the informal nature of communities of practice places great limitations in terms of management and control and that for their purpose—which is primarily to share tacit organisational knowledge and enhance organisational learning—it is fatalistic to try to impose and enforce control. Rather, these communities ought to be left alone to formulate their knowledge sharing activities, and management comes in to provide the support, both cognitive and practical in terms of resources, to ensure that time spent at work is productive, and the knowledge is well spread and used throughout. So, not only do we intend to identify a link between culture and communities of practice, but we will demonstrate that the former has great implications in the survival and success of the latter. A review of the most prolific literature is provided, followed by a debate about the relationship between these two distinct concepts, followed by our visions for the future.

EVOLUTION OF COMMUNITIES OF PRACTICE

Communities of practice have become a hot topic of discussion of late. Their benefit to organisational development and work is immense in that they facilitate organisational communication and collaborative work and support knowledge sharing, which is fast becoming the standard for achieving greater competi-

tive advantage. They bring together organisations that are both collocated and distributed and bridge the gap for distance work that is found particularly in the latter type, where there is a need for people to work in different locations, time zones, physical environments, and cultures.

Communities of practice have been described by Lave & Wenger (1991) as "as a set of relations among persons, activity, and world, over time and in relation with other tangential and overlapping communities of practice" (p. 7). It is within these communities that new members come and learn from the more experienced members by participating in the practice with which the community is concerned. At the beginning, new members tend to be peripheral participants, those who observe the more experienced members performing certain tasks and how they deal with problems that are encountered. After this initial period of observation and possibly making small contributions to the community, the member may move from the periphery to the center.

Moving from the periphery to the center is often related to the apprenticeship model, where knowledge is transferred within a certain setting to the apprentice by the master. Lave and Wenger (1991) reject other knowledge transfer models that isolate knowledge from practice, where there is no meaning to the learning due to the setting being isolated from the practice. It is through their concept of legitimate peripheral participation (LPP) that they see learning as an integral part of practice in the real world of work, rather than being situated where instances of practice are merely replicated. This is a view supported by Brown and Duguid (1991), where they say that "abstractions detached from practice distort or obscure intricacies of that practice" (p. 40).

According to Gamble and Blackwell (2001), communities of practice are a collection of individuals who have formed an informal relationship concerned with a practice that is related to their work within a common context. It is through the practice that "a community of practice develops a shared understanding of what it does, of how to do it, and how it relates to other communities and their practices" (Brown & Duguid, 1998, p. 25). Another common characteristic that can be found in these communities is the regular interaction with other members and communities; these interactions can last for undetermined periods of time. Furthermore, within these communities, members come together voluntarily with the desire to learn and contribute. Unlike teams, these communities do not have to answer to managers or have to deliver certain outputs and be accountable to management; they exist outside of the formal structure of the organisation.

Management faces the challenge of having to try and manage communities of practice. Within these, the values that members tend to adhere to are based on reciprocity through the high level of trust formed, the expertise of members determining the influence they have, and the shared responsibilities with the community. Where the values of the organisation differ from the community, that is, the emphasis on formal roles, a challenge is posed in terms of the viability of communities of practice. This is most evident from Orr's (1990) ethnographic study of Xerox customer service representatives, in which management at first was hostile to the informal communities of practice that had evolved between the service representatives; in this case, management acted by abolishing the community. However, later, when it was realised that this had impacted productivity in a negative way, leading to inefficiencies, the decision was reversed, and the service representatives were allowed to meet. The case highlights how management needs to constrain itself from exercising too much control.

At the core of communities of practice are the people. They drive the evolution and development of the community in different ways. It is not always easy to identify communities of practice within organisations, but they have been in existence in some form or another for a considerable period of time, and it is not necessary for these groupings of people who share a common interest to be labeled as a community of practice formally. This is not to say that management need not support these communities, as vital resources and facilities can still be provided. What is essential, is that management actively tries to identify these informal communities that may exist outside of the formal structure in order to help cultivate them. It is vital that communities of practice offer something in terms of value to its members on a regular basis in order to retain them. Membership to these communities should be controlled by the members, not management. If management was to try and control the membership, it is likely that the core people would leave or not be as active. This is also a reason why communities of practice that are developed by management tend to fail, as they are not organised and defined by the members who are engaged and at the heart of the actual practice of the community. It is of utmost importance that the community is able to determine the degree of intensity and interaction that takes place. Management must recognise that communities of practice do not have a set life span; some communities of practice last longer than others. Management needs to play the pivotal role of facilitating a conducive environment where different communities that are formed within the organisation can come together (Gamble & Blackwell, 2001).

ORGANISATIONAL CULTURE: MYTH, FAD, OR REALITY?

There have been many studies and investigations into the nature of organisational culture, resulting in various almost synonymous definitions (Brown & Duguid, 1998; Deal & Kennedy, 1982; Handy, 1991; Hatch, 1993; Martin, 2002; Ott, 1989; Sathe, 1985; Schein, 1992). Residing at the cognitive level of the organisation's structure, there is consensus in the notion that those things that constitute organisational culture are shared or common. Schein (1992) defined culture as "a pattern of shared basic assumptions that the group learned as it solved its problems of external adaptation and internal integration, that has worked well enough to be considered valid, and therefore, to be taught to new members as the correct way to perceive, think, and feel in relation to those problems". The application of this definition works well only when there is a group of people who, historically, have worked together long enough to develop a relationship that is based on interaction and behavior and is also pertinent to task. Schein goes further to use a typology to classify culture into different levels, ranging from the tangible and visible representations of group behavior to those that are embedded in cognitive mental models and which are manifested in action or behavior. In uncovering Schein's model of culture, we find that there are three distinct levels of culture, from the physical and tangible *artifacts*, to the *espoused values* that are adopted, to the intangible cognitive assets that lie in the *basic assumptions* that are gradually and naturally embedded in mental models and that lie at the core of behavior and practice at work.

Artifacts are those things that we can see and feel. Common, or shared, artifacts provide the resources for people to focus their collaborative activities and to obtain facilitative feedback from each other about the current state of the activity. They can be regarded as loci, or physical manifestations, of what is essentially purposeful social action. As they are man-made, physical objects, they have a structure that enables people to obtain certain kinds of "social and organisational" information from them (Rogers & Ellis, 1994). In terms of representing culture, they communicate meaning, intentionally and unintentionally, about "the way things work around here" (Deal & Kennedy, 1982). For example, the simple ways that furniture is placed, work areas are divided and dress codes are set and adhered to (or not) all convey an insight into how people in that particular organisation interact. Company documents such as reports, memos, and brochures also play a role in this; deciphering the language of an organisation is a crucial element of understanding its culture, a concept that is core to Ott's explanation of organisational culture. According to Ott (1989), every "culture, discipline, perspective, organisation and profession" (p. 20) builds its core concepts through a common language or jargon, and this jargon provides a medium through which the metaphors and symbols of culture are understood. Ott believes that language controls cognitive patterns, which in turn affect the way people think. It is not feasible to attempt to study and understand culture through artifacts alone. As there is possibility for misinterpretation and misunderstanding, it would be prudent to investigate the higher levels of Schein's framework too.

Espoused values are those aspects of shared understanding and behavior that come from "cognitive transformation". In simple terms, when a group adopts certain beliefs as the basis for structuring their work and they gradually begin to see and feel success because of it, they begin first as shared values or beliefs and then gradually become shared assumptions, the difference being that the latter is taken for granted and is naturally incorporated into work life, whereas the former is something that the group aspires to and uses to guide its way. This transformation comes from the social experience of being and working within the group over long periods of time. The manifestation of these values and beliefs becomes embedded in nature as people fall into patterns of practice and discipline; this is different from routine though in that, theoretically, one cannot fall out of this pattern of behavior because of where it resides in the cognitive memory, whereas routine is practical and can be changed. Each component of this level, though, is thoroughly connected to the highest level of basic assumptions. Schein describes this as the change that a person undergoes when a hypothesis or a belief becomes a reality. The beliefs become so far embedded in behavior that people assume that this is how things work naturally, compounding the investigation of the earlier level and its implications for practice and behavior. The fundamental difference between the values level and the assumptions level is that the latter is never questioned or dishonored. In effect, what happens is that the nature and essence of the organisation's culture reside firmly in the assumptions around which it works, but they manifest themselves in the revelation of values and beliefs and the artifacts that constitute the work environment.

There are, once again, varying definitions available as to what organisational culture really is. For example, Allen and Kraft (1982) define it as "norms". Davis (1984) identified *two* levels of organisational culture: "guiding beliefs and daily beliefs" (p. 000). Pacanowsky and O'Donnell-Trujillo (1984) take the view that "a culture is something an organisation *is*" (p. 000). These are some among many, but most have a similar underlying theme: organisational culture exists on an emotive and cognitive level and is embedded in social processes, each culture is unique, and it is a guiding and binding factor for work (Martin & Siehl, 1983; Ott, 1989).

LINKING THE TWO

What significance does this have for communities of practice? According to Schein (1992), the absence of a valid communication system undermines effective action, a conclusion he reached following a consultancy period in a small company. Imagine what can happen in large, dispersed organisations! Effective communication is what holds together the acts of internal integration and external adaptation as described in Schein's definition of organisational culture. The formation of common ground consisting of shared meanings and symbols to represent those meanings to the outside world is what takes organisational culture from concept to *perceived* reality, and, in turn, the culture supports the acts of work and communication. Thus:

...a society's culture consists of whatever it is one has to know or believe in order to operate in a manner acceptable to its members...culture is not a material phenomenon; it does not consist of things, behaviour, or emotions. It is rather an organisation of these things. It is the form of things that people have in mind, their models for perceiving, relating, and otherwise interpreting them. (Goodenough, 1957, p. 167)

It is here that organisational culture has great implications for the success of communities of practice.

You cannot manage knowledge; you can only manage the environment that leads to the knowledge being created and transferred. (Anonymous, from Kermally, 2002, p. 43)

During the review of the literature on these two distinct concepts, we were hard pressed to find anything that we felt adequately linked the two. The works of Swan, Scarborough, and Robertson (2002) and Brown and Duguid (1991) have come close to making distinct links, but few studies have actually delved deep into the nature of organisational culture for communities of practice; most of the literature is learning-oriented. Looking back at the formal definitions of communities of practice, it is evident that

these informal structures that arise of a common interest in a particular practice do not require formal procedures of management and control, rather the members should be left to their own devices, as they thrive best on working things out themselves. The two are sometimes brought together in practice and described in ethnographically-informed workplace studies focused on technology at work.

The focus on human aspects of work has motivated reflections not only on technology design, but also on the organisation of co-operative activities in the organisations that provided data for ethnographic fieldwork. These reflections led to the search for explanation of the changes in work practices that occurred as the result of unexpected impacts of technology in the workplace. (Rosenberg, 2000)

A case study from manufacturing illustrates how overt and covert organisational structures may facilitate the change in communities of practice, when the organisational culture is supportive. A large computer manufacturing organisation procured peripherals from a small company on a regular basis. One type of printer developed especially to fit the host system kept breaking down once installed at the customer sites. As an emergency measure, an ad-hoc team of engineers, developers, and system integrators from both companies was created to deal with this particular problem and left to develop their own forms of communication and collaboration. They created a community of practice so successful that it continued well beyond the printer problems, and they effectively became the troubleshooting "A-team". The small company was subsequently integrated, and the original A-Team formed the core of a new department, thus formalising the now successful original community. Several workplace studies investigated the interaction between overt and covert organisational structures (Hutchison & Rosenberg, 1994; Perry, Fruchter & Rosenberg, 1999; Plowman, Rogers & Ramage, 1995; Rogers & Ellis, 1994), as they became more and more visible due to the introduction of collaborative technologies into real-life organisations.

What has been gained from these studies is an appreciation of the importance of a supportive climate through the development of a working environment that on a cognitive level supports and advocates the creation of these communities. This climate needs to be developed at the senior level and filtered through the organisation; it is here that an organisation's cultural climate is to be put to use in a sense. Taking from Schein's work, with culture being based on values and beliefs, this idea that communities of practice can and should be developed for work has to be far enough embedded in working practice so that they become a natural and wholly integrated part of organisational life—already we're looking at the highest level of Schein's model, the basic assumptions which guide daily work. With these communities being so informal and therefore difficult to identify, the presence of this type of proactive culture can be subconsciously fruitful. Management does not need to explicitly specify that communities of practice should be used—it is certainly not something that can be worded into a mission statement. However, as was seen in the Xerox case, it is essential that communities be allowed to develop and create their own paths for work, and this can happen only if people feel they can. Swan et al. (2002) declared that organisational innovation must "support the development and circulation of knowledge within communities" (p. 482).

CONCLUSION AND FUTURE DIRECTIONS

The link is becoming clear. Clearly, the formation of communities of practice is neither something that can be planned for and structured, nor controlled and manipulated by outsiders; the purpose of these communities is to share knowledge and learn immediately, which may interact fruitfully with active or formal forms of control. Rapid advances in technology, specifically technologies that support collaborative work, are quickly exposing organisational boundaries. Consequently, the constant push for innovation and innovative measures at work is putting pressure on people at work to come up with ways to keep knowledge within the organisation yet use it as effectively as possible. Thus organisations will find that by supporting these informal structures and permitting this kind of free exchange, knowledge will flow more easily, practically, and effectively. "Knowledge-based information systems, office networking and groupware, and multimedia—and, more so, the confluence of the three technologies—are

rapidly changing the ways in which people—in particular, 'knowledge workers' (Drucker, 1969) in organisations—work with computers; together, they will have a far more significant impact on the way people work than have 'conventional' IT products (e.g., word processors, spreadsheets, databases)" (Hutchison & Rosenberg, 1994, p. 99).

At the core of the success of communities of practice lies the issue of management's willingness to give up a certain degree of control and permit free knowledge flow. Accepting that organisational culture develops in different ways in different organisations, one thing is constant: that culture is definitely based on certain principals that cannot be avoided, and although difficult to measure, is surely to impact organisational learning in positive ways if cultivated in the right way. Communities of practice thrive in lively, positive, proactive environments, where members are permitted to communicate as they please yet remain within the boundaries of organisational control, and culture can be the facilitator. As Hutchison and Rosenberg (1994) state, "culture, according to the definition we shall use, is then *not* the overt behaviour, nor is it the complex co-ordination of actions producing that behaviour. Rather, it is that knowledge which one must have internalized in order to generate that behaviour" (p. 112).

REFERENCES

Brown, J. S., & Duguid, P. (1991). Organisational learning and communities-of-practice: Toward a unified view of working, learning and innovation. *Organization Science, 2*, 40-57.

Brown, J. S., & Duguid, P. (1998). Organizing knowledge. *California Management Review, 40*(3), 90-111.

Deal, T., & Kennedy, A. (1982). *Corporate cultures: The rites and rituals of corporate life.* New York: Penguin.

Drucker, P. F. (1969). *The age of discontinuity: Guidelines to our changing society.* London: Heinemann.

Gamble, P. R., & Blackwell, J. (2001). *Knowledge management: A state of the art guide.* London: Kogan Page.

Goodenough, W. (1957). Cultural anthropology and linguistics. In P. Garvin (Ed.), *Report of the seventh annual round table meeting on linguistics and language study*. Washington, DC: Georgetown University Press.

Hutchison, C., & Rosenberg, D. (1994). The organization of organizations. In L. Wilcox (Ed.), *Journal of Information Technology, Special Issue on Organisational Perspectives on Collaborative Work, 9*, 99-117.

Kermally, S. (2002). *Effective knowledge management: A best practice blueprint*. Chichester, UK: John Wiley & Sons.

Lave, J., & Wenger, E. (1991). *Situated learning: Legitimate peripheral participation*. Cambridge, UK: Cambridge University Press.

Orr, J. E. (1990). Sharing knowledge, celebrating identity in war stories: Community memory in a service culture. In D. Middleton & D. Edwards (Eds.), *Collective remembering: Remembering in a society*. Thousand Oaks, CA: Sage.

Ott, J. S. (1989). *The organisational culture perspective*. CA: Brooks Cole.

Perry, M., Fruchter, R., & Rosenberg, D. (1999). *Co-ordinating distributed knowledge: A study into the use of an organisational memory, in cognition, technology and work*. London: Springer.

Plowman, L., Rogers, Y., & Ramage, M. (1995). What are workplace studies for? In H. Mamolin, Y. Sundblad, & K. Schmidt (Eds.), *Proceedings of the 4th European Conference on CSCW*. Kluwer Academic: Stockholm.

Rogers, Y., & Ellis, J. (1994). Distributed cognition: An alternative framework for analysing and explaining collaborative working. *Journal of Information Technology, 9*, 119-128.

Rosenberg, D. (2000). 3 Steps to ethnography: A discussion of inter-disciplinary contributions. In L. Pemberton (Ed.), *Communications in Design, 15*(1).

Schein, E. (1992). *Organisational culture and leadership* (2nd ed.). San Francisco: Jossey-Bass.

Swan, J., Scarborough, H., & Robertson, M. (2002). The construction of "communities of practice" in the management of innovation. *Management Learning, 33*, 477-496.

KEY TERMS

Community of Practice: An informal collective group of individuals bound by a common practice base engaged in knowledge sharing activities to add value to work.

Distributed Organisation: An organisation that works across physical boundaries and time zones with multiple sites or offices.

Knowledge: Thoughts and processes embedded in the minds, behavior, and understanding of people, which are used to enhance understanding of problems and issues at work. It is cognitive in nature and generally exists in two forms, tacit and explicit, and is viewed as an essential asset to organisations and whose efficient spread leads to greater competitive advantage.

Learning Organisation: An organisation which actively engages itself in promoting the development of knowledge assets by extracting, storing, and nurturing the knowledge that exists in individuals and systems.

Organisational Culture: The embodiment of shared understanding and common purpose, which serves to develop and sustain working practices within organisations and provide a binding factor for all employees within the same organisation, regardless of location, distance, time, or local environment.

Organisational Learning: The consistent and constant acquisition and transfer of organisational knowledge assets for increased competitive advantage.

Social Networks: Groups of people formed through establishment and fruition of common interests.

Values: The beliefs that people carry, formed of identity, national culture, behavior, and understanding that translate into organisational values at the corporate level, formed of common understanding at work.

Linking Small Business Networks with Innovation

Patrice Braun
University of Ballarat, Australia

INTRODUCTION AND BACKGROUND

Today, with an economy enabled and driven by connectivity, a fundamental shift in business models is occurring whereby information, knowledge, and relationships underpin competitive advantage. In order to compete in what some refer to as the networked economy, companies across the globe must use technology-mediated channels, create internal and external value, formulate technology convergent strategies, and organize resources around knowledge and relationships (Scott & Storper, 2003).

The rise of information and communication technologies (ICT) and electronic information networks has led firms of all sizes to implement more technology driven solutions for improved productivity and information flow. Malhotra (2000) identified three general information management (IM) developments that have revolutionized company information processes over the last 40 years. The first phase, the automation phase, increased company efficiency of operations. The second phase, the rationalization phase, streamlined those procedures by eliminating bottlenecks made apparent by the automation. The third phase, the business reengineering phase, radically redesigned information and knowledge management processes through technology-intensive implementation of procedures in workflows and work processes (Malhotra, 2000). Now we have reached a fourth phase, the knowledge creation and knowledge transfer phase, that, if possible, is even more closely associated with technology than business process reengineering.

With embedded knowledge flows and innovation linked to communities of practice as well as through linkages using technology, companies of all sizes have the potential to both collaborate and compete by taking advantage of connectivity and new relationships founded on the exchange and sharing of embedded knowledge. This article discusses how knowledge sharing environments such as communities of practice and virtual business communities can be important determinants of commercial viability and business success for small and medium sized enterprises (SMEs), provided that both the (virtual) environment and inter-firm relationships are conducive to information sharing and knowledge flows.

COMMUNITIES OF PRACTICE

Once the domain of special business units and cross-functional teams to perpetuate ideas and embed core competencies, a new form of collective community building has emerged through a spontaneous new knowledge exchange trend known as communities of practice, or CoPs. Burk (2000) simply calls CoPs expansions of one-on-one knowledge sharing. Theorists, Wenger, Snyder, and William (2000) describe them as informal groups of people who regularly share their expertise and experiences; are not formulated or controlled by management; set their own leadership; and follow their own agenda. Lave and Wenger (1991) were among the first to introduce CoPs as context-bound groups of workers who share knowledge around a particular practice.

In many ways, communities of practice are the Western adoption of the holistic Japanese approach outlined by Nonaka and Takeuchi (1995) in acknowledging the importance of tacit company knowledge and transforming it into explicit company assets. However, one of the central benefits of these self-constituting communities is that they sidestep the "ossifying tendencies of large companies and develop rich, fluid non-canonical worldviews to bridge the gap between their organisation's static canonical view and the challenge of changing practice" (Brown & Duguid, 1991, p. 50). This spontaneous think-tank mode of team building through face to face meetings, e-mail, knowledge sharing networks, intranets, and technology-mediated conferencing is an inherently innovative process and is proving to be a crucial aspect of organizational learning and innovation.

Initially, most communities of practice were internal company networking groups to foster shared learning and practices to encourage team-based incentives that directly influenced company profits (McDermott, 2004). Showing great promise in driving organizational learning and innovation, this form of knowledge creation is being adopted in both the public and private sectors, as it is considered the key to survival in the knowledge economy. Communities of practice have also flourished with members from different companies, as exemplified by the chief executive officers (CEOs) of different US companies who make up the Business Roundtable (Wenger et al., 2000).

The application of CoPs to an inter-firm context is in line with the new business models that are favored in the networked economy, in which connectivity, relationships, and knowledge sharing are key assets for competitive advantage (Soekijad, Huis in 't Veld & Enserink, 2004). Within the networked environment, the relationship between connectivity and companies should be seen as reciprocal, whereby ICT and related capabilities – such as virtual community environments like chat rooms and e-mail forums used for product development, product review, and other business information exchange—have a significant impact on how inter-organizational relationships are developed. Examples of new economy inter-firm knowledge sharing may be found in Internet-based companies such as Amazon and e-Toys, which have successfully redefined the value of knowledge assets by fostering information flows between organizations and industries in virtual community environments (Malhotra, 2000). It should be noted that, contrary to the latter example, there are many different forms of sociocultural communities of practice proliferating in the virtual environment with objectives other than competitive advantage, discussion of which falls outside the scope of this article.

While CoPs and other complementary inter-firm relationships have been the subject of considerable empirical research in large enterprises (e.g., McDermott, 2004, p. 624; Pfeffer, 2000, p. 358;

Soekijad et al., 2004, p. 623), studies on the role of CoPs with regard to small and medium sized enterprises (SME) are less abundant. Building on the concept that global positioning and competitive advantage for SME may be achieved through clustering or network building and collaborative knowledge creation, the rest of this article explores the potential of a timely synergy between connectivity and collaborative business models for SME in embracing knowledge community practices.

SMALL BUSINESS NETWORKS

As seen above, ICT and related capabilities such as virtual community environments can have a significant impact on how inter-organizational relationships are developed. However, the structure and culture of an existing network of organizations itself also seems to have considerable predictive power for the way in which the telecommunications network is developed, implemented, and used (Nouwens & Bouwman, 1995). Adoption of network structures and networked technologies by SME is generally related to the size and nature of SME and largely depends on their perception of affordability and business growth opportunities for their business (OECD, 2000).

New ways of doing business to achieve success in the techno-economic innovation paradigm bring to the fore ICT adoption and strategic planning issues. Research into the adoption of networked technologies by SME indicates that SME generally approach clustering and networked infrastructures such as the Internet with caution and still hesitate to invest their time and money in a rapidly changing economy (NOIE, 2000), nor do SME necessarily view the Internet as a vehicle to transform their individual business capability from a parochial to a networked or global level, which may be achieved through the set-up of electronic commerce (e-commerce) portals or other Web-enabled cluster structures (Murray & Trefts, 2000). The latter study cites lack of technology skills, lack of a strategic sense of how to move forward, and fear of competitor use of the Internet as significant barriers for uptake of networked technologies by SME. Therefore, creating network infrastructures and knowledge flows between small firms is contingent, not on adoption of ICT technology alone, but on economic and social contexts as well.

European studies on SME positioning in the networked economy point to SME networking and knowledge sharing as contingent on favorable economic conditions, for example, by providing government-sponsored external networks (Cooke & Wills, 1999; Fariselli et al., 1999). An Asian study similarly provides empirical evidence that successful SME collaboration needs to be underpinned by resources that provide SME with the tools to become global players (Konstadakopulos, 2000). The European studies on SME positioning in the new economy also associate social capital with enhanced business, knowledge, and innovation performance (Cooke & Wills, 1999; Fariselli et al., 1999). Social capital can be roughly understood to mean "the goodwill that is engendered by the fabric of social relations" (Adler & Kwon, 2002, p. 17). While social capital and embracing connectivity through public or private initiatives may facilitate the electronic linking of SME to one another for potential business-to-business (B2B) resource and information sharing, and help to reduce isolation of individual SME, there is another critical factor to consider in terms of knowledge exchange between SME, and that factor is trust.

There is a considerable body of literature on inter-firm trust. Trust and social capital are attributes not only of organizations but also of communities, industry networks, or even entire geographic regions, which can help expedite economic development and facilitate large-scale economic activities (Fukuyama, 1995). Trust between network partners is said to reduce fear of opportunistic behavior and improve collective learning and knowledge sharing. The trust may be historical and already exist between different firms, as illustrated above, or it can be built during the relational exchange (Gulati, 1995; Ring & Van de Ven, 1992). Some scholars argue that relationships do not necessarily have to be based on trust as long as systemic mechanisms are in place which allow stakeholders to have confidence that network partners will exhibit cooperative rather than opportunistic behavior and not take competitive advantage of knowledge-based exchanges (Das & Teng, 1997). In the aforementioned Asian example of SME collaboration, information sharing and learning was in fact taking place based on prior existence of trust and in an atmosphere of continued trust building between stakeholders (Konstadakopulos, 2000).

Corporate company members tend to join a community of practice or a community of interest for networking or learning purposes in their field, trusting other company members in the exchange of explicit and implicit company knowledge for the public good aspect and building of company assets. This type of collaborative learning and innovation in large companies is facilitated via the company infrastructure such as e-mail, company intranets, and other technology-mediated tools such as video-conferencing. Given the lack of networked infrastructure between SME and the frequent individualistic nature of small firms, SME are more likely to compete than collaborate in a knowledge exchange milieu. Thus, to approximate community of practice results, if indeed such results are desirable, SME networks are likely to require infrastructure and knowledge exchange conditions that engender trust (Braun, 2002).

FUTURE TRENDS FOR SME

With networking, collaborative information flows, and communal knowledge creation providing potentially beneficial economic outcomes for SME in today's economy, SME may be willing to engage in collaborative knowledge sharing relationships, provided that connectivity (infrastructure), network relationships, and trust conditions are present.

Figure 1 maps knowledge flows against limitations such as size, connectivity, and lack of interaction and suggests that fostering a culture of connectivity, cooperation, and trust building between SME to initiate and encourage community of practice type knowledge diffusion may offer a potential solution to possible loss of competitive advantage for SME in the networked economy.

The top of the knowledge flow model shows how a large company with high connectivity and an integrated infrastructure for information and knowledge exchange via communities of practice can lead to a high level of trust and subsequent innovation and competitive advantage. The bottom part of the knowledge flow model shows how an SME with low access to a networked infrastructure and a low level of knowledge exchange leads to a low level of trust in industry or alliance partnering, which in turn can lead to isolation and loss of competitive advantage for individual SME in the networked economy.

Figure 1. Knowledge flow model

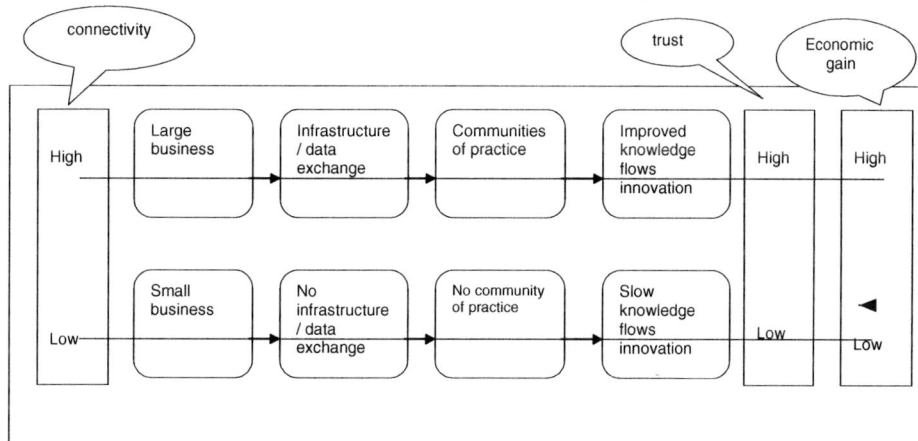

The model displays two extreme positions, for example, a large company with highly integrated connectivity vs. a small company with little or no connectivity. Alternative scenarios in terms of company size, connectivity, and level of interaction would impact on competitive advantage and the level of innovation; for example, a large company without an integrated infrastructure and a climate of data exchange would likely exhibit lower levels of knowledge flow and innovation. A small company with full connectivity (to the Internet) but without active participation in an online knowledge network would have competitive advantage over a small company with little (e-mail only) or no connectivity, while a high connectivity SME subscribing to a networked knowledge exchange culture would have competitive advantage over a connected yet network-isolated SME. Considering connectivity, local network conditions, the character of local practice, and the uniqueness of social capital within each region is pivotal in developing SME networks for competitive advantage.

In terms of connectivity, large firms with deep pockets have the ability to build their own integrated architecture, while SME access to a networked infrastructure may involve relying on the public purse. Where SME do not have a history of networking and knowledge sharing for the local or regional common good, a knowledge exchange culture will need to be cultivated and coupled to an economic framework that provides clear collaboration benefits and that is linked to tangible results for participating SME (Braun, 2003b).

In addition to offering a networked environment where SME can both collaborate and compete, a culture of trust needs to be fostered through personal interaction and strong relational or social cooperation between firms to learn critical information and know-how from alliance partners (Nonaka & Takeuchi, 1995). Established business networks and regional industry associations can play a key role in stimulating loose collaboration and knowledge exchange between SME.

Stronger ties could potentially be fostered through regional supplier and customer network linkages, as demonstrated by online companies such as Amazon.com. Such linkages can function as conduits for SME communities of practice. With networked technologies now able to provide centralized supplier and customer relations management functionalities, regional SME portals, SME intranets, and other interactive networked environments such as video and desktop conferencing may well provide the required interactive collaborative context to evolve into SME information and learning platforms (Lechner, Stanoevska-Slebeva & Tan, 2000; Stahl, 2000).

Joining the global market as a sole trader, let alone becoming an inter-firm network stakeholder may entail an enormous conceptual leap into the future for many SME owners and managers. Network novices will hence need substantial encouragement and support to make them willing to take the network plunge (Braun, 2003a). Creating awareness of networked opportunities, developing skills in using networked technologies, and providing educational processes around inter-firm relationship building may be beneficial to increase SME understanding of the potential of collaboration. Once individual SME and collective goal orientation is realized, participation in an SME community of practice or industry cooperative is likely to perpetuate trust between SME, which in turn would produce economically beneficial outcomes.

CONCLUSION

This article has discussed how collaborative relationships are becoming one of the most important determinants of commercial viability and business success in the networked economy. With embedded knowledge flows and innovation linked to communities of practice as well as through linkages using technology, SME have the potential to both collaborate and compete by taking advantage of connectivity and new relationships founded on the exchange and sharing of embedded knowledge.

There are obvious differences between large and small firms both in terms of availability of resources for connectivity and in terms of community of practice membership. Given the often independent nature of SME, exact emulation of a large company community of practice may not be feasible or even desirable, but a variety of models can be examined to identify appropriate connectivity-driven inter-firm network structures for SME and approximate desired community of practice results.

Having drawn on the inter-disciplinary fields of economic development, new business models, collaboration, and trust, it is suggested that overcoming potential loss of competitive advantage for SME in the networked economy will require strategic SME community building with connectivity, trust, and relational capital as pivotal building components. Taking communities of practice as the vehicle for knowledge sharing between SME, attention must be paid to local conditions, the concept of practice (Brown & Duguid, 2001), and how practice relates to trust. While this article demonstrates the importance of knowledge sharing in the networked economy, it also shows that the key to economic staying power is vested in people and their culture rather than in technology.

REFERENCES

Adler, P. S., & Kwon, S. W. (2002). Social capital: Prospects for a new concept. *Academy of Management Review, 27*(1), 17-41.

Braun, P. (2002). Digital knowledge networks: Linking communities of practice with innovation. *Journal of Business Strategies, 19*(1), 43-54.

Braun, P. (2003a, June 12-13). SME networks and innovation: Knowledge creation in small firms. *Proceedings of the 11th Annual International High Technology Small Firms Conference (HTSF).*

Braun, P. (2003b). Virtual SME networks: Pathways towards online collaboration. *Journal of New Business Ideas and Trends, 1*(2), 1-10.

Brown, J. S., & Duguid, P. (1991). Organizational learning and communities of practice: Toward a unified view of working, learning and innovation. *Organization Science, 2*, 40-57.

Brown, J. S., & Duguid, P. (2001). Knowledge and organization: A social-practice perspective. *Organization Science: A Journal of the Institute of Management Sciences, 12*(2), 198-214.

Burk, M. (2000). Communities of practice. *Public Roads, 63*(6), 18-22.

Cooke, P., & Wills, D. (1999). Small firms, social capital and enhancement of business performance through innovation programmes. *Small Business Economics, 13*, 219-234.

Das, T. K., & Teng, B. (1997). Sustaining strategic alliances: Options and guidelines. *Journal of General Management, 22*(4), 49-64.

Fariselli, P., Oughton, C., Picory, C., & Sugden, R. (1999). Electronic commerce and the future for SMEs in a global market-place: Networking and public policies. *Small Business Economics, 12*(3), 261-275.

Fukuyama, F. (1995). *Trust.* New York: Free Press.

Gulati, R. (1995). Does familiarity breed trust? The implications of repeated ties for contractual choice in alliances. *Academy of Management Journal, 38*, 85-112.

Konstadakopulos, D. (2000). Learning behaviour and cooperation of small high technology firms in the Asean region. *ASEAN Economic Bulletin, 17*(1), 48-62.

Lave, J., & Wenger, E. (1991). *Situated learning: Legitimate peripheral participation.* Cambridge: Cambridge University Press.

Lechner, U., Stanoevska-Slebeva, K., & Tan, Y. H. (2000). Communities & platforms. *Electronic Markets, 10*(4), 231.

Malhotra, Y. (2000). Knowledge management for e-business performance: Advancing information strategy to "Internet time". *Information Strategy, The Executive's Journal, 16*(4), 5-16.

McDermott, R. (2004). How to avoid a mid-life crisis in your CoPs. *KM Review, 7*(2), 10-13.

Murray, R., & Trefts, D. (2000). The IT imperative in business transformation. Retrieved May 6, 2005, from *http://www.brint.com/online/*

NOIE. (2000). *T@king the plunge 2000: Sink or swim.* Canberra: National Office for the Information Economy, Commonwealth of Australia.

Nonaka, I., & Takeuchi, H. (1995). *The knowledge-creating company: How Japanese companies create the dynamics of innovation.* New York: Oxford University Press.

Nouwens, J., & Bouwman, H. (1995). Living apart together in electronic commerce: The use of information and communication technology to create network organisations. Retrieved May 6, 2005, from *http://www.ascusc.org/jcmc/vol1/issue3/vol1no3.html*

OECD. (2000). A global marketplace for SMEs, progress report by the Industry Committee Working Party on SMEs. Retrieved May 6, 2005, from *http://www.oecd.org/EN*

Pfeffer, J., & Sutton, R. (2000). *The knowing-doing gap: How smart companies turn knowledge into action.* Boston: Harvard Business School Press.

Ring, P. S., & Van de Ven, A. (1992). Structuring cooperative relationships between organizations. *Strategic Management Journal, 13*(7), 483-498.

Scott, A. J., & Storper, M. (2003). Regions, globalization, development. *Regional Studies, 37,* 579-593.

Soekijad, M., Huis in 't Veld, M., & Enserink, B. (2004). Learning and knowledge processes in inter-organizational communities. *Knowledge and Process Management, 11*(1), 3-12.

Stahl, G. (2000). Collaborative information environments support knowledge construction by communities. *AI & Society, 14,* 71-97.

Wenger, E., Snyder, W. M., & William, M. (2000). Communities of practice: The organizational frontier. *Harvard Business Review, 78*(1), 139-146.

KEY TERMS

B2C: Business-to-consumer.

B2B: Business-to-business.

Cluster: A group of linked enterprises that share a common purpose of gaining competitive advantage and economies of scale.

Competitive Advantage: A condition which enables companies to operate in a more efficient or higher quality manner than the companies it competes with, which results in financial benefits.

Connectivity: The ability to link to the Internet via a computer.

E-Commerce: Connection, electronic data exchange, and transaction capability via the Internet.

Economies of Scale: Economies of scale refers to the notion of increased efficiency for the production and/or marketing of goods/products by pooling or sharing resources.

ICT: Information and communication technologies, including phone, fax, e-mail, the World Wide Web, and the Internet.

Portal: A Web site or service that provides access to a wide range of services.

SME: Small and medium sized enterprises. Refers to enterprises with a specific number of staff. A small size enterprise generally refers to firms with less than 20 employees.

Social Capital: Social capital refers to the collective value of all social ties and the inclinations that arise from these networks to do things for each other.

Web-Enabled: Business systems that are supported by Internet technologies.

The Living Tradition of "Yaren Talks" as an Indigenous Community of Practice in Today's Knowledge Society

Tunç Medeni
Japanese Advanced Institute of Science and Technology, Japan

INTRODUCTION

The "corporations" of craftsmen and guilds of artisans can be considered as the communities of practice of ancient times and the Middle Ages. The "Ahi (Brotherhood) Organization" in the Ottoman Empire is an example of these. Today, although the Ahi Organization itself has ceased to exist, the "Yaren (Friend) Talks" still continue in certain parts of Turkey as a living tradition of the Ahi principles and practices. Viewing the continuing custom of the Yaren Talks as a precious practice of society and a unique community of practice, in this short article, we will first provide some background information about this tradition and then address some related discussions, mainly about how to make this tradition continue. We believe that the Yaren Talks will provide a rich source that can be fruitful for improving the understanding and use of communities of practice in a knowledge society.

BACKGROUND

Yaren meetings assumed an important role in the members' life after work in the organization of Ahi.

Figure 1. Yarens in the Yaren room

Requiring its members to be honest and honorable, the Yaren organization did not allow any unethical behavior. It functioned as an educational institution, especially for the young boys in the community. The Yaren Talks also provided recreation for the members, with special events that included, among other activities, folk music and dance performances and the serving of special cuisine. The Yaren successfully played its role as an important social institution through the centuries in the Ottoman Empire and Turkey.

Çankiri, a small city in Anatolia, where the author spent his childhood, continues to organize the Yaren Talks every winter, holding onto the original principles and practices. Friends, who want to organize meetings in the cold winter months, gather and agree to establish an organization for the Yaren Talks. First, they choose the leaders of the organization, then select the other members, approved by these leaders, and decide on other issues necessary for the preparation of the meetings. The Yaren organization consists of weekly meetings, which are organized by one member in turn for an 11-week period. However, the influence of these special community events is much broader and remains much longer in the life of its members. Some explanation of this influence is as follows.

Both the members and guests invited by the organizer can participate in the Yaren Talks. The ongoing storytelling and conversation among these participants facilitate the flow of knowledge and information necessary to maintain social order in daily life. Moreover, in the Yaren meetings, any critical matter is judged privately among the members. Normally, all the routines in the meetings and in social life outside the meetings have certain rules, which the members are expected to obey. Moral and material punishments are available to discourage members from committing any immoral act or wrong

Figure 2. Examples of local plays performed in Yaren Talks

behavior; these can range from having to treat all the Yaren to a Turkish bath, meal, or shave to dismissal from the Yaren Talks, which is more severe than dismissal from one's homeland. On the other side, if a mistake is made within the community, first the members try to sort it out without intervention by the leaders or notice by others outside the community. Thus inside the Yaren community, members undergo an influential learning process that educates them as virtuous, responsible individuals for their community and society. Finally the unique theatrical performances during the Yaren Talks require participants to know the local plays so that the continuation of these customs can be maintained.

FUTURE TRENDS

As a living tradition and custodian of indigenous values and customs, the Yaren Talks still struggle to survive, having remained mainly as a volunteer effort. While the volunteer members try to guarantee the continuation and retain the originality of the Yaren Talks, they face the problem of finding funding to maintain their organization and develop their practices. The reinterpretation of the meaning and purpose of this traditional organization can be useful in attracting the attention that members seek. The classics of the Yaren Talks can be of benefit as sources of new synthesis and innovations, so that they fulfill their social function not only to educate community members, but also to share information and transfer knowledge according to the changing conditions of the community and society. In addition, some punishments and restrictions can be moder-

ated in order to enhance the role of the Talks in the society. With a careful and considerate reinterpretation, women might also attend certain meetings. On the other hand, if the necessary meaning reconstructions are not given the importance they deserve, the risk of getting lost in the various dilemmas posed by the need to adapt a precious tradition to the changing conditions of time may result in virtual extinction, and the Yaren Talks will only be seen in folklore exhibitions.

CONCLUSION

The Yaren Talks is an exciting example of community practice. Having lasted for centuries, they are continuing into the new millennium. The rich oral and performing arts customs and the interesting interaction opportunities available for the members' identity

Figure 3. Yaren assemblage

construction, learning, rewarding, and leadership are among the unique characteristics of the Yaren Talks that deserve to receive closer attention. Looking at the Ahi Organization and the Yaren Talks from the perspective of knowledge societies and communities of practice can help transfer this living tradition to the next generation, while it may also prove to be useful for the enrichment of this emerging and e-merging perspective itself.

ACKNOWLEDGMENT

This short article is dedicated to the living memory of Kemal Parilti, without whom this work could not be produced.

REFERENCES

Ayta, G. (1990). Çankiri Yaran Sohbetleri. *Ayane Kültür Edebiyati*, 3-36.

Baker, A.C., Jensen, P.J., & Kolb, D.A. (2002). Learning and conversation. In A.C. Baker et al. (Eds.), *Conversational learning: An experiential approach to knowledge creation* (pp. 1-14). Westport, CT: Quorum Books.

Lave, J., & Wenger, E. (1991). *Situated learning: Legitimate peripheral participation.* Cambridge, UK: Cambridge University Press.

OECD. (2000). *Knowledge management in the learning society.* Center for Educational Research and Innovation.

Tezcan, M. (1989). *Sosyal Degisme Sürecinde Çankiri Yaran Sohbetleri.* Ankara: Kültür Bakanligi Yayinlari.

Wenger, E. (1998). *Communities of practices: Learning, meaning, and identity.* Cambridge, UK: Cambridge University Press.

Wenger, E., & Synder, W.M. (2000). Communities of practice: The organization frontier. *Harvard Business Review,* (January-February), 139-145.

Yilmaz, A. (1985). *Folklor ve Halk Edebiyati Yönüyle Sosyal Teskilat Yeren.* Unpublished Master's Thesis, Ankara.

RESOURCES

Unfortunately, not much information about the Yaren Talks is available in English. We hope that the Internet sources below can provide some useful information for interested readers. See Table 1.

KEY TERMS

Ahi (Akhi) Association: Guilds in the Ottoman Empire that integrated vocational principles and practices with spiritual and moral ones, enabling them to

Table 1. Internet resources

Web Link	Author	Last Access	Explanation
http://www.discoverturkey.com/english/kultursanat/halk-eren.html	Ministry of Culture, Republic of Turkey	April 9, 2005	Ahi Organization, Turkish Humanism, and Dervishes[+]
http://coursesa.matrix.msu.edu/~fisher/hst373/readings/inalcik8.html	Halil İnalcık	April 9, 2005	Ottoman Economy, Government, and the Guilds[+]
http://steppenreiter.de/craftsman.htm	T.M.P Duggan, Turkish Daily News	April 9, 2005	A brief history of Ahi Associations[+]
http://www.sadibayram.com/?page=makaleler&mid=106&id=2	Sadi Bayram	April 9, 2005	Origins of Ahi Organization[+]
http://www.cankiri.gov.tr/ana/kultur/yaran.htm	Yaşar Ateşsoy	January 26, 2005	Detailed information about Yaren customs*
http://www.karatekin.net/yaren/	Organization of Karatekin	January 26, 2005	Slideshows with photographs about Yaren meetings*
http://www.cankirininsesi.com/kose_sozucar_faydasizlik.htm	A. Gazi Tekin	January 25, 2005	A critique of the Yaren Talks

fulfill an important role in economic, social, and cultural development.

Yaren (Yâren, Yaran) Talks: A social institution that has assumed responsibilities for educating members of the community, maintaining social security and order, and sustaining traditional arts and customs in Anatolia for centuries during the Ottoman Empire and Republic of Turkey periods.

Managing Complexity via Communities of Practice

Lai Ling Ng
Northumbria University, UK

Jon Pemberton
Northumbria University, UK

INTRODUCTION

Globalisation, liberalisation of trade, internationalisation of financial markets, and the information technology revolution are but some of the developments that organisations have had to contend with in the last few years. There are, therefore, huge challenges for business leaders in the wake of constantly shifting global competition and ever-increasing change, underpinned by complexity, unpredictability, instability, and ambiguity (Nixon, 2003).

In dynamic and complex environments, it is essential for organisations to continually create, validate, and apply new knowledge in the development of their products, processes, and services to ensure they add value (Bhatt, 2001). In essence, organisations seek to differentiate themselves on the basis of what they know, and managers of successful organisations are consistently searching for better ways to improve performance and results. Indeed, frequent disappointments with past management initiatives have motivated managers to gain new understanding into the underlying, but complex mechanisms, like knowledge, which govern an organisation's effectiveness (Wiig, 1997). Knowledge is, however, not a rigid structure that excludes what does not fit—it can deal with complexity in a complex way (Davenport & Prusak, 2000).

A variety of approaches to knowledge management (KM) exist, many relying heavily on technology. However, the focus of KM has moved from an early emphasis on technologies and databases to a keen appreciation of how deeply corporate knowledge is embedded in people's experience. Organisations have learned that technology is the easy part of supporting knowledge creation; the difficult aspect is working with people to improve collaboration and knowledge sharing (Allee, 2000). To sustain long-term competitive advantage, an organisation needs to ensure a fit between its technological and social systems. In effect, technologies can be used to increase the efficiency of the people and enhance the information flow within the organisation, while social systems facilitate better communications and understanding of complex issues by bringing multiple viewpoints to a variety of situations (Bhatt, 2001).

This socio-technical view of KM has spawned a number of initiatives in recent years embracing organisational, cultural, and individual issues (Pemberton & Stonehouse, 2000). One in particular, the notion of a community of practice (CoP), has played, and continues to play, a significant role in knowledge exchange and creation. Indeed, CoPs are KM's mechanism of choice and are a valuable means of unlocking this hidden treasure (McDermott, 2000). In this sense, CoPs have an important role in the management of complexity within organisations; this is the focus of this particular article.

BACKGROUND

Complexity refers to the degree to which the structure, behaviour, and application of an organisation is difficult to understand and validate due to its physical size, the intertwined relationships between its components, and the significant number of interactions required by its collaborating components to provide organisational capabilities. Furthermore, the complexity and volume of global trade today is unprecedented, with the number of global players, products, and distribution channels much greater than ever before. Information technology has caused these global elements to change rapidly, and the decline of centralised economies has created a more frenzied atmosphere within many organisations that feel compelled to bring new products and service to wider markets ever more quickly. This combination of global reach and speed compels organisations to ask themselves what they know, who knows what, and what they do not know (Prusak, 2001).

Strategies adopted by organisations that have previously placed emphasis on higher productivity via lean production methods, or shorter time-to-market through concurrent engineering, no longer provide differentiation, ensuring only survival, not growth (Rajan, Lank & Chapple, 1998). Thus, in the current climate, accelerated innovation by exploiting knowledge within the organisation is becoming the means by which superior performance ensues, with competitive success governed by an organisation's ability to develop new knowledge assets that create core competences to generate superior performance (Pemberton & Stonehouse, 2000). Moreover, the ability of individuals to apply their cognitive skills to extract and generate knowledge from existing sources of information has improved organisations' effectiveness to innovate and disseminate learning (McDermott, 1999).

Traditionally, the old adage 'knowledge = power' has been cherished by individuals for centuries, leading to the hoarding and protection by individuals of their knowledge assets. In an organisational context, however, the use of organisational knowledge relies on the sharing of this knowledge, and as it is shared, it multiplies (Allee, 2000). As a consequence, a radical rethink of basic business and economic models has catapulted the issue of knowledge to the top of the management agenda, and it has, arguably, become one of the most important factors of economic life (Stewart, 1997).

KM focuses on improving the means by which individuals' knowledge—and collectively held knowledge—is produced and integrated in organisations (Lesser & Storck 2001; Brown & Duguid, 2000, 1991; Scarbrough, 2003). Many researchers and practitioners in the knowledge arena argue that there is no knowledge outside of the individual, and they view externalised knowledge merely as information (Allee, 2000). The tacit knowledge that resides in the minds of individuals makes it difficult to share, and in this sense, the CoP has a significant

role to play in creating an environment that stimulates knowledge transfer and creation.

CoPs are physical or virtual groups of people who share a passion for something that they know how to do and who interact on a regular basis to learn how to do it better (Wenger, McDermott & Snyder, 2002). Essentially, they create, expand, and exchange knowledge and develop individual capabilities (Wenger & Snyder, 2000). They are formed through groups of many individuals who collaborate in the production of new knowledge of a mutually held kind. CoPs exist in a variety of forms, but they share a basic structure. A CoP is a unique combination of three fundamental elements: a domain of knowledge, which defines a set of issues; a community of people who care about this domain; and the shared practice that they are developing to be effective in their domain (Wenger et al., 2002).

This structure, outside the formal hierarchy of an organisation, provides the CoP with the potential to examine complex scenarios within organisations and provide a mechanism for dealing with it in an innovative and often unconventional manner.

CoPs and Complexity

In many organisations, CoPs are fostered to address untapped collaborative opportunities to serve as leverage to gain competitive advantage in today's complex business environment (Fontaine, 2001). The development and widespread adoption of global networks and communication protocols have not only made it possible, but also economically feasible to interconnect employees in large and geographically distributed organisations, allowing them to exchange information (Anand, Charles & William, 1998). This is a key process in the creation and management of collective knowledge, without which an organisation may not be able to extract this most valuable asset as a potential source of competitive advantage in complex environments. In this section, a number of pertinent themes of CoPs are introduced, many of which can assist in handling complexity.

Peer-to-Peer Help in Problem Solving

CoPs are informal networks that emerge of their own accord where members informally share knowledge in unstructured discussions. In this sense, CoPs further interact between members and identify whom participants can ask for help with a problem, thereby allowing them to ask questions so that peers can quickly comprehend and focus on the heart of a particular scenario. In more developed and established CoPs, exchange of ideas, building of skills, and the development of networks can take place via their own conferences and dinners, as well as the more usual chat rooms and informal social gatherings (Wenger & Snyder, 2000). Either way, these CoPs help to increase access to expertise to find answers, solve problems, and respond more quickly to complex situations in today's competitive environment.

Developing and Verifying Best Practices

A CoP does much more than work on specific problems—it is also an ideal forum for sharing and spreading best practices across an organisation. In the course of socialising, members develop a collective pool of practical knowledge that any one of them can draw upon (Wenger & Snyder, 2000). By exchanging their interpretations, members can build their own communities and share efficient techniques of working through different and complex situations. These members interact directly, use each other as sounding boards for new ideas, and help each other to learn, using one another as critical resources. As they interact with others, they are likely to understand and share their views of the same situation in a different light. This interaction process is helpful in developing a holistic view of the range of complex problems and situations, thereby facilitating the integration of a diverse body of knowledge in the organisations and, as a consequence, developing knowledge and verifying best practices (Bhatt, 2001).

Developing Capabilities to Thrive

With the knowledge and best practices CoPs are able to create, they are also able to facilitate the capture and re-use of existing knowledge assets and retention of organisational memory. Members engage themselves in conversations, experimentation, and shared experiences with other people who do what they do, and thus as they move beyond routine processes into more complex challenges, they rely heavily on their CoPs as their primary knowledge

resource (Allee, 2000). This is important especially in dynamic and complex environments where organisations face a series of unexpected problems and unforeseen situations that are difficult to control by one individual in the organisation. Yet, by coordinating the pattern of interaction between its members, technologies, and culture, organisations can work with complex and novel situations (Hutchins, 1991). Weick and Roberts (1993, p. 360) refer to these interaction patterns as the "collective mind" of the organisation. Through these meetings of minds, community members are able to find solutions that result in increased capability and improved performance that are critical for the organisation to thrive in complexity (Saint-Onge & Wallace, 2003).

Fostering Unexpected Ideas and Innovation

A conducive environment, where one solution leads to another, can generate innovative ideas. A key source of innovation is the close interaction with members in a CoP that stimulates more productive conversations. For example, innovative ideas do not appear from nothing; they are born in mindful, mind-opening, and productive dialogues. CoPs provide a safe environment to share problems and challenges, as well as test new ideas. In effect, members build upon one another's ideas in a high-trust vessel for exchange and will contribute significantly to elevate innovation of an organisation (Por, 2003b). In addition, participants of CoPs share their experiences and knowledge in free-flowing, creative ways that foster new approaches to problems (Wenger & Snyder, 2000). Hence, many of the small-scale, individually insignificant ideas scattered around an organisation add up to an enormous amount of knowledge creation of unexpected ideas and innovation that can help to solve complex problems more proficiently and effectively, or may lead to innovations in the marketplace.

Building a Learning Culture

Since CoPs typically house the valuable knowledge and practice in an organisation and where people really learn, they have the ability to build a robust learning culture on a small scale, without taking on the entire organisation. Having developed the ability to learn in a CoP, an organisation establishes a platform where collaborative problem solving and innovation are readily internalised in the manner in which individuals do their work (Por, 2003b). This fosters a culture in which innovation occurs naturally, almost as a by-product of how people perform their work. Over time, CoPs develop their own culture, and they can transform an organisation's culture through their collective influence on members and on other CoPs with whom they interact (Wenger et al., 2002). Indeed, CoPs can be instrumental to the cultural transformation of an organisation and facilitate the solution of complex problems.

Crossing Boundaries

CoPs are as diverse as the situations that give rise to them. A CoP can exist entirely within a business unit or stretch across divisional boundaries, and a community can even thrive with members from different companies (Wenger & Snyder, 2000). Yet, they are non-canonical and not always recognised by the organisation. Furthermore, they are more fluid and interpenetrative than bounded, often crossing the restrictive boundaries of the organisation and different time zones (Brown & Duguid, 1991). More often than not, knowledge has difficulty crossing boundaries of practice even within an organisation, but it flows easily within a practice, no matter what other boundaries exist (Wenger et al., 2002). A CoP makes knowledge "leaky"; this is essential since there is an increasing need to cross boundaries, as today's complex problems frequently require solutions that are not confined to any one practice or even to a single organisation.

Facilitating Change

Rigidity is a global dysfunction that occurs as a result of disorders at the level of core practices. These practices in the organisation are built over time, and gain value and momentum as problems are solved (Por, 2003a). The instability mentioned earlier in this article demonstrates that CoPs are adaptable by taking it on a path, whereby new interactions are explored so that it can eventually adapt to new goals and environmental changes (Huberman & Hogg,

1995). Only then can organisations possess the requisite flexibility to truly thrive on complexity. As companies try to keep pace with rapid changes in technology and cope with increasingly unstable business environments, CoPs can be a catalyst for change, helping organisations to evolve in significant ways by preparing them to operate successfully in the knowledge era (Saint-Onge & Wallace, 2003).

Driving Strategy

CoPs are informal networks that coexist within the formal structure of the organisation and serve many purposes, such as resolving the conflicting goals of the business, solving complex problems in more efficient ways, and furthering the interest of their members, to name a few. They thrive when the goals and needs of an organisation intersect with the passions and aspirations of its members (Wenger et al., 2002). With this convergence of purpose, organisations can be successful if there is a strong link between individual and organisational capabilities (Saint-Onge & Wallace, 2003). In spite of their lack of official recognition, CoPs' informal networks can provide effective ways of learning, a sharpened sense of belonging, and with the proper incentives, actually help to drive the strategic direction and enhance the productivity of the formal organisation (Huberman & Hogg, 1995).

FUTURE TRENDS: THE WAY AHEAD

The focus on knowledge is particularly acute in complex global environments where the development and delivery of timely and innovative products across heterogeneous cultures, localities, and markets is critical. In order to deal effectively with such challenges, organisations require knowledge to operate effectively across the temporal, geographic, political, and cultural boundaries routinely encountered. CoPs play a significant role in providing opportunities and overcoming the difficulties associated with sharing knowledge and transferring best practices within and across organisations—as such, they offer an alternative means by which complexity can be handled.

Cultivating a CoP is a practical way to manage knowledge as an asset, just as systematically as organisations manage other critical assets—they become the main focus and the primary means for realigning organisations to mirror the demands of an uncertain and unpredictable marketplace. The domain of CoPs' discussions may be their "war stories" and what is in the forefront of their work. As a by-product, as well as improving their capabilities in their field, these updates, ideas, and solutions for problem solving through learning can collaboratively add more value to the products and services that organisations offer to their customers. This sets up a mechanism by which employees exchange information, help each other in problem solving, and share new ideas to develop best practices and capabilities that facilitate organisations to thrive in complexity. Essentially, this is one of the key processes that enable organisations to leverage its most valuable asset.

In reality, successful organisations strive to produce an environment for encouraging broad participation between workers and managers. CoPs have a major role to play in this respect, as organisations can establish a shared goal where new knowledge is created and becomes necessary for continuous improvement and innovation. Additionally, the informal networks of a CoP also enable fragmented knowledge, often spanning cross-functional boundaries and embedded within individuals, to identify and share for the good of the organisation.

In a time of globalisation and disaggregation, organisational boundaries, a source of additional complexity in themselves, must be taken on board to avoid the problems they raise and take advantage of the opportunities they present. Effectively, CoPs can create links amongst individuals across these boundaries, and further afield, act as a vehicle for dealing with complex issues arising from them. In the process, CoPs create the potential for organised change far beyond an individual's capacity to change.

As a result, it is also important that a learning culture is cultivated for knowledge creation in organisations. For people to learn, a CoP can foster a learning culture for knowledge creation where members reinforce each other's perceptions and aspirations, and the potential is not just individual, but collective. As CoPs transform the organisations in which they thrive, knowledge becomes the central driver of an organisation's business and CoPs play an increasingly integral role here. When a CoP is aligned to the organisation's strategy, it begins to share knowledge for enhancing each other's learning capability and increasing the organisation's knowledge. Over time, if the collective influence of a CoP pervades the organisation, it is capable of transforming the culture of an organisation, and makes it more flexible and responsive to change.

For CoPs to contribute to an organisation, it is critical for management to acknowledge and support them as an important and complementary vehicle for the sharing of knowledge. Cultivating and encouraging CoPs in the long term helps to build both the communities and their shared practices, thus developing capabilities that are critical to the continuing success of the organisation to thrive in complexity.

CONCLUSION

Finally, as knowledge is the lifeblood of managing complexity, the future belongs to organisations that have learned to truly unleash the creative powers of a CoP. CoPs are an effective mechanism for nurturing collaborative culture, and escalating the speed and quality of knowledge diffusion. Their emergence in many organisations is symptomatic of a dynamic and forward-looking organisation. Indeed, it might be argued that where an organisation does not have a coherent strategy of actualising the potential of CoP for value creation, it may have negative consequences for its competitive strategy. The sooner an organisation recognises that fact, and its implications, the better the chances of dealing with innovation and creativity against a backdrop of ever-increasing complex business scenarios.

REFERENCES

Allee, V. (2000). Knowledge networks and communities of practice. *Journal of the Organisation Development Network, 32*(4). Retrieved March 10, 2004, from *http://www.odnetwork.org/odponline/vol32n4/knowledge_nets.html*

Anand, V., Charles, C.M., & William, H.G. (1998). An organisation memory approach to information management. *Academy of Management Review, 23*(4), 796-809.

Bhatt, G.D. (2001). Knowledge management in organisations: Examining the interaction between technologies, techniques and the people. *Journal of Knowledge Management, 5*(1), 68-75.

Brown, J.S., & Duguid, P. (1991). Organisational learning and communities of practice: Towards a unified view of working, learning and innovation. *Organisation Science, 2*(1), 40-57.

Brown, J.S., & Duguid, P. (2000). Balancing act: How to capture knowledge without killing it. *Harvard Business Review, 78*(3), 73-80.

Davenport, T.H., & Prusak, L. (2000). *Working knowledge: How organisations manage what they know.* Boston: Harvard Business School Press.

Fontaine, M. (2001). Keeping communities of practice afloat. *Knowledge Management Review, 4*(4), 16-21.

Huberman, B.A., & Hogg, T. (1995). Communities of practice: Performance and evolution. *Computational and Mathematical Organisation Theory, 1*, 73-92.

Hutchins, E. (1991). The social organisation of distributed cognition. In L.B. Resnick, J.M. Levine & S.D. Teasley (Eds.), *Perspectives on socially shared cognition* (pp. 283-307). Arlington, VA: American Psychological Association.

Lesser, E.L., & Storck, J. (2001). Communities of practice and organisational performance. *IBM Systems Journal, 40*(4), 831-841.

McDermott, R. (1999). Nurturing three-dimensional communities of practice: How to get the most out of human networks. *Knowledge Management Review, 2*(11), 26-29.

McDermott, R. (2000). Critical success factors in building communities of practice. *Knowledge Management Review, 3*(2), 5.

Nixon, B. (2003). Leading business transformation—learning by doing. *Industrial & Commercial Training, 35*(4), 163-167.

Pemberton, J., & Stonehouse, G. (2000). Organisational learning and knowledge assets—an essential partnership. *The Learning Organization, 7*(4), 184-194.

Por, G. (2003a). Building a case for communities of practice: What makes communities of practice an economic imperative? Retrieved January 20, 2004, from *http://www.co-i-l.com/coil/knowledge-garden/cop/thought.shtml*

Por, G. (2003b). Innovation and communities of practice. Retrieved March 20, 2004, from *http://www.co-i-l.com/coil/knowledge-garden/cop/thought.shtml*

Prusak, L. (2001). Where did knowledge management come from? Retrieved April 8, 2004, from *http://www.knowledgeboard.com/cgi-bin/item.cgi?id= 73527/*

Rajan, A., Lank, E., & Chapple, K. (1998). *Good practice in knowledge creation and exchange.* CREATE: Turnbridge Wells.

Saint-Onge, H., & Wallace, D. (2003). *Leveraging communities of practice for strategic advantage.* Boston: Butterworth-Heinemann.

Scarbrough, H. (2003). Why your employees don't share what they know. *Knowledge Management Review, 6*(2), 16-19.

Stewart, T. (1997). *Intellectual capital: The new wealth of organisations.* New York: Doubleday.

Weick, K.E., & Roberts, K.H. (1993). Collective mind in organisations: Heedful interrelating on flight decks. *Administrative Science Quarterly, 38*(3), 357-381.

Wenger, E.C., McDermott, R., & Snyder, W.M. (2002). *Cultivating communities of practice: A guide to managing knowledge.* Boston: Harvard Business School Press.

Wenger, E.C., & Snyder, W.M. (2000). Communities of practice: The organisational frontier. *Harvard Business Review,* (January/February), 139-145.

Wiig, K.M. (1997). Knowledge management: An introduction and perspective. *Journal of Knowledge Management, 1*(1), 6-14.

KEY TERMS

Community of Practice: A group of self-governing people with shared interests whose practice is aligned with strategic imperatives, helping each other to solve problems, share and benefit from each other's expertise, and are committed to jointly developing better practice.

Complexity: The degree to which the structure, behaviour, and application of an organisation is difficult to understand and validate due to its physical size, the intertwined relationships between its components, and the significant number of interactions required by its collaborating components to provide organisational capabilities.

Industrial Economy: A traditional economic perspective that presumes restricted expansion opportunities based on scarcity of physical resources, available labour, capital, and so forth.

Innovation: A process in which organisations create and define problems, actively developing new knowledge to solve them, and generating new products, processes, or services.

Knowledge: An organisational resource, either explicit or tacit, and often context specific, and the basis on which organisations create superior business performance.

Knowledge Domain: An area of knowledge a community agrees to learn about and advance, representing common ground and a sense of common identity that legitimises the community by affirming its purpose and value to members.

Knowledge Economy: A realm of economic activity encompassing factors forming the core of a high value-added economy where knowledge is a key to economic success and where skills and learning are valued and productively employed.

Knowledge Management: The systematic management and use of knowledge in an organisation by capitalising on individual, collective, and organisational capabilities and expertise.

Socio-Technical: The integration of social and technological systems to ensure optimal utilisation of information and knowledge-based resources.

M

Managing Intellectual Capital and Intellectual Property within Software Development Communities of Practice

Andy Williamson
Wairua Consulting Limited, New Zealand

David M. Kennedy
Hong Kong Institute of Education, Hong Kong

Ruth DeSouza
Wairua Consulting Limited, New Zealand

Carmel McNaught
Chinese University of Hong Kong, Hong Kong

INTRODUCTION

In this article, we will develop a framework for educational software development teams that recognizes the conflicts and tensions that exist between the different professional groups and will assist software teams to recognize the intellectual capital created by individuals and teams. We will do so by recognizing the inherent relationship between the tangible elements of intellectual property and the intangible organizational assets that form the basis of intellectual capital and by discussing how knowledge generated by a project team can become an explicit asset.

BACKGROUND

Universities are increasingly becoming developers of complex software-based applications. In-house development ranges from teaching aids and online learning resources to large information systems products that could ultimately become successful commercial ventures. Increased product complexity is easily recognized, yet research shows that the organizational aspects of a software development project are more likely to affect performance and outcomes than technical issues (Xia & Lee, 2004). Successful development and deployment of today's complex educational systems and environments comes with an imperative for an array of different and unique skill sets for the various stages of each project. One can view a software development team as a microcosm of the wider community of practice of software development professionals who work in information and knowledge management in higher education. As Wenger (1998) observes, such communities of practice are not random but constructed around required skills and through a process of negotiation based on mutuality and accountability.

Workforce mobility has increased: academic staff members regularly and easily move between institutions; development and design staff have many opportunities for contract-based work, move to other academic institutions or into the private sector. The ideas that lie behind a successful process or product are increasingly drawn from a wider pool of talent and, as people move around, these ideas are being taken with them and disseminated through informal and new work practices into a wider community of practice. How then does a team, formed to design and develop a technology-rich educational or systems environment, manage and control issues of intellectual capital and intellectual property such that all of those who contribute throughout the life of a project are acknowledged and rewarded fairly and appropriately for that contribution, even after they have left the project?

Team Formation and Relationships

Additional complexity leads to specialization (Jacobson, Booch & Rumbaugh, 1998). New ways of working bring with them a shift in power, where the academic expert will often lack the technical skills, time or resources to turn ideas into reality. Instead, they must rely on a team of experts from other disciplines to interpret their ideas, evolve them, and deliver the finished product. As complexity increases, communication between team members becomes paramount; specialist educational designers are required to translate pedagogy into functional specifications that can be understood by software developers and graphic designers. Modern software teams are project-based, where resources come and go as required.

Software development communities of practice exist within a larger organizational context. Roles and responsibilities will vary and are negotiated depending on the toolset and architecture used, the size of the project, and the culture of the organization (Phillips, 1997; Williamson et al., 2003). Project team members can be full- or part-time employees (academic or non-academic) or contractors retained specifically for the project. As such, these roles exhibit complex relationships and interfaces between each other and the project. In Figure 1, a range of typical roles and relationships found in a tertiary education software development project are shown.

During the various stages of the development process, various players move into prominent roles. One way to illustrate this shifting set of work responsibilities is to list the main players at each stage of the process. We will do this using the classic instructional systems design (ISD) model (Dick & Carey, 1990) as it is so well known. (There are many other models, many of which are discussed in Bannan-Ritland, 2003.) The key players at each stage of the ISD model are listed in Table 1. In reality, each team parcels out the work depending on the skill set of individuals in the team.

It is important to be aware of the different communities of practice that exist in this field and ensure that the role of individual team members is able to be

Figure 1. Intra-project relationships in software development teams (Williamson et al., 2003, p. 345)

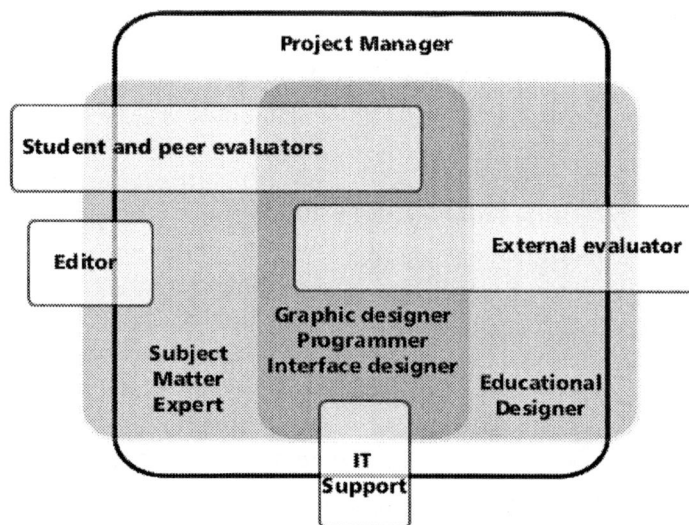

Table 1. Key players at each stage of the ISD model

Stage of the ISD model	Key players
Needs assessment	Subject matter expert
Analysis	Subject matter expert, Educational designer
Design	Subject matter expert, Educational designer, Project manager
Development	Project manager, Graphic designer, Programmer, Interface designer, Editor
Formative evaluation	Student and peer evaluators, Subject matter expert, Educational designer
Revision	Project manager, Graphic designer, Programmer, Interface designer, Editor
Implementation	Subject matter expert, IT support
Summative evaluation	External evaluator
Maintenance	IT support, Subject matter expert (Interface designer)

promoted appropriately. Professional recognition can come through either publication, a portfolio of work or through the finished product, and the importance of a successful project to the career development of individuals should not be underestimated. It is important to ensure that academic dissemination of successful projects through publication recognizes the contribution made by all team members, including the non-academic members. Many myths persist in relation to acknowledging the veracity of contribution with regard to educational software, and these often have the potential to leave team members feeling their effort and ideas have gone unrecognized and, at worst, feeling they have been exploited (Williamson et al., 2003). In the second half of this article, we will develop a framework that ensures appropriate outlets for reward and recognition of individual contributions within academic software development teams. Before doing this, we will define what is meant by intellectual capital and intellectual property.

Defining Intellectual Capital and Intellectual Property

Florida (2002) argues that the principal factors for successful software development are talent, knowledge, and intellectual capital (IC). The connection of new ideas and existing knowledge within an organizational context leads to the creation of IC. Stewart (1999) defines IC as the sum of everything everybody in a company, organization, or team knows and which provides some advantage over their competitors. Davidson and Voss (2001) agree, describing individual IC as "the sum of individual imagination, intelligence and ideas" (p. 60). They then extend this definition to encapsulate a model for organizational IC that is based on the talent of individuals (human capital), the knowledge that is captured within systems and processes (structural capital), and the characteristics of relationships with customers and suppliers (customer/supplier capital). Organizational IC comes from the "interplay of all three (structural capital augments the value of human capital, leading to an increase in customer/supplier capital)" (p. 61). In terms of this discussion as it relates to the appropriate recognition and acknowledgement of individual contributions within software development teams, human capital is our primary focus. Human capital is "what walks out of the

door at the end of the day" (p. 68); it is a vital intangible.

If IC is the intangible but invaluable contribution of human talent to a project, then Intellectual Property (IP) is a formal measurable subset. It is the tangible product that results from the idea. The UK Patent Office (United Kingdom Patent Office, n.d.) defines four formal types of IP:

- patents for inventions;
- trademarks for brands;
- designs for product appearance; and
- copyright for material (including software and multimedia).

This definition is then extended to cover a much broader and often more intangible grouping that extends to trade secrets, plant varieties, geographical indications, and performers rights. While many see copyright as a way of protecting IP, it is only a subset. Copyright provides recognition of their invention to the creators of software or multimedia products in order for them to be able to obtain economic rewards for their efforts (Macmillan, 2000). Historically, comparisons have been drawn between software development and the traditional arts, such comparisons reinforcing an argument that IP law is focused on the protection of software such that others are not able to modify the source product (White, 1997). It is important to note that copyright extends only to a tangible product, it does not lend protection to the more intangible areas of IC such as ideas and individual contribution. Since copyright has a primarily commercial imperative, it is a limited and perhaps inappropriate mechanism for acknowledging contribution. This is of greater importance in higher educational settings since copyright of educational materials can reside with the institution (particularly with off-campus courses), rather than the individual, and very few educational software products developed for specific content domains in higher education are ever commercialized (Alexander, McKenzie & Geissinger, 1998).

The Relationship between IC and IP

A relationship exists between the tangible elements of intellectual property and the three forms of

Figure 2. Intellectual capital and intellectual property (Williamson et al., 2002, p. 342)

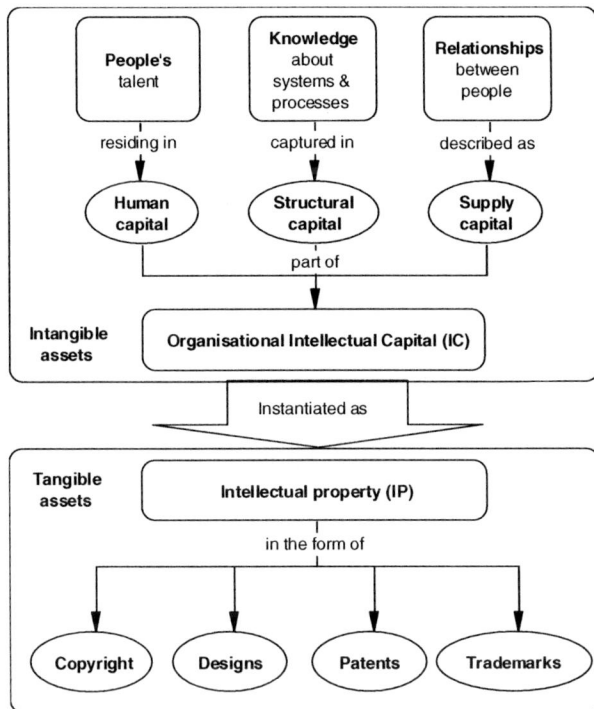

intellectual capital (the intangible organizational assets) discussed in the preceding section. These are shown in Figure 2.

IC/IP Management Framework

Having addressed the complexity of educational software development teams and defined IC and IP within an educational software context, we will now develop a framework that can be used to ensure proper recognition and reward for individual and collective ideas in such a setting.

Given the critical value of IC in software development (Florida, 2002), it is important that the processes used within educational software development are strengthened and formalized through the adoption of a strong project management framework. Project management is a key role in any project involving information and communication technologies and interactive multimedia software, and it requires specific skills and attributes. These include both the hard skills of contract negotiation, budgeting, scheduling, project definition, and scoping as well as the soft skills of human relations, team building, and facilitation

(Burdman, 2000; Schwalbe, 2000). Successful teams work well together because they have clear roles and relationships and because the terms of engagement within the team and with external parties are well defined, understood, and agreed by all. This provides a solid platform for the explicit incorporation of IC and IP policies into project documentation so that such issues can be considered early on, preferably during the project scoping phase.

A process and framework are required to recognize knowledge as it is created so that it becomes explicit. Without doing so, knowledge remains tacit and cannot be rewarded or acknowledged, that is, credited to the appropriate team members in the future. Extending this concept, knowledge that is explicit within the team can remain tacit beyond team boundaries if no process is in place to ensure appropriate recognition of contribution. It is, therefore, necessary for teams to negotiate clear, up-front delineation of roles, responsibility, and ownership of both tangible and intangible outputs from the project. This does not prejudge what that ownership might be, merely that the agreement takes place before the project commences. It is important to consider how IC/IP generated during the project's life will be disseminated, in what form, and by whom. Such a clear articulation of roles and responsibilities has the benefit of helping to make the process of dissemination more visible. By doing so, it is hopefully the case that team members will recognize the significance of the different sources of acknowledgment. This in turn will result in up-front agreement on potential opportunities for dissemination of original ideas among the team.

A seldom discussed aspect of the manner in which ideas might be disseminated (and credit obtained) is the potential synergy between individual team members. For example, among academic staff involved in the project, there is a possibility for cross-disciplinary publications.

This framework, shown in Figure 3, maps out two axes: the horizontal axis representing formal ownership of the tangible IP, the vertical axis representing a continuum of recognition for the IC generated during a project, ranging from no acknowledgment of individual effort and contribution to a full public acknowledgment. Intermediate steps include recognition at the team and institutional level.

Enacting the IC/IP Framework

Our discussion so far has shown that, regardless of the nature of the IP ownership, academics and professionals working in software development teams need appropriate recognition for their contributions, but certain factors can prevent this from happening. The challenge, therefore, is to identify a set of project attributes that can be used to inform project management practices such that institutions are cognizant of the need for appropriate recognition. In the following section, we identify seven key attributes of, or processes within, a successful project. The model is developed from a review of the authors' own experiences of software development teams where prob-

lems had occurred. This review led to the identification of which weaknesses in the process had resulted in these problems (Williamson et al., 2003). By ensuring that these seven attributes are recognized and actively negotiated by newly forming teams and enacted throughout the life of the team, this model can assist projects in identifying and filling gaps in the structure of development teams, hence future risk can be mitigated.

In essence, the IC embedded in the members of the project team is articulated in terms of the various IP contributions made by these team members. However, if the focus is exclusively on the tangible products of the process (e.g., software and papers) through only considering the IP, then the worth of

Figure 3. IP/IC management framework

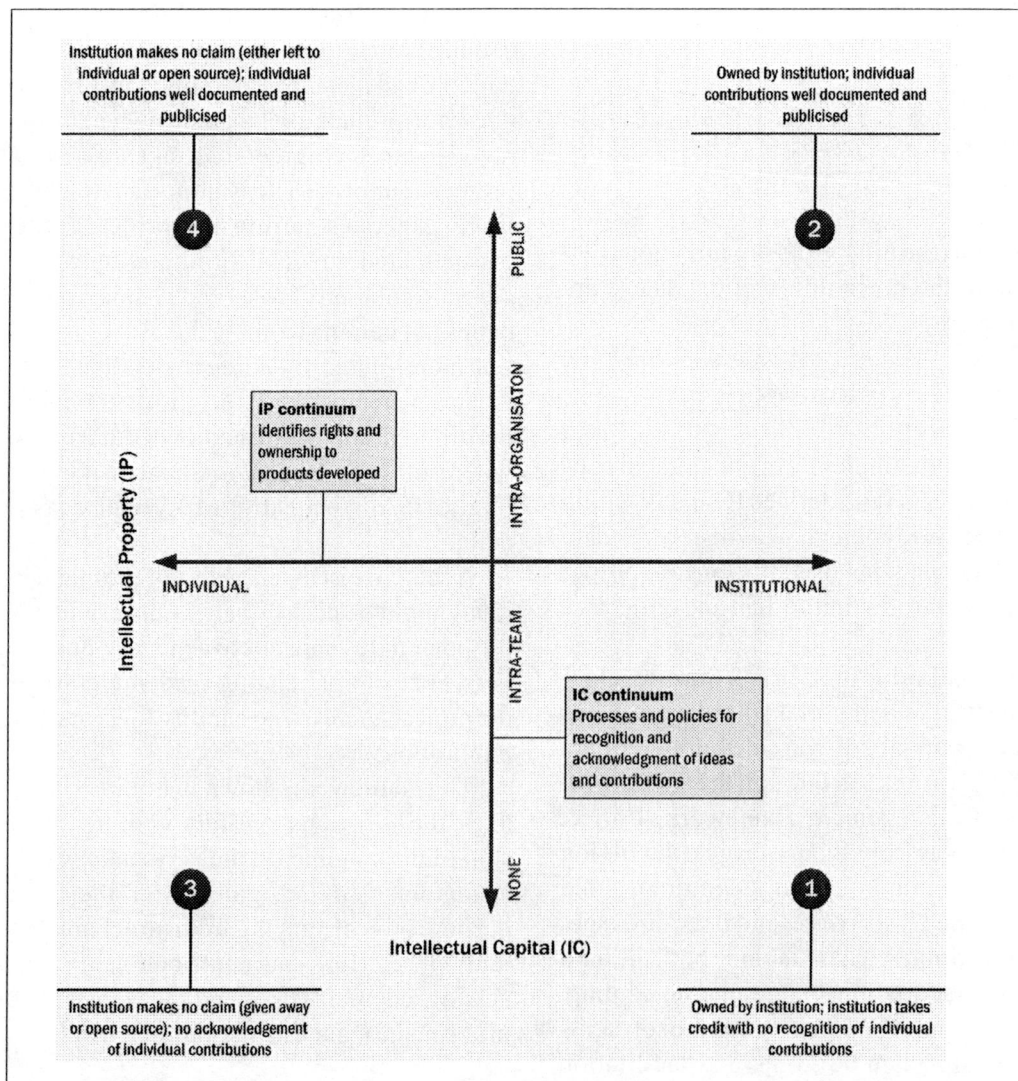

ideas (IC) can be underplayed, and their potential may not be realized. We are suggesting that explicit application of these seven guidelines can ensure more successful project outcomes and positive professional outcomes for all members of the team.

The nature of an effective community of practice for software development teams is discussed in terms of Figure 3. The two major axes and the four examples in Figure 3 are used to frame the seven attributes.

Intellectual Property: Individual Affirmation to Institutional Affirmation

Have an IP Acknowledgement Strategy

Highly successful projects exhibit a strong team dynamic which arises when the expertise and knowledge of individual team members can be communicated and shared with others. Part of this process involves ensuring that ideas are fairly acknowledged within and outside the team, whether by portfolio (graphic artists), publication (academics), or product (project managers and programmers).

Have an IP Review Strategy

It does not matter for the purposes of academic critical review whether the subject of study is a written paper, a software product, or a portfolio. Contribution from individual team members needs to be acknowledged through an inclusive authorship policy which is regularly revisited in team meetings. This process can strengthen collegiality and reinforce mutual valuing between team members.

Have a Strategy to Separate IP from IC

The IP might be owned by an organization or institution, but the IC remains with the individuals in the team. Formal acknowledgment of where the ownership of IP lies is important and needs to be negotiated ahead of the commencement of the project. In many higher education institutions, this has become standard practice and involves retaining a competitive advantage and protecting the resources produced by employees of the organization. There are risks associated with key project contributors leaving (for example, a lead programmer) and either taking intel-

lectual property with them or holding a software development team or institution to ransom by withholding access to code or other resources. In some organizations, the IC also remains with the organization via means of a nondisclosure agreement. Communities of practice might consider using a confidentiality agreement as part of a contract or offer of employment in order to keep this issue open and transparent.

Longevity Strategy: Ideas Remain

When a person leaves a team, they cede their IC to the project team or institution, and that contribution should continue to be recognized and acknowledged in project documentation, appropriate publications, and authorship in any finished product. In some projects, this may also involve ceding formal IP to the project (e.g., in the case of commercialization).

Intellectual Capital: No Formal Acknowledgment to Public Acknowledgment

Recognize the Emergent Nature of the Software Development Process and Its Impact on IC/IP for all Team Members

As software becomes more complex, it becomes less and less likely that the original academic imperative that led to the idea for the product will be instantiated in a form initially envisaged by the academic or the organizational unit that initiated the project. The development process and the end result will be strongly influenced by a wide range of individual and group contributions to the process and the product.

Ideas are Perishable

Software has a shelf-life, hence the IC that led to that product is also of limited use. The idea will become superseded and outdated as new ideas and new technologies emerge. For example, there are any number of commercial or free customizable online survey instruments (such as Survey Monkey, http://www.surveymonkey.com) that now exist. Learning Evaluation Online (LEO) was an early system that explored how customizable educational surveys could

Table 2. Implications for IP and IC

#	Scenario	Implications for IP and IC
1	IC and IP is owned by institution; institution takes credit with no contribution of individual creativity and effort.	This is a very poor scenario for developing the IC of an institution. Without affirmation, individuals will seek employment elsewhere and take their IC with them.
2	IP and IC are owned by the institution; individual contributions are well documented and publicized.	This is the scenario in a number of institutions worldwide, particularly those involved in distance education. This scenario is problematic when commercial aspects enter the situation as in the case of patents.
3	Institution makes no claim (software is given away or open source); no acknowledgment of individual contributions.	This is often the result of small student projects (although many institutions claim the IP of all undergraduate student work, not postgraduate) undertaken during a course of study). Most software of this type has a very limited life although there are some exceptions (Gunn, 1995).
4	Institution makes no claim (either left to the individual or open source community); individual contributions are well documented and publicized.	This applies to postgraduate work in universities. In many institutions, postgraduate (especially doctoral) students own their IP, and it is up to the student and supervisor to disseminate the details of the project. This aspect is changing as universities try to gain a competitive advantage, and many postgraduate students working in a large research department would do well to consider how the results of their studies might be retained, negotiating with the university in the early phases of the project. For example, some student projects (see moodle.org and moodle.com) have become very high profile products (Dougiamas & Taylor, 2003).

Table 3. Example of the application of the seven attributes of the IP/IP framework

#	Attribute	Enactment
1	Have an IP acknowledgment strategy	The acknowledgment of IP was never an issue within the project group. Individual contributions were always acknowledged by the core development team, becoming part of the documentation of the project. The extensive documentation ensured that no one was left off the credits on the *An@tomedia* Web site.
2	Have an IP review strategy	The existence of "prior art" was established in the early phases of the project. While the final product did not resemble the initial designs, it was always clear in the project meetings that the team was involved in the instantiation of the educational vision of the project leader (one of the core SMEs).
3	Have a strategy to separate IP from IC	The strategy to separate IP from IC was undertaken by the four principal authors as *An@atomedia* was moved from an interesting project to a commercial product. The documentation resulting from meetings included discussions of commercialization of *An@atomedia* and the associated need to separate IP from IC. The strategy adopted involved consultations with the university's legal advisors and the project team. The IP for the sale and commercial rights to *An@tomedia* were ceded to the four key authors by the other team members; however, the IC remained with members of the project team to use as they required.
4	Longevity strategy: Ideas remain	The credits list contains a list of all members who contributed to the project over a period of many years, including those individuals who either retired (in one case) or moved to other institutions (a number of people). It is possible for all members of the project team to include evidence of contributions to *An@atomedia* by reference to either the Web site or the CD-Roms (the form in which *An@tomedia* is published and sold).
5	Recognize the emergent nature of the software development process and its impact on IC/IP for all team members	The development of *An@tomedia* occurred over a considerable period of time. The genesis of some the clinical approaches adopted in the project were developed by the project leader and occurred well before *An@tomedia* commenced (Eizenberg, 1988, 1991). The use of technology followed as a consequence of the need to develop more effective and engaging approaches to the teaching of anatomy (Driver & Eizenberg, 1995). As the final design of the software emerged, it was always clear in meetings and associated documentation that other members of the project team were involved in the instantiation of prior concepts and developments in new and innovative ways, but the underlying concept derived from the earlier work in paper-based media.
6	Ideas are perishable	*An@atomedia* received a number of awards for innovation and excellence after the first release (see http://www.anatomedia.com/credits.shtml). However, as people come and go from the project, the initial ideas will be superseded or altered to reflect teaching evaluations, changes in the medical curriculum, and improvements in technology. Solutions developed in 1999 or 2000 may not be suitable in 2005. What was once a good idea may not be appropriate in the future, but the three major methods of affirmation remain—publication, portfolio, and vitae for all contributions.
7	Public acknowledgment of IP/IC requires the source material to be in the public domain	The *An@atomedia* Web site provides definitive acknowledgment of the specific contributions of individuals, including the evaluators, programmers, educational consultants, photographers, medical consultants, project managers, dissectors, illustrators, interface and graphic designers, and research assistants, to name a few.

be developed online using an entirely Web-based interface (Kennedy & Ip, 1998). At the time this was an innovative approach, but it has since been superseded by more robust software. Thus, the IC for LEO has long since expired. The idea behind LEO has been taken up by others and reproduced using different software code. The code is the instantiation of the idea and is the only part of the project subject to IP rights.

Public Acknowledgment of IP/IC Requires the Source Material to be in the Public Domain

Acknowledgment of unpublished work or work not publicly available is not sufficient to acknowledge IC and IP issues in a publication. In the case of academics where affirmation and professional career progress is at least partially a result of publication in accredited arenas such as books and journals, this is clearly not sustainable. Graphic artists, on the other hand, have their portfolios of work with iterations of visual designs that they take with them to the next project or job; and programmers have compilations of code: for these professionals, the publication is less important or substantive in career development. A key issue for an institution is providing the process by which academic publications can be developed without compromising the IP of the individual or trade advantages in the marketplace.

In summary, the implications for the four scenarios in Figure 3 are shown in Table 2.

THE FUTURE: APPLICATION OF THE SEVEN ATTRIBUTES OF THE IC/IP FRAMEWORK

In order to see how these attributes can be enacted in practice, the example of a major Australian multimedia project, *An@tomedia*, will be used. *An@tomedia* was designed to support problem-based learning (PBL) of anatomy in the Faculty of Medicine at the University of Melbourne (http://www.anatomedia.com). A number of academic evaluations on the role of *An@tomedia* in this PBL learning environment have been published (e.g., Kennedy, Eizenberg & Kennedy, 2000; Kennedy, Kennedy & Eizenberg, 2001).

The software has been successfully commercialized by the four subject matter experts (core SMEs or core authors) after other members of the development team ceded any personal commercial claims to the group by means of a legal document to that effect.

Affirmation and acknowledgment involving publishing for academic members (Kennedy et al., 2000; Kennedy et al., 2001), contributions to portfolios for non-academic members, and public acknowledgments in the *An@tomedia* Web site for every person who contributed in any significant way to the project were not affected by this written agreement. The public affirmation (particularly important for non-academic members of the project team) is illustrated by the observation made by a reviewer of *An@tomedia* in *The Lancet* (Marušiæ, 2004) where she mentions the extensive list of credits for all the members of the team (over 60) involved with the project. This process was accomplished quite simply because matters of IP had been previously discussed in the course of project meetings, and the "prior art" that existed and underpinned the educational approach was well known to all project members. Table 3 summarizes the way in which the seven attributes worked in this project.

CONCLUSION

While formalized tools exist for capturing IP generated during a project, most software development teams lack formal explicit processes for ensuring that the IC generated is accurately and adequately apportioned. This article has raised issues relating to how software development project teams are recognized for their contribution and a simple framework for measuring recognition of contribution has been presented. Seven key project attributes or processes have been identified to assist project teams develop an awareness of how project roles and structures can be negotiated so that tacit ideas and knowledge generated can become explicit. Such a model must recognize that the requirements for, and process of, recognition will differ within different multiskilled teams. The application of the framework to one major multimedia project has been discussed.

REFERENCES

Alexander, S., McKenzie, J., & Geissinger, H. (1998). *An evaluation of information technology projects in university learning*. Canberra: Australian Government Publishing Services, Department of Employment, Education and Training and Youth Affairs.

Bannan-Ritland, B. (n.d.). The role of design in research: The integrative learning design framework. *Educational Researcher, 32*(1), 21-24.

Burdman, J. (2000). *Collaborative Web development: Strategies and best practices for Web teams*. Reading, MA: Addison-Wesley.

Davidson, C., & Voss, P. (2001). *Knowledge management: An introduction to creating competitive advantage from intellectual material*. Auckland, NZ: Tandem.

Dick, W., & Carey, L. (1990). *The systematic design of instruction*. Glenview, IL: Foresman/ Little.

Dougiamas, M., & Taylor, P. (2003) Moodle: Using learning communities to create an open source course management system. In D. Lassner & C. McNaught (Eds), *ED-MEDIA 2003, Proceedings of the 15th Annual World Conference on Educational Multimedia, Hypermedia and Telecommunications* (pp. 171-178). Honolulu, Hawaii. Norfolk VA: Association for the Advancement of Computers in Education.

Driver, C., & Eizenberg, N. (1995). Constructing and deconstructing the human body: Applying interactive multimedia in the learning of anatomy. In J. M. Pearce, A. Ellis, C. McNaught, & G. Hart (Eds.), *Learning with technology: ASCILITE 95. Proceedings of the 12th Annual Conference of the Australian Society for Computers in Learning in Tertiary Education* (pp. 586-587). The University of Melbourne: The Science Multimedia Teaching Unit.

Eizenberg, N. (1988). Approaches to learning anatomy: Developing a program for pre-clinical medical students. In P. Ramsden (Ed.), *Improving learning: New perspectives* (pp. 178-198). London: Kogan Page.

Eizenberg, N. (1991). Action research in medical education: Improving teaching via investigating learning. In O. Zuber-Skerrit (Ed.), *Action research for change and development* (pp. 179-206). Avebury: Aldershot.

Florida, R. L. (2002). *The rise of the creative class: And how it's transforming work, leisure, community and everyday life*. New York: Basic Books.

Gunn, C. (1995). Useability and beyond: Evaluating educational effectiveness of computer-based learning. In G. Gibbs (Ed.), *Improving student learning through assessment and evaluation* (pp. 168-190). Oxford, UK: Oxford Centre for Staff Development.

Jacobson, I., Booch, G., & Rumbaugh, J. (1998). *The unified software development process*. Reading, MA: Addison-Wesley Longman.

Kennedy, D. M., Eizenberg, N., & Kennedy, G. (2000). An evaluation of the use of multiple perspectives in the design of computer-facilitated learning. *Australian Journal of Educational Technology, 16*(1), 13-25. Retrieved May 5, 2005, from *http://www.ascilite.org.au/ajet/ajet16/kennedy.html*

Kennedy, D. M., & Ip, A. (1998). *Learning Evaluation Online (LEO): A customizable Web-based evaluation tool*. In C. Alvegard (Ed.), *Computer aided learning and instruction in science and engineering. CALISCE'98 proceedings* (pp. 255-262). Goteborg: Chalmers University of Technology.

Kennedy, G. E., Kennedy, D. M., & Eizenberg, N. (2001). Integrating computer-facilitated learning resources into problem-based learning curricula. *Interactive Multimedia Electronic Journal of Computer-Enhanced Learning, 3*(1). Retrieved May 5, 2005, from *http://imej.wfu.edu/articles/2001/1/02/index.asp*

Macmillan, F. (2000). Intellectual property issues. In C. McNaught, P. Phillips, D. Rossiter, & J. Winn (Eds.), *Developing a framework for a usable and useful inventory of computer-facilitated learning and support materials in Australian universities. Evaluations and Investigations Program report 99/11* (pp. 189-205). Canberra: Higher Education Division Department of Employment, Education, Training and Youth Affairs.

Marušiæ, A. (2004). Media reviews: Interactive anatomy. *The Lancet, 363*(9404), 254.

Phillips, R. A. (1997). *A developer's handbook to interactive multimedia: A practical guide for educational applications.* London: Kogan Page.

Schwalbe, K. (2000). *Information technology project management.* Cambridge, MA: Course Technology.

Stewart, T. (1999). *Intellectual capital: The new wealth of organizations.* New York: Currency.

United Kingdom Patent Office. (n.d.). *What is intellectual property or IP?* Retrieved May 5, 2005, from *http://www.intellectual-property.gov.uk/std/faq/ question 1.htm*

Wenger, E. (1998). *Communities of practice: Learning, meaning and identity.* Cambridge, UK: Cambridge University Press.

White, J. A. D. (1997). Misuse or fair use: That is the software copyright question. *Berkeley Technology Law Journal, 12*(2). Retrieved May 5, 2005, from *http://www.law.berkeley.edu/journals/btlj/articles/ vol12/White/html/reader.html*

Williamson, A., Kennedy, D. M., McNaught, C., & DeSouza, R. (2003). Issues of intellectual capital and intellectual property in educational software development teams. *Australian Journal of Educational Technology, 19*(3), 339-355. Retrieved May 5, 2005, from *http://www.ascilite.org.au/ajet/ajet19/ williamson.html*

Xia, W., & Lee, G. (2004). Grasping the complexity of IS development projects. *Communications of the ACM, 47*(5), 69-74.

KEY TERMS

Educational Software Development Teams: Software development teams in university settings are multifaceted and multiskilled, requiring the skills of project managers, subject matter experts, educational designers, programmers, graphic designers, interface designers, IT support staff, editors, and evaluators. In many cases, one person can assume more than one role.

Intangible Assets: Higher education institutions are traditionally based upon ideas, one form of intangible assets. However, the modern university frequently seeks to differentiate itself from its competition. In this article, intangible assets include an investment or outcome enjoyed by the institution in knowledge-based resources and processes. Typically, these intangible assets are termed soft assets because they are not either infrastructure or equipment. Some examples are training programs, improvements to organizational communication flows, or new quality assurance systems.

Intellectual Capital (IC): IC is the sum of the individual imagination that, when aggregated, becomes everything everybody in an organization or team knows and which provides them with some advantage over their competitors. Organizational IC comes from the interplay of structural capital, which augments the value of human capital, leading to an increase in customer/supplier capital.

Intellectual Property (IP): IP is a formal measurable subset if IC; the tangible product that results from the idea and is represented and recognized through patents, trademarks, designs, and copyright (which includes software and multimedia). IP can also be extended to cover a much broader and often more intangible grouping that extends to trade secrets, plant varieties, geographical indications, and performers' rights.

Prior Art: In this article, the concept of "prior art" is used in a similar manner to that adopted by the patent office. However, higher education has a long tradition of valuing ideas, not just economic value (as in patent laws). The "prior art" in this instance refers to the intangible ideas (instantiated in the earlier publications) prior to the commencement of a project, such as the An@tomedia project. It was the basis of these earlier ideas that formed the nucleus of the design philosophy and influenced the manner in which the developers reached agreement on design decisions.

Project Framework: Negotiation of a model within the project to ensure that the contributions of all individuals (and their IC) are able to be appropriately recognized.

Recognition: Different professional groups look for and require different forms of recognition for their professional development. Where academic staff focus on publication, designers need to develop a portfolio of work, and software developers receive kudos and build a reputation based on a product that has been developed and the code therein.

Market of Resources as a Knowledge Management Enabler in VE

Maria Manuela Cunha
Polytechnic Institute of Cavado and Ave, Portugal

Goran D. Putnik
University of Minho, Portugal

INTRODUCTION

Knowledge is, undoubtedly, an indispensable asset for organizations to compete effectively (Alavi & Leidner, 2001; Murray, 2002).

New organizational models, such as the virtual enterprise (VE) model, characterized as dynamically reconfigurable information-based global networked structures, are emerging. New technological environments and solutions are being developed to support them, and the importance of knowledge and the capability of managing it by creating the organizational conditions that facilitate the generation, sharing, and application of knowledge are more and more critical.

In a global organization, as defended by Kluge, Stein, and Licht (2001), face-to-face relationships are not possible, giving rise to difficulties in accepting knowledge from outside. This applies more deeply in virtual enterprises (or in virtual organizations) in the interactions among the independent partners who tend more and more to fear the leakage of private knowledge. This situation promotes competition and rivalry and, as suggested by Prahalad and Hamel (1990), impedes collaboration and knowledge sharing, precisely two of the main underlying issues of this organizational model. A supporting environment, such as the market of resources proposed by the authors, is the way to assure effective knowledge management between the members of a virtual enterprise and business strategic alignment enabling the performance improvement of the VE.

In an environment to support VE integration, knowledge management is simultaneously a tool and an object. As a tool, knowledge management can be used by the market of resources to reduce transaction costs in VE integration and VE reconfiguration; as an object, knowledge must be protected and knowledge leakage prevented to assure trust and protection of VE participants.

The broker (an integrating element of the market of resources) is, besides other attributions, responsible for advising the VE owner in identifying and communicating the role of knowledge management within the VE business plan and for ensuring the permanent alignment between business strategy and knowledge strategy within the network of independent enterprises that constitute the VE. The broker must ensure that the global knowledge sharing is not threatened by deficient knowledge management procedures and, simultaneously, that any instance of the VE (as a reconfigurable network) at a given time, is able to respond to the market requirements with its maximum performance, that is, is business aligned.

In this article, we introduce the VE disabling factors and the functionalities for VE integration, briefly present the market of resources as an environment to support VE integration, assuring business alignment and knowledge management, identify the main strengths and problems associated with the implementation of knowledge management functions, and, finally, discuss the main opportunities associated to the implementation and exploitation of the market of resources.

BACKGROUND: VIRTUAL ENTERPRISE INTEGRATION

The virtual enterprise model can be viewed as a global networked and information-based organizational structure in dynamic adaptation (reconfiguration) to the market or business requirements. Virtual enterprises (in a broad sense) are defined as enterprises with integration and reconfiguration capability in useful time, integrated from independent enterprises, primitive or complex, with the aim of taking profit from a specific market opportunity (Byrne, 1993; Camarinha-Matos & Afsarmanesh, 1999; Cunha, Putnik & Ávila, 2000; Preiss, Goldman & Nagel, 1996; Putnik, 2000). After the conclusion of that opportunity, the virtual enterprise dissolves itself. During its lifetime, the VE changes its physical structure (reconfigures) to be permanently aligned with that market opportunity.

We designate, by *resource,* any function, service, or product provided by independent enterprises (*resources providers*), candidates to integrate a VE. The *resource* is a recursive construct; resources can be primitive or complex where a complex resource consists on a meaningful combination of primitive resources.

There are several factors determining the performance of the VE model. In the BM_Virtual Enterprise Architecture Reference Model (BM_VEARM) (Putnik, 2000), the author presents "fast adaptability" or "fast reconfigurability" as the most important characteristic for the competitive enterprise, enabling the agile alignment with the market.

In this section, we introduce the VE disabling factors, the tools proposed to overcome the disabling factors, and the functionalities required to efficiently implement this organizational model.

The Virtual Enterprise Disabling Factors

The main critical aspects associated with the recent concept of dynamically reconfigurable global networked structures; that is, the main factors against networking and reconfigurability dynamics are the *transaction costs* and the *leakage of private knowledge.*

In an ideal business environment, a firm makes an informed assessment of the relevant costs, benefits, and risks of outsourcing *vs.* internal procurement. If there exists a profitable opportunity to outsource a service or operation, the client and the suppliers enter into a contract with full knowledge of the nature of the work, signing a complete and explicit written agreement covering all aspects of the outsourced service and payments, eventually including contingency plans. But in most contractual relationships, things do not

happen this way; processes are much more complex than idealized.

In concrete, when integrating a VE rather than outsourcing a service or a set of simple products or operations, difficulties arise. Selection, negotiation, contractualization, and enforcement can be too complex and too delicate. There is a vast spectrum of available resources providers, each with different characteristics, leading to difficult selection and integration decisions.

The costs of outsourcing are composed of both the explicit cost of carrying out the transaction as well as hidden costs due to coordination difficulties and contractual risks. The major costs associated with outsourcing include (1) the transaction costs and (2) the leakage of private knowledge.

Transaction Costs

Transaction costs include the time and expense of negotiating, writing, and enforcing contracts. They include the adverse consequences of opportunistic behavior, as well as the costs of trying to prevent it. In the VE model, transaction costs are the firm (re)configuration costs, associated with partners search, selection, negotiation, and integration as well as permanent monitoring and the evaluation of the partnership performance (Cunha & Putnik, 2003a). If externalizing functions can involve high transaction costs, networking relies intensively on extending the enterprise boundaries, partnering functions, and the VE model is extremely dependent not only on networking but on dynamically reconfiguring. This way, the implementation of the VE concept requires tools to overcome the transaction costs barrier, and knowledge management is a tool, assured by the market of resources, supported by an intelligent knowledge base and by the human brokerage function.

Leakage of Private Knowledge

The *preservation of firm's knowledge* on organizational and management processes is the firm's competitive factor.

A firm's private knowledge is based on information that no one else knows and gives a firm an advantage in the market. Most of the time, this private knowledge is a core competitive advantage that distinguishes a firm from its competitors

(Prahalad & Hamel, 1990). It may concern production know-how, product design, or consumer information.

Networking or partitioning tasks between resource providers increases the risk of losing control of such type of information, which only through complete contractual agreements, could be safeguarded and, furthermore, through an environment assuring trust and accomplishment of the duty of seal. The implementation of the VE model enables the preservation of the firm's knowledge.

Overcoming the Disabling Factors

The implementation of networked and dynamically reconfigurable organizations requires the existence of tools and environments that overcome these two disabling factors, improving knowledge management and allowing dynamics as high as required to assure business alignment. The main tools suggested in the BM_VEARM (Putnik, 2000) for managing, controlling, and enabling networking and dynamics are:

- The *market of resources* is the environment for enabling and management of efficient configuration and assuring virtuality at low transaction costs and reduced risk of knowledge leakage, this last requiring particular attention by the definition of knowledge management procedures.
- The *broker* or organization configuration manager is the main agent of agility and virtuality, acting either between two operations of the VE (off-line reconfigurability, providing agility only) or online with the operation (online reconfigurability, providing virtuality and a higher level of agility).
- *Virtuality* that makes possible the transition from one physical structure (instance) to another in a way that the enterprise or process owner is not affected by the system reconfiguration and is not aware of the reconfiguration (the underlying service structure and reconfiguration process are hidden).

Functionalities for Virtual Enterprise Integration

The organizational challenges of (1) partitioning tasks among partners in the distributed networked environ-

ment so that they fit and take advantage of the different competencies in an VE, (2) integration of the same, (3) coordination and reconfigurability management in order to keep alignment with the market requirements are of main concern and can determine the success or failure of a VE project.

As discussed in (Cunha & Putnik, 2004, 2005), in order to achieve its maximum competitiveness, that is, to be competitive in delivery time, quality, and cost and to yield satisfactory profit margins, the implementation of the VE model requires a supporting environment assuring two main interrelated aspects (designated virtual enterprise requirements): (1) Reconfigurability dynamics (assuring fast transition between VE instantiations) and (2) Business alignment (maintaining the VE aligned with the market).

An environment designed to assure the two above-mentioned VE requirements should present as main characteristics the ability of (1) flexible and almost instantaneous access to the optimal resources to integrate in the enterprise, negotiation process between them, selection of the optimal combination and its integration; (2) design, negotiation, business management, and manufacturing management functions independently from the physical barrier of space; (3) minimization of the reconfiguration and integration time; and (4) managing knowledge within each instance of the integrated network.

The first characteristic implies the existence of a market of independent candidate resources for integrating a virtual enterprise. This market role is (a) to provide the environment and technology and the corresponding procedure protocols, that is, an *open system architecture* for the efficient access to resources, efficient negotiation between them, and its efficient integration and (b) to provide a domain for selection of participant resources providers in a VE, large enough to assure good options.

The second characteristic implies the utilization of advanced information and communication technologies to the operation of the market of independent resources, that is, technologies providing technical conditions to efficiently access to the globally distributed resources providers, efficient negotiation between them, and its efficient integration.

The third characteristic is necessary in order to provide flexibility, as high as possible, that is, reconfigurability as fast as possible.

The fourth characteristic is assured by the broker, supported by the market of resources knowledge base and information infrastructure by specific contractual agreements settled by the market of resources between the involved parties in a given VE instance and by a VE management regulation.

Any environment attempting to support the VE model, that is, to assure VE dynamic integration and business strategic alignment should implement a set of functionalities traducing the VE requirements, which, as proposed in Cunha, Putnik and Carvalho (2002), include:

- Responsiveness or almost real-time answer, as one instantaneous physical structure (or one instance) of a virtual enterprise may last (on the limit) only for a few days or even hours. It should be possible to reduce negotiation time and time-to-contract.
- The permanent alignment of the VE with the market (business) requirements, which can justify a dynamic process of VE performance evaluation and the analysis of reconfiguration opportunities.
- The ability to find the right potential partners and further efficient negotiation are essential; this should require a normalized description of products, operations, and services (resources) participating in the environment.
- Monitoring the performance of every integrated resource, increasing trust and the highest possible performance of the VE.
- Risk minimization through contractual agreements.
- Provision of knowledge in VE creation/reconfiguration through appropriate algorithms, artificial intelligence support, intelligent knowledge base, and brokerage systems.

MARKET OF RESOURCES: AN ENVIRONMENT FOR VIRTUAL ENTERPRISE INTEGRATION

Offer and demand are usually matched under several different circumstances from unregulated search to oriented search, from simple intermediation mecha-

nisms to the market mechanism, all of them with the possibility of being either manually performed or automated (Cunha, Putnik & Gunasekaran, 2002). A marketplace of resources providers will enable the matching between firms looking for potential partners for integration and firms offering their resources, facilitating VE integration, and offering to participants a larger number of business opportunities.

Several supporting infrastructures and applications must exist before we can take advantage of the VE organizational model such as electronic markets of resources providers, legal platforms, brokerage services, efficient and reliable global information systems, electronic contractualization and electronic negotiation systems, and software tools (Carvalho, Putnik & Cunha, 2002; Cunha, Putnik & Carvalho, 2002).

The authors have proposed the market of resources concept as an alternative to existing applications, which were developed to support isolated activities within supply chains such as partners search and selection, negotiation, and enterprise collaboration but without the purpose of responding to the VE requirements.

The market of resources is an institutionalized organizational framework and service assuring the accomplishment of the competitiveness requirements for VE reconfigurability dynamics, business alignment, quality assurance, trust and optimization in resources utilization, and quick response to market. It is an alternative to the dispersedly developed Internet-based solutions that can be used to support search and selection of partners to integrate in a given supply chain.

The operational aspect of the market of resources consists on an Internet-based intermediation service, mediating offer and demand of resources to dynamically integrate in a VE, assuring low transaction costs (demonstrated in (Cunha & Putnik, 2003a, b)) and the partners' knowledge preservation. Brokers act within the market of resources as intermediation agents for agility and virtuality (Ávila, Putnik & Cunha, 2002) and, simultaneously, as knowledge management promoters and consultants.

In this virtual environment, *offer* corresponds to *participants* (enterprises, resources providers) that make their *resources* available, as potential partners for VE integration, and *demand* corresponds to the *client* (or VE owner), the entity looking for a product, components, or operations (resources) to create/integrate/reconfigure a VE to satisfy the *customer*.

The service provided by the market of resources is supported by (1) a knowledge base of resources and results of the integration of resources in previous VE, (2) a normalized representation of information, (3) intelligent agent and algorithms, (4) a brokerage service, and (5) a regulation covering management of negotiation and integration processes. It is able to offer (1) knowledge for VE selection of resources, negotiation, and its integration; (2) specific functions of VE operation management; and (3) contracts and formalizing procedures to assure the accomplishment of commitments, responsibility, trust, and deontological aspects, envisaging the integrated VE accomplishes its objectives of answering to a market opportunity (Cunha, Putnik & Gunasekaran, 2002).

A comprehensive explanation of the market of resources structure and operation can be found in Cunha et al. (2005).

MARKET OF RESOURCES: AN ENVIRONMENT FOR KNOWLEDGE MANAGEMENT AND BUSINESS ALIGNMENT

In the context of VE integration, by aligning, we mean the actions to be undertaken to gain synergy between business, that is, the market opportunity (or business opportunity) and the provision of the required product with the required specifications at the required time with the lowest cost and the best possible return (financial or other) (Cunha & Putnik, 2005). In particular, we propose alignment strategies between business and the integration of resources in a VE to answer to a market opportunity, supported by the environment of a *market of resources*.

The Market of Resources Entities and Relationships

As introduced in Cunha and Putnik (2005), the entities present at the market of resources are:

1. **Clients (or VE owner):** Those looking for a product, components, or operations to integrate a VE, according to a VE project. Information considered relevant concerns the product to be produced and its process plan, the negotiation parameters, project constraints, and so forth.
2. **Virtual Enterprise:** The set of integrated resources providers respecting the VE project, able to answer to a market opportunity. The VE created/reconfigured is itself a complex entity, constituted by the client and the resources integrated to provide the operations to manufacture the product or its parts. The resulting VE is expected to produce the specified product, according to the process plan defined by the client, respecting all the project constraints. Information considered relevant concerns the network structure, dependencies between the resources providers, the contract and commitments between them and the client, and all the details in order to manage the process.
3. **Resources Providers (enterprises registered in the market of resources to specifically provide resources or add value to products or processes):** Resources providers are mapped into resources (products and operations). Information considered relevant concerns the enterprise, its structure, products/operations provided, conditions, and negotiation details. The same enterprise can be present in the market offering several resources.
4. **Products (we are including services in this entity):** Resources providers are mapped into components or parts of products. Information considered relevant concerns conditions in which resources providers provide each product or part, negotiation details, and availability.
5. **Operations:** Associated to each component of a product; elementary operations performed by resources providers while executing an operation on a specific product or part. Resources providers are mapped into operations, and operations are mapped into products. Information considered relevant concerns conditions under which resources providers provide each operation, negotiation details, information to allow further production control, and evaluation.

Knowledge Management and Business Alignment in Virtual Enterprise Integration

Business alignment in VE integration is complex and challenging, as alignment has to incorporate immaterial components in the relationships within the integration of resources. It is not just an internal strategy but a set of integrated and interrelated integration strategies that must be verified so that the integrated VE is able to meet the objective giving rise to the VE itself, that is, to meet the market requirements (Cunha et al., 2005). Business strategic alignment is a matter for knowledge management.

The introduction of the VE concept and of the corresponding supporting environments requires the definition of business alignment strategies. If organizations respond to market requirements (end-users requirements), traducing these requirements into a project of VE and pushing them along the process of selection and integration of resources providers under the format of a VE, the role of the market of resources is to assure the permanent (dynamic) alignment of the resulting VE with the market. The market of resources is the environment enabler for efficiency and effectiveness in the integration process, thus, generating, sharing, and using knowledge to strategically align the virtual enterprise with business.

Strategic alignment between business and VE integration involves a mix of dependencies between *market* requirements, *product* requirements, and *resource providers* requirements.

The market of resources must assure the client the alignment between the market and the resources providers selected and integrated in the VE. Also, the market of resources must try to assure that the client has correctly captured the market requirements, which is a task performed by a broker. This way, the process must align the client with the market (business) and then align the resources (by the selection and integration processes) with the client and business (Cunha & Putnik, 2005).

Integrating a VE corresponds to aligning the five entities of the previous section with business. The market of resources is expected to guide the client in aligning the VE with the market opportunity. The process consists on pushing downstream the market requirements.

The referential for alignment proposed in Cunha and Putnik (2005) considers:

1. **Market Alignment (alignment with customer or market requirements):** Before the creation of the VE, the client traduces the customer requirements into product specifications and projects the system of resources for the VE. The VE project consists on the generic identification of the characteristics of the resources that will accomplish the execution of the process plan to the required product, that is, the process plan that will allow the production of the product verifying the market requirements.

2. **Product/Service and Operations Alignment:** Aligning the product with the specifications, that is, with the market requirements. Operations provided by the selected resource providers must conduct to the desired product.

3. **Resources Providers Alignment:** Aligning resources providers with the market requirements involves the verification of which characteristics resources providers must assure so that the client can trust that the selected set of enterprises is able to be integrated in a VE able to produce the product that meets the requirements previously captured by the client (market requirements).

Resources providers requirements include economical, managerial, and organizational aspects.

These three sets of requirements for alignment are grouped in Table 1. In Cunha and Putnik (2005), the interested reader can find a development of these alignment strategies between business opportunities and the creation/reconfiguration of the VE that is expected to meet that opportunity.

KNOWLEDGE MANAGEMENT FUNCTIONS: STRENGTHS AND PROBLEMS

In this section, we highlight some strengths and problems associated with the implementation of knowledge management procedures and supporting technologies in the market of resources.

Table 1. Checklist of requirements to be considered in alignment (Cunha & Putnik, 2005)

Market Alignment	Product/Service/Operations Alignment	Resources Providers Alignment
– Price, cost, and profit – Quality – Quick response: the desired product, on time, in the required conditions – Transparency and legality – Trust and confidence – Correct capture of market /customer requirements	– Cost – Quality – Integrability – Interoperability between different providers – Standards	– Availability – Ability to meet product/service/operation requirements – Certification – Dependability – Flexibility – Responsiveness – Competitiveness and proactiveness – Past information of previous VE integrations

Knowledge Management Strengths

Knowledge management in the market of resources enables some functions such as (1) the ability to assure trust (given by the partnership performance monitoring and utilization of historical information in processes of search and selection) and responsiveness; (2) knowledge based guidance in VE design and integration (assured by the introduction of brokers); (3) electronic automated negotiation and contractualization; (4) performance evaluation of the VE participants; and (5) contract management and enforcement (based on performance evaluation of the VE participants).

Trust and Responsiveness

Trust is a major concern that any environment to support VE integration must assure. In the market of resources, trust is assured by a detailed regulation, enforcement procedures through contracts and safety mechanisms, duty of seal, and so forth. Responsiveness or almost real-time answer is essential. The market enables the reduction of the integration time and increases integration efficiency as demonstrated in Cunha and Putnik (2003a, b).

Knowledge-Based Guidance in VE Design and Integration

Brokerage implementation (human brokerage), search, and selection support algorithms and an efficient

organization of the market of resources intelligent knowledge base and business intelligence are on the origin of this knowledge-based guidance. The broker, supported by computer-aided tools, validates all the steps in the process of designing the VE project that is most suitable to achieve the VE underlying objectives, manages VE integration, and monitors VE operation.

Electronic Automated Negotiation

The market of resources service is designed to offer different processes of electronic negotiation (passive and active) and is supported by automated tools of search, selection, and negotiation, which can increase the performance of the process when the solution space size is high.

Performance Evaluation of the VE Participants

The requirement for permanent alignment of the VE with the market (business) asks for a dynamic process of VE performance evaluation and analysis of reconfiguration opportunities. To answer to this requirement, the market of resources offers procedures for performance monitoring and, through the broker allocated to a given VE project and using computer aided coordination mechanisms, is permanently monitoring the partnership and recording historical information to be used in the future. The market makes use of historical information of the behavior of the resources providers in previous integrations, in the search processes, to increase trust and achieve better results. This activity of monitoring the performance of every integrated resource increases trust and contributes to the highest possible performance of the VE.

Contracts Management and Enforcement

The market of resources offers mechanisms for contract generation, management, and enforcement. To reduce the contractualization time, the market of resources (empowered to represent the parties in the contract formalization), is able to perform almost real-time contractualization between the parties to integrate in the VE.

Knowledge Management Problems

Two of the main difficulties identified are related with (1) the difficulty of expressing the resources requirements by the VE owner who must be able to use a resources representation language to traduce the VE project and resources requirements in the knowledge base of the market of resources and (2) the necessity of implementing partnerships with other similar services in order to extend the coverage domain.

Difficulty in Expressing the Resources Characteristics

The efficiency of the service is dependent on the ability of representation and organization of the resources information in the market of resources intelligent knowledge base and the capture and translation of the requirements for resources selection and negotiation parameters. If the first is dependent solely of a unified representation language, the second requires also the ability of the VE client to translate the requirements for the VE project into this language, which is far more complex than describing the individual resources provided by resources providers. These functions are essential to knowledge organization, maintenance, and extraction.

The developments toward unified representation languages, such as the XML-based developments, represent a tremendous contribution that should help to overcome this problem.

Limitations in Coverage: Dependability on Similar Services

A project can touch many different areas, and our market is both vertical and horizontal (matricial) to allow a better coverage of domains of activity. To overcome the lack of coverage, it is necessary to establish partnerships with other similar markets so that the broker does not see its search space limited to the market of resources database, but this situation of partnering with similar services is constrained by the existence of unified representation languages. If this does not happen, translating software will be required to support interoperability between services, or the broker will have to know different representation languages in order to transport requests into other services.

OPPORTUNITIES FOR A MARKET OF RESOURCES AS AN ENVIRONMENT FOR KNOWLEDGE MANAGEMENT AND BUSINESS ALIGNMENT

Cost savings do not seem to be a major key driver for enterprises to use the market of resources. Rather, they should be interested in time and quality benefits, trust, dynamic reconfigurability, and so forth. Opportunities should come from technological developments, which will enable more efficiency in the implementation and from the current state of ICT investment and usage by the enterprises, which traduces the willingness to drive business online. But the main opportunity seems to come from the actual strong competition environment, which is expected to force companies to the adoption of VE models, and this shift may represent an opportunity for services as the one provided by the market of resources.

Emerging VE Organizational Model

As demonstrated in Cunha and Putnik (2003a, b), the market of resources environment is more efficient in coping with the VE model than the Internet-based traditional ways (e.g., WWW search using search engines, e-mail, etc.). With the predictable evolution of the organizational models, services as the one provided by the market of resources will appear as the previewed evolution toward a new generation of business-to-business electronic marketplaces and support services.

Technological Development

The rise of Internet-based business-to-business marketplaces is progressing rapidly. At the same time, we are assisting the fast appearing of networked enterprises, extended enterprises, and VE. However, the developments or solutions still do not respond to the VE model requirements. Several enabling technologies are suffering significant developments from electronic payment to security. Finally, the emerging standards for information representation will be a major requirement for efficiency and integrability in electronic business.

Investment in Information and Communication Technology

Several surveys (for example, Boston Consulting Group, 2002) suggest that there is a very favorable environment for the adoption and increased usage of new value-added services, as enterprises have invested in the enabling technology and are looking for reducing costs and increasing productivity, which means that it could be understood as potentiating the acceptability for the market of resources.

Competitive Pressures

We feel that enterprises of all sectors perceive the threat of competition and see, both in the emergent virtual enterprise organizational models and in the Internet-based applications, a possibility to improve productivity and reduce some type of costs. This is pushing traditional business to adopt business-to-business electronic commerce practices and represents an opportunity for the deployment of new applications, one of these, the market of resources.

At the same time, companies providing e-business services represent a competitive pressure toward the success of the market of resources (while competition is simultaneously a threat).

Technology Accessibility to Small and Medium Sized Enterprises

A key driver of growth of business-to-business electronic commerce will be the increased adoption of e-commerce initiatives by small- and mid-sized companies. Solutions, up until recently, associated with huge investments and dedicated to large companies are now accessible to small and medium enterprises.

CONCLUSION AND FUTURE TRENDS

This article intended to provide a better understanding of the environment supporting virtual enterprise integration from a knowledge management perspective. We have introduced the support to VE integration by the creation of a *market of resources* and introduced a referential for knowledge management to assure the alignment between business requirements and the integration of resources providers in a VE.

The development of environments to support the VE model in general are of increasing importance, and the market of resources intends to be a contribution toward that direction. However, it is an innovative approach when compared with the other developments that literature provides, which are not as integrated as the market of resources is and covering only aspects of the VE life cycle in a less dynamic approach to the virtual enterprise concept.

All the technologies and techniques necessary to support the several phases of the life cycle of a VE, as well as many valuable applications, already exist and some are in operation, but most of them fail to answer to the VE integration requirements, as they were not developed specifically to support this model. What we have designated as an adequate *environment* to support the requirements of the emerging VE paradigm is missing.

Simultaneously, besides the strengths identified for the implementation of knowledge management functions in the market of resources, some problems are also identified. The article also identifies some opportunities associated to the implementation and exploitation of the market of resources as an environment to support knowledge management in VE integration.

REFERENCES

Alavi, M., & Leidner, D. E. (2001). Knowledge management and knowledge management systems: Conceptual foundations and research issues. *MIS Quartely, 25*(1), 107-136.

Ávila, P., Putnik, G. D., & Cunha, M. M. (2002). Brokerage function in agile/virtual enterprise integration: A literature review. In L. M. Camarinha-Matos (Ed.), *Collaborative business ecosystems and virtual enterprises* (pp. 65-72). Boston: Kluwer Academic.

Boston Consulting Group. (2002, February). Company communication trends: Growth of new communication technologies demands rethinking of companies' internal communication strategies, a survey report. Retrieved May 8, 2005, from *http://www.bcg.com*

Byrne, J. A. (1993, February 8). The virtual corporation: The company of the future will be the ultimate in adaptability. *Business Week,* 98-103.

Camarinha-Matos, L. M., & Afsarmanesh, H. (1999). The virtual enterprise concept. In L. M. Camarinha-Matos & H. Afsarmanesh (Eds.), *Infrastructures for virtual enterprises* (pp. 3-14). Porto, Portugal: Kluwer Academic.

Carvalho, J. D. A., Putnik, G. D., & Cunha, M. M. (2002). Infrastructures for virtual enterprises. In A. Baykasoglu & T. Dereli (Eds.), *Proceedings of the 2nd International Conference on Responsive Manufacturing* (pp. 483-487). Gaziantep, Turkey.

Cunha, M. M., Putnik, G. D., & Ávila, P. (2000). Towards focused markets of resources for agile/virtual enterprise integration. In L. M. Camarinha-Matos, H. Afsarmanesh, & H. Erbe (Eds.), *Advances in networked enterprises: Virtual organisations, balanced automation, and systems integration* (pp. 15-24). Berlin: Kluwer Academic.

Cunha, M. M., Putnik, G. D., & Carvalho, J. D. (2002). Infrastructures to support virtual enterprise integration. In R. Hackney (Ed.), *Proceedings of 12th Annual BIT Conference - Business Information Technology Management: Semantic Futures.* Manchester, UK: The Manchester Metropolitan University (CD-ROM).

Cunha, M. M., Putnik, G. D., & Gunasekaran, A. (2002). Market of resources as an environment for agile/virtual enterprise dynamic integration and for business alignment. In O. Khalil & A. Gunasekaran (Eds.), *Knowledge and information technology management in the 21st century organisations: Human and social perspectives* (pp. 169-190). Hershey, PA: Idea Group Publishing.

Cunha, M. M., Putnik, G. D., Gunasekaran, A., & Ávila, P. (2005). Market of resources as a virtual enterprise integration enabler. In G. D. Putnik & M. M. Cunha (Eds.), *Virtual enterprise integration: Technological and organizational perspectives.* Hershey, PA: Idea Group Publishing.

Cunha, M. M., & Putnik, G. D. (2003a). Market of resources versus e-based traditional virtual enterprise integration - part I: A cost model definition. In A. Gunasekaran & G. D. Putnik (Eds.), *Proceedings of the 1st International Conference on Performance Measures, Benchmarking and Best Practices in New Economy.* Guimarães, Portugal.

Cunha, M. M., & Putnik, G. D. (2003b). Market of resources versus e-based traditional virtual enterprise integration: Part II: A comparative cost analysis. In A. Gunasekaran & G. D. Putnik (Eds.), *Proceedings of the First International Conference on Performance Measures, Benchmarking and Best Practices in New Economy.* Guimarães, Portugal.

Cunha, M. M., & Putnik, G. D. (2004). Trends and solutions in virtual enterprise integration. *Tekhné - Review of Politechnical Studies, 1*(1).

Cunha, M. M., & Putnik, G. D. (2005). Business alignment in agile/virtual enterprise integration. In M. Khosrow-Pour (Ed.), *Advanced topics in information resources management* (Vol. 4). Hershey, PA: Idea Group Publishing.

Kluge, J., Stein, W., & Licht, T. (2001). *Knowledge unpluged: The McKinsey & Company survey on knowledge management*. New York: Palgrave and McKinsey & Company.

Murray, P. (2002, March-April). Knowledge management as a sustained competitive advantage. *Ivey Business Journal*, 71-76.

Prahalad, C. K., & Hamel, G. (1990, May-June). The core competence of the corporation. *Harvard Business Review, 68*, 79-91.

Preiss, K., Goldman, S., & Nagel, R. (1996). *Cooperate to compete: Building agile business relationships*. New York: van Nostrand Reinhold.

Putnik, G. D. (2000). BM_Virtual Enterprise Architecture Reference model. In A. Gunasekaran (Ed.), *Agile manufacturing: 21ˢᵗ century manufacturing strategy* (pp. 73-93). UK: Elsevier Science.

KEY TERMS

BM_VEARM: BM_Virtual Enterprise Architecture Reference Model: BM_VEARM is a VE reference model conceived to enable the highest organisational/structural/reconfiguration and operational inter-enterprise dynamics of a VE, employing three main mechanisms for VE dynamic creation, reconfiguration, and operation: *Market of Resources*, *Broker*, and *Virtuality*. Additionally, BM_VEARM implies the highest level of integration and (geographic) distribution of VE elements (partners in the VE network).

Business Alignment: Actions to be undertaken by an organisation, to answer to a market opportunity with the provision of the required product, with the required specifications, at the required time, with the lowest cost, and with the best possible return.

Market of Resources: An institutionalized organisational framework and service assuring VE dynamic integration, reconfiguration, and business alignment. The operational aspect of the Market of Resources consists of an Internet-Based intermediation service, mediating offer, and demand of resources to dynamically integrate in a VE, assuring low transaction costs and partners' knowledge preservation. Brokers act within the Market of Resources as the intermediation agents for agility and virtuality and, simultaneously, as knowledge management promoters and consultants.

Virtual Enterprise: A dynamically reconfigurable global networked organisation, networked enterprise, or network of enterprises, sharing information, knowledge, skills, core competencies, market and other resources and processes, configured (or constituted) as a temporary alliance (or network) to meet a f(fast-changing) market window of opportunity, presenting as its main characteristics agility, virtuality, distributivity, and integrability (see Putnik, 2000). The factors against VE are transaction (reconfiguration) costs and preservation of enterprises', or firms' (partners in VE) knowledge.

Metaphors as Cognitive Devices in Communities of Practice

Iwan von Wartburg
University of Hamburg, Germany

INTRODUCTION

The role of language for knowledge creation in communities of practice (CoPs) and innovation teams has been stressed by the accounts of storytelling (Orr, 1996; Nonaka & Takeuchi, 1995). Stories work as metaphors connecting new problem situations with prior problem situations. They guide CoP members to arrive at new connections of prior unconnected knowledge domains within cognitive maps. Cognitive maps contain causal and temporal relations between cognitive concepts: "[Cognitive] maps portray causality, predicate logic, or sequences, all capture temporal relations: if this (in the now), then that (in the future)" (Weick, 1990, p. 1). New connections of knowledge domains brought about by metaphorical reasoning enable innovative problem solutions and serve as a 'platform' for new knowledge creation in the future. Thus, investigating metaphorical language usage promises to add value to the understanding of knowledge creation in CoPs.

BACKGROUND

Traditionally, the study of metaphors belongs within the study of rhetoric, linguistics, literature, cognitive psychology, and philosophy. Metaphors are "the outcome of a cognitive process that is in constant use—a process in which the literal meaning to a phrase or word is applied to a new context in a figurative sense" (Grant & Oswick, 1996, p. 1).

Metaphors are more than linguistic tools. Lackoff and Johnson (1980, pp. 5-7) state that metaphor is pervasive in everyday life, not just in language, but also in thought and action: "The essence of metaphor is understanding and experiencing one kind of thing in terms of another" (Lackoff & Johnson, 1980, p. 5). To speak metaphorically is to relate two entities (or terms) through the verb "to be" (or the copula

"is")—for instance, 'an organization is a machine' (Coyne, 1995). The consequences of such metaphorical utterances are of cognitive nature—metaphor is implicated in perception. During word processing for example, we actually see the computer screen as a sheet of paper. "Seeing as" is a fundamental act of perception (Goodman, 1978).

Accordingly, we are constantly engaged in metaphorical projections: we project one term, concept, or situation onto another (Coyne, 1995). A metaphor includes a primary and a secondary subject. In the metaphor 'producing an integrated circuit (IC) by using chemical vapor deposition (CVD) is building a complex labyrinth by using Lego toys', the primary subject is the 'CVD-Method' and the secondary subject is 'Lego toys.' The secondary subject is a whole system or a whole domain of elements in a cognitive map. Therefore, by relating a secondary subject domain to a primary subject domain, multiple comparisons, differences, and paradoxes can be discovered. However, it is important to stress that often the meaning of the secondary subject changes too. Thus, the knowledge subjects really interact in a sense that both concepts are given new or enriched meanings depending on context (Black, 1962).

FUTURE DEVELOPMENT AND CONCLUSION

Metaphor is a complex cognitive phenomenon that alters cognitive maps and therefore future action on the ground of a specific context. Metaphors are an "invitation to see the world anew" (Barret & Cooperrider, 1990, p. 219). Thus by using figurative speech in metaphorical statements, CoP members may generate knowledge that helps solve problems in actual practice.

Knowledge creation by further developing cognitive maps involves arriving at new classifications:

"Naming is always classifying, and mapping is essentially the same as naming" (Bateson, 1979, p. 30). However, making new classifications in maps always happens on the ground of what someone already knows: "You have to know something already in order to 'see' something different" (Weick, 1990, p. 2). Thus, the effectiveness of metaphorical statements is dependent on the amount of prior shared knowledge which can be activated. This emphasizes the importance of shared "absorptive capacity" between CoP members who interpret metaphors (Cohen & Levinthal, 1990). Therefore, it may be concluded that storytelling may not be facilitated independently from common actual practice in CoPs.

Acknowledging and revealing the central role of figurative speech in knowledge creation requires research that entails participation in discussions or a detailed analysis of content 'produced' during interactions. In this respect, open source software development CoPs seem to be an especially interesting research area: adaptive reuse of prior knowledge in software development—that is, the modified usage of existing code fragments for different software projects—may be conceived of as a metaphorical statement to a prior problem solution.

Furthermore, the role of metaphors as a language tool for socializing legitimated peripheral participating members in CoPs may be investigated. This could reveal whether core members use metaphors for knowledge creation and sharing from which outsiders probably cannot grab the whole associated meaning. Becoming a core member within a CoP as learning collective thus involves learning the associated figurative meaning of metaphorical statements uttered during actual practice.

REFERENCES

Barret, F.J., & Cooperrider, D.L. (1990). Generative metaphor intervention: A new behavioral approach for working with systems divided by conflict and caught in defensive perception. *Journal of Applied Behavioral Science, 26,* 219-239.

Bateson, G. (1979). *Mind and nature.* New York: Dutton.

Black, M. (1962). *Models and metaphors.* Ithaca, NY: Cornell University Press.

Cohen, W.M., & Levinthal, D.A (1990). Absorptive capacity: A new perspective on learning and innovation. *Administrative Science Quarterly, 35,* 128-152.

Coyne, R.D. (1995). *Designing information technology in the postmodern age.* Cambridge, MA: MIT Press.

Goodman, N. (1978). *Ways of worldmaking.* Hassocks, UK: Harvester Press.

Grant, D., & Oswick, C. (1996). Introduction: Getting the measure of metaphors. In D. Grant & C. Oswick (Eds.), *Metaphor and organizations* (pp. 1-20). London: Sage Publications.

Lackoff, G., & Johnson, M. (1980). *Metaphors we live by.* Chicago: The University of Chicago Press.

Nonaka, I., & Takeuchi, H. (1995). *The knowledge-creating company: How Japanese create the dynamics of innovation.* Oxford, UK: Oxford University Press.

Orr, J.E. (1996). *Talking about machines: An ethnography of a modern job.* Ithaca, NY: Cornell University Press.

Weick, K.E. (1990). Cartographic myths in organizations. In A.S. Huff (Ed.), *Mapping strategic thought* (pp. 1-10). Chichester, UK: John Wiley & Sons.

KEY TERMS

Cognitive Map: A collection of nodes linked by some edges (arcs). From a logical perspective, a node is a logical proposition and a link is an implication. Thus, cognitive maps consist of causal and temporal relations between cognitive concepts.

Metaphor: In language, a rhetorical trope where a comparison is made between two seemingly unrelated subjects. Typically, a first object is described as being a second object. In this way, the first object can be economically described because implicit and explicit attributes from the second object can be used to fill in the description of the first. (http://en.wikipedia.org/wiki/Metaphor)

Narrative Inquiry and Communities of Practice

M. Gordon Hunter
The University of Lethbridge, Canada

INTRODUCTION

The kind of research conducted by communities of practice may be related to preliminary investigations of subjects that may not have been extensively investigated in the past. Alternatively, communities of practice may be formed to expand previous investigations into relatively under-researched areas. In either case the approach to conducting these preliminary investigations may be qualitative in nature. An approach that facilitates qualitative investigations is *Narrative Inquiry*.

BACKGROUND

Narrative Inquiry documents "a segment of one's life that is of interest to the narrator and researcher" (Girden, 2001, p. 49) and includes "the symbolic presentation of a sequence of events connected by subject matter and related by time" (Scholes, 1981, p. 205). The documentation represents the story told by the research participant that is both contextually rich and temporally bounded. A story is contextually rich when it has been experienced firsthand by the research participant (Tulving, 1972; Swap, Leonard, Schields & Abrams, 2001). Stories that have a beginning and an end are temporally bounded and follow a chronological sequence of events (Bruner, 1990; Czarniawska-Joerges, 1995; Vendelo, 1998).

In order to add structure to a Narrative Inquiry interview, the Long Interview Technique (McCracken, 1988) may be incorporated. To begin, "grand tour" (McCracken, 1988) questions are addressed which are general and provide a context for the following more detailed discussion. Questions are posed during the interview which relate specifically to the area under investigation. In response to an answer, a "floating prompt" (McCracken, 1988) may be employed so that the researcher may delve into more detail. Near the end of the interview, "planned prompts" may be used to ensure issues are addressed which may have arisen from a literature search or previous interviews.

Data analysis involves searching for emerging themes, first within an interview and then across a series of interviews. The search for emerging themes is common practice in qualitative research (Miles & Huberman, 1994) and involves the interplay between both the data and the emerging themes. The process begins with a careful reading of the transcript, where noteworthy phrases or sentences are highlighted. Passages that seem conceptually linked are then considered together, and descriptions of the theme or pattern that the groupings share are developed. Subsequently, the data are reread to identify further evidence that supports or challenges the emerging themes. This second review process can lead to the identification of new themes, new classification of themes, or reclassification of data from one theme to another. Eventually, a consolidated interpretation of the data is achieved.

CONCLUSION

Narrative Inquiry allows research participants to relate stories of their own experiences. The Long Interview Technique provides structure to the process which identifies the underlying concepts of research participants' perspectives within the context of the individual. This approach should facilitate the type of preliminary investigations conducted by groups involved in communities of practice.

REFERENCES

Bruner, J. (1990). *Acts of meaning*. Cambridge, MA: Harvard University Press.

Czarniawska-Joerges, B. (1995). Narration or science? Collapsing the division in organization studies. *Organization*, 2(1), 11-33.

Girden, E.R. (2001). *Evaluating research articles— from start to finish* (2nd ed.). Thousand Oaks, CA: Sage Publications.

McCracken, G. (1988). *The Long Interview*. New York: Sage Publications.

Miles, M.B., & Huberman, A.M. (1994). *Qualitative data analysis: A new sourcebook of methods* (2nd ed.). Newbury Park, CA: Sage Publications.

Scholes, R. (1981). Language, narrative, and anti-narrative. In W. Mitchell (Ed.), *On narrativity* (pp. 200-208). Chicago: University of Chicago Press.

Swap, W., Leonard, D., Schields, M., & Abrams, L. (2001). Using mentoring and storytelling to transfer knowledge in the workplace. *Journal of Management Information Systems, 18*(1), 95-114.

Tulving, E. (1972). Episodic and semantic memory. In E. Tulving & W. Donaldson (Eds.), *Organization of memory* (pp. 381-404). New York: Academic Press.

Vendola, M.T. (1998). Narrating corporate reputation: Becoming legitimate through storytelling. *International Journal of Management and Organization, 28*(3), 120-137.

KEY TERMS

Contextually Rich: A story that has been experienced firsthand.

Narrative Inquiry: An approach to documenting a research participant's story about an area of interest.

Temporally Bound: A story that has a beginning, an end, and a chronological sequence.

Networks of People as an Emerging Business Model

Lesley Robinson
Lesley Robinson Company, UK

INTRODUCTION

Networking as a skill is becoming more and more important as traditional ways of doing business continue to change. Many organisations are moving from the industrial model of culture to a more "knowledge"-based culture, changing from having structured hierarchies to flatter structures with distributed responsibility. This has vast implications for how things get done. Instead of receiving instructions or being expected to work to a strict process, the knowledge-based organisations are giving people looser frameworks, and expect them to take responsibility for contributing ideas and sharing their knowledge.

EFFECTIVENESS THROUGH NETWORKING

The most effective way to work in these organisations is to build a network of contacts, colleagues, and teams. This networking approach is different from communities of practice where a group of people come together, formally or informally, to solve particular problems or discuss specific issues. Building a network is wider than just one specific focus; it is a new way of working and indeed a new way of thinking. This will give rise to many questions for organisations including structure, leadership, decision making, and much more. Many are not familiar in working in such an unstructured way.

Traditional communication techniques such as e-mail are also failing to deliver, as they dramatically overload people who have fallen into bad usage habits, thus restricting the techniques' effective use. The preferred way to communicate in the new 'networked organisations' is by using instant messaging and blogs, providing immediate business interactivity and truly engaging people.

Along with these developments, many organisations are downsizing and encouraging more virtual work-ing scenarios. This means that individuals are having to become more self-reliant and build up their networks for support and development purposes. The number of small businesses, independent workers, and those with portfolio careers is also growing, and they are starting to join and form their own online communities to share work, develop business ideas, and to gain profile. With this trend, the skills of networking become critical.

These new online businesses are becoming the 'new corporates', and moving from being efficient networkers to providing infrastructure and benefits to members of the network. An example of this is eBay, which has provided a very successful online world where you can make a living from buying and selling on the company's Web site; but eBay has also developed a huge amount of infrastructure around building a community. eBay promotes its community values as:

- We believe people are basically good.
- We believe everyone has something to contribute.
- We believe that an honest, open environment can bring out the best in people.
- We recognise and respect everyone as a unique individual.
- We encourage you to treat others the way you want to be treated.

Perhaps some of the remaining large corporates could learn from these values. eBay also runs workshops for members of its community, offers a forum facility, and even offers insurance to regular users.

THE FUTURE

Another example of a growing network is Ecademy. This is a business exchange that connects people to knowledge, contacts, support, and business. It is free to join Ecademy, and you can create a profile of yourself, read what is happening within Ecademy, search the site, and receive e-mail newsletters and updates. The idea of Ecademy is to build up a wide range of business contacts. For a small fee, £25 per annum, you can access and contribute to all areas of the site and build a network of up to 20 contacts. At this level you can also generate more awareness of yourself and your business through submitting content onto the Ecademy homepage. There are also a range of specific networking clubs you can join. For £120 per annum, you can have access to a growing list of premium Web site tools, but the ultimate level is the BlackStar level, which costs £2,500 per annum and is available to a limited audience. Benefits include personal introductions, mentoring and promotional opportunities, personal branding, networking tuition, online system training, and much more.

This is a huge support network backed up by regular events where it is not just a case of business cards flying around but real business gets done. They also publish a "Citizens Guide" on how to get the best out of the Ecademy community.

CONCLUSION

Examples like eBay and Ecademy are the beginning of many structured networks that set out to truly help people to do business, as well as provide a support network of value. All this without the company politics, but having true respect for the individual.

These trends are a serious challenge to the traditional corporate environment.

KEY TERM

Blog: A personal diary and a collaborative space. A breaking-news outlet, a collection of links, your own private thoughts. In simple terms, a blog is a Web site where you write material on an ongoing basis. New entries show up at the top, so your visitors can read what is new. Then they comment on it or link to it or e-mail you.

Observed Patterns of Dysfunctional Collaboration in Virtual Teams

Wing Lam
Universitas 21 Global, Singapore

Alton Chua
Nanyang Technological University, Singapore

Cecelia Lee
Universitas 21 Global, Singapore

INTRODUCTION

To collaborate is defined in the Wordsmyth (2002) dictionary as "to cooperate or work with someone else, especially on an artistic or intellectual project." The widespread adoption of the Internet and increasing sophistication of online communication tools have led to the emergence of collaboration in virtual teams in which members work with each other without the constraint of being physically together (Townsend, DeMarie & Hendrickson, 1996). Unlike traditional face-to-face teams, members of virtual teams may be geographically distributed, work in different time zones, and may never even meet face-to-face. Virtual teams therefore rely heavily on asynchronous (e.g., discussion boards, e-mail), and to some extent synchronous (e.g., videoconferencing, online chat, telephone) collaboration tools to support the interaction.

In the educational scene, many academic institutions are turning to the use of virtual teams to meet the growing demand for online education (Zhang & Nunamaker, 2003). Distance learners, who have limited face-to-face interaction opportunities, are organised into virtual teams to collaborate, solve problems, and conduct projects in much the same way as virtual teams in corporate organisations do. Apart from overcoming the barriers of space and time, virtual teams afford an environment conducive to peer-learning (Bailey & Luetkehans, 1998).

Although the dynamics of traditional face-to-face teams in the educational setting has been well studied (Slavin, 1989), the use of virtual teams raises new issues in relation to how the physical, temporal, and social separation of students affects the learning process. This article reports on the experiences of using virtual teams in an online university.

BACKGROUND: UNIVERSITAS 21 GLOBAL (U21G), ONLINE UNIVERSITY

Universitas 21 Global (U21G) is a pioneering online university formed from a joint venture between Universitas 21 (U21) and Thomson Learning. U21 is a network of 17 international universities that includes the National University of Singapore, Edinburgh University, McGill University, the University of Hong Kong, Melbourne University, and the University of Virginia. In August 2003, U21G launched its first academic programme, the MBA, which is delivered entirely online. There are neither physical classrooms nor the need for students to have face-to-face contact with other students or with their instructors. Instead, students are given access to a range of Web-based collaboration tools that include discussion forums, e-mail, and online chat that enable them to interact amongst themselves and their instructors. Given that U21G students may reside across the globe and study in different time zones, the learning approach is predominantly asynchronous to provide maximum flexibility. The global student base also affords a high level of cultural diversity.

Class sizes typically vary between 10 and 30 students. Students work in virtual teams, each comprising four or five members, on team assignments that usually revolve around the analysis of business cases. Since the team assignments contribute between 30% and 60% to a student's final mark, there is a strong incentive for students to participate to their fullest. The formation of teams is freely determined by the instructor. In fact, instructors at U21G are required to complete an online faculty training program that puts them through a similar kind of learning experience as a student would go through. This ensures that instructors are familiar with the learning opportunities afforded by the U21G pedagogy and are also sensitive to the type of problems a potential student may face.

While U21G's experience with virtual teams has generally been positive, there had been a number of teams in which dysfunctional collaborative behaviours were observed. This resulted in the poor quality of the work produced or the inability to complete the assignment on time. The following part of this article identifies four underlying reasons why these virtual teams were not as effective as they should be.

Dysfunctional Online Collaboration

Lack of Coordination

Some teams suffered from a general lack of coordination. They took a long time before getting started and generally experienced difficulties in meeting assignment deadlines. Members in such teams did not appear to have clearly assigned roles and responsibilities. Furthermore, the interaction between team members was observed to be ad hoc and irregular. One symptom of such teams is the uneven spread of effort throughout the time for which they had been given to complete the assignments.

Conversely, the more organised teams tended to reaffirm the overall goals and deliverables for the assignment early in the project lifecycle, identify the tasks needed to be completed, and divide responsibilities amongst themselves. Organized teams also tended to have an individual who assumed the role of an editor to assemble the documents produced by individual team members into a coherent whole. In some cases, the editor doubled as a project manager, reminding individual members of when their individual deadlines were due.

Minimal Social Exchanges

In teams that performed poorly, social exchanges amongst team members were observed to be minimal. Social exchanges, which include "idle banter" and "small-talk," are sine qua non to healthy, thriving teams. Kerr and Murthy (1994) suggest that the use of technology tools for collaboration tends to increase an individual's attention to the task, resulting in teams that tend to have fewer distractions and diversions than face-to-face teams. Furthermore, Warkentin, Sayeed, and Hightower (1997) explain that the difficulty in exchanging information has led virtual teams to lean towards task-oriented rather than social-emotional information. This slows the development of relational links among members.

Conversely, healthy teams were observed to be socially bonded through discussion threads and e-mail exchanges that transcend the scope of the assignments. They freely shared their academic and professional aspirations, and discussed cross-cultural culinary delights and vacation points. The use of emoticons, such as a colon followed by a right parenthesis, was also observed to be used rather liberally.

Lack of Deep and Active Discussion

Some teams appeared to adopt a 'get-it-over-with' mentality. Such teams were more pre-occupied with getting to the end of the assignment than relishing in the fullness of the educational insights that the assignment potentially had to offer. Hence, they did not fully benefit from a pedagogical standpoint. In such teams, the majority of the interaction was related to the division and completion of work activities rather than deep and active discussion about the problem at hand. Desanctis, Fayard, Roach, and Jiang (2003) describe deep discussion as one which involves challenging assumptions, reflecting the issue at hand and debating one's position. Deep discussions require students to critically analyse a problem and defend the appropriateness of potential solutions. Two possible reasons why these teams failed to engage in deep discussion include the fear of upsetting team harmony and not wishing to prolong the completion of a team assignment under the already time-pressuring conditions. Cultural factors may also play a part, where open discussion is seen as improper or alien behaviour.

In face-to-face teams, the immediate and responsive nature of exchanges between individuals induces a certain degree of spontaneity and vitality, or what one might call the 'heat' of discussion. In many virtual teams, however, heated discussion tended not to arise, and discussions were observed to be more clinical in nature. An explanation for this observation could be the predominant use of asynchronous collaboration tools in U21G. With asynchronous tools, the pressure of responding immediately to a question is lifted. A team member has time to mull over a message posted by another team member and formulate an appropriate response (Vonderwell, 2003). Koory (2003) notes that "written participation makes for a less spontaneous and sustained but more thoughtful and substantive class discussion than in a face-to-face situation." Hence, the act of having to communicate in a written form appears to suppress spontaneity.

Free-Riders and Easy-Riders

There were teams in which certain team members appeared to contribute very little to team activities, team discussion, and the creation of deliverables. This is not a phenomenon exclusive to virtual teams. The problem of 'free-riders' (Salomon & Globerson, 1989) exists in face-to-face teams as well. A milder form of free-riding, what the authors have termed 'easy-riding', was more commonly observed. An easy-rider is a student who makes a bare minimal contribution to the team. A similar concept known as 'social loafing' exists in the psychology literature which refers to the tendency of individuals to shirk when their lack of effort can be easily disguised within the activities of the team as a whole (Harkins & Szymanski, 1989). An easy-rider is a type of social loafer who may either face genuine difficulties coping with the study workload or is simply an indolent student.

FUTURE LESSONS

On the basis of the experience at U21G, several key lessons can be culled for institutions which intend to explore the use of virtual teams.

First is to provide clear guidelines on the roles for which each member is expected to play. In the absence of face-to-face interactions or prior engagement experience, members of a virtual team are unlikely to have a congruent view on how the team ought to function, and what is expected of each member. Ground rules¾for example, the style of writing, the tone of written text, and turnaround time for an e-mail¾could be spelt out to facilitate common understanding among members. Additionally, a private space can be set up to allow members to share thoughts outside the scope of assignment. Topics for social exchanges include family background and personal interests. Such measures, when implemented at the early stages of the team formation, would help develop a social context for trust and nurture the spirit of camaraderie which are essential ingredients to minimise a coordination problem, a problem found in virtual teams that are not sufficiently socialised.

Second is to demarcate intermediate milestones with achievable goals. Rather than presenting the question and expecting a report at the end of a time period, incremental goals coupled with the associated learning objectives could be sign-posted to help prod the virtual teams forward. In this way, any slippages along the way could be detected early, and appropriate actions can be promptly taken on the derailed team. Moreover, with the signposting of learning objectives, members become more perceptive of the knowledge and skills to be acquired. This measure is intended to sharpen the discourse among members of the virtual teams and hopefully promote deep discussion.

Third is to consider assessing not only the output of team assignments (i.e., the deliverables), but also the discussion that led to the creation of the output. Such discussions, which would be accessible on the discussion board, could be used as evidence that students had clearly discussed and debated issues surrounding the assignment. This would encourage students to engage in deeper and more active discussion.

Fourth is to explicitly provide students with exposure to different teams comprising members of varying communication style and performance. In the real world, virtual teams are unlikely to be perfectly formed and high performing. Hence, students have much to learn from situations where, for example, multiple members who have dominating personalities exist in the team. In addition, exposing students to teams where problems of conflict, coordination, and mistrust exist is a valuable learning experience for students. This is particularly so when students are asked, on hindsight, to reflect on how they solved or should have solved those problems.

Fifth is to investigate the use of media-rich collaboration tools that provide students with greater freedom of expression. At U21G, students rely primarily on text-based tools such as e-mail and chat. However, the use of voice and video-based collaboration tools would allow students to express a wider variety of emotions that are inherent in face-to-face interactions.

Finally we should introduce a peer appraisal system where each member confidentially appraises each other's performance. Members should be informed at the start of the assignment of the peer appraisal system. Dimensions in the appraisal system include the extent to which a member contributes to the team, the quality of the contribution, and the sense of responsibility demonstrated. This would help deter the problem of free and easy-riding.

CONCLUSION

As more universities, both online and traditional bricks-and-mortar, explore online education, virtual teams will become a more accepted way for students to work. The article has highlighted observations of dysfunctional collaborative behaviour in virtual teams in the asynchronous educational setting of U21G. The authors suspect such observations, while having drawn from the experience in U21G, are equally pertinent to virtual teams in other contexts. It would seem from an educator's perspective that one should not necessarily try to replicate a face-to-face team experience in a virtual environment. Rather, educators should consider carefully how they can best exploit the advantages that virtual teams have to offer, including the equal opportunity that all team members have to contribute, the emphasis on written communication, and the additional time students have for reflection.

REFERENCES

Bailey, M.L., & Luetkehans, L. (1998). Ten great tips for facilitating virtual learning team. *Proceedings of the Annual Conference on Distance Teaching and Learning,* Madison, WI.

Desanctis, G., Fayard, A., Roach, M., & Jiang, L. (2003). Learning in online forums. *European Management Journal, 21*(5), 565-577.

Harkins, S.G., & Szymanski, K. (1989). Social loafing and group evaluation. *Journal of Personality and Social Psychology, 56*(6), 934-941.

Kerr, D.S., & Murthy, U.S. (1994). Group decision support systems and cooperative learning in auditing: An experimental investigation. *Journal of Information Systems, 8*(2), 85-95.

Koory, M.A. (2003). Differences in learning outcomes for the online and F2F versions of "An Introduction to Shakespeare." *Journal of Asynchronous Learning Networks, 7*(2), 18-35.

Salomon, G., & Globerson, T. (1989). When teams do not function the way they ought to. *International Journal of Educational Research, 13*(1), 89-99.

Slavin, R.E. (1989). Research on cooperative learning: An international perspective. *Scandinavian Journal of Educational Research, 33*(4), 231-243.

Townsend, A.M., DeMarie, S.M., & Hendrickson, A.R. (1998). Virtual teams: Technology and the workplace of the future. *Academy of Management Executive, 12*(3), 17-29.

Vonderwell, S. (2003). An examination of asynchronous communication experiences and perspectives of students in an online course: A case study. *Internet and Higher Education, 6*, 77-90.

Warkentin, M.E., Sayeed, L., & Hightower, R. (1997). Virtual teams versus face-to-face teams: An exploratory study of a Web-based conference system. *Decision Sciences, 24*(4), 975-996.

Wordsmyth. (2002). Accessed January 27, 2005, from *http://www.wordsmyth.net/live/home.php?script =search&matchent=collaborating&matchtype=exact*

Zhang, D., & Nunamaker, J.F. (2003). Powering e-learning in the new millennium: An overview of e-learning and enabling technology. *Information Systems Frontiers*, 5(2), 207-218.

KEY TERMS

Asynchronous Communication: Communication between parties that does not require the parties to be available for communication all at the same time.

Deep Discussion: Discourse that involves critical analysis and debate.

Online Collaboration: The use of Internet-based communication tools to cooperate or work with someone else, especially on an intellectual project.

Peer Learning: The use of other students' experiences and knowledge as a primary resource in learning.

Social Loafing: The tendency of members of groups to put in less effort in group activities when group performance is measured than when individual performance is measured.

Socialisation: The process by which people become accustomed to the norms and roles that are necessary to function in a group setting.

Virtual Team: A group of remotely situated individuals who rely primarily on electronic communication to work together on group tasks.

Open Collectivism and Knowledge Communities in Japan

Hideo Yamazaki

Nomura Research Institute Ltd. Knowledge MBA School Tokyo, Japan

INTRODUCTION

The aim of this article is to introduce an Eastern CoPs' specific approach that is quite different from that of Western communities. In a collectivist prevalent societal type, the "sharing of feelings should come first, naturally followed by knowledge sharing" type of approach works very well even in a business environment.

One of Japan's traditional manufacturers has launched several interesting knowledge communities that are different from the accepted Western KM approach that emphasizes cost and effect straightforwardly. Their approach emphasized the generation of social networking on intranet first, and at a later stage, they proceeded to knowledge sharing through communities of practice.

This "go slow to go fast" approach may look like one of the typical and traditional Japanese management styles. However, in this approach, the culture of this company group steadily changes from introverted and closed to extroverted and open.

Their approach to build open and extroverted collectivism that is generated by knowledge communities could be one of the new management style prototypes of Japanese companies in the future.

BACKGROUND

QP Corporation is a top-class Japanese company, manufacturing and marketing mayonnaise and salad dressings in Japan. One of the group companies manufactures jam. QP is one of the top brands in the food industry. They have been in the food manufacturing business more than 80 years with 2,200 employees. The total number of employees including group companies is 6,000. This means that QP are typical of the traditional business group in Japan.

Their operation of knowledge communities has a 3.5-year history. In the past, their original KM approach failed when they built knowledge databases for knowledge sharing. This approach did not work well, let alone match the expectations of management. Therefore, this time, they decided against the mechanical approach such as building a knowledge database instead of building knowledge communities on their intranet was tried.

As far as the communities' approaches toward face-to-face activity, traditional socializing approaches such as company outings, pub drinking, and factory participation in local festivals where employees socialize with local people work very well in this company. And small-group employee activities for incremental improvement also work well. Therefore, the main activities of the revised QP KM plan focused on building knowledge communities on their intranet. The purposes was to change the company culture from a closed and introverted one to an open and extroverted one; promote employees to knowledge workers who are individually treasured; and promote workers into those who produce and share business ideas.

In the past, one of the features of Japanese companies was an introverted collectivism that supports the production of a quality product. However, in the 21st century of rapidly changing market environments, companies need fresh idea generation for new products and for new marketing, and it is clear that fresh ideas cannot emerge only between intimate colleagues. New ideas can come from anybody else inside and outside of organizations. Knowledge communities supply companies with "the strength of weak ties", backed by social networking that bridges silo type organizations.

SOCIAL NETWORKING BY KNOWLEDGE COMMUNITIES

KM Team Blogging

QP Corporation started building knowledge communities by using Lotus Notes software on their intranet. A centrally organized knowledge management team consisted of five employees. One of their roles is to behave as if they are company news reporters, and they compile different articles about the business and put them on the knowledge repository in Lotus Notes. They also provide company newsletters that brief all the employees of QP group by e-mail about these newly posted articles. After that, employees can voluntarily read some articles and post their comments. As far as posting comments are concerned, two different ways of posting were arranged by the KM team. One was posting for discussion on common forum, called the *tea lounge*, where a link from each article is prepared. Another way to comment is through the employee posting anonymously onto a bulletin board that is also linked to each article.

The category of these articles by the KM team varies from reports of shareholder meetings, introduction of new products, competitors' information to reports of management and employees' lives. As far as well visited articles are concerned, the articles relating to customers' voices are the most popular ones, and they convey customers' messages directly to all the employees. They sometimes include claims or request for product and services improvement.

The articles of customers' voices are compiled by the KM team, and the production process is the following: First, KM team contacts the call center to pick up customers' claims and needs. After that, the KM team contacts R&D or relevant manufacturing department for further information. Finally, compiled articles are posted on the bulletin board for employees' comments.

The same processes are repeated for the other categories of articles. They are regarded as KM team blogging, covering all the activities of companies. Interestingly, KM team blogging itself is declared as an informal message supply. It seems that currently, some articles are also voluntarily posted by local offices.

Employees' Regular Column Posting

In addition to the KM team article posting, individual employees' regular posting activities, called *my opinion,* are popular. *My opinion* is a weekly posted blogging by several employees. In *my opinion,* several volunteers such as young employees, experienced employees, men and women from varying divisions freely talk about their business and private life as either a diary or column, and the rest of the employees freely give their comments on them. In doing so, employees can get to know each other well, and they can share other employees' experiences and their way of thinking. For young employees, reading a senior's *my opinion* works as net-mentoring. Recently, female managers were employed in this company for the first time in its history. For these female employees, reading the female managers' *my opinion* is quite encouraging, and lots of them try to understand the way female managers think. It is a new form of tacit knowledge sharing in Japan. *My opinion* usually lasts for two to three months per employee. After that, they are replaced by different volunteers in rotation.

Business Volunteers' Recruiting

QP Corporation is a food manufacturer for consumers and businesses. Therefore, they sometimes recruit volunteers from employees to taste new products. For this purpose, *knowledge communities* are very useful and rewarding. When 10 to 20 employees are chosen for tasting volunteers, they sometimes voluntarily organize a community of practice on the intranet, and they regularly post their interesting reports for discussion about possible product improvements. They also sometimes recruit volunteer employees to distribute sample products in the employees' home neighborhood to pick up comments from their neighbors.

The recruiting and reporting by volunteers are all made known on *knowledge communities.* In particular, this reporting by volunteers is full of the voices of potential customers as well as their own feelings.

Usually, this type of business volunteers recruiting is requested from the KM team by official organizations such as marketing, and then the KM team coordinates between the *knowledge communities* and these divisions.

Business Q&A Exchange

The company recently started a business Q&A exchange. On the *knowledge communities*, knowledge seekers raise queries and knowledge givers post an answer. In this case, the KM team works as a facilitator.

COMMUNITIES OF PRACTICE AND FUTURE TRENDS

QP Corporation's Knowledge Community is a big community, called I-QP, covering all the group companies. As far as communities of practice are concerned, currently around 10 to 15 communities are operational. Among them, the working mother's community is the most fascinating. When it comes to the subject of raising knowledge workers, their current concern is how to raise the status of female workers. So far, the working mother's community is at an early stage, and the current central theme of discussion is baby food. However, this community has the potential to promote the improvement of female worker status in Japan.

As for business CoPs, the sales division has a community of practice, called *find*. In this CoP, all the sales staff are participants. All the noticed and found information in sales fields such as competitor's movement, distributor's movement, and popularity of new products is posted in this community of practice for knowledge sharing.

CONCLUSION

Throughout building knowledge communities, the QP organization has found that the employees have been gradually transforming themselves from "organization man" to self-supporting knowledge workers. Especially, the female workers have become more active and lively. However, from the building communities point of view, their approach to raise "open collectivism by knowledge workers" is still at an early stage and has a long journey ahead.

KEY TERMS

Blogging: Web logs, or blogs, are personal, online journals and one of the fastest growing trends on the Internet. Blogs are now considered one of the tools to maintain knowledge communities.

Business Q&A Exchange: One of the most popular knowledge exchanges in knowledge communities. The most refined software structure is called Q&A community.

Mechanical Approach: Means a (too) business process oriented approach such as building knowledge databases in rigid business process cycles.

Social Networking: Describes the process of connecting individuals via a personal network such as friends and acquaintances. In QP Corporation, social networking has been promoted for the foundation of knowledge exchange on intranet. Social networking nurtures a trusting atmosphere for knowledge communities on an intranet.

Strength of Weak Ties: In knowledge management, it means ideas can be shared not with intimate colleagues but with slight acquaintances.

Organisational Change Elements of Establishing, Facilitating, and Supporting CoPs

P.A.C. Smith
The Leadership Alliance Inc., Canada

INTRODUCTION

Although knowledge management (KM) is often proposed as a viable means to enhance business performance by facilitating knowledge creation and sharing, there is serious concern that it frequently fails to deliver on its promise (Despres & Chauvel, 2000; Fuller, 2001; Newell, Scarbrough, Swan & Hislop, 1999; Pietersen, 2001; Brown & Duguid, 2000; Storey & Barnett, 2000).

Smith and McLaughlin (2003) posit that KM's lacklustre performance can often be traced to non-rational emotion-based "people-factors" that negatively influence interpersonal relationships, and that are ignored during typical KM implementation. These authors argue that the success of any significant change initiative, including KM, will be critically dependent on understanding, and improving as necessary, the collaborative characteristics of the organisation's culture.

This article adopts the notion that effective KM is largely people-centric, and that communities of practice (CoPs), *when suitably grounded*, provide a practical framework for assisting in the development of appropriate "people-factors" and the nurturing of collaborative relationships. It builds on the work of Smith and McLaughlin (2003) by proposing an extension of their approach that helps ensure the presence of a truly collaborative culture in the target community.

BACKGROUND

Smith and McLaughlin (2003) describe in detail a number of practical remedial initiatives, including establishing CoPs, that may be undertaken to help "get the people factors right" when trying to ensure successful KM implementation. These initiatives are grounded in chaos theory and relate to three systemic "performance drivers":

1. **KM Focus:** A clear "who, what, where, when, and why" of the KM performance envisaged
2. **KM Resources:** The wherewithal to support KM Focus
3. **KM Will:** The intent to perform KM Focus

There are typically serious endemic barriers to optimising or even balancing these performance drivers. Four workforce development initiatives are recommended by Smith and McLaughlin (2003) to overcome these shortcomings:

1. community-wide collaborative development of a Vision *for the KM initiative* since this provides excellent Focus and Will for relationship-building through sharing of the individual yearnings of all employees;
2. management initiatives to address the physiological needs of individual employees (need for belongingness, esteem, and striving to be the best a person can be) based on Maslow's (1943) theory such that Will to form relationships is strengthened;
3. the nurturing of voluntary CoPs (Wenger, McDermott & Snyder, 2002) in order to promote formation of appropriate relationships based on conversations and activities of interdependent people in complex responsive processes (Stacey, 2001); and
4. introduction of CoP members and others to Action Learning methodology (Gaunt, 1991) as a means to:
    - enhance understanding of the "people-factors" that enhance or hinder relationship building, and provide participants with a process and the skills to further develop

their learning and collaborative capabilities; and

- improve the way people meet (and form relationships) by helping them become sensitised to the semiconscious and unconscious impulses that operate as individuals and groups struggle to come together.

Recent KM literature reflects this emphasis on the people-centric nature of KM implementation, particularly where knowledge is tacit and not easily shared (Hildreth, Kimble & Wright, 2000). Comments by authorities such as Wiig (2000; p. 4) are typical:

There are emerging realisations that to achieve the level of effective behaviour required for competitive excellence, the whole person must be considered. We must integrate cognition, motivation, personal satisfaction, feelings of security, and many other factors.

Wiig (2000, p. 14) cites a number of authors to support his contention that "overall KM will become more people-centric because it is the networking of competent and collaborating people that makes successful organisations." He goes on to say: "One key lesson to be learned is that we must adopt greater people-centric perspectives of knowledge...Technology only goes so far" (Wiig, 2000, p. 25).

Snowden (2000, pp. 237-238) notes that organisations:

...are gradually becoming aware that knowledge cannot be treated as an organisational asset without the active and voluntary participation of the communities that are its true owners. A shift to thinking of employees as volunteers requires a radical rethink of reward structures, organisational forms, and management attitudes.

Even where the KM focus is essentially technology based, the importance of people to the process is acknowledged. For example, Davenport and Prusak (1998, p. 129) wrote: "The roles of people in knowledge technologies are integral to their success."

As noted in the Introduction, this article adopts the notion that successful KM is largely people-centric, and that CoPs, *when appropriately grounded*, provide a practical framework for nurturing suitable relationships. Furthermore the article builds on the work of Smith and McLaughlin (2003) by proposing that there is a critical additional "fifth" development initiative that must be undertaken if a truly collaborative social fabric is to develop. This initiative involves the visualisation, optimisation, and utilisation of a variety of social networks across the organisation as the basis for establishing CoPs and other relevant groups (e.g., a KM Steering Committee). Issues that heighten the need for this initiative are presented in the next section. Identification of, and assessment of the influence of, the organisation's formal and informal opinion leaders is included in the initiative, as are efforts to involve them at all stages of KM design and implementation. The fifth initiative is based on Social Network Analysis (Wasserman & Faust, 1997), which is also described.

ISSUES CONCERNING CoP AND RELATIONSHIP BUILDING

As discussed above, success in the new knowledge economy, for a public or private organisation, is critically dependent on having an organisational culture that is characterised by ready and effective communications across voluntary collaborative partnership-networks of all kinds. It is no longer "*what you know*" or even "*who you know*" that leads to viability and well-being; it is "*who you know well enough to trust for advice, or have confidence in to get things done efficiently and effectively.*" In other words, the extent to which formal and informal conversations, storytelling, and interactions of all kinds can take place across stakeholder communities will be critical to learning and the widespread sharing/generation of knowledge (Stacey, 2001). The concept of *social capital* (SC) (Coleman, 1990; Burt, 1992; Putnam, 1993) is useful for representing the collaborative status of relationships across an organisation. Although there is no uniformly accepted definition of SC, its meaning in an organisational setting has been captured by Gabbay and Leenders (1999, p. 3): "The set of resources, tangible or virtual, that accrue to a corporate player through the player's social relationships, facilitating the attainment of goals."

Each individual's relationships with other individuals in an organisation form that individual's SC for better or worse; close relationships enhance SC, whereas distrust and lack of openness cause low SC (sometimes termed *social liability*). Furthermore, the SC of individuals aggregates into the SC of organisations. This is an important attribute since, as Burt (1992, p. 52) points out, a critical property of SC is that it creates opportunities for, or blocks, the transformation of human capital and financial capital into profit.

The formation of SC clearly depends on having positive individual attitudes with respect to forming and sustaining interpersonal relationships, and one might anticipate that nurturing SC could be fruitfully undertaken within a CoP framework (O'Donnell, Porter, McGuire, Garavan, Heffernan, & Cleary, 2003). Wenger, McDermott, and Snyder (2002, p. 4) have provided a widely accepted definition of CoPs as "groups of people who share a concern, a set of problems, or a passion about a topic, and who deepen their knowledge and expertise in this area by interacting on an ongoing basis." These authors add that "these people don't necessarily work together every day, but they meet because they find value in their interactions" (Wenger et al., 2002, p. 4), and go on to make it clear that in their view, the emphasis in CoPs is on "shared practice" where only behaviours and abilities *with respect to that practice* are enhanced (Wenger et al., 2002, pp. 41-44). Even though SC should be locally enhanced, this definition would seem to seriously constrain opportunities for overall relationship building, and be more conducive to development of "tight" cliques where group members become locked into like-minded close partnerships established in early community formation (Burt, 1992; Haythornthwaite, 1998).

There are issues related to whether more than a few viable CoPs can ever become established by simply allowing them to emerge as is normally recommended (Wenger et al., 2002; Saint-Onge & Wallace, 2003). For example, the formation of CoPs will be hampered where individuals lack networking skills, although workshops have been developed to address this issue (Smith & Godkewitsch, 2004). Indeed the stress of new networking cannot be overemphasised:

...our experiences of being and working in groups are often powerful and overwhelming. We experience the tension between the wish to join together and the wish to be separate; between the need for togetherness and belonging and the need for an independent identity. (Stokes, 1994, p. 19)

Even when CoPs do become established, authorities say little about how members really interact (Wenger et al., 2002; Saint-Onge & Wallace, 2003; Kimball & Ladd, 2004). If such groupings are to help nurture a broad-based collaborative organisational culture, members must focus on the attitudinal and behavioural nature of the various formal and informal group settings in which they meet. Smith and McLaughlin (2003) detail how effectively structuring such meetings provides a natural systemic way to shape the quality of interpersonal relationships through self-reflection, self-disclosure, and emotion, whilst energising individuals to act. These authors also indicate how these "meeting" issues may be explored through various group dynamics approaches (Egan, 1973; Nevis, 1987; Gabriel, 1999).

There are also issues concerning the true nature and extent of "sharing" relationships in CoPs. Individuals often resist sharing their knowledge in CoPs (Ciborra & Patriota, 1998), and knowledge is not shared easily even when an organisation makes a concerted effort to facilitate knowledge exchange (Szulanski, 1996). The success of knowledge sharing depends on the organisational KM system's social and technological attributes (Holsthouse, 1998), and on organisational culture (De Long & Fahey, 2000). Ardichvili, Page, and Wentling (2003, p. 29) report that employees in the virtual CoP they studied:

...view knowledge as a public good belonging to the whole organisation, [and] knowledge flows easily. However, even when individuals give the highest priority to the interests of the organisation and of their community, they tend to shy away from contributing knowledge for a variety of reasons. Specifically, employees hesitate to contribute out of fear of criticism, or of misleading the community members (not being

sure that their contributions are important, accurate, or relevant). To remove the identified barriers, there is a need for developing various types of trust, ranging from the knowledge-based to the institution-based trust.

Knowledge-based trust emerges on the basis of recurring social interactions between individuals, and is formed when the individuals get to know one another well, and are able to predict what to expect of one another, and how each will behave in a certain situation (Tschannen-Moran & Hoy, 2001). *Institution-based trust* is related to employees' trust across the whole organisation. Specifically, CoP members would need to have trust in the integrity of the organisation as a whole, and the competence of its members. This is based on the belief that necessary structures are in place to ensure trustworthy behaviour of individual members, and protect the members from negative consequences of administrative and procedural mistakes (McKnight, Cummings & Chervany, 1998). Institutional trust is enhanced by providing clear directions on what constitutes useful knowledge that can be posted on a CoP network, and by widely advertising examples of successful contributions by individuals. Clear communication is not enough; the organisation must demonstrate that it trusts its individual employees (DeLong & Fehey, 2000).

The work of Ardichvili, Page, and Wentling (2003) indicates that a virtual CoP, and indeed most CoPs, will function best when they: (a) are founded on, or have members that are drawn from, existing collaborative social networks; and (b) are part of an organisation that not only espouses trust in employees, but "walks the talk."

An organisation wishing to nurture KM and collaborative relationships through CoPs will almost certainly know the answer to (b), but they are not likely to know (a)—the patterns and nature of social networks in their organisation. This is because an organisation's social fabric is a complex mixture of closely-knit and more loosely woven formal, and informal, interpersonal and community relationships; the fabric may also display holes where community trust and collaborative knowledge sharing are absent.

Social Network Analysis (Wasserman & Faust, 1997) is therefore an important element in the "fifth"

development initiative because it makes possible the identification of the patterns and the nature of social networks in an organisation, and their existing or latent influential potential from a knowledge-trust standpoint. Given this insight, a CoP or other grouping may be encouraged to take root on one or more prior social networks where appropriate relationship capabilities and institutional-trust have already been demonstrated.

Social Network Analysis (SNA)

Special techniques are required to visualise the complexities of how people communicate and interact in social networks, and SNA provides this capability. Although it is a highly mathematical approach, a number of simplified descriptive texts exist, for example Scott (2000). SNA is a very rich theoretical methodology that is only recently emerging as a practical and dynamic approach to real organisational problems (Kilduff & Tsai, 2003). Because of its highly mathematical nature, computers are typically used for calculation and display (Borgatti, Everett & Freeman, 1999).

In practice, data regarding an attribute of interest are first collected from a target organisational population, or the whole organisation. The appropriate SNA is then applied to these data, and local interpretation of results undertaken. In this way key informal and formal players may be identified, the relationship networks visualised and compared to optimal patterns, and actions undertaken as necessary to realise the potential envisaged for the initiative at hand. In addition, the various influential network agents have recognisable characteristics that can be identified (e.g., individuals who link networks across organisational boundaries). Networks themselves may be characterised as displaying effective social communications and collaborative archetypes (Buchanan, 2002).

When mapped, "real" communications channels are distributed unevenly, since dense clusters tend to form around established relationships (e.g., existing CoPs). The strong ties formed in these clusters have many benefits, but it is also critical to have "weaker" links between clusters to ensure broad-based relationship building, the quick flow across the community of new ideas, and the timely awareness of new opportunities and challenges. For this reason, identification of weak ties and knowledge of their relationship utility (Kilduff & Tsai, 2003) are important aspects of the "fifth" development initiative. SNA is particularly necessary for pinpointing these weaker links, since such ties are often informal, having little obvious relationship to the official organisational-communications design.

FUTURE TRENDS

One may expect that SNA and related information will have increasing application in organisational optimisation in general, and in the development of SC in particular. It is anticipated that attempting to establish CoPs on a solid foundation of existing supportive relationships will also become a key concern with respect to KM design and implementation. Interpretation of an organisation's emergent social and communication patterns in dynamic and practical contexts (Kilduff & Tsai, 2003) is currently a hot topic that is expected to attract even more interest.

CONCLUSION

The emergence of a CoP is a necessary but insufficient condition when an organisation wishes to nurture a collaborative social fabric and optimise its SC. Practical activities to further nurture collaborative relationships within a CoP framework have been identified, and in particular, the importance of SNA as a precursor to viable CoP development has been explained.

REFERENCES

Ardichvili, A., Page, V., & Wentling, T. (2003). Motivation and barriers to participation in virtual knowledge-sharing communities of practice. *Journal of Knowledge Management, 7*(1), 64-77.

Borgatti, S.P., Everett, M.G., & Freeman, L.C. (1999). *UCINET 6.0 version 1.00.* Analytic Technologies, Natick, MA.

Brown, J., & Duguid, P. (2000). *The social life of information.* Boston: Harvard Business School Press.

Buchanan, M. (2002). *Nexus.* New York: Norton.

Burt, R.S. (1992). The social structure of competition. In N. Nohria & R.G. Eccles (Eds.), *Networks and organizations.* Boston: Harvard Business School Press.

Ciborra, C.U., & Patriotta, G. (1998). Groupware and teamwork in R&D: Limits to learning and innovation. *R&D Management, 28*(1), 43-52.

Coleman, J.S. (1990). *Foundations of social theory.* Cambridge, MA: Belknap Press.

Davenport, T.H., & Prusak, L. (1998). *How organisations manage what they know.* Boston: Harvard Business School Press.

DeLong, D., & Fehey, L. (2000). *Diagnosing cultural barriers to knowledge management. Academy of Management Executive, 14*(4), 113-127.

Despres, C., & Chauvel, D. (Eds.). (2000). *Knowledge horizons.* Boston: Butterworth-Heinemann.

Egan, G. (1973). *Face to face.* Monterey, CA: Brooks/Cole.

Fuller, S. (2001). *Knowledge management foundations.* Boston: Butterworth-Heinemann.

Gabbay, S.M., & Leenders, R.T.A.J. (1999). The structure of advantage and disadvantage. In R.T.A.J. Leenders & S.M. Gabbay (Eds.), *Corporate social capital and liability.* Boston: Kluwer Academic Publishers.

Gabriel, Y. (1999). *Organisations in depth.* London: Sage Publications.

Gaunt, R. (1991). *Personal and group development for managers: An integrated approach through action learning.* Harlow, Essex, UK: Longmans.

Haythornthwaite, C. (1988). A social network study of the growth of community among distance learners. *Information Research, 4*(1). Retrieved from *http://informationr.net/ir/4-1/paper49.html*

Hildreth, P., Kimble, C. & Wright, P. (2000). Communities of practice in the distributed international environment. *Journal of Knowledge Management, 4*(1), 27-38.

Holtshouse, D. (1998). Knowledge research issues. *California Management Review, 40*(3), 277-280.

Kilduff, M. & Tsai, W. (2003). *Social networks and organisations.* London: Sage Publications.

Kimball, L. & Ladd, A. (2004). Facilitator toolkit for building and sustaining virtual communities of practice. In P.M. Hildreth & C. Kimble (Eds.), *Knowledge networks: Innovation through communities of practice.* Hershey, PA: Idea Group Publishing.

Maslow, A. (1943). Theory of human motivation. *Psychological Review, 50,* 370-396.

McKnight, D.H., Cummings, L.L. & Chervany, N. (1998). Initial trust formation in new organisational relationships. *Academy of Management Review, 23*(3), 413-490.

Nevis, E.C. (1987). *Organisational consulting; A gestalt approach.* New York: Gardner.

Newell, S., Scarbrough, H., Swan, J. & Hislop, D. (1999). Intranets and knowledge management: Complex processes and ironic outcomes. *Proceedings 32nd Hawaii International Conference on Systems Sciences.*

O'Donnell, D., Porter, G., McGuire, D., Garavan, T.N., Heffernan, M. & Cleary, P. (2003). Creating intellectual capital: A Habermasian community of practice (CoP) introduction. *Journal of European Industrial Training, 27*(2/3/4), 80-87.

Pietersen, H.J. (2001). The IC/KM movement and human systems well-being. *Journal of Knowledge Management Practice, 2.* Retrieved from *http://www.tlainc.com/jkmpv2.htm*

Putnam, R.D. (1993). The prosperous community: Social capital and public life. *American Prospect, 13,* 35-42.

Saint-Onge, H. & Wallace, D. (2003). *Leveraging communities of practice for strategic advantage.* New York: Butterworth-Heinemann.

Scott, J. (2000). *Social network analysis.* London: Sage Publications.

Smith, P.A.C. & Godkewitsch, M. (2004). *Private communication.*

Smith, P.A.C. & McLaughlin, M. (2003, January 15-17). Succeeding with knowledge management: Getting the people-factors right. *Proceedings of the 6th World Congress on Intellectual Capital & Innovation.* McMaster University, Canada.

Snowden, D. (2000). The social ecology of knowledge management. In C. Despres & D. Chauvel (Eds.), *Knowledge horizons* (pp. 237-265). Boston: Butterworth-Heinemann.

Stacey, R.D. (2001). *Complex responsive processes in organisations.* London: Routledge.

Stokes, J. (1994). The unconscious at work in groups and teams: Contributions from the work of Wilfred Bion. In A. Obholzer & V.Z. Roberts (Eds.), *The unconscious at work: Individual and organisational stress in the human services* (pp. 19-27). London: Routledge.

Storey, J. & Barnett, E. (2000). Knowledge management initiatives: Learning from failure. *Journal of Knowledge Management, 4*(2), 145-156.

Szulanski, G. (1996). *Exploring internal stickiness: Impediments to the transfer of best practice within the firm. Strategic Management Journal, 17*(1), 27-44.

Tschannen-Moran, M. & Hoy, W.K. (2001). *A multidisciplinary analysis of the nature, meaning, and measurement of trust. Review of Educational Research, 70*(4), 547-593.

Wasserman, S. & Faust, K. (1997). *Social network analysis: Methods and applications.* Cambridge: Cambridge University Press.

Wenger, E., McDermott, R. & Snyder, W.M. (2002). *Cultivating communities of practice.* Cambridge, MA: Harvard Business School Press.

Wiig, K.M. (2000). Knowledge management: An emerging discipline rooted in a long history. In C. Despres & D. Chauvel (Eds.), *Knowledge horizons* (pp. 3-26). Boston: Butterworth-Heinemann.

KEY TERMS

Chaos Theory: A theory that deals with complex and dynamical arrangements of connections between elements forming a unified whole, the behaviour of which is simultaneously both unpredictable (chaotic) and patterned (orderly).

Human Capital: The attributes, competencies, and mindsets of the individuals that make up an organisation.

Institution-Based Trust: Trust formed when organisational members believe that their organisation as a whole has their best interests at heart and acts accordingly.

Knowledge-Based Trust: Trust that emerges on the basis of recurring social interactions between individuals, and is formed when the individuals get to know one another well.

Social Capital: The set of resources, tangible or virtual, that accrue to a corporate player through the player's social relationships, facilitating the attainment of goals.

Social Network: A set of nodes (persons, organisations, etc.) linked by a set of social relationships of a specified type (e.g., friendship).

Social Network Analysis: Data acquisition methods and computerised (typically) techniques that enable visualisation of social networks and articulation of their properties.

Promoting Participation in Communities of Practice

Carolyn W. Green
Texas A&M University - Kingsville, USA

Tracy A. Hurley
Texas A&M University - Kingsville, USA

INTRODUCTION AND BACKGROUND

One of the emerging themes in recent organization theory and strategic management research has been the central role that knowledge plays in organizational performance. Grant (2001), for example, looks at the advantages of a knowledge-based perspective in organization theory, focusing on knowledge as the critical resource in the production of goods and services. Similarly, Teece (2001) notes an "increasing recognition that the competitive advantage of firms depends on their ability to create, transfer, utilize and protect difficult to imitate knowledge assets" (p. 125). Nonaka, Toyama, and Konno (2001) claim that continuously creating knowledge is the reason for a firm's existence, noting widespread acceptance of the view that the ability to create and utilize knowledge is the most important source of a firm's sustainable competitive advantage. More recently, Simsek (2003), taking a knowledge-based view of the firm, has argued that firms with superior knowledge systems are better able to identify, take advantage of, and create information asymmetries in their competitive environments. Simsek's study found that knowledge-based capabilities were associated with more entrepreneurial activity, which was in turn related to higher levels of firm performance.

Interest in how knowledge affects organizational performance has also turned to a consideration of the role communities of practice play in increasing the knowledge-based capabilities of organizations. Brown and Duguid (2001), exploring contradictions associated with the tendency for knowledge to leak across organizational boundaries, focus on practice as the key to understanding the communities that connect professionals in their shared development of knowledge. They note:

...what individuals learn always and inevitably reflects the social context in which they learn it and in which they put it into practice. When learning a job is at issue, this context usually includes the firm as a whole, immediate colleagues, and the relevant discipline or profession (as well as idiosyncratic external social forces bearing on each individual). (p. 200)

It is in this social context that communal practice develops with those who are involved in the common pursuit of a profession or endeavor developing new knowledge and insights that are then shared among the members of the community. Regarding the organization as a "community of communities of practice", Brown and Duguid (1991, 2001, p. 203) suggest that it is in communities of practice that much of an organization's knowledge creation takes place. An organization's knowledge base extends beyond its boundaries, drawing on the knowledge of the communities of practice in which its members are involved. Brown and Duguid see the innovative capacity of an organization arising from its ability to coordinate the development and sharing of knowledge as its various communities of practice cooperate in carrying out value chain activities.

The learning required to become an experienced practitioner is not passive; it requires active participation in the community of practice in which the work is embedded (Argote, McEvily & Reagans, 2003). The success of these communities is largely determined by the willingness of its members to actively participate in knowledge generation and sharing (Ardichvili, Page & Wentling, 2003). Recent studies have focused on identifying conditions that either encourage or inhibit the knowledge creation and sharing activities that lie at the heart of a community of practice (e.g., Ardichvili et al., 2003; Argote et al., 2003; Bieber et al., 2002; Griffith, Sawyer & Neale, 2003; Kling & Courtright, 2003; Lee & Choi, 2003; Schlager & Fusco, 2003; Schwen & Hara, 2003). Among the factors commonly cited are organizational culture, norms, rewards, incentives, and technological support.

The purpose of this study was to examine the extent to which professionals engage in knowledge creation and sharing activities and to see whether supportive norms and access to information technology are associated with higher levels of activity. This was done by developing and testing a model relating the extent to which knowledge creation and sharing activities are performed in organizational settings, the extent to which these activities are supported by information technology, and the existence of norms that encourage information technology use. The study was designed to apply to a variety of organizational settings, including for-profit and non-profit organizations.

ORGANIZATIONAL CONDITIONS INFLUENCING KNOWLEDGE CREATION AND SHARING

Technological, cultural, and structural infrastructures have been identified as significant contextual elements that characterize and influence the environment in which knowledge management processes are embedded (Gold, Malhotra & Segars, 2001; Grover & Davenport, 2001). Each of these elements is expected to have an effect on the extent to which knowledge creation and sharing activities are carried out within the organization. These are very similar to the organizational factors identified in Leavitt's (1965) model of organizational change, which has been used in several studies of system development success and risk (e.g., Heeks, 2002; Lyytinen, Mathiassen & Ropponen, 1998; Van Offenbeek & Koopman, 1996). Leavitt identifies four factors that must be aligned to bring about change: technology, people, structure, and task. In Leavitt's model, the people factor focuses on attitudes and motivations that encourage the people in the organization to embrace a proposed change. These attitudes would include cultural factors like

norms. The infrastructures identified by Gold et al. (2001) fit under three of Leavitt's factors: technology, people, and structure. As noted in Lyytinen et al. (1988), Leavitt's model proposes that these factors are interrelated and must be reasonably congruent in order for the organization to function well. Part of the challenge of introducing organizational change is to identify a pattern of organizational factors that is congruent with the proposed change and then develop a transition plan that will minimize the risks associated with its implementation.

Technological infrastructure includes access to a comprehensive information and communication system that supports knowledge management activities (Gold et al., 2001). Teece (2001) notes that a "combination of IT and co-aligned organizational processes can significantly enhance learning and competitive advantage" (p. 130). Access to relevant information technology and higher levels of technology use would be expected to contribute to a higher level of knowledge management activity within the organization. Markus (2001) notes that knowledge reuse depends in part on the availability of information technology and repositories of knowledge. Wenger and Snyder (2000) see information technology as a vital part of the infrastructure that enables organizations to cultivate communities of practice. Studies of efforts to foster their development have found that the most effective approach has been to use information technology as a tool to strengthen and support already existing communities formed using traditional means of interaction (Kling & Courtright, 2003; Schlager & Fusco, 2003; Schwen & Hara, 2003). The development of virtual communities of practice—whether purely virtual or hybrids that have both traditional and virtual components—relies on information technology as an essential enabler (Bieber et al., 2002; Griffith et al., 2003; Lee & Choi, 2003).

Cultural infrastructure, which includes corporate vision and values, is also expected to have a significant effect on knowledge management activity (Gold et al., 2001). Norms promoting knowledge creation and sharing, including norms of reciprocity, are expected to affect the extent of knowledge creation and transfer (Alavi & Leidner, 2001; Argote et al., 2003; Markus, 2001) and to contribute to the development of vibrant communities of practice (Ardichvili et al., 2003). Active participation in virtual communities of practice also requires that members be comfortable with computer-mediated interaction (Ardichvili et al., 2003). Drawing from the existing TAM (technology acceptance model) and TPB (theory of planned behavior) research, norms encouraging technology use could also be important factors influencing knowledge management activity levels. Norms encouraging technology use would be expected to have a positive effect on technology use for knowledge management (Green, 1998; Taylor & Todd, 1995; Venkatesh & Morris, 2000), which in turn could result in higher levels of knowledge management activity. Trust—both competence-based and benevolence-based—is seen to be another necessary ingredient in active knowledge creation and sharing in communities of practice (Argote et al., 2003; Kling & Courtright, 2003; Lee & Choi, 2003; Levin, Cross & Abrams, 2002).

Structural infrastructure, which includes the organization's system of rewards and incentives (e.g., incentives to generate new knowledge and to share knowledge with others), is expected to have a significant effect on the extent of organizational knowledge management activity (Gold et al., 2001). This is consistent with observations made by others who have noted the need for organizational incentives to encourage knowledge management participation (Ba, Stallaert & Whinston, 2001; Hall, 2001; Markus, 2001; Stein & Zwass, 1995; Wenger & Snyder, 2000). Hall (2001) has identified extrinsic and intrinsic rewards that may be important in motivating knowledge management activity. These include economic rewards like salary increases and bonuses; access to information and knowledge; career advancement; job security; reputation enhancement; and personal satisfaction. Wenger and Snyder (2000) see reward structures and promotion systems as essential elements in organizations' efforts to foster the development of communities of practice.

Although these contextual elements represent distinct constructs, they are not assumed to be orthogonal. Consistent with Leavitt's (1965) model, interactions among the elements are expected. Development of an organizational culture that encourages knowledge management activity might, for example, lead to increased access to information technologies that support those activities. Similarly, investment in information technologies that support knowledge management might be accompanied by

concerted efforts to develop an organizational culture and organizational structures that would encourage its use.

FRAMEWORK FOR ANALYSIS OF KNOWLEDGE CREATION AND TRANSFER

In order to explore the relationship between organizational conditions and active participation in a community of practice, it is necessary to be able to identify and describe the types of knowledge creation and sharing activities that might be encountered in these settings. In their review of knowledge management and knowledge management systems research, Alavi and Leidner (2001) outline various types of activities that may be used to distribute and apply knowledge in organizations. The activities described in the framework include knowledge creation and transfer, which are the kinds of activities required for active participation in a community of practice. As an aid to identifying individual knowledge creation and transfer activities, a composite model was developed to depict the knowledge management elements found in Alavi and Leidner's framework (Figure 1)[1]. Each of the framework's elements is described in the following discussion.

Types of Knowledge

Effective management of organizational knowledge must be based in a rich understanding of the nature of knowledge at the individual, community of practice, and organizational levels. Knowledge management research has recognized that the development of an organizational knowledge base involves an interplay between individual knowledge creation and organizational knowledge transfer. Organizational knowledge can be thought of as collective knowledge consisting of the aggregation and integration of individual knowledge (Grant, 2001; Simon, 1991). Research focused on communities of practice offers an additional dimension with the organization viewed as a community of communities of practice (Brown & Duguid, 2001). From that perspective, organizational knowledge also aggregates and integrates the collective knowl-

Figure 1. Framework for analysis of knowledge management activities

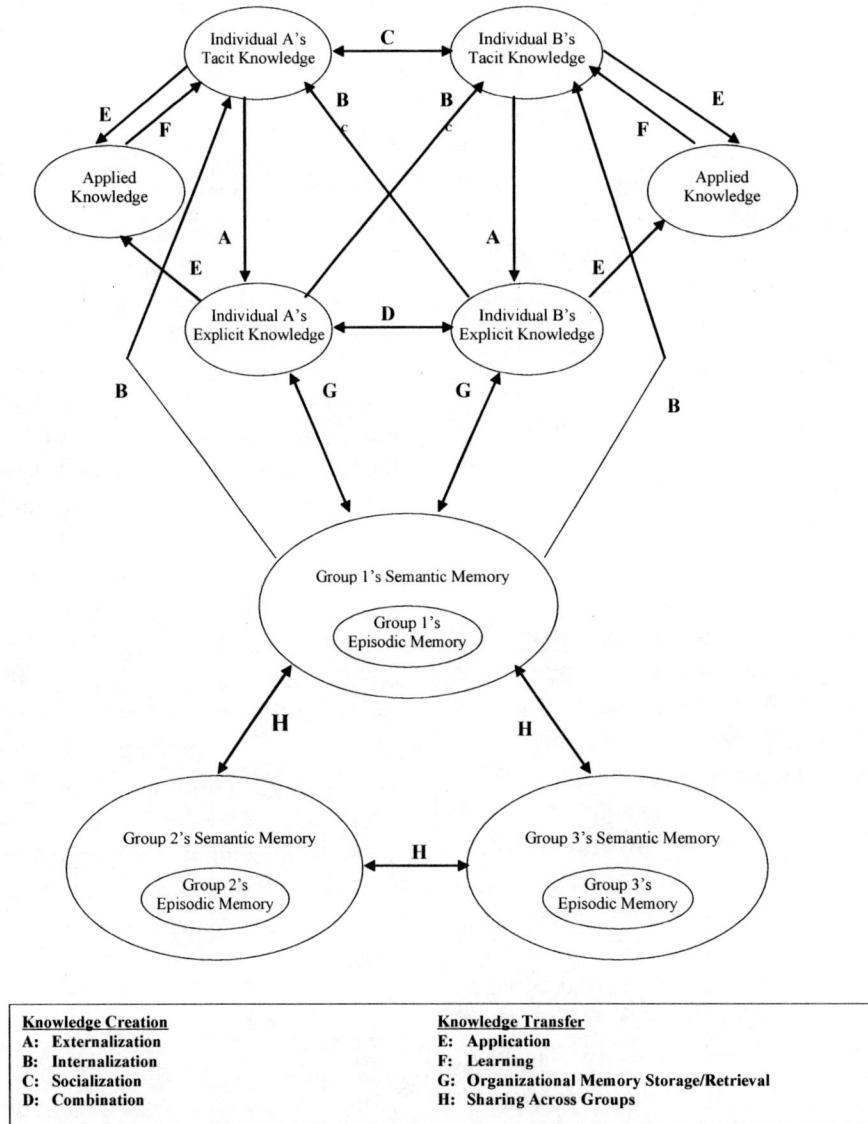

edge of its communities of practice. The framework depicted in Figure 1 recognizes and identifies different types of individual and organizational knowledge.

Studies of knowledge and knowledge management have identified two primary classes of individual knowledge: tacit and explicit. Tacit knowledge is defined as unformulated knowledge consisting of an individual's mental models (e.g., beliefs, paradigms, and mental maps) as well as know-how that may be applied to particular tasks or problems. Explicit knowledge is defined as public, objectified knowledge that has been articulated, codified, or

communicated to others, even in symbolic form (Alavi & Leidner, 2001; Nonaka & Konno, 1998; Polanyi, 1959). Knowledge management activities depicted in Figure 1 include processes by which tacit knowledge is made explicit and explicit knowledge is shared with others (activities A through D in Figure 1).

The framework depicted in Figure 1 also considers knowledge at the organizational level. Organizational knowledge is viewed as including written documents, structured information, codified knowledge, and documented procedures and processes, as well as tacit knowledge retained by individuals and networks of individuals who are part of the organization (Alavi & Leidner, 2001; Stein & Zwass, 1995; Tan et al., 1998). Organizational knowledge may also be thought of as organizational memory. Stein and Zwass (1995) note that organizational memory may be classified as either semantic or episodic. Semantic memory consists of generalized knowledge rather than memories of specific events. Episodic memory is context-specific, consisting of memories of individual experiences, including the time and context in which the events occurred (Alavi & Leidner, 2001; El Sawy, Gomes & Gonzales, 1996; Stein & Zwass, 1995, Tulving, 1983). Knowledge management activities depicted in Figure 1 include processes by which individual knowledge is made part of and enhanced by reference to semantic and episodic organizational memory (activities B, G, and H in Figure 1).

Knowledge Management Activities

Knowledge management itself has been conceptualized as consisting of those activities that result in the creation of knowledge and those activities that result in the transfer of explicit knowledge to others. The framework depicted in Figure 1 delineates four knowledge creation activities and four knowledge transfer activities described in the Alavi and Leidner framework and other knowledge management research. The knowledge creation activities are socialization, externalization, combination, and internalization. The knowledge transfer activities are learning, application, organizational memory storage/retrieval, and sharing across groups.

Knowledge Creation Activities

Externalization (Activity A) involves articulation of tacit knowledge and translation into forms that can be understood by others. This includes dialogue with others as well as other forms of communication that make use of words, concepts, figurative language, and visual aids. Internalization (Activity B) involves conversion of explicit knowledge into tacit knowledge. This requires the individual to identify what is personally relevant within the organization's knowledge base and put the knowledge into practice. Socialization (Activity C) involves sharing tacit knowledge through social interaction and shared experience. Socialization involves being in proximity, as

would be typical, for example, in an apprenticeship assignment. Combination (Activity D) involves converting explicit knowledge into more complex articulated knowledge through communication, diffusion, and systemization (Alavi & Leidner, 2001; Nonaka, 1994; Nonaka & Konno, 1998).

Knowledge Transfer Activities

Learning (Activity E) involves the knowledge transfer that occurs when individuals apply knowledge to a situation and develop new understandings by observing the results they achieve. Application (Activity F) includes integrating specialist's knowledge into the execution of organizational tasks in the form of rules and directives, sequences of tasks, organizational routines, and joint problem solving. Organizational memory storage/retrieval (Activity G) involves storage of and retrieval from explicit knowledge residing in forms like written documentation, electronic databases, e-mail messages, pictures, images, video, and music. Sharing across groups (Activity H) involves sharing group knowledge with other groups, whether internal or external to the organization, including importing information from external sources through dialogue, retrieval of written documentation, and access of external databases (Alavi & Leidner, 2001; Grant, 1996, 2001).

Research Model

The general model guiding the study is shown in Figure 2. The model was drawn from a review of prior research associated with organizational conditions influencing knowledge management activity as well as recent studies that delineate activities associated with participation in communities of practice. The model proposes that supportive technological, cultural, and structural infrastructures will be associated with higher levels of knowledge creation and transfer activity. Higher levels of knowledge creation and sharing are expected to have a positive effect on organizational performance (Grant, 2001; Nonaka et al., 2001; Simsek, 2003; Teece, 2001). The model also includes expected interactions among technological, cultural, and structural infrastructure factors.

Figure 3 shows the portion of the model that was tested in this study. The study's focus was on

Figure 2. A model of the effect of technological, cultural, and structural infrastructure on knowledge management activity and organizational performance

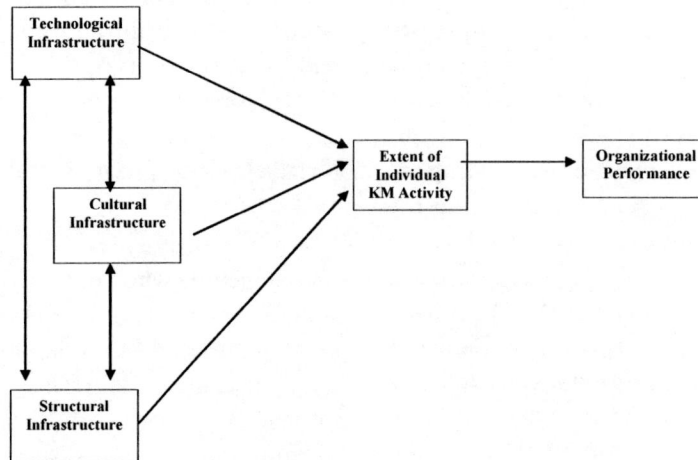

Figure 3. Effect of information technology use extent and norm on knowledge management activity

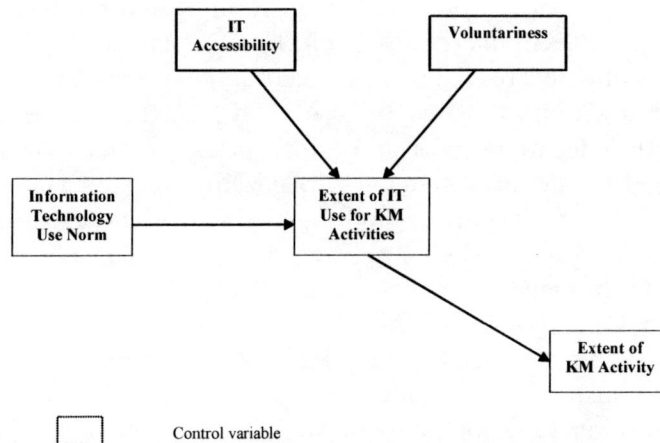

technological and cultural infrastructures that facilitate knowledge creation and transfer. The outcomes of interest were the extent to which information technologies were used in carrying out knowledge management activities and the extent to which knowledge management activities were performed. Information technology accessibility, voluntariness of use, and the extent of information technology use for knowledge management were used as measures of technological infrastructure. An information technology use norm was used as a measure of cultural infrastructure. The extent of knowledge management activity was represented by the extent to which activities described in the Alavi and Leidner (2001)

framework (Activities A-D and G-H, Figure 1) were performed by organization members.

Methodology

The data for the project were gathered via a survey questionnaire administered to 30 professionals who worked in country club management, real estate, and social services counseling. Of the 30 subjects included in the study, 19 worked in for-profit organizations, and 11 worked in non-profit organizations. Of the respondents, 20 were female, and 8 were male (2 did not respond to the question). Of the respondents, 6 were in their 20s, 10 in their 30s, 10 in their 40s, and 3 in their 50s (1 did not respond to the question). The survey instrument included questions designed to measure the dependent and independent variables described below.

Dependent Variable

Extent of knowledge management activity. The extent of the subjects' knowledge management activities was measured on a 5-point scale using 6 questions drawn from the knowledge creation and transfer activities (activities A-D and G-H shown in Figure 1) described in the framework depicted in Figure 1 ($\alpha = 0.79$ for knowledge creation activities; $\alpha = 0.71$ for knowledge transfer activities; $\alpha = 0.84$ for all knowledge management activity):

Knowledge Creation:

- I often document my ideas at work by writing them down for my own reference.
- I often incorporate my co-workers' written and/or documented ideas as a basis for forming or improving my own ideas or knowledge.
- I often use co-workers' ideas (information conveyed in conversation or informal communication) as a springboard to enhance my own ideas or knowledge.
- I often merge or incorporate information documented by my workgroup or department into my own knowledge base.

Knowledge Transfer:

- I often formally document what I have learned for work or job-process improvement so that others who may do my job in the future may learn from my experience.
- I often research organizational, workgroup, or industry records to enhance my workgroup's productivity.

Independent Variables

- **Information technology use norm.** The information technology use norm was measured using a series of questions based on the subjective norm measures developed by Ajzen and Fishbein (1980) and Venkatesh and Davis

(2000) ($\alpha = 0.88$). The subjective norm was measured on a 7-point scale using the following statements:

- People who influence my behavior think that I should use computer systems in my work.
- People who are important to me think that I should use computer systems in my work.
- My coworkers think that I should use computer systems in my work.
- The people with whom I work most closely think that I should use computer systems in my work.

- **Extent of information technology use for knowledge management.** The extent of the subjects' use of information technology for each area of knowledge management (i.e., activities A-D and G-H shown in Figure 1) was measured on a 5-point scale, with 1 = never, 2 = rarely, 3 = occasionally, 4 = frequently, and 5 = always ($\alpha = 0.88$).

Control Variables

- **Information technology accessibility.** Since use of information technology might be lessened if the technology were not accessible to knowledge workers, a measure of information technology accessibility was included in the study as a control variable ($\alpha = 0.71$). Accessibility was measured on a 7-point scale using the following statements:

  - I have very limited access to the computer systems that I would like to use to do my work.
  - I have adequate access to the computer systems that I would like to use do my work.
  - The computer systems available to me at work are adequate for performing my work in an effective manner.

- **Voluntariness of information technology use.** Management policies that make information system use mandatory may also have an influence on information technology use, causing employees to increase their usage despite their own preferences to the contrary. The voluntariness measure used by Venkatesh and Davis (2000) was used to control for an increase in use based on mandatory usage policies ($\alpha = 0.86$).

Results

The means and standard deviations of all variables included in the study are presented in Table 1. Information technology access, voluntariness of information technology use, and information technology use norm were measured on a 7-point scale. Subjects' responses indicated a moderate level of access to information technology ($M = 4.80$, $SD = 1.67$) and a low level of voluntariness of use ($M = 2.85$, $SD = 2.85$). Norms encouraging information technology use were prevalent ($M = 5.70$, $SD = 1.43$). Information technology use and knowledge management activity levels were measured on a 5-point scale. Means (with standard deviations in parentheses) for information technology use and the overall level of knowledge management activity were 3.51 (0.90) and 3.76 (0.71), respectively. Respondents reported higher levels of knowledge creation activity than knowledge transfer activity. The difference in the levels of knowledge creation

activity ($M = 3.96$, $SD = .59$) and knowledge transfer activity ($M = 3.35$, $SD = 1.09$) was statistically significant, $t(44.66) = 2.68$, $p = 0.01$, $d = 0.61$.

Correlations for all variables included in the regression analyses are shown in Table 1. The information technology use norm was significantly correlated with information technology accessibility ($r = 0.44$, $p < 0.05$) and voluntariness of use ($r = -0.37$, $p < 0.05$). This would be consistent with a commitment to provide information technology resources in organizations that want to develop an environment that encourages technology use. The inverse relationship between the information technology (IT) use norm and voluntariness of use may indicate that organizations lacking positive IT use norms turn to mandatory usage policies to achieve acceptable levels of use. As expected in the model, there was a significant relationship between IT usage and the IT use norm ($r = 0.45$, $p < 0.05$) and between IT usage and the extent of knowledge management activity ($r = 0.69$, $p < 0.01$).

Multiple regression analyses were used to test the expected relationships between the variables illustrated in Figure 3. Two regression models were tested. Results of the first regression model are reported in Table 2. The first model tested the effects of accessibility, voluntariness, and IT use norm on the extent of use of information technology. The result was a significant model ($F(2, 26) = 3.52$, $p < 0.05$) explaining 21% of the variance in IT usage

Table 1. Correlations and descriptive statistics (N = 30)

Variables	Mean	SD	Accessibility	Voluntariness	Norm	IT_Use
Accessibility	4.80	1.67				
Voluntariness	2.85	2.06	-.36			
Norm	5.70	1.43	.44[*]	-.37[**]		
IT_Use	3.51[1]	.90	-.03	-.01	.45[*]	
KM_Activity	3.76[1]	.71	-.33	-.10	.24	.69[**]
KC_Activity	3.96[1]	.59				
KT_Activity	3.35[1]	1.09				

[1]Based on a 5-point scale. All other variables are based on a 7-point scale.

[*]$p < .05$

[**]$p < .01$

Table 2. Summary of regression analysis for variables predicting extent of information technology use for knowledge management activity (N = 30)

Variable	B	SE B	β
Accessibility	-.28	.20	-.26
Voluntariness	.16	.24	.13
Norm	.58	.18	.62**

Note. R^2 = .29, adjusted R^2 = .21, $F(3, 26)$ = 3.52.

*$p < .05$

**$p < .01$

Table 3. Summary of regression analysis for variables predicting extent of knowledge management activity (N = 30)

Variable	B	SE B	β
Accessibility	-.34	.13	-.40*
Voluntariness	-.24	.14	-.23
Norm	.02	.13	.03
IT Use	.52	.12	.66**

*Note. R^2 = .62, adjusted R^2 = .56, $F(4, 25)$ = 10.05**.

*$p < .05$

**$p < .01$

(multiple R^2 = 0.29, adjusted R^2 = 0.21). The IT use norm was the only significant independent variable in the model (β = 0.62, $p < 0.01$). Results of the second regression model are reported in Table 3. The second model tested the effects of accessibility, voluntariness, IT use norm, and IT usage on the extent of knowledge management activity. The result was a significant model ($F (4, 25)$ = 10.05, $p <$ 0.01) explaining 56% of the variance in the extent of knowledge management activity (multiple R^2 = 0.62, adjusted R^2 = 0.56). As expected, IT usage had a significant effect on the extent of knowledge management activity (β = 0.66, $p < 0.01$). Accessibility was also a significant variable in the regression model (β = -0.40, $p < 0.05$).

CONCLUSION AND FUTURE TRENDS

As noted earlier, the success of a community of practice depends upon its members' active participation in knowledge generation and sharing. The focus of the data analysis in this study was on the extent to which professionals engage in knowledge creation and transfer activities, and whether supportive norms and access to information technology are associated with higher levels of activity. A framework for analysis of knowledge management activities (Figure 1) was adapted from Alavi and Leidner (2001). The activities described in the framework were used as the basis for measuring levels of knowledge creation and transfer activity. A general model of the relationships among technical, cultural, and structural infrastructures, and levels of knowledge management activities was also developed and used as a guide to the design of the study (Figure 2).

The study found moderate levels of knowledge creation and transfer activity among the respondents. Respondents' reported levels of knowledge creation activity were somewhat higher than their reported levels of knowledge transfer activities. The difference between the two activity levels was statistically significant. The relationships among supportive norms, access to information technology, and knowledge management activity levels were tested with the model depicted in Figure 3. Analysis of the subjects' responses supported the expected positive relationships between IT use norms and IT usage levels with IT use norms explaining 21% of the variance in IT usage levels. The results also supported the expectation of a positive relationship between IT usage levels and knowledge management activity levels. IT use norms and use of information technology were significant predictors of knowledge management activity levels, explaining 56% of the variance in knowledge management activity levels. The technologies used most commonly in the respondents' knowledge management activities were office productivity products (word processing, spreadsheet, and presentation software), e-mail, PDAs (personal data assistant devices), and the Internet.

Given the small size of the sample and the limited relationships explored in this study, caution is necessary in drawing conclusions from the analysis. Common method variance may also affect the reliability of the results obtained from the analysis. Additional research with a larger sample and additional sources of data would be needed to confirm the results found here.

The results of the study suggest that organizations that want to foster higher levels of knowledge creation and sharing should encourage the development of organizational norms favoring the use of information technology. Actually creating such an environment may require the use of rewards and incentives that reinforce the positive effects of supportive messages from management and the expenditure of funds to ensure technology accessibility. Efforts to encourage higher levels of partici-

pation in communities of practice within the organization would also benefit from a supportive information technology environment. Participation in virtual and hybrid communities of practice is especially dependent on information technology support. Supportive norms and access to information technology should contribute to participants' comfort with the computer-mediated communication required to participate in virtual communities of practice.

Further studies of technological, cultural, and structural influences on knowledge creation and sharing should be conducted in order to gain additional insight into the factors that encourage participation in communities of practice. Relevant technology issues might include the availability of knowledge repositories and the type of role played by the information technology (e.g., whether technology is used to lead the development of a community or is instead used to support already existing communities of practice). Cultural infrastructure influences might include norms of reciprocity and levels of trust among members of the community. Additional structural factors might include rewards, incentives, and promotion systems. Future studies should also explore the conditions in which increased knowledge management activity and participation in communities of practice have a positive effect on organizational performance.

REFERENCES

Ajzen, I., & Fishbein, M. (1980). *Understanding attitudes and predicting social behavior.* Englewood Cliffs, NJ: Prentice Hall.

Alavi, M., & Leidner, D. E. (2001). Review: Knowledge management and knowledge management systems: Conceptual foundations and research issues. *MIS Quarterly, 25*(1), 107-136.

Ardichvili, A., Page, V., & Wentling, T. (2003). Motivation and barriers to participation in virtual knowledge-sharing communities of practice. *Journal of Knowledge Management, 7*(1), 64-77.

Argote, L., McEvily, B., & Reagans, R. (2003). Managing knowledge in organizations: An integrative framework and review of emerging themes. *Management Science, 49*(4), 571-582.

Ba, S., Stallaert, J., & Whinston, A. B. (2001). Research commentary: Introducing a third dimension in information systems design: The case for incentive alignment. *Information Systems Research, 12*(3), 225-239.

Bieber, M., Engelbart, D., Furuta, R., Hiltz, S. R., Noll, J., Preece, J., et al. (2002). Toward virtual community knowledge evolution. *Journal of Management Information Systems, 18*(4), 11-35.

Brown, J. S., & Duguid, P. (1991). Organizational learning and communities-of-practice: Toward a unified view of working, learning, and innovation. *Organization Science, 2*(1), 40-57.

Brown, J. S., & Duguid, P. (2001). Knowledge and organization: A social practice perspective. *Organization Science, 12*(2), 198-213.

El Sawy, O. A., Gomes, G. M., & Gonzales, M. V. (1996). Preserving institutional memory: The management of history as an organization resource. *Academy of Management Best Paper Proceedings, 37*, 118-122.

Gold, A. H., Malhotra, A., & Segars, A. H. (2001). Knowledge management: An organizational capabilities perspective. *Journal of Management Systems, 18*(1), 185-214.

Grant, R. M. (1996). Toward a knowledge-based theory of the firm. *Strategic Management Journal, 17*, 109-122.

Grant, R. M. (2001). Knowledge and organization. In I. Nonaka & D. Teece (Eds.), *Managing industrial knowledge: Creation, transfer and utilization* (pp. 145-169). London: Sage.

Green, C. W. (1998). Normative influence on the acceptance of information technology: Measurement and effects. *Small Group Research, 29*(1), 85-123.

Griffith, T. L., Sawyer, J. E., & Neale, M. A. (2003). Virtualness and knowledge in teams: Managing the love triangle of organizations, individuals, and information technology. *MIS Quarterly, 27*(2), 265-287.

Grover, V., & Davenport, T. H. (2001). General perspectives on knowledge management: Fostering a research agenda. *Journal of Management Information Systems, 18*(1), 5-21.

Hall, H. (2001). Input-friendliness: Motivating knowledge sharing across intranets. *Journal of Information Science, 27*(2), 139-146.

Heeks, R. (2002). Information systems and developing countries. *The Information Society, 18*, 101-112.

Kling, R., & Courtright, C. (2003). Group behavior and learning in electronic forums: A sociotechnical approach. *The Information Society, 19*, 221-235.

Leavitt, H. J. (1965). Applying organizational change in industry: Structural, technological, and humanistic approaches. In J. G. March (Ed.), *Handbook of organizations.* Chicago: Rand McNally.

Lee, H., & Choi, B. (2003). Knowledge management enablers, processes, and organizational performance: An integrative view and empirical examination. *Journal of Management Information Systems, 20*(1), 179-228.

Levin, D. Z., Cross, R., & Abrams, L. C. (2002). The strength of weak ties you can trust: The mediating role of trust in effective knowledge transfer. *Academy of Management Proceedings.* MOC: D1-D6.

Lyytinen, K., Mathiassen, L., & Ropponen, J. (1998). Attention shaping and software risk: A categorical analysis of four classical risk management approaches. *Information Systems Research, 9*(3), 233-255.

Markus, M. L. (2001). Toward a theory of knowledge reuse: Types of knowledge reuse situations and factors in reuse success. *Journal of Management Information Systems, 18*(1), 57-93.

Nonaka, I. (1994). A dynamic theory of organizational knowledge creation. *Organization Science, 5*(1), 14-37.

Nonaka, I., & Konno, N. (1998). The concept of "Ba": Building a foundation for knowledge creation. *California Management Review, 40*(3), 40-54.

Nonaka, I., Toyama, R., & Konno, N. (2001). SECI, ba and leadership: A unified model of dynamic

P

knowledge creation. In I. Nonaka & D. Teece (Eds.), *Managing industrial knowledge: Creation, transfer and utilization* (pp. 13-43). London: Sage.

Polanyi, M. (1959). *The study of man.* Chicago: University of Chicago Press.

Schlager, M. S., & Fusco, J. (2003). Teacher professional development, technology, and communities of practice: Are we putting the cart before the horse? *The Information Society, 19*, 203-220.

Schwen, T. M., & Hara, N. (2003). Community of practice: A metaphor for online design? *The Information Society, 19*, 257-270.

Simon, H. A. (1991). Bounded rationality and organizational learning. *Organization Science, 2*(1), 125-134.

Simsek, Z. (2003). Toward a knowledge-based view of entrepreneurial initiatives and performance. *Academy of Management Proceedings, BPS*, J1-J6.

Stein, E. W., & Zwass, V. (1995). Actualizing organizational memory with information systems. *Information Systems Research, 6*(2), 85-117.

Tan, S. S., Teo, H. H., Tan, B. C., & Wei, K. K. (1998). Developing a preliminary framework for knowledge management in organizations. In E. Hoadley & I. Benbasat (Eds.), *Proceedings of the 4th Americas Conference on Information Systems* (pp. 629-631). Baltimore: Conference Proceedings.

Taylor, S., & Todd, P. A. (1995). Understanding information technology usage: A test of competing models. *Information Systems Research, 6*(2), 144-176.

Teece, D. J. (2001). Strategies for managing knowledge assets: The role of firm structure and industrial context. In I. Nonaka & D. Teece (Eds.), *Managing industrial knowledge: Creation, transfer and utilization* (pp. 125-144). London: Sage.

Tulving, E. (1983). *Elements of episodic memory.* Oxford, UK: Oxford University Press.

Van Offenbeek, M., & Koopman, P. (1996). Scenarios for system development: Matching context and strategy. *Behaviour & Information Technology, 15*(4), 250-265.

Venkatesh, V., & Davis, F. D. (2000). A theoretical extension of the technology acceptance model: Four longitudinal field studies. *Management Science, 46*(2), 186-204.

Venkatesh, V., & Morris, M. G. (2000). Why don't men ever stop to ask for directions? Gender, social influence, and their role in technology acceptance and usage behavior. *MIS Quarterly, 24*(1), 115-139.

Wenger, E. C., & Snyder, W. M. (2000). Communities of practice: The organizational frontier. *Harvard Business Review, 78*(1), 139-145.

KEY TERMS

Boundary: A systems concept, whereby all systems are held to have a boundary, and often judgments at the boundary will yield insightful results.

Sustainable Competitive Advantage: Those factors which enable an organization to maintain an advantage over competitors in the long term. Typically, it is contended that such advantage may be gained by the use of technology but cannot be sustained from this source.

ENDNOTE

[1] Alavi and Leidner (2001) use three figures to depict the framework. These have been consolidated into a single drawing in order to facilitate instrument development.

Psychoanalysis, Organisations, and Communities

Angela Lacerda Nobre
ESCE-IPS, Portugal

INTRODUCTION

This article explores the many hidden dimensions of human actions within the organisational environment. It considers the practice of the theory of psychodynamics and the role of consultants engaging with a client organisation. Creating shared meaning within a community is discussed within the context of situadedness, understanding, and discursiveness. A psychological framework, it is argued, enables a clearer understanding and facilitates development.

THE HIDDEN DIMENSION OF MEANING CREATION

A psychoanalytical approach to organisations highlights the fact that there are many hidden dimensions to humans' actions. The possibility of taking those dimensions into account, and of using them and exploring them in a positive and constructive way, is a critical move within current organisational environments, characterised by high levels of complexity.

The epistemology of knowledge and philosophy of science pose similar questions in the sense that they search for the sense-making process which legitimises specific theoretical approaches. This process may only be meaningful within a particular context, which in turn corresponds to a concrete community. Communities are not only the external arenas for social interaction, but they represent the identity and the social process through which knowledge is created, used, and shared. However, this bedrock function of communities is so subtle that it is almost invisible, with the consequence that its importance is often not acknowledged, and thus it risks remaining neglected.

Psychodynamics focuses on the inner structures which determine not only our actions, but also what we see, take into account, interpret, or take for granted. By gradually making these structures visible and explicit, it is possible to transform them through a developmental process that is characterised by continual reflection in action. Within a psychodynamic approach, both the inner structures and the developmental process are the result of a collective endeavour, as they are based upon specific and concrete relationships; that is, it is a social and relational based approach, both by focusing on current and previous relationships, as well as by co-constructing a new type of relationship with the psychoanalyst. This perspective is critical to the understanding of the fundamental role of communities as the ethos of relationships. If we take a reductionist perspective and state that knowledge is in people's heads, then we fail to take into account the critical issue of how it got there, and thus neglect the central importance of communities within an organisational setting.

When knowledge management stresses the distinction between tacit and explicit knowledge (Nonaka & Takeuchi, 1995), or when organisational learning focuses on double-loop learning processes as the need to question our deep held assumptions (Argyris, 1992; Argyris & Schon, 1978), both areas are highlighting the crucial issue of dealing both with the superficial and with the deep aspects of organisational life, with the visible and the invisible, the obvious and that which is subtle and volatile, the individual and the social, and the inner and the outer worlds.

BACKGROUND

There are further examples of management and organisational literature that aim at less direct, linear, and mechanistic approaches to organisational life. The interest in the concept and theory of communities of practice is probably the most striking example (Wenger, 1999; Wenger, McDermott & Snyder, 2002a). Equally paradigmatic is Stacey's work on learning and knowledge creation through the focus on the complex responsive processes of organisations (2001). In parallel, there is a vast array of theories and initiatives which pay tribute to the richness and complexity of organisational life. These include the importance of informal learning (Marsick & Watkins, 1990), the critical role of organisational meaning creation processes (Weick, 1995, 2001), the social aspects of information (Brown & Duguid, 2000), the question of the epistemology of knowledge (Von Krogh, 1995), and other organisational learning and knowledge management theories and approaches (Senge, 1990; Dixon, 2000; Davenport & Prusak, 1997).

In historical terms, the early works of the Tavistock Clinic, which was founded in 1920 in London, are an example of the boundary-crossing work of multidisciplinary teams that took a psychodynamic perspective. The socio-technical systems developed in the early 1950s within the same institution focused on self-regulating workgroups (Trist et al., 1963). The concepts of appreciative inquiry and of appreciative systems also pay a tribute to the reflection in action process, which was first developed by Vickers (1965, 1968; Chekland, 1999) and then further developed within the fields of soft systems theory and complex systems thinking. Lewin's (1951) action research methodology is another example of the need to constantly balance practice with a reflection over the same practice.

The works of Bakhtin and of his close circle of colleagues—works that were developed in the 1920s and the 1930s—are probably one of the richest examples of how a general theory of social philosophy may find its applications many decades later and in a wide range of areas (Brandist, 2002). The special meaning of dialogue, discourse, heteroglossia, or multiple voices is still being applied and developed within social science settings.

An application of a psychodynamic approach to organisational development may incorporate and integrate both earlier efforts to make sense of social interaction, as well as current theoretical approaches, such as those of knowledge management and of organisational learning. Information is not only that which we communicate, the content of our messages; it is also and most importantly that which inhabits our imaginary, and which constitutes our projection of the world, of ourselves, and of our place and role in that world. The better we understand this meaning creation process, the better able are we to use it in a productive, as well as constructive and gratifying way. Again and again, this is never an isolated process suspended in a vacuum. It is always the child of a community.

PRACTICE AND THEORY

Nietzsche, Marx, and Freud are known as the philosophers of the suspect, as their work radically questioned the assumed beliefs of their time. Nietzsche claimed the death of God, opening the way for postmodernism. Marx called attention to the importance of social determinations, in the sense that individuals are fundamentally affected by the social class to which they belong. And Freud explored the unconscious with a radical theory which put under question the rationalist idea of human beings as autonomous, independent, and rational individuals.

Psychoanalysis can be seen as belonging to the hermeneutics of suspicion *which, apart from Freud, is also represented by Marx and Nietzsche. All three of these had probed behind what they conceived as an illusory self-conscious to a deeper-lying, more unpleasant or 'shameful' one. In Freud it appeared as libido, in Marx as the economic interest, and in Nietzsche as the will to power.* (Alvesson & Sköldberg, 2000, p. 95)

Habermas' critical theory refers to Marx and Freud as examples of those working on an emancipatory mode (Alvesson & Sköldberg, 2000, p. 125), though Foucault points in the opposite direction, stating the importance of monitoring, disciplining, and control in modern discourses.

There are certainly anti-emancipatory elements in both the practice and ideology of psychoanalysis…the movement may have helped to build a therapeutic culture which makes people exaggeratedly and self-centredly preoccupied with their mental inadequacy and dependent on therapeutic authorities. (Alvesson & Sköldberg, 2000, p. 146)

The issue that is at stake when discussing a psychoanalytic practice as a form of critical reflection, and thus of transformational potential, or else as a form of further control, if we confront Habermas' and Foucault's perspectives, is the way that this practice is able to fully take and explore its own recreation capacity. This capacity, in turn, depends on the collective development of the community of practice of psychoanalysts, which pushes forward the creative and constructive function of their own collective practice. Psychoanalysts are themselves psychoanalysed, implying that there is an built-in mechanism for further development. It is precisely this characteristic that made the early work of the Tavistock clinic within the field of organisational consulting so powerful. The consulting team, whose members were all under psychoanalysis, had a shared understanding and a shared meaning on the potential benefit of a psychoanalytical approach so that they could put it in practice and create a new and innovative perspective on organisational reality.

Though the clinic had been founded in 1920, it was in the 1950s and 1960s that the combined consulting and research projects developed. These projects used the perspectives from both the social sciences and from psychoanalysis.

Together, these two perspectives provide us with tools not only for taking up a role as consultant, but also for thinking about our experiences as members of institutions. (Mosse, in Obholzer & Roberts, 1994, p. 5)

This comment is central to the point that this article wants to highlight, that the most effective initiatives at the organisational level must include in their practice an inside knowledge; that is, both the consultants and clients undergo the same process: the consultants practice what they preach, and it is this practice that may be learnt by and shared with clients.

As Freud and others discovered, there are hidden aspects of human mental life which, while

remaining hidden, nevertheless influence conscious processes...What was then required was interpretation of these symbolic expressions from the unconscious. (Halton, in Obholzer & Roberts, 1994, p. 11)

Though there is no exact parallel between individuals and institutions, psychoanalysis has contributed one way to approach thinking about what goes on in institutions.

[L]ooking at an institution through the spectrum of psychoanalytical concepts is a potentially creative activity which may help in understanding and dealing with certain issues. The psychoanalytical approach to consulting is not easy to describe. It involves understanding ideas developed in the context of individual therapy, as well as looking at institutions in terms of unconscious emotional processes. This may seem like a combination of the implausible with even more implausible or it may become an illuminating juxtaposition. (Halton, in Obholzer & Roberts, 1994, p. 11)

Halton refers to critical concepts and practices, such as the need to develop a listening position on the boundary between conscious and unconscious meanings, working simultaneously with problems at both levels. The avoidance of pain and the defence mechanisms against difficult emotions that are too threatening or too painful to acknowledge occurs both within individuals as well as within institutions. Some defence mechanisms are healthy, as they help the staff to cope with stress and develop through their work in the organisation. But some institutional defences, as some individual defences, can obstruct contact with reality, and in this way damage the staff and hinder the organisation in fulfilling its task and in adapting to changing circumstances. Denial is a central defence which consists of pushing certain thoughts, experiences, and feelings out of the conscious awareness because they have become too anxiety-provoking. This denial implies also a resistance to change which has to be worked analytically.

Halton refers to Melanie Klein's work with children and to the concepts of 'splitting'—creating stories where the good aspects are separated from the bad ones in order to gain relief from internal conflicts—and to 'projection'—which often accompanies splitting, and involves locating feelings in others rather than in oneself. Klein's conceptualisation of the unconscious world applies to adults and to organisations, and it is critical in understanding group processes (Sayers, 2000). Freud himself developed the study of unconscious processes in groups in early 1920, and many others followed, including Wilfred Bion, working first with the army during World War II, and later at the Tavistock clinic.

The experience of reading Bion's work conveys a sense of psychoanalysis as a process involving a movement toward infinite expansion of meaning. (Ogden, 2004, p. 285)

Bion identified two main tendencies in the life of a group, as the wish to face and work with reality, and the wish to evade it when it is painful or causes psychological conflict within or between group members; from here he established a framework for analysing some of the irrational features of unconscious group life.

A consultant engaging with a client organisation is engaging with a social system. The consultant must understand the organisation's description of itself and of its intended structure, and confront this against what may be observed in practice, regardless of what is claimed.

The institution is best served by a form of consultancy which does not have a preconceived idea of what the structure of the organisation should be on completion of the intervention...The outcome should be determined by a public process of striving towards understanding. With this comes the awareness that the task of monitoring and reviewing is never complete and needs to be supported in an ongoing way. The consultant who offers a psychodynamic understanding of institutional processes also brings a state of mind and a system of values that listens to people, encourages thought, and takes anxieties and resistance into account. At the end of the consultation, the organisation will, one hopes, have taken this stance into its culture: a new awareness of the potential risks to the work and *the workers as a result of stresses inherent in the organisation's tasks, together with greater clarity about how to proceed.* (Obholzer & Roberts, 1994, p. 210)

Robert de Board's book, *The Psychoanalysis of Organisations*, first published in 1978, with six editions until the 2002 one, refers to a wide range of authors and theories which influenced the study of groups, although the area of organisation theory itself only became established in the second quarter of the twentieth century. The importance of taking into account the early development, as well as the influences that affected organisation theory, relates to our capacity today to see the connections between traditional approaches and the relatively new as well as emergent theories and perspectives. It is this broad interpretation that enables us to acknowledge the rich potential of psychoanalysis over a century after its creation.

Today's work on communities of practice, organisational learning, knowledge management, social capital, human capital, trust, social networks, innovation, communication, organisational design, and organisational development is deeply rooted and grounded in the early development of social sciences. The early works of Max Weber, Kurt Lewin, S. Freud, F. Taylor, and H. Fayol created the bedrock basis for organisation studies. Group dynamics itself undergoes, still today, a process of exploring different theories, often conflicting ones, from psychoanalytic to empiricist, cognitivist, sociometric, systems theory, interaction theory, and field theory.

De Board, over 25 years ago, said: "It is ironic that as civilisation develops and the major killing diseases of the past are gradually eradicated, there is a steady rise in modern stress diseases such as coronaries, hypertension, and mental ill-health. It is salutary to realise..." that a quarter of the British population, at the time, had received prescriptions for some kind of mind-affecting drug (De Board, 1978 [2002], p. 112). He referred to Menzies' approach, as a psychoanalyst consultant working with client organisations, as 'socio-therapeutic', for its aim was to facilitate desired social change. The defence mechanisms of idealisation, splitting, denial, and projection are present in groups and organisations as social defence systems against anxiety.

[A] general symptom of stress manifests itself in an inability to concentrate on long-term plans and objectives...People tend to behave without considering the long-term effects of their action. (De Board, 1978 [2002], p. 113)

Within the field of organisational psychology, the issue of work-related stress captures significant attention: "[S]tress is possibly the most widely researched area of work and organisational psychology..." (Doyle, 2003, p. 114). However, most approaches to stress have a narrow individual focus.

Work and organisational psychology tends to focus on individuals and groups and to explore what goes on in organisations in terms of underlying psychological processes. So for instance, leadership is coming to be understood in terms of the social constructions of followers. In other words, followers interpret what is going on around them and come to a consensus about what that 'reality' is. Leaders are the people who influence these socially constructed interpretations through their exceptional social and communication skills. (Doyle, 2003, p. 12)

This comment illustrates how the rich and complex social process of creating shared meaning can be reduced and simplified into a reductive cognitivist focus. Leadership is critical, but it is one among many forces which shape a community.

Psychoanalysis is often regarded as one among the 500 different schools of psychotherapy (Roudinesco & Plon, 1997). In terms of therapeutic approaches, this may be true, though as a theory and knowledge field, the division lines are not so straightforward. Psychoanalysis may be understood as focusing on relationships, thus not on isolated, hermetic, and autonomous individuals. This justifies the importance of supervisory work among psychoanalysts (Wallerstein, 1999). Another way to interpret the distinction between psychoanalysis and other forms of psychotherapy is by relating psychoanalysis with the work of Wittgenstein (Bouverresse, 1995). Both Freud and Wittgenstein saw the therapeutic relationship and the use of free association as a way to construct meaning. Freud developed a theory, a technique, and a school to disseminate his findings. Though Wittgenstein was greatly impressed and influenced by Freud's work, and thought of his own work as a therapy, he was also critical of Freud and of the truth-seeking drive of traditional philosophy. Wittgenstein did not create any technique or systematised school, as his interest was to focus on understanding and constant and continual questioning.

The practice of psychoanalysis may be interpreted as this questioning, seeking of understanding, and meaning creation process, and it is in this sense that it may be distinct from more formal, prescriptive, or systematising forms of psychotherapy. This informal and unsystematic endeavour, the way that conversations within communities of practice take a similar path and pattern to free association, helps to recognise and to acknowledge the potential relevance of psychoanalysis to the development of community life. Wittgenstein's word games and Freud's free association technique are thus valuable contributions to a better understanding of the community-building process.

SITUATEDNESS, UNDERSTANDING, AND DISCURSIVENESS

Heidegger also had a conception of truth as a process of illumination and opening (Heidegger, 1953 [1996]; Guignon, 1983). An analogue of this process is what can be found in psychoanalysis.

It would be naïve to think that the goal of the psychoanalytic dialogue is to arrive at a 'correct representation' of the patient's mental state or of the precise sequence of events that led to his neurosis. On the contrary, its aim is to deepen, widen, enrich, and clarify his self-understanding, to allow him to see the broader range of connections, and to liberate him from pointless obsessions by making him more open toward the world...The language of 'disclosing', 'clearing', and 'lighting up' is much more appropriate here than that of 'correspondence'. Successful therapy is measured by its consequence for one's life. (Guignon, 1983, p. 250)

Probably one of the most interesting ways to grasp the complexity of creating shared meaning within a community is through Heidegger's thought.

If we are to understand the full import of Heidegger's conception of 'meaning', then, we must avoid seeing it as referring to something inner in any sense...Heidegger identifies three existentialia of what is called 'Being-in as such': situatedness, understanding, and discursiveness. Situatedness and understanding correspond to the passive and active aspects of Dasein's [existence's] formal determination...as situated, Dasein is disclosed as thrown into a definite range of pre-given, shared possibilities...As understanding, Dasein takes up the possibilities it discovers in its situatedness and projects itself onto some range of goals for its life as a whole...Meaning is that which makes possible that projection of possibilities in understanding...What is the source of this most primordial level of intelligibility? Heidegger says that it is 'discursiveness'. The concepts of 'discursiveness' and 'meaning' are closely related, so to clarify one is at the same time to illuminate the other. (Guignon, 1983, p. 111)

Heidegger's concepts allow for a rich interpretation of the critical role of community life for human beings' organisation within a society, a culture, and a civilisation. Life in the knowledge economy of the information age continues to be grounded in the same network of communities and of social and cultural embedded meaning creation processes.

To be Dasein is essentially to be a nexus of the socially constituted relations of a culture...Heidegger's phenomenology of everydayness works to counteract the tendency toward the displacement of meaning into subjectivity which began with the rise of modern science...Dilthey had already taken the first steps toward relocating meaning into the world in his attempt to grasp the human life-world...But, to the extent that he still conceives of meaning as a product of something inner which is expressed in the outer world, Dilthey remains ensnared in Cartesian dualism. By regarding the self as its meaningful expressions, Heidegger is able to fully break away with the Cartesian tradition. His conception of the intelligible world as a holistic field of meaningfulness undermines both the inclination to see reality as consisting solely of spatiotemporal particulars and the temptation to think of meanings and values as solely subjective. (Guignon, 1983, p. 110)

The importance of the concept of social subjectivity is acknowledged in the philosophical interpretation of communities. According to Gonçalves (1995), a community is a key ontological manifestation of the world and it has a decisive role in the development of the instance of being-in-the-world. A community may be understood from different perspectives, ranging from the psychological level to the sociological one. The deep soil of an institution, however, is its ontological grounding, in particular, and especially that of its community. When the presence of human beings is referred, it is not the individuals themselves, nor the sum of all individuals, that are being referred to, but rather the community itself whose statute largely overpasses the simple sociological horizon. The maximum degree of possibilities of being-in-the-world has in a community, thus interpreted, its fundamental interpreter and protagonist.

Thus an institution is never simply formed by the human beings who happen to be there at some point in time. An institution is, intrinsically and constitutively, a community of meaning. Apart from the human beings' dimension, which Gonçalves calls horizontal, there is the vertical dimension of the meanings which the tradition has accumulated over time and which are incorporated or stored in its memory. The current community, globally understood, represents the passing, yet essential, factor of an institution; the physical memory, and the norms and routines, preserve, develop, and transmit the marks of the unique style of reading and interpreting of an institution. The value of an institution is determined by its specific style, and it survives and develops when its community is constantly dynamised and energised by the recurrent movement between the vertical and the horizontal dimensions. Within this approach the concepts of action and of developmental action are central as they constitutively incorporate the meaning creation process implicit in all social and cultural activity.

This complex interpretation of organisational and community life may be incorporated into daily practice through an implementation of a psychoanalytical inspired framework. An organisation's strategy, mission statement, vision, culture, and structure have implicit a set of invisible patterns and assumptions which determine the organisation's members projections of their roles in the organisation, as well as the role of the organisation itself in the marketplace. The process of clarifying and sense-making of a psychoanalytic framework enables a clearer understanding, and facilitates the acknowledgement of the bottlenecks and barriers to further development. By focusing on the being-in-the-world instance, it avoids individualistic and reductive approaches, and brings forward the fundamental and constitutive role of communities within organisations.

THE FUTURE AND TRENDS: THE NEED FOR PSYCHOANALYTIC THEORY

In order to understand why the psychodynamic perspective is critical to communities of practice theory, it is necessary to follow a few reasoning steps. The core of the argument lies in the understanding of what both psychoanalytic theory and communities of practice theory are about. Firstly, communities of practice refer to groups of people that have a relationship that has a high degree of spontaneity and of informality. Wenger (2002b) stresses that a family that lives together in the same house cannot be immediately assumed to be a community. There need to be signs of 'community life', and these may be subjective and difficult to observe. It is not a black or white, or a yes or no question, but rather an issue of the quality and intensity of the community life. The quality of the community life within an organisational setting directly influences issues such as the organisational identity and the organisational meaning-making capacity. But why is psychoanalytic theory relevant? Because communities of practice, when compared with other organisational groups such as multidisciplinary teams, functional groups, or task-forces, have a much less degree of directiveness and have a less procedural and results-oriented nature. And this is precisely what makes them critical within contexts of high complexity where creativity and innovation are cru-

cial. There is an *emergence* character and a *self-directness* that critically distinguishes communities of practice as opposed to groups that are focused on explicit, measurable, and formal objectives.

Within any kind of group, there is the conjunction of two sorts of tensions: the tension related to external pressures and directions, on one side, and the tension related to the inner life of the individuals that form each group. The stronger and more explicit the external directives, the less the inner dynamism can manifest itself. Within very informal communities there is also a pressure from outside the group, but there is a wider margin for the personal and group dynamics to express themselves. Therefore, if psychoanalytic theory is relevant to the study of organisations and group behaviour in general, it is particularly more so within communities that have a freer and more loose structure, and where the complexity of human relationships is higher and more subtle.

The second aspect to highlight is that psychoanalytic theory is about the world of relationships. The inner subconscious world and the outer world of relationships is one and the same, as if the two sides of a mirror. "Freud stated that any treatment based upon an understanding and application of the concept of transference and resistance deserves to be called psychoanalysis" (Waldron, Scharf & Hurst, 2004, p. 444). Psychoanalytic theory claims that the way we relate to ourselves, to each other, and to the world is influenced by a few 'structural' relationships. These relationships start at childbirth, or even before, and they can be perpetually repeated or else be transformed into new relationships. More important than the primordial relationship with the parents or the child-carer is the capacity to form and create new relationships, as if life situations and specific protagonists could perform again the role of the mother or the father throughout life. Therefore, psychoanalytic theory is more about the future than the past and is more about creating new relationships—new dynamisms—than about exact and precise descriptions of objective relations. This being so, it implies that there can be relevant insights to be disclosed by the use of psychoanalytic theory in the processes of facilitating and nurturing communities of practice.

CONCLUSION

The two assumptions discussed above may be further strengthened by acknowledging the wealth of theoretical development that is presented from both sides of the equation. On one hand, from the communities of practice side, it is relevant to highlight the importance of the developments within management and organisation theory, such as the rise in interest in collaborative forms of work and learning, informal learning, reflexive practices, group identity, and meaning-making. The fields of organisational learning, knowledge management, and personal knowledge management witness these new developments. Chris Argyris, talking about organisational learning, refers to defensive mechanisms, and these are a critical component of psychoanalytical theory:

I'm interested in producing knowledge that is actionable...I have especially focused on the defensive routines of organisations that prevent actionability. And what I get from executives is a continual awareness of how important it is to overcome these defensive routines, especially if you are interested in changing organisations. (Argyris, 1998, p. 22)

On the other hand, psychoanalytic theory is one among several theoretical contributions that may have a profound impact in our understanding of community life. Aligned with psychodynamic theory, there is an array of social philosophy contributions that must be acknowledged and explored. These theories have the role of *mediators* in the process of grasping the complexity of human relationships. To mention just a few, hermeneutics, action philosophy, social semiotics, and critical realism correspond to different traditions that have in common a non-reductive or dualistic approach to reality and that may be of critical importance to the understanding of the role of communities within organisations.

REFERENCES

Alvesson, M., & Sköldberg, K. (2000). *Reflexive methodology.* London: Sage Publications.

Argyris, C. (1992). *On organisational learning.* Oxford, UK: Blackwell.

Argyris, C. (1998). An interview with C. Argyris. *Organisational Dynamics, 27*(2).

Argyris, C., & Schon, D. (1978). *Organisational learning.* Reading, MA: Addison-Wesley.

Board, R. (1978 [2002]). *The psychoanalysis of organisations.* London: Tavistock.

Bouveresse, J. (1995). *Wittgenstein reads Freud.* Princeton, NJ: Princeton University Press.

Brandist, C. (2002). *The Bakhtin circle: Philosophy, culture and politics.* London: Pluto.

Brown, J., & Duguid, P. (2000). *The social life of information.* Boston: Harvard Business School Press.

Chekland, P. (1999). *Soft systems methodology: A 30-year perspective.* West Sussex, UK: John Wiley & Sons.

Davenport, T., & Prusak, L. (1997). *Working knowledge.* Boston: Harvard Business School Press.

Dixon, N. (2000). *Common knowledge.* Boston: Harvard Business School Press.

Doyle, C. (2003). *Work and organisational psychology.* East Sussex, UK: Taylor & Francis.

Gonçalves, J. (1995). *Fazer filosofia: Como e onde?* (To do philosophy—how and where?). Braga, Portugal: Fac. Filosofia, U.C.P.

Guignon, C. (1983). *Heidegger and the problem of knowledge.* Indianapolis, IN: Hackett.

Halton, W. (1994). *Some unconscious aspects of organisational life: Contributions from psychoanalysis.* In A. Obholzer & V. Roberts (Eds.), *The unconscious at work.* London: Routledge.

Heidegger, M. (1953 [1996]). *Being and time.* New York: State University of New York.

Lave, J., & Wenger, E. (1991). *Situated learning: Legitimate and peripheral participation.* Cambridge, UK: Cambridge University Press.

Lewin, K. (1951). *Field theory in social science: Selected theoretical papers.* New York: Harper and Row.

Marsick, V., & Watkins, K. (1990). *Informal and incidental learning in the workplace*. London: Routledge.

Mosse, J. (1994). The institutional roots of consulting to institutions. In A. Obholzer & V. Roberts (Eds.), *The unconscious at work*. London: Routledge.

Nonaka, I., & Takeuchi, H. (1995). *The knowledge-creating company*. New York: Oxford University Press.

Obholzer, A., & Roberts, V. (1994). *The unconscious at work*. London: Routledge.

Ogden, T. (2004). An introduction to the reading of Bion. *The International Journal of Psychoanalysis, 85*(Part 2).

Roudinesco, E., & Plon, M. (1997). *Dictionnaire de la psychanalyse*. Paris: Fayard.

Sayers, J. (2000). *Kleinians: Psychoanalysis inside out*. Cambridge, UK: Blackwell.

Senge, P. (1990). *The fifth discipline: The art and practice of the learning organisation*. New York: Doubleday.

Stacey, R. (2001). *Complex responsive processes in organisations*. London: Routledge.

Thrist, E., Higgin, G., Murray, H., & Pollock, A. (1963). *Organisational choice*. London: Tavistock.

Vickers, G. (1965). *The art of judgement: A study of policy making*. London: Chapman and Hall.

Vickers, G. (1968). *Value systems and social process*. London: Tavistock.

Von Krogh, R. (1995). *Organisational epistemology*. New York: Palgrave Macmillan.

Waldron, R., Scharf, R., & Hurst, D. (2004). What happens in a psychoanalysis? *The International Journal of Psychoanalysis, 85*(Part 2), 443-446.

Wallerstein, R. (1999). The future of psychotherapy. In S. Schill & S. Lebovici (Eds.), *Challenge to psychoanalysis and psychotherapy*. London: Jessica Kingsley.

Weick, K. (1995). *Sense making in organisations*. Thousand Oaks, CA: Sage Publications.

Weick, K. (2001). *Making sense of the organisation*. Oxford, UK: Blackwell.

Wenger, E. (1999). *Communities of practice: Learning, meaning and identity*. Cambridge: Cambridge University Press.

Wenger, E. (2002b). International workshop. *Communities of practice dialogue*. Setubal, Portugal.

Wenger, E., McDermott, M., & Snyder, W. (2002a). *Cultivating communities of practice*. Boston: Harvard Business School Press.

KEY TERMS

Action: Has a signifying function and is characterised by an expansive mechanism which creates its own content and follows a result; this result—the world—belongs to the intentionality of the doing process, thus avoiding being closed into itself. This process, which occurs through action, is the de-centring process. Action is an organised whole, and it brings forth organised contents, always as a function of the ontological drive. Action means doing, in order to give more meaning, more signification. It is a process which generates a meaningful reality. All human collaboration is inherently this process of meaningful creation of reality. Humanism should not be reduced to the world of human beings, but considered as the humanism of the world, as the value of humans within the world process. Rationality is not a structure, a paradigm, nor a frame or a rigid law, permanently defined; it is the meaningful reality which emerges from action, which is the process of constituting an organised whole. Rationality cannot be reduced to mental schemes, as it is nurtured from a global rationality which arises from the structure of action.

Appreciative Systems: Developed by Vickers in the 1960s, the concepts of appreciative systems and of appreciative inquiry go beyond the paradigm of goal seeking to explain the processes of social activity, including decision making and action. Vickers criticised the reductionism of the perspective of focusing exclusively on goals, which he thought would be adequate to explain the 'behaviour of rats in mazes'. In order to describe and to explain the processes that characterise social systems, it is necessary to capture the establishing and modifying of relationships through time. Vickers also rejects the cybernetic paradigm where the course to be steered is available from outside the system, whereas systems of human activity themselves generate and regulate multiple and sometimes mutually inconsistent courses. Vickers' model is cyclical and starts with previous experiences which have created certain tacit patterns, standards, values, or norms; these lead to a readiness to notice certain features which determine which facts are relevant; the facts noticed are evaluated against the norms, leading both to regulatory action and to the modification of the norms so that future experiences may be evaluated differently. The organisation of this process is the appreciative system which creates an individual and a social appreciative world. The appreciative settings condition new experience but are also modified by the new experience. Since the state of an appreciative system is the function of its own history, this implies that they are learning systems, and for Vickers, learning is the most central and basic social process. Soft systems methodology and complex systems thinking have extended the use and notion of appreciative systems.

Concepts: Simultaneously a result and an agent; concepts are formed within the discourse, however they lack meaning if isolated from it. It is the concepts that bring density and relief to a discourse's content. A concept is an accumulation of meaning, and this meaning is produced within a discourse, through a metaphorisation process, constitutive of all natural language, and thus inherent to philosophy itself. The density and thickness of a text depends on the combination and hierarchisation of concepts. Philosophy is not a sophistication or a purification of concepts; it is discourse and text, where concepts have a key role, and can be searched for, never at the beginning but rather through the interpretation process itself.

Defence Mechanisms: Freud used this term in 1894 to classify the set of manifestations through which the ego protects itself from internal as well as external aggressions. As different branches of psychoanalysis developed, there has been a significant distinction between approaches which interpreted psychoanalysis as the effort to reinforce the ego and the conscious, through its adaptation to the external

environment, such as are examples ego psychology and self psychology, and strong critics of this approach, considered to be 'hygienic' and 'social orthopaedic' by Lacan, who develops a 'return to Freud' approach and thus a focus and preponderance of the unconscious and of the id. Lacan investigated the conditions of possibility of psychoanalysis and studied Heidegger's ontology and questioning process, and Saussure's and Levy-Strauss' works on symbolism, which inspired his notion of the unconscious organised as a language. The epicentre of this polemic is located around the question of whether defence mechanisms may be manipulated and indoctrinated in order to adapt to the demands of society, versus defence mechanisms that witness the huge complexity of the unconscious life that may be explored in order to create fuller meaning and further development. Defence concepts include projection, introjection, deflection, idealisation, splitting, and denial, and all have the common aim of overcoming anxiety. Groups and organisations develop their own defence mechanisms which may be explored through psychoanalytic-oriented consulting, aiming at social change.

Developmental Action: The action involved in the development of human beings and the action involved in the development of the world is inherently the same. It is not possible for human beings to manifest themselves unless in a constitutive relationship with the world. The world is, then, no longer the mere *object* of the consciousness of the *subject*.

Dynamic Perspective: The therapeutic approach which is based on a transference relationship between patient and therapist, and which includes psychoanalysis, but also nineteenth century approaches such as magnetism and hypnosis. Dynamic psychiatry is a term invented by historians in the early 1940s to describe the development of therapeutic methods interested in the psycho-genesis of mental illnesses. It takes from psychiatry its classifications and clinical approach; from psychology, the postulate of the dual reality of body and mind, and the proposal of the technique of observation of the subjects; and from the ancient tradition of witchcraft, the idea itself of transferential cure. Apart from the historiography context, within current use the term psychodynamic is directly linked to a psychoanalytical approach. The technical term of 'transference' belongs to psychoanalytical theory, as previously the term 'suggestion' was used. However, the historical interpretation of connecting psychoanalysis with anterior approaches risks reducing the rich and complex content of the specific use that 'transference' holds within psychoanalytic theory. Nevertheless, the full potential of psychoanalysis cannot be adequately understood without a reasonable understanding of its context and historical conditions for emergence.

Horizontal vs. Vertical Dimension: The horizontal dimension is represented by the current community at the institutional level. It is the dimension of structuralism which focuses on the 'parts' that make up a 'whole'. Science searches for total autonomy of its object, being thus insensitive to any referential—that is, to the vertical dimension. The horizontal dimension in philosophy represents the content and the ontological consistency of the de-centring in the world. Structuralism raises attention to the meaning of the content and to its relative autonomy. The horizontal dimension represents the presence of differences, and the signification power of the elements of the world, as a globallity. An empty world, void of differences, could never be a philosophical world. On the other hand, the vertical dimension is represented by the institutional memory. The vertical dimension is the attraction towards a referent. Philosophy is always facing a referent, never being closed within the game of its internal structures, though these are present and crucial in philosophy, where both vertical and horizontal dimensions are present.

Knowledge: Human life, as well as the philosophical exercise, are not a suffered climbing towards the unattainable mountain of truth. Human beings frequently face the question of truth as a victory over totality, the knowledge of everything, absolute knowledge—as if this total knowledge would be anterior and external to humans. The knowledge of being is one of the most radical constituent drives of humans, and it happens within the compromise of action, and never in the summing up, whether total or partial, of discreet units. At the origin of all knowledge, there is the drive to organise reality in order to optimise it. It is this drive that promotes the creation

of all knowledge and that also unifies it. All knowledge departs from cultural worlds; all knowledge is made possible through the action of human beings; and all knowledge is directed towards the widest horizon possible. The engine behind the global movement of knowledge is the ontological demand felt, in particular, by the areas that use natural language.

Philosophy: Two opposing interpretations: one more theoretical, contemplative, and interior to the human process of thinking; another one, which interprets philosophy as a doing, as a practice, and as an action. As philosophy is highly complex, both interpretations are possible, though the focus of the community of practice approach is on philosophy as action.

Psychoanalysis: A term created by Freud in 1896 to name a specific method of psychotherapy, which was inspired in the cathartic process, or treatment through speech, and was based in the exploration of the unconscious, through free association and interpretation. Catharses was already used by Aristotle, who was the son of a physician and thus was influenced by Hypocrites' thought. The idea was of a therapeutic process able to free individuals from oppressive experiences and to let the constraining element emerge. Freud developed a theory, a technique, and a school in order to formalise and perpetuate his approach. Freud also developed work on the unconsciousness of groups in the early 1920s. Though psychoanalysis started as an individual therapy, its insights, technique, and concepts have been extended to the domain of group analysis, family therapy, and organisational consulting. Its depth and breadth enable a high potential for application to the complex field of community life within organisational settings.

***Subject-Object* and *Being-in-the-World* Instances:** The *subject-object* instance refers to the closed relationship, unidirectional, between the researcher, the subject, and her object of research; the *being-in-the-world* instance, a concept of contemporary philosophy, precedes and includes the former one, and is ontologically rooted, thus bringing forth the concept of the world and the unity of action. Ontology is the manifestation of being, of all reality.

Psychologically Aware IT Workers

Eugene Kaluzniacky
University of Winnipeg, Canada

INTRODUCTION AND BACKGROUND: A NEED FOR DEEPER AWARENESS

Increasingly, information systems development has been recognized as a sociotechnical endeavor. We have seen calls in professional literature and at MIS conferences that IT workers must develop soft skills and emotional literacy and that they must learn to "grow their whole selves", to quote Extreme Programming initiator, Kent Beck. Many stories abound of system development efforts where the system was built right, but it was not the right system; that is, it did not satisfy the real needs of the users. As well, we see increasing reports of burnout among IT professionals, as the business and organizational environments look for an increasingly impactful role from newly developed systems. To quote Edward Yourdon (2002), "Burnout is still a topic that most senior managers would rather not confront, but it has become so prevalent and severe that some IT organizations have become almost completely dysfunctional" (cover page). The IT profession, no doubt, has been undergoing continual change in its orientation, methodologies, and technological tool sets. To deal with this constant change and increased expectations from IT, it has been proposed that the profession adopt an interdisciplinary, holistic approach to professional development, similar to that of an Olympic athlete who would consult the areas of physiology, psychology, nutrition, kinesiology, and so forth to enable optimum performance.

A powerful ally in the IT professional's quest for inner balance and resilience is *multidimensional psychological awareness*. IT work is done mostly with one's mind, one's psyche, and thus, a deeper awareness and understanding of one's own inner psychological dynamisms and those of one's co-workers is advisable. System developers can release previously pent-up, unavailable psychic energies for a more effective and less stressful work effort. Possible areas of psychological functioning that would warrant specific, concerted attention from the IT profession at this point in its evolution are personality typing systems, cognitive, creativity, and learning styles and also the reality of the "deepest inner self" (core or spirit).

Until now, focus on psychological factors in IT has been growing, albeit rather slowly. Groups of IT workers may have been introduced to a psychological perspective at a one-day seminar with no extensive, planned follow-up. A few organizations have taken their interests further and involved IT team members in exercises and assessments of the effect of their newly developed interests on their work performance and satisfaction. However, it may indeed be true that now the time has come for a concentrated effort in the IT profession to involve psychological factors in a more widespread, concerted, and thorough manner. The recent book, *Managing Psychological Factors in Information Systems Work: An Orientation to Emotional Intelligence* by Kaluzniacky (2004), provides a vision for such a possibility and issues a call to action. Psychological aspects promoted are the Myers-Briggs (MBTI) and Enneagram personality types; Kirton's Adaptor/Innovator cognitive styles, creativity styles as measured, for example, by the Creative Styles Inventory; four learning styles as defined by Kolb (1984); the deepest inner self as outlined by the PRH (1997); and Hoffman (1988) personal growth programs, and promoted by contemporary authors such as Borysenko (1990), Bedrij (1977), Dyer (1995), McGraw (2001), and Weizenbaum (1976). The book promotes professional acceptance and application individually and collectively and suggests specific areas for both academic research and development of materials (e.g., a full-scale methodology for psychological factors in IT) to facilitate such an acceptance within the IT profession.

A more concerted involvement of psychological factor awareness in specific aspects of IT work could give rise to Psych-factors Communities of

Practice. Such communities, physical and virtual, would communicate and collaborate on specifically applying areas of psychological awareness to certain IT tasks. A centralized Web site, such as the one for Kaluzniacky's book (currently at http://itwellness.ncf.ca), can act as a catalyst for such a potentially impactful movement. Awareness of specific psychological mechanisms, the emotions they generate in different situations, and the effect of such emotions on the psychic energy of the IT worker would give rise to the emotionally intelligent IT worker.

The following is a basic orientation to several significant psychological factors that could have a positive impact on IT work. Then the feasibility of IT Psych-factors communities of practice is promoted.

PERSONALITY FACTORS

Personality can be defined as "a complex set of relatively stable behavioral and emotional characteristics" of a person (Hohmann, 1997, p. 35). The Myers-Briggs Personality Type (Keirsey & Bates, 1978) approach to classifying personalities has been widely accepted and applied in a diversity of fields such as social work, counseling, career planning, and management. It assesses four different dimensions of a person:

1. **Introversion/extraversion:** relates to *how a person is oriented*, where he/she focuses more easily, within oneself or on other people and the surrounding environment. This dimension is coded I or E, respectively.
2. **Intuition/sensing:** relates to two different *ways of perceiving*, of taking in information. An intuitive person focuses on new possibilities, hidden meanings, and perceived patterns. A sensing person focuses on the real, tangible, and factual aspects. Thus, a sensing person can be described as being more practical, whereas an intuitive is more imaginary. This dimension is coded N for Intuitive and S for Sensing.
3. **Thinking/feeling:** relates to *how a person comes to conclusions*, how a person normally prefers to make judgments. A thinking person

employs logical analysis, using objective and impersonal criteria to make decisions. A feeling person, on the other hand, uses person-centered values and motives to make decisions. This dimension is coded T for Thinking and F for Feeling.

4. **Judging/perceiving:** relates to two essential attitudes of dealing with one's environment. A judging person prefers to make judgments, or comes to conclusions about what one encounters in one's outer environment. A perceiving person prefers to notice one's outer environment while not coming to conclusions or judgments about it. This dimension is coded J for Judging and P for Perceiving.

It can be hypothesized (Ferdinandi, 1994) that different personality factors would contribute in specific ways to different IS development tasks. Regarding the first Myers-Briggs dimension, tasks such as detailed data modeling, coding, quality assurance testing, and network design can lend themselves quite well to preferred introversion. However, extraverts can feel especially at home in requirements determination, joint application development, presentation to users/senior management, user training, and help desk activities, for example.

As for the second, sensing/intuition dimesnion, much of actual technology is practical, and activities such as system installation, detailed telecommunication design, physical data modeling, as well as programming, testing, activity scheduling, and detailed documentation would appeal to and energize the sensing person. Activities such as system planning, high-level business and data modeling, object modeling, and political positioning would be much more in the realm of intuitive types. Thus, there is considerable opportunity for both sensors and intuitives to find IT work appealing.

Third, considerable IS development activity, no doubt, involves the thinking function, whether practical thinking (as in telecommunication design or testing) or conceptual thinking (object modeling, system planning). Often, the thinking must be structured and yield specific deliverables that can execute on specific machines. But, how can feeling types find a home in IT work? Since they place considerable focus on harmony, feelers can be particularly sought after as group/team leaders, high-level business modelers or analysts, where considerable effective interaction with non-IT staff is essential. Feelers may become prominent IS "politicians" who can forge effective relationships with others in organizations. They can also contribute innovatively and effectively in development of training materials and in the training process itself. As systems move toward integration of a variety of communication modes through multimedia and Internet access, the contribution of artistically-minded feelers will be increasingly desirable.

Relating to the judging/perceiving component, computing itself is largely structured with emphasis on precision. Thus, procedural language programming, for example, would be ideal for a judging (structured) personality orientation as would be detailed telecommunication design. Yet, there certainly are activities in the development and maintenance of systems where too much structure and predictability would not be desirable. System planning and brainstorming, for example, thrive on flexibility and spontaneity. Business and data modeling for a new system also mandate adaptability and flexibility. Maintenance and help-desk work is often unpredictable and varied. These would likely be the domains of the perceiving (open-ended) worker. It is thus easy to see that consciously matching Myers-Briggs personalities with IT tasks can indeed increase effectiveness, motivation, and synergy.

While we can appreciate the insights provided by the Myers-Briggs system, we are equally aware that no one system can hope to address all aspects of personality. A noteworthy personality analysis tool that has achieved a significant presence in both personal growth and management applications is the *Enneagram system of personalities*. The essence is said to have descended from the ancient Sufis, and modern adaptations have been made by a variety of authors, including Riso and Hudson (1990), Palmer and Brown (1998), Rohr and Ebert (1990), and Goldberg (1996). Whereas MBTI attempts to explain *how* we function, the Enneagram focuses more on *why* we function in a particular way: what is the underlying emotion that guides the way we act? In this way, MBTI and the Enneagram can be viewed as complementary.

The Enneagram proposes nine different personalities, numbered consecutively from 1 to 9. Each type has an inner need for some factor to be "ok" for him/her to feel "ok". For example, the Perfectionist (1) can feel ok only if he/she is performing nearly perfectly—if he/she is doing the "right" thing. The Status Seeker (3) can feel ok only if his/her image in front of others is that of a successful, important person. The Knowledge Seeker (5) feels ok only if he/she has impartially observed, analyzed, and learned as much as possible about his/her environment. The Loyalist (6) can feel ok if she/he is obedient to external rules and dictations; she/he is often hesitant to act out of personal initiative.

From the Enneagram, we can appreciate much more clearly and directly how different people can be doing the same IT work but from different underlying motivations. Thus, they will recognize different challenges in the same IT work and will exhibit different reactions. Within the IT employee him/herself, the key question is how might his/her reaction drain his/her own inner mental and emotional energies, making such energies unavailable for solving the problem at hand. Between individuals, how might their differing reactions to a situation they face in common initiate caution, mistrust, and lack of synergy instead of the intended effective, synergetic communication? Whether alone or with others, a person who has not integrated the various perspectives of the Enneagram to at least a fair degree will experience a considerable amount of stifling energy blockage, underutilizing his/her potential and reducing his/her possible level of fulfillment.

COGNITIVE CONSIDERATIONS

While personality relates to one's behavior as a whole, *cognitive function* relates more explicitly to mental information processing. According to Hayes and Allinson (1998), cognitive style is " a person's preferred way of gathering, processing, and evaluating information." One main way of dichotomizing cognitive functioning is the *analytic, sequential* vs. *intuitive, wholistic* functioning. Some psychologists have referred to the former as "left-brain thinking" and the latter "right-brain thinking" (although other scholars may consider this an over-simplification). The former focuses on "trees", and the latter sees the "forest" in solving problems and coming to conclusions.

One of the main theories on cognitive styles, along with an instrument to evaluate the style, is the Kirton Adaptation/Innovation theory. This theory of cognitive strategy relates to the amount of structure a person feels appropriate within which to solve a problem or to embark on creativity. The Adaptor (left-brained) prefers to work within current paradigms, focusing on *doing things better*, while the Innovator (right-brained) prefers to "color outside the lines", constructing new paradigms, focusing on *doing things differently*. According to psychologist/consultant, Michael Kirton (1989), the Adaptor favors precision, reliability, efficiency, prudence, discipline, and conformity. The Innovator, on the other hand, cuts across and often invents new paradigms. The Innovator is more interdisciplinary, approaches tasks from unsuspected angles, and often treats accepted means with little regard.

The late researcher, Dan Couger (1996), applied Kirton's Adaptivity/Innovation Inventory (KAI) to a representative sample of IS professionals and found that most were inclined to the Adaptive rather than the Innovative style. However, a highly innovative employee working under an unaware adaptor may be an occasion for significant dysfunction. Yet, both styles have their definite place in IT. According to MIS researcher, Michael Epstein (1996), "the need to integrate the mechanistic logic of computers with the ambiguous nature of living human systems remain fundamental to the successful design and implementation of computer-based information systems."

THE DEEPEST INNER SELF

Another significant psychological dimension that could warrant consideration within IT is that of the "deepest inner self". This deepest self (inner core, center, being) can truly provide rejuvenated psychological energy in times of stress and change, and it can also provide stability and significant impetus to creative efforts. Thus, it is proposed here that conscious awareness of one's deepest self can indeed add a very important dimension to the work of an IT

professional, particularly one whose work involves human interaction. As well, the connection to one's inner self provides the theoretical and empirical basis for the notion of emotional intelligence.

For a majority of people, at least in today's Western world, their self-awareness seems to be largely limited to a one-tiered view. Here the mind (intellect), feelings (emotions), and the body (with its senses) are identified. The will is the force which channels one's consciousness to the various components during the day. However, a variety of personal growth programs and literature, some of which have received notable acceptance from professional circles, propose the two-tiered model of the human person. Here the three human dimensions are receiving energy and direction from the deepest self/ core energy/inner being, which operates on a separate, deeper level. Moreover, the substance of this core energy is *only positive*—there is nothing negative and nothing missing (for deep psychological wellness) in the inner being. The difficulty, though, lies in establishing and maintaining a strong enough connection between the three *human* and the *being* parts (see Hoffman, 1988; PRH, 1997).

A major contribution to such a new vision that can apply to IT professionals can be seen in the work of Gary Zukav, author of the bestseller, *The Seat of the Soul* (1990), and Linda Francis. In their collaborative book, *The Heart of the Soul* (2001), Zukav and Francis declare, "spiritual growth {awareness of the deep self} is now replacing survival as the central objective of the human experience" and:

...the old species explored the physical world and it created security by manipulating and controlling what it discovered. The new species creates security by looking inward to find the causes of insecurity and healing them...This is the path to authentic power....Authentic power is the alignment of your personality with your soul (i.e., deepest core energy). (p. 159)

Thus, the deepest self completes the earlier discussed psychological factors, providing a blueprint for an integrated whole. Cannot such a psychological integration be the basis for a strategic vision of the functional IT professional of the 21st century? If so, what is needed to implement such a strategy?

FEASIBILITY OF A PSYCH-FACTORS CoP

Twenty-five years ago, researchers Couger and Zawacki (1980) reported in their book that system development professionals, while having a lower-than-average need for socializing, had demonstrated a high need for growth. While, at that time, growth may have referred largely to the development of technical competence, today such an orientation may well be expanded to include a number of interdisciplinary, yet relevant concerns.

For the IT profession, as a whole, to begin accepting personality, cognition, and spiritual awareness as relevant to specific work tasks, considerable, frequent, and detailed communication will be essential. A Psych-Factors CoP, defined as "a worldwide collection of IT professionals who communicate regularly regarding growth in psychological awareness and how such growth explicitly enhances various facets of IT work" could be a welcome forum for such communication. Here, effects of such awareness on worker productivity and effectiveness could be discussed and compared.

Community members could share, for example, experiences of how a person with an analytic (left-brained) cognition and one with a heuristic (right-brained) approach have collaborated synergistically in the design of user-interfaces for a Web-based e-commerce system; how a predominately sensing analyst has complemented an intuitive one in eliciting user requirements; how a power seeker project manager has avoided near mutiny on a system development team by applying Enneagram wisdom; or how a loyal, burned-out employee had found unexpected psychological empowerment as a result of the Hoffman Quadrinity Process. Moreover, such a community of practice could provide valuable motivation and substance for MIS academics to research systematically relevant issues concerning personal psychology and IT.

CONCLUSION AND FUTURE TRENDS

Kent Beck (2000), the founder of Extreme Programming, has issued a call within the IT world for "growing the whole person". For the psychological emancipation of the IT professional, communication and community are indeed indispensable. With individual IT workers, employing IT departments, professional associations, university academics, and the user community to each do its part, the IT work world of the future, particularly through proliferating Psych-Factor communities of practice, will respond to its current challenges with integrated wisdom and fruitful, energizing innovation.

REFERENCES

Beck, K. (2000). *Extreme programming explained: Embrace change.* Reading, MA: Addison-Wesley.

Bedrij, O. (1977). *ONE.* London: Strawberry Hill Press.

Borysenko, J. (1990). *Guilt is the teacher, love is the lesson.* Warner.

Couger, J. D. (1996). *Creativity and innovation in information systems organizations.* Danvers, MA: Boyd & Fraser.

Couger, J. D., & Zawacki, R. (1980). *Motivating and managing computer personnel.* John Wiley & Sons.

Dyer, W. (1995). *Your sacred self.* London: Harper Paperbacks.

Epstein, M. (1996). *The role and worldview of systems designers: A multimethod study of information systems practitioners in the public sector.* University of Saskatchewan, College of Commerce.

Ferdinandi, P. (1994, July). Re-engineering with the right types. *Software Development.*

Goldberg, M. J. (1996). *Getting your boss's number.* New York: HarperBusiness.

Hayes, J., & Allinson, C. W. (1998). Cognitive style and the theory and practice of individual and collective learning in organizations. *Human Relations, 51,* 847-871

Hoffman, B. (1988). *No one is to blame.* Oakland, CA: Recycling Books.

Hohmann, L. (1997). *Journey of the software professional.* Englewood Cliffs, NJ: Prentice Hall.

Kaluzniacky, E. (2004). *Managing psychological aspects of information systems work: An orientation to emotional intelligence.* Hershey, PA: Information Science.

Keirsey, D., & Bates, M. (1978). *Please understand me.* Del Mar, CA: Prometheus Nemesis.

Kirton, M. J. (1989). (Ed.). *Adaptors and innovators: Styles of creativity and problem solving.* London: Routledge.

Kolb, D. A. (1984). *Experiential learning.* Englewood Cliffs, NJ: Prentice Hall.

McGraw, P. (2001). *Self matters.* New York: Free Press.

Palmer, H., & Brown, P. (1998). *The Enneagram Advantage: Using the nine personality types at work.* Nevada City, CA: Harmony Books.

PRH-International. (1997). *Persons and their growth.*

Riso, D. R., & Hudson, R. (1990). *The wisdom of the Enneagram.* New York: Bantam.

Rohr, R., & Ebert, A. (1990). *Discovering the Enneagram.* New York: Crossroad.

Weizenbaum, J. (1976). *Computer power and human reason: From judgement to calculation.* W.H. Freeman.

Yourdon, E. (2002, December). *Cutter IT Journal, 15*(12).

Zukav, G. (1990). *The seat of the soul.* New York: Fireside.

Zukav, G., & Francis, L. (2001). *The heart of the soul.* New York: Simon & Schuster.

KEY TERMS

Adaptive Cognition: A cognitive style that prefers to think sequentially and work within current paradigms.

Cognitive Style: A person's preferred way of gathering, processing, and evaluating information.

Deepest Inner Self: A psychological reality on a level deeper than (below) intellect, emotions and body, where rejuvenating psychological energy that is always only positive is found.

Enneagram Personality System: Identifies 9 personality types each based on a specific compulsion and underlying emotion; the enneagram diagram shows a different growth path for each of the nine types.

Innovative Cognition: A cognitive style that processes information, intuitively, integratively, and prefers to construct new paradigms.

Myers-Briggs Personality System: Identifies 16 personality types largely based on psychology of C. G. Jung and involving four dimensions: introvert/extravert, intuitive/sensing, thinking/feeling, and judging/perceiving.

Personality: A complex set of relatively stable behavioral and emotional characteristics of a person.

Psych-Factors CoP: A worldwide collection of IT professionals who communicate regularly regarding growth in psychological awareness and how such growth explicitly enhances various facets of IT work.

Quality of Knowledge in Virtual Entities

Cesar Analide
University of Minho, Portugal

Paulo Novais
University of Minho, Portugal

José Machado
University of Minho, Portugal

José Neves
University of Minho, Portugal

INTRODUCTION

The work done by some authors in the fields of computer science, artificial intelligence, and multi-agent systems foresees an approximation of these disciplines and those of the social sciences, namely, in the areas of anthropology, sociology, and psychology. Much of this work has been done in terms of the humanization of the behavior of virtual entities by expressing human-like feelings and emotions.

Some authors (e.g., Ortony, Clore & Collins, 1988; Picard, 1997) suggest lines of action considering ways to assign emotions to machines. Attitudes like cooperation, competition, socialization, and trust are explored in many different areas (Arthur, 1994; Challet & Zhang, 1998; Novais et al., 2004). Other authors (e.g., Bazzan et al., 2000; Castelfranchi, Rosis & Falcone, 1997) recognize the importance of modeling virtual entity mental states in an anthropopathic way.

Indeed, an important motivation to the development of this project comes from the author's work with artificial intelligence in the area of knowledge representation and reasoning, in terms of an extension to the language of logic programming, that is, the Extended Logic Programming (Alferes, Pereira & Przymusinski, 1998; Neves, 1984). On the other hand, the use of null values to deal with imperfect knowledge (Gelfond, 1994; Traylor & Gelfond, 1993) and the enforcement of exceptions to characterize the behavior of intelligent systems (Analide, 2004) is another justification for the adoption of these formalisms in this knowledge arena.

Knowledge representation, as a way to describe the real world based on mechanical, logical, or other means, will always be a function of the systems ability to describe the existent knowledge and their associated reasoning mechanisms. Indeed, in the conception of a knowledge representation system, it must be taken into attention different instances of knowledge:

- **The Existent Knowledge:** It will not be known in all its extension because it characterizes all the circumstances of the universe of discourse, known or unknown.
- **The Observed Knowledge:** Acquired by the experience, it must be noticed that it may depend upon the observer education, state of mind, and prejudices (to state a few).
- **The Represented Knowledge:** With respect to a certain objective, it may be irrelevant to represent a given set of data. This is the information that must be represented and understood.

In a classical logical theory, the proof of a question is made in terms of being true or false, or in terms of representing something about which one could not be conclusive. In spite of that, in a logic program, the answer to a question is only of two types: it can be true or false. This is due to the fact that a logic program shows some limitations in terms of knowledge representation. (It is not allowed explicit representation of negative information.) In addition, in terms of an operational semantics, it is applied the Closed World Assumption (CWA) to all the predicates.

The generality of the programs written in logic represents implicitly negative information, assuming the application of reasoning according to the CWA. An extension of a logic program may comprise negative information (Alferes et al., 1998; Neves, 1984), as well as directly describe the CWA for some predicates. Consequently, it is possible to distinguish three types of conclusions for a question: *true*, *false* or, when there is no information inferring one or another, the answer will be *unknown*.

In this work, the subject related with the qualitative knowledge is discussed behind the assumption that, when a system needs to reason about the real world, it must have the ability to infer upon imperfect knowledge. Hence, this knowledge imperfection may have an important role in the quality of the whole system when considered as a part of a wider community of virtual entities with a rich knowledge component, having sophisticated properties such as planning, reactivity, learning, cooperation, communication, and argumentation. Agent societies may mirror a great variety of human societies with emphasis on behavioral patterns and predefined roles of engagement and obligation.

PRELIMINARIES

Knowledge and belief are generally incomplete, contradictory, or error sensitive, being desirable to use formal tools to deal with the problems that arise from the use of incomplete, contradictory, ambiguous, imperfect, nebulous, or missing information. This work is supported by the developments in Analide (2004) where the representation of incomplete information and the reasoning based on partial assumptions is studied, using the representation of null values (Analide & Neves, 2000; Neves, 1984) to characterize abnormal or exceptional situations. The ELP language presents itself as a formal and flexible tool to obtain a solution for the problems just referred.

Null Values

The identification of null values emerges as a strategy for the enumeration of cases, for which one intends to distinguish between situations where the answers are *known* (true or false) or *unknown* (Analide & Neves, 2000; Traylor & Gelfond, 1993).

The representation of null values will be scoped by the ELP. In this work, two types of null values wil be considered: the first will allow the representation of unknown values, not necessarily from a given set of values, and the second will represent unknown values from a given set of possible values.

Consider the following as a case study to show some examples of how null values can be used to represent unknown situations. Consider the implementation of a time table to express the departure of trains through the predicate:

connect: City × Time

where the first argument denotes the city of departure and the second represents the time of arrival (e.g., connect(guimarães,17:00) denotes that the Guimarães's coming train is expect to arrive at 17 o'clock, Program 1).

Program 1. Extension of the predicate that describes arrivals at the train station:

connect(guimarães,17:00)
¬connect(C,T) ←
    not connect(C,T)

In Program 1, the symbol ¬ denotes the strong negation, denoting what should be interpreted as *false*, and the term *not* designates negation by failure.

Unknown

Following the example given by Program 1, one can admit that the connection from Oporto has not yet arrived. This situation will be represented by a null value of the type *unknown* that should allow the conclusion that the connection exists but to which it is not possible to be affirmative with respect to the arrival time (Program 2).

Program 2. Information about Oporto connection with an unknown delay:

connect(guimarães,17:00)
connect(oporto,⊥)
¬connect(C,T) ←
    not connect(C,T) ∧ not exception(connect(C,T))

exception(connect(C,T)) ←
    connect(C, ⊥)

Symbol ⊥ represents a null value of an undefined type, in the sense that it is a representation that assumes that any value is a potential solution but without given the clue to conclude about which value one is speaking about. Computationally, it is not possible to determine from the positive information, the arrival time of the Oporto's connection; by the description of the exception situation (fourth clause from Program 2, the closure of predicate connect), the possibility to be assumed as false any question on the specific time of arrival of that connection is discarded.

Unknown but Enumerated

Consider now the example in which the time of arrival of Lisbon's connection is foreseen as 18 o'clock but is 15 minutes delayed. It is not possible to be affirmative regarding the arrival at 18:00 or at 18:01 or even at 18:15. However, it is false that the train will arrive at 16:16 or at 17:59. This example suggests that the lack of knowledge may only be associated to an enumerated set of possible values.

Program 3. Representation of the connection with a 15-minute delay:

connect(guimarães,17:00)
connect(oporto,⊥)
¬connect(C,T) ←
    not connect(C,T) ∧ not exception(connect(C,T))
exception(connect(C,T)) ←
    connect(C, ⊥)
exception (connect(lisbon,T)) ←
    T ≥ 18:00 ∧ T ≤ 18:15

The exception occurs to the time interval 18:00...18:15. It is unknown that Lisbon's connection will arrive at 18:05 or at 18:10; it is false that it will arrive at 17:55 or at 18:20.

Interpretation of Null Values

To reason about the body of knowledge presented in a particular knowledge, set on the base of the formal-

ism referred to above, let us consider a procedure given in terms of the extension of a predicate called demo, using ELP as the logic programming language. Given a question, it returns a solution based on a set of assumptions. This meta-predicate will be defined as:

demo: Question × Answer

where Question denotes a theorem to be proved, and Answer denotes a truth value: True (T), False (F), or Unknown (U) (Program 4).

Program 4. Extension of meta-predicate demo:

demo(Q, T)   ← Q
demo(Q, F)   ← ¬Q
demo(Q, U)   ← not Q ∧ not ¬Q

The first clause of Program 4 sets a question to be answered with appeal to the knowledge base positive information; the second clause denotes the question is proved to be false with appeal to the negative information presented at the knowledge base level; the third clause stands for itself.

COMPUTING THE QUALITY OF KNOWLEDGE

Based on the assumptions presented previously, it is possible to establish mechanisms to analyze and process the information available in a way that turns feasible the study of the behavior of virtual entities in terms of its personification. Situations involving forgetfulness, remembrance, learning, or trust can be analyzed in the way proposed in this work; that is, the description of abnormal situations, declared as exceptions to a predicate extension, made one's goals possible.

Characterization of a Problem

Consider the following example, built up to illustrate the practical application of what is the main contribution of this work.

Program 5. Excerpt of an extended logic program, representing knowledge at a time t_i:

parent(carlos,joão)
¬parent(P,S) ←
    not parent(P,S) ∧ not exception(parent (P,S))

In Program 5, there is an axiom stating that Carlos is a parent of João. Assuming that this is all the knowledge available at instant t_i, the second clause of Program 5 enforces that all other situations where there is a lack of information and that are not being treated as exceptions must be considered false.

Suppose that, an instant later, t_j, the knowledge evolves in such a way that it may be represented as shown in Program 6.

Program 6. Knowledge base excerpt at instant t_j:

¬parent(P,S) ←
    not parent(P,S) ∧ not exception(parent (P,S))
exception(parent(carlos,joão))
exception(parent(luís,joão))
exception(parent(pedro,joão))

At a third instant of time, t_k, the knowledge base is shown as Program 7.

Program 7. Excerpt of the program that shows how the knowledge base evolves between instants t_j and t_k:

parent(⊥,joão)
¬parent(P,S) ←
    not parent(P,S) ∧ not exception(parent(P,S))
exception(parent(P,S)) ←
    parent(⊥,S)

Looking to the way the knowledge base evolved, between instants t_j to t_k, one may say that the information has been losing specificity. In the beginning, it was known that Carlos was a parent of João (t_i); after that, it was only known that the parent of João was Carlos, Luís, or Pedro (t_j); finally, in a third instant, the system only knows that João has a parent but cannot be conclusive about who such a person is; it is also not possible to state that João does not have a father.

Consequently, in terms of the temporal axis $t_i \rightarrow t_j \rightarrow t_k$, one may say that the knowledge evolution has taken a form of forgetfulness, leading to the emptying of the knowledge base knowledge. However, taking the knowledge evolution the other way around, that is, $t_k \rightarrow t_j \rightarrow t_i$, a similar analysis leads to the conclusion that the knowledge base learned something, showing that the knowledge base evolves in a way that secures its information.

The System Semantics

Last but not least, it is now possible to pay some attention to the human-like attributes to be represented at a system level, considering the ELP as the language to describe its knowledge bases or theories. Consequently, the objective here is to define those mechanisms that will allow the advent of computational agents at the system level with human-like properties and behaviors, making the way to a certain kind of personification of those computational entities.

Let us consider Program 5, referred to above, that describes the state of the system at instant t_i, where who is João's parent is questioned. In terms of the demo meta-predicate, one may have:

i. $\forall_{(P)}$: demo(parent(P,joão),T)?
                ∠ successful
    $\forall_{(P)}$: demo(parent(P,joão),F)?
                ∠ unsuccessful
    $\forall_{(P)}$: demo(parent(P,joão),U)?
                ∠ unsuccessful

This question is answered in terms of the knowledge base positive information that states that Carlos is João's parent. It is now possible to determine the amount and quality of the information that was used in this round. In other words, one intends to find the set of all the solutions that could contribute to solve the question referred to above, namely:

ii. $\forall_{(P,S)}$: findall(P,demo(parent(P,joão),T),S)?
                ∠ S = [carlos]

Let us now consider Program 6, referred to above, and in this context, endorse the same question as in (i). One may have:

iii. $\forall_{(P)}$: demo(parent(P,joão),T)?
                ∠ unsuccessful

$\forall_{(P)}$: demo(parent(P,joão),F)?
$\angle$ unsuccessful

$\forall_{(P)}$: demo(parent(P,joão),U)?
$\angle$ successful

That is, the question is solved but the answer is vague. This means that endorsing the question as in (ii) will give rise to an empty set of solutions when invoked in terms of the meta-predicate demo. One may have:

iv. $\forall_{(P,S)}$: findall(P,demo(parent(P,joão),U),S)?
$\angle$ S = [ ]

This situation denotes there are clauses defined as exceptions to the extension of predicate *parent*, allowing the solution to be unknown, U. One may now turn to the exceptions in order to evaluate the answer. One may have:

v. $\forall_{(P,S)}$: findall(P,exception(parent(P,joão)),S)?
$\angle$ S = [carlos, luís, pedro]

$\forall_{(S,N)}$: length(S,N)?
$\angle$ N = 3

In this case, attending to the fact that there are three exceptions to the predicate extension, the vagueness of the data is set to $1/3$.

Finally, let us consider the case described by Program 7, referred to above. By the application of the same procedures as in (i), one may have:

vi. $\forall_{(P)}$: demo(parent(P,joão),T)?
$\angle$ unsuccessful

$\forall_{(P)}$: demo(parent(P,joão),F)?
$\angle$ unsuccessful

$\forall_{(P)}$: demo(parent(P,joão),U)?
$\angle$ successful

That is, the solution to the question is undefined. In this case, acting as in (ii), one is presented with a specific result:

vii. $\forall_{(P,S)}$: findall(P,demo(parent(P,joão),U),S)?
$\angle$ S = [$\bot$]

$\forall_{(S,N)}$: lenght(S,N)?
$\angle$ N = ∞

That is, the evaluation of the truth value to assign to the solution falls back upon a mechanism that starts from an unlimited set of possible solutions. It is to be understood that the cardinality of such a set tends to be infinite.

CONCLUSION

The dissemination of computational systems with the ability to live in a virtual community is an open subject for discussion, attaining increasing relevance when the debate relates to the assessment of human-like attitudes and behaviors to virtual entities. The ability to represent and infer upon an entity body of knowledge built around the need to manage imperfect information is an advantage and a virtue explored in this work. This benefit is used to reason about the quality of the information that characterizes such body of knowledge. ELP proved to be an adequate tool for knowledge representation and reasoning, in particular, when one intends to endorse situations where the information is vague, imperfect, or incomplete, which is the case when representing properties and attitudes at the agent's level, only found in humans. The use of these techniques, in particular, in intelligent systems, are flexible to endorse problems where the knowledge of several agents has to be diffused and integrated, and the agents reason about the knowledge or the behavior of their peers in a competitive and/or collaborative way.

REFERENCES

Alferes, J., Pereira, L., & Przymusinski, T. (1998). Classical negation in nonmonotonic reasoning and logic programming. *Journal of Automated Reasoning, 20*, 107-142.

Analide, C. (2004). *Anthropopathy in virtual entities.* PhD Thesis, University of Minho, Department of Informatics.

Analide, C., & Neves, J. (2000, October). Representing incomplete knowledge. *Proceedings of the 1st CAPSI, Conference of the Portuguese Association of Information Systems*, University of Minho, Guimarães, Portugal.

Arthur, W. (1994). Inductive reasoning and bounded rationality (The El Farol Bar Problem). *Proceedings of the American Economics Review, 84* (p. 406).

Bazzan, A., Bordini, R., Vicari, R., & Wahle, J. (2000, November 19-22). Evolving populations of agents with personalities in the minority game. *Proceedings of the International Joint Conference: 7th IBERAMIA and 15th SBIA, Atibaia-SP,* Brazil.

Castelfranchi, C., Rosis, F., & Falcone, R. (1997, November 8-10). Social attitudes and personalities. In *Proceedings of the Agents, Socially Intelligent Agents, AAAI Fall Symposium Series,* MIT, Cambridge, Massachusetts.

Challet, D., & Zhang, Y. (1998). On the minority game: Analytical and numerical studies. *Physica A, 256,* 514.

Gelfond, M. (1994). Logic programming and reasoning with incomplete information. *Annals of Mathematics and Artificial Intelligence, 12,* 89-116.

Neves, J. (1984). A logic interpreter to handle time and negation in logic data bases. In R. L. muller and J. J. Pottmyer (Eds), *Proceedings of ACM'84 Annual Conference, The 5th Generation Challenge* (pp. 50-54).

Novais, P., Analide, C., Machado, J., & Neves, J. (2004, July). Reputation and trust in the context of logic based argumentation. *Proceedings of the ICKEDS 2004, 1st International Conference on Knowledge Engineering and Decision Support,* Oporto, Portugal.

Ortony, A., Clore, G. L., & Collins, A. (1988). *The cognitive structure of emotions.* Cambridge, UK: Cambridge University Press.

Picard, R. (1997). *Affective computing.* Cambridge, MA: MIT Press.

Traylor, B., & Gelfond, M. (1993). Representing null values in logic programming. *Proceedings of the ILPS'93 Workshop on Logic Programming with Incomplete Information* (pp. 35-47). Vancouver, Canada.

KEY TERMS

Anthropopathy: Attribution of human feelings to nonhumans (animals, objects or virtual entities).

Existent Knowledge: All the knowledge that characterizes a certain situation, entity, organization, and so on. Because it characterizes all the circumstances of that universe, identified or not, it will not be known in all its extension.

Imperfect Knowledge: The imperfection of knowledge has to do with its incompleteness (absence of a value), with its imprecision (lack of precision in a value) and iwth its uncertainty (doubt on the truth fulness of a fact).

Intelligent System: A conceptual system that learns during its existence and acts on the environment where he "lives." maybe, modifying it.

Negative Information: The kind of information about what one can be affirmative stating that it is false, with no doubt.

Null Values: The representation of null values emerges as a strategy for the enumeration of cases, for which one intends to distinguish between real facts of life and doubts or uncertainties.

Observed Knowledge: The knowledge that is acquired by the experience and/or by the interaction with the system. It must be noticed that it may depend upon many things, like the observer education or his state of mind, for example.

Positive Information: The kind of information about what one can be affirmative stating that it is true, with no doubt.

Quality of Knowledge: The quality of knowledge presented by an intelligent system is determined by an analysis of its knowledge body, by means of deciding about what kind of certainties and doubts it represents.

Represented Knowledge: The knowledge that must be represented and understood. It may be relevant, with respect to a certain objective, to represent a given set of data or not to represent another collection of information.

Virtual Community: A society of virtual entities, where they can interact with each other and with other objects, agents or physical entities.

Virtual Entity: An entity with a virtual life, generally supported by computational means and artifacts, with the ability to learn, communicate and interact with other entities (virtual or real). It is characterized by properties like anthropopathy and autonomy.

The Reformation of Communities of Practice

Aidan Pyke
Cork Institute of Technology, Ireland

INTRODUCTION

A *"knowledge set that distinguishes and provides a competitive advantage"* to a firm is considered to be that firm's *"core capability"* (Leonard-Barton, 1992, p.113). Firms cultivate groups or communities of practice to create, integrate, and disseminate organisational knowledge in particular fields of knowing (Brown & Duguid, 1991). Creating, sharing, and integrating knowledge is mission-critical to a firm because firm-specific advantage flows not from an organisation's resources per se, but from the *knowledge* that enables it to deploy such resources to leverage maximum benefit within its operational environment (Penrose, 1959).

Knowledge must be managed in a way that differs from the management of a firm's other resources (Spender, 1996). This requirement stems from the intrinsic qualities of knowledge; it emanates from individual thoughts, which shape and are in turn shaped by the social dynamic within an organisation, a fusion of cognitive and societal processes (Berger & Luckmann, 1967). The inherent qualities of knowledge affect its appropriability (Teece, 1998). Leonard and Sensiper (1998) assert that knowledge is a combination of tacit and explicit. Polanyi (1967) contends that explicit knowledge is knowledge in the abstract, while tacit knowledge incorporates experience and intuitive knowledge, which results from subconscious learning. Itami (1989) argues that it is impossible to separate the individual from the tacit knowledge that he or she possesses. The replication and transfer of knowledge is often impossible without the transfer of actors (Teece, 2000).

BACKGROUND

Many firms are confronted with a dilemma when they seek to transfer or replicate knowledge, as they are ignorant of the knowledge creating dynamics in operation within their own organisation, even though managers are cognisant of their exiting capabilities (Fransman, 1998). Lippmann and Rumelt (1982) use the term *causal ambiguity* to describe the ignorance of many firms concerning the interaction and combinations of resources, which producing their capabilities. They note that some resources are not tradable, as they are combined in ways that are difficult to reproduce because of causal ambiguity. This lack of clarity hampers the knowledge creation and integration processes. To circumvent this difficulty many organisations have increasingly opted to assign the task of knowledge creation and integration to communities of practice (Brown & Duguid, 1991).

The creation of a community of practice is a social and economic process that is nurtured through the ongoing interaction of individuals (Zuboff, 1988; Brown & Duguid, 1991). Nonaka and Takeuchi (1995) consider that individual intuition is a source of organisational knowledge. New members of the community of practice are *enculturated* in the values and norms of their communities (Brown, Collins & Duguid, 1989). Storytelling plays a unifying role in group formation, and indeed adopting the role of "storyteller" is a rite of passage for new group members (Orr, 1990). Storytelling also preserves group knowledge and supports and forms group values and beliefs (Jordan, 1989). Hence, organisational actors should be viewed as active participants in vibrant communities where social, cultural, and past forces contribute to a collective learning process aimed at satisfying individual and communal needs (Barney, 1986). An organisation's ability to learn and to cultivate its knowledge asset is therefore dependent upon its capability, through its evolving social character, to influence its employees' values, sense of self, and consequently their sense of community (Nelson & Winter, 1982; Spender, 1996).

Unfortunately, not all organisations posses this capability, therefore the creation of communities of practice is not without danger to the firm (Wetlaufer, 1999). Communities of practice can develop community-specific values, which may or may not be aligned with the values of the firm (Child, 1972).

Such communities of practice become enclaves of valuesvalues that are community centric rather than corporate centric; this has the potential to thwart the efforts of their organisations to develop firm-specific knowledge sets and capabilities (Butler & Pyke, 2003). The subversion of organisational structure and communities of practice by sectional interests is of particular concern to Child (1972). He notes that theoretical models of organisational structure draw attention to the:

...possible constraints upon the choice of effective structures but fail to consider the process of choice itself in which economic and administrative exigencies are weighed by the actors concerned, against the opportunities to operate a structure of their own and/or other organisational members' preferences. (p. 16)

Such preferences may emanate from the disaffected or those who resist change and wish to maintain the status quo because such communities can perceive the firm's activities and goals to be foreign and act to subvert them (Levinthal & March, 1993). They contend that members of a community who want existing traditions to prevail will subvert attempts to introduce new knowledge and practices. Orr (1990) states:

The process of working and learning creates a work situation which the workers value, and they resist having it disrupted by their employers through events such as reorganisation of work. This resistance can surprise employers who think of labour as a commodity to arrange to suit their ends. The problem for the workers is that this community which they have created was not part of the series of discrete employment agreements by which the employer populated the work place, nor is the role of the community in doing the work acknowledged. (p. 48)

A lack of alignment can be damaging to the organisation's efforts to create competencies and may in fact lead to the development of core rigidities, which "are but the flip side of core capabilities" (Leonard-Barton, 1995, p. 30). The firm must provide support that corresponds to the *real needs* of the community rather than just to the abstract expectations of the corporation (Brown & Duguid, 1991).

FUTURE TRENDS

The recommendations to corporate management concerning the governance of communities of practice are divergent, reflecting the uniqueness of each firm, the dynamics of its industry, and its constituent communities (Leonard-Barton, 1995). Nystrom and Starbuck (1984) and Schein (1990) argue that communities of practice must be allowed some liberty to free themselves from the absolute influence of received wisdom in order to develop and be innovative. Nonaka (1994) endorses this call, maintaining that firms create an environment or context conducive to the emergence of knowledge when they reduce control mechanisms and afford their employees a significant degree of latitude. Festinger (1957) contends that social communication and social influence processes should be interwoven with the processes of knowledge creation to achieve *dissonance reduction*. Ouchi (1979) identifies clan mechanisms as one of the three types of mechanisms of coordination, the others being market and bureaucratic mechanisms. Pugh (1969) contends that for the larger firm, more bureaucratic methods of control are appropriate to manage the increase in specialisation.

In benevolent environments—when firms are successful and profitable—control strategies may not produce the desired outcomes, as communal actors operating in successful firms do not have a strong incentive to change (Child, 1972; Miller & Minzberg, 1972). A successful firm may have difficulty in executing the necessary attitudinal transformations, and therefore are susceptible to stagnation and inertia (Leonard-Barton, 1995; see Selznick, 1957). But such a reformation or change of community credo can be facilitated by crises, upheavals, conflicts, and innovation (White, 1969; Nystrom, Hedberg & Starbuck, 1976; Bartunek, 1988). Crisis fuels and drives the process of change (Nystrom & Starbuck, 1984; Bartunek, 1988). Nonaka and Takeuchi (1995) promote the concept of creative chaos as a method of countering the threat of rigidities, because chaos can act as a disruptive force capable of altering the perspectives of communal actors. Creative chaos can thus act as a catalyst in the destruction of cognitive frameworks and undesirable routines. They describe creative chaos as an intentional creation of the organisation's leaders. Management nurtures creative chaos to promote a sense of crisis, which is intended to create new challenges and sunder the sense of inertia that may exist within the organisation.

CONCLUSION

The culture of the organisation acts as an incubator for the development of the individual's, the community's, and the firm's capabilities. A firm's organisational culture supports its strategic vision and promotes communication, learning, and capability development. The values of each of the organisational communities of practice must be aligned. This is vital for the creation of a firm that enthusiastically embraces innovation and respects the value of the group and the individual. Without the appropriate social infrastructure and the reshaping or renaissance of communities of practice, the creation and transfer of knowledge can become incestuous and introspective, and spawn a weakened firm that evaluates experiences and information in a proprietary way. Such a process would create or bolster rigidities leading ultimately to failure to satisfy the market.

REFERENCES

Barney, J.B. (1986). Organizational culture: Can it be a source of sustained competitive advantage? *Academy of Management Review*, *11*(3), 656-665.

Bartunek, J.K. (1988). The dynamics of personal and organizational reframing. In R.E. Quinn & K.S. Cameron (Eds.), *Paradox and transformation: Towards a theory of change in organ. and management* (pp. 137-162). Cambridge, MA: Ballinger.

Berger, P.L. & Luckmann, T. (1967). *The social construction of reality.* New York: Doubleday.

Boland, R.J. Jr., Ramkrishnan, V.T. & Dov. T. (1994). Designing information technology to support distributed cognition. *Organization Science*, 5(3), 456-475.

Brown, J.S., Collins, A. & Duguid, P. (1989). Enacting design. In P. Adler (Ed.), *Designing automation for usability.* New York: Review Press.

Brown, J.S. & Duguid, P. (1991). Organisational learning and communities of practice: Towards a unified view of working, learning and innovation. *Organisation Science*, *2*, 40-57.

Butler, T. & Pyke, A. (2003). Examining the influence of ERP systems on firm-specific knowledge and core capabilities: A case study of SAP implementation and use. *Proceedings of the 11th European Conference on Information Systems* (ECIS), Naples, Italy.

Child, J.L. (1972). Organizational structure, environment, and performance: The role of strategic choice. *Sociology, 6,* 1-22.

Festinger, L. (1957). *A theory of cognitive dissonance.* Standford, CA: Stanford University Press.

Fransman, M. (1998). Information, knowledge, vision, and theories of the firm. In G. Dosi, D.J. Teece & J. Chytry (Eds.), *Technology, organization, and competitiveness: Perspectives on industrial and corporate change* (pp. 147-192). New York: Oxford University Press.

Hedberg, B. (1981). How organisations learn and unlearn. In P.C. Nystrom & W.H. Starbuck (Eds.), *Handbook of organizational design.* London: Oxford University Press.

Itami, H. (1989). Mobilising invisible assets: The key for successful corporate strategy. In E. Punset & G. Sweeney (Eds.), *Information resources and corporate growth.* London: Punset and Sweeney.

Jordan, B. (1989). Cosmopolitical obstetrics: Some useful insights from the training of traditional midwives. *Social Science and Medicine, 28*(9), 925-944.

Leonard, D. & Sensiper, S. (1998). The role of tacit knowledge in group innovation. *California Management Review, 40*(3), 112-132.

Leonard-Barton, D. (1992). Core capabilities and core rigidities: A paradox in managing new product development. *Strategic Management Journal, 9,* 41-58.

Leonard-Barton, D. (1995). *Wellsprings of knowledge: Building and sustaining the sources of innovation.* Boston: Harvard Business School Press.

Levinthal, D.A. & March, J.G. (1993). The myopia of learning. *Strategic Management J., 14,* 92-112.

Lippmann, S. & Rumelt, R. (1982). Uncertain imitability: An analysis of interim differences in efficiency under competition. *Bell Journal of Economics, 13,* 418-453.

Miller, D. & Minzberg, H.L. (1972). *Strategy formulation in context.* Working Paper, McGill Univ.

Nelson, R.S. & Winter, S.G. (1982). *An evolutionary theory of economic change.* Cambridge, MA: Harvard University Press.

Nonaka, I. (1994). A dynamic theory of organizational knowledge creation. *Organizational Science, 5*(1), 14-37.

Nonaka, I. & Takeuchi, H. (1995). *The knowledge creating company: How Japanese companies create the dynamics of innovation.* New York: Oxford University Press.

Nystrom, P.C. & Starbuck, W.H. (1984). To avoid organizational crises, unlearn. *Organizational Dynamics, 12*(4), 53-65.

Nystrom, P.C., Hedberg, B.L. & Starbuck, W.H. (1976). Interacting processes as organization design." In R.H. Kilman, L.R. Pondy & D.P. Sleviin (Eds.), *Management of organization design* (vol.1, pp. 209-230). New York: Elsevier North-Holland.

Orr, J. (1990). *Talking about machines: An ethnography of a modern job.* PhD Thesis, Cornell University, USA.

Ouchi, W.G. (1979). A conceptual framework for the design of organizational control mechanisms. *Management Science, 25,* 833-848.

Penrose, E. (1959). *The theory of the growth of the firm.* London: Basil Blackwell.

Polanyi, M. (1967). *The tactic dimension.* Garden City, NY: Anchor Books.

Pugh, D.S. (1969). The context of organization structures. *Administrative Science Quarterly, 14,* 91-114.

Schein, E. (1990). Organizational culture. *American Psychologist, 45*(2), 109-119.

Selznick, P. (1957). *Leadership in administration: A sociological perspective.* New York: Harper and Row.

Spender, J.C. (1996). Making knowledge the basis of a dynamic theory of the firm. *Strategic Management Review, 17,* 45-62.

Teece, D.J. (1998). Capturing value from knowledge assets: The new economy, markets for know-how, and intangible assets. *California Management Reviews, 40*(3), 55-79.

Teece, D.J. (2000). Strategies for managing knowledge assets: The role of firm structure and industrial context. In I. Nonaka & D. Teece (Eds.), *Managing industrial knowledge: Creation, transfer, and utilization* (pp. 125-144). Thousand Oaks, CA: Sage.

Wetlaufer, S., (1999). Organizing for empowerment. *Interviews with CEOs, Harvard Business Review,* 33-63.

White, O.F. Jr. (1969). The dialectical organization: An alternative to bureaucracy. *Public Administration Review, 29,* 32-42.

Zuboff, S. (1988). *The sources of innovation.* New York: Oxford University Press.

KEY TERMS

Distributed Cognition: A term used to denote the herd instinct or an empathetic behaviour pattern that develops within any group or firm (see Boland, Ramkrishnan & Dov, 1994, p. 475).

Learning: A process of understanding, assimilating, and absorbing knowledge.

Organisational Routines: Actions, processes, or decisions that are enacted in response to an individual stimulus or to a series or combination of stimuli (Nelson & Winter, 1982).

Unlearning: The act of discarding outdated or undesirable knowledge sets. An individual, communal, or organisational inability to unlearn is deemed to be a fundamental weakness as it impairs the accumulation of new knowledge and innovation (see Hedberg, 1981).

The Role of Technology in Supporting Communities of Practice

Alton Chua

Nanyang Technological University, Singapore

INTRODUCTION

Most KM literature which investigates the role of technology in supporting communities of practice is oriented either technically or socially. Works that adopt a technical orientation tend to focus sharply on issues such as robustness, scalability, interoperability, and security. Furthermore, they invariably include a plethora of KM tools ranging from databases, portals, and search engines. The dynamics that occur in communities of practice, on the other hand, receive cursory treatment. Works that adopt a social orientation tend to delve into cognitive and social processes but are usually confined within the context of distributed communities. They discuss how technology is used to mitigate the geographical separation among members. However, beyond citing the generic capabilities of technology, such as enabling connection among members, holding electronic content, and providing search functions, the role of technology to meet the peculiar needs of communities of practice has rarely been expounded.

For this reason, this article seeks to clarify how technology can be used to support communities of practice. First, it develops a conceptual model which provides a parsimonious approach to unravel the nebular properties of communities of practice. Next, it explains the power and limitation of technology within the realm of knowledge management and proposes a suite of capabilities found in extant technology tools which supports communities of practice. Finally, this article briefly discusses emerging technologies that could be used to meet the rising needs of community members.

BACKGROUND: UNRAVELING COMMUNITIES OF PRACTICE

Communities of practice are commonly conceptualized as informal aggregations of members who are drawn by common interests to engage in sense-making activities through sharing, learning, and solving problems (Brown & Duguid, 1991; Lave & Wenger, 1991). Due to their spontaneity and richness of interaction among members, communities of practice have been widely acknowledged as the ideal social structure to support the sharing and development of knowledge (Lesser & Stork, 2001; Wenger et al., 2002).

Several studies have been done on communities of practice, for example, those involving Xerox technicians (Orr, 1996), flute makers (Cook & Yanow, 1993), military oilfield services engineers (Edmundson, 2001), and internationally distributed organization members (Hildreth et al., 2000). Communities of practice may differ in attributes such as size, lifespan, physical boundaries, and the degree of recognition in the organization. However, all communities of practice share certain common salient features. These features are included in a conceptual model illustrated in Figure 1. Even though in reality, the constituents in the model are amalgamated and not easily distinguishable, they have been identified discretely to facilitate analysis.

Shown at the core of the model are the *structural elements*. Structural elements distinguish communities of practice from all other types of groupings such as project teams, task forces, and interest

Figure 1. A conceptual model of communities of practice

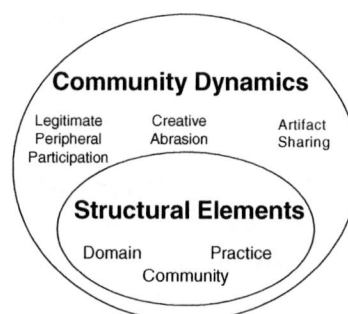

groups. The three structural elements are domain, community, and practice (Wenger et al., 2002). Domain refers to the sphere of knowledge related to a specific area of expertise held by members. Community denotes the sense of identity, confidence, and level of trust among members. Practice is the set of common tools, framework, methodologies, and vocabularies shared by members. Without the simultaneous presence of all three structural elements, a community of practice does not exist. For example, residents in the same neighborhood who have developed a sense of oneness but do not share expertise in a common domain are not regarded as a community of practice. Similarly, a community of practice does not come into existence simply by assembling a group of strangers who possess comparable levels of expertise in a given domain. There has to be a sufficient level of trust and confidence among them before they would consider themselves as part of the same community.

The outer layer in the model shows the *community dynamics*. Unlike structural elements which are defining features of communities of practice, community dynamics refer to the interaction and processes that occur in healthy and thriving communities. Community dynamics include legitimate peripheral participation (Lave & Wenger, 1991), creative abrasion (Leonard-Barton, 1995), and artifact sharing (Hildreth et al., 2000). Legitimate peripheral participation is a complex and composite process through which new members become matured members by acquiring knowledge from the group. It has three inseparable aspects, namely, legitimation, peripherality, and participation. Legitimacy concerns power and authority relations in the group, and it represents the degree of acceptance a member gains from the community. Peripherality refers to the individual's social rather than physical proximity in relation to the community. This in turn is dependent on the individual's history of participation in the group and the expectation of the individual's future interaction with others in the group (Lave & Wenger, 1991).

Creative abrasion, sometimes conceptualized as productive tension (Hirschhorn, 1997) or multi-faceted dialogue (Zárraga & García-Falcón, 2003), refers to the meeting of minds on common ground to explore and negotiate different opinions and, as a result, generates new ideas (Leonard-Barton, 1995).

It is seen in flourishing communities of practice whose members share heterogeneous experiences and perspectives but are bound by the spirit of community to confront a common challenge. Creative abrasion occurs only when there is a right balance between cohesiveness and diversity. Diversity without cohesiveness leads to disorder. On the other hand, cohesiveness without diversity results in group-think (Cohen, 1998).

Artifact sharing is a natural part of everyday work among members in the community of practice. Artifacts such as documents, charts, and images are used for planning, reflection, discussion of issues, and solving problems (Hildreth et al., 2000). They form part of the explicit corporate memory in the organization (Anand, Manz & Glick, 1998) and give permanence to the knowledge stewarded by the community.

TECHNOLOGY AND KNOWLEDGE MANAGEMENT

The use of technology in managing knowledge is not new. Intranets such as EPRINET used in the US electric utility industry (Mann et al., 1997) are based on early generations of networking and computer technology that sought to improve knowledge access. Mainframe computer technology was used to develop online conferencing and forums for collaboration and knowledge sharing (Foulger, 1991). In fact, for a long time, most organizations perceived technology as the cornerstone of all knowledge management initiatives, including developing communities of practice, mainly because technology represents a highly tangible and the easiest part of the implementation.

However, several researchers and practitioners have cautioned against the excessive focus on technology (Anand et al., 1998; Davenport & Prusak, 1999; Nonaka & Takeuchi, 1995). They argue that the success of a knowledge management initiative does not rest on the deployment of technology solutions alone. Research has revealed that some of the greatest difficulties in knowledge management include changing people's behavior, promoting a knowledge-friendly culture (Ruggles, 1998), and low absorptive capacity of the knowledge recipient

(Szulanski, 1996). Overcoming technological limitations, on the other hand, was typically a trivial issue (Ruggles, 1998).

It is important to recognize both the power and limitation of technology within the realm of knowledge management. Technology can be deployed to overcome the barriers of time and space. It can also be used to extend the reach and enhance the speed of knowledge transfer (Davenport & Prusak, 1999). Nonetheless, technology itself is not a compelling reason to cause people to develop a relationship of trust and the propensity to share knowledge with each other. Neither can technology substitute human social interaction in affording rich interactivity among individuals (Fahey & Prusak, 1998). For this reason, attempts to use technology to build communities of practice out of groups whose members neither shared a common domain nor had a feeling of communion invariably ended in failure (KcKinlay, 2002; Newell, 2001). Rather than to goad new communities of practice into existence, technology is most effective when used to support existing ones.

How Technology Supports Communities of Practice

With reference to the conceptual model presented in Figure 1, the role of technology is thus not to create structural elements but to enhance community dynamics, namely, legitimate peripheral participation, creative abrasion, and artifact sharing. Figure 2 shows a suite of capabilities found in extant technology tools that could be used for this purpose.

The Role of Technology in Supporting Legitimate peripheral Participation

Technology could support legitimate peripheral participation in two ways, namely, analyzing the partici-

pation pattern as well as enabling new and matured members in communities of practice to interact with one another. Technology solutions developed for the former purpose are known as social network analysis tools while those developed for the latter are known as collaboration tools.

Social network analysis tools seek to uncover the pattern of participation within communities of practice. A key feature of the tool is the ability to analyze the social network through a snapshot map, an example of which is shown in Figure 3. On the map, each community member is represented by a node, and the interactions among them are represented by lines. The attributes of the line, such as its thickness and color, denote the intensity of the participation. Various indices, such as network centrality and geodesic distance, are computed to determine the relative social position of members in the community. An example of a social network analysis tool is KNETMAP developed by Knetmap.

Collaboration tools provide a platform for community members to share knowledge with one another. The key features of collaboration tools include shared spaces, calendaring, workflow management services, synchronous communication such as net-based meeting and video conferencing, and asynchronous communication such as electronic discussion forums and e-mail. Some collaboration tools also incorporate peer-polling features so that the electronic entry posted by each member can be rated to reflect its quality. Examples of collaboration tools include the POLARIS project developed in-house by the University of Maastricht.

Figure 2. Capabilites of technology that support community dynamics

Capabilities of Technology	Community Dynamics
• Social network analysis • Electronic collaboration	Legitimate peripheral participation
• Idea generation • Simulation	Creative abrasion
• Content management • Concept mapping	Artifact sharing

Figure 3. An example of a social network

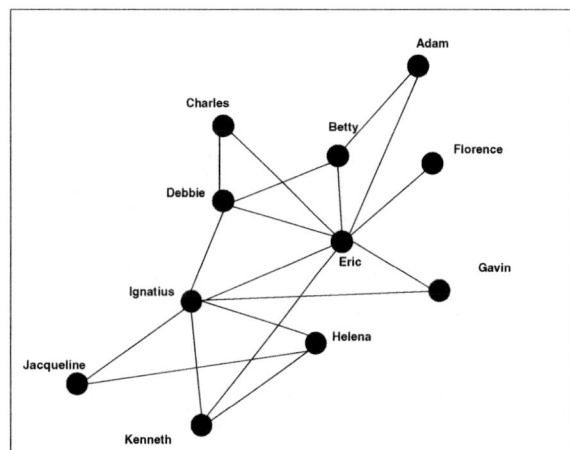

The Role of Technology in Supporting Creative Abrasion

Creative abrasion involves brainstorming, making improvements to initial ideas, and solving problems from multiple perspectives. Technology supports creative abrasion through exploitation and exploration of knowledge. Exploitation refers to the refinement of existing knowledge into new knowledge to achieve improvement in efficiency and effectiveness. Exploration refers to the creation of knowledge through discovery and experimentation (Manor & Schulz, 2001). Technology solutions that support exploitation possess idea generation capabilities. Idea generation stimulates thinking and association, which enables users to detect hidden patterns from mounds of data and discover relationships among entities. For example, data mining solutions, such as PolyAnalyst from Megaputer Intelligence and Clementine from SPSS, are designed to identify patterns from a sea of data.

Technology tools that support exploration possess simulation capabilities. Simulation goes beyond number crunching and allows qualitative and quantitative information to be modeled. By reflecting the subtleties of various systems, simulation solutions are able to identify potential scenarios and effectively communicate complex ideas through graphical representations, animations, and flow charts. Examples of simulation solutions include Powersim Studio Enterprise from Powersim, MindManager from Mindjet, and Witness from Lanner.

The Role of Technology in Supporting Artifact Sharing

Technology supports artifact sharing through content management and concept mapping. Content management establishes a structure to create and maintain different types of content in text, image, and video formats. It allows the content to be categorized and indexed to ease future searches. For example, one categorization approach is to tag content with attributes such as activities, themes, forms, types, and products (Heijst, Spek & Kruizinga, 1998). Concept mapping links several related concepts within a given theme or context. In so doing, it provides interdisciplinary perspectives and facili-

tates cross-referencing. For instance, a community member is able to locate relevant technical specifications when certain business information is retrieved. An example of technology designed for content management and concept mapping is SemioTagger from Entrevia.

Emerging Technologies and the Rising Needs of Community Members

Increasingly, community members' needs have broadened from the requirement to be connected electronically to the demand for multidirectional interaction and iterative negotiation processes. Thus, the trend in emerging technologies is toward creating social presence and supporting unstructured, rich content. Social presence is the degree of salience of the other person in the interaction and the consequent salience of the interpersonal relationships (Gunawardena & Zittle, 1996). For example, Babble, a tool currently under research, is designed specifically to make the presence and activities of others visible through an interface known as a social proxy (Thomas, Kellog & Erickson, 2001).

Emerging technologies with multimedia capabilities allow unstructured, rich content such as voice, images, and video to be transmitted, stored, and retrieved. While non-text documents are more difficult to be searched and browsed than text documents, improvements have been made to facilitate browsing through video documents. A technique known as summarization automatically produces a gallery of searchable still images from video documents (Lienhart et al., 1997). Work is also underway to improve the accuracy of automatic speech recognition that seeks to support speaker-independent recognition with unconstrained vocabulary. The aim is to produce text transcription from audio sources for indexing and searching purposes (Marwick, 2001).

CONCLUSION AND FUTURE TRENDS

In discussing the role of technology in support communities of practice, this article adopts a sociotechnical orientation. It argues that the effective use of technology is possible only if the nature of communities of practice and the capabilities of technology are jointly considered. For this reason, a conceptual model of communities of practice has been developed. Thereafter, the suite of capabilities found in extant technology tools that could specifically support the communities of practice is discussed.

For managers who wish to start communities of practice, this article serves as the warning against leaning toward a strictly technology-centric approach. Communities of practice cannot and should not be driven by technology alone. Instead, the role of technology is to support community dynamics rather than to create structural elements.

For knowledge management solutions vendors, this article calls for a greater inquiry into the nature of communities of practice. Instead of designing tools packed merely with more functions and features, the focus ought to first be on the understanding of the naturally-occurring processes and dynamics among community members. Tools can then be developed to aptly support these processes and dynamics. After all, technology is created for people, not people for technology.

REFERENCES

Anand, V., Manz, C. C., & Glick, W. H. (1998). An organizational memory approach to information management. *The Academy of Management Review, 23*(4), 796-809.

Brown, J. S. & Duguid, P. (1991). Organizational learning and communities of practice: Toward a unified view of working, learning and innovation. *Organization Science, 2*(1), 40-57.

Cohen, D. (1998). Toward a knowledge context: Report on the first annual U.C. Berkeley forum on knowledge and the firm. *California Management Review, 40*(3), 22-39.

Cook, S. D. N. & Yanow, D. (1993). Culture and organizational learning. *Journal of Management Inquiry, 2*, 373-390.

Davenport, T. H., & Prusak, L. (1999). *Working knowledge.* Boston: Harvard Business School Press.

Edmundson, H. (2001). Technical communities of practice at Schlumberger. *KM Review, 4*(2), 20-23.

Fahey, L., & Prusak, L. (1998). The eleven deadliest sins of knowledge management. *California Management Review, 40*(3), 265-276.

Foulger, D. A. (1991). *Medium as process: The structure, use and practice of computer conferencing on IBM's IBMPC Computer Conferencing facility.* Unpublished PhD thesis, Temple University, Department of Communications, Philadelphia,.

Gunawardena, C. N., & Zittle, R. (1996). An examination of teaching and learning processes in distance education and implications for designing instruction. In M. F. Beaudoin (Ed.), *Distance education symposium 3: Instruction, ACSDE Research Monograph, No. 12* (pp. 51-63). University Park, PA: Pennsylvania State University.

Heijst, J., Spek, V., & Kruizinga, L. (1998). The lessons learned cycle. In U. Borghoff & R. Pareschi (Eds.), *Information technology for knowledge management* (pp. 17-34). Berlin: Springer-Verlag.

Hildreth, P., Kimble, C., & Wright, P. (2000). communities of practice in the distributed international environment. *Journal of Knowledge Management, 4*(1), 27-37.

Hirschhorn, L. (1997). *Reworking authority: Leading and following in the post-modern organization.* Cambridge, MA: MIT Press.

Lave, J. & Wenger, E. (1991). (1991). *Situated learning: Legitimate peripheral participation.* Cambridge, UK: Cambridge University Press.

Leonard-Barton, D. (1995). *Wellsprings of knowledge: Building and sustaining the sources of innovation.* Boston: Harvard Business School Press.

Lesser, E. L. & Storck. (2001). communities of practice and organizational performance. *IBM Systems Journal, 40*(4), 931-841.

Lienhart, R., Silvia, P., & Wolfgang, E. (1997). Video abstracting. *Communications of the ACM, 40*(12), 54-62.

Mann, M. M., Rudman, R. L., Jenckes, T. A., & McNurlin, B. C. (1997). EPRINET: Leveraging

R

knowledge in the electric utility industry. L. Prusak (Ed.), *Knowledge in organizations* (pp. 73-97). Woburn, MA: Butterworth-Heinemann.

Manor, B., & Schulz, M. (2001, August). The uncertain relevance of newness: Organizational learning and knowledge flows. *Academy of Management Journal, 44*(4), 661-681.

Marwick, A. D. (2001). Knowledge management technology. *IBM Systems Journal, 40*(4), 814-830.

McKinlay, A. (2002). The limits of knowledge management. *New Technology, Work and Employment, 17*(2), 76-88.

Newell, S. (2001). From global knowledge management to internal electronic fences: Contradictory outcomes of intranet development. *British Journal of Management, 12*(2), 92-106.

Nonaka, I., & Takeuchi, H. (1995). *The knowledge creating company.* New York: Oxford University Press.

Orr, J. D. (1996). *Talking about machines: An ethnography of a modern job.* Ithaca, NY: Cornell University Press.

Ruggles, R. (1998). The state of the notion: Knowledge management in practice. *California Management Review, 40*(3), 80-89.

Szulanski, G. (1996). Exploring internal stickiness: Impediments to the transfer of best practice within the firm. *Strategic Management Journal, 17*, 27-43.

Thomas, J. C., Kellogg W. A., & Erickson, T. (2001). The knowledge management puzzle: Human and social factors in knowledge management. IBM Systems Journal, 40(4). Retrieved June 10, 2005, from *http://www.research.ibm.com/journal/sj/404/thomas.html*

Wenger, e. c., McDermott, R., & Snyder, W. M. (2002). Cultivating communities of practice. Boston: Harvard University Press.

Zárraga, C., & García-Falcón, J. M. (2003). Factors favoring knowledge management in work teams. *Journal of Knowledge Management, 7*(2), 81-96.

KEY TERMS

Artifact Sharing: A natural part of everyday work among community members where documents, charts, and images are shared for the purposes of planning, reflection, discussion, and solving problems.

Collaboration Tools: Technologies designed to allow users to share knowledge with one another. The key features include shared spaces, calendaring, workflow management services, synchronous and asynchronous communication.

Creative Abrasion: The meeting of minds on common ground to explore and negotiate different opinions and, as a result, generate new ideas.

Idea Generation Tools: Technologies designed to stimulate thinking and association. They enable users to detect hidden patterns from mounds of data and discover relationships among entities.

Legitimate Peripheral Participation: A complex and composite process through which new members become matured members by acquiring knowledge from the group.

Simulation Tools: Technologies designed to support the modeling of qualitative and quantitative information. The key feature includes the capability to identify potential scenarios and effectively communicate complex ideas through graphical representations, animations, and flow charts.

Social Network Analysis Tools: Software tools that are capable of uncovering the pattern of interaction among a group of users.

The Role of Trust in Virtual and Co-Located Communities of Practice

Jawwad Z. Raja
Royal Holloway University of London, UK

Anzela Huq
Royal Holloway University of London, UK

Duska Rosenberg
Royal Holloway University of London, UK

INTRODUCTION

Information and communication technologies (ICTs) have become popular in enabling organisations to work virtually, allowing them to organise and leverage their human assets in new ways. Numerous advantages are offered to organisations in the virtual world, including the ability to bridge time and space, and utilisation of distributed human resources without physical relocation of employees (Lipnack & Stamps, 2000). However, flexibility for organisations also comes with many challenges due to its own inherent characteristics. With the separation in time and space, possibly no history of working together, and communication options that are limited, working virtually can lead to undesired outcomes. There are many fundamental factors that not only drive knowledge sharing and transfer in virtual communities, but are believed to be important in their success and failure. One of these fundamental factors is trust. The literature on trust in co-located environments suggests that the establishment of trust is of importance in the working relationship (e.g., Mayer, Davis & Schoorman, 1995). Furthermore, it is argued that trust also leads to more open communication, cooperation, and a higher quality of decision making and risk taking (Lane & Bachmann, 2000). Lipnack and Stamps (2000) argue that the success of sharing and transferring knowledge virtually begins with trust, since trust functions as a mechanism to hold individuals together.

The aim of this article is to discuss types of trust and explain mechanisms of trust development in light of research on organisational dynamics. Although there is little standardisation in the trust literature, this article will attempt to critically assess contributions to the debate, illustrating points through references to communities of practice.

BACKGROUND

Communities of Practice and Trust

Wenger (1998) first coined the term "communities of practice," which refers to a theory that builds on learning as social participation. Within these communities, individuals actively participate in the practices of social communities and constructing identities in relation to these communities. These communities of practice do not have to be co-located or meet face-to-face at regular intervals, as through the use of technology they may just purely exist virtually. As long as the community has a common set of interests and is engaged in a work practice, it can be defined as a community. Furthermore, if the members of the community are learning and engaged in the practice, be it virtually or co-located, it can be defined as a community of practice. Thus, it is possible for communities of practice to function in distributed environments through the use of technology, though co-located contact may be essential for the further development of the community (Gamble & Blackwell, 2001).

Many organisations are beginning to realise that the knowledge their employees possess is their most valuable asset, but few firms have actually begun to actively manage their knowledge assets on a broad scale. Communities of practice offer organisations the opportunity, through the use of ICTs, to share knowledge in virtual environments. ICTs facilitate more than just information exchange between community members. Shared *cultural objects* can be created by communities through continuous communication in virtual environments, which can help bring about a sense of togetherness and trust among those who have never even met each other in the past—enabling a shared sense of belonging within communities (Brown & Duguid, 1991; Newell, Robertson, Scarbrough & Swan, 2002). It has been argued that initial trust has also been based on the competencies of community members and the fact that they are engaged in the same practice within the community (Boisot, 1995; Wenger, 2000).

Trust has been defined as "the expectation that a partner will not engage in opportunistic behaviour, even in the face of countervailing short-term opportunities and incentives" (Nooteboom & Six, 2003, p. 4). This definition emphasises that trust is one's own belief that the partner, with whom the relationship is taking place, will not fail to meet his/her expectations. Trust has been shown to be beneficial, as it enables cooperative behaviour (Gambetta, 1988), promotes adaptive organisational forms such as network relations (Miles & Snow, 1992), reduces harmful conflict, decreases transaction costs, facilitates rapid formulation of ad hoc workgroups (Meyerson, Weick & Kramer, 1996), and promotes the effective sharing of tacit knowledge. Furthermore, high levels of trust are also associated with a higher degree of economic performance. It is believed that trust provides organisations with greater flexibility, enables information exchange, increases product and technology quality, and expands productivity and profitability (Nooteboom, 2002). It also allows for the development of long-term relationships. Similarly, Bradach and Eccles (1988) emphasise that trust is "viewed as a precondition for superior performance and competitive success in [the] new business environment" (p. 1).

Trust builds up one's social capital, which includes "the features of social organisation such as networks, norms, and social trust that can facilitate coordination and cooperation for mutual benefit" (Putnam, 1995, p. 160). Fukuyama (1995) argues that social capital can only develop when individuals interact together in an organisation on the basis of trust, and the lack of such forms of trust are likely to lead people to maintain their relationships through other mechanisms, for instance through legal rules and regulations.

MAIN FOCUS: COGNITIVE AND AFFECTIVE ELEMENTS WITHIN TRUST

The rise of globalisation and the increased technological developments that have taken place in recent years have led to the emergence of virtual environments in organisations. Virtual environments enable people to communicate with each other across geographical boundaries through various technologies, such as, e-mail, videoconferencing, intranets, chat rooms, voicemail messages, and so on. In such environments, trust is a desirable prerequisite in order to build relationships between community members. Additionally, participating within a community of practice can be seen as a long- or short-term objective; it is more likely that trust will need to be developed more promptly where there are short-term objectives. In order to achieve this, members of communities need to be committed to their tasks, focus on regular communication, and display a sense of loyalty and understanding towards all community members.

In the context of communities of practice, trust should be represented as a multidimensional construct, with both cognitive (conditional) and affective (unconditional) elements (Lewis & Weigert, 1985; Jones & George, 1998). Research studies have discovered that trust develops from cognitive processes, referring to the calculative and rational characteristics demonstrated by trustees in newly formed relationships. These characteristics include integrity, perceived competence, responsibility, and professionalism (commonly characterised by affiliations with professional bodies such as the General Medical Council or the Law Society) (Mayer et al., 1995). The argument appears to be that cognitive trust fills the knowledge gap when beginning new relationships, which is vital for facilitating interaction where fellow trustees' values are unknown (Kanawattanachai & Yoo, 2002). Cognitive trust does not however explain the observations of Jarvenpaa and Leidner (1999), who demonstrated that teams show high levels of trust without having adequate information upon which to base judgements.

Thus, initial trust must be defined separately as trust that develops swiftly based on an individual's propensity to trust others (Meyerson et al., 1996). Broadly, swift trust is regarded as one of the most important trust types, without which individuals would not be able to enter productive new relationships. In studies of virtual teams in Sun Microsystems and Motorola, Lipnack and Stamps (1997) showed that this trust type could be an important predictor of future team performance. Indeed, teams identified with low levels of swift trust were rarely able to develop trust later or perform effectively. It does however limit the controllability of trust formation during development. As swift trust appears to relate to individual experiences, the ability to sustain trust levels relates more closely to personal factors, perhaps also explaining why swift trust has been identified as highly fragile. Supported by early research from Rempel, Holmes and Zanna (1985) which showed that initial trust can fall quickly over time, it is possible to conclude that swift and cognitive factors are not effective long-term methods of promoting cohesion within communities. While this could imply that cognitive trust is unimportant to the long-term success of communities of practice, it should be regarded as an important prerequisite for building long-term social relationships.

As communities of practice develop, it is important for them to build the strong relationships that are characteristic of affect-based trust. In the community of practice context, this trust is said to develop slowly, being linked to interpersonal relationships and far stronger motivations than are associated with cognitive measures. It is important to recognise the funda-

mental differences in personal and work relationships; however, it is possible to generalise that affect-based trust is less 'conditional' and goes far beyond cognitive or conditional factors. Although empirically affective-based trust is absent from most new relationships (Crisp & Jarvenpaa, 2000), affective trust is generally regarded as having greater longevity (Jones & George, 1998). Thus as this form of trust is a much stronger foundation for long-term effectiveness, it should be considered the long-term goal for development in communities of practice.

Trust can develop in different ways in a virtual context; that is to say, it falls into a three-fold typology. Firstly, trust may be developed when members of a community meet physically in a co-located environment and carry the relationship through to a virtual environment. The second way is where members interact virtually, due to physical boundaries, where trust develops over time. However, after a period of time, they may meet and focus on the practice in a co-located environment, which will further the relationship and help build affective-based trust. Thirdly, the community may interact entirely virtually, where members never meet physically in the same environment, though trust can still develop here, though this is unlikely to be unconditional trust.

FUTURE IMPLICATIONS

Virtual communities are more dependent on cognitive-based trust to both form and maintain swift trust. This

Table 1. Typology of trust development (Raja, Huq, & Rosenberg, 2004)

	Cognitive or/and Affective Elements	**Mechanisms Through Which Trust Is Developed**
Co-Located → Virtual	- Swift trust - Cognitive (conditional trust) - Affective (unconditional trust)	Background knowledge acquired through personal interaction at the time of meeting, past experiences, and reputation.
Virtual → Co-Located	- Swift trust - Cognitive (conditional trust) - Affective (unconditional trust)	Regular virtual interaction (i.e., e-mail based) and some background through daily knowledge acquired over time.
Virtual Only	- Swift trust - Cognitive (conditional trust)	Communication tools; norms and rules established through common purpose and reputation.

is because cognitive cues are easier to be read and interpreted in virtual environments. Members of non-virtual communities can base a large proportion of their judgements of other members on their physical contact and cues such as body language, whereas "it is more difficult to develop social relationships through computer-mediated communications due to the depersonalisation effect" (Kiezler et al., 1984). Given enough time, non-virtual communities can develop social relationships to help form affective-based trust, and it is in the interest of a non-virtual, as well as a virtual (if possible) community to build affective-based trust because it is associated with higher performance and well-being. A lack of working together, as well as no co-located communication, can create a sense of both physical and psychological distance among individuals (Huber, 2001). However, Iacono and Weisband (1997) "found that both the affective and cognitive elements of trust were prominent in the high-trust high performance virtual teams" (p. 34). This is because members of the community that bond on a social level, not just on a professional level, will be more likely to have more open communication, be able to compromise and cooperate, and members will be less likely to be susceptible to situations that may appeal to opportunism.

However, at the other end of the continuum, affective relationships can be too strong, and too much trust may be conferred, leaving the individual susceptible to the opportunism of other members. Given this vulnerability, perhaps the lack of affective trust and bonding within the virtual community is not altogether problematic. Personal relationships are not able to cloud judgements, and bad decisions are not made in order to protect friendships. By trusting only the basis of the quality of one's work and the aptitude of one's skills, members of the virtual community are better able to prevent personal feelings, positive or negative, from interfering with the promotion of community objectives.

This is not to say that problems do not exist within virtual environments. As Jarvenpaa and Leidner (1999) explain, "Dysfunctions such as low individual commitment, role overload, role ambiguity, absenteeism, and social loafing may be exaggerated in a virtual context" (p. 791). Nor can one overlook the value affect-based trust can have in organisations. As

Jones and George (1998) explain, it is this type of trust that inspires employees to perform behaviours that are strenuous, time consuming, or self-sacrificing.

CONCLUSION

People within communities of practice have shared social identities and interests which make the sharing of knowledge a central and significant activity. There exists a high level of trust between members in communities of practice that is caused by the sharing of experience of practice, which acts as the enabling mechanism between community members. This high level of trust helps overcome the barriers one may encounter when sharing knowledge in other organisational contexts. It has been argued that trust is easily developed in co-located environments as individuals interact directly and support their relationships through previous experiences, daily relations, and direct communication. Additionally, relationships can also develop in a virtual context where community members are dispersed geographically and whereby building trust can be challenging for community members, as they have to interact with each other under different conditions. Nevertheless, it is not impossible to sustain trust, as there are mechanisms, such as communication channels and norms, which support the development of relationships over time. Trust can be developed and maintained in both environments, but this will also depend on the willingness of members to work and interact with other members of the community. In addition, one cannot guarantee that trust will always develop, and that the various cognitive and affective elements will be present in the various circumstances, as social and cultural factors can also play a determining role.

In virtual environments, the task is to develop high levels of trust among people who typically do not have any sort of personal ties. Jarvenpaa and Leidner (1999) identify communication, strong initiative, and strong leadership as the essential qualities for trust in virtual environments. As shown in the results of the Jarvenpaa and Leider (1999) study, high trust levels can certainly be present in virtual environments, and such a state will lead to effective group work, cooperation, and innovation.

REFERENCES

Bazerman, M. (1998). *Judgement in managerial decision making.* New York: John Wiley & Sons.

Boisot, M. (1995). *Information space: A framework for learning in organizations, institutions and culture.* London: Blackwell.

Bradach, J.L., & Eccles, R.G. (1988). Market versus hierarchy: From ideal type to plural form. In W.R. Scott (Ed.), *Annual review of sociology* (vol. 15, pp. 97-118).

Brown, J.S., & Duguid, P. (1991). Organizational learning and communities of practice: Toward a unified view of working, learning and innovation. *Organization Science, 2,* 40-57.

Crisp, C.B., & Jarvenpaa, S.L. (2000). Trust over time in global virtual teams. *Proceedings of the Academy of Management Meeting,* Toronto.

Fukuyama, F. (1995). *Trust, the social virtues and the creation of prosperity.* New York: The Free Press.

Gabarro, J.J. (1978). The development of trust, influence, and expectations. In A.G. Athos & J.J. Gabarro (Eds.), *Interpersonal behavior: Communication and understanding in relationships.* Englewood Cliffs, NJ: Prentice-Hall.

Gambetta, D. (1988). Can we trust trust? In D. Gambetta (Ed.), *Trust: Making and breaking co-operative relations: 213237.* New York: Basil Blackwell.

Gamble, P.R., & Blackwell, J. (2001). *Knowledge management: A state-of-the-art guide.* London: Kogan Page.

Hardy, C., Phillips, N., & Lawrence, T. (1998). Distinguishing trust and power in interorganisational relations: Forms and facades of trust. In C. Lane & R. Bachmann (Eds.), *Trust within and between organizations.* Oxford: Oxford University Press.

Iacono, C.S., & Weisband, S. (1997). Developing trust in virtual teams. *Proceedings of the 13th Hawaii International Conference on System Sciences,* Hawaii.

Jarvenpaa, S.L., & Leidner, D.E. (1999). Communication and trust in global virtual teams. *Organization Science, 10*(6), 791-815.

Jones, G., & George, J. (1998). The experience and evolution of trust: Implications for cooperation and teamwork. *Academy of Management Review, 213,* 531-546.

Kanawattanachai, P., & Yoo, Y. (2002). Dynamic nature of trust in virtual teams. *The Journal of Strategic Information Systems, 11*(3-4), 187-213.

Kiesler, S., Siegel, J., & McGuire, T.W. (1984). Social psychology aspects of computer-mediated communication. *American Psychologist, 39*(10), 1123-1134.

Lane, C., & Bachmann, R. (1998). *Trust within and between organizations: Conceptual issues and empirical applications.* Oxford: Oxford University Press.

Lewis, J.D., & Weigert, A. (1985). Trust as a social reality. *Social Forces, 63*(4), 967-985.

Lipnack, J., & Stamps, J. (1997). *Virtual teams: Reaching across space, time and organizations with technology.* New York: John Wiley & Sons.

Lipnack, J., & Stamps, J. (2000). *Virtual teams: People working across boundaries with technology* (2nd ed.). New York: John Wiley & Sons.

Mayer, R., Davis, J., & Schoorman, F. (1995). An integrative model of organizational trust. *The Academy of Management Review, 20*(3), 709-719.

Meyerson, D., Weick, K.E., & Kramer, R.M. (1996). Swift trust and temporary groups. In R.M. Kramer & T.R. Tyler (Eds.), *Trust in organizations: Frontiers of theory and research.* New York: Sage Publications.

Miles, R.E., & Snow, C.C. (1992). Causes of failure in networked organisations. *California Management Review, 34*(4), 53-72.

Newell, S., Robertson, M., Scarbrough, H., & Swan, J. (2002). *Managing knowledge work.* Basingstoke: Palgrave.

Nooteboom, B. (2002). *Trust: Forms, foundations, functions, failures and figures.* Cheltenham: Edward Elgae.

Nooteboom, B., & Six, F. (2003). *The trust process within organizations: Empirical studies of the determinants and the process of trust development*. Cheltenham: Edward Elgar.

Putnam, R.D. (1995). Bowling alone: America's declining social capital. *Journal of Democracy, 6*, 65-78.

Rempel, J.K., Holmes, J.G., & Zanna, M.P. (1985). Trust in close relationships. *Journal of Personality and Social Psychology, 49*, 95-112.

Wenger, E. (1998). *Communities of practice: Learning, meaning, and identity*. Cambridge: Cambridge University Press.

Wenger, E. (2000). Communities of practice and social learning systems. *Organization, 7*(2), 225-246.

KEY TERMS

Co-Location: When members of a community work in the same physical space and time.

Computer-Mediated Communication: The facilitation and enabling of organisational discourse and interaction via computing technology, typically information and communication technologies (ICTs); particularly pertinent to virtual organisational work.

Knowledge: Thoughts and processes embedded in the minds, behaviour, and understanding of people, and which is used to enhance understanding of problems and issues at work; viewed as an essential asset to organisations and whose efficient spread leads to greater competitive advantage.

Social Capital: Resides in the structure of relationships between members of a community, and is managed equally by all members; property of the community.

Trust: A mechanism that enables cooperation and participation between members of a community in relation to the engagement of the practice.

Virtual Organisation: The translation of the traditional organisation to higher realms of work in terms of distance, time, space, and interaction; typically used to refer to organisations that engage in the use of ICTs to bridge gaps in these four dimensions.

Shaping Social Structure in Virtual Communities of Practice

Iwan von Wartburg
University of Hamburg, Germany

Thorsten Teichert
University of Hamburg, Germany

Katja Rost
University of Zurich, Switzerland

INTRODUCTION

Practice, that is, the execution of work relevant tasks, can take two forms: actual and espoused practice (Brown & Duguid, 1991). Espoused practice is formally and deliberately planned: formal organizational structuring, product manuals, error detection, and correction procedures represent just a few examples. Actual practice represents the solutions to problems and the execution of tasks as they really happened in a given context. Processes of knowledge generation and transfer are different for espoused or actual practice (Orr, 1996). While traditional modes of organizing work practice focus on espoused practice, newer organizational forms focus on actual practice: Communities of practice are groups of people bound together by shared expertise and passion for a joint enterprise on behalf of an organization (Wenger, 1998). To support effective work practices in an ever more distributed work environments, collocated CoPs are complemented by virtual communities of practice (VCoPs). Its members interact supported by collaborative technologies in order to bridge time and/or geographical distances. Toolkits of computer-mediated environments facilitate community building in addition to personal interaction (Hinds & Kiesler, 2002; Walther, 1995; Wellman et al., 1996).

There is a shared understanding that VCoPs are an especially effective organizational form for knowledge creation both within companies (Kogut & Metiu, 2001; Nahapiet & Ghoshal, 1998; von Krogh, Spaeth & Lakhani, 2003) and between companies (Constant, 1987; Vincenti, 1990). Therefore, VCoPs

are managerially desirable forms of virtual communities (Rheingold, 1993; Smith & Kollock, 1999; Wellman et al., 1996) in which learning in practice takes place; that is, professionals stick together because of exposure to common problems in the execution of real work. The "glue" which binds them together is a powerful mixture of shared expertise and experience, as well as the need to know what each other knows. Given that VCoPs offer such potential to enhance intellectual capital and to enrich social processes within companies, we look more closely at the social and knowledge generation processes within VCoPs from a *managerial point of view*. Viewed from this angle, VCoPs represent a difficult challenge for managers who want to profit from using them as an arena for desirable learning in practice. Although VCoPs are believed to be a desirable organizational form for knowledge generation, they are preferably modeled as a rather emergent phenomenon and believed to be only marginally manageable. Thus, on one hand, managers are urged to believe that VCoPs are something beneficial while, at the same time, they are told that VCoPs cannot be managed deliberately.

BACKGROUND

Studies of CoPs bring together studies from an ethnography of work (Orr, 1996) with theories of situated cognition (Lave, 1988, 1991; Lave & Wenger, 1991; Suchman, 1987). *In situ* learning context variables are becoming central research questions. A large part of the daily generation, application, and internalization of knowledge is achieved during learning in practice. If one wants to understand social learning processes, one has to analyze the contextual embeddedness of actors (Resnick, 1991). Learning in practice is delineated by the web of relationships between actors and takes place in a social and culturally constructed environment, the community. During processes of situated learning in VcoPs, knowledge is generated that cannot easily be articulated or captured. This sort of knowledge has been labeled *sticky* (Szulanski, 2003; von Hippel, 1994), *tacit* (Nonaka, 1994; Polanyi, 1967), and *declarative* (Cohen & Bacdayan, 1994). Individual and collective experiences as well as internalized work knowledge fall into this domain. VCoPs are arenas within which such social learning by doing is taking place (Lave, 1991; Levitt & March, 1988). Learning in practice not only enriches individual knowledge but also strengthens the identities and roles of actors within the learning community: Newcomers learn from old-timers by the legitimization to participate in certain activities as part of the practice in the community. New members first participate as peripheral community members. By continual learning and social identity as well as role building, they become core members This process has been termed *legitimate peripheral participation* (Lave & Wenger, 1991).

Based on the discussion of the learning processes in CoPs, we accept the basic proposition which legitimates the importance of CoPs from a managerial point of view: *VCoPs enhance the innovativeness and the productivity of individual actors and collectives beyond the degree of formal organizational structures.* Yet, albeit the acceptance of the basic propositions concerning relevancy of VcoPs, the question of manageability still remains.

KNOWLEDGE GENERATION AND DIFFUSION WITHIN VCoPs

In the treatment of VCoPs identity building, voluntarism, regularity, and experience through actual work practice are given central attention. This puts into perspective some prevailing and traditional instruments for learning and knowledge creation like seminaries, intranet manuals, e-learning modules, and workshops. These instruments, implemented as complements to formal job descriptions, are thought to be deliberately manageable. Because the environment of knowledge generation processes is assumed to be observable, foreseeable, and/or controllable, they can be planned and successfully implemented according to ex-ante considerations (Mintzberg & Waters, 1985). In contrast hereto, the central concern of knowledge creation processes within CoPs is to install learning as an integral part of everyday practice. Emphasis is placed on the rather informal nature of knowledge dynamics in CoPs (Orr, 1996). In this view, productive knowledge generation and exchange in VCoPs is a rather emergent phenomenon because the environment of knowledge generation processes is assumed to be unobservable, uncontrollable, and/or unforeseeable.

The view of knowledge generation in VCoPs as an emergent phenomenon should cause a natural resistance of managers to invest in VCoPs because of the uncontrollable and immeasurable character of possible outcomes. Given such a managerial reluctance against VCoPs, one may begin to question whether VCoPs will remain the praised form for knowledge creation in actual practice. It seems that common sense needs to be established that VCoPs are at least partially manageable. Most helpful in this respect are transformation models indicating whether there exist manageable rules and environments for CoPs which influence the possible outcomes in a desired direction depending on situational characteristics.

A conceptual framework reflecting this claim is presented in Figure 1. It delineates a causal chain between a construct called collective social structure of VCoPs (macro-level) as the starting point

Figure 1. Building blocks of the proposed transformational framework of VCoPs

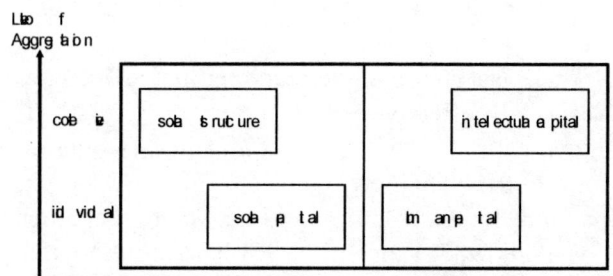

and the collective intellectual capital (macro-level) as the final dependent variable. Referring mainly to Nahapiet and Ghoshal (1998) as well as Adler and Kwon (2002), the causal relationship is conceptualized by introducing two intermediary variables on the individual actors' level, that is, social capital and human capital. Our conceptual model assumes a causal macro-micro-micro-macro sequence: the collective social structure of VCoPs delineates the social capital available for individual actors inside VCoPs. The latter enables the emergence of enhanced individual human capital by allowing members of VCoPs to make use of the social capital in social interactions. Finally, actual knowledge generating practices in VCoPs lead to the development of collective intellectual capital beyond and above simple aggregation of individual learning curves.

Details of the transformation framework are discussed elsewhere (Teichert, Rost & von Wartburg, 2003). Following, we will focus on *social structure* as the building block of the model which is assumed to be most adept for deliberate action in order to influence knowledge generation.

GAUGING SOCIAL STRUCTURE IN VCoPs

A specific kind of social structure needs to be kept alive in VCoPs in order for situated learning in actual practice to occur. It has been exemplified within the realm of open source software development communities (Kogut & Metiu, 2001; von Krogh et al., 2003). We develop our argument about social structure in VCoPs by *first* profiling five coordination types, that is, markets, social networks, hierarchies, expert cultures, and communities. Each coordination type represents a specific configuration of several coordination dimensions. *Second*, we discuss the proposed configuration of the coordination dimensions that make up the social structure in VCoPs.

As for the discussion of coordination types, we first refer to the economic distinction between markets, hierarchies, and networks (Adler & Kwon, 2002). In *markets*, resources, or goods and services, are exchanged spontaneously at arms length. *Hierarchy* rests on explicit and formal arrangements of power. Relationships between actors are governed by formal rules and contracts. Contracts remain

incomplete to a certain degree: not all relevant details can be anticipated and crafted in a contract *a priori*. *Networks*, referred to as *social networks*, consist of cooperative relations between actors. They represent a hybrid coordination type that lies between markets and hierarchies but still possesses an idiosyncratic quality (Powell, 1990). Coordination emerges spontaneously under the precondition that members believe there will be an opportunity to profit from other VCoP members in the future; that is, it is facilitated by mutual trust and a generalized norm of reciprocity (Gould, 1979).

The three coordination types discussed so far are incomplete when it comes to profiling the social structure in VCoPs. In particular, they miss specific aspects of specialized knowledge sharing and the specific problem-solving context rooted in actual practice. By relying on anthropological writings, two additional coordination types called communities and expert cultures (Weissbach, 2000). *Communities* are small groups of actors that get together on a temporal finite basis. They provide for normative and ideological security in cliques and friendship circles; they attract actors by what could be called "normative pull". In *expert cultures*, members are viewed as homogenous concerning the core knowledge and experience possessed by members. Exchange in expert cultures is guided by fixed rules derived from justified shared knowledge and expertise. Developing this deep expertise can be described as a long-term process of cognitive apprenticeship (Rogoff, 1990).

Based on previous research (Adler & Kwon, 2002; Weissbach, 2000; Wellman et al., 1996; Wenger, 1998), we propose a specific configuration of social structure in VCoPs for situated learning in practice to occur. It consists of specific parameter values of the coordination dimensions from basically three out of the five coordination types as presented in Table 1: expert culture, social networks, and communities. Specifically, we postulate the following characteristics of social structure in VCoPs:

- It refers to *expert culture* because the coordination in VCoPs is primarily achieved by common knowledge learned during prior problem-solving activities. The time horizon of exchange is long-term. Goods and services

S

are exchanged along specific but hard to codify rules guided by expertise. Access to the collective is generally open. However, rules exist which govern the qualification requirements of core members.

- It refers to *social networks* as there is shared believe in reciprocity, that one will also be able to reap a benefit from membership in the future. Thus, terms of exchange are rarely made explicit. Possible conflicts are solved by (re)negotiation. Arguments rooted in expertise will be most viable during these negotiations.
- It refers to *communities* because what is exchanged is viewed as valuable goods and services by members of the VCoPs. Members are therefore willing to trade obedience against such valuable knowledge for future work success.

As for possible influence tactics, managers seeking to implement VCoPs for difficult and knowledge intense work processes may gain some advice from Table 1. They might find that establishing and enforcing some basic and ballparking rules might enable situated learning in actual practice more than devising precise decision cut-off criteria, standardized conflict resolution processes, predetermined internal transfer prices, explicit group missions, and the like.

For example, since formal authority only coincidentally collapses with the kind of authority deployed for conflict resolution, they may explicitly and repeatedly claim that conflicts have to be negotiated on the basis of rational arguments rooted in deep expertise. Another example is ensuring the long-term character of VCoPs. Established on a long-term basis, VCoPs can provide knowledge guidance for members even when more formal organizational structures change. However, without the explicit concern of executives that VCoPs are long-term initiatives, norms of general reciprocity will hardly develop, and

Table 1. Coordination types, coordination dimensions: The proposed configuration of social structure in VCoPs

Coordination type	Community	Expert culture	Market	Social network	Hierarchy
What is exchanged?	Goods and services	Obedience for knowledge security	Goods and services	Favors and gifts	Obedience for material security
How is coordination achieved?	By norms, values, and ideology	By knowledge generation and transfer	By prices (money or barter)	By trust and generalized reciprocity	Formal rules
Are terms of exchange specific or diffuse?	Diffuse and spontaneous though grounded in norms and values	Specific along rules guided by expertise	Specific along the rules of relative prices	Diffuse (a favor will be returned at some point of time in the future)	Diffuse (not all issues can be specified ex ante)
Are terms of exchange made explicit?	Implicit (social inclusion by normative pull)	Implicit (social exclusion by means of exclusive expertise)	Explicit (social inclusion)	Implicit (limited social inclusion and exclusion)	Explicit (social exclusion by explicit employment contracts)
Is the exchange symmetrical?	Asymmetric (normative guidance)	Asymmetric (apprenticeship)	Symmetric (relative prices)	Symmetric (reciprocity)	Asymmetric (formal domination)
Actors are	Independent	Dependent	Independent	Interdependent	Dependent
How closed is the system?	Generally open	Open but rules for membership	Generally open	Limited and exclusive	Rules for formal membership
Time horizon?	Short-term	Long-term	Short-term	Mid-term	Long-term
How are conflicts solved?	Normative power	Expertise	Contracts and law	Negotiation	Formal authority

the willingness to contribute to the problem-solving strategies in VCoPs may be diminished.

Thus, by placing emphasis on influencing the social structure in VCoPs, the focus lies on a partially manageable aspect of VCoPs. The effects of influencing the social structure in VCoPs can have far reaching consequences which can be modeled as direct and indirect effects along a transformational framework like the one depicted in Figure 1.

CONCLUSION AND FUTURE TRENDS

We hypothesized that the desired learning processes in VCoPs can be described by a causal chain portrayed by the conceptual transformation framework summarized in Figure 1. Thus, a specific fabric of social structure will lead to the emergence of valuable individual social capital that in turn enhances the human capital available to actors within VCoPs. The interaction of these highly skilled and socially embedded individuals during actual practice will lead to an enrichment of intellectual capital at the collective level and therefore to an enhancement of collective intellectual capital.

We proposed social structure to be the most directly influenceable building block within the framework. VCoPs have been predominantly portrayed as a largely informal phenomenon that should be passively tolerated in order to tap their generative power as wellsprings of knowledge and diffusion structure. By focusing on the social structure and its components and by specifying a proposed configuration for these knowledge benefits to occur, we invite a step in the direction of more actively managing VCoPs in companies.

In order to stay in "fashion" in the future, the study of VCoPs needs to address more convincingly the question of manageability without falling victim to the illusion of controllability of knowledge generation and diffusion processes. In this respect, it will be helpful to map the social structure of social collectives studied along the essential coordination dimensions. By doing so, more insights about different kinds of (V)CoPs can be provided and terminological confusion about different (V)CoP labels can be reduced. In addition to such an existence analysis of different kinds of VCoPs, contingency analysis may yield insights about how situational characteristics influence the fabric of social structure in VCoPs. Finally, performing success factor research, one might test whether specific kinds of social structure entail more or less potential for success depending on situational characteristics.

REFERENCES

Adler, P. S., & Kwon, S. K. (2002). Social capital: Prospects for a new concept. *Academy of Management Review, 27*(1), 17-40.

Brown, J. S., & Duguid, P. (1991). Organizational learning and communities of practice: Toward a unified view of working, learning, and innovation. *Organization Science, 2*(1), 40-57.

Cohen, M., & Bacdayan, P. (1994). Organizational routines are stored as procedural memory: Evidence from a laboratory study. *Organization Science, 5*(4), 555-567.

Constant, E. W. (1987). The social locus of technological practice: Community, system, or organization? In W. E. Bijker, T. P. Hughes, & T. J. Pinch (Eds.), *The social construction of technical systems: New directions in the sociology and history of technology* (pp. 223-242). London: MIT Press.

Gould, S. (1979). An equity-exchange model of organizational involvement. *Academy of Management Review, 4*, 53-62.

Hinds, P., & Kiesler, S. (2002). *Distributed work.* Boston: MIT Press.

Kogut, B., & Metiu, A. (2001). Open source software development and distributed innovation. *Oxford Review of Econ-mic Policy, 17*(2), 248-64.

Lave, J. (1988). *Cognition in practice: Mind, mathematics, and culture in everyday life.* New York: Cambridge University Press.

Lave, J. (1991). Situating learning in communities of practice. In L. Resnick, J. Levine, & S. Teasley (Eds.), *Perspectives on socially shared cognition* (pp. 63-82). Washington, DC: American Psychological Association.

Lave, J., & Wenger, E. (1991). *Situated learning: Legitimate peripheral participation.* New York: Cambridge University Press.

Levitt, B., & March, J.G. (1988). Organizational learning. *Annual Review of Sociology, 14,* 319-340.

Mintzberg, H., & Waters, J. H. (1985). Of strategies, deliberate and emergent. *Strategic Management Journal, 6*(3), 257-272.

Nahapiet, J., & Ghoshal, S. (1998). Social capital, intellectual capital, and the organizational advantage. *Academy of Management Review, 23*(2), 242-266.

Nonaka, I. (1994). A dynamic theory of organisational knowledge creation. *Organization Science, 5*(1), 14-37.

Orr, J. E. (1996). *Talking about machines: An ethnography of a modern job.* Ithaca, NY: Cornell University Press.

Polanyi, M. (1967). *The tacit dimension.* London: Routledge and Kegan Paul.

Powell, W. W. (1990). Neither market nor hierarchy: Network forms of organization. *Research in Organizational Behavior, 12,* 295-336.

Resnick, L. (1991). Shared cognition: Thinking as a social practice. In L. Resnick, J. Levine, & S. Teasley (Eds.), *Perspectives on socially shared cognition* (pp. 1-20). Washington, DC: American Psychological Association.

Rheingold, H. (1993). *The virtual community: Homesteading on the electronic frontier.* New York: Harper Perennial.

Rogoff, B. (1990). *Apprenticeship in thinking: Cognitive development in social context.* New York: Oxford University Press.

Smith, M. A., & Kollock, P. (1999). *Communities in cyberspace.* New York: Routledge.

Suchman, L. (1987). *Plans and situated actions: The problem of human-machine communication.* New York: Cambridge University Press.

Szulanski, G. (2003). *Sticky knowledge. Barriers to knowing in the firm.* London: Sage.

Teichert, T., Rost, K., & von Wartburg, I. (2003). Aufbau von Sozial: Und Intellektuellem Kapital in Virtuellen Communities of Practice. In C. Herstatt & J. Sander (Eds.), *Produktentwicklung mit Virtuellen Communities* (pp. 249-276). Wiesbaden, Germany: Gabler.

Vincenti, W. G. (1990). *What engineers know and how they know it: Analytical studies from aeronautical history.* Baltimore: John Hopkins University Press.

von Hippel, E. (1994). Sticky information and the locus of problem solving: Implications for innovation. *Management Science, 40*(4), 429-439.

von Krogh, G., Spaeth, S., & Lakhani, K. R. (2003). Community, joining, and specialization in open source software innovation: A case study. *Research Policy, 32*(7), 1217-1241.

Walther, J. B. (1995) Relational aspects of computer-mediated communication: Experimental observations over time. *Organization Science, 6*(2), 186-203.

Weissbach, H.-J. (2000). Kulturelle und Sozialanthropologische Aspekte der Netzwerkforschung. In J. Weyer (Ed.), *Soziale Netzwerke: Konzepte und Methoden der Sozialwissenschaftlichen Netzwerkforschung* (pp. 255-284). München: Oldenbourg.

Wellman, B., Salaff, J., Dimitrova, D., Garton, L., Gulia, M., & Haythornthwayte, C. (1996). Computer networks as social networks: Collaborative work, telework, and virtual community. *Annual Review of Sociology, 22,* 213-238.

Wenger, E. C. (1998). *Communities of practice, learning, meaning and identity.* New York: Cambridge University Press.

KEY TERMS

Human Capital (Individual): Individual skills and capabilities that allow actors to act in new and innovative ways and to respond to new challenges with creative solutions.

Intellectual Capital (Collective): The aggregation of individual human capital in a sense that the aggregation is more than the sum of it parts, that is, encompassing organizational routines and capabilities.

Legitimate Peripheral Participation (LPP): A conceptual framework that recognizes that different kinds of learners can contribute to learning in VCoPs. As a consequence, novices are welcome to participate in the group even if their positions in the social network are peripheral and their contributions marginal. Over time, provided opportunities for situated learning, they are acquiring a growing body of knowledge and may become core members of the respective community.

Situated Learning: Learning that results from activities taking place in a real-life particular context. It is a synonym for apprenticeship and emphasizes the real-life learning aspects of knowledge generation.

Social Capital (Individual): Goodwill available to individuals which results from the structure (con-

figuration) and the relational content of the actor's social relations in a VCoP. The goodwill can be valuable in different respects, that is, can bring about information advantages, solidarity, and influence.

Social Structure (Collective): Members in VCoPs are considered a system organized by a characteristic pattern of relationships. The pattern is characterized by a specific configuration of coordination dimensions.

Virtual Communities of Practice (VCoPs): Communities of practice in which members interact supported by collaborative technologies in order to bridge time and/or geographical distances. VCoPs are partly self-organized, deliberate groups of people who share common practices, interests, or aims on behalf of the organization they belong to and want to advance their knowledge.

S

A Social Informatics Framework for Sustaining Virtual Communities of Practice

Umar Ruhi
Wilfrid Laurier University, Canada

INTRODUCTION: NASCENT PHENOMENON OF VIRTUAL COMMUNITIES OF PRACTICE

Reminiscent of the present-day Web vogue and the emergence of a myriad of e-enabled business models, virtual communities of practice are fast emerging as the next logical extension of traditional communities of practice. Virtual communities of practice exemplify the components of most contemporary communities of practice, which incorporate elements of physical social interactions, in combination with distributed virtual connections. These communities utilize technology applications to better manage their routine pursuits. More specifically, information and communication technologies (ICTs) are being used to facilitate the operations of a community of practice by providing tools for managing content (explicit knowledge) and a means for sharing expertise (tacit knowledge) through cooperation, coordination, and collaboration. The enabling technologies for institutionalizing a virtual community of practice range from simple user tools such as e-mail, teleconferencing, and groupware, to the more complex software applications, including group decision support systems (GDSSs) and corporate portals.

BACKGROUND: DRIVERS AND BARRIERS FOR VIRTUAL COMMUNITIES OF PRACTICE

Virtual communities proffer an inclusive embodiment of a technology platform that aims to provide an effective mechanism for enhancing the capabilities of traditional communities of practice. There are various benefits to institutionalizing communities of practice through the utilization of information and communication technologies. One of the main ad-

vantages of virtual communities is their ability to use networked technology, especially the Internet to establish links and form relationships across geographical barriers and time zones (Palloff & Pratt, 1999). Researchers and practitioners also recognize aspects of scalability and flexibility as important features of virtual communities enabled through Web-based applications. Squire and Johnson (2000) note that boundaries for virtual communities are relatively more "fluid" as compared to traditional communities, and this allows greater individual control over involvement in the community and its respective activities.

Based on the above-mentioned positive attributes of technology, an aspiring organization might be tempted to believe that an appropriate choice of information and communication technologies may be the sole basis for the success of a virtual community. However, several researchers and practitioners contend otherwise. The limitations of current technology in establishing communities of practice have been identified by various researchers (Haythornthwaite, Kazmer, Robins & Shoemaker, 2000; LeBaron, Pulkkinen, & Sconllin, 2000). Squire and Johnson (2000) affirm that among other things, Web-based environments run the risk of becoming impersonal without frequent contact between the participants. Some authors also cite the factors of trust and safety being major impediments in establishing a purely virtual community of practice. Although it is easier to establish trust in face-to-face interactions, communities that have their basis on the online medium exclusively often preclude that element (Palloff & Pratt, 1999). Palloff and Pratt also regard virtual communities to be more conducive to introvert members and hence believe that these communities may create a misbalance in the type of participation they pleat.

It is with these issues in the backdrop that a Social Informatics Framework is proposed as a useful lens for analysis of best practices in institutionalizing virtual communities of practice. As elaborated in the sections that follow, the framework allows for an all-encompassing treatise of cultural, organizational, and technical facets of virtual communities of practice.

VIRTUAL COMMUNITIES OF PRACTICE WITHIN A SOCIAL INFORMATICS FRAMEWORK

In order to achieve equilibrium between the use of technology and other organizational initiatives for institutionalizing virtual communities of practice, this article employs a social informatics framework to explicate a number of business recommendations. The field of social informatics yields best to this article's discussion since, by its very definition, the field entails an interdisciplinary study of the design, uses, and consequences of information technologies that takes into account their interaction with organizational and cultural contexts (Kling, 1999; Kling, Rosenbaum & Hert, 1998). Over the pervious 25 years, the body of research in social informatics has focused on socio-technical premises around the use of various technologies including the Internet, intranets, electronic forums, digital libraries, and electronic journals (Kling, 1996). As such, social informatics applies to a variety of ICTs, and it is logically instinctive to utilize relevant concepts from this discipline to theorize the multifaceted recommendations for establishing virtual communities of practice within organizations. Moreover, proffering the various recommendations for virtual communities constitutes a normative orientation in social informatics, whereby the discussion of best practices builds upon lessons learned from institutionalizing traditional communities of practices in the past. Figure 1 summarizes the general ideas of the social informatics framework by hinging virtual communities of practice as a technical facet within organizational and cultural viewpoints. In the past, such an orientation has been used by other socio-technical researchers (Kling, Crawford, Rosenbaum, Sawyer & Weisband, 2000).

From a technical standpoint, virtual communities offer an enabling platform for supporting knowledge management initiatives and organizational learning objectives within a firm's operational context. However, it should be realized that the deployment of virtual communities for such tasks requires organizational backing in the form of sponsorship and leadership initiatives, and also requires active and effective participation from core and peripheral community members. Researchers in the past have utilized similar approaches in discussing the critical success factors for traditional community-building initiatives within organizations (McDermott, 2001).

PHASED RECOMMENDATIONS FOR INSTITUTIONALIZING VIRTUAL COMMUNITIES OF PRACTICE

Any community of practice evolves overtime—it moves through various stages of development characterized by different kinds of prevailing activities and levels of interaction among the members. Hence, it is only natural to discuss specific recommendations for each stage in a community's evolution. In order to put forth the recommendations for commu-

Figure 1. Summary of the social informatics framework for virtual communities of practice (adapted from Pacey, 1983)

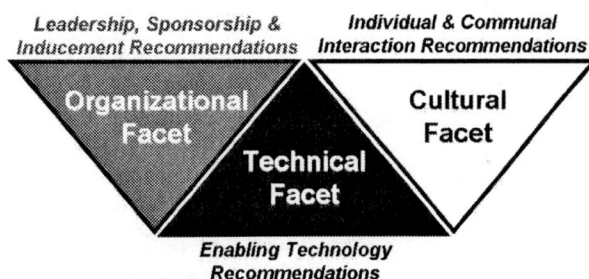

nities of practice within a social informatics context, this article adopts the *Community Evolution Model* developed by Gongla and Rizzuto (2001) in their discussion on the temporal stages of organizational community development. There are several reasons for selecting this particular model over other prevalent models such as the *Community Lifecycle Model* advocated by Wenger (1998) and its later adaptation by McDermott (2000). Firstly, the model proposed by Gongla and Rizzuto (2001) provides more inclusive coverage of the development points in the evolutionary phases of a community of practice. The model excludes the vanishing stages of a community's lifespan (referred to as the 'dispersed' and 'memorable' stages in Wenger's (1998) and McDermott's (2000) models). This is in line with the objective of the article and will allow the discussion to focus only on the formative and growth stages in a community's evolution. Secondly, the model also provides a good starting point to discuss technology applications in virtual communities of practice since it profiles the types of technologies that can advance traditional communities of practice to the next stages in their evolution. Table 1 summarizes the five stages in the Community Evolution Model by providing an overview of each stage and highlighting the fundamental underlying functions of the community at each stage. These underlying functions will form the basis of the cultural, organizational, and technical recommendations for various virtual communities of practice initiatives.

BEST PRACTICES MODEL FOR VIRTUAL COMMUNITIES OF PRACTICE

Based on the five stages that a traditional community of practice evolves through (see Table 1), this section aims to elaborate best practices for institutionalizing virtual communities of practice. More specifically, the best practices will lay the groundwork for an effective strategy based on a social informatics standpoint by addressing the technical, the organizational, and the cultural dimensions for virtual communities of practice. Figure 2 provides a pictorial summary of the best practices model. A discussion on the model follows.

The model is indicative of an increasing span of the community of practice as it evolves overtime through the various stages of its development. Within the social informatics context, the model addresses cultural, organizational, and technical recommendations progressively to facilitate the virtual community's advancement to the next phase in its evolution. These recommendations are elaborated further herewith.

Cultural Recommendations: Internal Processes

The cultural recommendations in the best practices model address the internal processes (Gongla & Rizzuto, 2001) in a community of practice. The

Table 1. Summary of the community evolution model (adapted from Gongla & Rizzuto, 2001)

Stage	Description	Fundamental Underlying Functions
Potential	The inception stage where the community is in its embryonic and foundational phase.	Connection
Building	The formalization stage where the community identifies its purpose and outlines its philosophy.	Memory & Context Creation
Engaged	The execution stage where the members start to affect new tactics that were shared across the community.	Access & Learning
Active	The realization stage where the members understand and appreciate the value and benefits of their collective efforts.	Collaboration
Adaptive	The profitization stage where the community and the sponsoring parties start utilization of new and shared knowledge for competitive advantage.	Innovation & Generation

Figure 2. Pictorial summary of the best practices model for virtual communities of practice

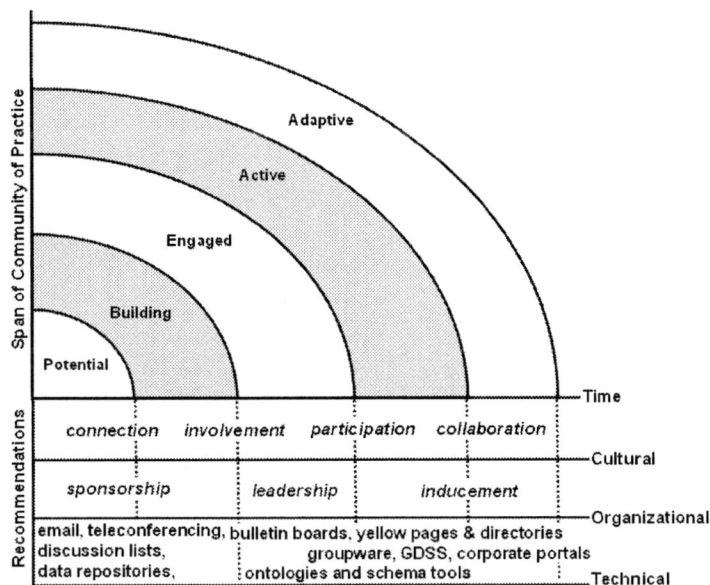

function of internal processes in a community of practice is to facilitate tacit knowledge sharing through community membership and member socialization. Internal processes within virtual communities of practice typify the manner in which tacit knowledge is shared among online participants through membership, socialization, and collaboration. As exposited in Figure 2, the recommendations along the cultural dimension of communities of practice support the involvement of the right people in the community, followed by their active participation in the community's activities and collaboration with other core and peripheral members. Lessons could be derived from studies by other theorists who have discussed socialization among members in a broad sense. For example, Cohen and Prusak (2001) talk about people as social individuals with their individual and group behaviors, as well as the larger organizational behavior influence vis-à-vis a community.

In order to get new members to start contributing towards the goals of the community and the sponsoring organization, social interaction and mutual engagement need to be encouraged organization-wide. In the early stages of a virtual community's development, thought leaders should try to provide information to prospective members through "one-way dissemination" (Rainey, 2002). The information provided should be persuasive enough to attract a high

number of professionals and experts to join the community. The objective at the early stages is to "connect" with individuals who can prove to be an asset to the community. Community leaders in this stage should also herald a sense of commonality to help foster feelings of mutual benefit among the early members.

Following the initial bonding between new members, people should be encouraged to take up more active roles in the virtual community. At this stage, members should begin to identify with everyone in the community through "two-way interactions" (Rainey, 2002). It is at this stage that members should be encouraged to share experiences and exchange tacit knowledge with one another. Organizational management should contribute some direction in defining the community's identity and assist with aspects of future planning.

Subsequent to these two-way interactions instigated and moderated by community leaders, the final stage of socialization dwells upon "multiple interactions" (Rainey, 2002) where members actively interact with one another directly without any intermediation from leaders. Leaders and managers need to encourage sharing work-related knowledge through experiential anecdotes and group stories from members. Furthermore, people should be convinced to contribute to the community without persuasion

(Gongla & Rizzuto, 2001). Once the community nears the active stage in its development, initiatives should be devised to allow "collaboration" to gain its full momentum by assisting new members to be absorbed into the community, and by effecting outward linkages with other communities and peripheral members.

Organizational Recommendations: External Processes

The organizational recommendations of the Best Practices Model address the external processes (Gongla & Rizzuto, 2001) in a community of practice. The function of external processes in a community of practice is to encourage and endorse member participation throughout the community via organizational sponsorship and ongoing leadership. As highlighted in Figure 2, these external process recommendations exhort the progressive use of sponsorship, leadership, and inducement for ensuring the efficient development of a virtual community of practice. Quite similar to traditional communities, at their inception, virtual communities lack the legitimacy and budgets of established organizational units. It is hence very important for leaders and managers who are interested in the community's development to support it by creating the initial executive buy-in for the community at the organization's top management levels. This patronage for a newfound commu-

nity is a crucial instigating point in the community's lifecycle as it affects the pace of the community's development overtime.

Once a community grows past its embryonic phase (i.e., the "potential" stage in the Community Evolution Model), the active members need to adopt diverse leadership styles to help the community advance to the engaged and active stages. Wenger (1998) outlines some of the leadership styles that can prove to be beneficial in developing communities of practice. The styles range from inspirational leadership to day-to-day leadership. Table 2 provides a summary of leadership styles that are relevant in a virtual community's development through its five stages of evolution, along with a list of sample initiatives that can be undertaken by leaders in a virtual community of practice.

In addition to organizational leadership, communities of practice also require encouragement and motivation from the organization's top management in the form of incentives and rewards. For example, the World Bank is well known for initiatives that fund positions for knowledge managers to assist in community development (Wenger & Snyder, 2000). Other organizations have developed promotion and recognition systems that formally acknowledge the work of community members. Variants of these programs can be established to foster an affirmative environment for virtual communities as well. For instance, community members could be rewarded based on

Table 2. Leadership styles through the five stages of a virtual community's evolution

Stages of Community Evolution	Suitable Leadership Styles	Examples of Initiatives in Virtual Communities of Practice
Potential	Inspirational leadership from thought leaders and recognized experts	Instigating of new forums for discussion and knowledge sharing.
Building	Interpersonal leadership from membership principals	Sending invitations to colleagues to join the community
Engaged	Day-to-day leadership from activity organizers	Moderating daily discussions and acting on member feedback for improvements
Active	Institutional leadership from executives	Publicizing success stories through member anecdotes
Adaptive	Boundary leadership from opportunists and futurists	Seeking new opportunities and synergies within the organization or across business partners

the quality of support they provide to incoming novice members in getting them acquainted; they can also be rewarded based on the number of problems they were able to resolve in the community, and the nature and value of the solution they have codified for permanent storage in the organization's explicit knowledge base.

It is worth mentioning once again that in providing external support to virtual communities of practice, leaders and managers need to avoid any tactics that may possibly sacrifice the community's self-organizing nature. In the words of Wenger and Snyder:

You can't tug on a cornstalk to make it grow faster or taller. You can however, till the soil, pull out the weeds, add water during dry spells, and ensure your plants have the proper nutrients. (Wenger & Snyder, 2000)

Both traditional and virtual communities are like these plants that can benefit from "cultivation" and "nurturing". Early organizational sponsorship, perpetual leadership, and continuous management support can provide the proper form of cultivation and nurturing for virtual communities of practice.

Technical Recommendations: Communication and Content Applications

Although new technology may well have been the genesis of a virtual community of practice, it should be evident by now that successful virtual communities require technology to be utilized in conjunction with suitable cultural and organizational directives. Technology in itself needs to be treated as an enabler for virtual communities of practice.

In order to describe the features and functions of technology that can prove to be favorable in the institutionalization of virtual communities of practice, this section categorizes technology functions under communication and content applications. At the simplest level, communication applications include tools such as e-mail, teleconferencing, and online discussion forums that can facilitate conversation and correspondence among community members in both synchronous and asynchronous ways. On the other hand, content applications include the

databases and document management systems that help store explicit organizational knowledge and make that knowledge accessible to core and peripheral community members. Finally, there are hybrid applications that comprise the dual aspects of communication and content applications—examples of such hybrid applications include groupware applications, group decision support systems (GDSSs), and corporate portals. By combining content storage and access functionality with the ability to communicate among group members, such applications enable a higher level of coordination within the community.

It should be noted that similar to the progressive recommendations for cultural and organizational processes, the recommendations for enabling technologies also represent a phased implementation plan that facilitates the community's development towards more advanced stages in its lifecycle. This is to say that the more complex technology applications such as GDSS and corporate should ideally be deployed at the later stages of the virtual community's development.

The communication and content technologies that support virtual communities of practice can also be related to the underlying fundamental community functions characterizing the five stages of community evolution (see Table 2). It can be said that a virtual community will start functioning by first utilizing simpler technologies and tools such as e-mail, teleconferencing, and discussion lists to enable "connection" among its members. At the same time, the community will establish its presence online through static Web pages to convey its purpose and intention to potential members ("context creation"). With the passage of time, it will enable "learning" by utilizing a multitude of communication tools to share and codify tacit knowledge, while organizing explicit knowledge for convenient "access" through databases and document management systems. Following that, a community will utilize expansive online directories and cross-functional tools for "collaboration" with new members and other external communities. Finally, as it proceeds towards the active and adaptive stages, a virtual community can proceed to utilize advanced coordination tools including GDSSs and corporate portals for "generating" new ideas and creating opportunities for "innovation."

CONCLUSION

Following the discussion of the Best Practices Model in this article, it should be reinforced that the recommendations provided herewith attempt to emphasize that virtual communities of practice require serious groundwork in the form of organizational, cultural, and technical initiatives for their successful institutionalization. Together, these initiatives provide the foundation for sponsorship, support, and coordination, which are essential effecting factors for any community of practice.

Furthermore, within a social informatics context, the Best Practices Model for institutionalizing virtual communities of practice can prove to be a useful conceptual vehicle for understanding:

1. a community's progression through various development stages;
2. the diverse sponsorship, leadership and inducement initiatives that are required at each stage;
3. the type of social interactions that need to be encouraged at each stage; and
4. the kind of technology tools and applications that can be utilized at each stage to accelerate the virtual community's advancement to the next levels.

REFERENCES

Cohen, D., & Prusak, L. (2001). *In good company: How social capital makes organizations work.* Boston: Harvard Business School Press.

Gongla, P., & Rizzuto, C.R. (2001). Evolving communities of practice: IBM Global Services experience. *IBM Systems Journal, 40*(4), 842-862.

Haythornthwaite, C., Kazmer, M., Robins, J., & Shoemaker, S. (2000). Community development among distance learners: Temporal and technological dimensions. *Journal of Computer Mediated Communication, 6*(1). Retrieved April 23, 2004, from *http://www.ascusc.org/jcmc/vol6/issue1/haythornthwaite.html*

Kling, R. (1999). What is social informatics and why does it matter? *D-Lib Magazine, 5*(1). Retrieved May 3, 2004, from *http://www.dlib.org/dlib/january99/kling/01kling.html*

Kling, R. (Ed.). (1996). *Computerization and controversy: Value conflicts and social choices* (2nd ed.). San Diego: Academic Press.

Kling, R., Crawford, H., Rosenbaum, H., Sawyer, S., & Weisband, S. (2000). *Learning from social informatics: Information and communication technologies in human contexts.* Bloomington, IN: Center for Social Informatics, Indiana University.

Kling, R., Rosenbaum, H., & Hert, C. (1998). Social informatics in information science: An introduction. *Journal of the American Society for Information Science, 49*(12), 1047-1052.

LeBaron, J., Pulkkinen, J., & Sconllin, P. (2000). Promoting cross border communication in an international Web-based graduate course. *Interactive Multimedia Electronic Journal of Computer Enhanced Learning, 2*(2). Retrieved May 3, 2004, from *http://imej.wfu.edu/articles/2000/2/01/index.asp*

McDermott, R. (2000). Community development as a natural step: Five stages of community development. *Knowledge Management Review, 3*(5), 16-19.

McDermott, R. (2001). Knowing in community: 10 critical success factors in building communities of practice. Retrieved from *http://www.co-i-l.com/coil/knowledge-garden/cop/knowing.shtml*

Pacey, A. (1983). *The culture of technology.* Oxford, UK: Basil Blackwell.

Palloff, R., & Pratt, K. (1999). *Building learning communities in cyberspace: Effective strategies for the online classroom.* San Francisco: Jossey-Bass.

Rainey, S. (2002, May). Building online professional networks: Three stages to success. Retrieved April 26, 2004, from *http://www.onlinecommunityreport.com/features/rainey*

Squire, K., & Johnson, C. (2000). Supporting distributed communities of practice with interactive television. *Educational Technology Research and Development, 48*(1), 23-43.

Wenger, E. (1998). Communities of practice: Learning as a social system. *The Systems Thinker, 9*(5), 1-5.

Wenger, E., & Snyder, W.M. (2000). Communities of practice: The organizational frontier. *Harvard Business Review, 78*(1), 139-145.

KEY TERMS

Community Moderators: Day-to-day leaders in virtual communities who control discussions in individual forums, point the community members in the right direction, and ensure that the social climate of the forum promotes participation and exchange among its members.

Corportal (Corporate Portal): An online corporate Web site based on the model of an enterprise information portal. An enterprise information portal acts as a single gateway to a company's information and knowledge base for employees, and sometimes its customers and business partners as well.

Group Decision Support System (GDSS): An interactive computer-based system that facilitates solutions to unstructured problems by decision makers working as a group. Among other features, the software package includes idea organizers, electronic brainstorming tools, questionnaire tools, and group dictionaries.

Groupware: Software applications that help people work together virtually while being physically located at a distance from one another. Groupware applications and services include the sharing of work schedules, event calendars, electronic meetings, shared databases, and group e-mail accounts.

ICTs (Information and Communication Technologies): An umbrella term that includes any communication devices, computing hardware, as well as software applications.

Normative Orientation in Social Informatics: Refers to research that aims to recommend alternatives for professionals who design, implement, use, or develop policy about ICTs (Kling et al., 2000).

Social Informatics: The field of interdisciplinary study of the design, uses, and consequences of information technologies that takes into account their interaction with institutional and cultural contexts (Kling, 1999).

Social Network Analysis and Communities of Practice

Dimitris Assimakopoulos
Grenoble Ecole de Management, France

Jie Yan
Grenoble Ecole de Management, France

INTRODUCTION

Social network analysis (Scott, 2000; Wasserman & Faust, 1994) is a relatively new theory and methodology that has found wide application in social science research. In the early 2000s, an increasing number of scholars have been interested in computerized Social Network Analysis (SNA) and have adopted social network theory and techniques to study communities of practice (CoPs). In this article, the authors introduce SNA from a historical perspective, compare SNA with non-network theories and methods, and introduce popular SNA software packages. With reference to recent empirical research, the authors discuss several areas in which SNA has been applied to CoP research.

BACKGROUND: SOCIAL NETWORK ANALYSIS

SNA is a set of theories and methods used to uncover, map, and analyze the underlying structure of relationships that bind people, and other human and non-human actors, together within and across groups, organizations, and communities. SNA has gained momentum since the 1950s as a result of increased use of mathematical, statistical, and computing methodologies in conjunction with social behaviour theory to model and explain patterns of social interaction within or across organizational settings (see, for example, Mitchell, 1969). With the rapid development of graph theory (Harary, 1969) and the dramatic increase in computing capabilities during recent decades, SNA has found important applications in modelling the spread of contagious disease, diffusion, and communication of technologi-

cal and administrative innovations, relational nature of competitive advantage, intra- and inter-organizational relations, the nature of social support, including CoPs, to name but a few areas of interest.

Generally, a social network refers to a social system constructed by a collection of actors, relations among these actors (e.g., friendship, knowledge exchange), and possible attributes (e.g., age, sex) for each actor. SNA assumes that the actors in a social system are not isolated but have linkages with others, each of whom in turn is linked to a few, some, or many others. How actors behave depends on how they are linked with others and where they are located in the whole network. In this sense, SNA seeks to describe, understand, and model the relationships among actors either at the personal level or at the group level, using metrics at the actor level (e.g., centrality) or the group level (e.g., density). The overall aim of SNA is therefore to analyse and explain how the relationships among sets of actors (i.e., dyads, triads, and larger subgroups) and the structure of relationships of a whole group influence individual behaviour, as well as the functioning of a group as a whole.

In SNA, the observed attributes of social actors are primarily understood in terms of their relational content. The unit of analysis is not, for example, an individual but a relation (e.g., advice) connecting two (or more) individuals. As a result, SNA focuses on a collection of individuals and the set of linkages among them with respect to one (or more) relations. The inclusion of information on relations among groups of actors is the fundamental feature differing SNA from other non-network methodologies such as statistical analysis with its primary focus on attribute data (e.g., age, sex) with no relational content (Scott, 2000).

As a result of this distinction, a non-network theory based on statistical analysis, as in neo-classical economic theory of self-interested profit maximizing actors, the units of analysis, regardless of whether they are people or organisations, are viewed as isolated, independent, and unrelated, and researchers implicitly assume that actors do not influence each other, as if they are undersocialised and disembedded from any kind of social system (Granovetter, 1985). However, from a SNA perspective, it is assumed that the economic behaviour of actors mainly arises from ongoing structural and relational workings of a social system (Wellman & Berkowitz, 1988). Researchers primarily focus on the properties of the network, community, or social system, rather than attributes of individual actors. The attribute data of individual actors only help to add additional explanatory power with statistical methods complementing SNA methods and metrics.

The relational data of SNA are measured in terms of two properties: whether a linkage is binary or valued, and whether it is directional or nondirectional. The directional valued data contains more information than the nondirectional binary data, which only presents the absence or presence of the linkage between two actors. Questionnaire survey is suggested to be the most pragmatic approach to collect relational data in organizational settings (Cross, Borgatti & Parker, 2002). Relational data are often stored in data matrices, called sociomatrices, suitable for electronic computing. Besides such measures as mean, median, standard deviation, and so forth borrowed from statistics, SNA provides a set of quantitative concepts and related techniques to analyze relational data. For example, the term *centrality* measures how critical an actor is in a network; *density* measures how closely a group of actors are connected, and *centralization* measures how variable or heterogeneous the actor centralities are (see also, Key Terms, at the end of this article). Many other advanced concepts, for example, structural equivalence, component, role, and position, are also widely used in SNA research.

Most of the mathematical calculations of SNA can be done by a variety of off-the-shelf computer SNA software packages. Scott (2000) suggests that GRADAP is good at handling fairly large data sets; STRUCTURE is slightly more user-friendly;

UCINET has many powerful features, and it is fast and efficient with a wide range of measures available; PAJEK is able to handle and analyze very large data sets. UCINET (Borgatti, Everett & Freeman, 2002) is the most widely used SNA software, and it is regarded by many researchers as the best for network analysis and visualization. It was developed by a group of network scholars at the University of California, Irvine (UCI). It is a general purpose and easy to use package covering a wide range of graph theoretical concepts, positional analysis, and multidimensional scaling (MDS) routines. Its functions for cohesion, components, centrality, subgroup, role, and position analysis are well designed and widely used.

CoP AS SOCIAL NETWORKS

A community of practice (CoP) is a collection of tightly intertwined interpersonal networks. As stated by Wenger, McDermott, and Snyder (2002), "the heart of a community is the web of relationships among community members, and much of the day-to-day interaction occurs in one-to-one exchanges" (p. 58). The CoP members are therefore linked by ongoing dyadic relations and the daily social interactions which form a well knit network of interpersonal relationships. Generally, interaction among CoP members is more intensive than in ordinary informal networks so that a CoP "could be in fact viewed as nodes of 'strong ties' in an interpersonal network" (Wenger, 1998, p. 283).

However, a CoP is more than just a set of interpersonal relationships. It is "not defined merely by who knows whom or who talks with whom in a network of interpersonal relations through which information flows" (Wenger, 1998, p. 74). The common practice gives the community members a knowledge domain, a shared identity, and cohesiveness to sustain interactions over time. CoPs are closely associated with a collective ongoing practice, in which interpersonal relationships are built up as a result of continuous knowledge generation and exchange. Thus, it is argued that every CoP consists of one (or many) social networks, but not every social network forms a CoP (Schenkel, Teigland & Borgatti, 2001). Moreover, CoPs are regarded as knowledge creation and sharing networks (Cross, Prusak &

Parker, 2001) while social networks can have a relational content based on other relational characteristics such as traditional friendship, kinship, or marriage relations.

APPLICATION OF SOCIAL NETWORK ANALYSIS IN CoP RESEARCH

Empirical CoP studies usually adopt an ethnographic research design and strategy backed up by participant observation and in-depth interviews, such as the investigations into midwives, tailors, butchers, and navy quartermasters (Lave & Wenger, 1991), photocopier machine repairmen (Orr, 1996), copier salespeople (Østerlund, 1996), and insurance claim processors (Wenger, 1998). In recent years, with the increasing need to conceptualize and measure the structural properties of CoPs, many scholars have made efforts to adopt SNA methodology in CoP research (Barab, Kling & Gray, 2004; Huysman, Wenger & Wulf, 2003). Basically, SNA concepts and techniques can be used to explore CoPs from a broad range of perspectives as discussed below.

Identifying CoPs

SNA has a powerful set of techniques to analyze complex social structures and identify CoPs (Parker, Cross & Walsh, 2001). CoPs can be identified following the three dimensions outlined by Lave and Wenger (1991), that is, joint enterprise, shared repertoire, and mutual engagement. The former two dimensions are concerned with what a group of people is doing and what is the common meaning of their shared activities, while the latter is directly associated with the frequency of interpersonal relationships. By surveying and questioning the channels and frequency of interpersonal communication, researchers measure the different levels of mutual engagement, and thus identify CoPs from interpersonal networks. Examples include Schenkel et al. (2001) measuring the help-seeking relations in an international construction project, Swarbrick's (2002) work on the personal exchange of information, Walsh and McGrath (2003) tracing the flow of business ideas within organizations, and Yan and Assimakopoulos' (2003) study of the advice seeking relations in a software engineering community.

Visualizing CoPs

SNA provides a set of techniques not only to analyze but also visualize social networks (Freeman, 2000; Freeman, Webster & Kirke, 1998). SNA often plots two- or three-dimensional maps of interpersonal communication using computer animated graph-making software such as Mage (Richardson & Presley, 2001), Krackplot (Krackhardt, Blythe & McGrath, 1995), and IKnow (Contractor, O'Keefe & Jones, 2001). Maps can be based on sociometric techniques such as structural equivalence and multi-dimensional scaling (for a conceptual outline, see Wasserman & Faust, 1994). Visual analysis can be made for various purposes, such as identifying CoPs, analyzing the structural properties of CoPs, distinguishing different roles actors play, tracing information flows, and so on. The visualization techniques are used in many of the cited empirical CoP studies.

Analyzing the Deep Structure of CoPs

As mentioned above, many existing empirical CoP studies are based on ethnographic case studies, describing the underlying structure and discussing the cognitive processes of how CoPs emerge and operate. For example, Schenkel et al. (2001) have used a set of concepts, including connectedness, graph-theoretic distance, density, core/periphery structure, and coreness to describe CoPs. Yan and Assimakopoulos (2003) also measure the intensity and density of advice-seeking relations in software engineering communities to increase our understanding of the complex relationships between CoPs and project teams. Some additional SNA concepts such as clusters, cliques, positions, and structural holes (Burt, 1992) have also been used in analyzing the deep structure of CoPs (Huysman et al., 2003).

Identifying Key Players of CoPs

The members of CoPs play different roles in knowledge creation and sharing processes. Wenger (1998) suggests a core-periphery structure, in which community members have different positions according to their levels of participation, that is, full participation (insider), legitimate peripherality, marginality, and full non-participation (outsider). Cross and

476

Prusak (2002) further identify four key players who have important roles in gathering and distributing knowledge including central connectors, boundary spanner, information brokers, and peripheral specialists. In small CoPs, the players can be located by visually analyzing the maps, while in large, complex communities, quantitative indicators can be calculated to distinguish the role each player holds. For example, as suggested by in Swarbrick (2002), central connectors are often those who have the highest score of degree centrality; information brokers and boundary spanners often have high scores of betweenness centrality and flow centrality (for more discussion of these concepts, see below and, also, Wasserman & Faust, 1994).

Managing Knowledge Activities

By tracing and visualizing the paths of knowledge flows, SNA has proven to be an invaluable tool to systematically assess and manage knowledge creation and exchange. SNA can serve to diagnose the health of informal CoPs and their relationship to formal work groups to find out bottlenecks of information flow, streamline reporting relations, and leverage overall efficiency of communication. Cross et al. (2002) have carried out consultancy projects for this purpose and suggested that SNA is effective in "promoting effective collaboration within a strategically important group; supporting critical junctures in networks that cross functional, hierarchical, or geographic boundaries; and ensuring integration within groups following strategic restructuring initiatives" (p. 28). Falkowski and his colleagues have also made use of SNA to help leverage innovation work in IBM (Falkowski & Krebs, 1999; Falkowski & Ray, 2000).

Studying the Evolution of CoPs

Currently, most of the empirical studies present a static picture of CoPs, reflecting the structural properties of the community at the time of field study. The evolution of CoPs is a topic worthy of further research from an SNA perspective (see, for example, Assimakopoulos, Everton & Kiyoteru's, 2003 study of the emergence of Silicon Valley's semiconductor community). If the process of network mapping is repeated in more than one cycle, it is possible

to get information about the emergence, growth, and structural change of CoPs over time. For example, by tracing how newcomers change their positions over time from peripheral to core, one can get a better understanding of how the legitimate peripheral participation process takes place (Lave & Wenger, 1991; Wenger, 1998). By analyzing how the structure of CoPs changes when organizational functional department and project teams are reorganized, for example, as a result of business process reengineering initiatives, it is possible to reveal more information about the relationship between CoPs and formal work groups (see also Yan & Assimakopoulos, 2003).

Studying Distributed CoPs

With recent trends of business and technology globalization, collaboration and networking among business partners who are not collocated have become increasingly common (Assimakopoulos, 2003). Information and knowledge are often required to be shared intensively across organizational and geographic boundaries. Sometimes these collaborations form the so-called distributed CoPs (Hildreth & Kimble, 2000) or virtual CoPs (Kimble, Hildreth & Wright, 2001). SNA is an appropriate approach to study CoPs whose members are not located in a single place. By tracing and measuring the boundary objects (Star & Griesemer, 1989) such as shared documents, e-mail exchanges, phone calls, and others, researchers can get a deep understanding of such boundary spanning CoPs.

Studying Internet-Based Virtual CoPs

Internet-mediated virtual communities are a recent phenomenon that have attracted the attention of many social researchers (Barab et al., 2004; Rheingold, 1993; Wellman, 1997). A group of people gathering and exchanging information on an Internet Web site or linked by group e-mails to discuss topics of common interest is termed a virtual community. Generally speaking, virtual communities are anonymous, and members are unknown to each other offline. In some cases, such virtual communities also form CoPs when the topics of common interest for the community are not personal like fishing or stamp collection but deal with ongoing work practices, such as software development (Assimakopoulos & Yan, 2004). Some researchers, for example, Mahony and Ferraro (2004) have studied Internet based CoPs for which are developed open source software using SNA techniques (see also Wasko & Teigland, 2002).

CONCLUSION AND FUTURE TRENDS

SNA is an innovative theory and methodology for knowledge management and organizational learning research, providing useful ways to study CoPs from novel perspectives. In particular, the quantitative techniques and visualization capabilities of SNA have demonstrated unique advantages in studying CoPs. The application of SNA concepts and methods has been used to identify CoPs from amorphous informal interpersonal networks, analyze the structural properties of CoPs, distinguish the different roles each actor plays, and manage how knowledge is created and shared within CoPs and between CoPs and formal work groups, such as project teams. SNA has also recently been used to study geographically distributed CoPs, Internet-mediated virtual CoPs, as well as the dynamics of structural change in CoPs. As a result of increasing globalization, the emergence of knowledge economies and societies and the distributed nature of technological and organizational innovations and associated CoPs, it is expected that much more empirical work will be undertaken in the near future for applying a SNA perspective in the study of the notion of CoPs.

REFERENCES

Assimakopoulos, D. (Ed.). (2003). Introduction: Collaboration and networking. *International Journal of Technology Management, 25*(1/2), 1-4.

Assimakopoulos, D., Everton, S., & Kiyoteru, T. (2003). The semiconductor community in Silicon Valley. *International Journal of Technology Management, 25*(1/2), 181-199.

Assimakopoulos, D., & Yan, J. (2004). The external linkages of community of practice. In Y. Hosni, R. Smith, & T. Khalil (Eds.), *Proceedings of the 13th International Conference on Management of Technology (IAMOT 2004).* (CD-ROM). Miami, FL: University of Miami.

Barab, S., Kling, R., & Gray, J. (Eds.). (2004). *Designing for virtual communities in the service of learning.* Cambridge: Cambridge University Press

Borgatti, S. P., Everett, M. G., & Freeman, L. C. (2002). *Ucinet 6.0 for Windows: Software for social network analysis.* Boston: Harvard Analytic Technologies.

Burt, R. (1992). *Structural holes.* Chicago: University of Chicago Press.

Contractor, N., O'Keefe, B., & Jones, P. (2001). *IKnow: Software for social network analysis.* University of Illinois at Urbana-Champaign.

Cross, R., Borgatti, S., & Parker, A. (2002). Making invisible work visible: Using social network analysis to support strategic collaboration. *California Management Review, 44*(2), 25-46.

Cross, R., Prusak, L., & Parker, A. (2001). Knowing what we know: Supporting knowledge creation and sharing in social networks. *Organizational Dynamics, 30*(2), 100-120.

Cross, R., & Prusak, L. (2002, June). The people who make organizations go or stop. *Harvard Business Review*, 104-112.

Freeman, L. C. (2000). Visualizing social networks. *Journal of Social Structure, 1*(1). Retrieved June 10, 2005, from *http://moreno.ss.uci.edu/freeman.pdf*

Freeman, L. C., Webster, C. M., & Kirke, D. M. (1998). Exploring social structure using dynamic three-dimensional color images. *Social Networks, 20*, 109-118.

Granovetter, M. (1985). Economic action and social structure: The problem of embeddedness. *American Journal of Sociology, 91*, 481-510.

Harary, F. (1969). *Graph theory.* Reading, MA: Addison-Wesley.

Hildreth, P., & Kimble, C. (2000). Communities of practice in the distributed international environment. *Journal of Knowledge Management, 4*(1), 27-38.

Huysman, M., Wenger, E., & Wulf, V. (Eds.). (2003). *Communities and technologies: Proceedings of the 1st International Conference on Communities and Technologies.* Dordrecht: Kluwer Academic.

Falkowski, G., & Krebs, V. (1999, February). Beyond the organization chart. *Across the Board Magazine.* Retrieved from *http://www.conferenceboard.org/publications/*

Falkowski, G., & Ray, R. (2000, October-December). Organizational knowledge mapping: Building a knowledge-based organization. *The Journal of International Association for Human Resource Information Management.*

Kimble, C., Hildreth, P., & Wright, P. (2001). Communities of practice: Going virtual. *Knowledge management and business model innovation* (pp. 220-234). Hershey, PA: Idea Group Publishing.

Krackhardt, D., Blythe, J., & McGrath, C. (1995). *KrackPlot.* Pittsburgh, PA: Carnegie-Mellon University.

Lave, J., & Wenger, E. (1991). *Situated learning: Legitimate peripheral participation.* Cambridge: Cambridge University Press.

Mahony, S., & Ferraro, F. (2004, January). Managing the boundary of an "Open Project" (Working Paper No. 537). University of Navarra, Barcelona, IESE Business School.

Mitchell, J. C. (1969). *Social networks in urban settings.* Manchester: Manchester University Press.

Orr, J. E. (1996). *Talking about machines: An ethnography of a modern job.* Ithaca, NY: Cornell University Press.

Østerlund, C. (1996). Learning across contexts: A field study of salespeople's learning at work. *Skriftserie for Psykologisk Institut, 21*(1).

Parker, A., Cross, R., & Walsh, D. (2001). Improving collaboration with social network analysis. *Knowledge Management Review, 4*(2), 24-30.

Rheingold, H. (1993). *The virtual community: Finding connection in a computerized world.* London: Minerva.

Richardson, D., & Presley, B. (2001). *MAGE 5.8..* Biochemistry Department, Durham, NC: Duke University.

Schenkel, A., Teigland, R., & Borgatti, S. P. (2001). Theorizing structural properties of communities of practice: A social network approach. *Proceedings of the Academy of Management Annual Conference.*

Scott, J. (2000). *Social network analysis: A handbook* (2nd ed.). London: Sage.

Star, S. L., & Griesemer, J. R. (1989). Institutional ecology "translations" and boundary objects: Amateurs and professional in Berkeley's Museum of Vertebrate Zoology, 1907-1930. *Social Studies of Science, 19*, 387-420.

Swarbrick, A. (2002). The knowledge network analysis report. Retrieved June 10, 2005, from *http://www.robotegg.com/projectsite/documents/*

Walsh, E., & McGrath, F. (2003, September). Identification and emergence of communities of practice: A social network analysis study. *Proceedings of the European Conference on Knowledge Management*, Oxford, UK.

Wasko, M. M., & Teigland, R. (2002). The provision of online public goods: Examining social structure in a network of practice. *Proceedings of the 23rd International Conference on Information Systems (ICIS),* Barcelona, Spain.

Wasserman, S., & Faust, K. (1994). *Social network analysis: Methods and application.* Cambridge: Cambridge University Press.

Wellman, B. (1997). An electronic group is virtually a social network. In S. Kiesler (Ed.), *Culture of the Internet* (pp. 179-205). Hillsdale, NJ: Lawrence Erlbaum.

Wellman, B., & Berkowitz, S. D. (Eds.). (1988). *Social structures: A network approach.* Cambridge: Cambridge University Press.

Wenger, E. (1998). *Communities of practice: Learning, meaning and identity.* Cambridge: Cambridge University Press.

S

Wenger, E., McDermott, R., & Snyder, W. M. (2002). *Cultivating communities of practice: A guide to managing knowledge*. Boston: Harvard Business School Press.

Yan, J., & Assimakopoulos, D. (2003). Knowledge sharing and advice seeking in a software engineering community. In L. M. Camarinha-Matos & H. A. Dordrecht (Eds.), *Processes and foundations for virtual organizations* (pp. 341-350). Dordrecht: Kluwer Academic.

KEY TERMS[1]

Betweenness Centrality: A measurement of centrality, indicating how powerful an actor is in terms of controlling information flow in a network. The idea here is that actors are central if they lie between other actors on the shortest paths connecting these actors.

Centrality: An index used to indicate how critical an actor is in a network. Degree is the most popular way to measure centrality; see also betweenness and closeness centrality below.

Centralization: An index at group level, measuring how variable or heterogeneous the actor centralities are. It records the extent to which a single actor has high centrality, and the other, low centrality.

Clique: A clique is a maximal complete subgroup in which all actors are directly connected to each other, and there are no other actors that are also directly connected to all members of the clique.

Closeness Centrality: A measurement of centrality, indicating how close an actor is to all other actors in a network. The idea here is that actors are central if they can quickly interact with all other actors in a network.

Degree: An index measured by the number of linkages incident with an actor.

Density: An index used to indicate how actors are closely or loosely connected in a network. It is measured by the proportion of possible lines that are actually present in a network.

Euclidean Distance: It is an index to measure structural similarity among actors of a network. The less two actors are structural equivalent, the larger the Euclidean distance between them.

Multi-Dimensional Scaling: A way to visualize Euclidean distances. Networks are often visualized by two-dimensional scaling in a graphical way with (x, y) coordinates, presenting a map of geometrical Euclidean distances among actors in a network.

Position: A position is a group of actors who have similar pattern of relations to all other actors in a network. People in similar social position are more likely to have similar profiles, that is, social activities, ties, and interactions, than people in different positions in a network.

Structural Equivalence: Two actors are structurally equivalent if they have mathematically identical connections (structural similarity) to and from all other actors in a network. The actors who occupy same social positions are said to be structural equivalent.

ENDNOTE

[1] The definitions of the terms above are based on Wasserman and Faust (1994).

Social Philosophy, Communities, and the Epistemic Shifts

Angela Lacerda Nobre
ESCE-IPS, Portugal

INTRODUCTION: DOMINANT AND PERIPHERICAL INFLUENCES

It is critical to distinguish between mainstream traditional management theory and the myriad of complementary approaches that have contributed to the development of alternative approaches to organisational and management theory. The dominant stream of management theory is still largely influenced by the command and control paradigm developed over a century ago by early theorists such as Weber, Taylor, and Fayol. Though the control paradigm today is closely connected to a technocratic and functionalistic perspective of management science, there is a growing awareness of the dangers of assuming a reductive and limited view of organisational complexity. In other words, it is important to recognise the role of bureaucratic, functional, and procedural-like aspects of organisational life, though it is critical to complement these perspectives with richer and more human-centred interpretations of organisational reality. This critical role is performed by, among others, communities of practice theory (Lave & Wenger, 1991; Wenger, 1999; Wenger, McDermott & Snyder, 2002; Brown & Duguid, 1991). In order to better understand the developments in terms of management thinking, it is relevant to revise the sequence of the different schools of thought that influenced the social sciences throughout the 20th century.

BACKGROUND

At the end of the 19th century, the spreading and dissemination of the use of electricity for industrial and domestic use implied profound changes in the way people lived and organised their lives. In parallel with this technologic development, there was a radical change in mentalities. The confident and self-assuring concept of human beings as rational, independent, and autonomous individuals had been the product of the Enlightenment movement of the 17th and 18th centuries, which followed from the 15th and 16th centuries' scientific revolution. Opposing the rationalist and utilitarist perspective, there emerged the idea that humans present different forms of rationality, some of which imply conflicting perspectives; that human beings are not that independent from the social structures in which they are immersed; and that there are hidden inner processes which undermine their apparent autonomy (Foucault, 1972). Following this process, holistic and systemic perspectives had to be incorporated and integrated into social sciences in order to acknowledge this complexity.

At the end of the 20th century, or rather throughout the whole century but more visible in the last quarter of the century, a similar move has occurred which has intensified the previous development. This development is still going through, and may go unnoticed if we fail to recognise the need for a change in perspective and of point of view. This change places its focus and its epicentre on the intrinsic and inherent nature of all human action and thought as socially embedded phenomenon. In order to grasp the importance of this change, it is critical to point out that this notion of social embeddedness has surpassed the traditional binary opposition between individual and social issues which still permeates current and mainstream management and organisational perspectives. Instead of opposing or separating psychological and sociological issues, it treats the individual and the collective, the internal and the external, the inner and the outer world as a unique single reality. In other words, it does not partition and divide, study each isolated part, and then take the result of this process for the whole. Rather, it takes the whole from the start.

Several contemporary organisational theories follow this pragmatist approach. Pragmatism was developed by Peirce who, at the end of the 19th century and together with Saussure, created the two large schools of semiotics that have been largely influential throughout the 20th century. Among these non-dualistic, post-cognitivist, and post-structuralist thinking approaches are Stacey's complex responsive processes (2001), Checkland's soft systems methodology (1984, 1999), Eijjnatten's chaordic systems thinking (2003), Alvesson and Skoldberg's reflexive methodology (2000), and Weick's organisational sense-making (1995, 2001).

MAIN FOCUS: THE EPISTEMIC SHIFTS

The importance of the concept of communities of practice at an organisational level is parallel to the growth in the interest of management approaches such as organisational learning and knowledge management. At a broader level, this development reflects the reactions from the management and organisational areas to the reality of the knowledge economy of the information age (Kearmally, 1999). This movement may be considered as the tip of an iceberg, as the culmination of a long process of development that is still going on.

Social philosophy is a valuable reading matrix for the interpretation of the current complexity of an organisation's environments. If we want to understand the work of Aristotle, we have first to grasp what issues and questions he was trying to answer—his context. In order to understand how to implement organisational practices such as collaborative work and learning, or knowledge creation and sharing, we first have to grasp the necessary conditions for them to work—that is, the relevant community of practice, of learning, or of interest. Communities refer to the form and context of human interaction, to the situated and embodied character of human action and thought. The community concept brings up the social, cultural, and historical underpinnings of individual embeddeness. The hidden part of the iceberg includes a myriad of threads which the philosophy of the social sciences may help to untangle.

Applying this line of reasoning to the field of communities of practice is equivalent to trying to unveil the hidden influences and underpinnings that condition its development, as well as its potential for action: the exploration of the full capacity of the communities of practice theory and practice within organisational settings, the thought-possibilities and action-possibilities of communities of practice as such.

The argument is that social sciences, as a whole, frame and condition the emergence of theories and concepts such as, for example, communities of practice theory. Foucault did a historical analysis of social sciences, or the "human sciences" as he called them (Foucault, 1970; Delanty, 2003; McHoul & Grace, 1993). He looked at the structure of knowledge of a time, at its way of establishing order. He starts long before the existence of the human sciences, and examines the development of the fields known in the seventeenth and eighteen centuries as general grammar, natural history, and the analysis of wealth. He considers the question of what marks the shift into the modern world and claims that before the 18th century, man did not exist. Of course human beings existed before that, and may even have looked at themselves as the centre of the universe. But they were central because God had made them that way. Man was then left with only himself at the centre, as the sole source of knowing, and thus turned to intense examination of what this knowing being was. The Human Sciences sprang up as old fields were re-examined, with a new notion of Man as both the object and the subject of study.

From empiricism and positivism, through rationalism, structuralism, interpretativism, pragmatism, post-structuralism, and post-modernism, the development of the philosophy of social sciences presents a broad array of trends and approaches. Though each one of them may still be present today through the influence it had in the development of specific knowledge areas, when taken as a whole it is possible to differentiate four epistemic shifts throughout the 20th century (Delanty & Strydom, 2003). The first and second shifts developed in the first part of the century, and the third and fourth on the second part. These divisions are not to be taken as once and for all changes, as each one of them still persists today. They mutually influenced and reacted against each other from the start.

The first is the *logical turn*, in early 1920s, originating in the work of the Vienna Circle, which reacted against 19th century's positivism and promoted logical positivism. Authors such as Schlick and Carnap are examples of this period. Wittgenstein travels at the time connected with Cambridge and Vienna thinkers.

The second shift is the *linguistic turn,* which developed out of the works of Saussure and Peirce at the turn of the century, and then it was further developed by Wittgenstein, Morris, and many others throughout the first half of the century. Levy-Strauss developed structuralism by applying Saussure's thought to culture in general. The interest in language included not only the syntax and semantic aspects, the structure and meaning, but also and above all the pragmatics of language use.

The third epistemic shift is the *context turn*, from post-war to the 1970s, and refers to an extension of the linguistic turn into a full historical-cultural revolution which radically contextualised science. Kuhn's work on the conflict of paradigms reflects this change. This development implied also a relativist turn. Further examples of this movement are the feminist standpoint epistemology, radical hermeneutics, constructivism, poststructuralism, and postmodernism, as well as the works of Foucault, Derrida, Rorty, and Bourdieu.

The fourth shift is the *knowledge turn* during the last quarter of the century; it is an attempt to deal with the problem of the inherent ambiguity and diffuseness of context. It is the turn towards knowledge in the discourses of the human and social sciences.

...knowledge recovers from the full implications of the historical-cultural turn...Knowledge is less about knowing reality than about emergent forms of the real and reflexive relation to the world in which reality is shaped.... (Delanty & Strydom, 2003, p. 10)

Apel, Habermas, and Fuller's work are examples of this shift. Delanty and Strydom call attention to the cognitive practices, structures, and processes that are constitutive of knowledge creation, that in turn occurs within research programs, traditions, and scientific communities. These authors refer that in parallel with the rise of cognitivism, and the interest

in reflexivity, new controversies emerged such as the constructivist versus realism dilemma.

From this broad spectrum of approaches related to the historical development of social philosophy, we will situate the emergence of the communities of practice theory.

FUTURE TRENDS AND INFLUENCES

Marginally ignorant and indifferent to the logical, linguist, context, and knowledge epistemic turns of the social sciences, the vast majority of dominant theories and approaches within the fields of management science and organisational theory grew out of the influence, and still remain today largely influenced, by positivist, Cartesian, rationalist, mechanist, and utilitarian philosophical approaches, which developed out of the European Enlightenment. The sophistication of management theory and practice often remains secured within the broad positivist umbrella. This influence is usually subtle and pervasive so that more often than not it remains invisible, thus unquestioned and uncriticised.

Under this perspective, none of the epistemic turns directly and fundamentally influenced the core and functionalistic approach of mainstream management science, as it still remains constitutively determined by pure rationalist influences. Simultaneously to this core-functionalistic management centre, there is a minority and peripheral development which has long invested in non-orthodox and non-traditional management and organisational studies, ranging from cultural and sociological analysis, to the study of organisational development. It is critical to acknowledge these contributions, though still today, they are far from attracting the attention that they deserve. Even within the relatively recent fields of organisational learning and knowledge management, there are still prescriptive, linear, reductionist, and immediatist trends. The holistic, complex, and interpretative perspective of organisational reality is more an ideal than a widely disseminated reality.

These comments are not a dualistic, black-and-white analysis, but rather they aim at calling attention to the immense importance and the critical role of the communities of practice concept. By definition, this concept is intrinsically and inherently non-positivist. It is dynamic, flexible, and open, because it does not stand on rigid rules, clear-cut definitions, quantitative rationale, and universal laws. Its focus is on the particular and not on the general, on the local and not on the universal, on quality and not on quantity. This implies that the theory of the communities of practice potentially embodies every single contribution of all different epistemic turns which occurred in the social sciences during the 20th century.

A second aspect of analysis and directly related to this discussion is what Delanty and Strydom (2003) referred to as the fourth epistemic shift, the knowledge turn. Under their view, there is a radical importance given to cognitivism which becomes as if the new centre of the universe. However, cognitivism is still under a Cartesian and dualistic influence, focusing on the passive observer; on the rational, autonomous, and independent subject;, and on the mind as the centre and single focus of all rationalisation processes. Here, once again, the communities of practice theory implies a 180-degree shift as it calls attention to the intrinsic social nature of knowledge creation, and to the cultural, social, and historical embeddedness of every single human activity, from the external, visible interaction to the internal and invisible thought processes.

CONCLUSION

This somewhat harsh and direct discussion needs a devil's advocate comment. Both positivist and cognitivist approaches represent extremely valuable contributions to humankind and to science in general. The extraordinary development of science and technology, as well as the direct gains obtained from cognitivist research, are overwhelmingly important. At a different level, and within the management field, the functionalistic and mechanistic approaches are crucial in defining routines and procedures. No organisation could survive if it completely ignored its basic and repetitive tasks, and its rigid and bureaucratic structures. The issue that we have been referring to goes beyond these contributions and relates to the urgent and drastic need to pay as much

attention to the visible, individual, immediate, repetitive, and standardised issues as to the invisible, collective, complex, subtle, and dynamic aspects of organisational life. Communities of practice theory corresponds to this effort which implies a hidden and subtle thought revolution. Referring again to the epistemic shifts, this thought revolution takes aspects of all previous contributions and integrates them within the fourth knowledge turn, which then becomes not solely restricted, reduced to, and limited by a cognitivist approach. And this is the greatest challenge and potential contribution that may be expected from communities of practice theory.

REFERENCES

Alvesson. M., & Sköldberg, K. (2000). *Reflexive methodology: New vistas for qualitative research.* London: Sage Publications.

Benton, T., & Craib, I. (2001). *Philosophy of social science.* London: Palgrave.

Brown, J., & Duguid, P. (2000). *The social life of information.* Boston: Harvard Business School Press.

Brown, J., & Duguid, P. (2001). Knowledge and organisation: A social-practice perspective. *Organisational Science, 12*(2), 198-213.

Checkland, P. (1984). *Systems thinking, systems practice.* West Sussex, UK: John Wiley & Sons.

Checkland, P. (1999). *Soft systems methodology: A 30-year perspective.* West Sussex, UK: John Wiley & Sons.

Delanty, G. (2003). Michel Foucault. In A. Elliot & L. Ray, L. (Eds.), *Key contemporary social theorists.* Oxford, UK: Blackwell.

Delanty, G., & Strydom, P. (2003). *Philosophies of social science.* Berkshire, UK: McGraw-Hill.

Eijjnatten, F. (2003). Chaordic systems thinking: Chaos and complexity to explain human performance management. *Proceedings of Business Excellence '03.*

Foucault, M. (1970). *The order of things: An archaeology of the human sciences.* London: Tavistock.

Foucault, M. (1972). *The archaeology of knowledge.* London: Tavistock.

Kearmally, S. (1999). *When economics means business.* London: Financial Times Management.

Lave, J., & Wenger, E. (1991). *Situated learning: Legitimate and peripheral participation.* Cambridge, MA: Cambridge University Press.

McHoul, A., & Grace, W. (1993). *A Foucault primer—discourse, power and the subject.* New York: New York University Press.

Stacey, R. (2001). *Complex responsive processes in organisations: Learning and knowledge creation.* London: Routledge.

Weick, K. (1995). *Sense making in organisations.* Thousand Oaks, CA: Sage Publications.

Weick, K. (2001). *Making sense of the organisation.* Oxford, UK: Blackwell.

Wenger, E. (1999). *Communities of practice: Learning, meaning and identity.* Cambridge, MA: Cambridge University Press.

Wenger, E., McDermott, M., & Snyder, W. (2002). *Cultivating communities of practice.* Boston: Harvard Business School Press.

KEY TERMS

Cognitivism and Post-Cognitivism: Cognitivism is a scientific branch of social sciences; it is also known as being part of cognitive science. Cognitivism focuses on cognition and on cognitive processes by following a perspective that is centred on the individual, on the idea of the mind, and on the neuro-physiology of brain processes. The Cartesian dualism and the radical distinction between individual and social processes, the rationalist and utilitarist perspectives that interpret the human subject as an independent and autonomous entity, and the results-oriented and objectives-centred approaches to human action may all come under the broad umbrella of cognitivism. Mainstream manage-ment theory is largely influenced by cognitivist approaches. Post-cognitivism does not deny the positive contributions brought by cognitivist thinking, though it highlights the need for further developments and for the exploration of alternative approaches. Communities OF Practice theory may be understood as being part of the post-cognitivism movement, which calls attention to the social embeddedness and embodiedness of all knowledge creation processes.

Epistemology and Epistemic Shifts: Epistemology corresponds to the philosophical branch that studies how knowledge fields develop. It corresponds to "knowledge about knowledge." As different epistemological traditions develop and mature, certain schools of thought may be identified. These alternative lines of thinking may develop into epistemic shifts which correspond to a homogeneous influence that spreads through several knowledge fields and across disciplinary borders.

Structuralism and Post-Structuralism: Structuralism, systems thinking, and single-text approaches may all be contrasted to post-structuralism, post-systems thinking, and multiple-texts approaches to reality. These different terms correspond to different knowledge fields and disciplines, though they share the same logic and rationale. Both structuralism and systems thinking developed in the post-war years, and in the 1950s, the former in philosophy and social sciences and the latter in computing science. Systems thinking has also been largely influential in management science throughout the second part of the 20th century. Structuralism developed in the '50s out of an interpretation of Saussure's semiotic theory, which had itself been developed at the turn of the nineteenth to the twentieth centuries. Its initial development is related to Levy-Strauss' work in anthropology, and it then rapidly spread to become one of the major influences in the thinking of the second part of the 20th century. The structuralist interpretation of Saussure's work highlighted the perspective on the analysis of language use in terms of a closed system, irrespective of its historical and social context and its cultural links. This process of analysis may be said to be part of a single-text approach so that reality itself or, in the case of anthropology, the living habits of a particular primitive tribe could be read and interpreted as a single text, as separate parts of a single and unified whole. Post-structuralism and post-systems thinking go one step ahead in this analysis by highlighting the importance of acknowledging the complexity and the multiple layer of influences brought by a historical, social, and cultural context. Therefore, these perspectives interpret reality as multifaceted and complex, in constant and dynamic transformation, unpredictable and uncontrollable, and a multiple-text analogue. The post-structuralism and post-systems thinking perspectives are becoming gradually acknowledged by management and organisational theoreticians, and they correspond to a critical contribution in terms of the understanding of organisational complexity. Communities of practice theory may be interpreted as a net contributor to this more complex approach, and post-structuralist movement by its social and cultural awareness to organisational reality.

Social Resistance in Virtual Communities

Rahul De'
Indian Institute of Management Bangalore, India

INTRODUCTION

The period from 1994, after the release of the Web browser, Mosaic, until the turn of the century saw the upsurge of what was termed e-commerce, which grew into a much-hyped and much-invested proposition that followed a predictable cycle of boom and then bust. Though the value propositions of e-commerce, as promised in business-to-business, business-to-consumer, and consumer-to-consumer models, survived, they drew much more attention from the media and publications than was, possibly, due to them. What was happening simultaneously with the business explosion of the Web was the alternative use of the Internet as an arena of dissent—as an organizing medium, as an activist space, and as a medium for counter-propaganda. These phenomena were not necessarily unnoticed or in any way secretive in nature, but they did not occupy the front pages of the media, and they did not attract investors. These phenomena were both defined and adopted by people in various capacities to advance a cause, an idea, or simply act.

There are 605.60 million users of the Internet worldwide, as estimated by the Scope Communications Group (http://www.nua.com), a Dublin-based company. Given that there are about 6.2 billion people in the world (Population Reference Bureau, http://www.prb.org) as a whole, the number of Internet users is about 9.6% of the total population. In comparison to television, where the estimates are around 4 billion viewers around the globe, the reach of the Internet seems to be small, but there remains a crucial and defining difference: the Internet enables users to participate in the content whereas television does not. Television and other media have tremendous reach but only as broadcast sources: a few control the content broadcast to many.

The phenomena of virtual communities on the Internet was recognized early in the 1990s and was defined as groups of people that communicate via the Internet. This is the broadest possible definition.

The Internet is a network of telecommunications networks, and its representation as a virtual community becomes possible as its members take for granted that the computer networks are also social networks spanning large distances (Wellman & Gulia, 1997). Aggregations of virtual communities form the society of the Internet, where the structure of this society is defined by the patterned organization of the network members and their relationships (Wellman, 1996). Defined in this manner, the Internet society is now amenable to analysis by sociological and political theories and constructs.

Various communities and groups have emerged in the society of the Internet. These communities are distinguished by their thematic content and the delivery mechanism they use. Free service providers, such as Yahoo! Groups, support thousands of informal groups with restricted or unrestricted access that define communities in the broadest sense. Other types of communities include chat rooms, multi-user gaming, *metaworlds*, blogs, and interactive video and voice (Wallace, 1999). Communities may form and disband easily on the Internet. The Internet is thus a virtual space that is not constituted by physical objects of land, bricks, cement, furniture, but of a collection of files, folders, and accounts. These digital assets are created as quickly as they are destroyed; what perpetuates them is the common interest of the community. Further, the members of this community may be widely dispersed geographically and so may the files and accounts of the community, their physical presence, and geographic location at any point of time, irrelevant to their functioning.

To understand and examine social resistance in virtual communities, the intent of this article has to first draw on the various themes in research literature that define and shape the contours of the discourse. We begin with a review of some of the literature on virtual communities. Social resistance, as understood by the acts of organizing, activism, and counter-propaganda, are then examined and their characteristics highlighted. We conclude with a summary of the basic ideas in this article.

BACKGROUND

Virtual communities on the Internet constitute an active area of research in both business and social science disciplines. The literature in information systems (IS) has studied various aspects of virtual communities quite extensively, including virtual teams, virtual organizations, virtual enterprises, and the issues related to information sharing, cooperation, collaboration, trust, and so forth. The literature in the social sciences has examined the very nature of virtual communities, their organization, scope, rationale for existence, and their implications for society. The principal distinction in the research is the understanding of communities on the basis of their observable properties, which may be engineered, as in the former stream of research vs. understanding the basis of their underlying sociology, as in the latter.

For example, in IS research, virtual communities are understood as social networks in a virtual space that bring people together for some purpose (Ridings, Gefen & Arinze, 2002). One binding force is trust, which is central to the sustenance of the group and its ability to generate participation and contribution. The motivation for the study is the insight that can be obtained on the digital economy and its consequent value in marketing campaigns and market research. (However, it will be noted that trust is not an essential binding force in all virtual communities. For example, in online gaming communities, trust is not an issue at all; the electronic environment of the game ensures that players play by the rules.) In another study, Raybourn, Kings, and Davies (2003) examine "cultural markers" used to define particular communities. The research identifies cultural cues that will help strangers identify others as also belonging to the same community to facilitate information sharing and communication. In both cases, the papers directly address issues related to retaining members and facilitating electronic interchanges for commercial benefit.

Virtual communities exist, and people flock to them because they derive economic benefit from such participation. Incentives for participation include gifts, public goods, and benefits derived from

reciprocity and sharing (Butler, 2001; Kollock, 1997). Participants gain from resources such as information, influence, and social support that are quantifiable as utilities or economic benefits.

Other research on virtual communities examines deeper propositions about society and the relations within societies that shape technology use or that are shaped by technology formations. With a view to examining emergence of newer forms of society, this research (Burkett, 2000; David, 2003; Stevenson, 2002) assumes that questions about the role, access, and impact of technologies need to be asked on an epistemological basis before the questions about the manipulations of technology for commercial reasons are addressed. The fundamental assumptions are: the move to the "information society" is not necessarily inevitable; the diffusion of technology is not also a metric of the advancement of a society; the information society is more democratic and participatory; and "given enough information we can solve all the world's problems"(Burkett, 2000, p. 680).

The phenomenon of social resistance on the Internet refers to a particular usage of technology that enables groups of people, loosely or densely bound, to actively engage against an overarching, hegemonic power (Conway, Combe & Crowther, 2003). Technology is manipulated and organized to support formations within groups and the emergent resources of such formations are then used in the resistance. The resistance may take many forms, those of organizing, activism, or of counter-propaganda, and in each case, technology plays a central role in coordination, information dissemination, and information gathering, among other uses.

SOCIAL RESISTANCE

When theorizing about the diffusion, spread, or acceptance of a technology in society, it is of interest to consider not only the extent of the diffusion but also to bear upon the changes that the technology introduces in the structures and functioning of the society. These changes may be studied in different ways. An interesting aspect of change introduced by technology in society has to do with how the unempowered, or marginal, can use technology to challenge dominant forces and the powerful (Conway et al., 2003; Kamat, 2002). Technology becomes a tool by which people can reinvent social relations or recast themselves in power relations; technology mediates the opposing dynamics in society and posits newer social formations. The following sections describe three radical uses of technology in society: one to organize resistance to war; another to actively challenge and shut down funding for hate groups; and a third as a counter-propaganda medium. These examples constitute brief case studies of the uses of technology; the general characteristics are obtained from the literature and other examples.

Organizing

On February 15, 2003, in over 600 cities around the world, millions of protestors took to the streets, marched, chanted, and waved banners against the impending bombing of Iraq by a U.S.-led coalition. Cities such as London, Berlin, Rome, Johannesburg, Kiev, Kuala Lumpur, Manila, and Sydney saw massive turnouts, and, in London, one commentator said that it was the largest protest of its kind in the history of Great Britain. Many have since stated that February 15 saw the largest coordinated people's protest of any kind in history. The coordination of the entire protest in 75 countries was managed through the Internet, through Web sites, e-mail, and list-servers. Although the message for the protests was carried forth in conferences (such as the World Social Forum and the Asia Social Forum), meetings, teach-ins, and local rallies, its worldwide coordination was effected through the Internet. Some characteristics of the organizing process are that although local government and elected officials were involved in individual capacity, there was no government-level official sanction or involvement in the organizing; the organizers put forth a common message, cutting across languages, cultures, and religious affiliations; there was an explicit agreement on the general character of the protests, emphasizing non-violent, civil disobedience techniques; the funding for the organizing was borne entirely by contributions of the people without any corporate or governmental involvement; the Internet formed the backbone for the organizing in that the plan was formalized over the Internet, but further communication at the local level was effected through community radio and other media that was available to those without Internet access.

This facet of the society of the Internet, of a space for organizing, contrasts sharply with the one written about in the mainstream media – where the Internet plays a role in leveraging business processes or in feeding content to passive consumers. Even though the protests were eventually unsuccessful, as the U.S.-led coalition did go to war against Iraq, the organizing of the protests represented a significant event in the global society of the Internet. The organizers relied on the openness, the English language (mainly), absence of censorship, the rapidly developed norms of organizing, the multiple formats of data exchange, and the widely available free software to successfully coordinate the rallies.

Activism

The Internet also provides a medium for direct action for a cause, in addition to its support capabilities. E-mail campaigns are widely known and used by people and organizations around the world. Some are designed to be irritants, as in the campaign by some students at the University of California, Berkeley, to jam the telephone lines of a notorious televangelist (who had been embroiled in tax fraud and sex scandals) by programming a computer dialer to incessantly dial the toll-free number, while others are directed at canvassing and lobbying public officials via e-mail, SMS, fax, and pager messages. Hacktivism is a term used to describe the clandestine use of computer hacking to promote a political cause (Manion & Goodrum, 2000). Hacktivists protest corporate control of the Internet through denial-of-service attacks on servers, as well as other forms of electronic protests that target Internet-based and accessible properties. Electronic civil disobedience follows a similar line of protests, where unfair or unjust laws are challenged and violated via electronic means.

The Internet provides a medium, perhaps unrivaled in history, where the rich, influential, and powerful have known and public addresses that people across the world can access to put across their message. Even though the messages may be scanned, filtered, pruned, and altered by layers of intermediaries, both human and nonhuman in nature, their impact is direct and tangible in many cases. This mode of activism has the following characteristics: the activists may be geographically distributed, where the target of their activism is a known entity with an address on the Internet; the actions may be performed by humans or by software agents acting on behalf of humans (as in the case of denial-of-service programs unleashed by hackers); the Internet permits both synchronous and asynchronous actions, where the action may be suited to the demands of the situation; and the Internet also permits anonymous action, where the activists are not obliged to leave a trail of their identity.

As an example of activism on the Internet, consider the case of the Campaign to Stop Funding Hate (CSFH), a loose coalition of students, professionals, workers, and artists who banded together on the Internet to stop the flow of dollars from nonresident Indians to sectarian hate groups such as the RSS (Rashtriya Swayamsevak Sangh or the National Self-Service Organization) and its affiliates in India. In October 2002, the CSFH released a document titled, "A Foreign Exchange of Hate: IDRF and the American Funding of Hindutva", simultaneously in India and on its Web site (http://www.stopfundinghate.org). Newspapers and the television media in India carried reports and interviews with the authors of the report that said that companies such as Cisco, Oracle, Sun, Microsoft, AOL, and so forth were used through the means of matching funds to contribute large sums of money to the IDRF (India Development and Relief Fund), a Maryland-based charity that was a front organization for the RSS and its affiliates. CSFH had used information available on the Web sites of the RSS organizations in the U.S., as well as public documents available from government resources, to conduct their research. After releasing the report, the CSFH set up petitions, letter campaigns, and an FAQ on their Web site to further inform the world media and donors about the activities of IDRF. The repercussions were immediate— many of the corporations stopped their contributions to IDRF, and many media channels carried reports on the exposure. CSFH sustained its efforts in publicizing the IDRF's activities, and stories appeared in major newspapers such as the *Financial Times* (report by Luce & Sevastopulo, February 20, 2003) and in the *Wall Street Journal* (report by Banks, February 18, 2003).

Counter-Propaganda

According to Webtster's Dictionary, propaganda means the methodical propagation of a particular doctrine or the material spread by the advocates of a doctrine. In lay terms, propaganda connotes a negative sense of forcing an idea or political philosophy on people by clever demagoguery, manipulation, or persuasion. Herman and Chomsky (1988) state that the mass media in the United States of America performs the role of inculcating individuals with values, beliefs, and codes of behavior by means of propaganda. When the media are controlled by those in power and those who have money, by their "propaganda model", they posit that news and information over the media are filtered to determine news fit to print, marginalize dissent, and allow government and dominant private interests to get their message out to the public. A similar view is held by Bagdikian (1997) who states that the major media in the U.S. are controlled by a handful of corporations whose interests are always positively presented and whose influence over citizens through control over the images and text of the media is more powerful than schools, religion, parents, and even the government itself.

The Internet is also a rich source of news and information, but it is not under the control of governments or wealthy private interests. (One can add as a caveat that the government of the U.S. funded the development of the Internet, still maintains the backbone, and through an act of legislation, made it open to the public.) In this respect, it is not party to the propaganda issued by the other media. As was pointed out earlier, the Internet enables users to post news and messages as much as to consume. In recent times, the Internet has become a *de facto* source for alternative news and viewpoints. Ninan (in *The Hindu*, March 2, 2003) notes that the Internet is now beginning to upstage the mainstream media, TV, and newspapers, as the main source of news for "news junkies" all over the world. The Internet reverses the propaganda intent of the controlled media in that alternative viewpoints can thrive and prosper. Mainstream media ventures such as CNN, BBC, and many others have set up Web sites that cater to this need of news seekers on the Internet, in many cases, to continue their propaganda efforts.

However. they remain far from dominating the content. As an example of counter-propaganda, consider the case of the EZLN in Mexico.

On January 1, 1994, an army of revolutionaries, known as the EZLN (Zapatista National Liberation Army), walked into San Cristobal in the Chiapas region of Mexico and took over its administration. The move had been carefully planned, and the leader Subcomandante Marcos had readied a slew of reports, articles, press releases, and news broadcasts that were released to all available media—television, newsprint, radio, and the Internet—in a coordinated and sequenced manner. The movement had begun and gained momentum as a reaction to the NAFTA (North American Free Trade Agreement) that the Mexican government had agreed to join and which subsequently resulted in massive displacement of labor, loss of livelihoods, loss of land to agribusinesses, and loss of resources to companies moving in from the U.S. The Mexican state's attempts to thwart the revolution and its moral support were defeated by the steady stream of communiqués and press releases from the EZLN. Marcos used the Internet, particularly e-mail, to broadcast interviews and opinions to people across the globe and particularly to those in Mexico who could not have received these through the state-controlled TV and co-opted newspapers. Public sympathy grew for Marcos' campaign even as the population was able to express its dislike of the corrupt and inefficient government. The communiqués and press releases by the Zapatistas may be termed *counter-information*, or counter-propaganda, as it was intended to directly countermand the propaganda of the state. The press releases over the Internet reached activists for human rights, anti-NAFTA organizations, and others through newsgroups and list-servers. The effect of these reports was that the repression of the Mexican government was often verified and publicly announced, much to the embarrassment of the state, and the awareness of the Zapatistas' ideology and sympathy for them grew around the world. In the words of Cleaver (1988), the Zapatistas formed an "electronic fabric of struggle to carry their revolution throughout Mexico and around the world."

The characteristics of counter-propaganda over the Internet are: counter-propaganda can be generated by those without the power of the state or private business and is not easily stifled; the content of counter-propaganda may be timed to the exigencies of the situation; and the messages may be targeted at like-minded and supportive groups around the world.

CONCLUSION AND FUTURE RESEARCH

The use of virtual communities for organizing, activism, and counter-propaganda on the Internet are facilities that have grown, not as a direct result of premeditated design by the users, but because of requirements that arose from the society at large. Those who look to the Internet for sustainable models of community or of commerce have to include in their analysis these serendipitous uses. Research shows that the social capital (Uslaner, 2000) of the Internet may or may not be increased by such deployments; however, these phenomena are important and have to be included in the study of virtual communities.

IS research relies on a fundamental epistemology of individuals as economic actors who act, mostly rationally, in their own interest and in conformity with the implicit goals of the organizations to which they belong. Within virtual communities, individuals acting as citizens (of their nation or of the world) need not act in a rational manner, as understood in the context of the organization, if they are participating in political organizing, activism, or counter-propaganda. Further research is required to generate an alternative set of hypotheses to understand their behavior in the larger context of the society. For example, the issues related to monitoring employee e-mail and Internet usage, a topic of much debate within the IS community, rely on assumptions regarding the individuals role and responsibilities within the organization. Monitoring of political activism or organizing requires a fresh set of decisions regarding what is to be considered private and what is a public, political space within organizations.

REFERENCES

Bagdikian, B. (1997). *The media monopoly.* Boston: Beacon Press.

Bank, D. (2003). companies face quandaries over matching-figt programs. *The Wall Street Journal,* February 18, 2003.

Burkett, I. (2000). Beyond the "information rich and poor": Futures understandings of inequalities in globalising informational economies. *Futures, 32,* 679-694.

Butler, B. S. (2001). Membership size, communication activity, and sustainability: A resource-based model of online social structures. *Information Systems Research, 12*(4), 346-362.

Cleaver, H. (1998). Zapatistas and the electronic fabric of struggle. Retrieved June 10, 2005, from *http://www.eco.utexas.edu/faculty/Cleaver/zaps.html*

Conway, S., Combe, I., & Crowther, D. (2003). Strategizing networks of power and influence: The Internet and the struggle over contested space. *Managerial Auditing Journal, 18*(3), 254-252.

David, M. (2003). The politics of communication: Information technology, local knowledge and social exclusion. *Telematics and Informatics, 20*(3), 235-253.

Herman, E. S., & Chomsky, N. (1988). *Manufacturing consent: The political economy of the mass media.* New York: Pantheon Books.

Kamat, S. (2002). *Development hegemony: NGOs and the state in India.* New Delhi: Oxford University Press.

Kollock, P. (1997). The economies of online cooperation: Gifts and public goods in cyberspace. In P. Kollock & M. Smith (Eds.), *Communities in cyberspace* (220-239). New York: Routledge.

Luce, E., & Sevastopulo, D. (2003). Blood money. *The Financial Times,* February, 20, 2003.

Manion, M., & Goodrum, A. (2000). Terrorism or civil disobedience: Toward a hacktivist ethic. *Computers and Society, 30*(2), 14-19.

Ninan, S. (2003). An advance from all directions. *The Hindu,* March 2, 2003.

Raybourn, E., Kings, N., & Davies, J. (2003). Adding cultural signposts in adaptive community-based virtual environments. *Interacting With Computers, 15*(1), 91-107.

Ridings, C., Gefen, D., & Arinze, B. (2002). Some antecendents and effects of trust in virtual communities. Strategic Information Systems, *11,* 271-295.

Stevenson, T. (2002). Communities of tomorrow. *Futures, 34,* 735-744.

Uslaner, E. (2000). Social capital and the net. *Communications of the ACM, 43*(12), 60-64.

Wallace, P. (1999). *The psychology of the Internet.* Cambridge, UK: Cambridge University Press.

Wellman, B. (1996). For a social network analysis of computer networks: A sociological perspective on collaborative work and virtual community. *Proceedings of the 1996 Conference on ACM SIGCPR/SIGMIS Conference,* Denver, CO (pp. 1-11). ACM Press.

Wellman, B., & Gulia, M. (1997). Net surfers don't ride alone: Virtual communities as communities. In P. Kollock & M. Smith (Eds.), *Communities and cyberspace* (pp. 331-366). New York: Routledge.

KEY TERMS

Boundary: A systems concept, whereby all systems are held to have a boundary, and often judgments at the boundary will yield insightful results.

Hackers: Computer experts who break into networks or computer systems with (usually) malicious intent.

Hacktivism: Clandestine use of hacking for the advance of political causes.

Organizing: The process of identifying, specifying, and assigning work, grouping work and resources into a structure, and establishing a chain of command between individuals and groups.

S

Propaganda: Methodical propagation of a doctrine or the material spread by the advocates of a doctrine.

Social Resistance: An organized struggle against a political ideology or process.

Virtual Communities: A term that describes the groups of people with shared interests that communicate socially via computer networks.

Sociotechnical Theory and Communities of Practice

Andrew Wenn
Victoria University, Australia

INTRODUCTION

Communities of practice (CoPs), by their nature, are social entities. Such communities may be large or small, geographically dispersed or located within a confined region. Essentially, communities of practice consist of members who chose to come together because they have a passionate dedication to sharing knowledge and a desire to develop their own and other's capabilities (Wenger & Snyder, 2000). No matter what type of CoP (collocated or virtual; intra or interorganizational) communication is one of the prime desiderata. Thus, it is highly likely that technology of some form will be involved. For instance, a virtual community of practice may use e-mail or a more sophisticated groupware application to keep in touch. CoPs within a knowledge management environment will certainly have access to technology.

To understand the workings of such communities requires a theory that enables us to deal at the levels of the individual, the group, and the larger world in which the community is embedded (Lave, 1988). Any such theory must be able to account for the role of technology within the community as well as its social aspects.

BACKGROUND

Sociotechnical Theory

As the compound word *sociotechnical* indicates, it is, according to Coakes (2002), a combination of two ideas or paradigms—the social and the technical. It is an attempt to provide a view of technology, organizations, and people that is more holistic and less biased than either could be on its own. Furthermore, "[s]ociotechnical perspectives can be characterised as holistic, and whilst not being panoptic in character, take a more encompassing view of the organization, its stakeholders in knowledge and

the environment in which it operates, than [many other perspectives] ...which are limited by their origins and paradigms" (p. 4).

Information systems, as with knowledge management systems, should not be seen as technology in isolation; they consist of humans, technological, and social artifacts linked in networks of relationships. These networks are called sociotechnical networks. The strong emphasis on the human or social side is considered an important factor in any information system (Clarke et al., 2003). There are several sociotechnical approaches that could be used. One, derived from the work of the Tavistock Institute (Coakes, 2002; Mumford, 2003), is concerned with highlighting the moral and ethical issues associated with the work environment and aims to enhance worker involvement in change within this environment. Mumford (2003), whose focus is on the application of sociotechnical principles to the systems design process, remarks that both the social (human) needs and the technical must be given equal weight where possible.

Another approach that is seeing increased use in information systems is that of Actor-Network Theory (ANT) (Callon, 1986; Law & Callon, 1992; Tatnall, 2003; Wenn, 2003). The emphasis of ANT is on the interplay between the social and the material, how they come together and are coproduced, and the relationships that develop between them (Callon, 1986). In ANT, the social and the technical are often called human and nonhuman actants, or more often, just actants (Latour, 1987). The term *actant* was deliberately chosen so that the social and the technical can be treated in equal fashion. It uses a semiotic approach whereby actors and actions are seen as network effects, and the relationships between actors are traced through the strategies, practices, and negotiations employed within the network. ANT does not seek explanatory factors for innovations but describes and constructs theories of actions that arise from technical and social negotiations. Unlike the approach of the Tavistock Institute, ANT does not specifically focus on the moral and ethical issues— it prefers neutrality—it does, however, enable us to see how networks of associations arise. Thus, it would seem to have much to offer for understanding the complexities of communities, both in their internal practices and the boundary work (Star & Griesemer, 1989) required from communities wishing to maintain their connections with the outside world (Wenger, 1998).

FUTURE TRENDS

As more is learned about the way knowledge claims are constructed, the methods by which local orderings shape such claims and how these are transmitted to other communities of practice either internal or external to the organization the more it becomes apparent that our understanding of such practices cannot be described by a hard and fast theory (Turnbull, 2000). Sociotechnical theories often assume that categories of social and technical are firmly fixed. One recent proposal is that sociotechnical approaches should also pay more attention to the way society and technology co-construct each other (Misa, Brey & Feenberg, 2003). Co-construction is the idea that technologies, society, and culture interact deeply and mutually affect each other (Misa, 2003). Another promising approach is that employing Foucault's (1986) concept of heterotopian sites recently employed by Liff and Steward (2003) to analyze the communities of users that gather at cybercafés.

CONCLUSION

Sociotechnical approaches such as that arising from the work of the Tavistock Group or Actor-Network Theory have much to offer when it comes to understanding communities of practice, particularly ones that arise in information or knowledge based organizations. In such situations, consideration needs to be given to how individuals, groups, and technological artifacts interact to the mutual benefit of all concerned. It is through an understanding of these interactions and the processes of co-construction that we will be able to make more productive use of communities of practice.

REFERENCES

Callon, M. (1986). Some elements of a sociology of translation: Domestication of the scallops and the fishermen of St Brieuc Bay. In J. Law (Ed.), *Power, action and belief* (pp. 197-234). London: Routledge and Kegan Paul.

Clarke, S., Coakes, E., Hunter, M. G., & Wenn, A. (Eds.). (2003). *Socio-technical and human cognition elements of information systems*. Hershey, PA: Information Science.

Coakes, E. (2002). Knowledge management: a sociotechnical perspective. In E. Coakes, D. Willis, & S. Clarke (Eds.), *Knowledge management in the sociotechnical world* (pp. 4-14). London: Springer-Verlag.

Foucault, M. (1986). Of other spaces. *Diacritics, 16*(1), 22-27.

Latour, B. (1987). *Science in action*. Cambridge, MA: Harvard University Press.

Lave, J. (1988). *Cognition in practice: Mind, mathematics, and culture in everyday life*. Cambridge, UK: Cambridge University Press.

Law, J., & Callon, M. (1992). The life and death of an aircraft: A network analysis of technical change. In W. E. Bijker (Ed.), *Shaping technologybuilding society: Studies in sociotechnical change* (pp. 21-52). Cambridge, MA: MIT Press.

Liff, S., & Steward, F. (2003). Shaping e-access in the cybercafé: Networks, boundaries and heterotopian innovation. *New Media & Society, 5*(3), 313-334.

Misa, T. J. (2003). The compelling tangle of modernity and technology. In T. J. Misa, P. Brey, & A. Feenberg (Eds.), *Modernity and technology* (pp. 1-30). Cambridge, MA: MIT Press.

Misa, T. J., Brey, P., & Feenberg, A. (Eds.). (2003). *Modernity and technology*. Cambridge, MA: MIT Press.

Mumford, E. (2003). *Redesigning human systems*. Hershey, PA: Information Science.

Star, S. L., & Griesemer, J. (1989). Institutional Ecology, "translations" and boundary objects:

S

Amateurs and professionals in Berkeley's Museum of Vertebrate Zoology, 1907-1939. *Social Studies of Science, 19*, 387-420.

Tatnall, A. (2003). Actor-network theory: As a socio-technical approach to information systems research. In S. Clarke, E. Coakes, M. G. Hunter, & A. Wenn (Eds.), *Socio-technical and human cognition elements of information systems* (pp. 266-283). Hershey, PA: Information Science.

Turnbull, D. (2000). *Masons, tricksters, and cartographers: Comparative studies in the sociology of scientific and indigenous knowledge.* Australia: Harwood Academic.

Wenger, E. C. (1998). *Communities of practice: Learning, meaning and identity.* Cambridge: Cambridge University Press.

Wenger, E. C., & Snyder, W. M. (2000). Communities of practice: The organizational frontier. *Harvard Business Review*, 139-145.

Wenn, A. (2003). Enrolling ANT or how one understanding of sociology of technology may be of interest to information systems practitioners. In M. G. Hunter & K. K. Dhanda (Eds.), *Information systems: The challenge of theory and practice.* Las Vegas: Information Institute.

KEY TERMS

Actants: A general term used to refer to both human and nonhuman artifacts that can be acted on or move the action onto some other. Actants are heterogeneous entities that form a network.

Co-Construction: Co-construction describes the way technologies, society, and culture mutually interact to shape the product of these interactions. It is a notion that accepts there will be varying degrees of uncertainty, resistance, ambiguity, accommodation, and enthusiasm in these encounters (Misa, 2003).

Heterotopian Site: Owing its existence to the work of Foucault (1986), a heterotopia is a site that offers "mixed joint experience" to actors situated in a single real space (p. 24). Such a site relates to others within a culture in a variety of ways such as reflection, inversion, contestation, or even contradiction (Liff & Steward, 2003).

Sociotechnical Networks: The networks of associations formed between human and nonhuman actants. The actants are persuaded or enrolled into the network through a variety of socially mediated actions.

Strategic Objectives of CoPs and Organizational Learning

Diane-Gabrielle Tremblay
University of Quebec in Montreal (Teluq-UQAM), Canada

INTRODUCTION

There is more and more interest in different forms of knowledge creation and management, and the conditions necessary to succeed in such initiatives from the point of view of individuals and organizations. A great deal of this interest stems from the fact that organizations expect substantial gains from knowledge. Knowledge management is seen in many organizations as a source of potential competitiveness and innovation. The concept of communities of practice stems from this interest, but is viewed as a specific form of knowledge development, in principle more centred on the individuals and their exchanges than on "management" by the firm, although the firm does seem to have a role to play in fostering such initiatives. Thus, the use of communities of practice has emerged as a way to develop collective skills and organizational learning, in order to foster innovation and success for the organization.

Organizational learning is part of a broader concern related to the development of collective skills. We know that a large proportion of effective relations within organizations are informal, a characteristic that relates to the concerns of the communities of practice, which are usually based on informal relations. Organizational learning goes beyond individual learning, which can lead to relatively permanent changes in the individual's behavior, because it results in the development of a knowledge base which could translate into a more significant change of another kind within the organization. The knowledge is disseminated throughout the organization, is transmissible between members, is subject to consensus, and is integrated into the work processes and the structures of the organization. From this perspective, organizational learning is closely linked with "meaningful" organizational processes, which are basically routines used by decision makers to detect certain problems, define priorities, find solutions,

and attempt to improve performance. In this article, we will present research results on some strategic objectives of CoPs and the attainment of these objectives, from the viewpoint of organizational learning.

BACKGROUND

The results presented in this article are derived from action research on a dozen communities of practice (CoPs) conducted under the aegis of the *Centre Francophone D'Informatisation des Organisations* (CEFRIO[1]). To date, a dozen CoPs have actively participated in the research, which was carried out from 2001 to 2003. One hundred and eighty (180) participants answered questionnaires on starting up a CoP, and slightly less than 100 participants answered evaluation questionnaires six months later. In addition, focus groups and recordings of critical incidents in each of the communities were also conducted so as to better understand the dynamics of each of the CoPs. We will focus on the aspects related to learning, paying particular attention to the conditions and challenges that emerge from our results.

Attainment of Objectives

Although the objectives of the communities of practice can differ (Jacob, Bareil, Bourhis, Dubé & Tremblay, 2003), they were mainly aimed at learning through exchange and collaboration in our cases. From this perspective, it is interesting to note how the objectives have evolved over time. When the communities were starting up, the objectives identified by the participants were usually related to exchange and sharing of information and knowledge, better utilisation of delocalized resources, as well as the creation of a collective memory—objectives which actually pertain to knowledge sharing.

However, after a few months of work in a virtual CoP, the achievement of objectives seemed to be uneven. In fact, although certain CoPs felt that they had achieved their objectives, as was the case of a CoP in the health sector (Tremblay, 2004a), this was not so true of other CoPs. Perhaps it was still too soon to assess the achievement of objectives since, unlike project teams or groups, CoPs are not supposed to have a specific schedule and they have to learn new operating modes in a short time.

Concerning the partial achievement of the objectives of CoPs, there were various possible reasons for this, including the frequent change of CoP leader, the loss of interest on the part of management or participants, or the lack of time for participation. However, it must be stressed that developing learning and experimenting with a new problem-solving approach, which were not always among the objectives considered to be the most important at first, seemed to have been relatively well achieved by a number of CoPs, and these forms of learning are greatly appreciated by the participants.

It must be stressed that all of the CoPs operated with a knowledge-sharing tele-software. The participants were either not very familiar with the software or had to more or less master it in a few months, depending on how easy or difficult it was for them to use this software and the time—which is generally limited—that they had. The use of software such as Knowledge Forum or Lotus Notes, which was different in each case, allowed CoP participants to exchange messages. These were then grouped together on a space, and could be reviewed and re-organized according to the themes discussed in the exchanges. In principle, this is how virtual (i.e., tele-working) communities must jointly develop knowledge.

We analysed the data on success or attainment of objectives according to various demographic variables, but only two (gender and age) came out significantly in some of the analyses. For various reasons, often lack of variance in the respondents, the other variables tested did not show up as significant: level of schooling, professional category, and language have however been tested and should eventually be the object of more analyses.

The success of the CoP was evaluated in different ways, amongst which was the attainment of the strategic and operational objectives of the CoP according to the demographic variables; as mentioned, analyses (ANOVA) revealed few significant links, except with gender and age, the latter which we highlight here.

In Table 2, all statements are significantly differentiated according to age. There are some differences with gender, but almost none with all other "demographic"[2] variables tested.

As concerns differences according to gender, in terms of strategic objectives, only the objective of valuing excellence presented a gendered difference (detailed tables available in Bourhis & Tremblay, 2004). For operational objectives as well, differences according to gender are not numerous, since only the objective of facilitating exchange and sharing of information was differentiated according to gender.

Success was measured in different ways, not only in terms of attaining objectives as shown in

Table 1. Links between the attainment of strategic objectives and age

AGE		Innovation was valued	Relation with client became better	Quality became better	Excellence was valued	Rationalization	Competencies were valued	Efficiency
Under 35 yrs	Mean	3.7500	3.1333	3.4118	3.5789	3.0625	3.4000	3.6111
	N	20	15	17	19	16	20	18
	StanDev	0.85070	0.74322	0.79521	0.83771	0.85391	0.82078	0.84984
35 - 49 yrs	Mean	3.6170	3.2162	3.3810	3.5625	3.0000	3.3846	3.3529
	N	47	37	42	48	24	39	34
	StanDev	0.87360	0.82108	0.79487	0.84818	0.88465	0.96287	0.84861
50 and over	Mean	2.7000	2.8571	3.1111	3.2000	2.0000	2.7500	2.2857
	N	10	7	9	10	6	8	7
	StanDev	1.33749	1.21499	1.45297	1.31656	1.09545	1.28174	0.95119
Total	Mean	3.5325	3.1525	3.3529	3.5195	2.8913	3.3134	3.3051
	N	77	59	68	77	46	67	59
	StanDev	0.98120	0.84718	0.89384	0.91206	0.94817	0.97248	0.93319

Table 2. Links between operational objectives and age

AGE		Facilitate exchange and sharing of information and knowledge	Experiment with a new approach to problem resolution	Better use of delocalized resources	Reduce workforce	Maximize working time	Reduce duplication	Stimulate creativity	Increase learning
Under 35 yrs	Mean	4.0500	3.9000	3.8421	2.3077	2.8889	3.2778	3.6842	3.9500
	N	20	20	19	13	18	18	19	20
	StanDev	0.82558	0.71818	0.89834	0.63043	0.75840	0.89479	0.74927	0.60481
35 - 49 yrs	Mean	3.8846	3.8000	3.8000	2.7059	3.1429	3.5000	3.6327	3.8462
	N	52	50	40	17	35	44	49	52
	StanDev	0.87792	0.80812	0.88289	1.15999	0.94380	0.95235	0.97241	0.82568
50 and over	Mean	3.0000	3.0000	2.7778	1.5000	2.1429	2.5714	2.7000	3.1111
	N	11	11	9	4	7	7	10	9
	StanDev	1.26491	1.34164	1.20185	1.00000	1.21499	1.51186	1.25167	1.36423
Total	Mean	3.8072	3.7160	3.6765	2.4118	2.9500	3.3478	3.5256	3.7901
	N	83	81	68	34	60	69	78	81
	StanDev	0.96850	0.91152	0.98407	1.01854	0.96419	1.02650	1.00291	0.87630

Tables 1 and 2, but also in terms of organizational learning and professional and personal enrichment. We only highlight the significant differences here.[3] We observed that success from the individual point of view is not strongly differentiated according to gender as concerns professional enrichment and satisfaction in participation, but women value more the personal enrichment they gained through the CoP. In other evaluations of success of the CoP, the numbers given by women are systematically superior to those of men, although not significantly.

Personal and professional enrichment as well as satisfaction were slightly differentiated according to professional category, but since there is little variance (most of the respondents are professionals), we do not show them here. As concerns measures of learning, it is differentiated according to gender, women indicating that they gained more professional and personal learning in this context than male participants. It is nevertheless an important dimension for all participants.

CONCLUSION

To conclude, a number of factors related to the conditions and challenges associated with CoPs are summarized in order to identify those that would help promote the wider use of these collaborative learning practices.

Table 3. Measures of success from the individual point of view, by gender

Gender		I found my participation in the CoP very enriching from a personal point of view.	I found my participation in the CoP very enriching from a professional point of view.	I am very satisfied of my participation in the CoP.	I contributed a lot to the CoP.
men	Mean	4.1765	4.6176	3.5000	3.0588
	N	34	34	34	34
	Standard Deviation	1.76619	1.68801	1.69223	1.73975
women	Mean	5.1176	5.2115	4.0769	3.7170
	N	51	52	52	53
	Standard Deviation	1.70466	1.69586	1.78057	1.85407
Total	Mean	4.7412	4.9767	3.8488	3.4598
	N	85	86	86	87
	Standard Deviation	1.78054	1.70795	1.75913	1.82874

Participants' commitment was considered to be a crucial factor in the success of CoPs (Tremblay, 2004). However, other factors can play a role in explaining the more mixed success of other cases: for example, the lack of dynamism on the part of the CoP leader, the frequent change of leaders, or the fact that some participants did not contribute much to the CoP although they maintained that they had learned a great deal by participating. These factors must be taken into account when developing learning through communities of practice.

To sum up, there are three major challenges related to the implementation of this new form of learning and training through CoPs. First, to motivate individuals to participate in the project or the joint enterprize; second, to find the means to sustain the interest of participants, but also of the organization which supports the learning project through the CoP; and third, to establish a form of recognition (not necessarily monetary) of the participation of individuals, especially if they are expected to devote their time to it.

As regards the organizational conditions to attain the objectives mentioned above, three major conditions of success of a CoP are retained. First, the organization that sponsors the CoP should assign a leader to it, and this person should not change too often. Second, participants must trust themselves as well as their colleagues so they can contribute actively to online exchanges without fearing that what they have written, which remains in the system, will be criticized. Lastly, participants should have enough time (ideally taken from working time, if the topic of learning is linked to work) in order to contribute and learn a great deal. We believe that if these conditions are not met, it will be hard to imagine that a CoP can be a valid means to develop forms of organizational learning through the exchanges and interactions between peers, as suggested by the authors of works on communities of practice.

On the other hand, although the CoP experiences were examined over a relatively short period of time (6 to 12 months), they seem to offer a promising course of action for organizational learning through peers, exchanges, and collaboration. However, it should not be forgotten that these experiences are not implemented in a vacuum, but in specific organizational contexts. The analysis shows that these contexts should be taken into account (hierarchical or non-hierarchical culture, habit of collaboration, as well as social relations of work between individuals) since they will have an impact on the participants' commitment and the level of success of CoP experiments.

In any case, although relatively new, this CoP formula offers interesting prospects for organizational learning, but we can see that it cannot be generalized without considering various dimensions: age, gender, commitment, and various characteristics of the community need to be taken into account, since they may have an impact on attainment of objectives of the community.

REFERENCES

Bourhis, A., & Tremblay, D.-G. (2004). *Les facteurs organisationnels de succes des communautes de pratique virtuelles.* Québec: Cefrio.

Jacob, R., Bareil, C., Bourhis, A., Dubé, L., & Tremblay, D.-G. (2003). Les communautés virtuelles de pratique: Levier de l'organisation apprenante. In G. Karnas, C. Vandenberghe & N. Delobbe (Eds.), *Bien-être au travail et transformation des organisations. Proceedings of the 12th World Congress on Work and Organisational Psychology* (pp. 181-192). Belgium: Presses Louvain.

Tremblay, D.-G. (2004). Communities of practice: Are the conditions for implementation the same for a virtual multi-organization community? *Proceedings of the 2004 National Business and Economics Society Conference.* USA: NBES.

Wenger, E. (1998). Communities of practice—learning as a social system. *Systems Thinker, 9*(5), June, 5.

Wenger, E., McDermott, R., & Snyder, W. (2002). *Cultivating communities of practice. A guide to managing knowledge.* Boston: Harvard Business School Press.

Wenger, E., & Snyder, W. (2000). Communities of practice: The organizational frontier. *Harvard Business Review, 78*(0-1), 139-145.

KEY TERMS

Operational Objectives: Short-term objectives, such as maximization of working time, reduction of duplication, stimulation of creativity, and so forth.

Organizational Learning: Goes beyond individual learning, because it results in the development of a knowledge basis which could translate into a significant change within the organization, and not only at the individual level.

Strategic Objectives: Long-term objectives, such as innovation, rationalization, efficiency, quality, and so forth.

Success of a Community of Practice: Can be measured in different ways, either in terms of specific objectives (innovation, rationalization, etc.) or in terms of organizational learning, and professional and personal enrichment.

ENDNOTES

[1] We would like to thank Cefrio for funding this research, which was conducted in partnership with six other colleagues who examined other aspects (communications, technology, etc.— see Jacob et al., 2003). The follow-up study of a dozen communities of practice in Québec organizations, entitled "modes de travail et modes de collaboration à l'ère d'Internet," along with other articles on this theme, can be found on the following sites: www.cefrio.qc.ca and www.teluq.uquebec.ca/chaireecosavoir. The data presented here is based on a research report written with Anne Bourhis (Bourhis & Tremblay, 2004).

[2] We used the basic demographic variables (age, gender, language), but also a few more (including familiarity with technology, professional category, level of schooling, presence of children, and civil status), to test relations. More detailed analysis will follow in the coming year.

[3] Again, for more detail, see Bourhis and Tremblay (2004).

Supporting Communities of Practice in the Electronic Commerce World

Charlene A. Dykman
University of St. Thomas, USA

INTRODUCTION AND BACKGROUND

The phrase *communities of practice* has entered the lexicon of our world today. It implies some sense of closeness, intimacy, and connection with people bound together through mutual interest in something (Wenger & Snyder, 2000). As the pace of change is increasing and technology and information overload is becoming an issue, the time available to nurture relationships in the real world is becoming threatened (Baker & Ward, 2002). At the same time, our need for answers and quick access to solutions through the exchange of knowledge and experience is growing exponentially. Knowledge management, as a practice, focuses on making effective use of the intellectual capital that is found in the network of relationships connected to a business or organization (Bate & Robert, 2002). This is a perfect match for the business that is engaged in e-commerce to provide its stakeholders with opportunities to build relationships, to find answers and solutions, to exchange knowledge, and to gain a sense of community deriving from the relationships with the business.

This article discusses the potential for small businesses to develop and nurture their virtual communities on the Internet. There is also discussion of the technical foundation needed to make this happen. A virtual community is "an electronic meeting place where a group of people gather to exchange ideas on a regular basis" (Powers, 1997, p. 52). Such communities "allow broad communities of interest (e.g., all stakeholders) to coalesce around specific products and services" (Nambisan, 2002, p. 392). A community of practice is not necessarily always a virtual community. But virtuality greatly increases the potential for development of such a community that can be of great benefit to all stakeholders in this relationship.

A virtual community represents more than just the activities involved in e-commerce or shopping online. Visiting a site and seeking information about a product may be the "portal" into involvement in a virtual community supported by that e-commerce retailer. Buttons and links to chat rooms, to Internet groups, and/or to similar retailers selling similar products are all part of the experience. Viewing the relationship with the business as the gateway to other relationships allows one to visualize the vast potential for community building through that gateway. If the business understands the potential benefits of providing this community building service, they will recognize the importance of sound technical infrastructure to support the efforts.

A well-designed technical infrastructure provides a strong foundation to ensure flexibility, scalability, and adaptability to meet the changing user requirements of a virtual community and address inherent economic fluctuations in the marketplace. Managing the IS resources, including hardware, software, data, procedures, and support personnel, is a difficult task in a virtual community because they are subject to over and underutilization based on market changes that are difficult to predict and control.

In his book, *How to Program a Virtual Community*, Powers (1997) defined five building blocks for a virtual community: inhabitants, places to see, things to do, a government, and an economy. Online members are inhabitants; often called avatars (or embodiments). A virtual community may have different places, locations, spaces, rooms, chat rooms, or even theme parks for its online members to visit. A community may also have different objects, props, and activities for online members, encouraging interaction among members who may be present at the same time.

The *governance* in a community includes support and trust in a democratic environment with commands and control, as well as rules and regulations that govern the cyber interaction. This includes monitoring online activities, managing the resources, and gathering the utilization and visitation statistics that are needed to make the optimal decisions to facilitate future growth. In fact, one of the downsides of the concept of communities of practice is the uncontrollable nature of the informal structure that develops (Wenger, McDermott & Snyder, 2002). Lastly, the economy is the exchange of things of value, from local and foreign currencies, objects, and games to banners, promotions, memberships, and events. The online community must have some kind of economic and/or knowledge exchange in order to maintain itself in cyberspace.

ONLINE COMMUNICATION PROCESSES

Technologies mirror modern society based on democracy, civil structure, culture, and education (Barber, 2001). Stakeholders expect open systems that have flexible and dynamic links within and outside their communities. The following sections discuss four different communication architectures: one-to-one, one-to-many, many-to-many, and peer-to-peer. Understanding the different nuances of these approaches to communication will help businesses to design their virtual community effectively.

One-to-One Communication

One-to-one communication in a virtual community is the digital interaction between a host and a member, or more commonly, between two members. This kind of relationship is direct, simple, and easy to understand and manage. A virtual community with many of these linkages will have a lot of individually driven communications and is often described as a private or closed system. It is a one-way communication at a specific instance of time. The host or members have direct control in this one-way, vertical, or horizontal communication. This is an isolated and non-interactive relationship in a computer-mediated environment and is probably not the best approach to building a virtual community.

One-to-Many Communication

The one-to-many architecture allows the host to distribute information effectively to all members in a group. This is usually a closed system, as members have to register to receive information or be on a mailing list obtained by the host site. A one-to-many virtual community is nearly always a vertical communication process. If the customers would like to send in feedback, a many-to-one communication process begins. Thus, one-to-many communications can be viewed either as a one-way or a two-way communication process. This method is highly centralized and often minimally interactive, and the host has maximum power in controlling information within the group. However, it does provide a cost-effective way to communicate with members of a community.

Many-to-Many Communication

The many-to-many communications architecture may better solve the problem of market fragmentation as it integrates many groups together at one place through common interests and linkages. Kozinets (1999) suggests that effective virtual communities are like "electronic tribes" structured around member interests. *Communities of commerce*, a phrase originated in 1995, are the Internet-based communication channels for suppliers and customers (Bressler & Grantham, 2000). A many-to-many communication structure is an ideal way for the online members to interact with others and an innovative design that accelerates communication velocity at low cost. For a large, fragmented, and unorganized group of vendors who are seeking to reach buyers in the same market, communities of commerce provide a many-to-many communication structure for them to meet electronically on the Internet. While the communities of commerce evolve, clusters of communicating groups are formed on the Internet, albeit with some limitations based on cultural and language differences. In this interactive, many-to-many community, members have to learn how to manage communication effectively to gain market recognition and other members' support.

Peer-to-Peer Communication

The latest design in virtual communities is a totally market-driven, peer-to-peer (P2P) communication network. According to the TechEncyclopedia (2001), a P2P network is "a communication network that allows all desktop and laptop computers in the network to act as servers and share their files with all other users on the network". This P2P community is an open and yet dynamic electronic exchange network, totally controlled by its members. It is a multiple version of the one-to-many network and the expansion of the many-to-many network. A peer-to-peer communication network tends to lack security, privacy, and trust. Members have less protection in cyberspace than in the real space because there are no rules or regulations and no consensual authority that monitors the activity. This type of self-governing is a concept very much in its infancy and complicated by multiple cultural frameworks or orientations easily engaged in a struggle to dominate.

Molitor (2001) describes five forces that are transforming this electronic communication: optical transmission, satellite communication, wireless and mobile communication, broadband digital technologies, and Internet resources. As we progress beyond the 21st century, virtual communities will enter a new digital age with revolutionary changes in real space markets and online environments.

INFORMATION ARCHITECTURE

Virtual communities of practice are sustainable only through well-designed information architecture integrating the five building blocks of management information systems. These are finance (including accounting), human resources, information services, manufacturing (including product development), and marketing. With respect to the architectural design, front-end and back-end information design are discussed. User interfaces and data flow are also addressed. It is important that a system or business analyst defines user requirements and functional specifications in this stage.

Front-End Information Design

The front-end information design focus is like home decorating and interior design for a house, placing various objects in an environment based on anticipated human behavior. The front-end information design provides the first layer of user interface for members in a virtual community. Front-end Web interfaces include Web content management, graphical user interfaces (GUI), and multimedia presenta-

tions. Designing how members will interact online involves designing and writing the Web site content. Based on the Activity Theory, described by Chaudhury, Mallick, and Rao (2001), a virtual community has the following elements: actors, tools, objects, processes, and outcomes. When developing Web content, managers must address each of these elements.

At the graphical user interface (GUI) level, managers should work with their Web designers to develop interactive features and tools based on user requirements with a goal of increasing site stickiness (the time spent that leads to loyalty). Well-designed GUI pages create a special mood or emotional feeling in the cyberspace at a particular time. Hence, understanding online members' psychological behavior, attitudes, and personality will help to create the right emotional appeal at the GUI level.

For business and news related Web sites, very often there are some multimedia presentations or audio-video files for download. The use of streaming media creates more excitement and higher stickiness than static forms of communication (Bressler & Grantham, 2000). However, this multimedia approach does rely on the users' bandwidth capacity. Additionally, member loyalty does not necessarily lead to higher profitability (Bughin & Zeisser, 2001).

Back-End Information Design

The back-end information design is like drawing the plans for a house and preparing for construction. The key is to determine how the front-end and back-end elements will interact and communicate with each other. Managers have to design how information will flow in and out of the Web site and to or from the server and desktop computers. Basically, there are three types of management systems to take care of the back-end information: customer relationship management, value chain management, and knowledge management. A customer relationship management (CRM) process deals with external information from customers or members. Through the virtual community based on a customer relationship management strategy, the organization can track customers' or members' online activities and search for new marketing promotional strategies for future campaigns. According to the Object Management Group (Haughey, 2001), the value chain is a "set of activities an organization performs to create and distribute its goods and services, including direct activities, such as procurement and production, and indirect activities, such as human resources and finance". When an organization can link the activities in its value chain in a cheaper and more efficient way, it will gain a competitive edge in the market. Knowledge management is an internal business process that positions an organization to store, retrieve, study, and learn from historical data. For example, NVST.com, established in 1995, is a virtual community for investors, entrepreneurs, and professional service providers to meet and do business online (Tudor, 1999). There are two databases: one is a contact database with free subscription, and the other is an investment opportunity database with a fee-based subscription. Managers can easily do data mining studies and study the differences in online user demographics. New insights from the KM system will help to improve the front-end information design for the virtual community

NETWORK INFRASTRUCTURE

Railsback (2001) says that the right mix of databases, application servers, data-mining tools, and customer relationship management solutions will be the foundation for a solid business infrastructure. For a small business virtual community, the infrastructure will be a little different depending on whether the online community is a business-to-business (B2B) or a business-to-consumer (B2C) network selling products or services.

Requirements for the Network Infrastructure

It is crucial for a small business trying to build a virtual community to have a reliable ISP as small businesses usually have limited resources to maintain their computing network in-house. A well-planned network infrastructure must be open, flexible, scaleable, robust, reliable, and secure. In the article, "Wish: An Entrepreneur's Dream", McClelland (2000) suggests that content providers,

payment providers, and consumers together create an efficient e-commerce environment on the Internet. There really is an emergent quality to these relationships in that the virtual community transcends its component parts.

In terms of the software and hardware components of a virtual community, Powers (1997) suggests three software components and a four-step process. The three software components are (1) the server and database, (2) the client browser, and (3) the network connection. The four steps of building a virtual community are finding the right machine, launching a server, using the client, and editing the online settings. A small business may need to work with an IT analyst to define specific user requirements to reach the ultimate marketing goals.

E-Business and Customer Databases

A virtual community requires a virtual office with robust and reliable functionalities in the areas of finance, human resources, information systems, manufacturing, and marketing functions. The traditional business relationship is about vendors targeting customers in a segment. Today, this is reversed as a virtual community customer seeks the most ideal vendor (Poynder, 1999). A virtual community interacts in an e-commerce area with a unique focus that integrates content and communication and provides access and convenience to a broad range of products and services. After many dotcom failures offering free content sites, the future trend will be to have broad and popular content sites that gain online exposure and generate revenue (Goldberg, 2001).

Web-Based Applications

Web-based applications must provide accurate and secure online and off-line communication between all Web pages in the virtual community and the application servers. A virtual community derived from e-business relationships needs Web based for content management, and Internet security to develop member's interests, drive online revenue streams, and develop online trust.

There are both free and paid content management tools. For a small business, services such as MSN community allow the host to choose to keep the community as an open or closed system with calendar, chat, message, and many other Web features, all at no cost on the Internet. Similarly, eProject is a free online project management application service provider (ASP) that members can use to check schedules, post information, and exchange files with other members.

Internet collaboration takes hard work, the right tools, patience, and the ability to develop Web design from scratch (Sherman, 2001). Online communities can also build in a 3D collaboration application (Miller, 2000). Although remote teams can collaborate and communicate in an open ISO standard, members may require accelerator video cards and intensive resources to implement a 3D collaboration. The building blocks for support of a virtual community are individually quite straightforward, but bringing them all together to effectively nurture a community of practice calls for technical, behavioral, and organizational understanding.

CONCLUSION AND FUTURE TRENDS

A virtual community of practice is an electronic meeting place where individual members can buy and sell products and exchange ideas and messages, thus facilitating relationships based on common interests. Members use that community as a platform to search for other sites and begin new relationships. Customer-centric virtual communities provide competitive products, services, and experiences for all members anywhere, anytime. Managing information systems in a virtual community is like managing the incoming and outgoing data flows from a Web site. It consists of five fundamental functions: finance, human resources, information systems, manufacturing, and marketing. The five building blocks for a virtual community are actors, places, objects, a government, and an outcome. These elements make a virtual community very attractive, interesting, informative, and interactive. Online members are motivated to make repeat visits and expand upon their transactions with the small business and its extension through the virtual community.

A peer-to-peer network provides open and dynamic access to all parties. The information architecture of a virtual community must integrate the front-end and back-end and be robust and reliable with well-defined user requirements and system applications. A mix of databases and Web-based applications, such as Web content management software and Internet security tools, enhances a Web site's functionality in support of a virtual community.

Managing information in a virtual community requires hard work, patience, commitment, and creativity. Small businesses should evaluate how Web design and Web content can be translated into an online environment that mirrors the real space environment. Understanding how a virtual community can satisfy the currently unfulfilled market needs will certainly create many new online opportunities for small businesses to generate revenue and gain a positive experience in the cyberworld in the future.

REFERENCES

Baker, P., & Ward, A. (2002). Bridging temporal and spatial "gaps". *Information, Communication, and Society, 5*(2), 207-224.

Barber, B. R. (2001). The uncertainty of digital politics. *Harvard International Review, 23*(1), 42-47.

Bate, S. P., & Robert, G. (2002). Knowledge management and communities of practice in the private sector: Lessons for modernizing the National Health Service in England and Wales. *Public Administration, 80*(4), 643-663.

Breidenbach, S. (2001, July 30). Peer-to-peer potential. *NetworkWorld, 18*(31), 44-46.

Bressler, S. E., & Grantham, C. E. (2000). Communities of commerce. Retrieved June 9, 2005, from *http://www.enen.com/new_enenbook.html*

Bughin, J., & Zeisser, M. (2001). The marketing scale effectiveness of virtual communities. *Electronic Markets, 11*(4), 258-262. Retrieved June 9, 2005, from *http://www.electronicmarkets.org*

Bughin, J., & Hagel, J. The operational performance of virtual communities: Towards a successful business model. *Electronic markets, 10*(4). Retrieved June 9, 2005, from *http://www.electronic markets.org*

Chaudhury, A., Mallick, D. N., & Rao, H. R. (2001, January). Web channels in e-commerce. *Communications of the ACM, 44*(1), 99-104.

Etzioni, A., & Etzioni, O. (1999). Face-to-face and computer-mediated communities, a comparative analysis. *Information Society, 15*(4), 241-248.

Fremaux, D. (2000, September). The next VAS generation. *Telecommunications, 34*(9), 113-119.

Goldberg, L. (2001). Survivors take slow road to Internet revolution. *Electronic Engineering Times,* 1192, p. 18.

Grant, R. (2001). Knowledge and organization. In J. Nonaka & D. Teece (Eds.), *Managing industrial knowledge creation transfer and utilization* (pp. 145-169). London: Sage.

Haughey, W. (1999, October 18). OMG: Strategic approach to value chain integration. Retrieved June 9, 2005, from *http://www.omg.org/news/about/strategy.htm*

Kozinets, R. V. (1999, June). E-tribalized marketing? The strategic implications of virtual communities of consumption [Abstract]. *European Management Journal, 17*(3), 252-264.

McClelland, S. (2000, May). Wish: An entrepreneur's dream. *Telecommunications, 34*(5), S10-S11.

Miller, S. K. (2000, July 10). 3-D collaboration shows promise. *InfoWorld, 22*(28), 61.

Molitor, G. T. T. (2001, September). 5 forces transforming communications. *The Futurist, 35*(5), 32-37.

Nambisan, S. (2002) Designing virtual community environments for new product development: Toward a theory. *Academy of Management Review, 27*, 392-413.

Powers, M. (1997). *How to program a virtual community: Attract new Web visitors and get them to stay!* Emeryville, CA: Macmillan Computer.

Quintas, P. (2002) Managing knowledge in the new century. In S. Little, P. Quintas, & T. Ray (Eds.), *Managing knowledge. An essential reader.* London: Sage.

Railsback, K. (2001, June 11). Analyst's top 5 picks. *InfoWorld, 23*(24), 51-58.

Sherman, E. (2001, June). The myths of Web design collaboration. *Electronic Business, 27*(6), S2-S6.

TechEncyclopedia. (2001). TechEncyclopedia: Define an IT term. Retrieved June 9, 2005, from *http://www.techweb.com/encyclopedia*

Tudor, J. D. (1999, April). A venture finance virtual community: NVST. *Database, 22*(2), 32-35.

Wenger, E., McDermott, R., & Snyder, W. M. (2002). *Cultivating communities of practice.* Cambridge, MA: Harvard Business School.

Wenger, E. C., & Snyder, W. N. (2000, January-February). Communities of practice: The organizational frontier. *Harvard Business Review, 78*(1), 139-145.

Wilkins, L., Swatman, P., & Castleman, T. (2002). Mustering consent: Government-sponsored virtual communities and the incentives for buy-in. *International Journal of Electronic Commerce, 7*(1), 121-134.

KEY TERMS

Electronic Commerce: All forms of trading across electronic rather than traditional physical media.

Information Architecture: Mostly, the technological architecture which acts as an enabling mechanism to the information system.

Information Systems: Normally taken to include all elements of information, encompassing both the technical and human aspects.

Tacit Knowledge in Communities of Practice

Michele Suzanne Zappavigna
University of Sydney, Australia

INTRODUCTION

Harnessing the tacit knowledge latent in communities of practice in organizations is a major impetus in knowledge management research and practice. The concept of *practice* itself is closely associated with activity that is below-view such as intuition and tacit knowing. Indeed, the features binding the members of the communities are often tacit in nature, including things such as rules of thumb, ideologies, embedded habits, or predispositions. Much research in knowledge management posits a dichotomy of doing and saying: what we can do is, in these frameworks, necessarily distinct from what we can say. Polanyi's (1966) idea that "we know more than we can tell" (p. 4) is often cited to affirm this differentiation. This article seeks to review this relation with tacit knowledge as a focus and suggests that the skillful practice of communities of practice is carried in the discourse which they produce. We adopt a functional approach to discourse, drawn from Systemic Functional Linguistics, that suggests a realization relationship between doing, meaning, and saying rather than a series of dichotomies involving these three semiotic modes. According to this view, what we can say embodies what we can mean which in turn embodies what we can do (Halliday, 1975). This approach is in accord with Wenger's (1998) opposition to formalist dichotomies when theorizing social action.

This entry is structured to present the potential of discourse analysis as an analytical tool to understand the tacit component of participation in communities of practice. The background section details the issues which theories of practice have raised for knowledge management and information systems research. We then review the analytical tools which the field of linguistics offers to uncover implicit knowledge and assumptions in communities (e.g., Iedema, 2003; Jorderns & Little, 2004; Zappavigna-Lee & Patrick, 2004; Zappavigna-Lee et al., 2003). We conclude by arguing that the nature of our skillful practice may be carried in language: we articulate what we know through patterns and features of language of which we are not consciously aware. Analysis of this kind of language aims to elicit implicit meaning and is allied with psychoanalytical methods that attempt to understand implicit aspects of social experience.

MAIN BACKGROUND

Communities of practice are communities which hold collective meanings. While part of this meaning may be explicitly defined via a description mechanism such as the title of an online forum, the social nature of these communities allows meanings that are implicit or even converted to coexist. This parallels research in psychology that claims that social behaviors are encoded automatically and without intention (Bargh, 1999). Wenger (1998) acknowledges that, within communities, the tacit and commonsensical is backgrounded but that "[c]ommunities of practice are the prime context in which we can work out common sense through mutual engagement" (p. 47). In the social sciences, identifying the implicit subject positions, which have been naturalized by culture, has been a focus (Bernstein, 1971; Bourdieu, 1990). This naturalization means that these positions remain below scrutiny. Bourdieu (1990) suggests that individuals internalize the cultural habitat in which they reside, their habitus. This means they form behavior dispositions and construe their experience in certain ways.

The acquisition of these structural constraints is a process of acculturation into specific socially-established groups or classes and "[a]gents to some extent fall into the practice that is theirs rather than freely choosing it or being impelled into it by mechanical constraints" (Bourdieu, 1990, p. 90). A similar implicit structuring of communities is also highlighted in Bernstein's (1971) concept of coding orientation as a concept for understanding "the relationship between a particular symbolic order and the structuring of experience" (p. 112). He applies this concept to educators in the domain of transmitted educational knowledge with a view to analyzing the different orientations to meaning making which people adopt and the different ways they construe the context of their meaning-making practices. Teachers are socialized into assimilating the code and "during this process, the teachers will internalize, as in all processes of socialization, the interpretative procedure of the code so that these become implicit guides which regulate and co-ordinate the behaviour of the individual teachers" (pp. 107-108).

While the social sciences have introduced such theories about the implicit regulation of experience at a macrolevel, determining appropriate microlevel analytical tools is integral to understanding practice in communities. When the object of study is the implicit components of that practice, that is, the tacit knowledge of community of practice members, finding an appropriate tool is challenging. While most studies adhere to Polanyi's (1966) position privileging practice over talk, we suggest that linguistics offers a range of analytical techniques which are of direct use. Polanyi's theory of tacit knowledge has been used in an argument for tacit knowledge as ineffable; however this view is incommensurable with a research tradition beginning with the Ancient Greeks. The psychoanalytical concept that we can learn about our own tacit knowledge by talking originates in the ideas of the great Greek philosophers, Plato and Aristotle, in their notions of introspection and peripateticism. Peripatetics walk, as was Aristotle's habit, and talk, often not knowing what they will say but nevertheless learning from that talk. Greek philosophical traditions are probably preceded by earlier interest in tacit knowledge, but our understanding of them is limited.

IS THE TACIT KNOWLEDGE HELD IN COMMUNITIES OF PRACTICE INEFFABLE?

In assessing the kind of analytical tools appropriate to understanding tacit knowledge in communities of practice, it is necessary to ask whether the presupposition that tacit knowledge cannot be articulated is a defining property of tacit knowledge or an artifact of our lens. The attribute most consistently ascribed to tacit knowledge in the range of disciplines in which it is theorized is ineffability (Baumard, 1999; Collins, 2001; Nonaka & Takeuchi, 1995; Reber, 1993). The

strong position is that tacit knowledge cannot be articulated in any linguistic form, while the weak position holds that it is difficult to articulate. Polanyi's (1966) widely cited suggestion that "we know more than we can tell" (p. 4) asserts the epistemological significance of tacit knowing in terms of its ineffability. In assessing this proposal, it is important to consider what it means to tell. If telling means making explicit codified artifacts that are directly transferred to the mind of the listener, then this kind of telling is not a possible means of exposing tacit knowledge. However, if we allow that telling involves processes of which the speaker is not necessarily aware and which are, in turn, subject to both unconscious and conscious interpretation by the listener, linguistic structure is reinstated as relevant to understanding tacit knowledge.

Thus, it appears that Polanyi's statement needs to be refined. We know more than we can tell only if we think about telling as making explicit knowledge. Such an assumption utilizes an impoverished model of communication. This model, often referred to as a mathematical model of communication, presupposes that meaning in communication is absolute and, as such, may be seamlessly transferred from the mind of the speaker to that of the listener. It applies what Reddy (1979) terms the conduit metaphor, that is, the notion that words are boxes with meanings inside that are unpacked by the person to which they are directed. Reddy (1979) argues that the metalingual resources of English privilege this kind of view, as the following examples suggest (p. 287):

Whenever you have a good idea practice capturing it in word.

You have to put each concept into words very carefully.

Just as in uttering the sentences above, we are unlikely to focus on the presuppositions about communication they presume. When we speak, that which we utter cannot be viewed as an overt object. We may well articulate what we know implicitly through patterns and features of language to which we do not directly attend. This is an argument that articulation does not produce a form that by definition is explicit, or in alternative terms, that articulation is not the equivalent of codification. Acknowledging this idea is a first step in overcoming what Byrd, Cossick, and Zmud (1992, p. 123) term between obstacles in communication, that is, the difficulties of interpretation of meaning which people have when they interact. This kind of obstacle often contributes to misalignments in the perception of systems analysts compared with users and knowledge engineers compared with experts.

FUTURE TRENDS: ANALYTICAL TOOLS FOR UNDERSTANDING TACIT KNOWLEDGE IN COMMUNITIES OF PRACTICE

Understanding the collective, implicit knowledge latent in communities of practice requires analytical tools which can address the macrolevel research directions suggested by social theorists such as Bourdieu (1990) and Bernstein (1971) as well as providing methods for corresponding microanalysis of the choices people make in communities. Systemic Functional Linguistics (SFL) addresses this goal as "it seeks to develop both a theory about social process AND an analytical methodology which permits the detailed and systematic description of language patterns"(Eggins, 1994, p. 23). Systemic Functional Linguistics (SFL) is a theory interested in describing language in terms of its semantic function in the social and cultural contexts within which it is put to use by speakers. In this way, it differs from the formal, syntactic approach of traditional grammars. Halliday (1998), a major figure in the development of SFL, describes language as a social semiotic. Butt (2000) provides a clear outline of what this means:

To say language is social implies that a community of speakers share knowledge about systems of sound and writing, about lexicogrammar, about meanings and about situations. To say that language is semiotic implies it is a system of signs which convey meaning about that culture, just as other sign systems such as dress and architecture are shared by a cultural group and constitute that culture. (p. 10)

Thus, the social semiotic perspective suggests that the relationship between language and meaning is not arbitrary. Due to this, SFL asks questions about how language is used by speakers and writers to make meanings in functional contexts and how is it organized to achieve this. It approaches instances of such meaning making as texts, that is, as units which have semantic significance.

Language enacts the social construction of knowledge in communities of practice. Specialist knowledge is realized within research communities, such as science, by the way in which they "utilize different resources from lexicogrammar, discourse semantics, register and genre" (Wignell, 1998, p. 297). For example, Halliday (1998) suggests the way the discourse of scientific communities operate to "regrammaticize" experience as technical knowledge through the distillation of technical meaning in grammatical metaphor:

In these discourses, the semiotic power of referring is being further exploited so as to create technical taxonomies: constructs of virtual objects that represent the distillation of experience (typically experience that has itself been enriched by design, in the form of experiment). The semiotic power of expanding—relating one process to another by a logical-semantic relation such as time—is being further exploited so as to create chains of reasoning: drawing conclusions from observation (often observation of experimental data) and construing a line of argument leading from one step to the next. (Halliday, 1998, p. 195)

This semiotic action centers on "a metaphoric transformation from a clausal to a nominal mode of construal" (Halliday, 1998, p. 195), that is, rendering processes as if they were static participants or objects. This nominal mode of construal allows communal bodies of meaning to be "packed-up" (Halliday, 1998) so that it may be referred to and shared in condensed and portable ways. Nominalization is the rendering of a process or happening as a noun. Members of communities who can trace the more congruent reading of the nominalization, that is, the meaning of the original process which has been condensed are able to access the shared understanding. Individuals outside the community who have not been socialized into this way of understanding the process will be semiotically excluded from the discourse. In this way nominalization allows speakers to refer to bodies of knowledge "by a kind of shorthand" that can include or exclude membership to a group (Melrose, 2003, p. 428). Iedema (2003) in studying the discourse of planning meetings in organizations notes the way in which the increasingly relational nature of participation in organizations, in which workers engage with many different communities, means that they are involved "in a struggle over what can be realized or expressed as if already taken-for-granted, what needs to be specified and particularized, and what is to remain silenced and invisible" (p. 95). He suggests the seminal role the nominal group plays in what he refers to as the *textualization* of work, that is, the realization of complex and competing meaning making. Identifying and unpacking nominalization will provide researchers with a rich insight into the realization of socially constructed shared meaning in communities of practice.

In addition to describing implicit practices, discourse analysis can also be used within communities to elicit or uncover the intricacies of those practices so that they are rendered explicit to participants in those communities. Indeed Eggins (1994) describes the aim of genre theory, an area in Systemic Functional Linguistic theory, as being "about bringing …unconscious cultural knowledge to consciousness by describing how we use language to do things (p. 46). An example of the application of this aim is seen in Jorderns and Littles' (2004) attempt to define a genre for ethical reasoning within the healthcare profession. This study details a spoken policy genre which characterizes the way in which medical practitioners explain their understanding of how to behave in particular scenarios in order to achieve highly specialized outcomes. Jorderns & Little argue that this is the unfolding of practical understanding in language, or in different terms, the realization of tacit knowledge in discourse.

A dialogic approach to understanding practical wisdom in speech is seen in the work of Zappavigna-Lee and Patrick (2004). This study presents a method for eliciting tacit knowledge from the language of interviewees through a process of directed interviews based on a linguistic model of tacit knowledge. This process centers upon the interviewer identifying semantic and grammatical features in the interviewee's language that suggest knowledge that the participant possesses which remains below-view. The knowledge is below-view in the sense that the linguistic choice that the interviewee has made indicates *under-representation*. Under-representation occurs when

components of knowledge are effaced in discourse as they have been automatized by the individual. For example, the agent in a clause may be omitted. In addition to simply being left out of discourse, the knowledge may be effaced through generalization that construes it as unavailable for deconstruction. For example, a verb may be nominalized, meaning that something that was a process with component steps is rendered as a static object. This means that there is less potential for these steps to be analyzed. Tacit knowledge is subsidiary in the sense that we do not attend to such obfuscation. The directed interview method entails:

1. Identifying the semantic feature that suggests knowledge that is under-represented in the interviewee's discourse and important for their current knowledge management task.
2. Asking a question that elicits a more delicate response from the interviewee and which prompts them to elaborate on this feature.

In applying this interview method in an ongoing case study in an Australian broadcasting organization, Zappavigna-Lee and Patrick (2004) demonstrate that identifying such features will contribute in eliciting a more delicate description of the interviewees' meaning than a strategy based solely on eliciting content. The description is more delicate not only in the sense of being more specific lexically, but in the sense of being increasingly precise lexicogrammatically.

Zappavigna-Lee and Patrick (2004) address the issue of nominalization, mentioned previously, as a fundamental way in which knowledge is under-represented in language. In a directed interview with a senior manager in the digital media division of the organization, the interviewer noted that the manager possessed tacit knowledge that was embedded in nominalizations. For example, the manager described his division as a "service area" that "provide[s] services including IT services". "Service" is the nominalization of a range of processes involving understanding, communicating, and delivering feedback to clients. An underlying component of these processes is negotiating shared cultural experience. Through elaborating this nominalization in the directed phase of the interview, the interviewer uncovered that the manager believed the greater the shared cultural experience and active cultural processes he was able to foster, the greater the shared knowledge and cohesion of his employees. This information was not elicited merely by asking the manager to be more specific, as this would simply have elicited a more detailed rendering of his explicit style. This may have merely produced a taxonomy of the IT services in the organization. Instead, embedded phenomena about the manager's beliefs and practices were uncovered by analyzing his implicit style. This phenomena is more specific in a particular way: it is the elaboration of parts of the interviewee's language of which they were not aware. This is an exercise below the surface of the text and below the content plane on which most interviews are conducted. As such, it involves a richer elicitation of the interviewee's experience. While this study addressed an individual manager's meaning making, there is direct potential for the interview technique to be applied in communities of practice in organizations as a way of demonstrating how members of these communities adopt similar discourse practices and of eliciting the tacit knowledge embedded in those practices.

CONCLUSION

Theorizing the social construction of knowledge in communities has been an ongoing research imperative in the social sciences. This domain has suggested the important role of social processes that are below-view in shaping the practice of members of groups. As researchers seek to understand these processes at a microlevel, we require tools to analyze the knowledge people implicitly assimilate as they are socialized into the practice of different communities. Due to the dominance of the perspective that the tacit knowledge these communities generate cannot be articulated, the role that existing techniques in fields such as functional linguistics might play has been largely overlooked. The tacit knowledge residing in communities of practice is potentially not as taciturn as we have assumed.

REFERENCES

Bargh, J. A. (1999). The unbearable automaticity of being. *American Psychologist, 54*(7), 462-479.

Baumard, P. (1999). *Tacit knowledge in organizations*. London: Sage.

Bernstein, B. B. (1971). *Class, codes and control*. London: Routledge and K. Paul.

Bourdieu, P. (1990). *In other words: Essays towards a reflexive sociology*. Stanford, CA: Stanford University Press.

Butt, D. (2000). *Using functional grammar: An explorer's guide.* Sydney: National Centre for English Language Teaching and Research Macquarie University.

Byrd, T. A., Cossick, K. L., & Zmud, R. W. (1992). A synthesis of research on requirements analysis and knowledge acquisition. *Management Information Systems Quarterly, 16*(1), 117-138.

Collins, H. M. (2001). Tacit knowledge, trust and the Q of Sapphire. *Social Studies of Science, 31*(1), 71-85.

Eggins, S. (1994). *An introduction to systemic functional linguistics*. London: Pinter.

Halliday, M. A. K. (1975). *Learning how to mean: Explorations in the development of language*. London: Edward Arnold.

Halliday, M. A. K. (1998). Things and relations: Regrammaticising experience as technical knowledge. In J. R. Martin & R. Veel (Eds.), *Reading science: Critical and functional perspectives on discourses of science* (pp. 185-235). London/New York: Routledge.

Iedema, R. (2003). *Discourses of post-bureaucratic organization*. Philadelphia: John Benjamins.

Jorderns, C. F. C., & Little, M. (2004). "In this scenario, I do this, for these reasons": Narrative, genre and ethical reasoning in the clinic. *Social Science and Medicine, 58*, 1635-1645.

Melrose, R. (2003). "Having things both ways": Grammatical metaphor in a systemic-functional model of language. In A. M. Simon-Vandenbergen, M. Taverniers & L. Ravelli (Eds.), *Grammatical metaphor: Views from systemic functional linguistics* (pp. 417-442). Amsterdam; Philadelphia: Benjamins.

Nonaka, I., & Takeuchi, H. (1995). *The knowledge-creating company: How Japanese companies create the dynamics of innovation.* New York: Oxford University Press.

Polanyi, M. (1966). *The tacit dimension.* London: Routledge and K. Paul.

Reber, A. S. (1993). *Implicit learning and tacit knowledge: An essay on the cognitive unconscious.* New York; Oxford: Oxford University Press and Clarendon Press.

Reddy, M. (1979). The conduit metaphor. In A. Ortony (Ed.), *Metaphor and thought* (pp. 164-201). Cambridge: Cambridge University Press.

Wenger, E. (1998). *Communities of practice: Learning, meaning, and identity.* Cambridge, UK; New York: Cambridge University Press.

Wignell, P. (1998). Technicality and abstraction in social science. In J. R. Martin & R. Veel (Eds.), *Reading science: Critical and functional perspectives on discourses of science* (pp. ix, 368). London; New York: Routledge.

Zappavigna-Lee, M., Patrick, J., Davis, J., & Stern, A. (2003). Assessing knowledge management services through discourse analysis. *Proceedings of the 7th Pacific Asia Conference on Information Systems*, Adelaide.

Zappavigna-Lee, M., & Patrick, J. (2004). Literacy, tacit knowledge and organisational learning. *Proceedings of the the 16th Euro-International Systemic Functional Linguistics Workshop*, Madrid, Spain.

KEY TERMS

Below-View: Elements of experience which are not available to direct inspection without applying some semiotic tool such as linguistic analysis.

Code Orientation: Similar to Bourdieu's (1990) conception of habitus. The code or framework which implicitly regulates the practice of members of communities, particularly institutionalized communities.

Discourse: Closely associated with the notion of a text, discourse refers to meaning making. This meaning making may be of many forms such as written, spoken, written to be spoken, and spoken to be written.

Discourse Analysis: Generally looks at aspects of texts above the clause or sentence level and approaches them in their social contexts rather than as isolated aspects of grammar.

Habitus: Bourdieu's (1990) term for the cultural context in which individuals reside and which influences their practice.

Nominalization: The rendering in discourse of a "goings-on", happening, or process as a noun.

Systemic Functional Linguistics: A theory of language which seeks to understand how speakers use language to make meaning. It is functional in the sense that it analyzes such language in its practical context, and it is systemic as it theorizes meaning as selection from system networks of choices.

Tacit: From the Latin tacitus, meaning silent.

Tacit Knowledge: Practical understanding possessed by individuals and shared in communities which is below-view in the sense that it is not subject to the conscious attention of those who hold it.

Under-Representation: Zappavigna-Lee and Patrick's (2004) term for a set of specific linguistic features that indicate the presence of tacit knowledge in an individual's talk.

Tacit-Explicit and Specific-General Knowledge Interactions in CoPs

Tunç Medeni
Japanese Advanced Institute of Science and Technology, Japan

INTRODUCTION

Over the last decade the fields of knowledge management and organizational learning have developed rapidly, showing increasing diversity and specialization in the academic literature. Ikujiro Nonaka has played a leading role in setting standards and earning academic legitimacy for the emergent field of "organizational knowledge management" (Easterby-Smith & Lyles, 2003). In the period 1995-2001, the book *The Knowledge-Creating Company* (Nonaka & Takeuchi 1995) was the most-cited knowledge management work from academic literature (Koenig & Srikantaiah, 2004). Interestingly, in this book and in following works, the authors themselves prefer to use the term "knowledge creation" rather than "knowledge management," later also dropping the term "organizational" from the initial proposition. Easterby-Smith and Lyles also state (2003, pp. 642-643) that in the field of organizational learning and knowledge management, among the topics of articles published in the last two years, "learning capabilities, experience, and absorptive capacity" is the largest category, including several articles that assess the impact of learning on performance. Seeming to be frequently interrelated, "organizational learning and knowledge management across boundaries," "knowledge creation and transfer," and "human resource management and human capital" are the next largest categories for articles. Communities of practice, socio-political processes, and the development of tacit knowledge or social identity are among the other topics frequently addressed in the literature, categorized in terms of "cognition, socio-political aspects, and tacitness."

Using the extant and emerging perspectives in knowledge management, organizational learning, and communities of practice literature, in the following sections of this short article, we will first discuss the importance of specific-general knowledge, and context for knowledge creation and management. Then we will introduce the conceptualization of "specific" and "general" knowledge interactions, and discuss a framework that proposes these interactions as contextual knowledge conversions for learning and practice. The following section will aim to contribute to the representation of our knowledge on these contextual knowledge interactions, using visualization tools like geometric figures. We will conclude our discussion by highlighting future research possibilities in the relevant research fields.

BACKGROUND

Specific-General Knowledge and Context for Knowledge Creation and Management

According to the organizational knowledge creation model of Nonaka and Takeuchi (1995), the continuous and dynamic interaction between tacit and explicit knowledge that happens at the individual, group, organizational, and inter-organizational levels can be significant for the sustainable development of any social setting. Nonaka and Takeuchi follow the distinction of Polanyi (1966) between tacit and explicit knowledge: Tacit knowledge is personal, context-specific, and therefore hard to formalize and communicate. Tacit-explicit knowledge interaction is identified as the epistemological aspect, while the interactions among the different levels (individual, group and organization, inter-organizational) correspond to the ontological aspect of the model. When the authors first introduced their model, at the epistemology level they identified four distinctive interactions between tacit and explicit knowledge: socialization, externalization, combination, and internalization. Socialization is the process of creating tacit knowledge from tacit knowledge, whereas externalization is that of articulating tacit knowledge into explicit concepts. Combination involves the process of systemizing concepts into an explicit knowledge system. Internalization is a process of embodying explicit knowledge into tacit knowledge.

Nonaka, Toyama, and Byosiere (2003) also suggest that at the foundation of their modeling lies *ba*: the context that knowledge needs in order to exist, in which it is shared, created, and utilized. Although the concept of *ba* shows similarities with that of communities of practice, especially highlighting the importance of context for learning and knowing, Nonaka et al. (2003) differentiate them according to the nature of the learning and participation that takes place within them. For instance, a community of practice is a place where members learn knowledge embedded in the community; *ba* is a place where new knowledge is created.

Real or virtual interactions among individuals or between individuals and their environments are key for the understanding of *ba* and knowledge creation. Especially, within the tacit knowledge conversions of socialization and externalization, a real *ba* where participants can interact face-to-face in the same time and space is essential (Umemoto, 2002). In general, with regard to the type of interaction (individual or collective) and the interaction medium (face-to-face contact or through "virtual" media) (von Krogh, Ichijo & Nonaka, 2000), four types of *ba* can be defined, corresponding roughly to socialization, externalization, combination, and internalization: originating *ba*, dialoguing *ba*, systemizing *ba*, and exercising *ba* (Umemoto, 2002).

Although initially knowledge creation and management was widely understood as simply the interaction between tacit and explicit knowledge, the type of interaction—individual (personal) or collective (group, social, societal)—is also increasingly being recognized as another dimension of knowledge interaction and conversion that parallels the tacit-explicit dimension of knowledge and knowledge interactions. For instance Wierzbicki (2004) sees socialization as the transition from personal tacit knowledge to group tacit knowledge; externalization, group tacit to group explicit knowledge; combination, group explicit to personal explicit knowledge; and internalization, personal explicit to personal tacit knowledge. In fact, tacitness relates to the transferability of knowledge, which also makes the location of knowledge an important issue (OECD, 2000).

According to von Krogh, Ichijo, and Nonaka (2001), knowledge can be observed and distinguished on two levels, individual and social. In addition, as recognized by various authors, there is "general knowledge," which is widely possessed by a large number of individuals and can be transferred easily among individuals, and "specific knowledge," which is idiosyncratic and narrowly possessed by a very limited number of individuals (Becerra-Fernandez, 2004). Whereas general knowledge is inexpensive to transfer, specific knowledge is expensive and costly (Jensen, 1998). Starting with global public knowledge, which is general and explicit, Stiglitz (2001) also analyzes the development of knowledge along two dimensions, general-local and implicit-codified. The description or classification of knowledge as a public or private good or asset retains an important place in the socio-economic modeling of knowledge (OECD, 2000). In order to redesign cross-cultural management, Holden (2002) discusses three domains of cultural knowledge as follows: general cultural knowledge, culture-specific knowledge, and cross-cultural know-how. While general cultural knowledge can be associated with explicit knowledge and cross-cultural know-how with tacit knowledge, culture-specific knowledge can be both tacit and explicit according to the convention. Gasson (2004) highlights the problems of managing and transferring local knowledge beyond its workgroup and specific context, and discusses the ways in which this distributed knowledge is managed, communicated, and translated across organizational boundaries. The shared explicit knowledge is transited into shared tacit knowledge, then to tacit distributed knowledge, and finally to explicit distributed knowledge.

Whether it is general, global, public, shared, common, collective, social, societal...or specific, idiosyncratic, local, private, distributed, individual, personal, and so forth, and although their units and levels of analysis differ, these various discussions all try to capture the same conceptual understanding about knowledge or knowing. However most of them remain as classifications of knowledge, rather than capturing the meaning of these knowledge types as knowledge interactions, and making use of these interactions for a better comprehension of knowledge creation.

"Specific" and "General" Knowledge Interactions

Nonaka and Takeuchi's (1995) generic knowledge conversion framework of socialization, externalization, combination, and internalization has become widely recognized in the academic and business worlds, and the organizational knowledge creation model has become known as the "SECI model." From this perspective, the main focus was on the tacit and explicit knowledge interactions as the epistemological aspect of the (organizational) knowledge creation model. There was no distinctive classification of knowledge interactions or conversions at what the authors called the ontology level, nor did the knowledge conversions of the SECI clearly identify the knowledge interactions at the individual, group, organizational, or social levels, and across these levels and entities. However, horizontal and vertical interactions among these levels and entities are just as important for organizational knowledge creation as the tacit-explicit knowledge interaction.

Nonaka and Takeuchi's classification of tacit and explicit knowledge interactions follows the logic that identifies *from which knowledge to which knowledge the knowledge interaction occurs*. The same distinction can be applied to the various individual, group, organization, and society levels. One way of doing this can be the realization of the knowledge dimension of general versus specific (or global versus local) knowledge besides that of tacit versus explicit (or implicit versus codified) knowledge. Specific knowledge can mean the knowledge of one specific individual, group, society, time, or location. General knowledge, on the other hand, can be understood as the knowledge that goes across or is shared by particular individuals, groups, societies, times, and locations.

These definitions can then be incorporated into the conceptualization of the knowledge creation model. Knowledge is now seen not only as tacit or explicit, but also specific or general. Like tacit-explicit knowledge, specific-general (local-global) knowledge is not a discrete dichotomy, for actually all knowledge is both specific and general to some degree. At the ontological level, the recognition of knowledge as both specific and general comprehensively takes into account individual, group, organization, society, collective, and locative perspectives. This is also a relative understanding: knowledge that is general for one level or entity can be specific for another. For a generic knowledge creation model, this relative understanding is important in order not to become limited to specific levels of analysis. For the application of the model then, these levels and entities can be made clear such as individual or group, social versus personal, inter-organizational or universal.

According to this model then, we can find four distinct, different knowledge interactions like those within tacit and explicit knowledge: local to local, local to global, global to global, and global to local. Together, the tacit-explicit and local-global (specific-general) knowledge interactions can nurture efficacious knowledge management and creation in institutions and socio-economic systems. We can reinterpret the SECI interactions in light of the knowledge conversions between specific and general knowledge, in addition to those of tacit and explicit knowledge. For an example interpretation, please see Table 1.

Specific-General Knowledge Conversions as Contextual Interactions (for Learning and Practice)

All these specific-general interactions highlight the importance of the context within which knowledge creation takes place. Moreover, they are able to direct our attention to the interactions across boundaries, which are very crucial to understand if we

Table 1. Tacit-explicit and general-specific knowledge conversions in SECI interactions

Knowledge Interaction	Tacit-Explicit Knowledge Conversion	Specific-General Knowledge Conversions such as
Socialization	Tacit to Tacit	General to Specific
Externalization	Tacit to Explicit	Specific to Specific
Combination	Explicit to Explicit	Specific to General
Internalization	Explicit to Tacit	General to General

want to improve our knowledge about knowledge, and its management and creation. This is the same as when we discuss learning through communities of practice. Wenger (1998) points out that knowing in practice involves an interaction between the local and global. Explaining organizational knowledge creation as simplified tacit-explicit knowledge conversions does not highlight the significance of this cross-boundary interaction, often placing too much emphasis on solutions to problems that cannot take advantage of the "landscape of practices" (p. 140). Since the processes of knowledge and learning are based and situated on the context, knowledge interactions on and across boundaries becomes essential.

In recognition of these cross-context transactions in knowledge transfer, in the academic and business literature, the "tacit-explicit"-ness and "specific-general"-ness of knowledge have noticeably been used together. For instance, Sudarshan (2000) concludes his discussion of the governance of economic and social development by claiming that a revival of concern for the role of local and tacit knowledge in development transformation should emerge as a global issue. He gives the example of India, which strongly rejected such local knowledge in favor of a world view derived from within the boundaries of codified and general knowledge, based on "colonial rule, with its accompanying territorial annexation and the need to create an elite with a shared world view and language...To revive the missing dimensions is a challenging, though not impossible, task" (p. 101).

Burton-Jones also contends (1999) that the commercial application of IT has had a 'leveling' effect, "reducing the specific to the general, the idiosyncratic to the standard, tacit knowledge to explicit knowledge, scarce goods and services to commodities." One reason for this is that businesses have historically "opted to use IT to control and standardize their operations." Another major reason is the tendency by firms to adopt standard software packages. For instance, SAP can provide "in detailed, explicit terms, a set of best practices for operating many routine functions of a business." Thus the companies can choose "to rely on externally provided knowledge of best practices, rather than have to develop and maintain that knowledge themselves...in doing so they may lose the benefits

of firm-specific or tacit knowledge embedded in internal processes and procedures which are replaced by the standard system" (pp. 9-10). The same argument can also hold for other types of systemic interventions into institutions, including consultancy, management learning, and training and development activities.

To an important extent the literature has come to recognize the specific-general dimension of knowledge together with that of tacit-explicit. When it comes to the recognition of specific-general knowledge interactions alongside tacit-explicit ones, the existing knowledge base does not provide much beyond the works that we have discussed above or similar articles. However, the intermingling of these two types of interaction is very important for knowledge creation—identifying these interactive processes of knowing beside the classification of different knowledge types deserves special attention. This is especially true when we approach the concepts of knowledge, learning, action, and participation as contextual interactions within *ba* or communities of practice. For instance, originating *ba* through face-to-face interactions offers a context for socialization that yields "sympathized knowledge" (Nonaka & Takeuchi, 1995, p. 72), as important tacit knowledge can be shared among individuals. Empathizing with others, an individual transcends the boundary between self and others (Umemoto, 2002).

In Nonaka and Takeuchi's original model of organizational knowledge creation (1995), the starting point of knowledge creation, socialization, corresponds in practice to not only the sharing of tacit knowledge among the team members, but also to learning through apprenticeship. What is called "learning by doing" matches with internalization, although this interaction and the resulting tacit knowledge has also the meaning of feedback that can initiate a new knowledge spiral. These three concepts—sharing of tacit knowledge, learning by doing, and apprenticeship—have been incorporated into the conceptualizations of situated learning, legitimate peripheral participation, and communities of practice, while the term 'socialization' is understood with regard to its meaning in sociology literature (Lave & Wenger, 1991; Wenger, 1998).

On the other side, what is implicitly evident in the conceptualization of socialization as field-building tacit knowledge interaction in Nonaka and Takeuchi's original model (1995), is that a shared context exists among the participants of the knowledge interaction process. The participants share a background of common knowledge, which enables them to communicate without using explicit terms. As in the case of apprenticeship, newcomers can learn first by watching their seniors and then by practicing themselves. In the case of knowledge sharing among group members, ways of tacit interaction can be enough. Even conversation, which can be seen as a means of explicit knowledge interaction, actually incorporates important tacit and explicit elements together (Baker, Jensen & Kolb, 2002). For all these cases, in order to exchange and share tacit knowledge, participants should also share a common base of mutually understandable and usable knowledge, which enables them to transfer knowledge with tacit terms. Penrose (1989), for instance, exploits this tacit common base to explain the implicit understanding and communication between two mathematicians, even if they cannot express or do not understand each other explicitly. This common ground can start to develop with the start of the participant's acceptance in a community of practice. Then the members gradually learn and develop their knowledge with their practice and participation in the community.

In fact, the dynamic interaction between tacit-explicit and specific-general knowledge can be matched well with this mutual existence of learning and practice, or the duality of participation and reification in communities of practice (Wenger, 1998). Nonaka et al. (2003) try to differentiate *ba* from communities of practice according to the nature of the learning and participation that takes place. Nevertheless, both communities of practice (CoPs) and *ba* highlight the importance of context for learning and knowing, and the conceptualizations of SECI and *ba* can be integrated and contribute to our understanding of communities of practice. The conversions of knowledge can be enabled in the communities of practice, which can lead to the learning of existing knowledge or the generation of new knowledge, together with the development of community and its members. Nermian Al-Ali suggests (2003), for instance, that CoP strategies can be applied for creating new knowledge as innovations through the transfer of mainly tacit general knowledge.

Representation of our Knowledge on Contextual Knowledge Interactions

To represent our knowledge about communities of practices metaphorically, we can also use the geometric characteristics of an ellipse. Forming around the interplay and the duality of concepts like practice and learning, communities of practice can be visualized as an ellipse that is drawn over two center points of practice (or action) and learning (or knowing) (see Figure 1).

In this model we use the geometric characteristics of an ellipse to develop the existing models about learning, learning by doing (or doing by learning), and communities of practice. The ellipse has two fixed points, which are labeled practice (or action) and learning (or knowing) in our model. Thus we can suggest that our community of practice forms around the concepts of both practice and learning. In the ellipse, the sum of the distances between any point on the plane curve and the fixed points is constant. In our CoP ellipse, this also means that whatever is done in the community of practice always consists of some action and some learning, although their ratio could be different. Using other characteristics of ellipse, such as eccentricity, we can use this kind of approach to help us solve some problems related to the different taxonomies of learning and action, which can also be a topic of another article.

Secondly, we can fit the SECI interactions onto this ellipse of community of practice, benefiting from the attachment of tacit knowledge with *bodily experienced practice* and explicit knowledge with *learnt theory in mind* (Nonaka & Takeuchi, 1995). However, by no means do we aim to propose a taxonomy, which could end up with another dichotomy. On the contrary, this modeling suggests that both tacit and explicit knowledge are bound to be together. Then,

Figure 1. Community of practice ellipse

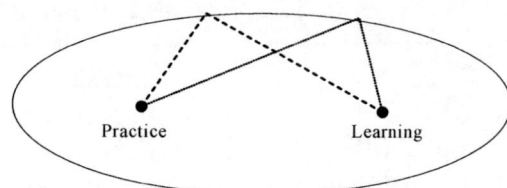

in the ellipse of the communities of practice, the conversion of knowledge can be enabled, which can lead to the generation of knowledge and development of the CoP (see Figure 2).

Then the specific and general knowledge conversions identify the interactions that go beyond individual entities, passing across boundaries. We distinguish these interactions in Table 2.

As suggested earlier, these interactions have relative meaning depending on the positioning of the entity besides other entities and levels (see Figure 3).

It is also possible to reinterpret and visualize SECI conversions as contextual, transboundary interactions as in Figure 4.

The conceptualization of tacit-explicit and specific-general knowledge interactions in a context for learning, participation, and knowledge creation highlights the existence of various processes as knowledge conversions or transactions. These processes of tacit-explicit and specific-general knowledge interactions, rather than the so-called "SECI" labels, can then be used to explain important mechanisms and dynamics in organizational knowledge creation and the communities of practice. For instance, a discourse can be seen as making any tacit knowledge explicit, which can only be truly understood with the specific knowledge of the context of that discourse. Relevant explanations can be made for other processes like reflection, justification, empa-

Figure 2. SECI interactions in the community of practice

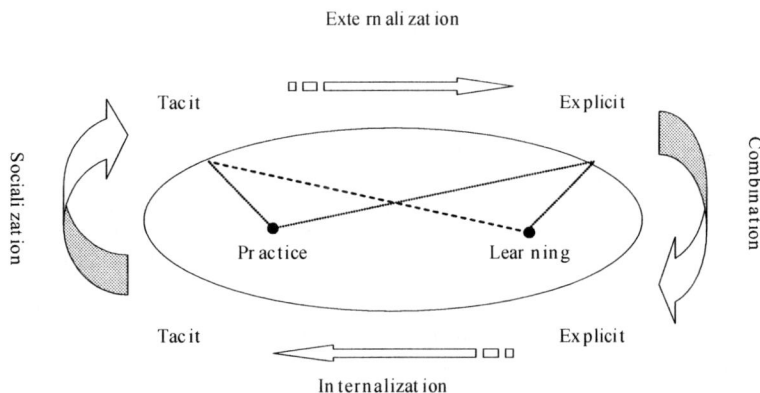

Table 2. Local-global (specific-general) interactions

From To	Local	Global
Local	LOCAL	GLOCAL
Global	LOBAL	GLOBAL

Figure 3. Specific-general knowledge interactions within and across contextual entities and levels

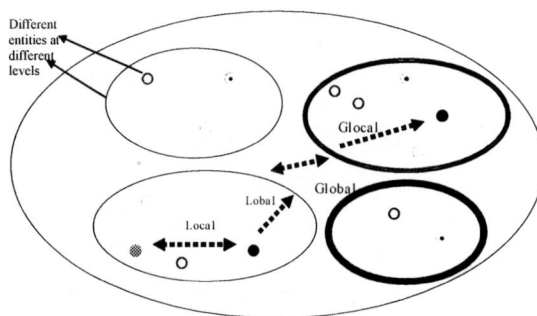

Figure 4. SECI conversions as contextual transboundary interactions

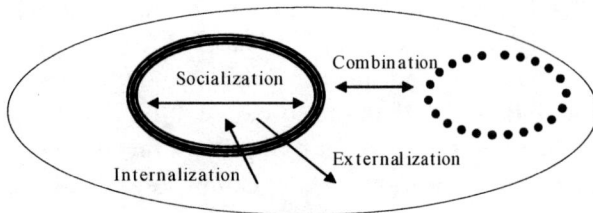

thy, and different ways of learning. Even the tacit-explicit knowledge interaction within the knowledge conversion process itself can be better understood with the recognition of specific and general knowledge interactions. What makes these delicate knowledge interactions possible is the awareness of all participants of the similarities and differences among each other. Candid conversations in informal settings make community members aware of each other's background and personality characteristics, and then they can properly position themselves, adjusting their own attitudes and behaviors for the benefit of all individuals and the community.

FUTURE RESEARCH TRENDS

Distinguishing specific-general knowledge beside tacit and explicit underlines the necessity of a better articulated generic model of knowledge creation, which should aim to incorporate conceptualizations like context (*ba* or CoP), learning and action—knowledge and knowing (Cook & Brown, 1999); integrate different levels of analysis at individual, group, organization, and society level; and look more carefully to the interactions among various entities at these different levels. According to this, the knowledge interactions in a group can also be understood as knowledge interactions among individuals within the context of the group environment, and the knowledge interactions in an organization as knowledge interactions among individuals and groups within the context of the organizational environment. Moreover they are all under the influence of local, global, 'glocal', and 'lobal' dynamics coming from different levels and entities within and out of their own boundaries. The same applies to communities of

practice and the community members, as well.

According to the findings of Easterby-Smith and Lyles (2003) about recent and future research trends, the following are likely to have the greatest impact on knowledge management and organizational learning in the next five years:

- research methods and measures of organizational learning/knowledge management;
- learning across boundaries;
- cognitive, socio-political aspects and tacitness, which also includes CoPs; and
- strategy, technology, and competitive advantage.

Among these, "learning across boundaries" has already become popular and is likely to continue its influence, in sharp contrast to the research on "learning capabilities, experience, and absorptive capacity." Less work has been done on the latter two areas; this is likely to change in the near future (pp. 645-646).

CONCLUSION

These findings point out that research on learning across boundaries and CoPs has a profound place in the literature of organizational development, and of knowledge creation and management. Studies on contextual specific-general knowledge interactions and CoPs, which can theoretically and practically identify and analyze the existing numerous interactions, knowledge transfer, and transitions, can contribute significantly to the knowledge-base of this emerging literature. The obverse is also true, that studies on organizational knowledge management can be used to further our understanding about CoPs and learning interactions across boundaries with regard to these communities and practices.

REFERENCES

Baker, A.C., Jensen, P.J., & Kolb, D.A. (2002). Learning and conversation. In A.C. Baker et al. (Eds.), *Conversational learning: An experiential approach to knowledge creation* (pp. 1-14). Westport, CT: Quorum Books.

Becerra-Fernandez, I., et al. (2004). *Knowledge management: Challenges, solutions, and technologies.* Singapore: Pearson Prentice-Hall.

Burton-Jones, A. (1999). *Knowledge capitalism: Business, work, and learning in the new economy.* Oxford, UK: Oxford University Press.

Cook, S.D.N., & Brown, J.S. (1999). Bridging epistemologies: The generative dance between organizational knowledge and organizational knowing. *Organizational Science, 10*(4), 381-400.

Easterby-Smith, M., & Lyles, M.A. (2003). Introduction: Watersheds of organizational learning and knowledge management. In M. Easterby-Smith & M.A. Lyles (Eds.), *Handbook of organizational learning and knowledge management* (pp. 1-16). Oxford, UK: Blackwell.

Easterby-Smith, M., & Lyles, M.A. (2003). Organizational learning and knowledge management: Agendas for future research. In M. Easterby-Smith & M.A. Lyles (Eds.), *Handbook of organizational learning and knowledge management* (pp. 1-16). Oxford, UK: Blackwell.

Gasson, S. (2004). The management of distributed organizational knowledge. *Proceedings of the 37th Hawaii International Conference on Systems Science,* University of Hawaii at Manoa, USA.

Holden, N. (2002). *Cross-cultural management: A knowledge management perspective.* Singapore: Pearson Education.

Jensen, M.C. (1998). *Foundations of organizational strategy.* London: Harvard University Press.

Lave, J., & Wenger, E. (1991). *Situated learning: Legitimate peripheral participation.* Cambridge, UK: Cambridge University Press.

Nermien, A.-A. (2003). *Comprehensive intellectual capital management.* Hoboken, NJ: John Wiley & Sons.

Nonaka, I., Toyama, R., & Byosiere, P. (2003). A theory of organizational knowledge creation: Understanding the dynamic process of creating knowledge. In M. Dierkes et al. (Eds.), *Handbook of organizational learning and knowledge* (pp. 491-517). Oxford, UK: Oxford University Press.

Nonaka, I., & Takeuchi, H. (1995). *The knowledge-creating company: How Japanese companies create the dynamics of innovation.* Oxford, UK: Oxford University Press.

OECD. (2000). *Knowledge management in the learning society.* Center for Educational Research and Innovation. Paris: OECD.

Penrose, R. (1989). *The emperor's new mind: Concerning computers, minds, and laws of physics.* Oxford, UK: Oxford University Press.

Polanyi, M. (1966). *The tacit dimension.* London: Routledge and Kegan Paul.

Ponzi, L.J. (2004). Knowledge management: Birth of a discipline. In M.E.D. Koening & T.K. Srikantaiah (Eds.), *Knowledge management lessons learned: What works and what doesn't* (pp. 9-28). Medford, NJ: American Society for Information Science and Technology.

Stiglitz, J. (2001). *Joseph Stiglitz and the World Bank: The rebel within.* London: Anthem Press.

Sudarchan, R.M. (2000). New partnership in research: Activist and think tanks. An illustration from the NCAER in New Delhi. In D. Stone (Ed.), *Banking on knowledge: The genesis of the global development network* (pp. 87-103). London: Routledge.

Umemoto, K. (2002). Managing existing knowledge is not enough: Knowledge management theory and practice in Japan. In C.W. Choo & N. Bontis (Eds.), *The strategic management of intellectual capital and organizational knowledge* (pp. 463-476). Oxford, UK: Oxford University Press.

Von Krogh, G., Ichijo, K., & Nonaka, I. (2000). *Enabling knowledge creation: How to unlock the mystery of tacit knowledge and release the power of innovation.* New York: Oxford University Press.

Von Krogh, G., Ichijo, K., & Nonaka, I. (2001). Emergence of "ba": A conceptual framework for the continuous and self-transcending process of knowledge creation. In I. Nonaka & T. Nishiguchi (Eds.), *Knowledge emergence: Social, technical, and evolutionary dimensions of knowledge cre-*

ation (pp. 13-30). Oxford, UK: Oxford University Press.

Wenger, E. (1998). *Communities of practices: Learning, meaning, and identity.* Cambridge, UK: Cambridge University Press.

Wierzbicki, A. (2004). Knowledge theory at the beginning of the civilization era of informational and knowledge economy. *Proceedings of the 1ˢᵗ International Symposium on Knowledge Management for Strategic Creation of Technology,* Ishikawa High-Tech Exchange Center, JAIST.

KEY TERMS

Ba: A physical, virtual, or mental context that enables effective knowledge creation, based on Japanese idea of "place."

Explicit Knowledge: Relatively objective, codified knowledge that is transmittable in formal, systematic language like written documents or spoken sentences.

General Knowledge: The contextual knowledge that goes across or is shared by particular individuals, groups, societies, times, and locations. General knowledge can be tacit or explicit.

Global (General) Knowledge Interaction: Knowledge interaction, which occurs among entities that are relatively at the same global level, such as the development and use of international standards among nations and institutions.

Global Knowledge Interaction: Knowledge interaction from the global knowledge level to that of local knowledge; for instance, tailoring the product of a multinational company, or the development project of an international aid organization, with respect to the unique characteristics of a region or community.

Knowledge Creation: The continuous and dynamic interaction between tacit-explicit and specific-general knowledge that happens at different individual, group, organizational, and social levels, and within and among entities.

Local Knowledge Interaction: Knowledge interaction from the local knowledge level to that of global, such as the promotion of local indigenous values to be appreciated at the national or international level.

Local (Specific) Knowledge Interaction: Knowledge interaction that occurs among entities that are relatively within the same local level, such as knowledge sharing activities among the members of a community.

Specific Knowledge: The contextual knowledge of one specific individual, group, society, time, or location. It can be tacit or explicit.

Tacit Knowledge: Relatively subjective, personal knowledge, which is hard to formalize and communicate.

Teamwork Issues in Virtual Teams

Diane-Gabrielle Tremblay
University of Quebec in Montreal (Teluq-UQAM), Canada

INTRODUCTION AND BACKGROUND

Over the past few decades, there has been much interest in various forms of participation in the workplace and in its impacts on learning from work for individuals and organisations. Teamwork has been the object of much attention in labour economics, in sociology of work, as well as in human resources management (Tremblay, Rolland & Davel, 2000; Davel, Gomez da Silva, Rolland & Tremblay, 2001). Collaborative work and learning have also been the object of much attention in HRM and organisational learning debates, as well as in education circles (Henri & Lundgren, 2000). Much of this interest stems from gains that organisations can expect to obtain from interaction between workers in terms of quality of products, innovation, productivity, and the like. Knowledge management has also spurred interest in recent years, partly on the basis of these expected gains from a better management of the knowledge hidden within organisations. More recently, the concept of communities of practice has been put forward as a form of knowledge management which paves the way to attainment of the various organisational objectives: productivity, quality, innovation, and so forth. In our view, this last concept is closely related to teamwork issues, and we will show how in the following pages.

Teamwork and Learning through Interaction at Work

Teamwork is a flexible configuration that can be adapted to many production and organisational contexts. Its diversity and conceptual polysemy (Durand et al., 1999; Salerno, 1999) are due to the different theoretical approaches to groups in organisations, but also to the different societal contexts that are, to some extent, transforming the theoretical model (Tremblay & Rolland, 1998). Moreover, it should be recognised that its polysemy stems from the fact that this expression is used to describe diverse realities and, in particular, teams functioning at different hierarchical levels. Management teams, production teams, support staff teams, project teams, continuous improvement groups, and client service teams are but a few illustrations of the variety of groups that firms use in their day-to-day operation (Hackman, 1990; Cohen & Bailey, 1997), and we could add to this list communities of practice, since their objectives are often similar.

Forms of Teamwork

In the late 1970s, interest in teams became identified with the *quality of worklife* movement which favoured the transformation of the workplace through labour-management cooperation, as well as the development of knowledge through interactions at work facilitated by the creation of semi-autonomous groups of production workers. Individual satisfaction as well as organisational advantages were the objective of this configuration of work, as is sometimes the case with communities of practice.

It should be pointed out that even if the establishment, operation, and social relations within the work team are far from homogeneous and uniform (Lévesque & Côté, 1999), many authors are in agreement about the core of team-based work organisation; in our view, this can be adapted to the communities of practice context.

Thus, to make up a team, members must have a minimum of: (a) task interdependency among members, (b) shared responsibilities, (c) team identity, and (d) power to manage the relationship between the team and the organisation (Hackman, 1987; Guzzo & Dickson, 1996; Sundstrom, De Meuse & Futrell, 1990; Cohen & Bailey, 1997; Savoie & Mendes, 1993). These elements appear interesting, and in our view, they could be transposed to CoP experiments and other forms of collaborative work and learning through interacting.

This vision can be used to distinguish teamwork from the Taylorist and Fordist systems of work organisation. Teamwork allows members to achieve a level of multi-skilling, to share information, and to be more responsible for quality and productivity (Marx, 1998), as well as providing less rigid and disciplinary supervision. Even when supervisors tend to change their hierarchical role in order to become facilitators, coordinators, or even resource persons, firms do not always eliminate certain forms of control such as performance indicators (Salerno, 1999).

The Distribution of Responsibilities in the Context of Teamwork

The involvement expected of workers in firms that are structured into teams goes far beyond the simple execution of predetermined tasks, which was the norm in the Taylorist and Fordist systems. Workers grouped into teams are, in principle, given the incentive to manage their unit in addition to accomplishing their work. In other words, teams (usually referred to as autonomous and semi-autonomous) should determine not only when and how to accomplish the work assigned to them, but sometimes also the work pace.

According to Marchington (1992), teamwork is the most advanced form of the reconfiguration of tasks and responsibilities, since it allows for an extension of responsibilities that is both horizontal (workers execute more tasks at the same level) and vertical (workers are made responsible for more tasks that previously came under other hierarchical levels, that is, under foremen and supervisors), and leads to learning on the job that is more complete than in traditional contexts of work. Thus, teamwork includes not only the delegation of tasks, but sometimes also the transfer of part of the control over tasks within the team.

Unions often maintain that responsibilities are assumed in various ways and at different stages when carrying out tasks. According to them, in any teamwork there are two types of tasks that are absolutely essential and inextricably linked, that is, technical tasks and social tasks. Technical tasks are those directly related to work execution and production. They concern the definition of production goals, the planning of activities and establishment of deadlines, the choice and examination of material means, the assessment of staffing needs, the definition and allocation of tasks between team members, the development of work schedules, the evaluation of costs and preparation of budgets, and the evaluation of results.

Social tasks include the exercise of leadership, training of members, health and safety, specific programs, the definition of communication channels, and team meetings. They have a decisive influence on the quality of life within the team and make the concrete expression of the values shared by its members possible. They also make trust possible between members as well as with the team leader. Autonomy will increase over time, depending on the evolution and maturity of the team, the dynamics of the relationships between teams, and the agreed-upon rules in the collective agreement (Tremblay, Rolland & Davel, 2000).

All this is also observed in the development and analysis of CoPs, but it is the process of fostering team responsibility or interaction which appears to be most challenging.

The Process of Fostering Team Responsibility and Interaction

Even though teamwork obviously requires the transfer of responsibilities to teams, this transfer alone does not explain the involvement and interaction between team members. According to a number of authors, the effectiveness of teams and their willingness to interact with each other and undertake new responsibilities are influenced by a whole set of factors. Savoie and Beaudin (1995) link the effectiveness of team interaction to functional components such as: (a) interdependency in terms of the environment (feedback from clients, supervisors, team mission, inter-team coordination, management support); (b) task interdependency of team members (skills development) and consequences (sanctions based on results); and (c) the quality of transactions between team members (interpersonal relations, production energy, shared effectiveness, and group cohesion).

Some authors underline that the process of fostering team interaction will achieve the objectives of increased productivity, flexibility, and effectiveness as soon as teams enjoy conditions that are conducive to decision making and collective learning (Edmondson, 1999). These conditions will allow teams to become truly committed to the new responsibilities or activities that they have been given.

Indeed, for some authors (Guzzo & Shea, 1992; Grant, Bélanger & Lévesque, 1997; Tremblay, Rolland & Davel, 2000), the level of team interaction and responsibility varies according to the degree of autonomy that they have been given. More traditional structures will give work teams powers that are less extensive, and interaction will thus be limited. Thus, for many authors, the degree of autonomy and types of responsibilities given to teams appear to evolve according to their maturity (Roy, 1999; Roy et al., 1998), since learning the team decision-making process requires time, experience of life as a team, and a degree of social cohesion (McGrath, 1991). According to this vision, the decision-making autonomy of teams follows an evolutionary process that develops in parallel with group maturity. This process is also seen as characteristic of the life of communities of practice, as is presented in the work of Wenger, McDermott, and Snyder (2002).

The most detailed model of the evolution of communities of practice was presented by Wenger et al. (2002), who define five stages (see Figure 1). At the beginning, the community is an informal network, a potential community. It then unites itself and acquires maturity, and then momentum, and becomes productive (Gongla & Rizzuto, 2001; Mitchell, 2002) until at some point, an event makes it essential for the community to change or renew itself.

According to Savoie and Beaudin (1995), the process of increasing team responsibilities is directly related to the interpersonal relationships between team members. It is presumed that effectiveness and involvement are supported more when team members help each other or have appropriate and enriching social interactions. This process of interaction refers to behaviours and reactions of team members regarding the exchange of information, expression of feelings, and formation of coalitions (Guzzo & Shea, 1992).

Thus, in addition to being a source of solidarity and social cohesion (Hodson, 1997), the quality of interaction within the team is fundamental to understanding the affective and behavioural consequences of forming a team or community. All these elements are surely important in the implementation of a CoP, and this is why we paid attention to the social relations between participants in the CoP.

Communities of Practice

Communities of practice have raised more and more interest over recent years. We will first present the definition of the concept, recall a few elements highlighted by other researchers as impacts or benefits expected from these communities of practice (CoPs), before we highlight the benefits as well as individual and organizational advantages and disadvantages of communities of practice, and link these to the context or conditions that appear to favour such benefits or advantages, as well as link them to teamwork.

The term *communities of practice* was first used by Wenger and Lave (1991). Many different views and definitions have been presented since, but most refer to the importance of sharing information within a small group, as well as the value of

Figure 1. Stages of development of a community (adapted from Wenger et al., 2002, p. 69, and Bourhis & Tremblay, 2004)

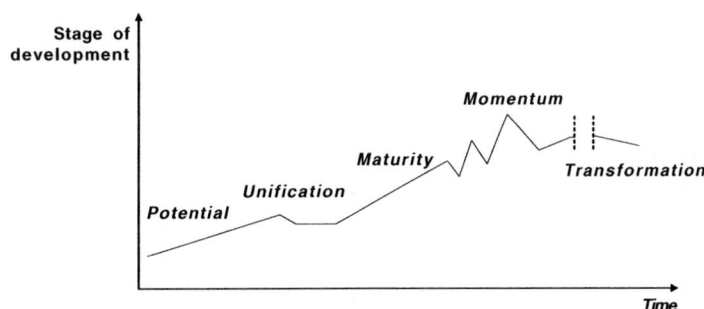

informal learning for a group and for an organization as a whole, bearing in this some similarity with the teamwork literature. A few definitions of communities of practice are presented in Mitchell (2002):

Communities of practice are groups of people who share a concern, a set of problems, or a passion about a topic, and who deepen their knowledge and expertise in this area by interacting on an ongoing basis. (Wenger et al., 2002, p. 4, quoted in Mitchell, 2002, p. 12)

A group whose members regularly engage in sharing and learning, based on their common interests. (Lesser & Stork, 2001, p. 831, quoted in Mitchell, 2002, p. 12)

The main elements stressed here are the sharing of a concern, a set of problems, the ongoing interaction between the group, the ongoing sharing and learning, bearing again some similarity with the teamwork literature. As we will see later, these definitions correspond to the type of community we studied, while other definitions insist on an informal dimension, which was absent from our case study.

Indeed, more conventional definitions of communities of practice exist, which refer to a more informal group, whereas the communities we studied, and the one presented here, are structured by an organization and much more formal. Here are a few other definitions, centred on the informal dimension:

Groups of people informally bound together by shared expertise and passion for a joint enterprise. (Wenger & Snyder, 2000, p. 139)

Informal clusters and networks of employees who work together—sharing knowledge, solving common problems and exchanging insights, stories and frustrations. (Lesser & Prusak, in Lesser et al., 2000, p. 831, quoted in Mitchell, 2002, pp. 11-12)

Over recent years, more and more interest has been placed in communities that work from a distance, although sharing a project, and it is this type of community that caught our interest. From an empirical point of view, these communities of practice are very similar to teams working from a distance,

although some theoretical differences have been highlighted by some authors. They are seen as a group that has a common mission, that has a common task, and must deliver a product based on the regular exchanges and information sharing within the group, as defined in McDermott (1999). Work teams usually have a predetermined goal and schedule, often very clearly defined tasks, and their activity is usually centred on their work tasks and done during working hours; often work teams disintegrate once the objective is attained, but in the manufacturing sector, they often remain to assume general work tasks collectively (Tremblay & Rolland, 1998). Also, work teams are often characterised by a strong division of labour, whereas communities theoretically imply more direct cooperation between the members (Tremblay, Rolland & Davel, 2000). Communities of practice are seen as having wider and less defined objectives, as not having a specific schedule and dates for attaining the various objectives (contrarily to work tasks), and usually go on for quite some time (indeterminate often), although this is not always the case.

As indicated in much of the literature on work teams as well as communities of practice, working *together* as a group usually requires some preconditions, the main one appearing to be trust in other members of the group. This is all the more important in a context of communities of practice, since members of the community are expected to share tacit knowledge, to construct collectively new knowledge, and possibly new products or services (McDermott, 1999, 2001; Wenger & Snyder, 2000; Adams & Freeman, 2000; Deloitte Research, 2001). It is precisely because of this trust element that many authors recommend that virtual communities of practice be developed on the basis of existing informal groups, groups that share values and already trust each other. This is however often not possible in firms and is why many virtual communities of practice are designed without taking this element into account. This of course represents an additional challenge for CoPs—that is, when previous acquaintance and trust of members has to be developed within the CoP.

Amongst the other main prerequisites often mentioned in the CoP literature are the importance of the leader or animator of the community, the interest and motivation of individuals to work together as a group, and the support received from the organisation: support and legitimisation of the group on the part of the immediate superior or higher levels of hierarchy, financial, or non-monetary rewards for the participants and the like (Wenger et al., 2002). Available technology and technological support are sometimes mentioned, but most research seems to indicate that the human resources and organisational challenges are more important, and that technology plays a more limited role in the success or failure of communities of practice (Bourhis & Tremblay, 2004).

The CoPs are seen as ways of delivering the following benefits, according to Mitchell (2002): the informal dissemination of valuable information, improvements in productivity, and fostering of innovation and the reinforcement of strategic direction of the organisation that is responsible for the CoP and supports it.

CONCLUSION AND FUTURE TRENDS

In our view, much of the literature on communities of practice should take into account all the work that has been done on teamwork in order to better understand the social relations of production which are at play in such a context. While much of the literature on communities of practice is in the management field, and often tends to be very normative in essence, the analytical work done on teamwork since the '90s has much to bring to the analysis and to the development of such communities.

The analysis of teamwork has shown that the process of fostering collective responsibility in a community or group, like all processes of organisational innovation, is not a simple linear process of transferring responsibilities or tasks. On the contrary, it refers to a set of dimensions such as task interdependency, interdependency with regard to the organisational environment, type of supervision, interpersonal relations between members, degree of autonomy, availability of resources, management support, organisational structure, and a whole set of variables related to the context in which the team or community evolves. These elements are crucial in the development of communities of practice, as we observed in empirical research on communities of practice (Tremblay, 2004; Bourhis & Tremblay, 2004).

REFERENCES

Bourhis, A., & Tremblay, D.-G. (2004). *Les facteurs organisationnels de succès des communautés de pratique virtuelles.* Québec: Cefrio.

Cohen, S.G., & Bailey, D.E. (1997). What makes teams work: Group Effectiveness research from the shop floor to the executive suite. *Journal of Management, 23*(3), 239-290.

Davel, E., Gomez da Silva, J.R., Rolland, D., & Tremblay, D.-G. (2001). Comunicaçao e competências no trabalho em equipe. *Dans Arche Interdisciplinar, 9*(28), 39-59.

Deloitte Research. (2001). *Collaborative knowledge networks: Driving workforce performance through Web-enabled communities.* Retrieved from *http://www.deloitte.com/dtt/research/0,1015, sid%253D6975%2526cid%253D12924,00.html*

Durand, J.-P., Stewart, P., & Castillo, J.J. (Ed.) (1999). *Teamwork in the automobile industry: Radical change or passing fashion?* London: Macmillan Business Press.

Guzzo, R.A., & Shea, F.P. (1992). Group performance and intergroup relations in organizations. In M.D. Dunnette & L.M. Hought (Eds.), *Handbook of industrial and organizational psychology* (vol. 3). Palo Alto, CA: Consulting Psychologists Press.

Hackman, J.R. (1987). The design of work teams. In J.W. Lorsch (Ed.), *Handbook of organizational behavior.* Englewood Cliffs, NJ: Prentice-Hall.

Henri, F., & Lundgren-Cayrol, K. (2000). *L'apprentissage collaboratif: Essai de définition.* Télé-Université.

Henriksson, K. (2000). *When communities of practice came to town: On culture and contradiction in emerging theories of organizational learning.* Retrieved from *http://www.google.ca/search? hl=fr&q=Henriksson%2C+K.+%282000%29*

Hildreth, P., Kimble, C., & Wright, P. (2000). Communities of practice in the distributed international environment. *Journal of Knowledge Management, 4*(1), 27-37.

T

Hodgson, R. (1997). Group relations at work: Solidarity, conflict and relations with management. *Work and Occupations, 24*(4), 426-452).

McDermott, R. (1999). Nurturing three-dimensional communities of practice: How to get the most out of human networks. *Knowledge Management Review,* 11, Fall, 26-29.

McDermott, R. (2000). Knowing in community: Ten critical success factors in building communities of practice. Retrieved from *http://www.co-i-l.com/ coil/iknowledge*

Mitchell, J. (2002). *The potential for communities of practice*. Sydney, Australia: John Mitchell and Associates.

Roy, M. (1999). Les équipes semi-autonomes au Québec et la transformation des organisations. *Gestion-Revue Internationale de Gestion, 24*(3), 76-85.

Salerno, M.s. (1999). *Projeto de Organizações Integradas e Flexíveis. Processos, Grupos e Gestão Democrática via Espaços de Comunicação-Negociaação*. Sao Paulo: Atlas.

Shea, G.P., & Guzzo, R.A. (1987). Group effectiveness: What really matters? *Sloan Management Review, 28*(3), Spring, 25-32.

Sundstrom, E., Demeuse, K.P., & Futrell, D. (1990). Work teams: Applications and effectiveness. *American Psychologist, 45,* 120-133.

Tremblay, D.-G., Rolland, D., & Davel, E. (2000). Travail en équipe, compétences et pratiques de sélection au Québec. In L. Cadin (Ed.), *Internationalisation de la GRH?* CD-Rom des Actes du XI Congrès de l'Association francophone de Gestion des ressources humaines. Also Research Note on *www.teluq.uquebec.ca/chaireecosavoir*

Tremblay, D.-G., & Rolland, D. (1998). *Gestion des ressources humaines: Typologies et comparaisons interantionales*. Québec: Presses de L'Université du Québec.

Tremblay, D.-G., & Rolland, D. (2000). Labour regime and industrialisation in the knowledge economy; the Japanese model and its possible hybridisation in other countries. *Labour and Management in Development Journal, 7.* Brisbane: The Australian National University. Retrieved from *http://www.ncdsnet.anu.edu.au*

Wenger, E. (1998). Communities of practice—learning as a social system. *Systems Thinker, 9*(5), June, 5.

Wenger, E., McDermott, R., & Snyder, W. (2002). *Cultivating communities of practice. A guide to managing knowledge*. Boston: Harvard Business School Press.

Wenger, E., & Snyder, W. (2000). Communities of practice: The organisational frontier. *Harvard Business Review, 78*(0-1), 139-145.

KEY TERMS

Effectiveness of Team Interaction: Effectiveness of team interaction is linked to functional components such as (a) interdependency in terms of the environment, (b) task interdependency of team members (skills development and consequences (sanctions based on results) and (c) the quality of transactions between team members (interpersonal relations, production energy, shared effectiveness and group cohesion.

Intentional Community of Practice: Community that is created by an organization rather than being an informal cluster or network of employees who share a passion, who share knowledge or work together to solve problems.

Team: To make up a team, members must have a minimum of (a) task interdependency among members, (b) shared responsibilites, (c) team identity, and (d) power to manage the relationship between the team and the organization.

Virtual Community of Practice: Informal clusters of employees who work together—sharing knowledge, solving common problems, and exchanging insights, stories, and frustrations—and who do this working from a distance, rather than face-to-face.

Team-Work Issues in Virtual Teams

Barbara J. Cargill
Swinburne University of Technology, Australia

INTRODUCTION

There has been much written about virtual organisations and virtual teams in the last five years. We have begun to research the shift in work organisation paradigms and structures, translating much of what we knew already about workgroups and teams in conventional workplaces into the new contexts, and adding some new issues and understandings into the mix. We may need to translate a little further to come to grips with the 'virtual teams' that are actually communities of practice (CoPs).

BACKGROUND

We have long known that all work groups are characterised and defined by their shared sense of time and task, if not place. Without the task and the interaction around that task, we have only a semi-random collection of individuals who happen to be in a particular space (temporally and/or geographically). A team is a very particular kind of group, where systematic contributions of different kinds from each of the members are required to complete the task (Tyson, 1998).

A workgroup may be less dependent on such contribution from each of the members and have some tolerance for members to be more passive and low-participative, provided they do not actually obstruct the progress on the shared task. Virtual work teams have been researched rather more than the more voluntary network of a CoP, and we are thus still learning how a CoP might manifest team issues, especially since some such communities are not virtual. The CoP does have a learning and sharing task, but it is often loosely defined, and so group cohesion is a more difficult process. Where the CoP is connected electronically rather than geographically co-located, then the dynamics are even more uncertain.

We have also known for a long time that when groups first form, they are typified by *ambiguity*. Members will ask themselves and perhaps each other whether they really wish to be in the group, whether they will get enough out of it, whether it will be worth their invested time, what it is all about anyway, if this CoP will really be able to share its collective knowledge, and so forth.

As the group begins to come together around the shared task, members may also sense a great deal of *ambivalence*. Becoming part of a group also means having to relinquish some aspects of one's own individuality. Will members feel that they must go too much with the demands and directions of others? Will they really be able to be heard and to also draw from the group? Will they retain their individual identity and sense of self-worth, or will the group diminish them in some way? Can the group become a close-knit, fully functioning team without the individuals' identities disappearing entirely? These are not foolish concerns and tend to simmer in each member's mind at various points of a team's formation and life.

As with all small groups, leadership issues will surface irrespective of what formal leadership arrangements and protocols have been put in place. The issue of *authority* pervades group interaction, whether they are face-to-face or remote. Members ask themselves who is in charge of this group, and test how the leader(s) will respond to certain interactions and challenges. Even when leadership and authority issues appear resolved comfortably, they are nonetheless present and potentially volatile if the task suddenly makes new demands on the group or team.

Dynamics of ambivalence, ambiguity, and authority inevitably generate some levels of *anxiety* in a group, and these will manifest in a variety of ways, ranging from withdrawal or tentativeness through to aggression and other dysfunctional behaviours (Wells, 1982; DeBoard, 1978; Hirschhorn, 1991). In a CoP, many of these issues will be amplified by the factors of loose structure, low positional authority, and the potential for lack of task clarity. If a virtual team is ever to become an effective learning and sharing group, a true CoP, *then the early formative stages are vital and must be given special attention.* Hoefling (2001), Nemiro (2004), Pauleen (2004), Haywood (1998), and many others agree that the start-up, team-building stages of forming a virtual team are worthy of special attention, since if the cohesiveness of the virtual team is not established then, the group or CoP may never be fully functioning. In a CoP, given the voluntary nature of the exercise, these start-up issues require special attention from whomever holds the instigating leadership 'torch'. The protocols of communication, the clarification of the learning tasks, and so forth are all laid down in these early stages, or the loose connection of the virtual group rapidly become disconnected entirely.

FUTURE AND CONCLUSION

In essence, the key issues for virtual teams pivot around cohesion, communication, and enabling processes and technology. Within these broad bands are embedded issues of possible cultural diversity, trust and interpersonal connectedness, and creation of a working culture that both encourages creativity and yet permits honest assessment and reflection on progress and outcomes. Since the 'virtual' element reduces visual communication and pares down normal interpersonal cues, it is clear that all aspects of the team life are potentially more fragile and at risk of faltering. Accordingly, leadership and facilitation, however informal or shifting, are pivotal ingredients in the success of the virtual team.

REFERENCES

DeBoard, R. (1978). *The psychoanalysis of organizations—a psychoanalytic approach to behaviour in groups and organizations.* London: Routledge.

Haywood, M. (1998). *Managing virtual teams: Practical techniques for high-technology project managers.* Boston: Artech House.

Hirschhorn, L. (1991). *Managing in the new team environment: Skills, tools and methods.* Reading, MA: Addison-Wesley.

Hoefling, T. (2001). *Working virtually: Managing people: Managing people for successful virtual teams and organizations.* Sterling, VA: Stylus Publishing.

Nemiro, J. (2004). *Creativity in virtual teams: Key components for success.* San Francisco: Jossey-Bass.

Pauleen, D. (Ed.). (2004). *Virtual teams: Projects, protocols and processes.* Hershey, PA: Idea Group Publishing.

Tyson, T. (1998). *Working with groups* (2nd ed.). South Yarra, Australia: Macmillan.

Wells, L. Jr. (1982). *The four 'A's in groups.* Unpublished Address, Melbourne.

KEY TERMS

Group: A number of individuals who interact for purposes of addressing some shared purpose or task. This differentiates human collectives from random clusters of people who happen to be co-located, or who have no purpose which requires interaction.

Team: A defined group that intentionally combines its members' various skills and knowledge to undertake a shared task that requires them to coordinate their various efforts for satisfactory completion. Teams will normally have a defined leadership and authority structure, or define their own in order to address the task.

Virtual Organisation: A commercial or non-profit enterprise that does not exist as a tangible entity at a single location, nor even at multiple sites, but comprises networked nodes and individuals who are connected solely for the purpose of pursuing that enterprise.

Virtual Team: A work unit that is created for the purpose of contributing some function, project, or other output to an organisation's mission, but which does not exist in a particular location or place. Remotely located team members are obliged to interact with each other via codified and acknowledged authority structures, each contributing to the team output by means of electronic and/or telephonic linkages. These connect the team and enable the shared task to be progressed. Virtual teams are often multi-location and international in nature.

Technical Issues Facing Work Groups, Teams, and Knowledge Networks

Lynda R. Louis
Xavier University of Louisiana, USA

INTRODUCTION

Communities of practice have become prevalent in today's business and educational environments. Loosely defined, communities of practice (CoPs) are informal groups within or across organizations. These groups are viewed to have a common set of information needs or problems. Davies, Duke, and Sure (2003) suggest that while not a typical organizational unit, these informal, work-oriented networks share a common agenda, interests, or issues. Millen and Fontaine (2003) assert that these communities are defined by a common disciplinary background, similar work activities and tools, and have shared stories, contexts, and values. Burk (2000) says CoPs are expansions of one-on-one knowledge sharing. This article discusses first the background to such issues as team formation, member participation, and sustained membership; it then considers future trends in knowledge capture and sharing, and the support such communities require.

BACKGROUND

The primary advantage of a CoP is knowledge sharing, and facilitating learning and organizational memory (Gallivan, 2000). Hara and Kling (2002) say the CoP provides an informal learning environment where both novices and expert members of the community may interact, share their experiences, and learn from each other. Wenger, McDermott, and Snyder (2002) assert that these communities thrive because they deliver value to their organization, to the teams on which the members serve, and to the community members themselves.

Even though CoPs are not formal parts of an organization in the same sense as a workgroup, a work team, or a traditional knowledge network, research reveals that many of the same technical issues abound for these communities. All have been well documented from the CoP perspective since the term has become so widespread in both the corporate and educational arenas. For this discussion, these issues are grouped into the following categories: team membership, knowledge management, communication mechanisms, and community support. Team membership encompasses such topics as team formation, member participation, and sustained membership. Knowledge management includes topics such as knowledge creation, knowledge capture, knowledge sharing, and knowledge management (KM). Communication mechanisms concerns the ways in which members communicate and the tools management may provide to facilitate that communication. Support centers on what support tools and mechanisms are employed from an information technology (IT) perspective. Each of these issues presents challenges and opportunities for both the corporate and the academic communities.

Community Formation

Communities of practice generally form through an informal network. The term was coined when it was noted that copy machine technicians often shared much of their informal discussion in natural settings during social interactions, often around the water cooler. Orr (1996) reported that these technicians shared experiences associated with repairing machines that was not part of the standard documentation used to service the devices.

CoPs form as needed to meet the knowledge needs of its members. Membership is not fixed and provides a vehicle for constant change. However, researchers acknowledge that in order to be successful and alive, members must be active participants of the community.

Sustained Membership

Since membership in a CoP is usually voluntary, sustained membership is of concern to many corporations that provide support for these communities. Ferran-Urdaneta (1999) asserts that the community outlives its members. Burk (2000) says that these communities can be impeded by turnover and restructuring.

Member Participation

Since membership is voluntary and not part of an assigned workgroup, CoP participation is not mandatory and does not absorb great deals of energy or attention of any one individual member. However, with the new tools that are being deployed to capture the knowledge of the members, some corporations are contemplating ways to make participation a requirement. Kankanhalli, Tanudidjaja, Sutanto, and Tan (2003) suggest that participation be made part of an employee performance appraisal. However, this raises the issue of whether membership is then considered 'voluntary'. Lesser and Storck (2001) discussed the implications of CoPs on organizational performance. They argued that the social capital in CoPs lead to behavioral changes, which in turn influence business performance. Corporations must find a way to encourage the continued participation in the community and not threaten the existence of the CoP or the performance of the business.

Rogers (2000) conducted a study of an online (virtual) community of practice to address the issue of fostering coherence in a virtual community. His study concluded that interactions in this community demonstrated characteristics of mutual engagement, joint enterprise, and shared repertoire. These are explained in the "Key Terms" section.

FUTURE TRENDS

Knowledge Creation

Hara and Kling (2002) assert that there are three types of socially constructed knowledge that members must learn: cultural knowledge and two types of subject-matter knowledge (either book knowledge or practical knowledge). The challenge here is to know what type of knowledge works best with the community so that the community meets its objective and commitment to the group, and brings value to the organization.

Knowledge Capture, Sharing, and Management

Researchers acknowledge that to be beneficial to corporations, the knowledge that is shared between members of CoP needs to be captured and preserved. As such, many knowledge management tools have been deployed to assist the members in documenting and reusing that knowledge. Attention must be given to ensure that the information technology and knowledge management systems enable CoPs to flourish (Agresti, 2003).

The KM tools allow for the economic reuse of the knowledge (Kankanhalli et al., 2003). The tools should allow queries and solutions to be posted efficiently and provide discussion threads. The majority of the knowledge capture tools include the computer-supported cooperative work (CSCW) tools used for collaborations (Birnholtz & Bietz, 2003; Cluts, 2003).

Communication Mechanisms

The primary mechanism for communication between members of a CoP has moved from the initial face-to-face social environment to electronic media, as the value of CoPs has been recognized within the organization. As such, companies are now providing Internet, intranet, and Web-based availability to members using a special purpose collaborative environment. Also, knowledge management tools are being deployed to assist with the capture and reuse of the knowledge that the communities hold.

Community Support

It has become apparent by most corporations that support the activities of CoPs that more than word of mouth support is needed. As such, most corporations are provided support from their information technology (IT) organizations. This support must come with committed funding and tools. Training to use the technology tools to communicate is also critical (Gallivan, 2000). As noted above the main tool used for communication between members is the electronic media.

Another facet of IT support comes in the form of knowledge management (KM) systems that many companies now employ to capture the knowledge of the CoP members. Eales (2004) presents a collaborative approach to user support.

CONCLUSION

Communities of practice are here to stay. They have been recognized as a vital part of the organization. As with the formal groups within the organization (the teams, workgroups, and knowledge networks), CoPs have many technical issues that have emerged. These issues have been well documented over the years and will continue to be subject to much research, debate, and discussion. While these issues—membership (formation, sustaining, participation), knowledge management (capture, sharing, and management), communication mechanisms, and support—are not new for any corporation, these issues must continue to receive the support of management in order for the communities of practice to continue to provide the necessary benefits to both its members and to the business.

Research abounds on these areas that mainly address the 'work environment'. There is evidence of research that deals with knowledge management and user training for both the traditional workgroups and CoPs. Academics are challenged to incorporate this knowledge of the functional CoP into the curricula so that students get a complete picture of the work environment. This will allow them to gain a full understanding of the synergies of the workforce, and the value of human knowledge and the role this knowledge plays in promoting the success of businesses.

REFERENCES

Agresti, W.W. (2003). Tailoring IT support to communities of practice. *IT Pro, 5*(6), 24-28.

Birnholtz, J.P., & Bietz, M.J. (2003, November 2-12). Data at work: Supporting sharing in science and engineering. *Proceedings of GROUP '03* (pp. 339-348).

Burk, M. (2000). Communities of practice. *Public Roads, 63*(6). Retrieved from *http://www.tfhrc.gov/pubrds/mayjun00/commprac.htm*

Cluts, M.M. (2003, November 2-12). The evolution of artifacts in cooperative work: Constructing meaning through activity. *Proceedings of GROUP '03* (pp. 144-152).

Davies, J., Duke, A., & Sure, Y. (2003, October 23-25). OntoShare—a knowledge management environment for virtual communities of practice. *Proceedings of K-CAP '03* (pp. 20-27).

Eales, R.T.J. (2004). A knowledge management approach to user support. *Proceedings of AUIC2004, Conferences in Research and Practice in Information Technology, 28* (pp. 33-38).

Ferran-Urdaneta, C. (1999). Teams or communities? Organizational structures for knowledge management. *Proceedings of SIGCPR '99* (pp. 128-134).

Gallivan, M.J. (2000). Examining workgroup influence on technology usage: A community of practice prospective. *Proceedings of SIGCPR 2000* (pp. 54-66).

Hara, N., & Kling, R. (2002). IT supports for communities of practice: An empirically based framework. White Paper retrieved from *http://www.slis.indiana.edu/CSI/WP/WP02-02B.html*

Kankanhalli, A., Tanudidjaja, F., Suntanto, J., & Tan, B.C.Y. (2003). The role of IT in successful knowledge management initiatives. *Communications of the ACM, 46*(9), 69-73.

Lesser, E.L., & Storck, J. (2001). Communities of practice and organizational performance. *IBM Systems Journal, 40,* 831-841.

Lueg, C. (2000, September 11-13). Where is the action in virtual communities of practice? *Proceedings of the Workshop on Communication and Cooperation in Knowledge Communities at the*

German Conference on Computer-Supported Cooperative Work (D-CSCW), Munich, Germany. Retrieved from *http://www-staff.it.uts.edu.au/~lueg/papers/commdcscw00.pdf*

Millen, D.R., & Fontaine, M.A. (2003, November 2-12). Improving individual and organizational performance through communities of practice. *Proceedings of GROUP'03* (pp. 205-211).

Orr, J.E. (1996). *Talking about machine: An ethnography of a modern job.* Ithaca, NY: Cornell University Press.

Preece, J. (2004). Etiquette, empathy and trust in communities of practice: Stepping stones to social capital. *Journal of Universal Computer Science.* Retrieved from *http://www.ifsm.umbc.edu/~preece/Papers/Tacit_Know_COPs.pdf*

Rogers, J. (2000). Communities of practice: A framework for fostering coherence in virtual learning communities. *Educational Technology and Society, 3*(3). Retrieved from *http://ifets.ieee.org/periodical/vol_3_2000/e01.html*

Wenger, E. (2004). Communities of practice—a brief introduction. Retrieved from *http://www.ewenger.com/theory/communities_of_practice_intro.htm*

Wenger, E., McDermott, R., & Snyder, W.M. (2002, March 25). Seven principles for cultivating communities of practice. *Harvard Business School Working Knowledge.* Retrieved from *http://www.askmecorp.com/pdf/7Principles_CoP.pdf*

KEY TERMS

Collaborative Support System: Network-based system designed to facilitate and augment the collaborative support for a solution of computer-related problems. Collaborative support is that support given by work colleagues and friends, usually operating in an informal environment (Eales, 2004).

Community of Practice: Informal group within or across an organization, whose members share a common set of information needs or problems. They are defined by a common set of problems and share a common agenda, interest, or issues (Agresti, 2003; Burk, 2000; Davies, Duke & Sure, 2003; Ferran-Urdaneta, 1999; Lesser & Storck, 2001; Millen & Fountaine, 2003; Preece, 2004). In some organizations, CoPs may be referred to as learning networks, thematic groups, or tech clubs (Wenger, 2004).

Community Characteristics: Mutual engagement—refers to fact that members are engaged in a common negotiated activity. Without mutual engagement the community is likely to represent a network of individuals or individual groups (Rogers, 2000).

Computer-Supported Cooperative Work (CSCW): The study of how people work together using computer technology. Typical CSCW tools/applications include e-mail, awareness and notification systems, videoconferencing, chat systems, multiplayer games, and real-time shared applications (such as collaborative writing or drawing). These software tools are used in CoP knowledge capture/storage/sharing (Birnholtz & Bietz, 2003; Cluts, 2003).

Distributed Community of Practice: A group whose members are distributed (possibly over countries), work is done in a core group, members meet face-to-face occasionally, and communication is maintained via electronic media. Its members are interacting with the real world, and learning takes place in the real world (Lueg, 2000).

Joint Enterprise: Allows the community to extend the boundaries and interpretations beyond those that were created. By sharing a common goal, members of the community negotiate their situations in their reactions to them. With joint enterprise it is more likely that the community will sustain its validity.

Shared Repertoire: Refers to the fact that there is a pool of resources that members share, contribute to, and renew. These resources can be physical (e-mail, word processors, common textbook, etc.) or intangible (common discourse, common means, or methodology for accomplishing tasks).

Virtual Community of Practice: Group of people that relies primarily on networked communication media to communicate and connect in order to: discuss problems and issues associated with their profession; share documents, solutions, or best practices; collaborate on projects; plan for face-to-face meetings, or continue relationships and work beyond face-to-face events (www.educause.edu/vcop).

Work Teams/Groups: Formed by management and report to a boss. These have defined member-ship, deadlines and schedules, and specific deliverables (Burk, 2000). They are usually small in number; membership is fairly stable; and the members must trust each other, coordinate the work amongst themselves, understand each other's importance, and hold each other accountable (Ferran-Urdaneta, 1999).

Training and Articulating Public Agencies in Argentina

Graciela Mónica Falivene
INAP, Argentina

Ester Kaufman
INAP, Argentina

INTRODUCTION

The deep institutional crisis the Argentine state has gone through has led the Heads of the Public Management Training Program (Instituto Nacional de Administración Pública (INAP)) to question the usual way of facing reforms and training policies. Thus, the design of the activities within this program has been carried out from a knowledge management perspective. According to this approach, professional state forums (PSFs) were created in 2002. These PSFs' practices followed the CoPs and networks conception; their main goals were strengthening public organizations and upgrading professional standards.

The PSFs involve public executives who focus on specific practices. These PSFs can cut across agencies or function at an internal level. Their members essentially seek to develop their competencies in the practice considered. In order to recognize the PSFs/CoPs, previous analysis was done regarding the way each process or task is performed in each agency.

BACKGROUND

Context

The reform conceived for the Argentine State in the 1990s was partially implemented—only to privatizations and downsizing in public agencies. Despite official announcements, less attention was given to the training and to the recruitment of professionals. Instead, recruitment was organized mainly around political "cronyism." Likewise, the use of resources was hindered by political mismanagement.

Ironically, this limited "reform" seemed to be working by the mid-1990s, fed by an overabundance of foreign funding (by multilateral organizations) and political oversight with regards to the public debt that was being accrued.

As a consequence of the political crisis (December, 2001), INAP came forward with the following diagnosis:

1. The rule was that public institutions were fragmented and isolated due to the constant changes of organizational structures encouraged by international agencies.
2. Within this lack of institutional framework, it was too difficult to establish the kind of competencies that were required.
3. Up until December 2001, INAP only offered "packaged courses," which did not regard the agencies' needs.

In response to the existing crisis, INAP developed a program whose strategic goals were:

1. to foster articulation at an institutional level;
2. to promote organizational learning;
3. to encourage CoPs; and
4. to recognize, sustain, and build knowledge as a policy and practice for modernizing the state.

Knowledge Management and CoPs

INAP has adopted Logan's (1997) definition, which suggests that:

Knowledge management is the organizational activity of creating the social environment and technical infrastructure so that knowledge can be accessed, shared, and created. (p. 23)

Nonaka and Takeuchi (1995) posit that organizational knowledge is created through a continuous and dynamic interpersonal interaction between tacit and explicit knowledge. While tacit knowledge is personal and difficult to communicate, explicit knowledge is transmittable in systematic language. In an interview (Scharmer, 1996), Nonaka stated:

This interaction between the two types of knowledge brings about what we call four modes of knowledge conversion—that is, socialization (from individual tacit knowledge to group tacit knowledge), externalization (from tacit knowledge to explicit knowledge), combination (from separate explicit knowledge to systemic explicit knowledge), and internalization (from explicit knowledge to tacit knowledge). (p. 4)

INAP developed two types of practices:

1. *CoP forums* (turning tacit knowledge into explicit knowledge); and
2. *good practices* research (mainly technological), generating knowledge databases for CoP members to share.

The conception of CoPs describes groups of people who share a concern or a passion for something they do and who interact regularly to learn how to do it better (Lave & Wenger, 1991; Cohendet & Creplet, 2001). CoPs constitute an effective way of dealing with unusual problems, sharing knowledge beyond the border of traditional structures (Tuomi, 1999). In Argentina the chaotic condition of government structures, added to the weak administrative career, played a strong role in choosing this strategy.

Bureaucratic Models and CoPs

Through the PSFs, members of public agencies tried to solve some of the increasing bureaucratic drawbacks, such as:

1. lack of shared effort and a tendency towards fragmentation,
2. reduced capacity to integrate innovation,
3. censorship related to such innovation,
4. self-centeredness and isolation, and

5. low self-esteem and low social standing of the public officials.

Some current government policies encourage CoPs as strategic lines of action as in Canada (www.communities-collectivites.gc.ca), the United States (www.gsa.gov/collab), Australia (www.agimo.gov.au), and other countries. These policies are oriented towards reinforcing federal policies, as well as supporting government structures in complex processes.

Referring to U.S. experiences, Snyder and Souza Briggs (2003) state:

For a variety of reasons, the federal government is uniquely positioned to help foster the evolution of nation-scale community-of-practice networks through five principal mechanisms: (1) leveraging infrastructure efficiency; (2) promoting agency learning and alignment; (3) diffusing learning and innovation across states and nations; (4) establishing standards for measuring performance outcomes; and (5) modeling an approach for diffusing ideas and methods that can be used at state and local levels. (p. 64)

The developments implemented by the above-mentioned administrations have key government authorities as their sponsors. Such was the case with Al Gore in the U.S. when he was the vice-president of that country ("Reinventing Government," 1998). Conversely, in Argentina, this initiative stems from the Public Management Undersecretariat and INAP (a mere public office depending on the latter). In view of this categorization, it is difficult to undertake a political strategy as strong as in the previously mentioned countries.

As the PSFs started growing, conflicts arose with the existing political authorities. Both sectors had different logical ways of thinking as well as different focuses of interest. As far as CoPs were concerned, their goal was to solve daily issues associated to their needs or professional practices. The CoPs within the state were driven by a technical rationality aimed at solving difficulties in management. By contrast, for most political authorities, their main interest lay in their party relationships, political commitments, and state agenda. With respect to the informal structure of CoPs, governmental control was limited, causing certain uneasiness in political authorities.

In other countries there are also misunderstandings. Snyder and Souza Briggs (2003) state:

There are several ways to address these concerns: by seeing the emergence of CoPs as an evolutionary process, not a cataclysmic revolution; by distinguishing the knowledge-building and knowledge-sharing functions of these communities with the primarily transactional focus of product- and service-delivery units; and by understanding that collaborative, boundary-crossing networks need not mark the loss of government's public-service identity and influence, but rather serve as an expansion of both. (p. 51)

DEVELOPMENT OF CoPs RESULTING FROM PSFs

During the crisis, the INAP has set up many PSFs focused on overcoming bureaucratic limitations. The INAP involved different public sectors to create PSFs/CoPs. This gave rise to forums through which organizational knowledge could be accessed to generate competencies and thus overcome the crisis. Thus, cross-agencies forums were created with reference to the following activities: human resources, file management, documentation center, IT areas, front desks, budget management, statistics, international cooperation units, and so forth.

Such PSFs have been extremely useful inasmuch as they have enabled the practitioners in several areas to reach compromises, to agree on different ways of working, and to reinforce the power of the public institutions.

The following are two paradigmatic examples of the Argentine experience:

1. The IT Professional Forum (ITPF): A CoP to Learn and Innovate for Developing the E-Government

The ITPF became a cross-public agencies CoP that involves the IT professionals of the public adminis-

tration (Kaufman, 2004, pp. 151-187). The ITPF generated a true CoP that set its own rules and innovations and proved to be autonomous. It may be supposed that this phenomenon took place due to the central importance acquired by the most basic aspects of technology into the crisis. Technology enabled the government to overcome the communication problem through the use of e-mails, thus solving crucial problems, such as the lack of paper or ink for printers. IT provided a solution for these shortages in an informal way, insuring an adequate number of functioning computers—something that could not be taken for granted at the time.

In order to set up this forum, representatives of more than 100 National Public Administration organizations (APN) were invited to participate. Ninety organizations and 200 technicians worked in different processes within the forum. Surprisingly, the ITPF current sustainability can be attributed to lack of resources and the sector's need for participation.

In the past, IT experts from international agencies and consulting companies were sought after for the design and implementation of technological policies. But when the crisis set in, the government was left without economic resources to keep up such expenditures. As a result, the local IT staff started to assemble the forum invited by INAP and the IT National Office.

The ITPF organizes its practice articulating several work meetings. The meetings are plenary, thematic, or group meetings. The global structure of the activities depends on a "Core Group" open to all members and made up of its coordinator.

Participants of the plenary meetings are experts in the IT national areas, although over time professionals from other local governments as well as from the academic sector have joined in. The average turnout is 100 people, with monthly meetings throughout 2002-2004. At present, face-to-face meetings are less frequent due to the widespread growth of the virtual forum.

The working groups are fully dedicated to the development of leading issues, such as Free Software, Cross Agencies Applications, Web Site, Software Licenses, Training in LINUX, Interoperability, Computing Crime, and so forth.

The Forum's strengths can be found:

1. in the steady attendance of its members, considering that the initiatives are generated by the permanent or quasi-permanent staff;
2. in the certainty that the work produced will be implemented, given that the professionals who present the innovations implement them;
3. in the legitimacy of its productions due to the general consensus regarding their suitability; and
4. in its transparency and responsibility as a result of a periodic and steady collective control of the initiatives, processes, and products.

The challenges the forum has to confront are:

1. its continuity beyond changes in government;
2. the need for greater resources, considering the extraordinary growth of its activities; and
3. the channeling of foreign funding destined to information technologies for their use and control via the forum.

The ITPF has come a long way. Since its inception, its members were against interacting with other systems, actors, or forums. Nowadays, they are starting to change this attitude of "isolation" by interacting with legislators, scholars, lawyers, human resources directors, and front desk chiefs, as well as members of other forums. For example:

1. **The Front Desk Forum** is developing a network together with the ITPF in order to agree on interoperability follow-up filing systems.
2. **The Center of Documentations Forum** is transferring to the ITPF tools for KM documents (which the ITPF used for its Web site). On the other hand, the ITPF is training the Center of Documentation Forum members to preserve digital documents.

The evolution of the ITPF is also reflected in the language its members have been adopting. Rather than just using IT jargon, they have integrated IT terms to an interlinguistic field. Through their own experience at the ITPF, its members have learned to work in networks, to perceive the environment, communicate skills, and so forth. They confront the challenge of bridging the gaps with non-IT areas,

taking into account that the latter are constantly producing data that needs to be standardized in order to be included in IT systems and to feed new developments. The ITPF has contributed to collective knowledge about the culture of organizations in connection with information and technology, as well as the implicit hierarchies and their informal structure.

INAP's strategy to bridge the gaps with non-IT areas was to foster border meetings, where some of the members of two or more CoPs get together in order to boost an interchange of practices and to trigger thinking processes into the community itself or in the "border practices." For example, as we previously mentioned, the ITPF needed to acquire competencies developed by librarians in order to be able to classify their innovations. The Forum of Documents Center, in turn, needed to incorporate competencies from the former in order to work with digital documents and to be able to deal with them.

2. Forum of Document Information Centers: A Community of Learning Outlook

According to Lave and Wenger (1991), the analysis unit is moving from the individual level to the community levels, where learning consists of developing an identity as a member of a community in order to attain learning skills as part of the same process. Learning is no longer the acquisition of knowledge by individuals. It becomes a process of social participation on which the nature of the situation has a significant impact. This is called process of legitimate peripheral participation because the new participant moves from the periphery of the community towards its center. The activity becomes the link between the individual and the community, where individual practices are legitimized. From the perspective of this social learning theory, meaning and identities are set up through interaction. Likewise, the making of these meanings and identities is influenced by the context. In CoPs there is no division between the development of identity and the development of knowledge.

The Forum of Documents Information Centers became a cohesive CoP because of four joint causes:

1. the need to train the people in charge,
2. the lack of a permanent and suitable environment in which activity-related issues could be discussed,
3. the need to reengineer services due to the rise of ICT, and
4. the need to face the crisis.

The process started with a definition of the specific objectives of the forum. On the basis of this input, a Tentative Training Agenda was set up and agreed upon in a workshop. As a complementary activity a presentation was organized on "Typologies of Training Activities in Organizations for the Maintenance and Development of Competencies."

In these activities discussions arose regarding the different purposes of training. The forum took into account Le Boterf's (1991) formulation of four complementary forms of training:

1. training associated with maintaining the existing competencies,
2. training related to the solution to a specific problem,
3. training linked to projects of change, and
4. training connected to the foreseeable evolution of the profession.

The members of the forum used the input resulting from the workshop and several meetings to create a "Course of Specialization in Management of Documents Centers" (CEDID)—their own postgraduate development. The expected result was to keep, improve, and develop competencies for the management and transformation of each working unit. This application and evaluation task consisted of an analysis of the problems faced by the unit or area and the development of a project to overcome these problems. This project called for the application of the knowledge acquired and the techniques learned in informal courses about quality management, project technologies, and organizational tools. The forum members had to take advantage of the contributions made by others in a cross-transference learning situation. The permanent forum identified good practices shared on different working days. Some of them dealt with such issues as virtual libraries, documents preservation, and setting quality

models, among others (Trouvé & García Costa, 2004).

FUTURE TRENDS

CoPs constitute an effective way of solving unusual problems, sharing knowledge beyond traditional structural borders through the coexistence of informal integration models and bureaucratic models. Within these interrelationships, formal structures can be fed by the production generated in turn by CoP members. The different CoPs intertwine in a blurred way and cut across the organizational arena (Tuomi, 1999, p. 398). These CoPs also contribute to feeding some teams that are constituted for specific government projects, recognizing that:

*   formally managed projects work best when:
    1. problems can be clearly defined;
    2. reliable, quantifiable measurements are established; and
    3. an authority structure is in place to ensure that project results get implemented.
*   communities are most effective when:
    1. problems are complex and dynamic or very situation-specific;
    2. measures require stories to link cause and effect; and
    3. authority is decentralized and depends more on professionals' intrinsic commitment to getting results (versus extrinsic appraisals and incentives). (Snyder & Wenger, 2003)

The IBM Center for the Business of Government has performed case studies, led by Snyder and Souza Briggs (2003), which reveal the strategic relevancy the U.S. federal government has placed on the development of CoPs to support a variety of state-related issues, such as children's health, highway controls, antiterrorism, e-government, and so forth.

Australia has also included CoPs in their experiences. The Australian Government Information Management Office (AGIMO, www.agimo.gov.ar) states:

CoPs are practical vehicles for sharing and building knowledge and promoting better practice...In this spirit, AGIMO's role is that of a catalyst and facilitator, providing initial structure, while encouraging ownership and engagement by community of practice members. Facilitation of the CoPs is shared with other government agencies.

CONCLUSION

The CoPs may help government:

- **To strengthen a weakened state:** Historically, the traditional structures of government have failed to provide integral answers, above all in Third-World countries such as Argentina. This seems also to be the case in First-World countries, as seen from the above-mentioned examples. Traditional structures have very sharp internal boundaries marking isolated compartments. Therefore, it is hard for them to incorporate functions and actors to interact with different contexts, even when a strong political environment encourages this.

- **To achieve professional standards in the career of public official:** In Argentina, reform projects with foreign funding have always failed because they are not committed to permanent institutional processes or with the way new behavior becomes institutionalized. Instead, they emphasize explicit learning through the incorporation of a sustained technological advance by means of IT tools. Thus, their failure emphasizes the difficulty of managing tacit knowledge, ignoring the strategic relevance of the latter. By contrast, CoPs facilitate the conversion learning process. PSF members go through the stage of socialization, externalization, combination, and internalization. Likewise, these are the mechanisms through which the knowledge acquired by each member spreads within his/her own organization. This approach turns obsolete the previous "adult education professor" model. The responsibility lies with professionals themselves, since they are aware of the changes they want to achieve, the goals to reach, and the necessary learning processes.

REFERENCES

Cohendet, R., & Creplet, F. (2001). CoPs and epistemic communities: A renewed approach of organizational learning within the firm. Retrieved November 22, 2004, from *http://www-eco.enst-bretagne.fr/Etudes_projets/RNTL/workshop1/dupouet.pdf*

Kaufman, E. (2004). E-gobierno en Argentina: Crisis, burocracia y redes. In R. Araya & M. Porrúa (Ed.), *América Latina puntogob* (pp.151-187). FLACSO, Santiago de Chile & OEA.

Lave, J., & Wenger, E. (1991). *Situated learning-legitimate peripheral participation.* New York: Cambridge University Press.

Le Boterf, G. (1991). *Cómo invertir en formación. Proceedings of Gestión 2000* (pp. 64-70). Barcelona, Spain.

Logan, L.K. (1997). *The fifth language: Learning a living in the computer age* (p. 23). Toronto: Stoddart.

Nonaka, I., & Takeuchi, H. (1999). *La organización creadora de conocimiento: Cómo las compañías Japonesas crean la dinámica de la innovación* (pp. 61-103). Mexico DF: Oxford Press. English version: Nonaka, I. & Takeuchi, H. (1995). *The knowledge creating company: How Japanese companies create the dynamics of innovation.* New York: Oxford University Press.

Scharmer, C. (1996, February 23). Knowledge has to do with truth, goodness, and beauty. Conversation with Professor Ikujiro Nonaka, Tokyo. Retrieved September 15, 2004, from *http://www.dialogonleadership.org/Nonaka-1996.pdf*

Snyder, W.M., & Souza Briggs, X. (2003, November). *Communities of practice: A new tool for government managers.* Collaboration Series. IBM Center for the Business of Government. Retrieved December, 22, 2004 from *http://www.business ofgovernment.org/pdfs/Snyder_report.pdf*

Snyder, W.M., & Wenger, E. (2003) *Communities of practice in government: The case for sponsorship.* Report to the CIO Council of the U.S. Federal Government. Retrieved December 30, 2004, from *http://www.ewenger.com/pub/pubusfedcio download.htm*

Trouvé, A., & García Acosta, A. (2004, November 3). Proyecto padrinazgo de publicaciones periódicas Argentinas (4p-AR): Reconocimiento de las competencias desplegadas en el trabajo en redes. *Proceedings of the CLAD Workshop*, Madrid, Spain. Retrieved January 5, 2005, from *http://www. clad.org.ve/fulltext/0049739.pdf*

Tuomi, I. (1999). Organizing for strategic knowledge creation. In *Corporate knowledge: Theory and practice of intelligent organizations* (Chapter 14, pp. 398-410). Retrieved May 29, 2003, from *http://www.jrc.es/~tuomiil/articles/Organizing ForStrategicKnowledge CreationCh14.pdf*

KEY TERMS

Border Meetings: Bring together some of the members of two or more CoPs in order to foster an interchange of practices and to trigger thinking processes into the community itself or in the "border practices."

Explicit Knowledge: Knowledge codified into a formal and systematized language. It may be transmitted, preserved, retrieved, and combined.

Good Practice: Institutional practice or design of a project, geared towards solving a problem, achieving a goal, improving the process, better rendering a service, or the upgrading of productivity, quality, or organizational effectiveness.

Institutional Strengthening: Awareness and effective use of human and technological available resources in all their scope to accomplish institutional goals.

Professional State Forums (PSFs): Involve the public executives who focus on specific practices within the state. These PSFs can cut across agencies or function at an internal level.

Tacit Knowledge: Personal knowledge which includes cognitive technical and attitude elements. This knowledge coincides with the person's competencies.

Training Knowledge Management Approach: Develops real and virtual training space to facilitate the knowledge conversion cycle.

T

Transfer of Information and Knowledge in the Project Management

Jerzy Kisielnicki
Warsaw University, Poland

INTRODUCTION AND BACKGROUND

Success and failure in information technology (IT) projects depend on many factors. Based on the analysis of literature as well as the author's research and experience, we can build a working hypothesis of a significant influence of the communication system on a final project outcome in the context of:

- Communication between the project team and the outside world (users, suppliers, other project teams, etc.)
- Communication within a project team

In project management literature, communication occupies a significant position (Candle & Yeates, 2003; Maylor, 2003). Most research projects, however, are focused on the analysis of communication between the project team and the outside world while communication within the project team seems to take a second place. From the literature dealing with building effective project teams, research carried out by Mullins (2001) deserves a closer look. Mullins researched the key contradiction within a project team; he discovered that project leaders demand from their team members the willingness to compromise and subordinate while at the same time they promote individualism and want to foster creativity. Chaffe (2001), on the other hand, concluded that most people during their professional career lose both their creativity and individualism and prefer to conform to the existing standards. This is the very reason why some leaders prefer to build their teams from young people knowing that they lack experience. By doing that, they realize they increase the risk of not achieving their goals. Therefore, the IT leaders need to combine these conflicting trends and build the project team to ensure the overall success of the project. Adair (1999) indicates three criteria that need to be taken into consideration when evaluating potential team members: competence, motivation, and personal traits.

The subject of this article is to prove the hypothesis that the communication system within the team significantly influences the its effectiveness. The key question that needs to be answered is: what conditions does the project leader need to create in order to maximize the positive and minimize the negative effects of teamwork?

While at first glance this hypothesis might seem obvious, detailed analysis does not lead to decisive conclusions. While executing the project, teams could use different communication methods to both define the project tasks as well as evaluate results. The effectiveness of various communication methods can be very different; therefore, we want to prove the hypothesis that:

THE NETWORK COMMUNICATION SYSTEM PROVIDES THE MOST EFFECTIVE FRAMEWORK FOR THE MANAGEMENT OF THE INFORMATION TECHNOLOGY PROJECTS

Network communication system is a system where communication between all team members is direct and cross-divisional. In such system, the role of a project leader is not only to build the seamless flow of information between the team members but also to build trust between them. A simple network communication system is illustrated in Figure 1.

During my professional career in IT, I went through all steps of a corporate ladder, from a systems analyst to a senior project manager in charge of large software delivery projects. I researched the effectiveness of many IT projects but did not investigate large projects from other industries, for example, construc-

Figure 1. Simple network communication system

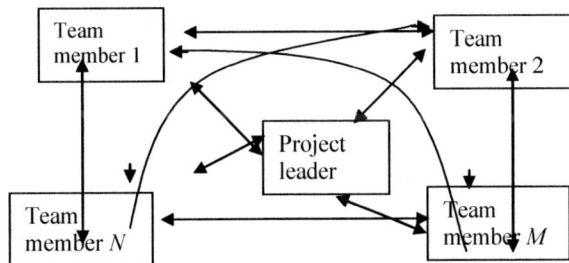

tion. Despite that, the results of my research can be adapted to any other industry, since the primary focus was on the internal project communication, which is generic rather than industry specific.

My research was related to the application of the Transactive Memory Systems (TMS), conducted primarily by Hollingshead (1998) analyzed the TMS as a structure to deliver data, information, and knowledge to the project teams. The research of 69 project teams indicated that team stability, team member familiarity, and interpersonal trust all have a positive impact on the TMS. Consequently, the TMS has a positive influence on team learning, speed-to-market, and a success of new products (Akgün et al., 2004).

RESEARCH DESCRIPTION: ANALYSIS OF COMMUNICATION SYSTEMS

The analysis of communication systems was based on 22 IT projects carried out between 1995 and 2002. The author actively participated in 12 of these projects; the information about the remaining 10 projects was based on project documentation as well as interviews with project participants. The main difficulty in the research is the fact that all projects are unique. (The ideal research would require an experiment where the same team would carry out the same projects at the same time with the only difference the communication method). Therefore, the conclusions of this research are based on estimates.

The majority of the projects included in the research targeted the business process improvement of the large organizations through the use of information technology. The project range was quite broad: implementation of IT in accounting for the major textile factory, improvement of the existing IT appli-

cation in the insurance and pension institution, implementation of MRP II/ERP in a pharmaceutical company, application of IT in a municipality of a large municipality, strategic application of a new IT for a national bank, application of IT to improve the management of a large top-security penitentiary, and application of IT for education (use of information technology program for senior executives), and so forth.

These projects represent a very diverse group of IT implementations; 18 of these projects were business applications for various industries, and 4 were for non-profit organizations. Success was defined based on schedule, cost, and scope; the project was considered successful if a variance at project completion for these three metrics was 10% or less. Despite the fact that 15 of these projects were classified as successes, during their implementation, the teams had to overcome significant problems.

The size of project teams in each of these projects was 20 people or more. The teams were cross-divisional; they included both IT personnel as well as industry specialists. The selection of such teams allowed the author to research a group that both required at least a three-level communication and could not be managed by one person. In such a project team, level one consisted of system analysts designing a system; level two consisted of operational managers or team leaders; and level three was a project leader accountable for the entire project. To complement the standard communication channels (i.e., project leader to team leader to system analyst), the author researched communication channels between project leader and systems analyst and between system analysts themselves.

The author searched for answers to the following questions:

1. How effective are main communication channels within a project team? Did the team members receive adequate information and knowledge from other team members?
2. What project management methods would ensure a seamless information flow within a project team?
3. What communication system is recommended for implementation of IT projects?

In the context of this research, effective communication is measured by earlier defined project success criteria.

The method of research is asymmetrical; the focus is on identification of causes of failure while a success is treated as a given. The methods of analysis are:

*   Review of project initial documentation (preliminary analysis, business case, application specifications, etc.) and project progress documentation (schedule, budget, delivered scope).
*   Questionnaires for both project managers and project team members.
*   Author's notes from the project meetings where the team discussed project issues, risks, and solutions.

The information from the project meetings was the key source for the analysis while project documentation and questionnaires provided the necessary background and were used for further result verification and diagnosis. Project documents and questionnaire results indicated there was a problem while the discussions were a source of recommended solutions. In most cases, the discussions were within the project team with participation of specialists from other project teams or from user groups. Each significant deviation from budget, schedule, or scope was presented and discussed. Project documents and questionnaire results would then help verify if decisions made by the group were effective. One of the key questions from the questionnaire was: Would you like to work with the same team on the next project?

Occasionally, the author used the experiment where he would pass specific information to one team member or a group and measure the time it would take for this information to reach all project members. In such an experiment, the author would send an e-mail and check when the e-mail is read, monitor the usage of project database, and monitor the usage of Internet. The analysis also included the understanding and usefulness of the information as perceived by the project team in the context of project scope, schedule, and budget. The results showed that there are two categories of roadblocks:

*   Communication roadblocks caused by external factors like delay in supply of required technology, project financial issues, incomplete documentation supplied by users, change in regulations, strategic organization changes with the organization on the receiving end of the project, unplanned absence of a team member, and so forth.
*   Communication roadblocks caused by internal factors like insufficient communication, lack of knowledge and experience in carrying out the project, personal conflicts within the team, errors in project managements, and so forth.

While external factors listed above affect the project in general, internal factors were strongly related to the flow of information within the team.

The communication system within the team was evaluated using the following criteria:

*   How significant was cost, schedule, and scope variance at project completion?
*   How effective was risk management process?
*   How effective was a conflict resolution process?
*   Were team members willing to cooperate and share knowledge?
*   Were the team members willing to work together on the next project?

Considering the scope of this article, the author presents only the most important facets of the research.

MAIN COMMUNICATION SYSTEMS AND THEIR ELEMENTS

The research includes two communication systems used by project teams:

Figure 2. The traditional three-level communication system

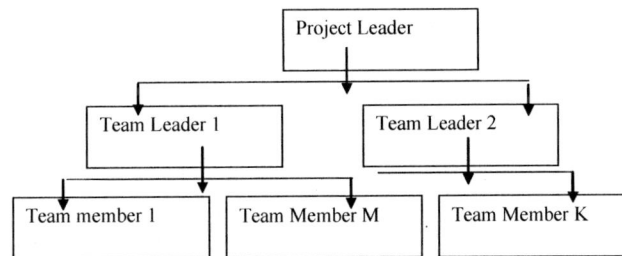

The network-based system presented in its simplified form in Figure 1. In reality, the network system is more complex, since besides the project leader, there are also team leaders accountable for delivery of portions of the overall solution. Fourteen projects selected for this research followed such structure and used the network-based communication method. The diagram depicting communication channels in such structure is presented in Figure 4.

The traditional, hierarchical communication system is depicted in its basic form in Figure 2. Eight projects selected for this research used the hierarchical communication system.

Regardless of the communication method, all projects were using various aspects of information technology to provide a business solution: computer aided system engineering (CASE), databases, Internet, and e-mail, or online cooperation.

All communication systems within the project team include basic elements presented on Figure 3.

Communication systems depicted on Figures 1 and 2 consist of "bricks" presented on Figure 3. The communication system is effective only if all individual bricks function properly. The information flow between these individual elements—regardless of the used technology—is deformed due to various disturbances caused by:

- **Technology:** Hardware and software cannot transfer the contents and/or form of the information.
- **Semantics:** The recipient cannot read or interpret received information.
- **Pragmatism:** Delivered information does not add anything new to the recipient's knowledge,

and, consequently, the effort to receive information was wasted.

- Analysis of communication systems covered by this research proved that out of 14 projects using network communication systems, 11 (80%) were successful. Out of eight projects using traditional hierarchical communication systems, 4 (50%) were successful. These four projects were MRPII/ERP-like package implementation projects.

As stated earlier, the communication system is not the only project success factor. However, the answers to the question quoted earlier—Would you like to work with the same team on the next project?—were symptomatic.

- Among team members operating within the hierarchical communication system, between 60% and 70% of managers provided the positive answers while only 30% of system analysts provided a positive answer.
- Among the team members operating within the network communication system, between 70% and 80% provided a positive answer, and there was no difference between the management team and systems analysts.

In addition, the number and magnitude of project issues were much smaller in a project using the network communication as compared to the projects using the hierarchical communication.

The results of research on the speed of information flow proved that, in the network system, information flow was 30% faster than in the hierarchical systems. This research also proved the principles of management system design presented by M. Ham-

Figure 3. Basic elements of communication systems

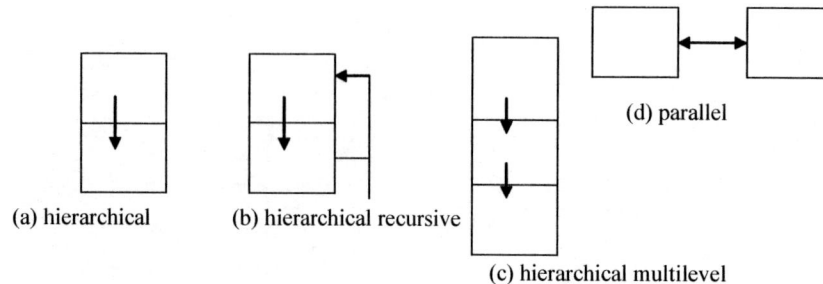

(a) hierarchical   (b) hierarchical recursive

(d) parallel

(c) hierarchical multilevel

mer in his business process reengineering method. According to Hammer (1995), it is critical to eliminate the middle man in order to improve the effectiveness of communication. Other scientists also confirmed these principles in their research (Kisielnicki, 2002).

Communication systems presented in Figures 2, 3a, and 3c are the least desirable and not recommended as in these systems where a team member only receives directives. Such situation in reality cannot and does not exist; there is always an exchange of information where the team member at least informs a project manager about progress of the project tasks. However, as stated in works (Grochowski & Kisielnicki, 2000), in the hierarchical relationship, team members reluctantly inform the project leaders about the project progress even though they consider it their duty. It seems that the reason of such behavior is psychological; when asked why team members withhold information from project managers, the answers were ambiguous. Also, discussions carried out within project teams did not bring about a conclusive answer. However, during one-on-one conversations, it became clear that team members perceive a project leader as a competitor; the typical answer was: If he is a project leader and receives higher salary, I will not advise him— it is up to him to make a decision.

Morgan (1986), in his work on different organizations, writes that the hierarchy is a source of various conflicts between people. These conflicts are not about solving business problems; they are about people's position in organizational hierarchy. Based on observations, we can say that the situation is different when team members cooperate with each other, and each individual's performance evaluation is driven by the evaluation of the final project outcome. In such environment, cooperation becomes a

necessity, and knowledge transfer between team members is always significant. The leader's influence should focus on fostering, promoting, and demanding—when necessary—knowledge transfer between the team members. Savatera (1998) writes, "Greek preferred to solve issues with his equal rather than receive a solution from his Master; to make mistakes on his own behalf rather than to follow orders". The author believes that people carrying out IT projects these days are such contemporary Greeks.

There are two categories of IT projects:

- Package implementation projects (for example, implementation of MRPII/ERP) where creativity is not as important as following standards and proven procedures. However, this approach is often criticized since, in reality, the business processes and overall business environment change, and package implementation projects do need to adopt the standard application to these changes.
- Projects that deliver new and unique applications where team members need to use creativity to a certain degree.

Therefore, (understanding the limitation pointed out above) the communication system presented in Figure 3b is effective in package implementation projects where it is critical that the system delivery procedures are followed. For projects delivering new applications, the communication pattern presented in Figure 3d is more appropriate.

In a hierarchical communication system, presented in Figure 2, the majority of elements are as presented in Figures 3a, 3b, and 3c. In a network communication system, the majority of elements are as presented in Figure 3d.

THE NETWORK COMMUNICATION SYSTEM AND ITS EVOLUTION: COMPARATIVE ANALYSIS

In reality, the network communication system depicted in Figure 1 is used for project teams consisting of five to seven people. For larger teams, this model takes on a more complex form, presented in Figure 4. This diagram represents a modification of the network communication system presented by Mintzberg and Van der Heyden (1999).

The network communication system (presented on Figure 4) is therefore recommended for implementation of complex IT systems. This system has been proven in several IT implementations projects; it was well received by the team members, and, most importantly, it was proven effective.

The network communication structure presented in Figure 4 has the following key characteristics:

1. **Division of the project team into smaller teams happens dynamically during the project using two techniques:** PERT combined with the Critical Path Method (CPM) as well as Management by Objectives (MBO). These techniques are supplemented with the analysis of skills and personality traits of the individual team members. (Team building methods will be a subject of a separate article.)

2. **The network communication system is based on direct reports. The only person responsible for the entire project is a project leader. Team leaders have dual responsibilities:** they are both team leaders and team members (system analyst, business analyst, etc.). During the project, after teams have completed their tasks, they were reorganized; the team leaders as well as team assignments would change.

A colloquium on *Participant-Centered Learning* organized by Harvard Business School in 2002 followed a very similar pattern; during discussions on various case studies, both team leaders and team members would periodically change. During the entire session, the author was a team leader only once, and, every week, he was working in a different team. All participants accepted this method as obvious and natural. Also, in the researched IT projects, the team accepted the changes in team leaders. These changes were introduced and explained at the beginning of the project. The financial aspect of the team leader position was such that the position of a leader required additional effort as well as different skills and was considered recognition. However, it did not trigger additional compensation. While changes in the team leader assignments worked well, reassignments to different teams did not. The reasons were twofold:

Figure 4. Organizational structure and communication flow in a network-based project team

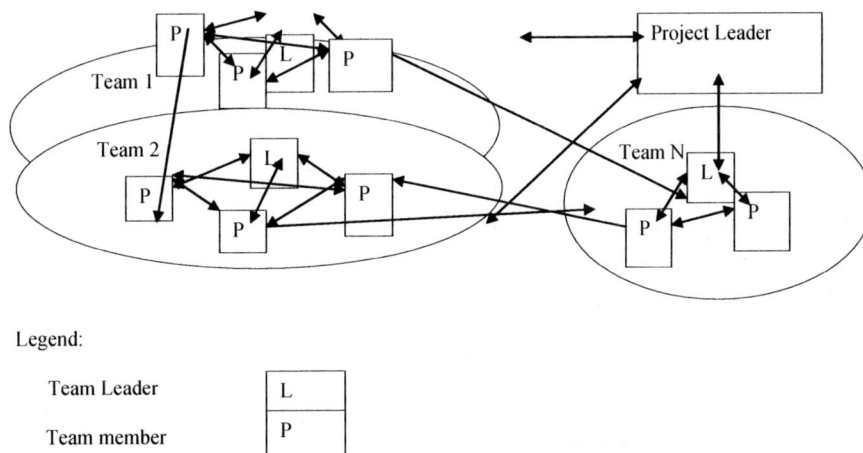

Legend:

Team Leader	L
Team member	P

- **Schedule:** Different teams finished their deliverables in different times.
- **Personal relationships** created during the project between the individual team members. This was the significant element supporting a strong communication within the individual team.

Participation of team members from one team in achieving tasks of other team. For example, selected group of more experienced team members would spend 20% to 30% of their time assisting in completion tasks from another group. This arrangement builds the relationship between the project participants and facilitates the flow of information as well as knowledge transfer. Methods PERT/CPM, as well as MBO, help decide which teams should share resources in this manner.

CONCLUSION AND FUTURE TRENDS

Research on effectiveness of both communication systems indicates that the network communication systems are superior to the hierarchical system in the following aspects:

Progress Monitoring

Possible deviations from scope, schedule, and budget were communicated earlier in the network system than in a hierarchical system thus allowing for earlier intervention. This was due to the fact that all team members in the network system (TMS) felt responsible for the project success.

Cooperation and Knowledge Transfer

There was strong cooperation as well as knowledge and information transfer between team members; there were no artificial barriers (i.e., manager against worker). Each team member was or could be a team leader depending on the need and situation.

Problem Solving

There were fewer conflicts within the network structure. The problems that did occur were less intense,

and they were resolved faster than within the hierarchical structure.

The communication system (TMS) in a network structure, to be effective, requires several conditions be met. The most important one is the competence of individual team members and their willingness to cooperate. This system is difficult for so called individualists as well people preparing for a project management career path. In the recommended system, career path leads toward professional development but does not provide a stepping stone from a system analyst position to a project leader position. It is also a system difficult for the project manager whose responsibility stretches from hiring and organizing the team members to creating an atmosphere conducive to open communication and cooperation. Compared to the hierarchical system, project leaders of network organizations need to delegate more of their duties to the teams while they retain full accountability of overall project success. For this very reason, many project leaders prefer the hierarchical system as easier to execute (it allows them to rely on formal authority) and to enforce the timeliness of delivery even though they fully understand its limitations.

These conclusions, however, still do not provide a decisive answer to the following questions:

- Which of these two systems is effective for all IT projects?
- What is the efficiency of replacing the hierarchical system with a network system?

Each business process needs to be both effective and efficient, and an information system delivery process is no exception. There are many contributing factors that influence both its efficiency and effectiveness. Therefore, in conclusion, the author would like to point out that the communication system, however critical to project success, is only one of these factors. Additional influence comes from the team make-up as well as motivation techniques. The hiring and team building has been briefly discussed already. The effective motivation system, while critical to the overall communication strategy within the project team, is a separate topic. The effective motivation system also depends on the organizational culture, overall state of economy (the job market, in particular) as well as the country itself; different motivation systems will be effective in India, UK, Poland, or the United States. The communication system remains a key component in building effective teams since it is independent from team make-up and utilized motivation techniques.

REFERENCES

Adair, J. (1999). *Decision making and problem solving (management shapers)*. London: Chartered Institute of Personnel and Development (CIPD).

Akgün, A. E., Keskin, H., Byrne, J., & Imamoglu, S. Z. (2004). Knowledge networks in new product development projects: A transactive memory perspective. *Information and Management*. Available online February 2005.

Candle, J., & Yeates, D. (2001). *Project management for information systems*. London: Prentice Hall.

Chaffee, J. (2001). *The thinker's way: 8 steps to a richer life*. Boston: Little Brown.

Grochowski, L., & Kisielnicki, J. (2000). Reengineering in upgrading of public administration: Modelling and design. *International Journal of Services Technology and Management, 14*, 331.

Hammer, M. (1995). *The reengineering revolution*. New York: HarperBusiness.

Hollingshead, A (1998). Retrieval processes in transactive memory systems. *Journal of Personality and Social Psychology, 74*, 659-671.

Kisielnicki, J. (2002). IT in improvement of public administration. In M. Khosrow-Pour (Ed.), *Cases on information technology*. Hershey, PA: Idea Group Publishing

Maylor, H. (2003). *Project management*. London: Prentice Hall.

Mintzberg, H., & Van der Heyden, L. (1999, September-October). Organigraphs: Drawing how companies really work. *Harvard Business Review*, p. 87.

Moreland, R. L., & Myaskovsky, L. (2000). Exploring the perfromance benefits of group training: Transactive memory or improved communication? *Organizational Behavior and Human Decision Processes, 82*, 117-133.

Morgan, G. (1986). *Image of organization*. London: Sage.

Mullins, L. (2001). *Management and organizational behavior*. London: Pitman.

Savatera, F. (1998). *Polityka para amador*. Editorial Ariel S.A., Barcelona.

Wegner, D. M. (1986). Transactive memory: A contemporary analysis of the group mind. In B. Mullen, G. R. Goethals (Eds.), *Theories of group behavior* (pp. 185-208). New York: Springer-Verlag.

Wegner, D. M. (1995). A computer network model of human transactive memory. *Social Cognition, 13*, 319-339.

Wegner, D. M., Guiliano, T., & Hertel, P. T. (1985). Cognitive interdependence in close relationships. In W. J. Ickes (Ed.), *Compatible and incompatible relationships* (pp. 253-276). New York: Springer-Verlag.

KEY TERMS

Pragmatism: Having a basis in practice but little explicit or implicit theoretical underpinning.

Semantics: In this article, the sense of this is that information should be in a form which is intelligible to those receiving it.

Trust in Virtual Teams

Christopher Lettl
Berlin University of Technology, Germany

Katja Zboralski
Berlin University of Technology, Germany

Hans Georg Gemünden
Berlin University of Technology, Germany

INTRODUCTION

In the past few decades, organizations have faced radical changes of their business environment. In order to meet the challenges of increasing global competition in a knowledge-based economy, traditional work forms have partly been replaced and complemented by more flexible organizational structures. Thereby, advances in information and communication technologies (ICTs) have created the means for interacting across boundaries both in space and time (Picot, Reichwald & Wigand, 2001; Townsend, DeMarie & Hendrickson, 1998). In this context, virtual teams have increasingly gained attention in theory and practice alike (Ahuja, Galetta & Carley, 2003; Kelley, 2001; Kirkman, Rosen, Gibson & Tesluk, 2002). This new organizational form aims to leverage advantages of the traditional team-based work structure while at the same time coping with the challenges of decentralization and geographical dispersion.

Traditional co-located teams have been studied by researchers of many disciplines, such as sociology, psychology, and business studies. Thereby, each discipline has its own focus. Consequently, there is an abundance of theories and no common definition of the term *team* (Stock, 2004). Generally, a team in any organization can be defined as a social system of three or more people, whose members perceive themselves and are perceived by others as team members, and whose members collaborate on a common temporary task (Guzzo & Shea, 1992; Hackman, 1987; Hoegl & Gemuenden, 2001).

Regarding *virtual teams,* this definition has to be extended by the issues of communication modes and location. Hence, in this article, a virtual team is defined as a social system characterized by context, identity, and common contemporary task, and whose members rarely meet in person, but rather communicate primarily through ICTs, as they are geographically dispersed (Lipnack & Stamps, 2000; Lurey & Raisinghani, 2001; Maznevski & Chudoba, 2000).

With respect to the term *trust,* it has to be taken into account that this construct can be viewed from a rational or social perspective. While the rational perspective centers on the calculus of self-interest—for example, decrease in transaction cost due to less self-protecting actions—the social perspective centers on moral duty (Jarvenpaa, Knoll & Leidner, 1998). Taking an integrated view of both perspectives, the definition from Mayer, Davis, and Schoormann (1995) is adopted:

The willingness of a party to be vulnerable to the actions of another party based on the expectation that the other will perform a particular action important to the trustor, irrespective of the ability to monitor or control that other party. (p. 712)

FACTORS THAT SUPPORT TRUST IN A VIRTUAL TEAM

Research on traditional work teams has outlined trust as a critical antecedent for team performance (Costa, Roe & Taillieu, 2001; Erdem & Ozen, 2003; Politis, 2003). Establishing trust is therefore fundamental for the formation, growth, and performance of any work team. For a team to operate effectively, its members need to be sure that everyone will fulfill his or her obligations and behave in a consistent manner. The teams studied, however, mainly worked together on a face-to-face basis, an interaction pattern which has been shown to support the development of mutual trust (Hallowell, 1999; Madhavan & Grover, 1998; Nooteboom & Six, 2003). In virtual teams, however, traditional ways of establishing bonds and socializing are absent or at best limited. Therefore, the emergence of trust in a virtual environment is difficult (DeSanctis & Monge, 1999; Handy, 1995; Holton, 2001). This raises the following questions: (1) Can trust be developed in a virtual setting at all? (2) Which mechanisms support the emergence of trust within a team of individuals working across distance, time zones, and cultures?

To analyze whether trust can emerge in a virtual team setting at all and how it might evolve, one needs to understand the antecedents of trust in dyadic relationships first. In such relationships trust arises from the attributes that are associated with a trustee and a trustor (Mayer et al., 1995). The trustee's attributes are his or her perceived *ability* (set of skills that enables a trustee to be perceived as competent in a specific domain), *benevolence* (willingness to do good to the trustor beyond egocentric profit motives), and *integrity* (adherence to a set of principles thought to make the trustee reliable). On the side of the trustor, his or her *propensity to trust* is considered to be the key attribute. Looking at the team level, the development of 'collective' trust is more complex than 'dyadic' trust as there are multiple trustees with a different set of attributes (Cummings & Bromiley, 1996). Conceptually, the trustor needs information cues to assess the ability, benevolence, and integrity of the trustee. Co-located team members can exchange this information by face-to-face interactions. This type of interaction enables the trustor to gain rich, non-verbal information cues about a trustee based on several dimensions such as looks, gesture, facial expressions, and behavior.

In virtual teams, non-verbal information cues are rare. Therefore, virtual team members need to rely almost exclusively on written information cues such as e-mails. Can virtual team members develop trust in such an environment? Empirical research shows that trust does indeed exist in virtual teams, but it develops in a very different way than in a conventional team setting (Henttonen & Blomqist, 2004; Jarvenpaa et al., 1998; Jarvenpaa & Leidner, 1999).

Past studies on traditional co-located teams reveal that trust tends to evolve in three stages. Firstly, deterrence-based trust is developed as team members simply comply as they fear sanctions. Secondly, knowledge-based trust emerges as each team member becomes more familiar with each other and, thereby, is able to predict the behavior of other team members. Finally, identification-based trust is built on empathy and shared values (Lewicki & Bunker, 1995; Rosseau, Sitkin, Burt & Camerer, 1998). In virtual teams, however, this three-stage pattern could not be observed. Jarvenpaa and Leidner (1999) found that a form of swift, action-based trust evolved between the team members. It is based on the expert reputation that team members have from the beginning of the project and the willingness to trust in this expert knowledge. Other sources for swift trust in a virtual context are the identification of team members with the organization, a previous common work history, and the personal fit of team members (Henttonen & Blomqist, 2004).

Instead of evolving slowly through stages, trust in virtual teams tends to be established right at the outset. Thereby, the first messages "appeared to set the tone for how the team interrelated" (Jarvenpaa & Leidner, 1999, p. 810). Starting the cooperation with confidence, optimism, and a propensity to initiate or respond to electronic communication seems to be an essential basis for trust building in virtual settings. To maintain trust communication should focus on the project and related tasks; social communication complementing task communication can strengthen trust. To summarize, we conclude that trust in principle can evolve in virtual environments, although it is a different kind of trust.

But through which mechanisms does trust emerge in virtual teams? As the assessment of ability, integrity, and benevolence of virtual team members is rather problematic, any mean that contributes to an enhanced information base of each team member can be regarded as important enablers of trust in a virtual environment. In this context, three categories of factors and mechanisms can be distinguished: team process factors, individual characteristics of team members, and the technical infrastructure.

Team process factors refer to mechanisms that focus on increasing communication and commitment within the virtual team. Team-building exercises are examples of such mechanism (Jarvenpaa et al., 1998). Hereby, members are encouraged to exchange information about themselves that is relevant for assessing one another's project-related skills, their motivations to contribute to the team, and their work habits believed to be supportive for team success. As observed by researchers (Jarvenpaa & Leidner, 1999; Lipnack & Stamps, 2000), especially the first interactions of the virtual team members are decisive for the emergence of trust in a virtual team. Consequently, the very early phases of team formation are crucial. Other team process factors are the setting of clear goals and the communication of feedback on a regular basis. These process factors increase the commitment towards the team and ultimately lead to a set of collectively shared rules. Hence, team identity emerges which in turn enhances trust (Govindarajan & Gupta, 2001; Holton, 2001).

With respect to *individual factors*, several characteristics of members have been outlined to be critical for the formation of trust in virtual teams. Members need to have the discipline to stick to the rules and norms that the team has agreed upon in the team-building process (Ardichvili, Page & Wentling, 2003; Govindarajan & Gupta, 2001). In addition, communicating via various virtual channels such as e-mail, teleconferences, and videoconferences requires the competence to use these instruments in an appropriate way (Straus & Olivera, 2000; Wong & Burton, 2000). Furthermore, team members need to have social competencies. Working together with members from different backgrounds and/or cultures on a virtual basis requires empathy and individual tolerance for unfamiliar communication and working styles, for example, patterns of problem solving and diverse perceptions of time (Saunders, van Slike & Vogel, 2004). The bundle of these skills and qualifications can be considered as a 'virtual competence'.

Besides those rather 'human factors', the *technical infrastructure* is an important prerequisite of trust in the virtual world. Hereby, the selection and effective utilization of communication tools designed to initiate and develop meaningful dialogue among virtual team members is essential (Hinds & Kiesler, 1995; Wiesenfeld, Raghuram & Garud, 1999). Regardless of technological advances in virtual communications, it has to be considered, however, that ICTs are only a necessary, but not a sufficient condition for trust development. They do not compensate for team

processes and individual characteristics that are critical for the emergence of trust in the virtual context (Handy, 1995).

FUTURE RESEARCH

There are a variety of questions that need to be addressed by future research. With regard to individual characteristics of team members, the personality traits may be an important factor for trust and success in virtual teams; for example, introverts who prefer to process information internally and who express themselves in writing may be better suited for communication in a virtual team than extroverts (Geber, 1995). As virtual teams very often include team members from different cultures, research also needs to address the influence of cultural factors on trust building. In addition, future research needs to develop a better understanding of how conflicts are resolved in virtual teams and which coordination mechanisms are most effective (Montoya-Weiss & Song, 2001). Furthermore, the impact of various communication modes and intensities on the emergence of trust in virtual teams is an important research question. Hereby, further insights are needed regarding a possible mix of diverse virtual and personal communication modes best suited to build up trust in various phases of the team process (Kirkman, Rosen, Tesluk & Gibson, 2004). With respect to appropriate empirical research methodologies, new techniques are required to improve the understanding of various mechanisms and critical success factors of virtual teams—for example, the participative observation of researchers who become 'members' of a virtual team. This approach allows an in-depth understanding of the processes in virtual teams which cannot be gained by survey research approaches alone.

CONCLUSION

Virtual team work has become a crucial part in the present economy. Advances of ICT enable the transfer of information across continents, time zones, and organizational boundaries. Despite these technological developments, 'human factors' should not be neglected in an increasing virtual environment.

Trust is not just important in conventional teams; it has been emphasized to be particularly crucial in virtual teams. Empirical research reveals that trust can be established in virtual teams, but the supportive mechanisms differ from those patterns in teams that cooperate on a face-to-face basis. Particularly, the early phases of team formation and collaboration are crucial for trust building in a virtual context. Furthermore, it has been shown that the kind of trust emerging in virtual teams is different.

Management needs to take these specialties of trust formation in virtual teams into account by establishing an appropriate environment. As a 'virtual competence' of team members is essential for the emergence of virtual trust, team members should be trained accordingly. Hereby, the cultural, social, and communication competencies are enhanced, and team members are sensitized for the problems that can occur in a virtual team context. Additionally, management needs to provide the adequate technical infrastructure that facilitates various modes of communication.

REFERENCES

Ahuja, M.K., Galetta, D.F., & Carley, K.M. (2003). Individual centrality and performance in virtual R&D groups: An empirical study. *Management Science, 49*(1), 21-38.

Ardichvili, A., Page, V., & Wentling, T. (2003). Motivation and barriers to participation in virtual knowledge-sharing communities of practice. *Journal of Knowledge Management, 7*(1), 64-77.

Costa, A.C., Roe, R.A., & Taillieu, T. (2001). Trust within teams: The relation with performance effectiveness. *European Journal of Work and Organizational Psychology, 10*(3), 225-244.

Cummings, L.L., & Bromiley, P. (1996). The Organizational Trust Inventory (OTI): Development and validation. In R.M. Kramer & T.R. Tyler (Eds.), *Trust in organizations: Frontiers of theory and research* (pp. 302-330). Thousand Oaks, CA: Sage Publications.

DeSanctis, G., & Monge, P. (1999). Introduction to the special issue: Communication processes for virtual organizations. *Organization Science, 10*(6), 693-703.

Erdem, F., & Ozen, J. (2003). Cognitive and effective dimensions of trust in developing team performance. *Team Performance Management, 9*(5/6), 131-136.

Geber, B. (1995). Virtual teams. *Training, 32*(4), 36-41.

Govindarajan, V., & Gupta, A.K. (2001). Building an effective global business team. *MIT Sloan Management Review,* (Summer), 63-71.

Guzzo, R.A., & Shea, G. (1992). Group performance and intergroup relations in organizations. In M.D. Dunnette & L.M. Hough (Eds.), *Handbook of industrial and organizational psychology* (2nd ed., vol. 3, pp. 269-313). Palo Alto, CA: Consulting Psychologists Press.

Hackman, J.R. (1987). The design of work teams. In J.W. Lorsch (Ed.), *Handbook of organizational behavior* (pp. 315-342). Englewood Cliffs, NJ: Prentice-Hall.

Hallowell, E. (1999). The human moment at work. *Harvard Business Review, 77*(1), 58-65.

Handy, C. (1995). Trust and the virtual organization. *Harvard Business Review, 73*(3), 40-50.

Henttonen, K., & Blomquist, K. (2004, April 2-3). Communicating trust across distance: Empirical study on trust in relationship development through technology-mediated communication of two virtual teams in the ICT sector. *Proceedings of the 5th European Conference on Organizational Knowledge, Learning and Capabilities,* Innsbruck, Austria.

Hinds, P., & Kiesler, S. (1995). Communication across boundaries: Work, structure, and use of communication technologies in a large organization. *Organization Science, 6*(4), 373-393.

Hoegl, M., & Gemuenden, H.G. (2001). Teamwork quality and the success of innovative projects: A theoretical concept and empirical evidence. *Organization Science, 12*(4), 435-449.

Holton, J.A. (2001). Building trust and collaboration in a virtual team. *Team Performance Management, 7*(3/4), 36-47.

Jarvenpaa, S.L., Knoll, K., & Leidner, D.E. (1998). Is anybody out there? Antecedents of trust in global virtual teams. *Journal of Management Information Systems, 14*(4), 29-64.

Jarvenpaa, S.L., & Leidner, D.E. (1999). Communication and trust in global virtual teams. *Organization Science, 10*(6), 791-815.

Kelley, E. (2001). Keys to effective virtual global teams. *Academy of Management Executive, 15*(2), 132-133.

Kirkman, B.L., Rosen, B., Gibson, C., & Tesluk, P. (2002). Five challenges to virtual team success: Lessons from Sabre, Inc. *Academy of Management Executive, 16*(3), 67-79.

Kirkman, B.L., Rosen, B., Tesluk, P.E., & Gibson, C.B. (2004). The impact of team empowerment on virtual team performance: The moderating role of face-to-face interaction. *Academy of Management Journal, 47*(2), 175-192.

Lewicki, R.J., & Bunker, B.B. (1995). Developing and maintaining trust in work relationships. In T.R. Tyler (Ed.), *Trust in organizations* (pp. 114-139). Thousand Oaks, CA: Sage Publications.

Lipnack, J., & Stamps, J. (2000). *Virtual teams. People working across boundaries with technology* (2nd ed.). New York: John Wiley & Sons.

Lurey, J.S., & Raisinghani, M.S. (2001). An empirical study of best practices in virtual teams. *Information & Management, 38*, 523-544.

Madhavan, R., & Grover, R. (1998). From embedded knowledge to embodied knowledge: New product development as knowledge management. *Journal of Marketing, 62*(4), 1-12.

Mayer, R.C., Davis, J.H., & Schoormann, F.D. (1995). An integrative model of organizational trust. *Academy of Management Review, 20*(3), 709-734.

Maznevski, M.L., & Chudoba, K.M. (2000). Bridging space over time: Global virtual team dynamics and effectiveness. *Organization Science, 11*(5), 473-492.

Montoya-Weiss, M.M., & Song, M. (2001). Getting together: Temporal coordination and conflict management in global virtual teams. *Academy of Management Journal, 44*(6), 1251-1262.

Nooteboom, B., & Six, F. (Eds.). (2003). *The trust process in organizations: Empirical studies of the determinants and the process of trust development.* Cheltenham: Edward Elgar.

Picot, A., Reichwald, R., & Wigand, R.T. (2001). *Die grenzenlose unternehmung: Information, organisation und management* (4th ed.). Wiesbaden: Gabler.

Politis, J.D. (2003). The connection between trust and knowledge management: What are its implications for team performance? *Journal of Knowledge Management, 7*(5), 55-66.

Rosseau, D.M., Sitkin, B., Burt, R.S., & Camerer, C. (1998). Not so different after all: A cross-discipline view of trust (introduction to special topic forum). *Academy of Management Review, 23*(3), 393-404.

Saunders, C., van Slike, C., & Vogel, D.R. (2004). My time or yours? MANAGING time visions in global virtual teams. *Academy of Management Executive, 18*(1), 19-31.

Stock, R. (2004). Drivers of team performance: What do we know and what have we still to learn? *Schmalenbach Business Review, 56*(July), 274-306.

Straus, S.G., & Olivera, F. (2000). Knowledge acquisition in virtual teams. *Research on Managing Groups and Teams, 3*, 257-282.

Townsend, A.M., DeMarie, S.M., & Hendrickson, A.R. (1998). Virtual teams: Technology and the workplace of the future. *Academy of Management Executive, 12*(3), 17-29.

Wiesenfeld, B.M., Raghuram, S., & Garud, R. (1999). Communication patterns as determinants of organizational identification in a virtual organization. *Organization Science, 10*(6), 777-790.

Wong, S.-S., & Burton, R.M. (2000). Virtual teams: What are their characteristics, and impact on team performance? *Computational & Mathematical Organization Theory, 6*, 339-360.

KEY TERMS

Deterrence-Based Trust: The first stage of trust is based on the consistency of behavior. It develops as team members simply comply as they fear sanctions and damage of the relationship.

Identification-Based Trust: The third stage of trust is built on empathy and shared values; team members can completely rely on each other.

ICTs (Information and Communication Technologies): A diverse set of technological tools and resources used to produce, store, process, disseminate, and exchange information and, thereby, aid communication.

Knowledge-Based Trust: The second stage of trust is grounded on the other's predictability and emerges as team members become more familiar with each other.

Team: A social system of three or more people, whose members perceive themselves and are perceived by others as team members, and whose members collaborate on a common temporary task.

Trust: The willingness of a party to be vulnerable to the actions of another party based on the expectation that the other will perform a particular action important to the trustor, irrespective of the ability to monitor or control that other party.

Virtual Team: A social system of three or more people, whose members perceive themselves and are perceived by other as team members, whose members collaborate on a common temporary task, and whose members rarely meet in person, but communicate mainly through telecommunication and information technologies.

T

Understanding Communities of Practice to Support Collaborative Research

José Córdoba
University of Hull, UK

Wendy Robson
University of Hull, UK

INTRODUCTION

In this article, we present an overview of the relationship between communities of practice and collaborative research. We relate this to an observation of how information technologies could be used as a tool for cultivating research communities of practice. Our aim is to explore similarities between research as an activity developed and supported by communities and the concepts that have been developed by advocates of communities of practice. We focus our attention on research in the social sciences and how it can be improved with the support of community-enabling technologies.

BACKGROUND: RESEARCH AS COLLABORATIVE ACTIVITY

There is little doubt about the transformation of research activity in the social sciences in the past few years. Traditional ways of doing research are still prominent, but changes are happening. Research processes, in which concepts and theories from single disciplines are used to inform new developments, are being replaced by more socially distributed activities where we understand that knowledge is generated by a variety of actors whose competencies and skills are needed for the delivery of final products. This means that the activities of research are developed by different people from different institutions, people who sometimes compete with each other and, at other times, collaborate for the achievement of a common objective. For example, the role of universities as knowledge generating organizations is being replaced by that of a coordinating role in which they work together with business and government orga-

nizations in addressing problems of the communities where they are based (Brulin, 2001). This does not mean that universities do not any longer engage in traditional research activities; they do. In addition, they are becoming closely linked to the regions where they are based while they are collaborating with others outside their own institutional or traditional academic boundaries to conduct research.

This emerging picture also shows that research can also be developed by geographically dispersed actors, involving researchers from different parts of the globe. Their encounters and interactions generate new sets of methodologies, concepts, and theories which are used and assessed in a particular context of application but which could also be used to address new problems in other geographical and institutional realms. The opportunities given by information technologies enable online communication between people and lead them to create, share, and exchange their knowledge to generate new possibilities, opportunities, and initiatives.

In short, we see an emerging type of research that is collaborative. It can be seen as a continuous dialogue between a variety of social actors who are concerned not only with participating in producing research outcomes, but also benefiting from them (Nowotny, Scott & Gibbons, 2003). The gathering of different stakeholders to do research also entails using and applying various sources of knowledge that have differing criteria for assessment. In this context, researchers become social activists who need to maintain and develop communication with other professionals and stakeholders (Callon, Law & Rip, 1986). They share interests and problems they face in their own disciplines, as well as needing to develop new competencies, methodologies, and tools to address societal problems.

This type of research very much resembles a new mode of research, or Mode 2 (Gibbons et al., 1994; Nowotny et al., 2003), which entails a high degree of collaboration between actors and is different from traditional or single disciplinary research (Mode 1). As pressures mount on researchers to deliver excellence, collaboration has also been seen as the medium and outcome of success. It enables the creation of appropriate conditions for the transference of knowledge and its use where it is needed (Commission of the European Communities, 2000). Despite the need for researchers to produce tangible outcomes on time, advocates of Mode 2 research also raise the importance of establishing and nurturing appropriate environments that support the continuous generation of ideas and knowledge, as well as their social production and distribution (Nowotny et al., 2003). With time, these environments could allow for the development of relations of trust between research actors, which may in turn lead them to generate further possibilities for collaboration (Department of Trade and Industry, 1990).

Therefore, research as a collaborative activity for knowledge creation requires developing continuous interactions between stakeholders. This and other issues have been addressed and explored in the theory of communities of practice (CoPs). We now turn our attention to it.

COMMUNITIES OF PRACTICE

Wenger (1998) and Wenger, McDermott, and Snyder (2002) provide the core ideas for the theory of communities of practice (CoPs). According to them, a community of practice is a group of people who share a concern, a set of problems, or a passion about a topic and who deepen their knowledge and expertise in this area by interacting on an ongoing basis. Over time, they develop a unique perspective on their topic as well as a body of common knowledge, practices, and approaches (Wenger et al., 2002, p. 4). Communities of practice do not emerge by the imposition of rules of interaction or formal structures on individuals. They are the result of continuous processes of learning in which individuals engage and sustain through time.

Communities of practice are a vehicle for the creation and dissemination of knowledge. When a community of practice is acknowledged (i.e., across organizational units or organizations), it permits people to acquire skills and expertise that otherwise would be difficult to acquire because of institutional hierarchies or constraints. Individuals find in communities of practice meaning to what they do, which confers on them an identity beyond performing tasks at a particular place. People are able to learn from others who are experts in particular topics, complement their expertise, and perform better in their daily activities.

A community of practice is different from other forms of organization in which multiple individuals participate. There is a core set of interests that is maintained and developed. Moreover, there is a general concern in advancing knowledge in a particular topic, which could go beyond simply accomplishing a set of objectives. Wenger et al. (2002) provide three features that characterize a community of practice:

- **A domain:** It defines a set of issues and questions (resolved or open) about which it is worth caring. A domain embraces an understanding of what matters to people of the community and therefore guides its inquiries. It keeps members of the community together and gives them opportunities for exchanging and creating knowledge. A domain does not only include a set of specific tasks or relevant problems at a particular time. It also encompasses the reasons why knowledge of a topic is relevant.

- **A community:** The community binds together by building and sustaining relationships and generates a sense of belonging and mutual commitment. To build a community, members must interact regularly on issues important to their domain. A community requires sharing norms and values; it requires establishing and maintaining trust and open communication about different issues. It also gives a notion of "us" and "others" and of boundary between them. A sense of community helps members to deal with emerging conflicts and differences.

- **A practice:** It consists of knowledge that people share about a particular topic and creates a common foundation to support collaboration. This knowledge can be explicit and tacit. It includes tangible outcomes and supporting elements of collaboration (i.e., books, Web sites, theories, models, etc.) as well as ways of be-having and approaching problems and opportu-nities. Successful practice in a community de-pends on a balance between joint activities of exploration and the production of "things" that show the outcomes of collaboration.

For individuals, participation in communities of practice is voluntary and requires them to be willing to contribute to and benefit from it. Sustaining a commu-nity of practice could be a lengthy but rewarding process. Success requires the collaboration of those individuals who care about a domain and want to see it developed. Participation embodies the use and advancing of practices and their sharing within an environment appropriate for learning. Self-interest is also served by this participation as value to the indi-vidual comes from membership of a successful com-munity of practice.

It is interesting to note that the three elements of CoPs discussed above show that relationships be-tween people develop as a *continuous process* based on sharing common concerns, respecting diversity, and having implicit and explicit understandings of knowledge areas. It is also our intention to draw attention to the correspondence between the nature of collaborative research and the characteristics of a community of practice. Because of this correspon-dence, we can use the ideas of communities of practice to apply to collaborative research. In particular, under-standing the elements of CoPs could guide managing and resourcing collaborative research activities. The following ideas provide some guidelines about devel-oping collaborative research as communities of prac-tice:

- The recognition and valuing of a *domain* of knowledge (i.e., research) refers to the possibil-ity of broadening research efforts and extending them beyond the achievement of specific tasks (i.e., project ideas, proposals, joint publications, discussions, etc.). The identification of a domain is essential to sustain collaboration over long periods of time. Researchers willing to engage in collaborative research could organize their inter-ests into broad areas to comprise their source of inspiration and to include the outcomes of their efforts. Timescales to develop a domain could be considered in relation to research pressures and

the need to establish solid social relationships between researchers.
- The existence of a sense of community allows sharing particular values and norms to inform collaborative research. If this is followed, col-laborative research should pay attention not only to processes of completion in which in-tended outcomes are defined and produced, but also to processes of collaboration in which shared values and interests are created, ex-plored, demonstrated, and sustained (i.e., why research is important, what research is for, who is to benefit, etc.). One without the other would not be collaborative research, nor would it be a community of practice.
- At the practice level, attention should be paid to the need for researchers to recognize their obligation to reflect on what constitutes their shared practices. This applies whether those practices are explicit (i.e., research outcomes, methods, approaches) or tacit (i.e., paradigms and styles).
- Communication between researchers is essen-tial to guarantee the development of domain, community, and practice. In this respect, it would be possible to use communication tools to support interaction, including bridging the gaps that could arise, and enhancing interaction even where there are no gaps. For example, facilitating communication that otherwise is difficult for researchers that reside in different geographical locations or offering new models or modes of communication.

DEVELOPING COMMUNITIES OF COLLABORATIVE RESEARCH

With these ideas in mind, we suggest that it is possible to draw upon the ideas of CoPs to improve collabo-rative research. To initiate and sustain collaborative research work at two levels is needed: a *community-oriented level and a practice-oriented level*. The interaction between these two also needs attention: a *domain-oriented integration*. The *community-oriented level* implies building and sustaining rela-tionships between researchers and creating warmth, friendliness, and trust to work together. This layer has a focus on sharing, socialization, building relationships, warmth, and exploration. Those are the attributes of an effective, sustainable community. The initiation of this layer could begin by defining or identifying some particular areas of interest between researchers and creating a sense of working together on them. This identification could be informed by particular observations of and conversations about what researchers do. Fostering a sense of transdisciplinary work could help in identifying common problems, issues, and dilemmas. Further work on this level could be developed by activities like meeting regularly (we will discuss the extended meaning of meeting later when we discuss the role of ICT) to share research agendas and issues; inviting people to join existing networks; exchanging ideas about methodologies, methods, and tools; and disseminating information about opportunities. In this way, steps are taken toward establishing informal networks of researchers who could meet regularly to define and refine issues of concern.

The *practice-oriented layer* requires developing joint activities that aim at producing tangible outputs through time. This is a way of directing the activities of a community of researchers and, at the same time, legitimizing it (Wenger et al., 2002). Practice can be developed by initiating a joint task which requires commitment and participation for its completion. The focus of this layer should be on achievement, efficiency, development, and practicality. An obvious example of activity on this level is a research project which brings different partners together to produce deliverables (Somekh & Pearson, 2002). Other examples could include joint writing for research funding or publication; organizing and participating in conferences of common interest; and visiting institutions for specific purposes like seminars or courses. These examples are artifacts also serving as "boundary objects" (Star & Griesemer, 1989; Wenger, 1998) defining the nature of practice to observers and so contributing to generate the sense of boundary necessary for the community. For this layer of practice, it is important to continuously have something "in the pile to do" and complement this with more informal activities and deadlines in which people could share a sense of identity with others. Collaborative research as a community of practice needs these boundary objects to continuously provide the "nexus of perspectives" that Wenger (1998) discusses.

To sustain collaborative research, community and the practice layers need to complement each other: Opportunities for joint research delivery or production (practice) help research communities to steer their activities. At the same time, creating and maintaining a community constitutes a source for new ideas and knowledge which could be applied to new research situations. Through time, working on the above two layers could also generate possibilities for creating and sustaining a *domain-oriented integration* that will ensure that communities have shared enterprises. In the case of collaborative research, this is the difference between a single project and a sustained relationship in which finishing a specific effort does not end the enterprise. It takes time and effort to build true collaborative research where there is a commitment to partnership rather than temporary alliances (Department of Trade and Industry, 1990). Wenger (1998) also argues that the development of this domain is a process, not a static agreement. This is how collaborative research differs from a single research project. An adequate way to begin the definition of research domain(s) for a community could be developing shared understandings of common research interests that later could inform the definition of core practice elements of collaborative research like methodologies, frameworks, concepts, and so forth.

These ideas might not be different from what researchers could already be doing. We suggest paying equal attention to these layers and advance on these directions as a way of transcending the boundaries of one's own research interests. Doing so is the first step toward beginning collaborative research and composing communities to support it.

THE ROLE(S) OF INFORMATION AND COMMUNICATION TECHNOLOGIES (ICT)

As well as the challenges noted previously, it takes additional time to develop collaborative research when researchers are beginning to collaborate with each other in a shared effort or are not physically together (Somekh & Pearson, 2002). To complement the development of the above layers, we see the design and use of information and communication

Table 1. Support of ICT for layers of collaborative research as communities of practice

Element of Community of Practice	Layer of Collaborative Research	Purposes of ICT
Community	*Community-oriented:* Creation and maintenance of relationships	Provide efficient communication between actors; sharing of stories, metaphors, and mental models (Schwen & Hara, 2003) **Examples:** Discussion groups, electronic forums, Web sites, user content management tools, interactive whiteboards, simulation, etc.
Practice	*Practice-oriented:* Achievement of outcomes	Exchange and dissemination of information; collaborative production of documents (Somekh & Pearson, 2002) **Examples:** E-mail, whiteboard systems, computer supported cooperative work (CSCW); shared diary systems; electronic submission systems; document imaging systems; multiuser editing software, etc.
Domain	*Integration-oriented:* Building a sense of identity across activities and time	Creation of common knowledge, experience and heuristics. **Examples:** Knowledge management forms, data stores and tools, data mining, software agents, distributed systems; version management software; archives, etc.

technologies (ICT) as vital to sustain collaborative efforts. It enables communication between researchers where it is physically difficult and allows models of interaction not possible in the real world. It is unrealistic to expect that the deployment of ICT will generate a community of practice. In fact, ICT often constitutes more of an impediment for the formation and sustaining of communities of practice than an adequate support (Schwen & Hara, 2003). However, it is possible to provide ICT tools that facilitate rather than create, organize, or impose a community. In this respect, researchers and designers of collaborative efforts could foster the collective creation of knowledge, the negotiation of meaning, and the evolution of the community (Wenger et al., 2002) as well as the achievement of more practice (research-oriented) goals.

A summary of some examples of how ICT could support the formation of communities of practice in collaborative research is presented in Table 1.

As seen in the Table 1, ICT can be chosen, designed, and used to support the development of each layer, community, practice, and domain layers, at least by providing minimal (Wenger, 1998) but adequate conditions for interaction in these layers. This use could provide a balance and complement to face-to-face interaction between members. This means that wherever possible, it could help individuals to communicate and develop their collaboration. However, technology mediated interaction does not substitute for physical interaction, and efforts should be made to develop continuous and regular encounters between researchers. As well as this, minimal support ICT, as always, offers the possibility to interact and produce in ways that better the physical world. This could be the case of communication without barriers and constraints of time zones, number of participants, language, record keeping, pattern and rule recognition, and visual representation.

We suggest that ICT tools usage should be *blended* to provide support to both community and practice oriented layers. By blended, we mean a mixture of ICT and face-to-face that combines the best of each so that ICT is in balance with face-to-face interactions. The nature of this blend depends on how a community decides to develop their work. Although it has been argued in this article that developing a community-oriented layer requires establishing social relationships, this does not mean this should only be done physically (Wenger, 1998). In physically distributed or closed research environments, ICT can contribute to provide a "social gluing" element for collaboration. In addition, ICT expands the spectrum of communication, models, learning opportunities, and the nexus of perspectives of communities.

Therefore, we suggest that the use and blending of ICT be defined by the different needs that research communities have at a particular time in relation to their development. Attention should be paid, though, to avoid depending only on an electronically linked community which could easily slip to the back of people's minds (Somekh & Pearson, 2002). The needs for using ICT in collaborative research could be defined by:

- Where face-to-face engagement builds *community-oriented layer*
- Where technology supports *community-oriented layer*
- Where face-to-face achieves *practice-oriented layer*
- Where technology constructs and disseminates *practice-oriented layer*
- Where face-to-face learning generates *domain-oriented integration*
- Where technology synthesizes *domain-oriented integration*

These needs could help those engaged or willing to engage in collaborative research to define and review periodically their priorities for the use of ICT in collaborative research.

Further research is needed to provide methodologies that help constructing the blend between ICT and face-to-face along the above lines, in particular, to support the formation and development of communities of many forms: one-to-one, groups, subgroups, dispersed, or joined. These more sophisticated forms will be more important in collaborative research as dynamic and evolving activity as we see it developing in the next few years.

CONCLUSION AND FUTURE TRENDS

In this article, we have provided an understanding of collaborative research as a social activity that should be supported by the use of concepts of the theory of communities of practice (CoPs). With these ideas, we have argued that collaborative research could be developed and sustained by working simultaneously on three aspects of activity: *A community-oriented layer; a practice-oriented layer; and a domain-oriented integration*. All aspects are important to sustain social relationships between researchers and create opportunities for collaborative research. The blended use of ICT and face-to-face can support activity and should be directed to provide tools for researchers to develop and maintain their communities of research. Technology should be defined by the needs that research groups have through their collaborations and how they need to balance it with face-to-face interactions. In this way, we hope to have contributed to the understanding of how communities of practice address the importance of collaborative research.

REFERENCES

Brulin, G. (2001). The third task of universities or how to get universities to serve their communities. In P. Reason & H. Bradbury (Eds.), *Handbook of action research: Participative inquiry and practice* (pp. 440-446). London: Sage.

Callon, M., Law, J., & Rip, A. (1986). *Mapping the dynamics of science and technology: Sociology of science in the real world*. London: Macmillan Press.

Commission of the European Communities. (2000). *Communication from the commission to the council, the European parliament, the economic and social committee and the committee of the regions: Towards a European research area*. Brussels: Commission of the European Communities.

Department of Trade and Industry. (1990). *Collaboration between business and higher education*. London: Council for Industry and Higher Education.

Gibbons, M., Limoges, C., Nowotny, H., Schwartzman, S., Scott, P., & Trow, M. (1994). *The new production of knowledge: The dynamics of science and research in contemporary societies*. London: Sage.

Nowotny, H., Scott, P., & Gibbons, M. (2003). "Mode 2" revisited: The new production of knowledge. *Minerva, 41*, 179-194.

Schwen, T., & Hara, N. (2003). Community of practice: A metaphor for online design? *The Information Society, 19*, 257-270.

Somekh, B., & Pearson, M. (2002). Intercultural learning arising from Pan-European collaboration: A community of practice with a "hole in the middle". *British Educational Research Journal, 28*(4), 485-502.

Star, L., & Griesemer, J. (1989). Institutional ecology, "translations" and boundary objects: Amateurs and professionals in Berkeley's Museum of Vertebrate Zoology, 1907-1939. *Social Studies of Science, 19*, 387-420.

Wenger, E. (1998). *Communities of practice: Learning, meaning and identity*. Cambridge: Cambridge University Press.

Wenger, E., McDermott, R., & Snyder, W. M. (2002). *A guide to managing knowledge: Cultivating communities of practice*. Boston: Harvard Business School Press.

KEY TERMS

Blend: A judicious mixture of face-to-face and computer-mediated interactions to facilitate collaboration.

Collaborative Research: Type of research that is developed by individuals who belong to different academic or practical disciplines, in which there isare a variety of purposes, methods, and outcomes.

Community: A group in which a way of talking about mutual purposes and participating are recognised as worth pursuing.

Community of Practice: Group of people who share a set of concerns and sustain their collective actions through their participation and generation of new knowledge.

Information Technologies: Computer and information based systems which are mediated through electronic communication and integrated into the activities of individuals and groups.

Layer: A set of practices that characteriszes activities of a community. These practices are not homogenous, although they share common features.

Mode 1 Research: Research that is conducted by those belonging to a single discipline, aiming to solve questions within this discipline. The outcomes will be used by the discipline and may not transcend a particular context of application.

Mode 2 Research: Research that is developed with the participation of different disciplines, with the aim of solving a problem that has a particular context and requires the use of a variety of methods, and sources of knowledge. The outcomes of this research will be used in other contexts as reference.

Practice: Shared historical and social resources and perspectives that help sustaining a community through time.

Research: Systematic endeavour to find answers to a problem, with the help of scientific methods or critical investigation.

Use and Methods of Social Network Analysis in Knowledge Management

Tobias Müeller-Prothmann
Free University Berlin, Germany

INTRODUCTION

Whilst the primary importance of informal communities of practice and knowledge networks in innovation and knowledge management is widely accepted (see Armbrecht et al., 2001; Brown & Duguid, 1991; Collinson & Gregson, 2003; Jain & Triandis, 1990; Lesser, 2001; Liyanage, Greenfied & Don, 1999; Nahapiet & Ghoshal, 1998; Nohria & Eccles, 1992; Wenger, 1999; Zanfei, 2000), there is less agreement on the most appropriate method for their empirical study and theoretical analysis. In this article it is argued that social network analysis (SNA) is a highly effective tool for the analysis of knowledge networks, as well as for the identification and implementation of practical methods in knowledge management and innovation.

Social network analysis is a sociological method to undertake empirical analysis of the structural patterns of social relationships in networks (see, e.g., Scott, 1991; Wasserman & Faust, 1994; Wellman & Berkowitz, 1988). This article aims at demonstrating how it can be used to identify, visualize, and analyze the informal personal networks that exist within and between organizations according to structure, content, and context of knowledge flows. It will explore the benefits of social network analysis as a strategic tool on the example of expert localization and knowledge transfer, and also point to the limits of the method.

BACKGROUND

Words have meanings: some words, however, also have a 'feel'. The word 'community' is one of them. It feels good: whatever the word 'community' may mean, it is good 'to have a community', 'to be in a community'. (Bauman, 2001, p. 1)

The term "community" is widely used, yet imprecisely defined in the sociological literature. Whilst there is consensus that community is a fundamental unit of social organization, there is little agreement on how best to describe it as a sociological entity (see Poplin, 1979, pp. 11-12). The fact that the term "community" refers to different things, depending upon who is using it and upon the context in which it is used, can render it useless for scientific purposes (see Poplin, 1979, p. 4). Nevertheless, the use of the community concept, or community "metaphor," is flourishing in the social sciences, as well as in political debates and management strategies. One of the foremost applications of the term is in the domain of knowledge communities or communities of practice.

One alternative approach is to view communities as networks. Drawing on the methods and tools of sociometry, the development of formal approaches to social networks began with Moreno (1934), and was systematized and fundamentally elaborated by means of graph theory (König, 1936) through Cartwright and Harary (1956). The breakthrough of social network analysis as a method of structural analysis was reached in the 1960s by White and his Harvard colleagues (see Scott, 1991, pp. 33-38; for a review of the large number of applications of social network analysis, see, e.g., Wellmann & Berkowitz, 1988).

A conceptualization of communities as social networks was outlined by Poplin (1979) in his analysis of community literature as a "network of interaction" (pp. 14-18). In Poplin's view, there is at least one major advantage in conceptualizing communities in this way: "It serves well as a tool by which to describe systematically the interrelationships of the various units that compose the community. This alone can help increase our understanding of community structure and process" (p. 16). Poplin's perspective helps us to build the case of communities of practice as social networks. In doing so, it provides us with both a unit of analysis and the means to develop and employ an empirical method and practical tool, that of social network analysis. The provision of a conceptual framework and powerful tool for the analysis of informal social structures is emphasized here as its major advantages.

USE

Informal knowledge networks are not a new invention in the knowledge management literature. Crane (1972), for example, published her widely recognized study on the diffusion of knowledge in scientific communities. Even earlier, the classic Hawthorne studies included in their principal report of 1939 various sociograms that the research team saw as reflecting the "informal organization" of a bank's wiring room (as opposed to the formal organization depicted by the organization chart) (see Roethlisberger & Dickson, 1947, pp. 500-548). Whether speaking about communities of practice, knowledge communities, or knowledge networks, all these concepts have a common core that can be subsumed under the "social capital" construct. Burt (2000) elaborates upon this point and suggests that the social capital concept is essentially "a metaphor about advantage" (p. 2), that is, the better the social connections between people, the higher the collective and individual returns for them. Cross, Parker, and Borgatti (2002) describe this advantage of connection as "who you know has a significant impact on what you come to know" (p. 2). From here, we can identify the logical underpinning of social network analysis as the empirical study of connections between individuals within communities.

Social network analysis uses several techniques to empirically identify underlying patterns of social structure. It then compares these individual patterns with their influence on specific network behavior variables and performance outcomes. From a knowledge management perspective, social network analysis helps us identify basic network properties, positions of network members, characteristics of relations, cohesive sub-groups, and bottlenecks of knowledge flows. By pointing to who shares knowledge with whom, social network analysis shows us the informal relations within and between organizations (see Figure 1). In doing so, it allows the researcher to identify and maybe influence a network's and its members' ability to create and to share knowledge.

Although social network analysis must always begin with some initial populations, one important advantage of the method is that it does not view formal boundaries (such as departments) as truly social boundaries. Rather, it traces social relationships wherever they may exist and extend. (Laumann, Marsden & Prensky, 1989, discuss the boundary specification problem in network analysis at length.) In this way, we can identify the following core knowledge management applications from social network analysis that are explained in the following example:

Figure 1. Formal vs. informal in a petroleum organisation (IBM Institute for Business Value, 2002)[1]

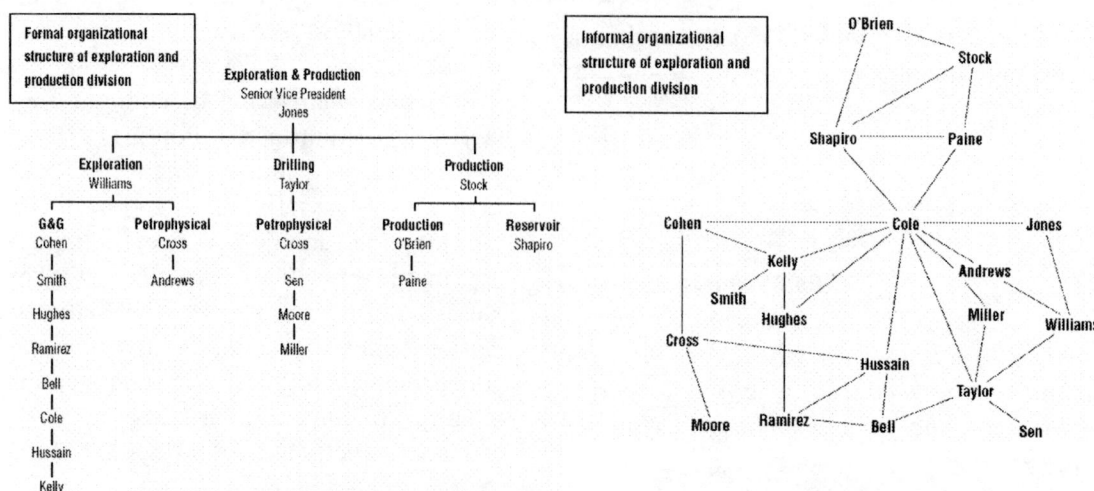

- identification of personal expertise and knowledge,
- research into the transfer and sustainable conservation of tacit knowledge, and
- discovery of opportunities to improve communication processes and communication efficiency.

Other knowledge management applications from social network analysis that go far beyond the scope of this article include studies into the development of core competencies (like leadership development), the identification and support of communities of practice, approaches for the harmonization of knowledge networks (for example after mergers and acquisitions), and the sustainable management of relationships between distributed sites and external partners.

METHODS AND MEASURES

Social network analysis perceives social structure as the pattern organization of network members and

their relationships. Network data are defined by the members of the network and their relationships. (Note: Social network analysts talk of "actors" rather than of "members.") Using graph theory, a sociogram consists of "nodes" (or points), representing individual network members, and "ties" (or lines), representing the connections between the members (relations); these graphs clearly record and visualize social relationships (see Figure 2). Another advocated means to represent information about social networks is in matrices. In their simplest form, network data consist of a square matrix, the rows of the array represent the persons, the columns of the array represent the same set of persons, and the elements represent the ties between the persons (so-called "adjacency matrix"—see Figure 3). Matrices are also used as data input for social network analysis processing. (For an introduction to graph theory and the use of matrices in social network analysis, see, e.g., Scott, 1991, pp. 39-65, or Hanneman, 2001, pp. 2-4 & pp. 26-36.)

As this article serves only as an introduction to social network analysis in knowledge management, what follows is a short guide to the analytical concepts and measures that sit at the heart of the technique and are of primary importance for a pragmatic adaptation as a method and practical tool in knowledge management. For a more comprehensive introduction to social network analysis see Scott (1991) or Hanneman (2001).

- **Knowledge Flows Within Large and Small Networks—Size:** Size is a basic property of a network—sharing knowledge between all members of a large network (say for example, between a total of 25,000 members of a whole business unit) would be extremely difficult

Figure 2. A graph of social relationships

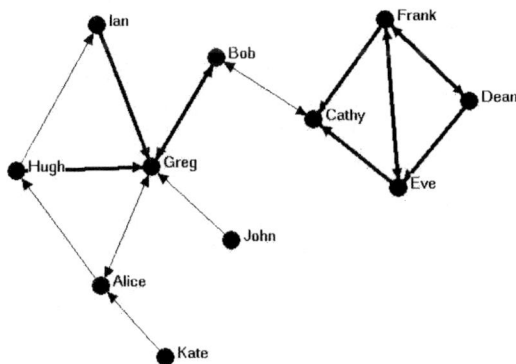

Figure 3. Adjacency matrix of social relationships as visualized in Figure 2

	Alice	Bob	Cathy	Dean	Eve	Frank	Greg	Hugh	Ian	John	Kate
Alice	0	0	0	0	0	0	1	1	0	0	0
Bob	0	0	1	0	0	0	2	0	0	0	0
Cathy	0	1	0	0	0	0	0	0	0	0	0
Dean	0	0	0	0	2	2	0	0	0	0	0
Eve	0	0	2	0	0	2	0	0	0	0	0
Frank	0	0	2	2	2	0	0	0	0	0	0
Greg	1	2	0	0	0	0	0	0	0	0	0
Hugh	0	0	0	0	0	0	2	0	1	0	0
Ian	0	0	0	0	0	0	2	0	0	0	0
John	0	0	0	0	0	0	1	0	0	0	0
Kate	1	0	0	0	0	0	0	0	0	0	0

compared to sharing knowledge between all members of a small network (say for example, between a total of 11 members—as presented in Figures 2 and 3).

- **Linkages of Networks—Density:** Density describes the global level of linkage of a network. Even if fully saturated networks are empirically rare (where all possible ties are actually present), measures of density look at "how closely a network is to realizing this potential" (Hanneman, 2001, p. 41). As a measure that is especially relevant for the case of communities of practice, density describes the overall linkage between the community members.

- **Expertise and Power—Degree Centrality:** Degree centrality is a measure of the incoming and outgoing connections held by an individual network member. Incoming connections (in-degree) define the popularity of a member; those with many ties are members who are considered particularly prominent or—in the case of knowledge networks—have high levels of expertise (like Greg in Figures 2 and 3 for example). Out-degree defines the number of outgoing connections or the power of a member; having a high out-degree, a person is considered as particularly influential in the network (like Frank in Figures 2 and 3). Insufficient member links (as well as links between sub-groups—see below) might indicate the potential resources of network members that are not used. Excessive linkages might indicate the stress and overload of individual members. Degree centrality is a measure that helps to purposefully support individual members of a community of practice.

- **Integration or Isolation—Closeness Centrality:** While degree centrality is a measure of the immediate ties of a network member, closeness centrality (as well as betweenness centrality) measures the reachability of members. This is achieved by including indirect ties. Closeness centrality focuses on the distance of a member to all others in the network through means of geodesic distance. It determines a member's integration in the network (in Figures 2 and 3, Greg displays high in-closeness and Frank high out-closeness). Thus, high close-

ness centrality indicates the greater autonomy of an individual person, since he or she is able to reach the other members easily. Low closeness centrality on the other hand indicates higher individual member dependency, that is, the willingness of other members to gain access to the network's resources. By determining the average closeness centrality of a network, the relative isolation or integration of persons can be identified. People who are not well integrated into a group could represent untapped skills. They may be highly expert people who are not being utilized appropriately (see Cross et al., 2002, p. 6).

- **Knowledge Brokers and Gatekeepers—Betweenness Centrality:** Betweenness centrality is a measure of the extent that a network member's position falls on the geodesic paths between other members of a network. Thus, it determines whether an actor plays a (relatively) important role as a broker or gatekeeper of knowledge flows, with a high potential of control on the indirect relations of the other members (like Greg and Bob in Figures 2 and 3).

- **Strength and Weakness of Ties—Multiplexity:** Network members may maintain a tie based on one relationship type only (a narrowly specialized relationship, for example, sharing news on only one topic of research). Alternatively, they may maintain a variety of relations—broadly multiplex relationships, for example, sharing information, working together in different projects, and playing golf together. The latter are known as multiplex ties (the example in Figures 2 and 3 shows only two strengths of ties: 1 or 2). On the one hand, multiplex (strong) relationships share more intimate, voluntary, supportive, and durable ties (see Wellman & Wortley, 1990), and thus, form a solid basis for trust. On the other hand, most people only share a small number of strong relationships, so that especially weak ties are warranty for access to a large variety of resources (Granovetter's (1973) popular "strength of weak ties"). With regard to communities of practice, the importance of multiplex relationships gives reason for various kinds of community-building activities.

- **Sub-Cultures and Clusters of Expertise—Sub-Groups and Cliques:** Sub-sets of members can build dense connections and develop cohesive sub-groups of the network (like Cathy, Dean, Eve, and Frank in Figures 2 and 3). These are known as cliques and clusters (Watts & Strogatz, 1998; Roethlisberger & Dickson, 1947, already wrote about "cliques" in their 1939 report, pp. 508-510).[2] Cliques or clusters are of special interest to network analysts as they are important for understanding the behavior of the whole network. For example, organizational sub-groups or cliques can develop their own "sub"-cultures and attitudes toward other groups (Cross et al., 2002, p. 6). They can also gain influence over the overall network. Exploitation and integration of the sub-groups' potential resources can be a critical factor to failure or success of a community of practice.

- **Bottlenecks and Knowledge Gaps—Cut-Points and Structural Holes:** Often, networks are not only clustered into cohesive sub-groups, but are also split into loosely coupled components. In this case, not all possible connections are present: there are structural holes (Burt, 1992). Persons of pivotal significance in holding components together are called cut-points or bridges: central nodes that provide the only connection between different parts of the network (like Bob in Figures 2 and 3). Cut-points build bridges between sub-groups that would otherwise have been cut-off and split into separate, unconnected components. They represent the network's bottlenecks and are critical to the knowledge flow of a network. Yet too many links can lead to inefficiency of knowledge exchange. Generally speaking, links between sub-groups (for example, between members of different departments) must be coordinated effectively and efficiently (see for example the role of hubs described below).

- **Enablers of Effective Knowledge Transfer—Hubs:** As networks are clustered, some members are important as simultaneous actors in many clusters. These are known as hubs (Kleinberg, 1999; Rosen, 2000). As Barabási (2003) puts it, these persons "have played in very different genres during their careers" (p. 61). They can effectively link different sub-groups of the network and can facilitate knowledge flows between different departments or to external organizations.

EXAMPLE

Consider for example that you are working in a research and service organization.[3] You have to acquire funding for new research projects, and you know that programs and financing are available from the EU to help you. You have an idea for a new project, but you do not know how best to prepare a project proposal. However, there are other people in your organization who have successfully acquired EU funding for their projects. The question is: How do you acquire the knowledge to know who knows about developing a winning proposal for an EU project acquisition? Who are the experts in your organization, and who do you need to know and contact outside of your organization to assist you (at the European Commission, for example)?

- First, you will have to define the knowledge domain that is relevant. This is what you did already: you want to learn about experts for the acquisition of EU projects within the organization and then contact appropriate persons external to your organization.

- Second, you will need to formulate questions that address your goal adequately and operationalize the survey. Typical questions might be: Who do you know who is an expert about project acquisition? Who do you know who is currently working on an EU project or has just finished one? Whom did you help with regard to project acquisitions? Additionally, your data collection would be best complemented by analyzing other available documents and resources such as project documentations, team meetings, conferences, and so forth.

- Third, after encoding the data you perform procedures of formal network analysis and visualization—for example, using the popular and very sophisticated software for network analysis called UCINET (see Borgatti, Everett & Freeman, 2002).

- Fourth, you will need to analyze your findings from the previous step and interpret the results.
- Finally, you will try to sustainably improve knowledge flows with regard to EU project acquisition. Indeed, you might perform a social network analysis on the same topic on a frequent basis to evaluate success or failure of your interventions.

Based upon the results and their interpretation, interventions (like workshops, dialogues, establishment of communities of practice) are recommended to improve knowledge flows (examples of interventions are outlined in more detail in Müeller-Prothmann & Finke, 2004, pp. 697-698). These can be designed to foster knowledge communication, strengthen relationships within the network, build relationships to other networks, and develop strategies for the creation of flourishing knowledge environments and for sustainable knowledge transfer. One very basic intervention would be to simply ask people to spend five minutes on their network visualizations and "to identify what they 'see' in the map, the structural issues impeding or facilitating group effectiveness, and the performance implications for the group" (Cross et al., 2002, p. 11).

This example shows that social network analysis provides a method to trace knowledge flows, analyze network structures and personal expertise, with additional value to a simple knowledge map or yellow pages. On the one hand, social network analysis tries to identify knowledge flows wherever they may go. On the other hand, social network analysis gives us a detailed picture of actors' positions, the characteristics of their connections, and the overall structure of relationships. Social network analysis provides us with a well-elaborated set of methods and measures for various applications in knowledge management that especially help us to foster the development of communities of practice.

FUTURE TRENDS

The application of social network analysis as illustrated in the example above affects all four dimensions that Cross et al. (2002, p. 7) found to be critical for a relationship to be effective in terms of knowledge creation and use. These are: (1) knowing what someone knows, (2) gaining timely access to that person, (3) creating viable knowledge through cognitive engagement, and (4) learning from a safe relationship. Despite the assertions of Cross et al., the use of social network analysis as a practical tool in knowledge management has been limited. To become more widely adopted in practice, the sophisticated scientific methods of social network analysis will need to be pragmatically adapted to suit practical needs.

The future focus of research must be put at two distinct levels of analysis, the individual networker and the organizational level. For both levels, the challenge will be to develop new methods in social network analysis that deliver practical value to knowledge management and provide for new models of interpretation and intervention. This includes methods for: (1) the clear definition of network members and network boundaries (for example based upon focused knowledge domains), (2) faster data collection, (3) efficient analysis and interpretation of results, and (4) effective intervention to improve knowledge sharing, knowledge flows, and network structure based on the given results.

In addition, the success or failure of social network analysis as a valuable knowledge management method depends on the successful integration of specific organizational conditions and requirements into the methodological process. This demands research activities on the cultural factors that influence network structure and performance. Then, insights from the application of social network analysis could provide the basis to develop measures for assessing the contribution of knowledge communities and networks to overall organizational performance and innovation. Especially, it should be strongly considered to integrate basic measures of social network analysis into a knowledge-orientated, balanced score card, and to expand techniques of social network analysis for examination and support of mobile networking and collaboration; without doubt, these technologies will gain influence as communication tools in distributed communities of practice and raise demands for thorough explorations.

CONCLUSION

Social network analysis is a powerful empirical tool for assessing network ties and structure and their influence on knowledge creativity, sharing, and innovation. The brief sociological discussion about the use of the term "community" early in this article shows that the development of a network perspective from a theoretical view on communities provides an adequate analytical framework for communities of practice, knowledge communities, and other social constructs subject to the knowledge management debate. As introduced in this article, social network analysis has practical application beyond a narrow theoretical perspective. By focusing on the social aspects of knowledge management in a methodically rigorous manner, the technique has much potential.

Of course, the limitations of social network analysis cannot be neglected. Since social network analysis is based on the study of bilateral interactions, it provides a merely descriptive picture of structures and positions. Further, aspects and characteristics of communities and social networks, like shared identity or shared norms of the network members for example, cannot be covered through social network analysis in a strict sense. Therefore, knowledge management processes focusing on communication structures and its related aspects, like advice and support networks, knowledge flows, and communication efficiency for example, are best explored through methods of social network analysis. These fields of application could make use of social network analysis as a powerful tool, as well as approaches for sustainable support of communities of practice.

Nevertheless, the use of social network analysis in knowledge management may be limited in environments characterized by high social complexity and a large variety of organizational constraints. Thus, a widespread adoption of social network analysis as a method and tool for knowledge management will depend on evidence provided by further research, case studies, and practical implementation in organizational business strategies in the various fields of application as proposed here.

ACKNOWLEDGMENTS

I am very grateful to Chris Lawer, Cranfield School of Management and The OMC Group, England, for critical and fruitful comments, and his valuable contribution in reviewing and correcting a draft version of this article. I am also very grateful for the unbureaucratic permission by Judith A. Quillard from the IBM Institute for Business Value, Cambridge, Massachusetts, and Rob Cross, assistant professor at the University of Virginia's McIntire School, to publish the network example presented in Figure 1.

REFERENCES

Armbrecht, F.M.R. Jr., Chapas, R.B., Chappelow, C.C., Farris, G.F., Friga, P.N., Hartz, C.A., McIlvaine, M.E., Postle, S.R., & Whitwell, G.E. (2001). Knowledge management in research and development. *Research Technology Management, 44*(4), 28-48.

Barabási, A.-L. (2003). *Linked. How everything is connected to everything else and what it means for business, science, and everyday life*. New York: Plume.

Bauman, Z. (2001). *Community. Seeking safety in an insecure world*. Cambridge, UK: Polity Press.

Borgatti, S.P., Everett, M.G., & Freeman, L.C. (2002). *Ucinet 6 for Windows*. Harvard: Analytic Technologies. Retrieved from *http://www.analytictech.com*

Brown, J.S., & Duguid, P. (1991). Organizational learning and communities of practice: Toward a unified view of working, learning, and innovation. *Organization Science, 2*(1), 40-57.

Burt, R.S. (1992). *Structural holes. The social structure of competition*. Cambridge, MA: Harvard University Press.

Burt, R.S. (2000). The network structure of social capital. Pre-print for a chapter in R.I. Sutton & B.M. Staw (Eds.), *Research in organisational behaviour* (vol. 22). Greenwich, CT: JAI Press. Unpublished manuscript.

Cartwright, D., & Harary, F. (1956). Structural balance: A generalization of Heider's theory. *Psychological Review, 63*(5), 277-293.

Collinson, S., & Gregson, G. (2003). Knowledge networks for new technology-based firms: An international comparison of local entrepreneurship promotion. *R&D Management, 33*(2), 189-208.

Crane, D. (1972). *Invisible colleges. Diffusion of knowledge in scientific communities*. Chicago; London: The University of Chicago Press.

Cross, R., Parker, A., & Borgatti, S.P. (2002). A bird's-eye view: Using social network analysis to improve knowledge creation and sharing. Retrieved January 14, 2003, from *http://www-1.ibm.com/services/strategy/files/IBM_Consulting_A_birds_eye_view.pdf*

Granovetter, M.S. (1973). The strength of weak ties. *American Journal of Sociology, 78*(6), 1360-1380.

Hanneman, R.A. (2001). *Introduction to social network methods*. University of California, Riverside, Department of Sociology. Retrieved January 29, 2004, from *http://faculty.ucr.edu/~hanneman/SOC157/NETTEXT.PDF*

Jain, R.K., & Triandis, H.C. (1990). *Management of research and development organizations. Managing the unmanageable*. New York: John Wiley & Sons.

Jansen, D. (1999). *Einführung in Die Netzwerkanalyse. Grundlagen, Methoden, Anwendungen*. Opladen: Leske und Budrich.

Kleinberg, J.M. (1999). Authoritative sources in a hyperlinked environment. *Journal of the ACM, 46*(5), 604-632.

König, D. (1936). *Theorie der endlichen und unendlichen graphen*. Leipzig: Akademische Verlagsgesellschaft.

Laumann, E.O., Marsden, P.V., & Prensky, D. (1989). The boundary specification problem in network analysis. In L.C. Freeman, D.R. White &

A.K. Romney (Eds.), *Research methods in social network analysis* (pp. 18-34). Fairfax, VA: George Mason University Press.

Lesser, E.L. (2001). Communities of practice and organizational performance. *IBM Systems Journal, 40*(4), 831-841.

Liyanage, S., Greenfied, P.F., & Don, R. (1999). Towards a fourth generation R&D management model: Research networks in knowledge management. *International Journal of Technology Management, 18*(3/4), 372-394.

Moreno, J.L. (1934). *Who shall survive? A new approach to the problem of human interrelations*. Washington, DC: Nervous and Mental Disease Publishing Company.

Müeller-Prothmann, T., & Finke, I. (2004). SELaKT: Social network analysis as a method for expert localisation and sustainable knowledge transfer. *Journal of Universal Computer Science, 10*(6), 691-701.

Nahapiet, J., & Ghoshal, S. (1998). Social capital, intellectual capital, and the organizational advantage. *Academy of Management Review, 40*(2), 242-266.

Nohria, N., & Eccles, R.G. (1992). Face-to-face: Making network organizations work. In N. Nohria & R.G. Eccles (Eds.), *Network and organizations* (pp. 288-308). Boston: Harvard Business School Press.

Poplin, D.E. (1979). *Communities. A survey of theories and methods of research* (2nd ed.). New York: Macmillan.

Roethlisberger, F.J., & Dickson, W.J. (1947). *Management and the worker. An account of a research program conducted by the Western Electric Company, Hawthorne Works, Chicago*. Cambridge, MA: Harvard University Press.

Rosen, E. (2000). *The anatomy of buzz. How to create word-of-mouth marketing*. New York: Doubleday.

Scott, J. (1991). *Social network analysis. A handbook*. London: Sage Publications.

U

Wasserman, S., & Faust, K. (1994). *Social network analysis: Methods and applications*. Cambridge, MA: Cambridge University Press.

Watts, D.J., & Strogatz, S.H. (1998). Collective dynamics of 'small-world' networks. *Nature, 393*(4), 440-442.

Wellman, B. (Ed.). (1988). *Social structures: A network approach*. Cambridge, MA: Cambridge University Press.

Wellman, B., & Wortley, S. (1990). Different strokes from different folks: Community ties and social support. *American Journal of Sociology, 96*(3), 558-588.

Wenger, E. (1999). *Communities of practice: Learning, meaning, and identity*. Cambridge, MA: Cambridge University Press.

Zanfei, A. (2000). Transnational firms and the changing organisation of innovative activities. *Cambridge Journal of Economics, 24,* 515-542.

KEY TERMS

Betweenness Centrality: Betweenness centrality is the sum of all probabilities that a member n_i lies on the path between other pairs of members:

$$C_B(n_i) = \sum_{j<k}^{n} g_{jk}(n_i) / g_{jk} \quad (i \neq j \neq k)$$

Closeness Centrality: The sum of all geodesic distances for each member is the farness of the member from all others. Taking the reciprocal, this measure is converted into a measure of closeness, called closeness centrality of member n_i:

$$C_C(n_i) = \left[\sum_{j=1}^{g} d(n_i, n_j) \right]^1 \quad (i \neq j)$$

Degree Centrality: Degree centrality counts incoming and outgoing connections of an individual network member n_i:

$$d(n_i) = \sum_{j=1}^{g} x_{ij} = \sum_{i=1}^{g} x_{ij} = x_{i.} \quad (i \neq j)$$

For non-symmetric data the in-degree of a member n_i is the number of ties received by this member:

$$d_I(n_i) = \sum_{j=1}^{g} x_{ji} = x_{.i} \quad (i \neq j)$$

Out-degree is the number of ties initiated by member n_i:

$$d_O(n_i) = \sum_{j=1}^{g} x_{ij} = x_{i.} \quad (i \neq j)$$

In case of symmetric data, in-degree and out-degree are identical.

Density: The density of a network is the total number of ties divided by the total number of possible ties:

$$\Delta = \frac{\sum_{i=1}^{g} \sum_{j=1}^{g} x_{ij}}{g(g-1)} \quad (i \neq j)$$

In a network the number of unique ordered pairs is derived from $g(g-1)$. (If not indicated otherwise, all formal notations given follow Wasserman & Faust, 1994.)

Ego-Centred Network Analysis: Social network analysts basically distinguish between ego-centred and whole network analysis. Ego-centred network analysis starts from a defined set of persons who are asked questions in order to generate a list of members of their personal network. This approach to generate a list of network members is called "snow-balling"; for example, a person may be asked to name the people that they turn to for advice about a specific work problem.

Geodesic Distance: Geodesic distance d(i,j) indicates how many intermediary persons are on the shortest path from member n_i to member n_j.

Graph: Formally, graphs are defined a set of actors (g-nodes) and a set of their defined relations (l-lines). The set of actors is defined by the nodes $\{n_1, n_2, n_3, \ldots n_g\}$.

Multiplexity: Network multiplexity is the relation between the number of actual multiplex ties and the number of potential multiplex ties in a network:

$$M = \frac{\sum_{i=1}^{g} \sum_{j=1}^{g} x_{ij(m)}}{g(g-1)} \quad (i \neq j) \text{ (see Jansen, 1999, pp. 104-105)}$$

Relations: Relations between network members can be undirected or directed and/or valued. They can be represented in an undirected or a directed graph and/or valued graph and their matrices, respectively (Figures 2 and 3 show a directed valued graph and matrix).

Size: The size of a network is indexed by counting its members (nodes).

Whole Network Analysis: In a whole network analysis, the network members are completely defined from the beginning and people are asked to identify their individual connections of some specific content to the other members.

ENDNOTES

[1] The example is taken from Cross et al. (2002, p. 4), with kind permission by Rob Cross and the IBM Institute for Business Value, Cambridge, Massachusetts; names have been changed at the request of the company.

[2] For a definition and formal description of the concepts outlined above, see the Key Terms section below. The description of formal approaches to analyse clusters, structural holes, and hubs would exceed the scope of this article; for a comprehensive introduction, see for example, Scott (1991) or Jansen (1999).

[3] The example given in this section is inspired by a study on expert networks about the acquisition of EU projects in a research and service organisation; see also Müeller-Prothmann and Finke (2004).

Using Emerging Technologies for Effective Pedagogy in Management Education

Sunil Hazari
University of West Georgia, USA

INTRODUCTION

The past decade has brought tremendous changes to higher education. Technology components that supplement teaching and learning are integrated into programs and courses in most universities. Tools, such as course management systems, portals, PDAs, wireless technology, and Web services are used to create virtual communities that provide an interactive platform for learning. Previous research (Alavi, 1994; Lake, 1998; Yip, 2004) has shown that technology-based instruction results in positive learning outcomes. Colleges and universities are trying to understand this phenomenon of digital education and restructure themselves to take advantage of emerging technologies so that students can be prepared to be leaders and managers, who not only realize the benefits of using collaborative tools in virtual space, but also have competencies in using these tools effectively. In addition, because emerging technologies make it possible to extend physical boundaries of a university, new markets could bring additional revenues and expand access to programs nationally as well as globally.

Management education with its use of problem-based learning and case study approach has been a leading candidate for integrating technology tools for scholarship and research. Business schools have been under constant pressure to provide students the skills and experience needed to effectively use emerging technologies (Alavi, Wheeler & Valacich, 1995; Hildebrand, 1995) that are used by businesses to gain competitive advantage (Leidner & Jarvenpaa, 1993). Webster and Hackley (1997) have identified previous studies of business schools adopting computer mediated distance learning for business cases and simulations. A strong community of practice (CoP) is critical for building collaboration between faculty in universities that may be separated by space but connected using networks that can be leveraged to extend programs and provide faculty partnerships and foster student scholarship. CoP can foster the spirit of discussion and collaboration. Brown and Duguid (2000) have defined CoPs as groups of people who share a common vision or passion and work closely together within the context of a particular practice or field of study (Garrison, Hawes & Kanuka, 2003). CoP has also been defined as a group of people who share a common concern, set of problems, or interest in a topic, whose members come together to fulfill both personal and group goals. The main goals of a CoP are to generate knowledge, contribute to identification of effective practices, and definition of underlying principles. CoPs also help create common vocabularies and conceptual frameworks (NLII Virtual Communities, 2003). There are several tools that try to address virtual collaboration, but very few tools are used effectively. The purpose of this article is to look at three tools: portals, course management systems, and videoconferencing to explore how CoP can thrive by use of these tools.

The Internet has quickly evolved from merely a distribution channel to an interactive environment for collaborative learning. In what can be considered a partial response to Frost and Fukami's (1997) challenge to the profession to think in deep ways about management education and teaching, faculty have realized the tremendous potential of actively engaging students in the online environment. Students have also appreciated the benefits and convenience of e-learning. The technology component is now integrated with almost every functional area of business education. As an example of a classroom, students in a supply chain course can discuss implications of global partnerships between suppliers and manufacturers, review best practices in supply chain management, and learn from case studies of international corporations. Lectures and discussions using streaming video and tools such as whiteboards, chat forums, and interactive audio can be used to explore cultural diversity and international business culture. To make this happen, costs associated with providing resources should be realized and budgeted. Business schools committed to research, student learning, and effective teaching have developed strategic plans that underscore significant investments in IT infrastructure, software development (such as portal technology, course management tools, and Web services).

The technology infrastructure and services that are deployed in business schools should provide a strong base for teaching, research, and community outreach and foster cross-disciplinary collaboration between units. The interaction may not necessarily be only among students, but can extend to professional exchange of ideas between students and faculty at one institution with a group of professionals at any institution. An example of a discipline-based CoP is the information systems professional Web site and listserv: ISWORLD. The Web site (http://www.isworld.org) provides members an opportunity to stay current on the happenings in information systems teaching, research, and service. An active listserv generates daily informational posts that include conference announcements, book/journal publications, position announcements, sharing of ideas on resources, offers of collaborative opportunities, further inquiry into research and systems implementation questions, and challenges to commonly believed hypotheses. The daily interaction, postings, responses, and response to responses provide constant and relevant information to list subscribers and also provide them an opportunity to stay active and participate in discussions.

The above example shows the use of a listserv via e-mail. In the past few years, more interactive environments have emerged to support communities of practice. Characteristics of three representative technologies (portals, course management systems, and videoconferencing) that can be considered to have potential to improve management education are given below.

PORTALS

Davydov (2001) defines a portal as "an entry point or originating web site for combining a fusion of content and information dissemination services" (p. 57). A corporate portal often includes customizable start pages to help users locate information tailored to their need as well as access to applications and business intelligence tools, data warehousing, collaborative and workflow systems, EAI tools, Web publishing, and personalization tools. In educational institutions, EduPortals are developed to provide easy access to academic and administrative resources and services. The main goal of EduPortals is to connect the institution's internal and external constituents to campus resources using a personalized interface. Some of the challenges in building such portals have been the integration of university directory services, single sign-on procedures, aggregation, organization, and delivery of information from multiple sources. Technical as well as organizational problems must be overcome to deploy university-wide portals that will be integrated with daily tasks of faculty, staff, and students. From a user perspective, the objective of an EduPortal should be to provide an attractive, easy to use gateway to navigate through the network of both public and private information, services, and business functions of the school and the university. It should provide a secure infrastructure to present Web-based applications and information to the university community. The portal should focus on tools for collaboration, research, and personal productivity. For future growth, today's portal should take into account scalability, integration of legacy systems, and future enterprise-wide system compatibility. As a specific management education example, benefits of portals to faculty and students can be single sign-on access to e-mail and academic calendars, browser based access to networked files from remote locations, collaboration that provides the ability to do audio/video/text chat, whiteboard, file transfer, and application sharing, profile messaging (faculty, staff, student, or a combination) to receive department, course, or club messages. The single sign-on feature, although difficult to implement in most cases due to integration issues with legacy systems, provides a convenient access to course materials along with the ability to set personal bookmarks and browse Web resources. EduPortals can also integrate with course management systems described below.

COURSE MANAGEMENT SYSTEMS

Several methods exist for developing and presenting Web courses. Although the basic framework of Web documents is built on HTML tags, today it is not necessary to know programming/markup languages such as HTML and PERL to develop course materials for the Web. This is due to development of software applications called Web course tools or course management systems that have a "shell" in which documents such as syllabi, schedules, instructor information, lecture materials, case studies, and so forth can be uploaded by faculty for distribution to students and interactions within the online course environment. Course management systems also allow interactivity that engages students with not only instructor developed course materials, but also provides access to other students, professionals, multimedia elements, and programs for interaction (Hazari, 1998). Developing courses using integrated features of course management systems offers advantages of a single authentication scheme, directory structure, consistent interface, and a simple way to publish and update content. HTML layout editors that were initially used to create course content produced static material which was rigid and nonconforming to different learning environments.

The new generation of Web course development tools provides features that allow instructors to adapt course components according to learning outcomes of the course. Use of such tools can promote collaborative learning, enhance critical thinking skills, and give every student an equal opportunity to participate in classroom discussions. Use of Web course development tools can piggy-back on huge investments higher education institutions have made in not only installing the hardware and software but also planning the network infrastructure to link offices, libraries, classrooms, and student dormitories for local, wide area, and Internet connectivity. Course management systems have been considered the academic equivalent of ERP Systems (Morgan, 2003). With proper implementation and sound pedagogical design, Web-based instruction can create meaningful enterprise learning environments by empowering faculty to engage students in active application of knowledge, concepts, and provide an opportunity to control pace and monitor learning which will help them grow and evolve as the course progresses.

VIDEOCONFERENCING

Course delivery and interaction can be offered in a rich interactive environment that goes beyond a purely text-based approach of previous generation distance education tools. One of the tools that have been increasingly popular in industry as well as in higher education is the use of videoconferencing for distance learning. Leveraging the power of Internet, business schools are looking to expand their programs nationally and internationally. A powerful technology tool that can be an enabler for expanding programs is videoconference technology. Videoconferencing allows synchronous (live) two-way communication using video and sound. A point-to-point (two-person) video conferencing system works like a videophone, where each party has a video camera, microphone, and speakers attached to their computers. Multipoint videoconferencing allows three or more participants to be present in a virtual conference room and communicate as if they were sitting next to each other. From a research viewpoint, use of videoconferencing has been studied in education and industry in relation to privacy, communication media choice, and systems analysis and design (Webster, 1998).

In general, distance education integrates various communication technologies to bring together students, faculty, and guest speakers by using communication that is mediated by technology. Multiple sites can be connected using videoconferencing. This offers an opportunity for students and faculty to interact in real time with participants at different sites by using audio and video data. Ancillary material such as videotapes, whiteboards, and slides can be shared over videoconferencing links. The use of videoconferencing follows defined standards for video compression and audio coding to allow systems from different vendors to communicate with each other using global standards. In business and industry, most conference rooms are ISDN and IP videoconferencing ready, and the sessions can be recorded to provide an archive for future use. Examples of educational videoconferencing session are the possibility of senior administrators in different institutions (national and international) collaborating to discuss joint projects and broadcast of classroom lectures and discussion to participants in the industry who in real time are able to interact the instructor and students. Within a community of practice, guest speakers from the industry can also be invited to discuss current projects and management concepts, thereby providing an interactive environment for students to engage in a dialogue with industry personnel.

FUTURE TRENDS: RECOMMENDATIONS FOR EFFECTIVE PEDAGOGY

A mixed-model approach that combines traditional teaching with use of technology tools can offer a faculty-moderated active learning environment and prepare business management students to seamlessly integrate information technology in their work environment. To achieve this, faculty must be trained to teach using technology. Communities of practice that discuss use of technology in teaching using any of the three tools mentioned above would find applications in improving teaching and learning among faculty and students. Traditional approaches to teaching must be reengineered to repurpose existing courses and related pedagogies. Effective diffusion of technology into practice of teaching is a critical requirement for management education. Team teaching within the school or by establishing external partnerships with global universities can provide value-added instruction to students that go beyond the constraints of a local geographic area. Forming communities of practice can also provide students and faculty a vision that encourages collaboration, experimentation, and broader learning. Another strategy would be to select faculty and give incentives to demonstrate best practices in each area, so the value of using the existing state-of-the-art infrastructure can be demonstrated to other faculty. With the right strategy that emphasizes technology as an integral

part of teaching, learning, and research, as well as partnerships with industry, faculty can be recognized as leaders in management education for having successfully addressed and integrated issues that pertain to delivery of education in the digital economy.

CONCLUSION

The Internet has drastically changed the way in which students and professionals interact in a dynamic environment. Communication and collaboration tools have evolved to support communities of practice that share a common goal. Use of a multimedia environment such as text, video, audio, as well as applications such as teleconferencing make it possible to disseminate and discuss information without the constraints of time and space boundaries. A thriving community of practice that uses tools such as portals, course management systems, and videoconferencing adds value to shared discussions by informing participants of diverse views and offering a common platform for exchange of ideas.

REFERENCES

Alavi, M. (1994). Computer-mediated collaborative learning: An empirical evaluation. *MIS Quarterly, 18*(2), 159-174.

Alavi, M., Wheeler, B. C., & Valacich, J. S. (1995). Using IT to reengineer business education: An exploratory investigation of collaborative telelearning. *MIS Quarterly, 19*(3), 293-312.

Brown, J. S., & Duguid, P. (2000). *The social life of information*. Boston: Harvard Business School Press.

Davydov, M. (2001). *Corporate portals and e-business integration*. New York: McGraw-Hill.

Frost, P. J., & Fukami, C. V. (1997). Teaching effectiveness in the organizational sciences. *Academy of Management Review, 40*(6), 1271-1281.

Garrison, D., Hawes, D., & Kanuka, H. (2003). Knowlege management and e-learning. Retrieved June 10, 2005, from *http://commons.ucalgary.ca*

Hazari, S. I. (1998). Evaluation and selection of Web course management tools. Retrieved June 10, 2005, from *http://www.sunilhazari.com/education/webct*

Hildebrand, J. E. (1995). Videoconferencing in the business curriculum. *Journal of Business and Technical Communication, 9,* 228-240.

Lake, M. (1998). Training teachers for distance education programs: Using authentic and meaningful contexts. *International Journal of Educational Telecommunications, 4*(2), 147-170.

Leidner, D. E., & Jarvenpaa, S. L. (1993). The information age confronts education: Case studies on electronic classrooms. *Information Systems Research, 4*(1), 25-54.

Morgan, G. (2003). *Faculty use of course management systems* (vol. 2). Boulder, CO: Educause Center for Applied Research.

NLII Virtual Communities. (2003). Retrieved June 10, 2005, from *http://www.educause.edu/576*

Webster, J. (1998). Desktop videoconferencing: Experiences of complete users, wary users, and non-users. *MIS Quarterly, 22*(1), 257-286.

Webster, J., & Hackley, P. (1997). Teaching effectiveness in technology-mediated distance learning. *The Academy of Management Journal, 40*(6), 1282-1309.

Yip, W. (2004). Achieving positive learning outcomes through problem-based learning, Web-based support and action project learning. *Proceedings of Second International Conference on Information Technology, China.* Retrieved June 10, 2005, from *http://charybdis.mit.csu.edu.au/icita/2004/papers/38-6.pdf*

KEY TERMS

Course Management Systems: Integrated course environments (e.g., WebCT, Blackboard) that include components such as e-mail, discussion group, chat, grade book for delivery, and management of instruction.

Distance Education: Instructional delivery to students at remote sites. Students communicate in this environment using electronic mail and discussion forums.

E-Learning: Synonymous with Web-based learning, it is instruction delivered over the Internet/intranet using a Web browser.

EduPortal: A portal geared toward education that provides single sign-on access to academic and administrative resources for students, staff, and faculty.

PDA: Personal Digital Assistant. A handheld device that combines computing, telephone/fax, Internet, and networking features. A typical PDA can function as a cellular phone, fax sender, Web browser, and personal organizer.

Portals: A Web site or service that offers a broad array of resources and services such as e-mail, forums, search engines, and personalized information.

Videoconferencing: A conference between two or more participants at different sites by using computer networks to transmit audio and video data.

Web Services: Integration of Web-based applications using common standards over an Internet protocol backbone.

U

Virtual Role–Playing Communities, "Wold" and World

Tunç Medeni
Japanese Advanced Institute of Science and Technology, Japan

Tolga İ. Medeni
Ankara, Turkey

INTRODUCTION

The increasing popularity of online role-playing games, and the virtual communities they create, are attracting much attention from business and academics. These virtual communities and environments provide invaluable opportunities for researchers to investigate various social and psychological aspects and issues. In this short article, we would like to share our views on some of these issues and opportunities in establishing a community identity with various virtual and real aspects, and transferring knowledge between these two aspects.

BACKGROUND

A fantasy role-playing game is one which allows a number of players to assume the roles of imaginary characters and operate with some degree of freedom in an imaginary environment. (Lortz, 1979, p. 36). Gaming fantasy combines the expressive freedom of fantasy with the structure characteristics of games. Fantasy role-playing gamers, the party of players and game masters, create their own cultural systems, generating identities and meanings in complex social worlds (Fine, 2002).

Online role-playing games have developed entire virtual worlds and communities with a sense of purpose, a shared history, and complex social interactions (Powasek, 2001; Murray, 2004). While the primary purpose of these (online) gaming communities is to have fun, nestled within them are important social and psychological phenomena such as identity construction, storytelling, learning, leadership, cooperation, and competition, which can/should be seen from the viewpoint of communities of practice. Moreover, phenomena emerging in these online communities can also be investigated almost entirely by unconventional online research methods. For example, our work on the Woldian games includes participation as an online member and the use of entirely information and communication technology (ICT) tools to research the emergence of interesting phenomena in the community.

The fantasy world of Wold, which began as a homemade local campaign in 1985, now has become an online community with almost 100 active members, who asynchronously interact with each other by posting on various boards for gaming and chatting within the community Web site, www.woldiangames.com. By paying the utmost attention to maintaining its free and volunteer nature, blending veteran players with new recruits, the Woldian world has achieved a lot as a community. When the life companion of one member passed away, an hour of silence was conducted on the players' chatting board to show their grief. The Woldian world also has made good use of its online environment by archiving all the games and chats, and providing virtual facilities for learning, research, and development.

Online fantasy role-playing communities like Woldian games present interestingly complex cases in the sense that they are build upon three different dimensions: (1) the real world, (2) a fantasy world, and (3) an online, virtual world, which blend with each other in an interesting manner. Even the term "virtual" means the mental/fantasy and the online/computerized aspect together. While both the fantasy and virtual world would share the common denominator of non-reality, the online dimension provides the environment that makes this mixing of reality and non-reality possible in an unprecedented way. The various types of knowledge transfer in the form of identity and experience that occur between the different real, virtual, and mental dimensions are highlighted here.

As a representational format, ICT operates as a sensual masking. The character chosen by a role-player can be seen as a mask acting as a metaphor for the person. Just as the transfer of knowledge from IT to production (the CAD and 3D modeling systems) materially affects our architectural world, the roles played in virtual worlds can have an important practical impact on our personal identities and personalities. Besides, the two terms—object and metaphor—inform each other. Having been involved in fantasy role-playing, it becomes difficult to think of fantasy creatures without seeing them as a particular form of computer user (Wiszniewski & Coyne, 2002). In fact, the self-images that people create in virtual communities show that there is a reflection of the person in the fantasy character that is created and played. These reflections may be physical, but also may be aspects of a person's personality. Some players almost self-consciously construct a persona that is completely opposite to the one they project in real life. The virtual environment provides a filter and can be used as a way to express a different side of personalities, escape the social constraints of real life, or experiment and find out what kind of person one wants to be in real life (Twist, 2004). Furthermore, in the fantasy role-playing games, personal rivalries can also be masked as role rivalries, for example, when determining who will be leading others (Fine 2002). In Wold, in one case, players threatened to leave the game, having developed hard feelings about the leader's role in the game.

FUTURE TRENDS AND CONCLUSION

As Puwasek (2001) and Twist (2004) point out, tens of millions of people worldwide interact in online games, and that number is growing. Role-playing games have developed entire virtual worlds and communities with complex social elements and interactions aside from the aspect of gaming. This can become so complex that sometimes the boundary between the virtual fantasy and real worlds becomes indistinct. The more technological or administrative control players have over their fantasy characters, the more likely the character is some sort of reflection of the real person, a creation of a new reality in a mirror world.

We would like to conclude our article with these suggestions for further analysis and research into online fantasy role-playing communities:

- Various studies have already been conducted to understand these virtual fantasy realities. Much of the interest here has been given to real-time games. However, asynchronous communication methods, such as the one in the Woldian campaigns and chats, also deserve special attention, since this method is specifically regarded as being open to discussion promoting the development of understanding, even through disagreements (Joinson, 2003), and better enabling to players to manage their own time, in general.
- The Japanese concept "ba" that addresses the (1) real, (2) mental, and (3) virtual (ICT using) contexts for knowledge creation (Von Krogh, Ichijo & Nonaka, 2000) can match well with the fantasy, online, and real aspects of the online fantasy role-playing communities, making them good examples of ba that exemplify its philosophical underpinnings.

Finally, studies to make sense of these special communities of practice are important, not only because they increasingly are becoming part of our lives, but also because analogies can be drawn for our default lives from the results of these studies. What could make the difference for our lives can be not only the similarities, but also the differences between these two modes of living.

REFERENCES

Anderson, T., & Kanuka, H. (2003). *E-research*. Boston: Pearson Education.

Fine, G.A. (2002). Shared fantasy: Role-playing games as social worlds. Chicago: University of Chicago Press.

Joinson, N.A. (2003). *Understanding the psychology of Internet behavior: Virtual worlds, real lives.* New York: Palgrave Macmillan.

Lortz, S.L. (1979). Role-playing. *Different Worlds, 1,* 36-41.

Murray, J. (2004). *From game-story to cyberdrama.* Retrieved February 2, 2005, from *http://www. electronicbookreview.com/v3/servlet/ebr?essay_ id=murray &command=view_essay*

Powazek, D.M. (2001). *Design for community: The art of connecting real people in virtual places.* Indianapolis: New Riders Press.

Squire, K., Scheckler, R., & Barab, S. (n.d.). Retrieved February 2, 2005, from *http://inkido.indiana. edu/onlinecom/*

Turkle, S. (1997). *Life on the screen: Identity in the age of the Internet.* New York: Simon & Schuster.

Twist, J. (2004). Retrieved February 2, 2005, from *http://news.bbc.co.uk/1/hi/technology/3683260. stm*

Von Krogh, G., Ichijo, K., & Nonaka, I. (2000). *Enabling knowledge creation: How to unlock the mystery of tacit knowledge and release the power of innovation.* Oxford, UK: Oxford University Press.

Wenger, E. (1998) *Communities of practices: Learning, meaning, and identity.* Cambridge, UK: Cambridge University Press.

Wenger, E., & Synder, W.M. (2000). Communities of practice: The organization frontier. *Harvard Business Review,* (January-February), 139-145.

Wiszniewski, D., & Coyne, R. (2002). Mask and identity: The hermeneustics of self-construction in the information age. In K.A. Renninger (Ed.), *Building virtual communities: Learning and change in cyberspace (Learning in doing: Social, cognitive & computational perspectives).* Cambridge, UK: Cambridge University Press.

KEY TERMS

(Fantasy) Role-Playing (FRP): Like being and playing in an improvisational drama or free-form theatre, in which the participants (actors) adopt imaginary characters, or parts, that have personalities, motivations, and backgrounds different from their own.

(Fantasy) Role-Playing Game (RPG): A type of game which allows players to role-play imaginary characters in an imaginary setting. Usually, role-players engage in cooperatively creating a story, each restricting themselves to the character they themselves introduced to the story.

Online (Fantasy) Role-Playing Game: A role-playing activity in which, rather than meeting in real life and using items like pens, pencils, dice, and so forth, players can benefit from ICT tools and interact with each other by using either asynchronous methods like posting on boards or sending e-mails, or synchronous ones that are evolving from text-based non-graphical into graphical forms.

Virtual Teaming

Brenda Elshaw
IBM, UK

INTRODUCTION

Given the ongoing advances in technology and the consequent changes in the work environment with the introduction of a mobile workforce, it is inevitable that more activities will be undertaken by virtual teams.

As with any team, a virtual team is a group of people who share a common objective and combine to provide a variety of different and complementary skills in order to achieve that objective. Unlike traditional teams, they are not collocated and can be working from a variety of different geographical locations, which can be either office or home-based. Often, they work across different time zones, adding to the challenges involved in successfully bringing a team together.

BACKGROUND: THE STAGES OF TEAM DEVELOPMENT

Tuckman (1965) identified four stages of team development which detailed a process that all teams need to work through:

1. **Forming:** When teams first come together, knowing little about each other and the project.
2. **Storming:** When roles and responsibilities are being clarified, and team members are striving to establish their position within the team.
3. **Norming:** When trust is established between team members and communication becomes more open and honest.
4. **Performing:** When the team members undertake the tasks at hand.

Virtual teams go through the same stages, but this needs to be facilitated by a strong and competent team leader. Steps 1 and 2 are essential for trust to be established between the team members; therefore, team building sessions need to be scheduled to allow this to happen. Time has to be built into the schedule to enable the members to get to know each other and their respective roles within the team.

BUILDING A VIRTUAL TEAM

Bringing together a virtual team requires strong management, communication, and facilitation skills. Often, the team leader is the one common link in bringing the team together and ensuring that team objectives are successfully met.

Most traditional teams come together initially via a face-to-face kick-off meeting, where they get to know each other and agree on their "team charter". For the best results, this should still be done for all virtual teams as people tend to build trust faster once they have met their teammates. It is far more comfortable to meet virtually when you already know the person involved compared to dealing with just a faceless voice at the other end of a telephone call. This forming stage often continues beyond the initial formal meeting with the session extending to a team dinner or sociable drinks.

However, it is not always possible to bring the team together for an initial meeting, and this has to be carried out via other means. Where this is conducted remotely, for example, by means of a conference call, video conference, or e-meeting, the team leader should ensure that each team member is involved and gets an opportunity to introduce him/herself.

Team members should be encouraged to provide profile information in advance of the meeting, covering not only their contact details and skills, but also some personal facts and interests. Contact details should also include a photograph – putting a face to a voice often helps get over the problem of speaking to someone who you have never met.

It is recommended that, even if the team members only meet virtually, the team leader makes every effort to meet the team members face-to-face when first enrolling them into the team.

LEADING A VIRTUAL TEAM

When participating in a virtual team, it is essential that the team members get the opportunity to communicate and share ideas on a regular basis. Not only does the team leader need to be in regular contact with each team member to understand how they are performing, but the team members need to be proactively in regular contact with each other.

One advantage that team members have when they work in the same location is the opportunity of meeting informally over lunch or coffee, encouraging them to share information, report on progress and discuss problems, issues, and so forth. As this is not an option for a virtual team, communication has to be more formal and regularly scheduled. Processes need to be put in place to define the different communication methods and how and when to use them.

During formal team meetings, the team leader should ensure that each participant in the call reports on their progress. With large teams, this may mean splitting the team down into smaller, more manageable subgroups. A regular meeting schedule involving everyone at some level ensures that team members do not become isolated. This is a good forum at which to identify and share "quick wins" with team members to keep the momentum going. In addition to the formal meetings, the team leader should ensure that he/she has regular contact with individuals and subgroups outside the meeting schedule.

TOOLS AND TECHNOLOGY

A wide range of tools and technologies exist to support virtual teams. The level of functionality and sophistication of the tool set available to individual team members, irrespective of their location, is increasing rapidly, allowing them to easily and transparently communicate with each other. However, this puts an increasing reliance on the availability of a supporting infrastructure, from both an IT user support and training perspective.

Some of the tools available are:

- **Broadband:** The increasing coverage of broadband services allows mobile workers to have a permanent connection to office systems, irre-

spective of their location. This is particularly useful for home-based team members.

- **Internet/Intranet:** The increasing use of the Internet and provision of in-house intranets allows dispersed users to have access to common systems and information sources.

- **Company Portals:** The growth of role-based, personalized portals enables team members to log into their systems and obtain a common desktop from a variety of different locations. It also ensures that new team members can be set up with a common team work environment quickly and easily.

- **E-mail:** Team members can easily communicate via e-mail. Documents can also be passed between team members using this mechanism, but this could lead to problems with duplicate copies at different versions. It can also discourage sharing by allowing team members to store exclusive copies of documents in mail files. For that reason, the use of file servers or Web places for team documents should be encouraged.

- **Videoconferencing:** The use of videoconferencing allows team members to meet face-to-face from a distance. This works best when team members are based in office locations that have easy access to videoconferencing equipment, although it usually places restrictions on the number of participants and requires advance scheduling to ensure that the facility is simultaneously available from all locations.

- **Instant Messaging:** The use of instant messaging allows a dispersed team to remain in contact with each other. It is an excellent resource for quick, ad-hoc interchanges between individual team members and also for synchronous team discussions. It can reduce the feeling of isolation felt by some team members and can assist in making them feel part of a team. It can also help them identify when other team members are online, particularly when working across time zones.

- **Web Place:** A project Web place can provide the team with a shared workspace structured for their specific needs. It contains structured spaces for them to store content, discussion areas for them to communicate irrespective of location or time zone, and can also provide project-based facilities such as team diaries. It reduces the need for large documents to be passed around via e-mail and ensures that documents are easily accessible by team members. Access can be controlled by the team leader to ensure that content is accessible only by those team members requiring it.

- **E-meetings:** The availability of Web-based e-meeting tools allows teams to have scheduled and ad hoc meetings via nothing more sophisticated than a Web browser and a telephone. Most systems have the ability to white board ideas and share documents, allowing control to be passed between team members when required. This means that team presentations, demonstrations, and so forth can be seen by the team, and documents can be worked on collaborativel—regardless of location.

WORKING WITH THE RIGHT TOOL SET

The tools required to support the team in their activities should be selected once the requirements of the team have been identified. At no time should the team members feel that the technology is driving what they do. An effective technology is one that supports the team activities seamlessly and transparently. There should never be any doubt as to which tool should be used when.

As the team develops their team charter, they should define what the communication and storage requirements are to support them at each stage. Part of the charter should detail the rules of etiquette regarding usage of the systems. For example, items to be agreed could include:

- when to use e-mail as opposed to Web places;
- when to be available via instant messaging;
- how often to go into systems to check for new information;
- how long team members have to respond to questions; and so forth.

ADVANTAGES OF VIRTUAL TEAMS

- Virtual teams allow the team leader to bring together the best people for the job, regardless of their location and organization.
- Time spent by team members is used more effectively as less time is lost traveling.
- Travel and subsistence costs are reduced.
- Team members can work to their own schedules, allowing for people in different time zones to work together for a common cause. This can also allow for teams to be represented 24 hours a day.
- Team members get to know and work with colleagues outside of their immediate work groups. This can increase their personal networks and encourage the flow of knowledge and experience across boundaries.

EFFECTIVE TEAMING

As Opper and Fersko-Weiss (1992) stated of team behavior, "For the…team, the shift is from a mentality of working alone to one of being in a state of collaboration." Deborah Harrington-Mackin (1994) identified one of the attributes of a non-effective team—when the team has "focused on task activities to the exclusion of work on team member relationships."

One of the great strengths of a team is the ability of its members to work together and build on each other's ideas. In many respects, the techniques and tools required to achieve this for a virtual team are different to those of a more traditional, collocated team and can certainly prove more challenging. It is easy to promote an environment where each team member works in isolation rather than as part of a team. A skillful team leader is required to encourage an environment where team members build the relationships with each other to give them the impetus to collaborate on activities. This may mean that time has to be allowed for some team communications focused solely on building team trust rather than producing deliverables.

CONCLUSION AND FUTURE TRENDS

Virtual teams are becoming the norm in many organizations. While some members struggle with the concept of being part of a dispersed team, many find that the advantages outweigh the difficulties. This is particularly true of the younger generation of employees who have grown up with and are more accepting of Web technologies.

For the organization, the ability to boost productivity by making best use of their expertise, wherever it resides, is of great benefit. For the employee, virtual teams open up opportunities to work within different teams, enabling them to increase skills and expertise by contact with a wider community. In turn, the organization gains a more skilled and satisfied workforce.

Virtual teams require a set of tools, selected and configured to meet their needs. The tool set used should be supportive rather than intrusive, responding to the varying needs of the team.

They also need to work within a framework of clearly defined processes and communication channels, facilitated by a competent and experienced team leader. When successfully achieved, a virtual team can bring all the benefits usually associated with team working, plus added benefits that only a virtual team can bring.

FURTHER READING

Buckman, R. H. (2004). *Building a knowledge driven organization.* McGraw-Hill.

Charles Steinfield Computers. (2002, March). *Realizing the benefits of virtual teams, 35.*

Fisher, K., & Fisher, M. D. (2001). *The distance manager: A hands-on guide to managing off-site employees and virtual teams.* New York: McGraw-Hill.

Godar, S. H., & Ferris, S. P. (Eds.). (2004). *Virtual and collaborative teams: Process, technologies and practice.* Hershey, PA: Idea Group Publishing.

Lipnack, J., & Stamps, J. (1997). *Virtual teams: Reaching across space, time and organizations with technology.* New York: John Wiley & Sons.

Nash, S. (1999). *Turning team performance inside out: Team types and temperaments for high impact results.* Palo Alto, CA: Davies-Black.

Pauleen, D. (2004). *Virtual teams: Projects, protocols & processes.* Hershey, PA: Idea Group Publishing.

REFERENCES

Harrington-Mackin, D. (1994). *The team building toolkit: Tips, tactics and rules for effective workplace teams.* AMACOM.

Opper, S., & Fersko-Weiss, H. (1992). *Technology for teams: Enhancing productivity in networked organizations.* New York: Van Nostrand Reinhold.

KEY TERMS

E-Meeting: An electronically-facilitated meeting allowing participants to share and work on documents remotely.

Instant Messaging: A computer application that allows two or more users to communicate with each other in real time via typed messages.

Portal: A computer desktop environment that provides organized aggregated access to the applications, systems and websites used by a community member, based on their role within the community.

Team Charter: A document created when a project team is formed that details the team's objectives and rules of engagement including items such as the mission statement, team roles and responsibilities and deliverables to be produced.

Virtual Team: A group of people brought together from different locations who do not meet face to face but work as a team.

Virtual Teams and Communities of Practice

M. Gordon Hunter
The University of Lethbridge, Canada

INTRODUCTION

Interdisciplinary research is being supported by universities and funding agencies, which in turn require a collaborative approach by researchers with complimentary yet different sets of expertise. Communities of practice are also facilitated by a collaborative approach, with groups of researchers investigating an area of common interest.

It is important to note that collaborative research is not an extension of the single researcher approach. Goode (1973) originally suggested that collaborative research may be depicted as a delicate balance of collegiality and bureaucracy. Bradley (1982) supported this idea and further suggested that to increase the probability of group success, it is important to reach, as early as possible, a mutually acceptable and explicit agreement about group members' responsibilities.

BACKGROUND

McGrath, Arrow, and Berdahl (2000) present a very thorough review of research about groups. Their Theory of Groups suggests "that groups are complex, adaptive, and dynamic systems" (p. 97). Their definition relates to systems theory, and in this light they consider groups to be open, complex systems interacting with other systems through fuzzy dynamic boundaries.

An important factor that contributes to the complexity of groups is location of individual team members. That is, if team members are geographically dispersed, the group dynamics (interaction for instance) will be affected. Barczak and McDonough (2003) suggest that the challenge for leaders of geographically dispersed teams is to integrate and coordinate team members. The importance of communication becomes important. Kayworth and Leidner (2002) also determined that leadership becomes even more important when physical separa-

tion is introduced into group projects. They found that effective leaders were able to display empathy and assert authority in dealing with team members. Also, as above, they were good communicators able to define the roles of team members and provide useful feedback on performance.

Another important factor in collaborative research relates to the background of the individual researchers. Gelfand, Meyers, and Ross (2002) determined that indigenous researchers were able to approach the investigations with more of an understanding of the culture. Korabik, Lero, and Ayman (2003) addressed the issues of *emic* and *etic* (Pike, 1954; Berry, 1990; Headland, Pike & Harris, 1990) approaches in a large-scale international study. An emic approach suggests a framework developed from within a culture and based upon criteria from that culture. An etic approach develops a universal framework by assessing and comparing universal criteria. Triandis (1972) has suggested an extension of these concepts with the term "pseudo-etic" which employs criteria from a limited number of cultures to develop a universal framework. The most effective research approach will be one that incorporates both emic and etic elements. As Early and Mosakowski (1995) suggest, "the most useful approach...is to focus on the pseudo-etic approach to develop quasi-universal constructs which may be subsequently challenged to more universal tests of validity" (Earley & Mosakowski, 1995, p. 9). This may be accomplished by starting from an emic base and then conducting emic studies in other cultures from which it might be possible to evolve an etic model of universal constructs based on the similarities and differences which emerge from the emic data.

CONCLUSION

The above factors represent important considerations for communities of practice involved in research situations requiring the establishment of virtual teams.

REFERENCES

Barczak, G., & McDonough, E.F. III. (2003). Leading global product development teams. *Research Technology Management, 46*(6), 14-18.

Berry, J.W. (1990). Imposed-etics, emics and derived-etics: Their conceptual and operational status in cross-cultural psychology. In T.N. Headland, K. L. Pike &M. Harris (Eds.), *Emics and etics: The insider/outsider debate.* Newbury Park, CA: Sage Publications.

Bradley, R.T. (1982). Ethical problems in team research: A structural analysis and an agenda for resolution. *The American Sociologist, 17*(May), 87-94.

Earley, P.C., & Mosakowski, E. (1995). A framework for understanding experimental research in an international and intercultural context. In B.J. Punnett & O. Shenkar (Eds.), *Handbook of international management research.* Malden, MA: Blackwell.

Gelfand, M. J., Meyers, M.K., & Ross, K.E. (1997). Methodological issues in cross-cultural organizational research. In S. Rogelberg (Ed.), *Handbook of research methods in industrial/organizational psychology.* Malden, MA: Blackwell.

Goode, W.J. (1973). The theoretical limits of professionalisation. In W.J. Goode (Ed.), *Explorations in social theory* (pp. 341-382). London: Oxford University Press.

Headland, T.N., Pike, K.L., & Harris, M. (Eds.). (1990). *Emics and etics: The insider/outsider debate.* Newbury Park, CA: Sage Publications.

Kayworth, T.R., & Leidner, D.E. (2002). Leadership effectiveness in global virtual teams. *Journal of Management Information Systems, 18*(3), 7-40.

Korabik, K., Lero, D.S., & Ayman, R. (2003). A multi-level approach to cross cultural work-family research: A micro and macro perspective. *3*(3), 289-303.

McGrath, J.E., Arrow, H., & Berdahl, J.L. (2000). The study of groups: Past, present, and future. *Personality and Social Psychology Review, 4*(1), 95-105.

Pike, R. (1954). Language in relation to a united theory of the structure of human behavior. *Proceedings of the Summer Institute of Linguistics*, Glendale AZ.

Triandis, H.C. (1972). *Analysis of subjective culture.* New York: Wiley Interscience.

KEY TERMS

Emic: A framework developed from constructs identified from within a specific culture.

Etic: A universal framework developed from universal constructs.

What Organisational Development Theory Can Contribute to Our Understanding of Communities of Practice

W

Jim Grieves
University of Hull, UK

INTRODUCTION AND BACKGROUND

Organisational development (OD) is an approach to developing organisations through the application of behavioural science knowledge, practices, and processes. Essentially, OD enables organisations to achieve effectiveness through careful analysis and diagnostic techniques as well as through carefully considered intervention strategies. Although some of its earlier planned change practice was adopted by approaches to quality management and business excellence in the late 1980s, much of this adaptation is generally regarded as overly mechanistic and formulaic. Indeed, as social science disciplines developed, corresponding changes occurred to OD methodology. In this regard, while OD can be regarded as an attempt to improve the total organisational system, it has moved beyond its earlier functionalist and behaviouristic assumptions to embrace critiques of the planned change process. OD should therefore be regarded today as "an evolving mixture of science and art" (Cummings & Huse, 1989, p.1) that integrates strategy, structure, and process in the pursuit of organisational change.

As organisational development matured over the past 20 years, it came to focus increasingly on organisational learning. Its main contribution to organisational learning is recognition that the quality of the diagnosis, interpretive judgments, and the sensitivity of the change agent to the nature of the intervention is much more important than the mechanistic application of planned change programmes. In order to explain this further, it would be useful first to say a few words about communities of practice and then, second, to illustrate some issues linking organisational development to the process of organisational learning.

Communities of practice can be described as informal groups or networks of people who share similar interests and objectives. The identification and development of tacit and formal knowledge is therefore the central activity of a community of practice. This informality of practice is generated by a group of people who are motivated to acquire and share knowledge in relation to an agreed objective. Once this has been applied to organisations, social networks (and I include virtual networks in this definition), geographical and spatial communities, then we can begin to get a feeling for the types of interactions that are now likely to be generated. This, of course, has increased exponentially with the use of modern communication technology and the World Wide Web, in particular.

While shared experiences and insights into best practice are essential to the activities of a community of practice, it is the desire to share a similar problem focus that brings us close to the heart of organisational development. While we can agree with those authors who argue that a community of practice is a knowledge exchange mechanism through informal learning (Lave & Wenger, 1991; Wenger, 1998), this in itself would not make a CoP a satisfactory mechanism for an organisational development intervention because an effective diagnostic framework would require a methodological approach to the identification of an agreed problem. It is in this sense that we need to argue for the application of an OD methodology.

Others (e.g., Boud & Middleton, 2003) have argued that, since communities of practice depend on learning, the outcome of any shared activity must require further skills which include mastery of organisational processes, negotiating the political, and dealing with the atypical (that is, having the flexibility to solve problems without resorting to mechanistic or formulaic approaches). Such arguments, of course, lend further weight to a more disciplined approach to the activities of a community of practice.

ORGANISATIONAL DEVELOPMENT AND THE IMPLICATIONS FOR COMMUNITIES OF PRACTICE

While the older definition of organisational development in the late 1960s and 1970s was primarily concerned with planned, organisation-wide change programmes that were managed from the top in order to increase organisational effectiveness and health, by the 1990s, it was possible to identify a different focus in relation to personal development and organisational learning and analyses informed by newer methodological approaches such as symbolic interactionism and discourse analysis. Thus, contemporary OD has developed a mature perspective for managing change involving stakeholders and collaborative action (Grieves, 2003).

A Methodology for Transformation

According to Wenger, McDermott and Snyder (2002), communities of practice, unlike organisational teams, do not necessarily require a tangible result to their activities. For example, communities of practice may focus on the clarification or the development of knowledge which is intangible and difficult to measure. This is not a totally convincing argument, however, if one is concerned about the dynamics of the CoP. This is because a CoP, like a team, needs to energise, motivate, and build cohesion.

The argument here is that communities of practice need to become more robust by adopting the main ingredients of a contemporary organisational development perspective. These are illustrated in Table 1.

In order to explain the main characteristics of an OD approach, it should be clear that the *methodology* is essentially that of action research which requires the systematic collection of data in relation to organisational problems. The *approach* adopted, although implicit in the CoP idea, needs to recognise the value and importance of the joint diagnostic relationship unfettered by status or functional position. In relation to this, the *interests* of CoP members are essentially pluralistic and should be driven by the members themselves rather than by the activities of a transformational leader. The word

Table 1. Characteristics of organisational development (Modified from Grieves, 2003)

Methodology	Action Research
Approach	Joint diagnostic involving stakeholders
Interests	Pluralist
Development	Personal and organizational learning
Culture	As analytical tool
Values	Promotes humanistic values
Mode of intervention	Process focused

development is also important since a CoP should be essentially concerned with the development of the organisation through its people. As a result, individual and organisational learning characterises the enterprise of a CoP. *Culture* is also central because, as a CoP develops the dynamics and strains of the network need to be made transparent and articulated at various stages of its development from birth to closure. Finally, *values* are critical, and, in this sense, I have to suggest that the adoption of humanistic values implicit in the OD approach become central to the very existence of a CoP.

Contemporary organisational development is of benefit to communities of practice because by adopting a more rigorous approach, communities of practice would provide a dynamic process of modifying group behaviour by defining the nature of the problem (even if this is simply the generation of knowledge) and by clarifying any proposed initiative and *intervention strategy*. Furthermore, in many cases, it will be necessary to identify the critical processes that either inhibit or progress some form of organisational transformation. It is also likely that, in many situations, individual behaviours may need to be made transparent by articulating the dynamics of the group or network because these will impact on organisational outcomes.

The adoption of process consultation requires the relationship with the client system to be clearly defined and articulated. Useful texts on this are provided by Schein (see, for example, 1987, 1988, 1997). A community of practice is required to engage organisational members in a diagnostic relationship which, by implication, means using appropriate methods for data collection. In addition, great care is required with the type of intervention. It is essential that the community of practice and the client system are seen as partners in the process where some type of transformation is likely to result from knowledge generation to organisation analysis and organisational redesign.

It is equally important to recognise that, once the client/consultancy relationship is established, a *modus operandi* may well include the need to address social dynamics. A good deal has been written on this over the years in relation to the systems/psychodynamic approach which emerged from the work of social scientists at the Tavistock Institute in the 1950s and 1960s. More recently, the work of Gabriel (1999) provides some interesting insights into individual and group dynamics. However, concern is expressed at this stage in relation to the level at which a community of practice should operate. Without such experience and expertise, caution must be exercised, as Harrison (1996) observed, in relation to the depth of the intervention carried out.

CONCLUSION AND FUTURE TRENDS: HOW ORGANISATIONAL DEVELOPMENT CAN ASSIST COMMUNITIES OF PRACTICE TO DEVELOP A STRATEGIC FOCUS

Communities of practice are the antithesis of bureaucracy since they depend on relationships which establish a common boundary and develop shared norms and standards. They have also been referred to as "revolutionary cells" or "change communities" (Ward, 2000). The approach indicated above (Table 1) will challenge bureaucratic barriers, but CoP members should nevertheless be mindful of the barriers they may face. These include:

1. differences in the extent to which members display tolerance to various pressures including the use of time;
2. differing levels of competence and cultural and value differences within the CoP; and
3. the operational pressures of working within inflexible, hierarchical structures.

According to Wenger, McDermott, and Snyder (2002), communities of practice can drive strategy, start new lines of business, solve problems quickly,

transfer best practices, develop professional skills, and help in recruiting and retaining talent. However, the argument proposed here is that these things are unlikely to happen without an awareness of the organisation's strategic configuration and decision-making practices. This refers, first, to the location of the organisation in relation to its position in the industry, marketplace, and competitors; second, to the structures and systems through which the key decisions are made and determine the degree of centralisation standardisation of products and services; and third, to the performance standards, methods, and control systems required to achieve the desired products or services.

One of the greatest dangers that communities of practice can face is an organisation which still engages in a form of strategic planning requiring the collection of large quantities of data to be regularly collected, processed, and cascaded downward before action can be taken. In other words, "the strategic level of the organisation cannot act like an all-seeing central planner because threats will emerge which have to be dealt with incrementally in ways not originally foreseen by the strategy" (Boisot, 1995, p. 33). Because most organisations operate today in turbulent environments, communities of practice must find a solution to Ashby's (1956) "law of requisite variety" which states that for an organisation to survive and prosper, its rate of learning must at least match the rate of change in its environment. Communities of practice can therefore, if they are effectively equipped, provide a solution to this problem by:

1. establishing a clear focus for their activities;
2. motivate members and determine the course of action; and
3. ensure that members have the capabilities and competencies to progress effectively toward their final goal.

If operated successfully using ideas from contemporary organisational development, communities of practice can avoid what Argyris (1997) referred to as "defensive reasoning and the doom loop" (p. 201) through which people commit themselves in principle but not in practice.

REFERENCES

Argyris, C. (1997). Teaching smart people how to learn. In C. A. Carnal (Ed), *Strategic change* (pp. 201-215). Oxford: Butterworth-Heinemann.

Ashby, W. R. (1956). *An introduction to cybernetics.* London: Methuen.

Boisot, M. (1995). Preparing for turbulence. In B. Garratt (Ed.), *Developing strategic thought* (pp. 29-45). Maidenhead, UK: McGraw-Hill.

Boud, D., & Middleton, H. (2003). Learning from others at work: Communities of practice and informal learning. *The Journal of Workplace Learning, 15*(5), 194-202.

Cummings, T.G., & Huse, E.F. (1989). *Organisation development and change* (4th ed.). St. Paul, MN: West.

Cummings, T. G., & Worley, C. G. (1993). *Organisation development and change* (5th ed.). St. Paul, MN: West.

Gabriel, Y. (1999). *Organisations in depth.* London: Sage.

Grieves, J. (2003). *Strategic human resource development.* London: Sage.

Harrison, R. (1996). Choosing the depth of organisational intervention. *Collected Papers of Roger Harrison.* Maidenhead: McGraw-Hill.

Lave, J., & Wenger, E. (1991). *Situated learning: Legitimate peripheral participation.* Cambridge: Cambridge University Press.

Schein, E. H. (1987). *Process consultation (Vol. 1).* Reading, MA: Addison-Wesley.

Schein, E. H. (1988). *Process consultation (Vol. 2).* Reading, MA: Addison-Wesley.

Schein, E. H. (1997). Process consultation principles in respect of clients. *Journal of Organizational Change Management, 10*(3), 3-38.

Ward, A. (2000). Getting strategic value from constellations of communities. *Strategy & Leadership, 28*(2), 4-9.

Wenger, E. (1998). *Communities of practice: Learning, meaning and identity.* New York: Cambridge University Press.

Wenger, E., McDermott, R., & Snyder, W. M. (2002). *Cultivating communities of practice.* Boston: Harvard Business School Press.

KEY TERMS

Communities of Practice: Although it is common to refer to informal groups or networks of people who share similar interests and objectives, CoPs have been seen as an alternative to teamwork where a variety of problems may be better considered through knowledge shared by loose coalitions of people who develop their own tacit knowledge and methods for doing things. This is more common among certain professions such as lawyers, barristers, GPs, academics, and so forth whose conduct is regulated by professional associations and who share a similarity of attitudes and conventions.

Depth of the Intervention: For Andrew Harrison, intervention strategies range from deep to surface level. Deep interventions are those which act on emotional involvement. These require a high level of behavioural knowledge and skill as well as a sensitivity to the client's needs. Furthermore, there are clearly ethical issues which require the willing participation of a client.

Intervention Strategy: Refers to specific approaches to developing the organisation through its people. While obvious intervention strategies may include techniques such as team development, self-directed learning, training approaches to personal growth and empowerment, T-groups, force field analysis, and organisational learning, a more methodological approach is identified by Blake and Mouton. In their Strategies of Consultation, they identify five types of intervention (acceptant; catalytic; confrontation; prescriptive; principles models and theories) that may be applied at five levels (individual; team; inter-group; organisation; society). The important point about this schema is that it forces the consultant to think carefully about the purpose of the intervention.

Joint Diagnostic Relationship: As a result of the methods used, the approach of OD is informed by a joint diagnostic relationship between the OD consultant, or change agent, and various stakeholders in the organisation. This enables the problem to be understood from multiple perspectives.

Organisational Development: A professional approach to organisational action guided by careful diagnosis based on social scientific enquiry. The period between the 1960s and late 1970s was highly experimental and established the principles of OD for much of the twentieth century. By the end of the twentieth century, new approaches to organisational development had emerged as a result of critiques of functionalist methods and behaviourism, in particular, and also because of Morgan's book, Images of Organisation, which gave rise to multiple diagnoses. Increasingly, organisational learning became a focus for OD activities.

Process Consultation: The skills we can identify as characteristic of the ideal OD professional result from specific training in three obvious skills: process consultation; diagnostic ability; and research methods. Process consultation is a method that enables the OD practitioner to engage organisation members in diagnosis and by collecting and analysing information through a variety of methodological techniques, including the use of statistics, survey techniques, and force field analysis.

Social Networks: The decision to act on any form of knowledge involves an intervention. Thus, to implement any change through a CoP is a micro-political process that operates through internal and external social networks. This may be geographical and/or spatial communities.

Working and Learning in Interdisciplinary Project Communities

Patrick S.W. Fong
The Hong Kong Polytechnic University, Hong Kong

INTRODUCTION AND BACKGROUND: CREATING KNOWLEDGE IN INTERDISCIPLINARY PROJECT TEAM SITUATIONS

Designing a product or service does not form a complete and coherent body of knowledge that can be precisely documented or even articulated by a single individual. Rather, it is a form of knowing that exists only through the interaction among various collective actors (Gherardi & Nicolini, 2000). Existing literature (Kanter, 1988; Nonaka, 1994) has highlighted a need for the development of a diverse workforce if knowledge creation is to be promoted and sustained. This literature suggests that a diverse set of resources (experts with different backgrounds and abilities) provides a broad knowledge base at the individual level, offering greater potential for knowledge creation.

Sahlin-Andersson (1998) viewed projects as local arenas for knowledge creation, as individuals possessing different experience and skills work together to solve a common task within a limited timeframe. Through collaboration, new technical knowledge and knowledge for organizing the project are developed over time. March et al. (1991) argued that organizations learn from experience to improve future performance. By the same token, projects can be used as a medium for organizational learning, where knowledge and experience gained in one project can be transferred and utilized in the next. This strategy does not aim solely to save time and money, but also to avoid "reinventing the wheel", which is something that occurs frequently in every new project. Penrose (1959) argued that utilizing and employing experience and the knowledge thus created makes an organization grow.

Conceptually, a team can be viewed as a socially constructed phenomenon or linking mechanism that integrates individuals and organizations (Horvath et al., 1996). A multidisciplinary team is defined by Nonaka and Takeuchi (1995) as "a self-managed, self-organised team in which members from various functional departments, and/or areas of expertise, work together to accomplish a common goal" (p. 85). The primary goal of the multidisciplinary composition (see Figure 1) is to marry diverse bodies of knowledge in a way that forces out a synergistic knowledge outcome that is innovative, contextualized, difficult to imitate, and, as such, has strategic value. For the most part, project team tasks are nonrepetitive in nature and involve the application of considerable knowledge, judgment, and expertise.

The advantage of adopting multidisciplinary project teams is that they are quicker in integrating the expert knowledge of different functions, for example, design, construction, property management, marketing, and so forth. Cross-functional project teams with mutual accountability and collective work products have been found to decrease development time and increase product quality (Van de Ven, 1986; Wheel-

Figure 1. A multidisciplinary composition of team members with diverse knowledge, judgment, and expertise

wright & Clark, 1992). Multidisciplinary project teams create a "task culture", facilitating close linkages and direct personal contacts between different functions (Cohen & Levinthal, 1990). These close connections are necessary, as new product development by its very nature includes uncertainty about the potential market response and about new technology (Henke, Krachenberg & Lyons, 1993). The multidisciplinary project team can be viewed as an unusual team arrangement primarily because it is composed of professionals from various disciplines who take pride in their fields of expertise. They are committed to the basic assumptions of their paradigms, and they perceive their roles in the team as representing their knowledge bases in the best possible way.

KNOWLEDGE SHARING IN PROJECT TEAMS

To enhance competitiveness and meet organizational goals, organizations need to ensure that people share both tacit and explicit knowledge. The increased sharing of knowledge raises the likelihood of new knowledge being created, tending to support valuable innovation (Nonaka & Takeuchi, 1995). Though organizations can codify some of the knowledge people use, it is easy to find cases or examples that do not fit the codified knowledge of the organization. This unarticulated knowledge requires communication among people in the organization. Orr (1996) found that photocopier technicians often searched for solutions beyond their manuals. He explained that "the expertise vital to such contingent and extemporaneous practice cannot be easily codified" (p. 2). When documentation proves insufficient, people need to access each other's experience to solve more difficult problems. Orr showed how technicians sometimes use narrative to recount each other's experience. Technicians might use breakfast or lunch meetings to share knowledge. Other accounts of knowledge sharing demonstrate how workers use computer-mediated communication. For example, Constant, Sproull, and Kiesler (1996) showed how people use a computer-mediated network to seek help and advice. Similarly, Hargadon and Sutton (1997) explained how product designers search for knowledge

by sending out pleas for help via electronic mail. In both cases, communication is the key to sharing knowledge.

Knowledge sharing relies on reaching a shared understanding of the underlying knowledge, in terms of not just the content but also the context of the knowledge, or "Ba", to use Nonaka and Konno's (1998) term. Exchanging information represents only a partial view of knowledge sharing activity. The essence lies in unveiling and synthesizing paradigmatic differences through social interaction.

Many definitions of the word *paradigm* exist. Neufeldt and Guralnik (1988) defined it first as "a pattern, example, or model" and second as "an overall concept accepted by most people in an intellectual community...because of its effectiveness in explaining a complex process, idea, or set of data" (p. 979). Kuhn (1970, p. 181), who popularized the term, provided two definitions for a paradigm. In the primary sense of the word, a paradigm is a "disciplinary matrix", the ordered elements of which are held by the practitioners of a discipline. According to this definition, a paradigm includes symbolic generalizations (laws and definitions), shared beliefs, and shared values. In an alternate use, Kuhn (1970, p. 187) defined paradigms in a more circumscribed manner as "exemplars" or "shared examples". More recent work by Boland and Tenkasi (1995) indicated the use of the concept of "perspective taking" and "perspective making" to resolve paradigmatic differences through appreciating individuals' different paradigms. By synthesizing the various definitions and insights, a paradigm as used in this chapter, is defined as a team perspective or belief which is collectively constructed and accepted by members of the team. This definition reflects the perspective of social construction as well as the opportunity for paradigmatic differences to be resolved through social interaction between members in collective settings, such as teams or organizations.

Knowledge sharing is not constrained to exchanges among and across the employees of a company. It can occur between employees and customers, or between organizations or firms in entirely different industries (von Hippel, 1988). Some of the very important knowledge identified in a survey among knowledge-intensive businesses includes customer, competitor, and product knowledge (Skyrme & Amidon, 1997). The more knowledge is shared about the needs of current and potential customers among project team members, the better they may understand realistic customer requirements. With such knowledge, greater value for customers may be created because the resultant products may better satisfy customer needs and expectations. Accordingly, they may have a better chance of success in the marketplace. In the same vein, shared competitor knowledge could be helpful in developing products ahead of market requirements (getting products to market ahead of competitors, developing products on schedule). It could yield high value to customers (extending a product's success in the marketplace), possibly improving product performance (better overall product performance than that of competitors). In addition, shared product knowledge (product advantages, disadvantages, strengths, history, and technologies) may be important in improving development productivity (reducing development costs) and production costs (reducing overall production costs).

It is clear that sharing diverse knowledge can enhance problem solving as well as create the culture required for knowledge creation. Communication is the key to knowledge sharing. Knowledge sharing is regarded as a combination of processes sharing and using knowledge directly without language (socialization) and with language (externalization).

KNOWLEDGE INTEGRATION IN TEAM SITUATIONS

More information and knowledge are not always the answer. What may be needed is to better integrate the information and knowledge already available within the team. According to Weick (1995):

more information will not help them. What will help them is a setting where they can argue, using rich data pulled from a variety of media, to construct fresh frameworks of action-outcome linkages that include their multiple interpretations. The variety of data needed to pull off this difficult task is most available in variants of the face to face meeting. (p. 86)

In a new product context, Hayes, Wheelwright, and Clark (1988) suggested that members of new product development teams should have a basic knowledge of other functions in addition to an in-depth knowledge

of their own specialty. It is suggested that "specialists are inventors; generalists are innovators" (Galbraith, 1982, p. 22) and that people who are willing to cross functional or other boundaries are likely to be more innovative (Kirton, 1988), or to be able to resolve conflicts because of their ability to see both sides (Gregory, 1983). Nonaka and Takeuchi (1995) alluded to generalism when they talked about Japanese firms' support of "information redundancy" or knowledge overlap between people. Although the use of the term *redundancy* might seem to denote inefficiency, it might in fact turn out to be effective in an innovation situation. This positive effect of knowledge overlap may explain the positive association between the use of job rotation and new product success found in a number of product development studies (Souder, 1981; Wiebecke, Tschirky & Ulich, 1987). Wiebecke et al. (1987) proposed that job rotation promotes an understanding of the work of other functions and facilitated cross-functional "bilingualism". Souder (1981) found that in all the cases where "equal partners" harmony, associated with product success, was attained between marketing and research and development (R&D), the marketing personnel were all technically trained, most having worked in R&D previously. Cross-functional skills learned in job rotation may facilitate the combination of existing knowledge to produce new knowledge. Having considered the importance of integrating knowledge in team situations, the following section will focus on tensions in the process.

Following the above discussion, knowledge integration is defined as a collective process of synthesizing different knowledge and paradigms through the social interaction of team members/stakeholders in order to facilitate the construction of new knowledge or combine existing knowledge.

A project in which a multidisciplinary team is involved can be described as a transformation process, superimposed on the regular or cycled activities of an organization (Beale & Freeman, 1991). In this regard, a project becomes part of a wider venture (Beale & Freeman, 1991), the first part of which is the production of a product or service followed by an operating cycle. The project therefore takes place within a complex corporate, legal, financial, and regulatory environment (Fox, 1984). This environment leads to a number of parties having a stake in the project, from internal departments to external regulatory bodies and customers, since the project decisions have a potential impact on all stakeholders (Cleland, 1986).

As Grant (1996) indicated, competitive advantage does not evolve from knowledge, per se, but from the integration of such knowledge as facilitates the construction of new knowledge. The diversity of the specialized knowledge involved in the integration process determines its difficulty. Hence, the uniqueness of multidisciplinary teamwork is in its potential to integrate different bodies of knowledge into a new synergy. From an organizational standpoint, the prime purpose of the multidisciplinary team is to function as a knowledgeable entity engaged in creating new knowledge. In other words, the function of a project team is to convert knowledge inputs into new products and processes, bringing together participants with expertise in the right specialized knowledge domains and skills necessary to integrate and coordinate the knowledge of diverse participants. With these paradigms of divergent thinking, basic assumptions, and the "professional egos" (Dougherty, 1992) held by team members, this difference of opinion is likely to challenge invalid assumptions and bring more information and knowledge to bear on issues; it may also neutralize tendencies toward "groupthink" (Janis, 1982).

LEARNING AND PROJECT COMMUNITIES

The metaphor of projects as learning experiments for the company embraces an awareness of the importance of both exploration and exploitation of knowledge in organizations (Burgelman, 1991; March, 1991). To see an individual project as an experiment means that new knowledge is created and explored among project participants. The project knowledge and experience gained from earlier or current projects can be used to create new knowledge to suit current situations or problems. Projects, as a form of organizing work, can be one way to explore new knowledge, project-related as well as operational. During participation in a project, team members, through their engagement in the learning process, gain new experience and knowledge that could be used to solve

problems. Furthermore, this knowledge and experience could be useful for other projects. In that sense, a project can be viewed as a learning experiment for the companies involved (Drew & Smith, 1995).

In project-oriented companies, learning from projects is the key to building strategic competitive advantage. During a project's existence, a number of decisions are made. Every decision involves a degree of uncertainty. Packendorff (1995), for example, argues that the problems or mistakes that cause this uncertainty are often of a similar character. Yet it is not clear whether this is a global generalization or whether it depends on the sector or stage of an industry's life cycle. Nevertheless, experience to date has shown that once experience is gained in a project, knowledge is created that may be reapplicable. The basic hypothesis of the project learning approach is that learning from projects can reduce the uncertainties that might lead to inefficiencies. The use of project experiences and their integration into the organization to expand the body of knowledge are important and valuable cornerstones in a project learning approach. Ensuring that people pass on their experience to others is one of the greatest challenges for an organization and its organizational memory (Morris, 1994). However, learning and projects are not a natural combination (Bartezzaghi, Corso & Verganti, 1997) since conflicts of a basic logical character are involved. These conflicts comprise the time aspect, the task orientation, the team structure, and the transitional culture of projects (Lundin & Söderholm, 1995).

To carry out their project work effectively, project team members need to develop the ability to manage across boundaries. If learning is assumed to be social, learning is engagement in practice and dealing with boundaries (Wenger, 1998). Project-based organizations offer an excellent opportunity to engage in learning and to acquire reflective habits that transcend the boundaries of projects. Learning is supported not only by the nature of single projects, but also by the web of relationships that is created in project management organizations.

Membership in projects is temporary and thus offers individuals the opportunity to belong to multiple communities. In project-based organizations, there are a large number of weak ties that help diffuse knowledge and practices (Granovetter, 1973). In the majority of organizations, project members maintain their links with their primary organizations (to which they will return upon the completion of the project). Membership in multiple existing teams contributes to the creation of informal webs of people who act as knowledge brokers (Wenger, 1998). Project-based organizations thus enable the continuous building and cultivation of relationships, nurturing the development of communities of practice (Brown & Duguid, 1999). Communities of practice are natural internal mechanisms where ideas and practices spread in work settings, although they tend to exist outside the boundaries of the formal hierarchy (Wenger & Snyder, 2000). Project-based organizations may grow into constellations of interrelated communities of practice, offering a web of mutual support for cultivating reflective practices. When projects share members, they are bound together and become embedded in the same social network (Granovetter, 1973). The recursive interaction among projects creates social networks of mutual assistance. Project-based learning looks to augment the natural workings of such social networks and communities of practice as already exist.

When a project is completed, the members either return to their functional units or organizations or move on to the next project, which makes project teams unique from any other organizational arrangement. In addition, it is not uncommon for individual team members to be members of several teams simultaneously (Henke et al., 1993).

CONCLUSION AND FUTURE TRENDS

Designing a service or product requires the collaborative interaction of individuals from different professional backgrounds. Their diverse expertise represents different interests and issues. These different experiences, mental models, and motivations can be expressed only partly in explicit language. Thus, socialization is a valuable mode of sharing knowledge in teams without language through imitation, observation, and sharing experience face-to-face. Nonaka (1994) emphasized that socialization was also an important way to further trust between partners. Saint-Onge (1996) referred to socialization as a way of creating a sufficient level of congruence to enable individuals to understand each other and work together toward their common goals from different

perspectives. Social constructionists regard language as coordination of action (Burr, 1995) and therefore a fundamental tool in knowledge creation. The commonly employed tool in externalization is dialogue. Dialogue triggers the unconscious elements of knowing and not-knowing, as well as revealing gaps in knowledge compared to what the community knows (Ayas, 1996).

An important aspect of knowledge integration is the willingness to combine knowledge from within and outside the team. The more differentiated the knowledge inputs needed in a task, the higher the knowledge diversity and the greater the scope for knowledge integration. Design, involving art, engineering, finance, and business, is a process of knowledge integration, and a product's design emerges from the collaboration of project participants and stakeholders. Leonard-Barton (1995) viewed the creation of new knowledge as occurring by combining previously unconnected elements or by developing ways of combining elements previously associated.

In project-intensive companies, learning from projects is the key to building strategic competitive advantage. During a project's existence, a number of decisions are made. Every decision involves a degree of uncertainty. However, the problems or mistakes that cause this uncertainty are often of a similar character (Packendorff, 1995). Penrose (1959) argued that utilizing and employing the experiences and knowledge created makes an organization grow. Takeuchi and Nonaka (1989) found that learning could potentially occur within a project team along two dimensions: across different levels (individual, team, and organizational) and across multiple functions or disciplines.

Project team members have to incorporate new knowledge into their understanding in order to solve the technical challenges they face. Thus, learning is inherent in the work they do (Mohrman, Mohrman, & Cohen, 1995). In the role of reflecting expert (Schön, 1987), one is expected, like the technical-rational practitioner, to "know one's business", to possess the relevant know-how. However, one need not know everything, let alone have all the answers. One recognizes that others, too, possess relevant knowledge and that people can learn from each other, gaining insights that result in good solutions. New learning is created through the transformation of experiences, but that learning is not leveraged before an understanding of the experience and task is established (Kolb, 1984). Learning has to be linked to a change in an individual's interpretation of events and actions (von Krogh & Roos, 1996).

Membership in projects is temporary and thus offers individuals the opportunity to belong to multiple communities. In project-based organizations, there are a large number of weak ties that help diffuse knowledge and practices (Granovetter, 1973). Cross-team learning or inter-team learning can occur when teams share their internal approaches with one another. Collective learning can be considered a vital mechanism (Huber, 1991) and a final product of knowledge creation (Senge, 1990).

REFERENCES

Ayas, K. (1996). Design for learning and innovation. *Long Range Planning, 29*(6), 898-901.

Bartezzaghi, E., Corso, M., & Verganti, R. (1997). Continuous improvement and inter-project learning in new product development. *International Journal of Technology Management, 14*(1), 116-138.

Beale, P., & Freeman, M. (1991). Successful project execution: A model. *Project Management Journal, 22*(4), 23-29.

Boland, R., & Tenkasi, R. (1995). Perspective making and perspective taking in communities of knowing. *Organization Science, 6*(4), 350-372.

Brown, J. S., & Duguid, P. (1999). Organizing knowledge. *Reflections: The SoL Journal, 1*(2), 28-44.

Burgelman, R. A. (1991). Intraorganizational ecology of strategy making and organizational adaptation: Theory and field research. *Organization Science, 2*, 239-262.

Burr, V. (1995). *An introduction to social constructionism.* London: Routledge.

Cleland, D. I. (1986). Project stakeholder management. *Project Management Journal, 17*(4), 36-44.

Cohen, W. M., & Levinthal, D. A. (1990). Absorptive capacity: A new perspective on learning and innovation. *Administrative Science Quarterly, 35*(1), 128-152.

Constant, D., Sproull, L., & Kiesler, S. (1996). The kindness of strangers: The usefulness of electronic weak ties for technical advice. *Organization Science, 7*(2), 119-135.

Dougherty, D. (1992). Interpretive barriers to successful product innovation in large firms. *Organization Science, 3*(2), 179-202.

Drew, S. A. W., & Smith, P. A. C. (1995). The learning organization: "Change proofing" and strategy. *The Learning Organization, 2*(1), 4-14.

Fox, R. (1984). Evaluating management of large complex projects. *Technology in Society, 6*, 129-139.

Galbraith, J. R. (1982). Designing the innovating organization. *Organizational Dynamics, 11*(2), 3-24.

Gherardi, S., & Nicolini, D. (2000). The organizational learning of safety in communities of practice. *Journal of Management Inquiry, 9*(1), 7-18.

Granovetter, M. S. (1973). The strength of weak ties. *American Journal of Sociology, 78*, 1360-1380.

Grant, R. M. (1996). Prospering in dynamically-competitive environment: Organizational capability as knowledge integration. *Organization Science, 7*(4), 375-387.

Gregory, K. L. (1983). Native-view paradigms: Multiple cultures and culture conflicts in organizations. *Administrative Science Quarterly, 28*(3), 359-376.

Hargadon, A., & Sutton, R. (1997). Technology brokering and innovation in a product development firm. *Administrative Science Quarterly, 42*(4), 716-749.

Hayes, R. H., Wheelwright, S. C. and Clark, K. B. (1988). *Dynamic manufacturing.* New York: Free Press.

Henke, J., Krachenberg, R., & Lyons, T. (1993). Perspective: Cross-functional teams: Good concept, poor implementation! *Journal of Product Innovation Management, 10*(3), 216-229.

Horvath, L., Callahan, J. L., Croswell, C., & Mukri, G. (1996). Team sensemaking: An imperative for individual and organizational learning. *Proceedings of the Academy of Human Resource Development Conference.*

Huber, G. P. (1991). Organizational learning: The contributing processes and the literatures. *Organization Science, 2*(1), 88-115.

Janis, I. L. (1982). *Groupthink: Psychological studies of policy decisions and fiascoes.* Boston: Houghton Mifflin.

Kanter, R. M. (1988). When a thousand flowers bloom: Structural, collective, and conditions for innovation in organizations. In B. Staw & L. L. Cummings (Eds.), *Research in organizational behaviour* (pp. 169-212). Greenwich, CT: JAI Press.

Kirton, M. J. (1988). Adaptors and innovators: Problem solvers in organizations. In K. Grønhaug & G. Kaufmann (Eds.), *Innovation: A cross-disciplinary perspective* (pp. 65-85). Oslo: Norwegian University Press.

Kolb, D. A. (1984). *Experiential learning: Experience as the source of learning and development.* Englewood Cliffs, NJ: Prentice Hall.

Kuhn, T. (1970). *The structure of scientific revolutions.* Chicago: University of Chicago Press.

Leonard-Barton, D. (1995). *Wellsprings of knowledge: Building and sustaining the sources of innovation.* Boston: Harvard Business School Press.

Lundin, R. A. & Söderholm, A. (1995). A theory of the temporary organization. *Scandinavian Journal of Management, 11*(4), 437-455.

March, J. (1991). Exploration and exploitation in organizational learning. *Organization Science, 2*(1), 71-86.

March, J., Sproull, L., & Tamuz, M. (1991). Learning from samples of one or fewer. *Organization Science, 2*(1), 1-13.

Mohrman, S. A., Mohrman, A. M., Jr., & Cohen, S. G. (1995). Organizing knowledge work system. In M. M. Beyerlein, D. A. Johnson, & S. T. Beyerlein (Eds.), *Advances in interdisciplinary studies of work teams* (pp. 61-91). Greenwich, CT: JAI Press.

Morris, P. (1994). *The management of projects.* London: Thomas Telford.

Neufeldt, V., & Guralnik, D. B. (Eds.). (1988). *Webster's new world dictionary of American English.* New York: Webster's New World.

Nonaka, I. (1994). A dynamic theory of organizational knowledge creation. *Organization Science, 5*(1), 14-37.

Nonaka, I., & Konno, N. (1998). The concept of "Ba": Building a foundation for knowledge creation. *California Management Review, 40*(3), 40-54.

Nonaka, I., & Takeuchi, H. (1995). *The knowledge-creating company: How Japanese companies create the dynamics of innovation.* Oxford: Oxford University Press.

Orr, J. E. (1996). *Talking about machines. An ethnography of a modern job.* Ithaca, NY: ILR Press.

Packendorff, J. (1995). Inquiring into the temporary organization: New directions for project management research. *Scandinavian Journal of Management, 11*(4), 319-333.

Penrose, E. T. (1959). *The theory of the growth of the firm.* New York: John Wiley & Sons.

Sahlin-Andersson, K. (1998). The social construction of projects. A case study of organizing of an extraordinary building project: The Stockholm Globe Arena. In N. Brunsson & J. P. Olsen (Eds.), *Organizing organizations* (pp. 89-106). Bergen-Sandviken: Fagbokforlaget.

Saint-Onge, H. (1996). Tacit knowledge: The key to the strategic alignment of intellectual capital. *Planning Review, 24*(2), 10-14.

Schön, D. A. (1987). *Educating the reflective practitioner: Towards a new design for teaching and learning in the professions.* San Francisco: Jossey-Bass.

Senge, P. M. (1990). *The fifth discipline: The art and practice of the learning organization.* New York: Currency Doubleday.

Skyrme, D., & Amidon, D. M. (1997). *Creating the knowledge-based business.* London: Business Intelligence.

Souder, W. E. (1981). Disharmony between R&D and marketing. *Industrial Marketing Management, 10*, 67-73.

Takeuchi, H., & Nonaka, I. (1989). The new new product development game. *Managing Projects and Programs* (pp. 85-103). Boston: Harvard Business School Press.

Van de Ven, A. H. (1986). Central problems in the management of innovation. *Management Science, 32*(5), 590-607.

von Hippel, E. (1988). *The sources of innovation.* New York: Oxford University Press.

von Krogh, G., & Roos, J. (1996). Imitation of knowledge: A sociology of knowledge perspective. In G. von Krogh & J. Roos (Eds.), *Managing knowledge: Perspectives on cooperation and competition* (pp. 32-54). London: Sage.

Weick, K. E. (1995). *Sensemaking in organizations.* Thousand Oaks, CA: Sage.

Wenger, E. (1998). *Communities of practice: Learning, meaning and identity.* Cambridge, UK: Cambridge University Press.

Wenger, E., & Snyder, W. (2000). Communities of practice: The organizational frontier. *Harvard Business Review, 78*(1), 139-145.

Wheelwright, S. C., & Clark, K. B. (1992). *Revolutionizing product development: Quantum leaps in speed, efficiency, and quality.* New York: Free Press.

Wiebecke, G., Tschirky, H., & Ulich, E. (1987). *Cultural differences at the R&D/marketing interface: Explaining inter-divisional communication barriers. Proceedings of the IEEE Conference on Management and Technology,* Atlanta, Georgia.

KEY TERMS

Ba: A Japanese word which can be translated into the English word *place*. *Ba* can mean a common context, platform, or space for developing collaborative relationships which can be physical, virtual, mental, or any combination of these.

Experience: Experience is a subset of tacit knowledge, but not all experiences are tacit. It can be either obtained through repeatedly performing a task in a similar way or through experimentation with new approaches to complete a task.

Knowledge Sharing: People share their experiences, knowledge, and insights through explicit or implicit means.

Learning: Knowledge or experience gained from working on a task, situation, or problem, which can be applied in future situations. Learning can happen in conscious or subconscious ways.

Multidisciplinary Project Team: A team of people who possess diverse skills, knowledge, and experience, who join together to work on a project which has a limited duration. Team members will disband upon the completion of the project.

Project: An assignment or task that needs to be tackled by a group of people. Projects usually have clearly set objectives, a fixed timescale, and limited resources.

Social Construction Perspective: Knowledge is a set of shared beliefs, constructed through social interactions and embedded within the social contexts in which knowledge is created.

Task Culture: This is characterized by an emphasis on problem solving by a team of experts. Teams are formed to deal with particular problems or projects. Once the task is completed or the project is over, the team will disband. Here the culture is one which attaches importance to knowledge and expertise.

Team Paradigm: A team perspective or belief which is collectively constructed and accepted by members of the team.

Index of Key Terms

Index

A

academic institutions 392
action learning (AL) 30
action research 590
actual practice 459
advance supply systems 295
AL (action learning) 30
ambiguity 357
anthropopathic 436
application 239
artifact 342
    sharing 450
artificial intelligence 436
asynchronous 258
autonomous communities 129

B

ba 515
behavioral pattern discovery 280
best practices 66, 92, 278
boundaries 237, 558, 597
boundary 12, 33, 194
    paradox, the 240
bureaucracy 587
business
    environment 93, 397, 552
    performance 400
    processes 293

C

capabilities 407
case study 157, 575
change programme 35
civil infrastructure systems 286
classifications 21
co-located environments 453

codification 239
codify 595
coercive power 56
cognition 459
cognitive 432
    model 234
    processes 476
CoI (communities of implementation) 35
collaborate 392
collaboration 466
    tools 392
collaborative
    approach 587
    culture 400
    relationships 400
    research 560, 558
    work 340
collective
    learning 115
    process 210
    responsibility 35
collectivist 397
collocated 494
commercial sector 301
common
    characteristics 60
    domain 210
    interest 210
communal
    common ground 78
    knowledge space 246
    learning 328
communication 185, 340
    model 236
    options 453
    technologies 141
communicative action theory 51